Rainer Pöttgen, Thomas Jüstel, Cristian A. Strassert (Eds.)

Rare Earth Chemistry

Also of interest

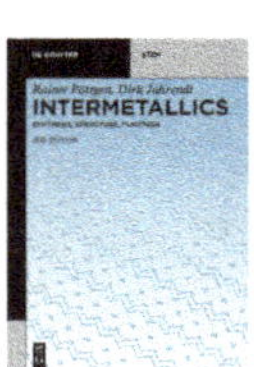

Intermetallics.
Synthesis, Structure, Function
Rainer Pöttgen, Dirk Johrendt, 2019
ISBN 978-3-11-063580-5, e-ISBN 978-3-11-063672-7

Handbook of Rare Earth Elements.
Analytics
Alfred Golloch (Ed.), 2017
ISBN 978-3-11-036523-8, e-ISBN 978-3-11-036508-5

Chemistry of the Non-Metals.
Syntheses – Structures – Bonding – Applications
Ralf Steudel, 2020
ISBN 978-3-11-057805-8, e-ISBN 978-3-11-057806-5

Elemental Analysis.
An Introduction to Modern Spectrometric Techniques
G. Schlemmer, L. Balcaen, J. L. Todolí,
M W. Hinds, 2019
ISBN 978-3-11-050107-0, e-ISBN 978-3-11-050108-7

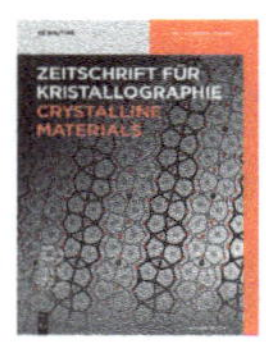

Zeitschrift für Kristallographie – Crystalline Materials
Rainer Pöttgen (Editor-in-Chief)
ISSN 2194-4946, e-ISSN 2196-7105

Rare Earth Chemistry

Edited by
Rainer Pöttgen, Thomas Jüstel, Cristian A. Strassert

DE GRUYTER

Editors
Prof. Dr. Rainer Pöttgen
Institut für Anorganische und Analytische Chemie
Westfälische Wilhelms-Universität Münster
Corrensstrasse 30
48149 Münster
Germany
pottgen@uni-muenster.de

Prof. Dr. Cristian A. Strassert
Institut für Anorganische und Analytische Chemie
Westfälische Wilhelms-Universität Münster
Corrensstrasse 30
48149 Münster
Germany
cstra_01@uni-muenster.de

Prof. Dr. Thomas Jüstel
Fachhochschule Münster
FB Chemieingenieurwesen
Stegerwaldstraße 39
48565 Steinfurt
Germany
tj@fh-muenster.de

ISBN 978-3-11-065360-1
e-ISBN (PDF) 978-3-11-065492-9
e-ISBN (EPUB) 978-3-11-065372-4

Library of Congress Control Number: 2020935460

Bibliographic information published by the Deutsche Nationalbibliothek
The Deutsche Nationalbibliothek lists this publication in the Deutsche Nationalbibliografie; detailed bibliographic data are available on the Internet at http://dnb.dnb.de.

Cover image: Dr. Danuta Dutczak
Typesetting: Integra Software Services Pvt. Ltd.
Printing and binding: CPI books GmbH, Leck

www.degruyter.com

Preface

The rare earth elements together with scandium, yttrium and lanthanum fill 17 places in the periodic table. Besides the transition metals, they comprise one of the large groups of elements for functional materials in inorganic, solid state, coordination and organometallic chemistry. The rare earth elements are not as rare as their name might imply, and we all commonly use rare earth-based materials in daily life. Typical applications are all kinds of luminescent materials for LED lighting, small permanent magnets in headsets, large magnets in wind turbines, rare earth-based alloys, metal-hydride batteries for hybrid engines, gadolinium contrast agents in tomography, laser crystals, lambda oxygen sensors in exhaust gas technology or LCD screens.

Besides these few examples, the rare earth elements dominate the fields of materials research and catalysis. Meanwhile, about 80% of the current material-based patents at least partly include a rare earth-based function. The gross market for modern functional materials based on rare earth elements includes: (i) metallurgy including alloys and permanent magnetic materials, (ii) glass and ceramics including polishing materials, (iii) chemicals and catalysts and (iv) phosphors and electronics. The production of and demand for rare earth metals have shown strong increases since many years, and one can without doubt count rare earth-based industry among the basic future technologies.

Although of extraordinary technological importance, rare earth chemistry is still poorly treated in most inorganic chemistry textbooks. In view of the high impact on the gross domestic products, basic knowledge on the rare earth elements and rare earth-based materials should be a prerequisite for a chemist, physicist and materials scientist. Over the last forty years, high-quality review articles on the many facets of rare earth functionalities have been published in the famous *Handbook on the Physics and Chemistry of Rare Earths* [1]. The present initiative is not just another compilation on rare earth functionalities and rare earth materials. It is meant as an introduction to this broad field on the level of senior undergraduate and postgraduate students. An inevitable prerequisite for the study of this book is the knowledge of some basic chemistry and physics.

This textbook covers fours main chapters: (i) *The Elements*, (ii) *Reactivity and Compounds*, (iii) *Characterization and Properties* and (iv) *Materials and Applications*, addressing many important aspects of rare earth chemistry. Due to the enormous combinatorial variety of the elements, it is impossible to know and consider all classes of rare earth compounds. Nevertheless, we hope that we have made a good compromise and covered most of the basic knowledge on rare earth materials. The chapters are written by leading scientists who are all experts in the topics of their individual contributions. They present a broad literature overview. The interested reader can use this secondary literature (books and review articles) for further in-depth information.

https://doi.org/10.1515/9783110654929-202

Such a book project is not realizable without the help of colleagues and coworkers. We thank Gudrun Lübbering for continuous help with literature search. We are especially grateful to our colleagues for their immediate consents to write up a chapter. It is always challenging to compile a concise Table of Contents and find suitable coauthors. We are indebted to the editorial and production staff of de Gruyter. Our particular thanks go to Kristin Berber-Nerlinger and Dr. Vivien Schubert for their continuous support during conception, writing and producing the present book.

Münster, Steinfurt, 01 May 2020
Rainer Pöttgen, Thomas Jüstel,
Cristian A. Strassert

[1] J.-C. G. Bünzli, V. K. Pecharsky (Eds.), Handbook on the Physics and Chemistry of Rare Earths, Volumes 1–54, North-Holland, Elsevier, Amsterdam, 1978–2020.

This book contains a token, pointing to:

recommended literature for further reading;
i. e. relevant text books, review articles or important original articles

Contents

List of contributors

Armbrüster, Prof. Dr. Marc
Faculty of Natural Sciences,
Institute of Chemistry
Materials for Innovative Energy Concepts
Technische Universität Chemnitz
09107 Chemnitz
Germany
E-mail: marc.armbruester@chemie.
tu-chemnitz.de

Bassil, Dr. Bassem S.
University of Balamand
Faculty of Arts and Sciences
P.O. Box 100
1300 Tripoli
Lebanon
E-mail: bbassil@jacobs-alumni.de

Baur, Dr. Florian
Fachbereich Chemieingenieurwesen
Fachhochschule Münster
Stegerwaldstrasse 39
48565 Steinfurt
Germany
E-mail: florian.baur@fh-muenster.de

Bertau, Prof. Dr. Martin
Institut für Technische Chemie
TU Bergakademie Freiberg
Leipziger Straße 29
09599 Freiberg
Germany
E-mail: Martin.Bertau@chemie.tu-freiberg.de

Bredol, Prof. Dr. Michael
Fachbereich Chemieingenieurwesen
Fachhochschule Münster
Stegerwaldstrasse 39
48565 Steinfurt
Germany
E-mail: bredol@fh-muenster.de

Brik, Prof. Dr. Mikhail G.
Institute of Physics
University of Tartu
Werner Ostwald Street 1
Tartu 50411
Estonia
E-mail: mikhail.brik@ut.ee

Bünzli, Prof. Dr. Jean-Claude G.
Swiss Federal Institute of Technology
Lausanne (EPFL)
Institute of Chemical Sciences and
Engineering BCH 4104
1015 Lausanne
Switzerland
E-mail: jean-claude.bunzli@epfl.ch

Charbonnière, Prof. Dr. Loïc
Institut Pluridisciplinaire Hubert Curien
CNRS DELEGATION REGIONALE ALSACE
IPHC
23 rue du Loess
BP 28
67037 STRASBOURG CEDEX 2
France
E-mail: l.charbonn@unistra.fr

Daumann, Jun.-Prof. Lena J.
Department Chemie und Biochemie
Ludwig-Maximilians-Universität München
Butenandtstrasse 5–13 (Haus D)
81377 München
Germany
E-mail: lena.daumann@cup.lmu.de

Dupré, Dr. Klaus
EOT GmbH (formerly FEE GmbH)
Struthstrasse 2
55743 Idar-Oberstein
Germany
E-Mail: KDupre@eotech-de.com

https://doi.org/10.1515/9783110654929-204

Eckert, Prof. Dr. Hellmut
Institut für Physikalische Chemie
Westfälische Wilhelms-Universität Münster
Corrensstrasse 30
48149 Münster
Germany
E-mail: eckerth@uni-muenster.de

Faust, PD Dr. Andreas
European Institute for Molecular Imaging (EIMI)
Waldeyerstraße 15
48149 Münster
Germany
E-Mail: faustan@uni-muenster.de

Funke, M. Sc. Sabrina
Institut für Anorganische und Analytische Chemie
Westfälische Wilhelms-Universität Münster
Corrensstrasse 30
48149 Münster
Germany
E-mail: sabrina.funke@uni-muenster.de

Gross, Andreas
EOT GmbH (formerly FEE GmbH)
Struthstrasse 2
55743 Idar-Oberstein
Germany
E-Mail: AGross@eotech-de.com

Hildebrandt, Prof. Dr. Niko
Institut de Biologie Intégrative de la Cellule (I2BC)
Bâtiment 21
Avenue de la Terrasse
91190 GIF-SUR-YVETTE
France
E-mail: niko.hildebrandt@u-psud.fr

Janka, PD Dr. Oliver
Universität des Saarlandes
Anorganische Festkörperchemie
Campus C4 1, Raum 4.03
66123 Saarbrücken
Germany
E-mail: oliver.janka@uni-saarland.de

Johrendt, Prof. Dr. Dirk
Department Chemie
Ludwig-Maximilians-Universität München
Butenandtstraße 5-13 (Haus D)
81377 München
Germany
E-mail: johrendt@lmu.de

Jüstel, Prof. Dr. Thomas
Fachbereich Chemieingenieurwesen
Fachhochschule Münster
Stegerwaldstraße 39
48565 Steinfurt
Germany
E-mail: tj@fh-muenster.de

Karst, Prof. Dr. Uwe
Institut für Anorganische und Analytische Chemie
Westfälische Wilhelms-Universität Münster
Corrensstrasse 30
48149 Münster
Germany
E-mail: uk@uni-muenster.de

Kögerler, Prof. Dr. Paul
Institut für Anorganische Chemie
RWTH Aachen University
Landoltweg 1
52074 Aachen
Germany
E-mail: paul.koegerler@ac.rwth-aachen.de

Kohlmann, Prof. Dr. Holger
Institut für Anorganische Chemie
Universität Leipzig
Johannisallee 29
04103 Leipzig
Germany
E-mail: holger.kohlmann@uni-leipzig.de

Kortz, Prof. Dr. Ulrich
Jacobs University
Department of Life Sciences and Chemistry
Campus Ring 1
28759 Bremen
Germany
E-mail: u.kortz@jacobs-university.de

Kremer, Dr. Reinhard K.
MPI für Festkörperforschung
Heisenbergstrasse 1
70569 Stuttgart
Germany
E-mail: rekre@fkf.mpg.de

Kynast, Prof. Dr. Ulrich
Fachbereich Chemieingenieurwesen
Fachhochschule Münster
Stegerwaldstrasse 39
48565 Steinfurt
Germany
E-mail: uk@fh-muenster.de

Leker, Prof. Dr. Jens
Institut für betriebswirtschaftliches
Management im Fachbereich Chemie und
Pharmazie
Universität Münster
Leonardo-Campus 1
48149 Münster
Germany
E-mail: leker@uni-muenster.de

Lezhnina, Dr. Marina M.
Fachbereich Chemieingenieurwesen
Fachhochschule Münster
Stegerwaldstrasse 39
48565 Steinfurt
Germany
E-mail: marina@fh-muenster.de

Liebald, Dr. Christoph
EOT GmbH (formerly FEE GmbH)
Struthstrasse 2
55743 Idar-Oberstein
Germany
E-Mail: CLiebald@eotech-de.com

Lorenz, Dr. Tom
Institut für Technische Chemie
TU Bergakademie Freiberg
Leipziger Straße 29
09599 Freiberg
Germany
E-mail: tom.lorenz@chemie.tu-freiberg.de

Möckel, Dr. Robert
Helmholtz-Institut Freiberg für
Ressourcentechnologie am HZDR
Chemnitzer Straße 40
09599 Freiberg
Germany
E-mail: r.moeckel@hzdr.de

Mudryk, Prof. Dr. Yaroslav
Ames Laboratory
Iowa State University
Ames, IA 50011-2416
USA
E-mail: slavkomk@ameslab.gov

Nocton, Dr. Grégory
Ecole polytechnique
Laboratoire de Chimie Moléculaire
Route de Saclay
91128 Palaiseau
France
E-mail: gregory.nocton@polytechnique.edu

Pecharsky, Prof. Dr. Vitalij K.
Ames Laboratory
Iowa State University
Ames, IA 50011-2416
USA
E-mail: vitkp@ameslab.gov

Peltz, Dr. Mark
EOT GmbH (formerly FEE GmbH)
Struthstrasse 2
55743 Idar-Oberstein
Germany
E-Mail: MPeltz@eotech-de.com

Pöttgen, Prof. Dr. Rainer
Institut für Anorganische und Analytische
Chemie
Westfälische Wilhelms-Universität Münster
Corrensstrasse 30
48149 Münster
Germany
E-mail: pottgen@uni-muenster.de

Pues, M.Sc. Patrick
Fachbereich Chemieingenieurwesen
Fachhochschule Münster
Stegerwaldstrasse 39
48565 Steinfurt
Germany
E-Mail: pues@fh-muenster.de

Rambeck, Prof. Dr. Walter
Veterinärwissenschaftliches Department
Lehrstuhl für Tierernährung und Diätetik
Schönleutnerstraße 8
85764 Oberschleißheim
Germany
E-mail: rambeck@lmu.de

Redling, Dr. Kerstin
Veterinärwissenschaftliches Department
Lehrstuhl für Tierernährung und Diätetik
Schönleutnerstraße 8
85764 Oberschleißheim
Germany
E-mail: kerstin@redling.com

Ronda, Prof. Dr. Cees
Research Group Oral Healthcare
Philips Group Innovation, Research
High Tech Campus 11, ground floor
5656 AE Eindhoven
The Netherlands
E-Mail: cees.ronda@philips.com

Rytz, Dr. Daniel
EOT GmbH (formerly FEE GmbH)
Struthstrasse 2
55743 Idar-Oberstein
Germany
E-Mail: rytz@eotech.com or rytz@fee-io.de

Schlatt, M. Sc. Lukas
Institut für Anorganische und Analytische Chemie
Westfälische Wilhelms-Universität Münster
Corrensstrasse 30
48149 Münster
Germany
E-Mail: l_schl21@uni-muenster.de

Schwung, Dr. Sebastian
EOT GmbH (formerly FEE GmbH)
Struthstrasse 2
55743 Idar-Oberstein
Germany
E-mail: SSchwung@eotech-de.com

Seitz, Prof. Dr. Michael
Institut für Anorganische Chemie
Universität Tübingen
Auf der Morgenstelle 18 (A-Bau)
72076 Tübingen
Germany
E-mail: michael.seitz@uni-tuebingen.de

Simon, Prof. Dr. h.c. mult. Arndt
Max-Planck-Institut für Festkörperforschung
Heisenbergstraße 1
70569 Stuttgart
Germany
E-mail: A.Simon@fkf.mpg.de

Sperling, Dr. Michael
Institut für Anorganische und Analytische Chemie
Westfälische Wilhelms-Universität Münster
Corrensstrasse 30
48149 Münster
Germany
E-mail: ms@speciation.net

Srivastava, Dr. Alok M.
GE Global Research
One Research Circle, Niskayuna
New York 12309
USA
E-mail: srivastava@ge.com

Stöwe, Prof. Dr. Klaus
Faculty of Natural Sciences
Institute of Chemistry, Chemical Technology
Technische Universität Chemnitz
09107 Chemnitz
Germany
E-mail: klaus.stoewe@chemie.tu-chemnitz.de

Strassert, Prof. Dr. Cristian A.
Institut für Anorganische und Analytische Chemie
Westfälische Wilhelms-Universität Münster
Corrensstrasse 30
48149 Münster
Germany
E-mail: cstra_01@uni-muenster.de

Van Leusen, Jan
Institut für Anorganische Chemie
RWTH Aachen University
Landoltweg 1
52074 Aachen
Germany
E-mail: jan.vanleusen@ac.rwth-aachen.de

von Bremen-Kühne, M. Sc. Maximilian
Institut für Anorganische und Analytische Chemie
Westfälische Wilhelms-Universität Münster
Corrensstrasse 30
48149 Münster
Germany
E-mail: ms.vbk@uni-muenster.de

Walter, Prof. Dr. Marc D.
Technische Universität Braunschweig
Institut für Anorganische und Analytische Chemie
Hagenring 30
38106 Braunschweig
Germany
E-mail: mwalter@tu-bs.de

Wentker, Dr. Marc
Institut für betriebswirtschaftliches Management im Fachbereich Chemie und Pharmazie
Universität Münster
Leonardo-Campus 1
48149 Münster
Germany
E-mail: marc.wentker@uni-muenster.de

Werner, Prof. Dr. Jan
Hochschule Koblenz, University of Applied Sciences
WesterWaldCampus Höhr-Grenzhausen
Fachbereich Bauen-Kunst-Werkstoffe
Fachrichtung Werkstofftechnik Glas und Keramik
Rheinstraße 56
56203 Höhr-Grenzhausen
Germany
E-mail: jan.werner@fgk-keramik.de

Wesemann, Dr. Volker
EOT GmbH (formerly FEE GmbH)
Struthstrasse 2
55743 Idar-Oberstein
Germany
E-Mail: VWesemann@eotech-de.com

Wickleder, Prof. Dr. Mathias S.
Department Chemie
Universität zu Köln
Greinstrasse 6
50939 Köln
Germany
E-mail: mwickled@uni-koeln.de

Zimmermann, Prof. Dr. Jörg
Abteilungsleitung Energiematerialien
Fraunhofer-Projektgruppe IWKS
Rodenbacher Chaussee 4
63457 Hanau
Germany
E-mail: joerg.zimmermann@isc.fraunhofer.de

1 The elements

Jean-Claude G. Bünzli

1.1 Discovery of the rare-earth elements

1.1.1 The protagonists

The rare earths (*REs*) are a homogeneous group of 17 elements. According to the IUPAC nomenclature rules, they correspond to the elements 21 (Sc), 39 (Y) and 57–71 (Ln = La-Lu). The Ln subgroup should be called "lanthanoids" but "lanthanides" (corresponding in principle to Ce-Lu) is still the most used designation for these metallic elements and their compounds. In the long form of the periodic table, Ln elements are inserted between Ba (56) and Hf (72) with Sc, Y and Lu, forming a column to the left of Ti, Zr and Hf (Figure 1.1.1).

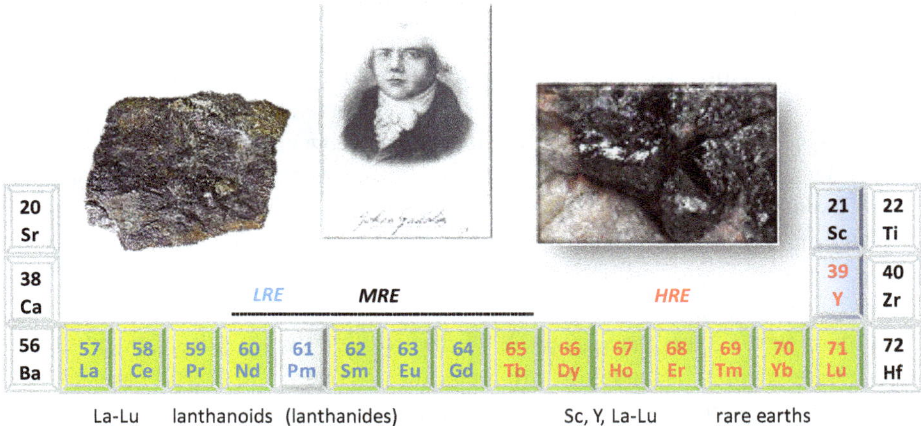

Figure 1.1.1: Rare-earth elements with the two generic minerals, cerite (left) and gadolinite (right) and the reproduction of a photograph of Johan Gadolin (middle, © 1910, Acta Societatis Scientarium Fennicae). Element 61, first artificially produced, is only found as minute traces in some uranium ores.

When it comes to subdividing the lanthanides, chemists rely on the electronic structure (see Section 1.6) of the most common oxidation state, the trivalent ions Ln^{III}, $[Xe]4f^n$: the lighter lanthanides (*LREs*) are those which have no paired 4f electrons ($n = 0–7$, La-Gd), while the heavier lanthanides (*HREs*) correspond to $n = 8–14$, Dy-Lu. Geochemists use different definitions, excluding Eu which has "anomalous" properties as compared to the *LRE*s, leaving it alone in a special group. In metallurgy and industry, *LREs* correspond to La-Nd (also called ceric *RE*s), middle *RE*s (*MREs*) to either Sm-Gd or Sm-Dy, and *HREs* to Dy-Lu or Ho-Lu; finally yttric *RE*s are those from Sm to Lu, plus Y. Fortunately, everybody agrees with the nonlanthanoid

https://doi.org/10.1515/9783110654929-001

elements: yttrium has chemical properties very similar to Dy-Ho, so it is included in *HREs*, while scandium has geochemical and chemical behaviors so different from of all the other *RE*s that it is not listed in any of these groups.

1.1.2 The search for *RE*s: an intricate problem

The discovery of *RE* elements has been an extraordinary endeavor spanning over a century, from 1794 (Y) to 1907 (Lu) for the naturally occurring elements and extending to 1945 for the radioactive Pm. This quest has been a tremendous success for crystal and analytical chemistry at first and, later, in the mid-1850s, for spectroscopy, since pure elements had to be separated from intricate mixtures: indeed, due to the similar chemical properties of *RE*s (with the exception of Sc), *RE* minerals contain several of these elements, usually almost the entire *LRE* or *HRE* series. It is also a tortuous story reflecting both on the little amount of information available at that time on chemical bonding and element systematics, as well as on the unsophisticated methods of analysis available. Different compositions of inhomogeneous minerals, depending on where they were found, increased the difficulty. To this should be added the hefty competition and rivalry between scientists, mirroring the race of nations for gaining supremacy in science and technology. The thrilling story of the isolation of *RE* elements is, therefore, full of incorrect claims and heated disputes among would-be discoverers, whilst reflecting the developments in separation, analytical, and spectroscopic techniques which took place during the nineteenth century [1, 2].

Before starting the description of *RE* discovery, a note on the concept of element discovery may be useful. In principle, the discoverer should prepare the new element in its elemental form, that is metallic form for *RE*s. But these elements are highly electropositive and their metallic form is difficult to obtain unless special reduction techniques are used, that were not always available at the time when the new *RE* elements were separated and identified. In particular, electrochemical methods only started to be mastered with Sir Humphry Davy's (1778–1829) experiments in the first decade of the nineteenth century. As a consequence, the discovery of a new *RE* was often attributed to the scientist who was able to produce a pure oxide of the element. These oxides are termed "earths" and their denomination takes the ending "a" instead of "um" for the element: yttria for yttrium oxide, terbia for terbium oxide and so on. Moreover, it is also noteworthy that element names were sometimes changed or interchanged, for instance, erbium and terbium, adding to the confusion. The denomination "rare earths" is not quite correct since the abundance of some *RE* elements is larger than several other technological metals such as silver or indium and comparable to that of lead. *RE* resources are also found in many places around the world in Europe, Asia, Australia, and North and

South Americas. The term “rare earths” stems from the fact that while the *RE* content in specific minerals is relatively large (35% to 75%), these minerals are rather dispersed into the rocks or sands in which they are found.

Although it is claimed that more than 270 minerals contain *RE* elements, two minerals have played a pivotal role in the discovery of the 4f-elements: gadolinite (previously ytterbite, $RE_2FeBe_2Si_2O_{10}$) containing mainly Y and the heavier *RE*s (Eu-Lu) and cerite, a complex silicate of composition $(Ca, Ce)_{10}(Fe^{3+})(SiO_4)_6(OH,F)_5$, essentially composed of La, Ce, Nd and smaller quantities of Sm-Eu. These two minerals mirror the subdivision of *RE*s into heavier (yttric) and lighter (ceric) *RE*s. A third mineral, samarskite, appearing in several chemical forms, $(Y,Fe^{3+},Fe^{2+},U,Th,Ca)_2(Nb,Ta)_2O_8$ and $(Yb,Y,U,Th,Ca,Fe)(Nb,Ta)_2O_8$ for instance, also played a decisive role in the discovery of several *RE* elements, including samarium.

1.1.3 The first period: yttrium and cerium

In 1787, Carl Axel Arrhenius (1757–1824), an artillery lieutenant in the Swedish army, chemist and an amateur mineralogist noticed and collected a black rock in a quartz and fluorspar (fluorite) quarry near Ytterby in Sweden, on Resarö island, about 20 km northeast from Stockholm and named it ytterbite. At first sight, the rock resembled asphalt or coal but was remarkably heavy. Bengt Reinhold Geijer (1758–1815), a geologist at the Swedish Royal Mint, examined the “black stone,” measured its density (4.22 g cm^{-3}), noticed it was diamagnetic and published its findings in 1788 as a short letter to the editor in a German journal [3]. He (wrongly) suspected that it contained the recently discovered heavy element tungsten (1783). Chemical analysis had to wait because there was a break in the succession of Swedish analytical chemists after the passing away of Torbern Olof Bergman (1735–1784) and Carl Wilhelm Scheele (1742–1786). Johann Gadolin (1760–1852), a Finnish chemist and an advocate of Finnish independence from Sweden, was appointed at the University of Åbo (presently Turku, Finland) in 1785 but during 1786–1788, he was completing his education by a study tour in Central and Western Europe working with various renowned chemists. He started to study the Arrhenius’ ytterbite sample only in 1792, separating it and obtaining a white compound that he recognized being a new earth. The paper relating the experiments conducted for separating and isolating yttria with the help of wet chemistry and calcination with the “blow pipe” appeared in 1794 in Swedish [4] and in 1796 in German [5] (the manuscript was sent to the Editor Crell on July 3, 1794, but arrived only on March 26, 1796 . . .). A part of these experiments are schematized in Figure 1.1.2. Johan Gadolin also found that yttrium can be precipitated by the addition of oxalate, a reaction still in use today in the chemical analysis and separation of *RE*s. Anders Gustaf Ekeberg (1767–1813)

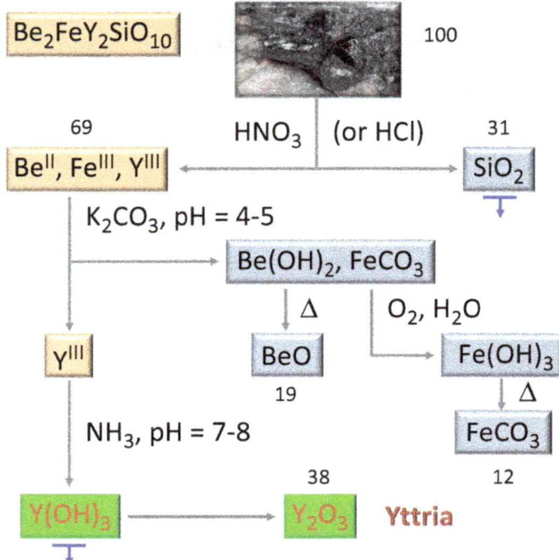

Figure 1.1.2: Analytical scheme summarizing some of the experiments conducted by Johan Gadolin on gadolinite [5]. Numbers represent the relative proportions of the isolated compounds. Please note that beryllium was not known at that time so that corresponding chemicals were taken as being aluminum compounds.

confirmed Gadolin's investigation in 1797, named the new earth yttria and proposed the name gadolinite for the mineral, which at the time turned out to be the only mineral bearing the name of a living scientist.

Cerium could have been discovered before yttrium: in 1751, Axel Fredrik Cronstedt (1722–1765) found two minerals with exceptional heavy densities near Nya Bastnäs, close to Riddarhyttan in Västmanland country, Sweden, where iron was mined since Roman times. He called them "tung sten" (heavy stones). Noticing that they were different minerals interspersed in the stones, he suspected there might be new earths in them. This suspicion was confirmed by Scheele who isolated tungstic acid (H_2WO_4) in 1781 from one of the minerals, which later tuned out to be calcium tungstate (scheelite).

The second mineral was subjected to many analyses by various scientists with contradictory results. However, in 1803, the young Jöns Jacob Berzelius (1779–1848), educated as a medical doctor and working as an assistant in the Chemistry department at the Collegium Medicum in Stockholm (presently the Karolinska Institute), visited Wilhelm Hisinger (1766–1852), an iron-works industrialist in Västmanland, for discussing the recent results obtained on different minerals. They came to the conclusion that the second stone from Bastnäs did not contain tungstic acid and despite similarities in its properties with those of gadolinite, there were some differences. So they correctly inferred the presence of another new element. They initially named it bastium (from Bastnäs) but soon changed the name into cerium, after a recently discovered asteroid, Ceres. The mineral was consequently named cerite. The full paper was sent to a new German journal edited by Adolph Ferdinand Gehlen (1775–1815) [6]. Upon receiving the manuscript, the editor informed the authors that he had just received another contribution from Martin Heinrich Klaproth (1743–1817) from Berlin on the analysis of a mineral from Bastnäs, that he had named ochroït, and that the author had concluded to the presence of a new earth. Klaproth's contribution was published in the issue 3 of the second volume [7] and Hisinger and Berzelius contribution's in issue 4 of the same year. There had been some preliminary disclosures of the experiments and animated discussions with other scientists. A priority dispute arose that turned to the advantage of the Swedes because they were referring to a new metallic element, while Klaproth was mentioning a new "earth" and did not name it. It is noteworthy that Hisinger and Berzelius paper was translated and published in several other journals in Swedish, French and English.

1.1.4 The second period: essential contributions from Carl Gustaf Mosander (1797–1858)

After the excitement stirred by the discovery of yttrium and cerium, none bothered much about these elements over the next 20 years or so until Mosander, who started his career as a pharmacist apprentice in Stockholm and became Berzelius' assistant, decided to reinvestigate the two oxides. In particular, one of his first objectives was to produce metallic cerium. After several attempts, he succeeded by starting with cerium sulfide, heating it in a tube with chlorine and then passing potassium vapor in a hydrogen atmosphere over the formed cerium chloride. The result was potassium chloride and the elemental cerium which violently dissolved in water with the evolution of hydrogen. Mosander then studied the reaction of cerium with various chemicals and slowly became convinced that the initial cerium oxide contained another element. Mosander was a very conscientious, but a slow and reserved researcher and at the instigation of Berzelius, he finally announced in 1839 that the initial ceria did contain about 40% of another, more basic, oxide from an element he named lanthanum, from the Greek word *lanthano* meaning "escaping notice" or "hidden," as suggested by Berzelius. Using a similar reaction as for cerium, he also obtained the elemental lanthanum. Furthermore, in the course of fractional precipitation and crystallization of the two components of cerite: ceria and lanthana, involving tens of steps, Mosander noticed that the two oxides were colored in amethyst-rose, lanthana being less tainted than ceria. From this observation, he correctly concluded that a third element might be present and presuming it was the least alkaline one compared to La and Ce, he finally succeeded in separating its sulfate by fractional crystallization in 1842. The corresponding oxide was named didymia (Di), from the Greek *didymos*, twin, because it was closely associated with lanthana and ceria. All these experiments took several years but then it was firmly established that lanthana is a white powder while ceria is yellowish and didymia purple.

In parallel, Mosander and Berzelius had been interested in finding out if yttria would also contain new elements, because some preparations were rather yellowish. Berzelius added ammonia to a solution of yttrium nitrate which resulted in precipitating several compounds, first beryllium oxide, then a yellowish substance, and finally whitish yttria. Mosander duplicated the experiment and found that the residue of the yellowish precipitate contained two fractions, a colorless one and a pale amethyst one. He repeated the experiment with oxalate: the first precipitate was again almost colorless and he named it erbium while the second amethyst-colored precipitate was named terbium. Mosander's findings were reported in 1842 and in 1843. He proved (as suspected by other scientists) that gadolinite contained at least three elements, similarly to cerite (Figure 1.1.3). However, if erbium could be readily confirmed, several other researchers questioned the existence of terbium, stirring heated debates. During this long process, another name was proposed for terbium, mosandrium, and stunningly the names erbium and terbium were inverted

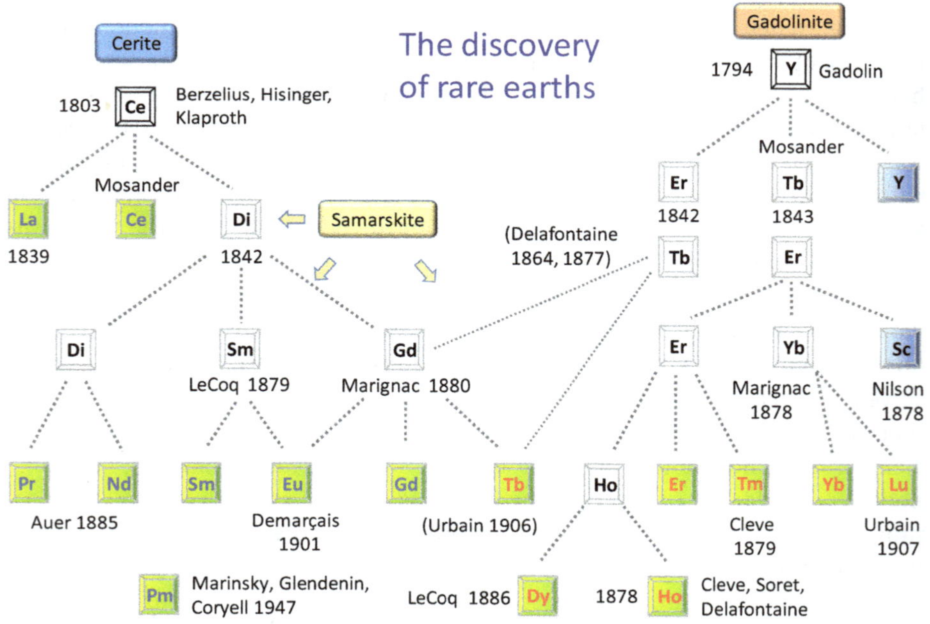

Figure 1.1.3: Simplified genealogy of the discovery of *RE*s. Adapted from ref. [1]. Official discovery dates are given as presently accepted. Important contributions are indicated within parentheses. The symbol Di stands for didymium.

by Marc Delafontaine (1838–1911) in 1864. Delafontaine confirmed this inversion when he unequivocally proved the existence of Y, Tb and Er by spectral analysis in 1877. The debate about terbium (ex erbium) was definitively closed when its full 194-line optical spectrum was published by Henry E. Roscoe (1833–1915) in 1882 [8].

1.1.5 The third period: the playground of spectroscopy

Similarly to what happened after the initial discoveries of yttrium and cerium, one had to wait 35 years before new *RE* elements were uncovered (Figure 1.1.3 and Table 1.1.1). One reason for this gap is that the attention of chemists was more focused on organic than inorganic chemistry. *RE*s were still studied but the work was mainly done with cerium, lanthanum and didymium, although questions about the existence of terbium/erbium stirred quite a bit of investigations. But then, new findings happened at an accelerated pace: six new elements were identified in the three years between 1878 and 1880 and three more in the period 1885–1886. Meanwhile, chemical research had also progressed enormously thanks to the opening of new research and education centers around Europe as well as in the fastly developing chemical industry. Accurate atomic weights were now at hand, owing to the hard work of J. J. Berzelius, Jean Servais Stas (1813–1891) and Jean Charles Galissard de Marignac (1817–1894), and due to the availability of a workable periodic table of the elements (D. Mendeleev, 1869, L. Meyer, 1870). Furthermore, analytical chemistry was benefitting from a major tool, spectroscopy, and the blow pipe was gradually losing its importance. On the other hand, analytical separations remained work-intensive with fractional crystallization and precipitation being the main techniques.

In 1875, Robert Wilhelm Bunsen (1811–1899) published his legendary spectroscopic investigations in which he mentioned having certified the presence of Ce, La, Di, and Er, and that he did not find Tb. At the University of Geneva, Switzerland, Delafontaine started to analyze samarskite, a mineral found in Ural mountains and named after the officer in charge of the mine, Vasilii Samarsky-Bykhovets (1803–1870). In 1877, he found terbium in this mineral but he also suspected the presence of a new element, which he called philippium (Pp) in 1878. At the same time and at the same university, Jacques-Louis Soret (1827–1890) was reanalyzing gadolinite and concluded that an additional oxide was present to which he gave the name "earth X," while Marignac separated ytterbium from erbium, also in 1878. Similar analyses were performed in parallel at the University of Uppsala. Lars Fredrik Nilson (1840–1899) separated scandium from erbia in 1879 thanks to an elaborate 13-step fractionation starting from the nitrates. He determined its molecular weight and demonstrated that scandium corresponds to an element predicted by Mendeleev, eka-boron. Next, Per Teodor Cleve (1840–1905)

Table 1.1.1: Chronology of the discovery of the rare-earth elements. Data from ref. [9].

Symbol	Z	Name	Year	Discoverer(s). Place(s) of discovery. Origin of name
Y	39	yttrium	1794	J. Gadolin, Åbo, Sweden (Today Turku, Finland). From Ytterby
Ce	58	cerium	1803	J. J. Berzelius and W. Hisinger, Västmanland, Sweden. From asteroid Ceres, discovered in 1801. M. H. Klaproth (Germany)
La	57	lanthanum	1839	C. G. Mosander, Stockholm, Sweden. From *lanthano*, "hidden" in Greek
Er	68	erbium	1842	C. G. Mosander, Stockholm, Sweden. From Ytterby
Tb	65	terbium	1843	C. G. Mosander, Stockholm, Sweden. From Ytterby
Ho	67	holmium	1878	P. T. Cleve, Uppsala, Sweden; J.-L. Soret and M. Delafontaine, Geneva, Switzerland. From *Holmia*, Latin word for Stockholm
Yb	70	ytterbium	1878	J. C. Galissard de Marignac, Geneva, Switzerland. From Ytterby
Sc	21	scandium	1879	L. F. Nilson, Uppsala, Sweden. From *Scandia*, Latin word for Scandinavia
Tm	69	thulium	1879	P. T. Cleve, Uppsala, Sweden. From Thule, ancient name for Scandinavia
Sm	62	samarium	1879	P. E. Lecoq de Boisbaudran, Paris, France. From samarskite mineral, found in Ural and named from the director of the mine, colonel Vasilii Samarsky-Bykhovets
Gd	64	gadolinium	1880	J. C. Galissard de Marignac, Geneva, Switzerland. In honor of Johan Gadolin
Pr	59	praseodymium	1885	C. Auer von Welsbach, Vienna, Austria. From Greek words *prasios* (green) and *didymos* (twin)
Nd	60	neodymium	1885	C. Auer von Welsbach, Vienna, Austria. From Greek words *neos* (new) and *didymos* (twin)
Dy	66	dysprosium	1886	P. E. Lecoq de Boisbaudran, Paris, France. From Greek word *dysprositos,* "difficult to access"
Eu	63	europium	1901	E. A. Demarçay, Paris, France. From Europe
Lu	71	lutetium	1907	G. Urbain, Paris, France. From *Lutetia*, Latin word for Paris
Pm	61	promethium	1945	J. A. Marinsky, L. E. Glendenin, C. D. Coryell, Oak Ridge, USA. From *Prometheus* who stole fire from the gods to give it to mankind; later it was found as minute traces in some uranium ores.

succeeded, in the same year, in separating erbia into three components, erbium and two other elements he named holmium (from the Latin word *Holmia* for Stockholm) and thulium (from *Thule*, ancient name for Scandinavia). These separations were achieved by fractional crystallizations with nitrate and sulfate. In view of the small concentrations of the *RE* elements, large initial batches of mineral had to be treated. It turned out that holmium was identical to Pp and "earth X." Hence, the discovery of holmium is credited to the three scientists and the discovery date set to 1878, although 1879 is commonly mentioned too.

Spectral analysis was a dedicated tool for proving the purity of the samples and also for suggesting the presence of new elements. A good example is the fractionation of didymium. In 1878, Delafontaine noted that didymium was probably not homogeneous since samples separated from cerite and from samarskite did not have the same absorption spectra. This suspicion was proved correct by Paul Émile Lecoq de Boisbaudran (1838–1912), an eminent spectroscopist, who measured the absorption and emission spectra of most elements known around 1870. He built the first database of atomic spectra [10] and discovered gallium in 1875. In 1879, he reported having found a new faint spectroscopic line at 400 nm in samarskite that was not present in cerite; he assigned it to a novel element. The double salt of the latter with potassium sulfate precipitated with Di but the oxalate separated before Di. Hence, Lecoq de Boisbaudran developed a complicated but a working separation process and named the element after the mineral, samarium. Analysis of samarskite and the newly discovered samarium was also performed by Marignac in 1880. He separated it first by precipitation with potassium sulfate, then used the "Delafontaine" formate method and finally applied the separation method with the oxalates, obtaining a substance that differed in many ways from the behavior of other lanthanides. He named it Yα. Delafontaine studied the new substance carefully, showed it could also be obtained from cerite and that it corresponded to decipium, an element he had discovered earlier. The current accepted name of gadolinium was given by Marignac in 1886, who was credited for its discovery.

Enters now a famous name in rare-earth science and technology, the Austrian Carl Auer von Welsbach (1858–1929). This extraordinary entrepreneur is the founder of the rare-earth industry with the incandescent mantle for gas lamps, ThO_2: Ce(1%), in 1891 and still in use today, as well as the flint stones based on mischmetal (30% Fe + a mixture of metallic *RE*, essentially Ce and La) in 1903. He is also the founder of two companies that still exist today, Treibacher Industrie A.G. (1898) and Osram (1906), a contraction of osmium and wolfram (German name of tungsten), in view of the alloy he used to replace carbon filaments in Edison electric bulbs. As a doctoral student of Bunsen in Heidelberg, he started to study *RE*s towards the end of the 1870s. After completing his PhD in 1882, he went back to Vienna and continued his research on the separation of these elements. He was convinced that didymium was not homogeneous (as many other scientists thought) and developed new separation methods based on fractional crystallization. He was then able to separate didymium into two fractions in 1885 by a painstaking procedure of more than 100 separations, each step taking 1–2 days. The color and spectra of the various fractions were different and proved the presence of two new elements, praseo(di)dymium (green didymium) and pink neo(di)dymium (new didymium). It is noteworthy that Mosander's name for didymium ("twins") was therefore fully justified! The syllable "di" was rapidly omitted in the names of the new elements for the sake of simplicity.

The following year, in 1886, Lecoq de Boisbaudran, who had just confirmed the discoveries of holmium and thulium by spectroscopy, developed an intricate method for separating gadolinite. The method involved 32 precipitations with ammonia followed by 26 precipitations with oxalate, each fraction characterized by its optical spectra (absorption and emission). He came to the conclusion that holmium was not homogeneous and in 1886, he identified another element which he called dysprosium ("difficult to access" in Greek). The element was accepted by chemists without any discussion or challenge and this discovery ended the long and complex history of the fractionation of erbium.

As for many newly discovered elements and particularly when it comes to *RE*s, the homogeneity of samarium (after separation from gadolinium) was questioned and triggered detailed studies. Cleve spent a long time studying this element and finally wrote a monograph in 1885 in which he stated that, indeed it was a homogeneous new element. But the following year, the French chemist Eugène Anatole Demarçay (1852–1903) announced that he most probably identified a new *RE* in samarium, judging from optical spectra. He then tried to separate it from a large quantity of samarium by tedious successive recrystallizations in nitric acid. In 1896, he isolated a more or less colorless compound that he named Σ, because his main proof was still spectral evidence. Lecoq de Boisbaudran also saw new lines in the spectrum of samarium and was convinced that they were from a new element. In England, Sir William Crookes (1832–1919), the inventor of the vacuum cathode-ray tube bearing his name, a gifted spectroscopist and the designer of UV-blocking sunglasses consisting of cerium-doped optical glass, also saw an "anomalous line" in samarium samples. His thinking was that the corresponding element was different from those identified by the French scientists. Finally, in 1901, after further purifications, Demarçay obtained a pure potassium double nitrate and published a short paper in which he described the new element as being between samarium and gadolinium, with atomic weight 151, and which he called europium. Georges Urbain (1872–1938) confirmed the finding in 1904 after developing a new separation method making use of a "separating element," namely bismuth. Bismuth nitrate frequently crystallizes between two rare-earth nitrates, making their separation easy. Urbain then entered into a fierce dispute with Auer von Welsbach about the last element of the series. In 1905, Auer proposed a new separation method based on double oxalates and applied it to ytterbium. He suspected this element consisted of two components. He repeated the experiment starting with 500 kg (!) of yttrium-group oxalate and announced in 1907 in an Austrian journal that, indeed, there was a new element, cassiopeium. In parallel, Urbain applied his own separation method and reached the same conclusion, proposing lutetium as the name for the new element. Although Urbain's manuscript was published after Auer's one, it was sent to the editor about 40 days earlier and, therefore, he won the priority dispute. German chemists did not accept this verdict and kept the name cassiopeium until 1945.

What about the missing element 61? Its existence, as well as that of three other elements ($Z = 43, 72, 75$), was postulated by Henry Moseley (1887–1915, killed at the battle of Gallipoli in Turkey). Moseley studied the X-ray spectroscopy of the elements and stated that the square of the frequency of the K_α line is proportional to the atomic number, a concept he invented. After a couple of false discoveries, promethium was found in fission products of uranium in 1945 by the team of Jacob A. Marinsky (1918–2005) at Oak Ridge National Laboratory. This research was classified and this finding was only published in 1947. This explains why this date is often retained as the discovery date. Later, it turned out that traces of promethium can be found in nature, in some uranium-containing ores. Only minute quantities of promethium are available and hence its properties are not well known. For instance, an accurate value of its ionization energy (5.58188 eV) could be determined only in 2019 [11].

1.1.6 Concluding remarks

There are two lessons to learn from this fascinating story of the discovery of rare-earth elements that may be useful to young researchers. The first one is that many of the major discoveries have been achieved thanks to an acute sense of observation: a slight change in color, a marginally different chemical reactivity or solubility or a faint line in an optical spectrum carefully described in the laboratory notebook were hints that something unusual was happening. This, combined with inventive thinking and very patient and arduous work, was required to achieve one's goal and gain recognition. The second one is that the best attitude for a scientist is, in addition to rigor, modesty and openness to discussion; a scientist should keep geopolitical factors and nationalistic feelings away from scientific investigations.

References

[1] F. Szabadvary, The History of the Discovery and Separation of the Rare Earths, in K. A. Gschneidner Jr. and L. Eyring, Handbook on the Physics and Chemistry of Rare Earths, North Holland Publ. Co., Amsterdam, 1988, Vol. 11, Ch. 73, pp 33–80.
[2] J. L. Marshall, Discovery of the Elements, a Search for the Fundamental Principles of the Universe, Pearson Custom Publishing, Boston, 2002.
[3] B. R. Geijer, Crells Chem. Ann. 1788, 229.
[4] J. Gadolin, Kungl. Sven. Vetenskapsak. Handl. 1794, Ser. 2, 15, 137.
[5] J. Gadolin, Crells Chem. Ann. 1796, 313.
[6] W. Hisinger, J. J. Berzelius, N. Allgem. J. Chem. (Gehlen J. Chem.) 1803, 2, 397.
[7] M. H. Klaproth, N. Allgem. J. Chem. (Gehlen J. Chem.) 1803, 2, 303.
[8] H. E. Roscoe, A. Schuster, J. Chem. Soc., Trans. 1882, 41, 283.
[9] J. Emsley, The Elements, Clarendon Press, Oxford, 1998.
[10] P. E. LeCoq de Boisbaudran, Spectres Lumineux, Spectres Prismatiques Et En Longueurs D'Ondes Destinés Aux Recherches De Chimie Minérale, Gauthier-Villars, Paris, 1874.
[11] D. Studer, S. Heinitz, R. Heinke, P. Naubereit, R. Dressler, C. Guerrero, U. Köster, D. Schumann, K. Wendt, Phys. Rev. A 2019, 99, 062513.

Annex. Important personalities in the discovery of rare earths

C.A. Arrhenius
gadolinite

B. R. Geijer
gadolinite

J. Gadolin
yttrium

A. G. Ekeberg
(yttria)

A. F. Cronstedt
cerite

J.J. Berzelius
cerium

W. Hisinger
cerium

M.H. Klaproth
cerium

C.G. Mosander
lanthanum, erbium, terbium

H.E. Roscoe
(Tb spectrum)

M. Delafontaine
holmium

J.-L. Soret
holmium

P. T.Cleve
holmium
thulium

J. Galissart de Marignac
gadolinium, ytterbium

L. F. Nilson
scandium

P.E. Lecoq de Boibaudran
samarium
dysprosium

C. Auer von Welsbach
neodymium,
praseodymium

E.A. Demarçay
europium

G. Urbain
lutetium

J.A. Marinsky
promethium

Photographic credits: Royal Swedish Archives: 1, 2, 4, 6, 7, 9, 16; Archetron (creative commons): 5; Wikipedia commons: 8, 10–13, 15–18; Abe Books (U.K.): 14; L'Usine Nouvelle: 19.

Tom Lorenz, Martin Bertau, Robert Möckel

1.2 Rare-earth minerals and rare-earth mining

Within the first 150 years of their known existence, from 1787 until the 1940s, rare earths were solely of scientific importance. There were no applications involving these elements, and since rare earths were found only in rare minerals in low concentrations, there was no mining either. This changed during the Manhattan project (1942–46) and the upcoming nuclear technology, as rare earths gained first attention as fission products of U. Due to their capability to absorb neutrons (^{157}Gd, ^{147}Sm), increasingly effective separation techniques had been developed, for example, ion exchange and solvent extraction [1]. In the mid-1980s, the first rare-earth-containing magnets were available to the market; it expanded their application massively as most electric devices contain at least one element of this group. Since then, rare earths have become increasingly important for many high-tech products, and this trend will certainly continue.

1.2.1 Economic importance and applications

Nowadays, rare-earth magnets can be found in all electric engines, speakers, wind turbines, smartphones and hard disk drives. Beside magnets, rare earths constitute an essential component of many high-tech products, such as metal alloys, illuminants, glasses, ceramics, polishing agents and catalysts (Figure 1.2.1). Among others, they are also used in medical diagnostics, for example as contrast agents in magnetic resonance imaging. It is due to their unique properties that they can hardly be substituted by other elements.

Between 2006 and 2012, the demand increased in all seven fields. In case of magnets, twice the amount of rare earths had been consumed in 2012. The trend changed within recent years, as five fields of application recorded lower demands. Polishing agents kept their level at approximately 22,000 t (2017). Only magnetic alloys experienced a positive trend, which is expected to be continued until 2020 [3]. The particular rare earths are distributed unevenly among the seven fields. Consequently, every element is influenced to a different degree by the trends of the market. Table 1.2.1 provides an overview of the most important applications for every element.

Rare earths that serve as components of magnetic alloys especially benefit from increases in demand. This applies in particular for Nd and Sm, because both metals act as main components of the most common and strongest permanent magnets (FeNdB and SmCo alloys). Additives like Pr, Tb and Dy are also affected

https://doi.org/10.1515/9783110654929-002

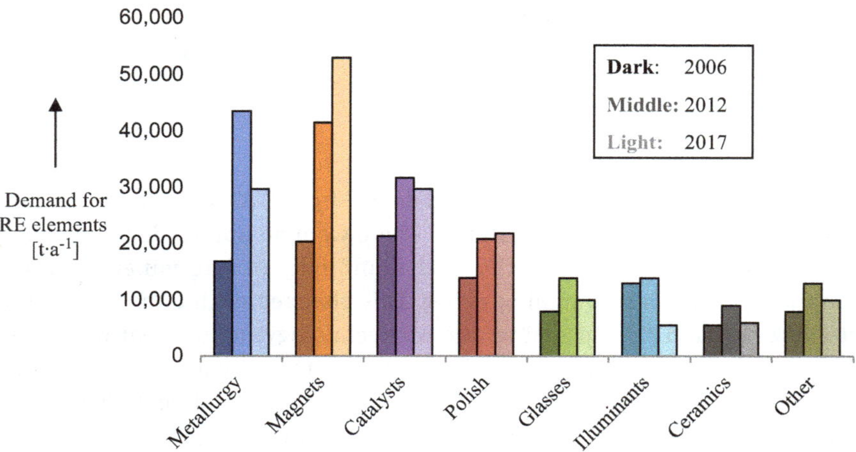

Figure 1.2.1: Fields of application and demand for *RE*s in 2006, 2012 and 2017 (annual average of [2]).

as they achieve contents of up to 8.2 wt%. Rare earths that are used as phosphors in illuminants will probably suffer from decreases in demand, because fluorescence lamps are increasingly replaced by light-emitting diodes (LEDs). NiMH accumulators, in which the anode is made of a rare-earth alloy, will also decrease in demand as Li batteries in hybrid electric vehicles (HEVs) replace them. For polishing agents and catalysts, the demand will remain roughly the same. Despite all market trends, all 17 elements have one feature in common: their demand is solely secured by primary raw materials. Since the 1980s, China has been the largest producer of rare earths in the world. Figure 1.2.2 provides an overview of how production capacities have changed over the years.

From 1994 until 2005, the Chinese share of world annual production increased from 47.5% to 96.9%, as many companies withdrew from the market. Meanwhile, China had significantly increased its production capacity, especially between 2002 (75,000 $t{\cdot}a^{-1}$) and 2005 (119,000 $t{\cdot}a^{-1}$). This led to an almost-monopoly position. In 2009, China produced approximately 120,000 t of rare-earth oxides, whereas the world's reserves had been estimated to be 99 million tons, of which 36.4% were located in China [9]. The situation has changed only slightly, since. In 2018, production amounted to 170,000 t while resources have increased to 116 million tons, due to newly discovered deposits (Figures 1.2.2 and 1.2.3) [10]. China's share in rare-earth production capacity decreased to 71.4%, which, however, still constitutes the predominant part. Nevertheless, there is a difference of three orders of magnitude between rare-earth reserves (116 million tons) and production (180 kt). By hypothesizing static consumption, it becomes obvious that rare-earth supply should be secure for centuries to come, yet rare-earth metals are considered critical.

Table 1.2.1: Application of all 17 *RE* elements in the fields of (a) metallurgy, (b) magnets, (c) catalysts, (d) polish, (e) glasses, (f) illuminants and (g) ceramics. [4–7].

Element	Application Examples	a	b	c	d	e	f	g
Scandium	Light alloys with Al and Mg *Traces*: catalysts, high pressure vapor lamps	x		x		x		
Yttrium	Alloys for electrodes, Y_2O_3:Eu-phosphor *Traces*: FeNdB magnets, YAG laser	x	x				x	x
Lanthanum	$LaCo_5$ magnets, glasses with high index of refraction, polish, catalyst additive in FCC process, LAP-phosphor	x	x	x	x	x	x	
Cerium	Additive for Al und Fe alloys, polish for glasses, support for automotive catalysts	x			x	x	x	x
Praseodymium	UV-absorbing glasses, Pr-Fe-Co magnets		x			x		
Neodymium	FeNdB magnets, colored glasses (violet)		x	x		x		
Promethium	*Traces*: permanent phosphor (^{147}Pm), radio nuclide batteries (^{147}Pm)						x	
Samarium	$SmCo_5/Sm_2Co_{17}$ magnets, IR-absorbing glasses, catalysts for hydrogenation		x	x		x		
Europium	Phosphors (SCAP, BAM, Y_2O_3:Eu), safety color (50 € banknote)						x	
Gadolinium	Gd-Fe-Co alloys, MRT contrast agent *Traces*: CRT phosphor (Gd, Ce, Tb)MgB_5O_{10}	x					x	
Terbium	FeNdB magnets, Tb-Fe-Co alloys, phosphor, safety color (100 € banknote)	x	x					
Dysprosium	FeNdB magnets, Tb-Dy alloys, special glasses	x	x			x		
Holmium	Ho-Fe-Ni-Co alloys for magnetic bubble memory, special glasses *Traces*: doping in YAG laser	x				x		
Erbium	IR-absorbing glasses. *Traces*. Er:YAG laser					x		
Thulium	*Traces*: activator in phosphors, special glasses					x	x	
Ytterbium	$Yb_2Co_{13}Fe_3Mn$ magnets, steel additive *Traces*: Yb:YAG laser	x	x					
Lutetium	*Traces*: Scintillator material, Lu:YAG for LEDs						x	

More than availability of physical resources, uneven allocation of production capacities causes the actual risk of supply. Hence, China's local trade policy is capable of largely affecting the global market, which explains what happened between 2009 and 2011: Due to increased demand, China started to reduce export contingents, in steps, from 50,145 t of rare-earth metals (2009) to 31,130 t (2012) [12]. Simultaneously, environmental issues related to illegal mining were tackled by licensing rare-earth exports and levying taxes [13, 14]. Packey and Kingsnorth estimated that 40-50% of China's rare-earth production originated from illegal mining, at this time [15]. Therefore, the Chinese Ministry of Trade reduced the

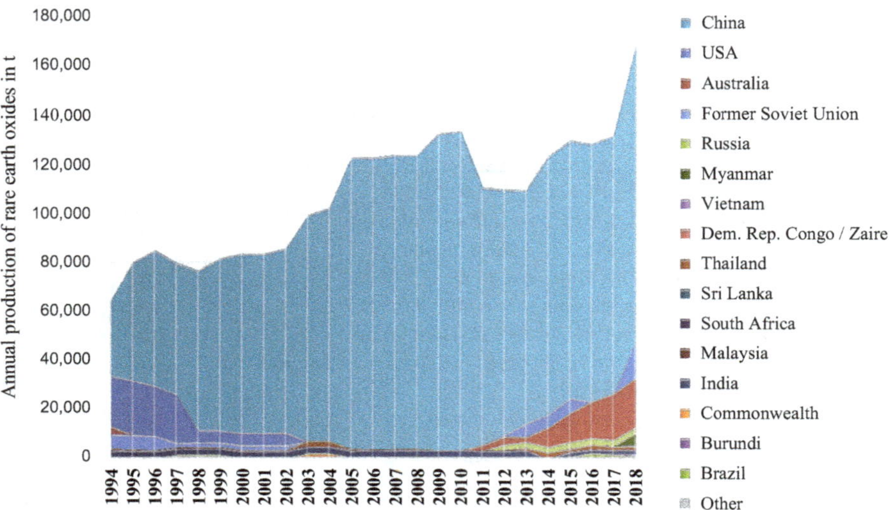

Figure 1.2.2: Annual rare-earth production in tons per year from 1994 until 2018 [8–11].

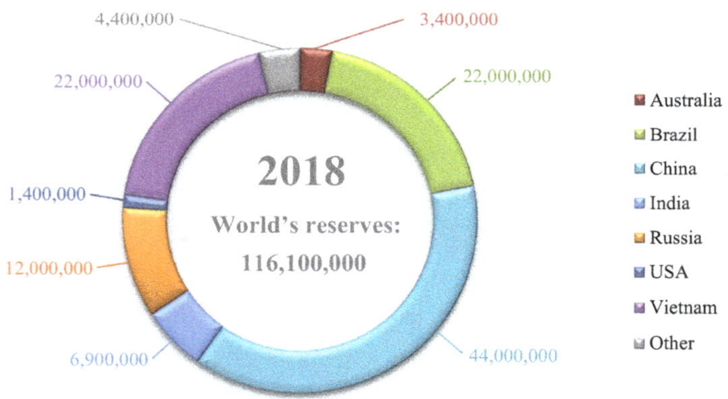

Figure 1.2.3: Estimated rare-earth reserves in 2018 (in tons of rare-earth oxides) [11].

number of companies that were allowed to export rare-earth products to 28 in 2013 [16]. Companies, which did not entirely comply with environmental regulations lost parts of their granted export contingents [14]. Consequently, exports could not even meet the reduced contingent set by the Chinese government. For all these reasons, rare-earth commodity prices rose rapidly reaching their all-time maximum in 2011 (Figure 1.2.4).

In particular, Nd, Dy, Tb and Eu were affected by the price rise (Figure 1.2.4). These elements are essential for producing permanent magnets, phosphors and

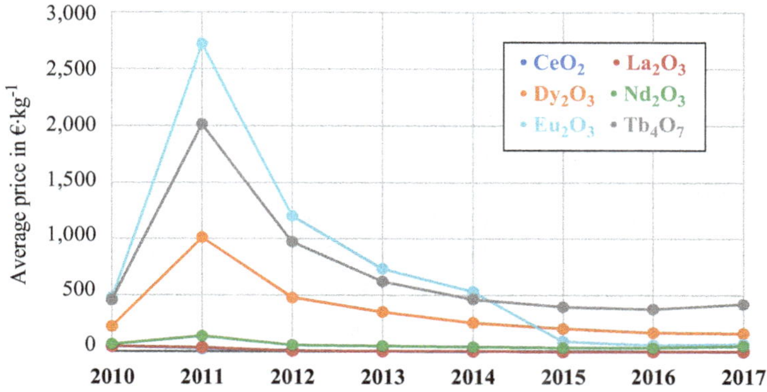

Figure 1.2.4: Average *RE* oxide prices from 2010 to 2017 [10, 14].

special lenses in high-tech products. In order to reduce the dependency on rare-earth imports, companies and countries started extensive R&D programs focusing on exploration, substitution and recycling.

As a result, numerous processes and techniques have been developed allowing for recycling rare earths from different waste streams such as spent $Fe_{14}Nd_2B$ magnets, fluorescence lamps or NiMH accumulators [17, 18–20]. Despite the fact that all processes focus on a more sustainable use of rare-earth elements, almost none of them has so far been realized on industrial scale. In the course of falling commodity prices, almost all recycling processes relying on end-of-life (EOL) waste became uneconomical. Compounds like Y_2O_3, Eu_2O_3 and Dy_2O_3 lost more than 70% of their value until 2013. The same happened to primary-rare-earth production. In 2010, Molycorp reopened the Mountain Pass mine in California (USA) to exploit the local rare-earth-rich bastnäsite and monazite deposits. Five years later, Molycorp had to stop mining and went bankrupt, as the winning process had already become unprofitable. Without Mountain Pass, the USA were taken off the list of the rare-earth producing countries. In 2017, Mountain Pass was reopened again. The company, MP Materials, that was founded by an American-Chinese consortium, runs the mine. Consequently, rare-earth production in the USA continued in 2018. Only a few exceptions managed to implement their recycling processes on the market. For instance, in 2009, Umicore and Solvay started a joint venture in which Solvay recovered rare-earth oxides from slags that Umicore obtained as by-product in the process of recovering Ni and Co from NiMH accumulators [19, 20]. Nevertheless, the original aim of being less dependent on rare-earth imports from China has been missed throughout the world. The development of economic processes has become much more difficult, as they have to compete with large-scale production facilities mining rare earths in open-pit mines, while struggling with commodity prices that have returned to the level of 2010 (Figure 1.2.4).

1.2.2 Primary raw materials

Contrary to what their name suggests, rare earths are by far more common than their discoverers had expected. The content of La (39 mg kg^{-1}) in the Earth's crust is about twice as high as the content of Li (20 mg kg^{-1}) and about 4 orders of magnitude higher than gold [21]. In the course of the last century, large deposits that could secure global demand for many centuries were discovered. What is comparatively rare is a concentration up to an economic feasible level. Most common minerals incorporate a certain level of rare-earth elements – the most prominent of these might be quartz, which can, although considered as one of the purest minerals, contain up to a few hundred ppm of rare earths [22]. For ore and mineral examinations, the material is compared to several normalization standards. Mostly the C1 chondrite standard is used to represent matter from an early, undifferentiated solar nebular. Therefore, these normalized patterns reveal an enrichment (ratio of an element in a rock to C1) or a depletion (ratio <1), compared to the C1. Comparing these patterns gives hints on the genesis of a certain mineral, for example, on the aforementioned quartz [22]. On the other hand, it also reveals important information about the ratio of heavy-rare-earth elements (HREEs) to light-rare-earth elements (LREEs), that is of crucial economic interest.

There are about 300 known minerals that contain significant rare earth amounts – meaning that the rare-earth elements compellingly appear in the mineral's composition [23]. Some of these minerals are rather rare (this time for real) and were found in single locations only; others may occur more frequently and in relevant amounts, forming deposits. Among these are the minerals bastnäsite, monazite and xenotime and a few others of rather minor importance (see Table 1.2.2). Due to the fact that it is not a mineral in sensu strictu, the so-called ion adsorption clays (IACs) are missing in the table, although they probably play the most important role in rare-earth production. They contain rather low total contents of rare earths (typically <0.5% total rare-earth oxide, TREO), but since the rare-earth ions are adsorbed on the surface of (clay) minerals like kaolinite and muscovite, they can be extracted easily. Consequently,

Table 1.2.2: The economically most important rare-earth minerals.

Mineral name	Formula
Bastnäsite	$(La,Ce,\mathit{RE})[CO_3](OH,F)$
Monazite	$(La,Ce,\mathit{RE})[SiO_4]$
Xenotime	$(Y,\mathit{RE})[SiO_4]$
Eudialyte	$Na_4(Ca,Ce)_2(Fe,Mn,Y)Zr[Si_8O_{22}](OH,Cl)_2$
Churchite	$Y[PO_4]_2 \times H_2O$
Loparite	$(Ce,Na,Ca)_2(Ti,Nb)_2O_6$
Fergusonite	$YNbO_4$
Allanite	$(Ce,Ca,Y)_2(Al,Fe)_3(SiO_4)_3(OH)$

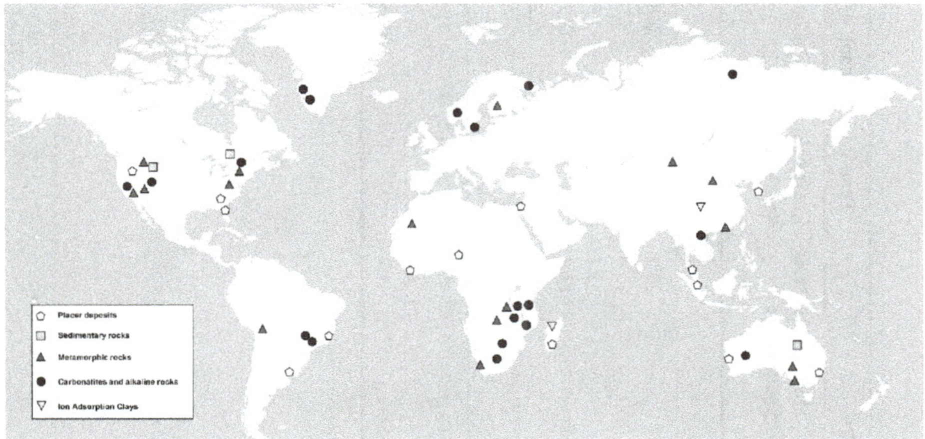

Figure 1.2.5: Locations of rare-earth deposits (modified from [7, 24]).

they are of high economic interest. Figure 1.2.5 shows where the most important deposits in the world are located. Rare earths are predominately mined from four different minerals, namely, loparite, monazite, bastnäsite and xenotime (Table 1.2.3).

Monazite and xenotime are minerals consisting of rare-earth phosphates ($REPO_4$). Monazite mainly contains light-rare earths (Sc, La-Eu), while xenotime consists of heavy-rare-earth phosphates (Y, Gd-Lu). Bastnäsite $RE_2(CO_3)_3(OH,F)$ is a rare-earth carbonate with a high content of fluoride. Loparite $(RE,Na,Ca)(Ti,Nb)O_3$ is an oxide mineral, which belongs to the group of niobates [25]. Bastnäsite and loparite are, like monazite, a source material for light rare earths. Since the 1980s, ion-adsorbing clays have completed the list [5]. Nowadays, heavy rare earths are mainly mined from these clays. Furthermore, they are also an important source for Sm, Pr and Nd (Table 1.2.3) [26]. The main deposits are located in Inner Mongolia, where ores are enriched with light rare earths, and Longnan, where clays particularly contain heavy rare earths. Thus, China possesses the most economically important deposits in the world [6]. Other extensive deposits are situated in Russia, Brazil, India, the USA and Australia. Furthermore, there are large reserves in Greenland (Kvanefjeld and Kringlerne) and the Tomtor region in Siberia (Figure 1.2.5). They cannot be mined profitably, because the infrastructure is missing and the climate is harsh (permafrost soil).

The rare-earth elements in these minerals differ in terms of total contents and their ratios as mentioned above, driven by fractionation within the rare-earth series. Xenotime, for example, reveals a higher HREE:LREE ratio than monazite and bastnäsite, represented by a positive slope in the normalized diagrams (Figure 1.2.6). The HREE are therefore underrepresented in most of the minerals (and deposits), resulting in higher demand and consequently higher prices, (Figure 1.2.6), as they are

Table 1.2.3: Rare-earth contents of ores from selected deposits [5, 6].

	Bastnäsite		Clays		Monazite			Loparite	Xenotime
Rare-earth oxides	Bayun Obo, China	Mountain Pass, USA	Xunwu, China	Lognan, China	Mt. Weld, Australia	Guandong, China	Steenkampskraal, South Africa	Lovozersky, Russland	Malaysia
LREO*									
La_2O_3	23.0	33.2	43.3	1.8	25.5	23.0	20.7	28.0	0.5
CeO_2	50.0	49.1	2.4	0.4	46.7	42.7	45.8	57.5	5.0
Pr_6O_{11}	6.2	4.3	9.0	0.7	5.3	4.1	5.1	3.8	0.7
Nd_2O_3	18.5	12.0	31.7	3.0	18.6	17.0	17.7	8.8	2.2
Eu_2O_3	0.2	0.1	0.5	0.1	0.6	0.1	0.1	0.1	0.2
HREO*									
Tb_4O_7	0.1	<0.1	<0.1	1.3	0.1	0.7	0.2	0.1	1.0
Dy_2O_3	0.1	<0.1	<0.1	6.7	0.2	0.8	1.0	0.1	8.7
Y_2O_3	<0.1	0.1	8.0	65.0	0.4	2.4	4.2	<0.1	60.8

*Light-rare-earth oxides (LREO) and heavy-rare-earth oxides (HREO)

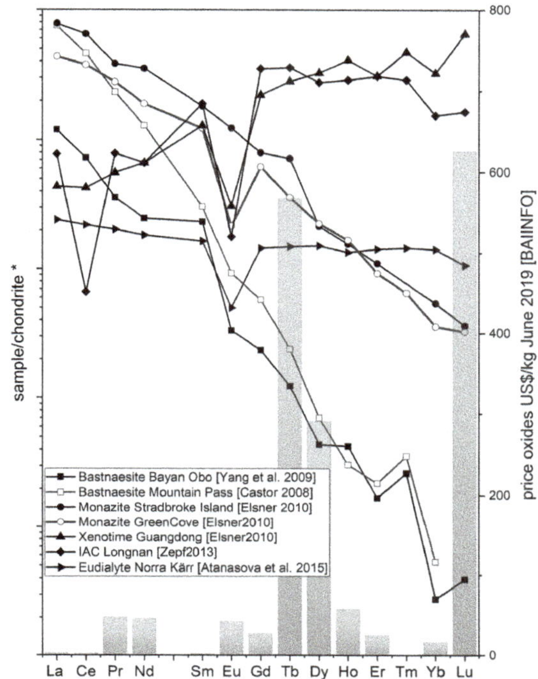

Figure 1.2.6: Rare-earth minerals from different deposits showing different HREE:LREE ratios. The HREE with lower numbers of total production are therefore the most expensive ones (*please note that absolute values can be ignored, as displayed values do not represent total contents) [27].

currently irreplaceable in some technologic applications (e.g., magnet production, fiber optics and phosphors).

Although there is no general consensus to be found in the literature, rare-earth deposits (or occurrences) can be classified roughly into three main categories: (1) primary, magmatic deposits with initially high contents; (2) secondary deposits with an enrichment due to “erosive processes” and (3) metamorphic deposits with an enrichment due to “additional magmatic” processes (Figure 1.2.7). Of course, this classification is not conclusive, and since no two rare-earth deposits are alike, a more detailed view on each single occurrence is necessary to categorize it sufficiently.

Primary deposits are basically magmatic (igneous) rocks, with initially high amounts of rare-earth elements. Usually, these rocks are formed out of highly evolved, specialized melts from both mantle and crustal sources (Figure 1.2.8A). Carbonatites, rocks with more than 50% carbonate mineral content, are probably the most important ones, although alkaline to peralkaline rocks (where $Al_2O_3 < Na_2O + K_2O$) can also contain significant rare earth amounts.

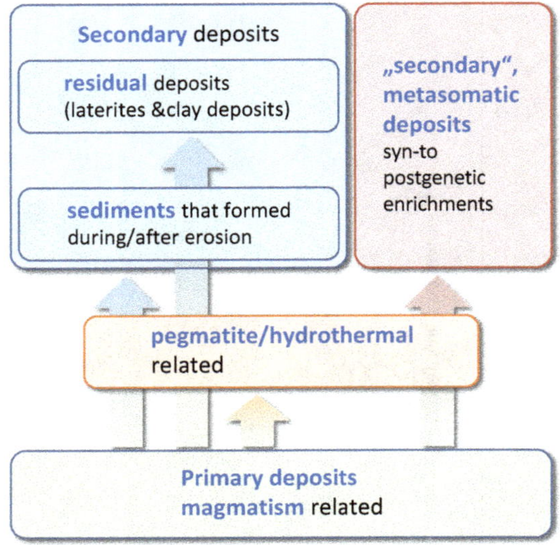

Figure 1.2.7: Classification of natural rare-earth deposits.

These enrichments are uneconomic to exploit (an exception is carbonatites with their economically high-rare-earth contents). Typical minerals in these rocks are the *RE*-(fluoro)carbonates (bastnäsite and usually, to a much lesser amount, parisite and synchisite), monazite, apatite and loparite. The carbonatites typically show an enrichment in LREE. The carbonatitic Mountain Pass deposit, California, USA, is a well-known rare-earth deposit, which has been mined during various times. Other examples are the Lofdal complex (Namibia), Mount Weld (Australia) and the Bayan Obo district (China). There is currently only one noncarbonatitic magmatic mine producing rare earths, that is in the Lovozero syenitic complex (Russia), although other occurrences are under investigation (e.g., Norra Kärr, Sweden; Illimaussaq, Greenland) [28]. Within these rocks, the mineral eudialyte is a typical rare-earth carrier, with a higher HREE:LREE ratio than, for example, bastnäsite (Figure 1.2.6). Related to Al_2O_3-undersaturated rocks are pegmatites with partly very high contents of rare-earth minerals, of minor economic interest though, since the sizes of these bodies are typically small. Carbonatitic vein structures are typically small in tonnage as well, but may have high grades. Examples are Steenkampskraal, South Africa, and the Gakara deposit in Burundi [28].

A so-called metasomatic influence may play an important role in the aforementioned occurrences (Figure 1.2.8 B). It is characterized by an alteration and/or metamorphism by magmatic fluids that can happen after or during the emplacement of the initial magmatic body, called post- or syngenetic, respectively. Those fluids may not only increase the REE significantly, but also the other elements like Zr, Nb, Hf and

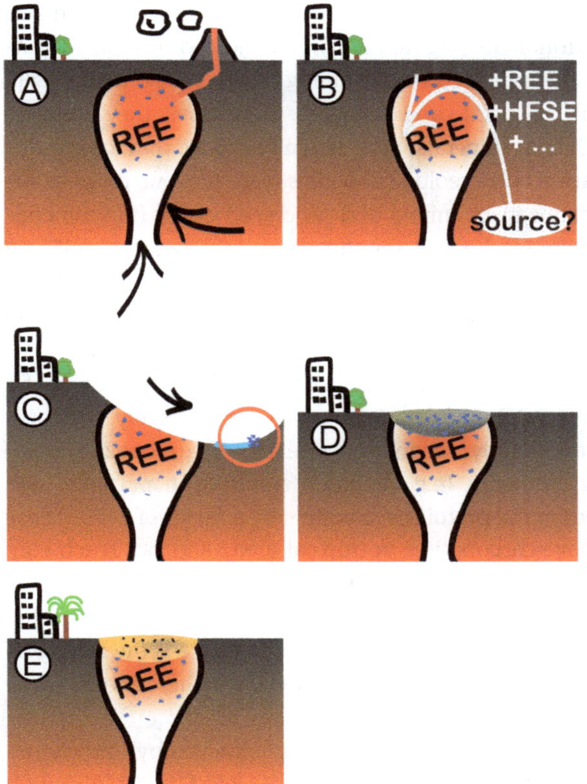

Figure 1.2.8: Geologic mechanisms that result in the formation of rare-earth deposits, which can be classified into (A) magmatic, (B) metasomatic/metamorphic, (C) (fluvial) placer, (D) lateritic and (E) ion adsorption clay deposits.

even U and Th. These elements belong to the so-called high field strength elements (HFSE) and some of them are of additional economic interest. Minerals that appear in such environments are zirconium-silicates like zircon and gittinsite, as well as oxides like pyrochlore. Also, the HREE:LREE ratio may change during the metamorphism. In many cases, which is unfortunate for every mineral processing engineer, the mineralogy changes likewise, increasing the complexity by adding or transforming REE-minerals, increasing the number of valuable minerals and changing grain sizes and liberation. The processing costs are therefore high, which is the reason why none of these occurrences are currently mined. The Strange Lake deposit, Canada and the Khalzan Buregte area (Mongolia) are typical examples for these kinds of deposits [29]. Of course, the metamorphic impact is hard to quantify, and sometimes the genesis of deposits is still under discussion.

Placers, belonging to the secondary deposits are natural, gravimetric enrichments in fluvial, marine and aerial environments (Figure 1.2.8 C). Well-known placers include enrichments of Au, Sn and diamonds; rare-earth-containing minerals also tend to separate gravimetrically, for example, monazite and xenotime. Sometimes, they occur together with other (industrial) heavy minerals, like ilmenite, rutile and cassiterite. Primary sources of the placers are eroded geological bodies (e.g., igneous rocks as described above). In general, placers can be mined and processed easily (the minerals are already separated), but since monazite and xenotime may incorporate significant amounts of U and Th, their contents are high in placers as well, and that may be a major drawback. REE-enriched placers are known to occur in India, Australia and Turkey [28].

Lateritic rare-earth deposits are basically the product of chemical weathering of a rare-earth rich protolith (Figure 1.2.8 D). Resistant minerals are relatively enriched, compared to the initial rocks. Additionally, secondary minerals form (monazite, xenotime) at the expense of primary minerals (e.g., apatite), resulting in partially high grades (up to 25% TREO), exceeding the protoliths' contents by a factor of 5–20. These laterites form predominantly above carbonatites in tropical areas and can reach considerably large thicknesses. The Mt. Weld carbonatite (the protolith), for example, is covered by an up to 70 m thick weathering zone, enriched in REE bearing monazite, churchite and other secondary minerals [30]. Another typical example is the Dong Pao district in Vietnam, with more than 60 known lenses of heavily weathered rock, seven of which are currently of economic interest [31].

The so-called IAC deposits evolve in humid, subtropical climates by (bio)chemical weathering of predominantly granitic protoliths (Figure 1.2.8 E) [32, 33]. Although the mechanisms of formation are not totally clear, rare-earth minerals are naturally leached, and the rare-earth cations are adsorbed onto the surface of minerals from the clay fraction, namely, kaolinite, halloysite, illite and mica. Despite the fact that the contents are relatively low (<0.5% TREO), IACs play a major role, mainly due to two facts: (1) REEs can be easily extracted by leaching, in many cases even in-situ heap leaching and (2) the high percentage of HREE, matching major industrial demand [32, 33]. IACs are currently mined solely in southern China, but other deposits are under development, e.g., in Madagascar and Brazil.

In addition to rare-earth minerals, ores contain gangue and accessory minerals, that consist mostly of silicates and oxides, for example, zircon ($ZrSiO_4$), rutile (TiO_2), ilmenite ($FeTiO_3$), cassiterite (SnO_2), and carbonate minerals like calcite ($CaCO_3$) and dolomite ($CaMg(CO_3)_2$). Furthermore, U and Th usually emerge as impurities. Monazite from Australia typically contains 4–8wt% Th and 0.01–0.1 wt% U. The same applies for deposits in India (8–10 wt% Th), the USA (0.02 wt% Th, 0.002 wt% U) or Greenland [5, 24, 34]. Both elements require additional treatment, as they have to be separated from rare-earth elements by precipitation and have to be treated as hazardous waste.

1.2.3 Secondary raw materials

End-of-life products (EOL): Disused consumer goods that are inoperative.
End-of-use products (EOU): Functional consumer goods that have been replaced by new goods, for example smartphones, TVs and Tablet-PCs.
Production waste: Unavoidable by-product that has to be disposed or recycled if possible, for example, indium tin oxide in the production of touchscreens

Any waste material that enables the production (recycling) of a raw material is considered a secondary raw material. In principle, potential feedstock for future rare-earth recycling comprises many different waste materials, due to their versatile applications. Rare-earth compounds can be found in fluorescence lamp scraps, spent speakers, broken wind turbines, disassembled electric engines, disposed polishing agents or deactivated catalysts from styrene production. In general, there are three different kinds of waste that are considered secondary raw materials: EOL products, end-of-use (EOU) products and production waste. The third is predominantly recycled during normal operation at the production site. EOL- and EOU products are usually disposed of according to regulations specified by law, that does not distinguish between these kinds of waste. Therefore, both generally contribute to the same waste stream. The ideal waste material should provide high-rare-earth contents and low amounts of impurities. It should be easy to collect and available in large quantities. Frankly, there is no waste material satisfying all these criteria. Secondary raw materials usually amount de-centralized. It depends on how effectively waste collection is organized if large amounts are available. Furthermore, most waste materials contain rare earths in quantities that are, by far, too small for efficient recycling, for example, YAG-lasers, LEDs, liquid crystal displays (LCD) or catalysts. It is conceivable that these rare earths can only be recycled as by-products, while focusing on different valuable metals. Even if the waste material contains components providing high-rare -earth contents, disassembling is often too expensive. Instead of sorting, most waste materials are crushed. This procedure significantly reduces the chances of rare earths being recycled economically. NiMH accumulators and rare-earth magnets are particularly affected by this problem. While hypothesizing sufficient separation, the next challenge already awaits, as secondary raw materials usually contain impurities that may largely affect process economics. For instance, $Fe_{14}Nd_2B$ magnets consist of >60 wt% Fe. Therefore, when $Fe_{14}Nd_2B$ magnets are introduced to acid leaching, Fe and rare earths are dissolved at once. The impurity, Fe, causes acid consumption to increase by up to five times. Compared to primary production and the rare earth-rich ore concentrates utilized therein, recycling experiences considerable disadvantages when processes involve exactly the same steps of treatment. Adjustments are inevitable to compensate for the obstacles mentioned. Numerous scientific articles deal with the topic of improving process economics of rare-earth recycling. However, in the course of commodity prices dropping back to the levels they were in 2010, most recycling approaches relying on EOL material became uneconomical and, consequently, rare-earth recycling rates of EOL waste still remain below 1% [35]. Recycling strategies have to make a compromise as industrial processes can only take effect when there is more sustainable resource management and as long as they are economical. Recent approaches consequently focus on secondary raw materials with the highest rare-earth contents available. Four waste materials, in particular, are currently under close examination.

(i) $Fe_{14}Nd_2B$ magnets contain up to 35 wt% rare earths. As permanent magnets in wind turbines and electric vehicles, rare earths are essential to clean energy technologies [36]. Rotor segments of only one 3MW wind power plant use not less than 500 kg of these magnets (Figure 1.2.9) [37]. $Fe_{14}Nd_2B$ alloys are also preferred for hard-disc drives, microphones and speakers, because of their high magnetic energy densities $(BH)_{max}$ amounting to ~ 450 kJ·m^{-3}. Based on rare-earth demand (Figure 1.2.1) and by estimating that rare-earth content achieves an average of 30 wt%, current annual production roughly amounts to 180,000t of rare-earth magnets.

Figure 1.2.9: Spent $Fe_{14}Nd_2B$ magnets (5 × 3 × 2 cm^3) originating form a wind turbine.

(ii) $SmCo_5$ and **Sm_2Co_{17} magnets** are second to $Fe_{14}Nd_2B$ in magnetic energy densities (BH_{max}=160–260 kJ·m^{-3}) and costs of production, but high Curie temperatures of 700–850 °C allow their application above 240 °C [6, 38]. Subsequently, they are used in special applications like electric engines in motor sports. SmCo alloys consist of ~33% rare-earth metals, among which, Sm is the main component. However, less than 2% of spent rare-earth magnets consist of SmCo alloys, which currently amount to ~2,000–3,000 t·a^{-1} [12]. It is unlikely that the quantity stated would justify rare-earth recycling. Moreover, SmCo alloys are not collected separately. They typically appear as trace metals in $Fe_{14}Nd_2B$ magnet scrap. A combined treatment of all rare-earth-containing magnets, therefore, seems to be the best option. In this regard, SmCo recycling shares the fate of $Fe_{14}Nd_2B$ magnets. Besides Sm and Nd, additional rare earths can be part of these magnets. Pr sometimes substitutes Nd, while Dy and Tb improve temperature stability of $Fe_{14}Nd_2B$ magnets. For SmCo alloys, Gd and Ce increase temperature stability and corrosion resistance. Thus, magnet scraps always contain a mixture of rare earths.

(iii) Three-band phosphors are a kind of luminescent material used in fluorescent lamps. Figure 1.2.10 depicts its basic structure. As a white layer on the inside of the glass mantle, it is their function to transfer UV light emitted by the Hg-plasma inside into visible light. The phrase "three-band" refers to the fact that there is always a mixture of phosphors with either a blue, green, or a red emission spectrum. The combination of phosphors and their respective spectra gives the color of light and, eventually, of the fluorescent lamp. Table 1.2.4 provides an overview of the most common three-band phosphors.

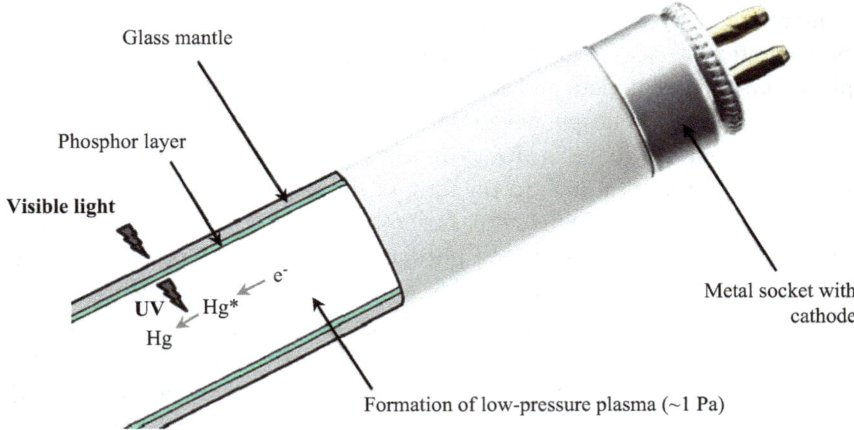

Figure 1.2.10: Structure and function of fluorescence lamps.

In general, phosphors consist of two parts: the grid and at least one activator that adjusts the emission spectrum. Activators are usually used in small amounts, since grids are only doped with them. In three-band phosphors, rare earths mostly serve as activators that would require small amounts of rare earths. Y and La are exceptions, because both act as the grid of a red and a green phosphor. Y, especially, is used in large quantities, since there is only one red phosphor available. It is for this reason

Table 1.2.4: Selected three-band fluorescent materials.

Fluorescent material		Emission color
$Y_2O_3:Eu^{3+}$	(YOE)	Red
$La(PO_4):Ce^{3+},Tb^{3+}$	(LAP)	Green
$(Ce^{3+}, Tb^{3+})MgAl_{11}O_{19}$	(CAT)	Green
$(Gd^{3+}, Ce^{3+}, Tb^{3+})MgB_5O_{10}$	(CBT)	Green
$BaMgAl_{10}O_{17}:Eu^{2+}$	(BAM)	Blue
$(Ca, Sr, Ba)_5(PO_4)_3Cl:Eu^{2+}$	(ScAp)	Blue

that three-band phosphors are much more expensive than the commonly used halophosphate phosphor $Ca_5(PO_4)_3(F,Cl):Sb^{3+},Mn^{2+}$. For decades, the halophosphate has been the dominant phosphor in fluorescence lamps. The situation changed in 2009, when the EU passed the regulation EG 245 introducing higher standards for light quality [39]. The rare-earth-free halophosphate phosphor is unable to meet the new requirements. In order to comply with the new regulation, suppliers of fluorescence lamps, like Osram and Philips, were obliged to use three-band phosphors. At first, the change affected the composition of production waste, the rare-earth content of which increased considerably. Though EOL fluorescent lamp scraps amount to much lower contents, their rare-earth contents are increasing and achieve partially more than 7 wt%. EOL wastes always contain small amounts of Hg, which has to be separated by vacuum distillation. These wastes are always a mixture of different phosphors, pieces of glass and some metal parts originating from the sockets. The content of three-band phosphors varies along a wide range. The same applies to the rare-earth content. All six REEs always appear together: 70–75% thereof account for Y that occurs jointly with Eu^{3+}, in the form of the red-emitting fluorescent substance $Y_2O_3:Eu^{3+}$. Gd (2–3%), Ce (8–10%), Eu^{2+} (4–6%) and Tb (4–6%) are essential components of green- and blue-emitting phosphors. La (5–10%) solely appears as green-emitting $LaPO_4:Ce^{3+},Tb^{3+}$ [18, 40]. As for availability, fluorescence wastes are second to none. Hg requires special treatment for disposal, and fluorescence lamps are therefore collected separately at recycling depots. The disadvantage of the need for special treatment proves to be an advantage, in terms of availability. EOL phosphors appear as white, fine-grained powder, because the phosphor layer is blown out of the glass mantle after the sockets have been cut off. Production waste usually amounts as aqueous slurry, which originates from impregnating the inside of the glass mantle with the phosphor layer. A white powder is gained when water is removed by drying. The main problem of fluorescence wastes is low-rare-earth commodity prices. Rare earths are mostly traded as oxides. The average price for 1 kg Y_2O_3 is 3.3 US$ (purity 99%; FOB) [41]. Prices of La_2O_3 (2.3 US$·kg^{-1}) and CeO_4 (2.2 US$·kg^{-1}) are even lower [41]. Consequently, ~85% of the rare earths found in fluorescence wastes are of small value. The remaining 15% comprise Tb, Eu and Gd which are of higher worth, e.g., Eu_2O_3 (56.4 US$·kg^{-1}) and Tb_4O_7 (458.6 US$·kg^{-1}) [41]. However, these three rare earths are part of green and blue phosphors that are highly resistant to acids, rendering their recovery rather expensive. Consequently, rare-earth recovery from phosphor wastes solely employs the best waste material available: production waste. For EOL waste, recycling processes have to find new ways to improve process economics [42].

(iv) Last but not the least, **NiMH accumulators** are the fourth promising material for rare-earth recycling. In 2017, ~2,023 t of these accumulators were sold in Germany [43]. La, Ce, Pr and Nd are part of their anodes and appear either as type AB_5 alloys or seldom as type AB_2 (A = *RE* and B = Ni, Co, Al). The mixture of rare-earth metals (Mischmetall) employed for manufacturing the anodes consists of 50–55 wt.% Ce, 18–28 wt% La and 12–18 wt% Nd [44]. Sometimes, Pr is also present. In general, NiMH accumulators comprise nine different parts (Figure 1.2.11) [45]. The anode readily achieves rare-earth contents of 30 wt%, but the rare-earth content of the waste depends strongly on whether NiMH accumulators are separated from battery waste and how far they are disassembled [44]. If the whole accumulator is crushed, this content decreases to 2.5–10 wt% [12, 46]. The latter is state of the art, since accumulator recycling focuses more on recovering Ni and Co for steel production than on recovering rare-earth metals [12]. Collecting the anodes is therefore still a problem, when rare-earth recovery targets NiMH accumulators. One would suggest that a combined recycling of Ni, Co and rare earths should be favored, then. Another reason why recycling processes have to provide a combined approach is the fact that commodity prices for La and Ce are comparably low among rare earths (cf. three-band phosphors). Nd_2O_3 generates revenues of 50.7 US$·kg^{-1} (purity 99.5%, FOB), but the Nd content of the anode is quite low compared to La and Ce. Therefore, Umicore and Solvay Rare Earths adopted the combined approach and extended Umicore's existing battery recycling [19, 20]. NiMH accumulators (Figure 1.2.11) and Li-ion batteries – both containing Ni and Co – are introduced to pyrometallurgical treatment to recover both metals for steel industry. Umicore's smelter in Hoboken (Belgium) is able to utilize ~7,000 t battery waste per year [19]. Ignoble metals like rare earths leave the process with the slag, and serve as starting material for Solvay's rare-earth recovery. The following process comprises hydrometallurgical steps of treatment (including acid leaching and solvent extraction) in separating rare earths from each other. Though rare earths are already recycled in industrial scale that way; still, there are several disadvantages to the process. Anode alloys, are, by far, a more attractive starting material than slags that consist of slag formers and all ignoble metals found in batteries and accumulators introduced to the process. The powdery alloy can readily be accommodated by sieving the accumulator scrap, but that requires a separate sorting of the battery waste. The anode powder also contains significantly lesser impurities. Therefore, the slag needs more steps of treatment and larger quantities of chemicals than those required by anode material.

The alternative approach is switching the order of recovery, that is, rare earths are recycled before Ni and Co. This requires sorted NiMH accumulators as available

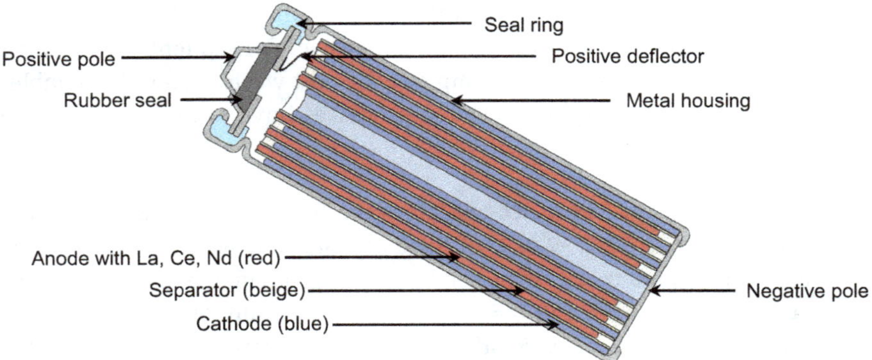

Figure 1.2.11: Structure of a AA-sized NiMH accumulator (cross section, modified according [45]).

waste stream. Rare-earth recycling would undoubtedly benefit chemical consumption, but it also has to fulfil new requirements related to Ni and Co recovery. While separating both metals from rare earths, for example, by precipitation, the Ni-Co precipitate has to be free of sulfur, chlorine, nitrogen and phosphorus; this process may have been quite complicated in times when mineral acids (HCl, H_2SO_4, H_3PO_4 and HNO_3) were used for digestion. This problem is often underestimated, as current recycling approaches are mainly focused on rare-earth recovery. However, process development must provide an answer to the entire recycling problem (holistic approach) and not just for a specific valuable material.

The future trend of NiMH accumulators depends largely on the production of HEVs. Approximately 85% of the HEVs built in 2016 were equipped with NiMH accumulators, and 90% of all electric cars built worldwide were HEVs [47]. What is, on the one hand, a highly available waste material will have to deal with serious problems on the other, in future. Forecasts show a negative trend for the market share of NiMH accumulators in HEVs, as Li-ion batteries gain in importance [47, 48]. If reality meets the predictions, the market share in HEVs will decrease stepwise from 85% to 65% in 2020, and to 10% in 2025 [47]. At least, the demand for NiMH accumulators in non-HEV related applications like consumer batteries will roughly stay constant [48]. Overall, market trends are mostly likely to affect the availability of this waste material negatively.

References

[1] E. R. Tompkins, J. X. Khym, W. E. Cohn, J. Am. Chem. Soc. 1947, 2769.

[2] a) H. Wilken, Verfügbarkeit von Seltenen Erden. Vortrag auf dem 21. Dechema Kolloquium, Rostock, 2016; b) statista, "Verwendung von Seltenen Erden nach Einsatzbereichen in den Jahren 2006 und 2012 (in Tonnen)", can be found under http://de.statista.com/statistik/daten/studie/246406/umfrage/verwendung-von-seltenen-erden-nach-einsatzbereichen/, 2016.

[3] statista, "Weltweite Nachfrageentwicklung von Seltenen Erden nach ausgewählten Anwendungsgebieten im Zeitraum der Jahre 2015–2020", can be found under https://de.statista.com/statistik/daten/studie/209229/umfrage/weltweite-nachfrageentwicklung-von-seltenen-erden-nach-anwendungsgebieten-bis-2020/, 2016.
[4] a) R. Dittmeyer, W. Keim, G. Krevsa, A. Oberholz (Eds.) Winnacker-Küchler Chemische Technik. Prozesse und Produkte, Band 6: Metalle, Wiley-VCH, Weinheim, 2006. ISBN: 3527307710; b) T. Jüstel, "Seltene Erden – Vorkommen und Anwendung", can be found under https://www.fh-muenster.de/ciw/downloads/personal/juestel/juestel/Seltene_Erden-Vorkommen_und_Anwendungen-1.pdf, 2012.
[5] R. E. Kirk, D. F. Othmer, J. I. Kroschwitz, M. Howe-Grant, Encyclopedia of Chemical Technology, Wiley, New York, Chichester, 1998. ISBN: 0471527041
[6] P. Kausch, M. Bertau, J. Gutzmer, J. Matschullat (Eds.) Strategische Rohstoffe – Risikovorsorge, Springer, Berlin, Heidelberg, 2014. ISBN: 978-3-642-39703-5
[7] Ullmann's Encyclopedia of Industrial Chemistry, Wiley-VCH, Chichester, 2010. ISBN: 3527306730
[8] a) J. B. Hedrick, Mineral Commodity Summaries 1996, Rare Earths, 1996; b) J. B. Hedrick, Mineral Commodity Summaries 1997, Rare Earths, 1997; c) J. B. Hedrick, Mineral Commodity Summaries 1998, Rare Earths, 1998; d) J. B. Hedrick, Mineral Commodity Summaries 1999, Rare Earths, 1999; e) J. B. Hedrick, Mineral Commodity Summaries 2000, Rare Earths, 2000; f) J. B. Hedrick, Mineral Commodity Summaries 2001, Rare Earths, 2001; g) J. B. Hedrick, Mineral Commodity Summaries 2002, Rare Earths, 2002; h) J. B. Hedrick, Mineral Commodity Summaries 2003, Rare Earths, 2003; i) J. B. Hedrick, Mineral Commodity Summaries 2005, Rare Earths, 2005; j) J. B. Hedrick, Mineral Commodity Summaries 2006, Rare Earths, 2006; k) J. B. Hedrick, Mineral Commodity Summaries 2007, Rare Earths, 2007; l) J. B. Hedrick, Mineral Commodity Summaries 2008, Rare Earths, 2008; m) J. B. Hedrick, Mineral Commodity Summaries 2009, Rare Earths, 2009; n) J. B. Hedrick, Mineral Commodity Summaries 2010, Rare Earths, 2010; o) D. J. Cordier, Mineral Commodity Summaries 2011, Rare Earths, 2011; p) D. J. Cordier, Mineral Commodity Summaries 2012, Rare Earths, 2012; q) J. Gambogi, Mineral Commodity Summaries 2013, Rare Earths, 2013; r) J. Gambogi, Mineral Commodity Summaries 2014, Rare Earths, 2014; s) J. Gambogi, Mineral Commodity Summaries 2015, Rare Earths, 2015; t) J. Gambogi, Mineral Commodity Summaries 2016, Rare Earths, 2016; u) J. Gambogi, Mineral Commodity Summaries 2017, Rare Earths, 2017.
[9] J. B. Hedrick, Mineral Commodity Summaries 2010, Rare Earths, 2011.
[10] J. Gambogi, Mineral Commodity Summaries 2018, Rare Earths, 2018.
[11] J. Gambogi, Mineral Commodity Summaries 2019, Rare Earths, 2019.
[12] K. Binnemans, P. T. Jones, B. Blanpain, T. van Gerven, Y. Yang, A. Walton, M. Buchert, J. Cleaner Prod. 2013, 51, 1.
[13] J. Gambogi, 2010 Minerals Yearbook – Rare Earths. [Advanced Release], 2012.
[14] J. Gambogi, 2011 Minerals Yearbook – Rare Earths. [Advanced Release], 2013.
[15] a) D. J. Packey, Rare Earths: Diversification is the Key to Sustainable Supply. Vortrag am Helmholtz-Institut Freiberg, 2017; b) D. J. Packey, D. Kingsnorth, Resour. Policy 2016, 48, 112.
[16] J. Gambogi, 2013 Minerals Yearbook – Rare Earths. [Advanced Release], 2016.

[17] a) T. Elwert, Dissertation, TU Clausthal, Clausthal, 2015; b) T. Elwert, D. Goldmann, F. Schmidt, R. Stollmaier, World Metall.-Erzmet. 2013, 66, 209; c) T. Lorenz, Dissertation, TU Bergakademie Freiberg, Freiberg, 2018; d) T. Lorenz, P. Fröhlich, M. Bertau, Recycling von Seltenen Erden aus Leuchtstoffen. Alternative Recyclingstrategien, Vortag auf der Tagung Aufbereitung und Recycling 2014, Freiberg, 2014; e) T. Lorenz, P. Fröhlich, M. Bertau, Chem. Ing. Tech. 2017, 66, 209; f) T. Huckenbeck, R. Otto, E. Haucke, WO2012143240 A2, 2012; g) J.-

J. Braconnier, A. Rollat, US8501124 B2, 2013; h) J.-J. Braconnier, A. Rollat, US9102998 B2, 2015.

[18] T. Lorenz, K. Golon, P. Fröhlich, M. Bertau, Chem. Ing. Tech. 2015, 87, 1373.
[19] J. Tytgat, Umicore Battery Recycling. Recycling of NiMH and Li-ion Batteries. Vortrag auf dem Green Cars Initiative PPP Expert Workshop, Brüssel, 2011.

[20] U. Bast, R. Blank, M. Buchert, T. Elwert, F. Finsterwalder, G. Hörnig, T. Klier, S. Langkau, F. Marscheider-Weidemann, J.-O. Müller et al., Recycling von Komponenten und strategischen Metallen aus elektrischen Fahrantrieben (MORE). Abschlussvorhaben zum Verbundvorhaben, 2015.
[21] K. Hans Wedepohl, Geochim. Cosmochim. Acta 1995, 59, 1217.
[22] J. Götze, R. Möckel, Quartz: Deposits, Mineralogy and Analytics, Springer Berlin Heidelberg, Berlin, Heidelberg, 2012. ISBN: 978-3-642-22160-6.
[23] A. R. Chakhmouradian, F. Wall, Elements 2012, 8, 333.
[24] H. Elsner, H. Sievers, M. Szurlies, H. Wilken, "Das Mineralische Rohstoffpotenzial der Arktis", can be found under http://www.bgr.bund.de/DE/Gemeinsames/Produkte/Downloads/Commodity_Top_News/Rohstoffwirtschaft/41_mineralisches-rohstoffpotenzial-arktis.pdf;jsessionid=3FF4D488FDC0CFC1712A597314E0A796.1_cid321?__blob=publicationFile&v=6, 2014.
[25] A. M. Clark, M. H. Hey, Hey's Mineral Index. Mineral Species, Varieties, and Synonyms, Chapman & Hall, London, 1993. ISBN: 0412399504

[26] C. K. Gupta, N. Krishnamurthy, Extractive Metallurgy of Rare Earths, CRC Press, Boca Raton, 2005. ISBN: 9780415333405
[27] a) X.-Y. Yang, W.-D. Sun, Y.-X. Zhang, Y.-F. Zheng, Geochim. Cosmochim. Acta 2009, 73, 1417; b) S. B. Castor, Can. Mineral. 2008, 46, 779; c) V. Zepf, Rare Earth Elements. A New Approach to the Nexus of Supply, Demand and Use: Exemplified along the Use of Neodymium in Permanent Magnets, Springer, Berlin, Heidelberg, 2013. ISBN: 978-3642354571; d) H. Elsner, "Heavy Minerals of Economic Importance", can be found under https://www.bgr.bund.de/DE/Themen/Min_rohstoffe/Downloads/heavy-minerals-economic-importance.pdf;jsessionid=B08C86F582CFE6217C2CE0223ADBA393.1_cid292?__blob=publicationFile&v=5, 2010.
[28] K. M. Goodenough, F. Wall, D. Merriman, Nat. Resour. Res. 2018, 27, 201.
[29] a) S. Salvi, A. Williamsjones, Lithos 2006, 91, 19; b) U. Kempe, R. Möckel, T. Graupner, J. Kynicky, E. Dombon, Ore Geol. Rev. 2015, 64, 602.
[30] S. Jaireth, D. M. Hoatson, Y. Miezitis, Ore Geol. Rev. 2014, 62, 72.
[31] N. D. Chau, P. Jadwiga, P. Adam, D. van Hao, K. Le Phon, J. Paweł, Vietnam J. Earth Sci. 2017, 39.
[32] G. A. Moldoveanu, V. G. Papangelakis, Mineral. Mag. 2016, 80, 63.
[33] P. L. Verplanck, M. W. Hitzman, Ore Geol. Rev. 2016, 18, 1.
[34] B. Achzet, Empirische Analyse von preis- und verfügbarkeitsbeeinflussenden Indikatoren unter Berücksichtigung der Kritikalität von Rohstoffen, Dissertation Universität Augsburg, Disserta Verl., Hamburg, 2012. ISBN: 3954250926.
[35] a) T. E. Graedel, J. Allwood, J.-P. Birat, M. Buchert, C. Hagelüken, B. K. Reck, S. F. Sibley, G. Sonnemann, Recycling Rates of Metals. A Status Report, United Nations Environment Programme, Nairobi, Kenya, 2011. ISBN: 9789280731613; b) T. E. Graedel, J. Allwood, J.-P. Birat, M. Buchert, C. Hagelüken, B. K. Reck, S. F. Sibley, G. Sonnemann, J. Ind. Ecol. 2011, 15, 355.
[36] K. Habib, H. Wenzel, J. Cleaner Prod. 2014, 84, 348.
[37] M. Mocker, J. Aigner, S. Kroop, R. Lohmeyer, M. Franke, Chem. Ing. Tech. 2015, 87, 439.
[38] O. Gutfleisch, M. A. Willard, E. Bruck, C. H. Chen, S. G. Sankar, J. P. Liu, Adv. Mater. 2011, 23, 821.

[39] EU-Verordnung (EG) Nr. 245/2009. EG 245/2009, 2009.
[40] a) T. Lorenz, P. Fröhlich, M. Bertau, DE 102014224015, 2014; b) T. Lorenz, P. Fröhlich, M. Bertau, DE102014206223 A1, 2014.
[41] Deutsche Rohstoffagentur DERA, Volatilitätsmonitor Oktober 2018, Berlin, 2018.
[42] S. Hopfe, K. Flemming, F. Lehmann, R. Möckel, S. Kutschke, K. Pollmann, Waste Manage. 2017, 62, 211.
[43] N. Hüsgen, "Erfolgskontrolle 2017 der Stiftung GRS Batterien", can be found under http://www.grs-batterien.de/fileadmin/user_upload/Download/Wissenswertes/Infomaterial_2018/GRS_Erfolgskontrolle2017Web.pdf, 2018.
[44] L. Pietrelli, B. Bellomo, D. Fontana, M. R. Montereali, Hydrometallurgy 2002, 66, 135.
[45] J. A. S. Tenório, D. C. R. Espinosa, J. Power Sources 2002, 108, 70.
[46] V. Innocenzi, F. Vegliò, J. Power Sources 2012, 211, 184.
[47] B. Zhou, Z. Li, C. Chen, Minerals 2017, 7, 203.
[48] C. Pillot, Battery Market Development for Consumer Electronics, Automotive, and Industrial: Materials, Requirements and Trends, Vortrag auf der Tagung Batteries 2014, Nizza, 2014.

Tom Lorenz, Martin Bertau

1.3 Rare earth resources and processing

Noble metal: Metals that are stable under atmospheric conditions. Within the electrochemical series, they have a positive potential. Typical examples are Cu (+0.35 V, Pt: (+1.20 V) or Au (+1.50 V). They are nobler than hydrogen (±0.00 V)
Ignoble metal: Metals with a negative electrochemical potential, such as Zn (−0.76 V) or Li (−3.04 V). Naturally, they appear in oxidized form in compounds (e.g., salts). These metals react with O_2 (+1.23 V), sometimes even with N_2 from air.
(Standard potentials at 25 °C; 101.3 kPa; pH = 0; ion activity 1 mol/l)

In the course of chemical processes, it is essential to decide which part(s) of the raw material should be recovered, what the end product would be and which components or by-products should be disposed or introduced to different processes. Depending on the raw material, different steps of processing become necessary. In this regard, Table 1.3.1 illustrates the fundamental steps in the treatment of primary and secondary raw materials as well as of noble and ignoble metals. Both raw materials serve as source for valuable metals, but while ores are usually mined at one location, wastes have to be collected at productions sites, landfills or recycling depots. For both, the next step involves the concentration of the product, aiming at the production of an intermediate with high contents of valuable metals. Herein, mechanical techniques (shredding, grinding) are employed, beside physical (magnetic and density separation) and chemical methods (acid leaching, solvent extraction).

Noble metals are easier to reduce than the impurities accompanying them. Therefore, reduction is commonly carried out before refining. The fine purification of the raw metal that is obtained removes all the remaining impurities to yield high-purity metals. For noble metals, like Cu, refining consists of, for example, an additional electrolysis converting the crude metal to one with the desired level of purity. The opposite order of treatment applies to ignoble metals (e.g., Zn, Ni, Fe). Impurities are often nobler than the target metal. In order to prevent contamination during reduction, they have to be refined first. Impurities are removed from the intermediates, for example, by solvent extraction or ion exchange. Later, the respective metals are produced via smelt electrolysis, carbothermic or metallothermic reduction [1, 2]. Since rare earths have negative electrochemical potentials between −1.991 V (Eu^{3+}/Eu^{0}) and −2.379 V (La^{3+}/La^{0}), all elements of the group are among the most ignoble metals. Consequently, the second and fourth columns in Table 1.3.1 are applicable to their production.

https://doi.org/10.1515/9783110654929-003

Table 1.3.1: Fundamental principles of treatment according to raw material and metal.

Primary raw materials		Secondary raw materials	
Noble metals	**Ignoble metals**	**Noble metals**	**Ignoble metals**
Mining	Mining	Collection	Collection
Concentration	Concentration	Concentration	Concentration
Reduction	Refining	Reduction	Refining
Refining	Reduction	Refining	Reduction
Products			

1.3.1 Mining and collection

As stated in Table 1.3.1, rare earth production from primary raw materials involves mining as the first step of the process chain. They are mined in open pit mines due to the large deposits just below the surface. Mining by underground methods is too expensive. Deep-set deposits such as the rare earth-bearing magnetite-Fe orebody of the Pea Ridge mine (Missouri, USA) are not exploited with regard to rare earths. The same applies to reserves in Siberia or Greenland, where mining is still prevented by harsh climatic conditions and a lack of infrastructure.

Rare earths are divided into light (Sc, La–Eu) and heavy rare earth elements (Y, Gd-Lu), due to slight differences in their chemical properties. The composition of rare earth ores reflects these differences. Consequently, deposits of the two groups can also be divided into light and heavy rare earth deposits. There are six important deposits where light rare earths are mined, namely Bayan Obo, Dalucao, Weishan and Maoniuping in China, Mountain Pass in the USA and Mt. Weld in Australia [3]. In all of them, rare earth ores consist of monazite and bastnäsite (cf. Table 1.2.3). For heavy rare earths, the most important mines are located in the south of China, namely in Lognan, Xunwu, Changting, Chongzou and Xinfen [3]. The ion-adsorbing clays originating from this region secure the world's supply of heavy rare earths. Moreover, clay from Xunwu also achieves Nd_2O_3 contents above 30 wt%, which is the highest Nd concentration in all of China.

Overall, the world's largest rare earth mine is located in Bayan Obo in China (Figure 1.3.1). This particular mining operation comprises several open pit mines, two of which are more than 1 km in diameter. Figure 1.3.1 depicts how the areas of open pit mines, tailings and tailing ponds expanded during operations between 2001 and 2006. The mining is accompanied by massive environmental problems as highly acidic wastewater and sludge are disposed of with little or without any kind of treatment. The U and Th content of the ores makes more than 160 million tons of

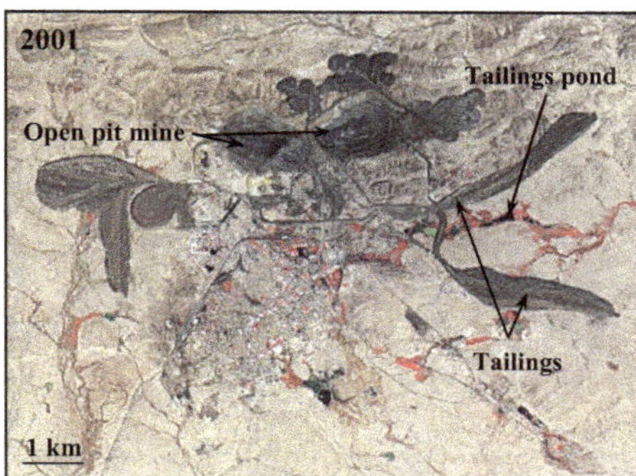

Figure 1.3.1: Satellite images of the world's largest rare earth mine in Bayan Obo (Inner Mongolia, China) from 2001 and 2006 [6].

waste and excavated soil radioactive [4]. This waste material is disposed of, almost untreated, on the production site, which leads to severe soil and water contamination in the Baotou region. Strict government regulations tried to solve the environmental problems in the late 2000s, albeit with limited success. Moreover, regulations were the main cause for the significant increase in rare earths commodity prices in 2011. However, the environmental issues related to rare earth mining and processing are not limited to China alone. Rare earth ores mined in the Mt. Weld deposit in Australia are transported to Malaysia for processing where environmental standards and production costs are much lower [4]. As acidic and radioactive waste is dumped on the production site without special care, rare earth processing leads to a contamination of soil, water and organisms similar to that in China. Malaysia has revised its policy toward *RE* processing dumps. Lynas Corp. (AUS) has to remediate their tailings until 2023 in order to preserve their authorization to process *RE* ores [5].

Instead of mining, secondary raw materials need to be collected before they can be processed (cf. Table 1.3.1). Waste materials are usually disposed of decentralized and in rather small amounts, which makes waste collection as well as separate sorting vital steps in all recycling processes. A closer look on the availability, impurities and accessibility of potential starting materials often reveals problems that can completely compromise any recycling effort. If, for example, only small quantities are available, an industrial process cannot be operated. As mentioned above, four waste materials are the most promising in terms of future rare earth recycling, namely $Fe_{14}Nd_2B$ magnets, SmCo magnets, three-band phosphors, and NiMH accumulators. Table 1.3.2 provides an overview of the most important facts about each of the four secondary raw materials.

With regard to available amounts, rare earth magnets are superior to phosphors and accumulators. About 300,000 t of end-of-life waste will be prospectively generated in 2020, containing between 15 and 33 wt% of rare earth metals [7]. Though

Table 1.3.2: Comparison of starting materials for rare earth recycling.

Material	Kind of waste	Appearance	REE content** [wt %]	REE stocks 2020*** [t waste•a^{-1}]	Separate sorting
$Fe_{14}Nd_2B$ magnets	EoL; EoU*	Magnet scrap	15–33	300,000	Partially
	Production	Powder, slurry, chips	15–33	No data	Yes
SmCo magnets	EoL; EoU	Magnet scrap	30–33	6,000	No
	Production	Powder, slurry, chips	30–33	No data	Yes
Three-band phosphors	EoL; EoU	White powder	0–8	25,000	Yes
	Production	Slurry	15–28	No data	Yes
Anodes of NiMH accumulators	EoL; EoU	Scrap or slag	2.5–10	50,000	No

*EoL, end-of-life products; EoU: end-of-use products.
*Rare earth element content estimated from various samples analyzed at the TU Bergakademie Freiberg.
**Rare earth element stocks estimation from [7].

this amount is more than sufficient for rare earth recycling, current approaches have to face several challenges. First, spent magnets are part of many different products – mostly electric devices – that have to be collected and dismantled separately to give the actual starting material. To tackle the collection problem, Bast et al. examined how magnets can be removed undamaged from rotor segments of electric engines [8]. In 2010, Hitachi announced the first automated process that separates magnets from hard-disc drives to replace manual disassembling [9]. The next problem arises as soon as magnets are separated from electronic waste, because all kinds of magnets are collected mutually. This also applies to SmCo magnets since there is no separate sorting of these alloys. In this regard, Schmidt examined how rare earth magnets can be separated from rare earth-free ferrite magnets by sorting them according to their magnetic flux density. This worked quite well as 90% of the magnets were sorted properly [10]. Only production waste achieves high recycling rates but pyrometallurgical processes used for this purpose are unsuitable for end-of-life waste due to their inability to separate impurities [9, 11]. End-of-life recycling will consequently rely on hydrometallurgical processes based on processes already used for primary production. Though magnet collection remains an issue and current recycling rates are still <1%, magnets are highly promising as starting material due to their large quantity and an increasing demand for rare earth magnets [7, 12].

Why does *RE* ore processing produce radioactive waste? The chemistry of lanthanides (*RE* metals) and actinides (U, Th and others) is very similar. The reason is that in both groups, inner *f*-orbitals are filled while the outer valence shell, which is relevant for the chemical properties, remains equal. In the course of ore formation, particularly in magmatic ores, there are certain minerals, the structure of which is robust enough to tolerate foreign ions (diadochy). Upon magma cooling, these minerals "collect" magma incompatible elements from the melt. Incompatible cations may be boron, lanthanides and actinides, anions phosphate and halogens.

The forecast states another rare earth element stock of 25,000 t for three-band phosphors. This amount is significantly lower than for magnets, but three-band phosphors have a special advantage when it comes to waste collection. The toxic impurity, Hg, is what labels these phosphors a hazardous waste but it is also the reason why three-band phosphors is the best material considering selective waste collection. Hg needs to be removed and, therefore, fluorescence lamp scraps are collected as a waste stream and introduced into a distillation process. This procedure yields all phosphors in form of relatively pure white powder. This powder is readily available for rare earth recycling. On the other hand, phosphor waste is always a mixture of many different compounds and provides comparably low rare earth contents, about 0–8 wt%. With Y, La and Ce, phosphors consist mainly of rare earths of small value, which is why recycling processes will most likely focus on production waste instead of fluorescence lamp scraps [13]. In future, rare earth demand for fluorescence lamps will most likely decrease because LED are increasingly replacing these lamps. LEDs also contain rare earths like Y and Ce, but in significantly lower concentrations. This trend will negatively affect the availability of the waste material.

NiMH accumulators will approximately amount to 50,000 t in 2020, which is twice the amount of lamp phosphors but they are commonly collected along with other kinds of primary and secondary batteries. Separate sorting is carried out only to a limited extent since Ni and Co are targeted for recycling. For instance, Umicore utilizes Li-ion batteries and NiMH accumulators in their process. After crushing, accumulator scraps contain between 2.5 and 10 wt% rare earth metals, which might be a starting point for future rare earth recycling. To the present day, battery recycling usually prevents rare earth recycling because of low commodity prices and slags with low rare earth contents. Nevertheless, Umicore and Solvay have managed to merge their processes to conduct rare earth recycling from battery waste on industrial scale. The overall demand for NiMH accumulators is expected to decrease mainly due to Li-ion batteries replacing these accumulators in HEVs.

1.3.2 Concentration

According to Table 1.3.1, primary and secondary raw materials can follow the same fundamental route of treatment once mining and waste collection provide sufficient material to justify industrial processing. These steps are (i) concentration, (ii) refining and (iii) reduction. The methods used to eventually produce rare earth metals differ, especially, at the beginning of the process. Once the rare earths are separated from all impurities, there is no difference in treatment between the two kinds of raw materials. For information about how secondary raw materials are processed, we refer to chapter 1.5. The aim of the concentration steps is the extraction of rare earths from raw material. The obtained intermediate consists of rare earths but is not entirely free of impurities. The purification of this intermediate takes place through various refining steps that will be explained in the next part.

Typically, ore processing starts with mechanical and physical steps of treatment to produce an ore concentrate for further treatment [14]. The ores are crushed, grinded and introduced to different types of density separation, for example vibrating tables. Siliceous impurities have a relatively low density of ~2.6 $g \cdot cm^{-3}$ and can be easily separated from rare earth minerals. The denser monazite (4.6–5.4 $g \cdot cm^{-3}$), for example, remains in the heavy fraction with zircon ZrO_2, rutile TiO_2, ilmenite $FeTiO_3$ and cassiterite SnO_2 [14]. Fe-bearing accessory minerals like ilmenite are removed with the aid of strong magnetic fields. For nonmagnetic minerals, flotation is used. The product is an ore concentrate providing high rare earth contents. Monazite concentrates contain up to 95 wt% rare earth oxides [15]. Bastnäsite concentrates from Mountain Pass, USA, achieve 68–72 wt% rare earth oxides [14]. The order of the steps and which steps are necessary depends on the respective ore. Figure 1.3.2 depicts, for example, how the production of ore concentrates is carried out with the Bayan Obo ore in China.

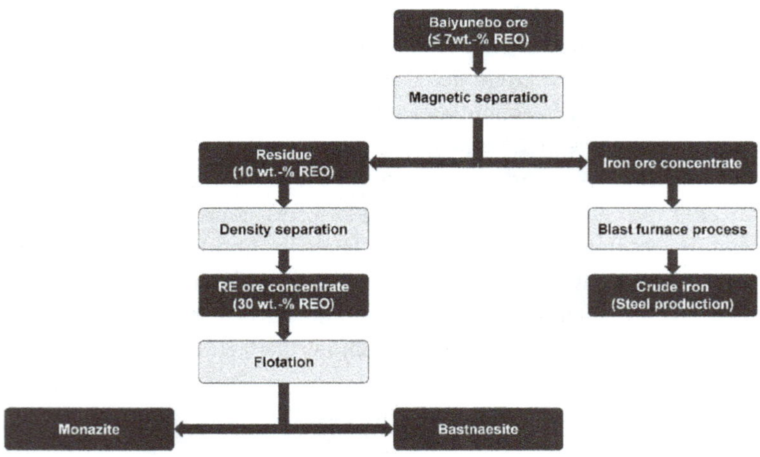

Figure 1.3.2: Mechanical and physical steps to produce the ore concentrate (modified from [14]).

The following treatment steps involve hydrometallurgical methods, which are applied to primary and secondary raw materials in a very similar manner. The hydrometallurgical treatment of the ore concentrates has two objectives: Rare earths are to be separated from the ore and impurities such as Th and U are to be removed before the rare earth elements are separated from each other. Depending on the ore, the impurities and the rare earth content, a variety of different digestions and purification steps are used (Figure 1.3.3).

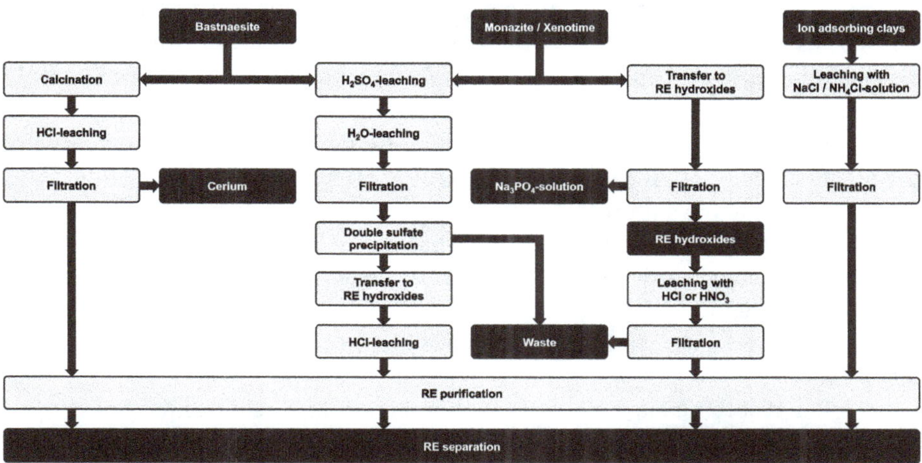

Figure 1.3.3: Simplified scheme of the hydrometallurgical treatment of the most important rare earth minerals [2, 14–16].

For the phosphate minerals monazite and xenotime, two main digestion procedures are favored. Sulfuric acid is used for digestion when the rare earth content is low and the concentration of silicate impurities is high. This way, the impurities are not dissolved. The rare earth phosphates are converted into the sulfates by utilizing concentrated sulfuric acid at 150 °C. The formed rare earth sulfates cannot dissolve under these conditions. The first aim comprises the separation of the phosphate which would form insoluble salts with rare earths. With sulfuric acid, the phosphate reacts to an impure phosphoric acid, which is separated by filtration [17]. In the subsequent leaching step with cold H_2O, the rare earth sulfates dissolve. By adding Na_2SO_4 to the solution, the precipitation of the light rare earths as double sulfates $RE_2(SO_4)_3 \cdot Na_2SO_4 \cdot n\ H_2O$ can be initiated at this point. Heavy rare earth elements do not form insoluble double sulfates with alkali metals. The double sulfates are separated by filtration and then transferred to their respective hydroxides with $NaOH_{(aq)}$. For further processing, the light rare earth hydroxides are again dissolved by adding hydrochloric acid. This preseparation technique produces two solutions, one with the light and the other with the heavy rare earths.

In the case of ores rich in rare earths and low in silicates, however, the alkali route is preferred to the sulfuric acid route. The first step is leaching with 50% $NaOH_{(aq)}$, whereby, a Na_3PO_4 solution is formed as a tradable by-product while the rare earths are uniformly transferred into the hydroxides [14, 16]. To separate these from the hydroxides of Th^{4+} and Fe^{3+}, which are also insoluble, the rare earths are selectively dissolved with hydrochloric or nitric acid at pH values of 3–4 in the subsequent leaching step [2, 14, 15]. Since the radioactive radium is formed from Th by α decay, $^{228}Ra^{2+}$ is co-precipitated by the addition of Ba^{2+} and $SO_4{}^{2-}$ ions. The precipitate, which consists of $RaSO_4$ and $BaSO_4$, is disposed of together with $Th(OH)_4$ as radioactive waste. Alternatively, Th and rare earth hydroxides can be dissolved in nitric acid and separated via solvent extraction. This is how Solvay in La Rochelle separates Th from monazite sands [14]. Especially, U- and Th-rich deposits require additional process steps which burden the economic efficiency of the processes.

In addition to the sulfuric acid and alkali procedure, hydrochloric acid can also be used for the digestion of bastnäsite. The bastnäsite is first calcined to convert the carbonates in metal oxides and gaseous CO_2. The obtained rare earths oxides can then be dissolved in hydrochloric acid at 95 °C. The insoluble CeO_2 constitutes an exception and is separated from the dissolved rare earths in the subsequent filtration step.

For ion-adsorbing clays, the digestion in Figure 1.3.3 requires neither mineral acids nor strong bases. The rare earths are adsorbed on the surface of the clays and can be desorbed with a $NaCl/NH_4Cl$ solution. Apart from ion exchange and the hydrochloric acid process, according to which Molycorp (Mountain Pass, USA) had processed bastnäsite ores (Figure 1.3.3), the hydrometallurgical digestion of rare earths is very chemical-intensive. This applies, in particular, to the multiple changes between strongly acidic and strongly alkaline pH values. These changes result from the fact that with monazite and xenotime, the phosphate has to be first separated. In this regard, the processing of clays and bastnäsite is comparatively simple but here too, the impurities, especially U and Th, must be removed.

In addition to the methods described, there are many processes for all kinds of rare earth minerals. For monazite and bastnäsite ores, for example, there are processes that conduct an alkaline digestion step with a 65% Na_2CO_3 solution or a basic salt fusion with solid Na_2CO_3 or NaOH [1]. Alternative to wet-chemical leaching, bastnäsite, xenotime, monazite or gadolinite $(Y,Ce,Nd)_2Fe^{2+}Be_2O_2(SiO_4)_2$ can be directly chlorinated with carbon and chlorine gas at 1,000–1,200 °C according to the Goldschmidt process [1]. For the digestion of the oxide mineral euxenite $(Y,Ca,Ce,U,Th)(Nb,Ta,Ti)_2O_6$, hydrofluoric acid is first used to remove Nb, Ta and Ti. Instead, acid salt fusion employing $KHSO_4$ can be used to avoid strongly acidic wastewater. Eventually, all types of digestion finally result in one (or more) aqueous rare earth concentrate.

1.3.3 Refining and separation

From the digestion solution, rare earths are separated from each other which is costly due to their similar chemical and physical properties. Rare earth elements differ mainly in the occupation of the energetically low 4f orbitals. With increasing atomic number, the seven 4f orbitals are occupied with more electrons until the maximum of 14 is achieved (Lu). However, chemical bonds are exclusively formed via the energetically higher $5d^1$ and $6s^2$ orbitals which are occupied by a maximum of three electrons. The main oxidation state +3 is thus the same for all 17 rare earths and all elements, therefore, react in very similar manner. Rare earth separation is, accordingly, difficult and laborious. The oxidation states +2 and +4 can only be achieved when either (i) the positive charge of the nucleus is still comparatively low (light lanthanides), (ii) the resulting 4f orbitals would be half- or unoccupied or (iii) the oxidation state can be stabilized within a chemical complex. [18] In the aqueous solutions, however, only Eu^{2+} (half-occupied $4f^7$) and Ce^{4+} (unoccupied $4f^0$) are stable ions. Both elements can thus be separated individually from all trivalent rare earths. Eu^{2+} is usually precipitated selectively as $EuSO_4$, and CeO_2 is the only rare earth oxide resistant to hydrochloric acid. Consequently, CeO_2 remains undissolved by leaching steps utilizing this acid (Figure 1.3.3). There are three basic procedures employed to separate all 17 rare earth elements from each other: (i) fractional precipitation (hydroxides, double sulfates) or crystallization (double nitrates), (ii) ion exchange and (iii) solvent extraction [2, 16, 19].

Fractional precipitation and **crystallization** are only of historical importance for rare earth separation [20]. By repeating dissolution and crystallization steps a hundred times with the rare earth nitrates or hydroxides, the smallest differences in solubility were used to separate rare earth elements from each other. At present, double sulfate precipitation is still being used to precipitate light from heavy rare earths in one step [2, 16, 21, 22]. On the other hand, this preseparation requires an additional effort as the double sulfates obtained must be transferred into the hydroxides before they can be dissolved again in hydrochloric acid. Subsequently, light rare earths are separated from each other by solvent extraction.

In the 1950s, fractional separation was replaced by **ion exchange** due to the immense effort involved. Among all separation techniques, this method achieves the highest purity of rare earths (>99.9999%) but is significantly more expensive than extractive separation because of the costly recovery of the resin and the low rare earth concentrations in the eluate ($2–10\ g \cdot L^{-1}$) [2]. Strongly acidic cation exchange resins such as polystyrene-divinylbenzene copolymer or phenol-formaldehyde resins with sulfonic acid groups are usually used as stationary phase [14, 23, 24]. The adsorption of rare earths is not selective. Only by the second step, the elution, the rare earths are separated. This is achieved by eluting with concentrated salt solutions (often NH_4Cl) containing different complexing agents, for example, citric acid, lactic acid, HEDTA and EDTA. Subsequently, rare earths are separated on 15–34 exchanger columns due to slightly different complex equilibria [14, 23]. In the process, NH_4Cl regenerates the ion exchanger. More selective ion exchangers, in which the rare earths adsorb less well with increasing ion radius due to steric distortion of an organic lattice structure, are still the subject of current research [25].

Since the 1960s, **solvent extraction** has predominantly been used for rare earth separation. The organic phase mostly consists of kerosene, an extraction agent, and a modifier (e.g., 1-decanol) to accelerate phase separation [22, 26]. Chelating ligands commonly serve as extraction agents, for example, carboxylic acids (EDTA) or phosphoric acids and esters (e.g., TBP, DEHPA, PC-88A, Cyanex 302) [1, 2, 26, 27]. Both extraction and reextraction are carried out in countercurrent mixer-settler plants and are strongly dependent on the pH value of the aqueous phase.

With the pH value decreasing gradually, the light rare earths are re-extracted before the heavy rare earths (Figure 1.3.4) [14, 16, 21, 22]. Washing steps are additionally conducted between every extraction step. Complete separation of all 17 rare earth elements subsequently requires numerous repetitions of extraction, washing and re-extraction steps, which makes rare earth separation an expensive and laborious procedure. Usually, several groups of mixer-settler units with 10–100 extraction stages each are employed for this purpose. The Solvay plant in La Rochelle has ~1,000 of these units [14]. The purity of the products is usually lower than that of the ion exchangers but exceeds >99.99% [23]. Since rare earths and the compounds thereof are traded almost exclusively in the purity range between 99 and 99.99%, both separation methods are sufficient. In contrast to ion exchangers, solvent extraction is operated continuously and achieves up to 10–50 times higher rare earth concentrations in the extract, which makes solvent extraction the method of choice for separation [2].

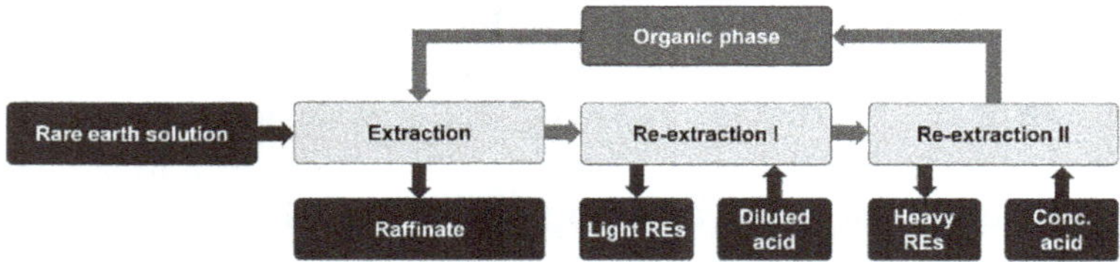

Figure 1.3.4: Schematic separation of rare earths via solvent extraction [2, 8, 16].

References

[1] C. K. Gupta, N. Krishnamurthy, Extractive metallurgy of rare earths, CRC Press, Boca Raton, Fla, 2005. ISBN: 9780415333405

[2] I. McGill, "Rare Earth Elements" in Ullmann's Encyclopedia of Industrial Chemistry, Wiley-VCH, Chichester, 2005. DOI: 10.1002/14356007.a01 001

[3] P. Friedrichs, Dissertation, Rheinisch-Westfälische Technische Hochschule Aachen, Aachen, 2017.

[4] L. Marschall, Seltene Erden. Im Spannungsfeld zwischen "schmutziger" Herstellung & "sauberer" Anwendung in Umwelttechnologien, München, 2018.
[5] Atomic Energy Licensing Board of Malaysia, Pemindahanproses "Cracking and Leaching" Keluar Negara Dan Pembinaan PDF. Antarasyarat Lesen Operasi Lynas, Dengkil, 2019.
[6] NASA, "Rare Earth in Bayan Obo", can be found under https://earthobservatory.nasa.gov/images/77723/rare-earth-in-bayan-obo, 2012.
[7] K. Binnemans, P. T. Jones, B. Blanpain, T. van Gerven, Y. Yang, A. Walton, M. Buchert, J. Cleaner Prod. 2013, 51, 1.
[8] U. Bast, R. Blank, M. Buchert, T. Elwert, F. Finsterwalder, G. Hörnig, T. Klier, S. Langkau, F. Marscheider-Weidemann, J.-O. Müller et al., Recycling von Komponenten und strategischen Metallen aus elektrischen Fahrantrieben (MORE). Abschlussvorhaben zum Verbundvorhaben, 2015.
[9] Pressemitteilung, "Hitachi Develops Recycling Technologies for Rare Earth Metals", can be found under http://www.hitachi.com/New/cnews/101206.pdf, 2010.
[10] M. Schmidt, Untersuchungen zur Charakterisierung der Wertstoffinhalte von Permanentmagneten. Diplomarbeit an der TU Bergakademie Freiberg, Freiberg, 2016.
[11] a) K. Hirota, T. Minowa, US6960240 B2, 2002; b) Y. Hirose, T. Go, M. Renda, N. Kawamura, A. Otaki, WO2000039514 A1, 2000; c) H. Onishi, T. Terada, Y. Yamagata, F. Yamashita, EP1096517 B1, 2005.
[12] a) statista, "Weltweite Nachfrageentwicklung von Seltenen Erden nach ausgewählten Anwendungsgebieten im Zeitraum der Jahre 2015-2020", can be found under https://de.statista.com/statistik/daten/studie/209229/umfrage/weltweite-nachfrageentwicklung-von-seltenen-erden-nach-anwendungsgebieten-bis-2020/, 2016; b) T. E. Graedel, J. Allwood, J.-P. Birat, M. Buchert, C. Hagelüken, B. K. Reck, S. F. Sibley, G. Sonnemann, Recycling rates of metals. A status report, United Nations Environment Programme, Nairobi, Kenya, 2011. ISBN: 9789280731613
[13] a) T. Lorenz, P. Fröhlich, M. Bertau, DE102014206223 (A1), 2014; b) T. Lorenz, P. Fröhlich, M. Bertau, Recycling von Seltenen Erden aus Leuchtstoffen. Alternative Recyclingstrategien. Vortag auf der Tagung Aufbereitung und Recycling 2014, Freiberg, 2014.
[14] R. Dittmeyer, W. Keim, G. Krevsa, A. Oberholz (Eds.) Winnacker-Küchler Chemische Technik. Prozesse und Produkte. Band 6: Metalle, Wiley-VCH, Weinheim, 2006. ISBN: 3527307710
[15] P. Kausch, M. Bertau, J. Gutzmer, J. Matschullat (Eds.) Strategische Rohstoffe – Risikovorsorge, Springer Berlin Heidelberg, Berlin, Heidelberg, 2014. ISBN: 978-3-642-39703-5
[16] R. E. Kirk, D. F. Othmer, J. I. Kroschwitz, M. Howe-Grant, Encyclopedia of chemical technology, Wiley, New York, Chichester, 1998. ISBN: 0471527041
[17] F. H. Firsching, S. N. Brune, J. Chem. Eng. Data 1991, 36, 93.
[18] A. F. Holleman, E. Wiberg, N. Wiberg, Lehrbuch der anorganischen Chemie, De Gruyter, Berlin, 2007. ISBN: 3110177706
[19] F. Xie, T. A. Zhang, D. Dreisinger, F. Doyle, Miner. Eng. 2014, 56, 10.
[20] J. Herzfeld, O. Korn, Chemie der seltenen Erden, Salzwasser Verl., Paderborn, 2011. ISBN: 9783864443978
[21] T. Elwert, D. Goldmann, F. Schmidt, R. Stollmaier, World Metall.-Erzmet. 2013, 66, 209.
[22] T. Elwert, Dissertation, TU Clausthal, Clausthal, 2015.
[23] F. Habashi, Can. Metall. Q. 2013, 52, 224.
[24] S. Barany, V. Strelko, Adsorption 2013, 19, 769.
[25] a) Y. Tasaki-Handa, Y. Abe, K. Ooi, M. Tanaka, A. Wakisaka, Dalton Trans. 2014, 43, 1791; b) M. Tian, Q. Jia, W. Liao, J. Rare Earths 2013, 31, 604.
[26] N. V. Thakur, Miner. Process. Extr. Metall. Rev. 2000, 21, 277.
[27] C. Yan, J. Jia, C. Liao, S. Wu, G. Xu, Tinshhua Sci. Technol. 2006, 11, 241.

Tom Lorenz, Martin Bertau

1.4 Rare earth metal production and refinement

Although rare earths are chemically very similar to each other, three different methods have to be employed to reduce every element to the metallic state. These three methods are smelt electrolysis of rare earth chlorides, an oxide–fluoride mixture or metallothermic reduction. An electrolysis of an aqueous solution like the one that is conducted for Cu production is impossible for this group of ignoble elements. Rare earth metals react with water in a very exothermic reaction, due to their extremely low standard electrode potential (from Eu^{3+}/Eu^{0} −1.991 V till La^{3+}/La^{0} −2.379 V) [1]. Consequently, reduction is realized using pyrometallurgic methods.

1.4.1 Smelt electrolysis and metallothermic reduction

Accordingly, electrolysis is conducted in melts of molten salts at 700–1,000 °C [2 to 4]. The crucibles are made of graphite or refractory metals such as W and Ta. The anodes consist of graphite, and cathodes are mostly made of W or Mo. The materials of the electrolysis cell have to be chosen carefully, because they inevitably contribute to the impurities found in the resulting rare earth metals [5]. Depending on the method, rare earth oxides, chlorides and/or fluorides are used as starting material. For rare earth chlorides, a molten mixture consisting of NaCl, $CaCl_2$, $BaCl_2$ and NaF serves as electrolyte. Such melts are mainly used for light rare earth elements with comparatively low melting points ($\ll$1,300 °C). The heavy rare earths, whose metals have melting temperatures >1,300 °C, cannot be separated that way, because the electrolyte mixture already evaporates at these temperatures [4]. Heavy rare earths are instead reduced in a melt containing LiF and BaF_2 [2–5]. Table 1.4.1 provides an overview of the rare earths and the methods by which they are reduced.

The four elements Sm, Eu, Tm and Yb are exceptions, since they also exist as RE^{2+} ions in molten salts. They cannot be further reduced by smelt electrolysis. Instead, these rare earths are reduced via metallothermic reduction at 1,000–1,600 °C. The oxides or fluorides of the four elements are mixed with a cheaper but nobler rare earth metal, such as La, Ce or misch metal. After heating to the desired temperature, La_2O_3 is formed, while Sm, Eu, Tm and Yb are reduced to their metal state. The metals obtained are then separated from La_2O_3 by distillation [5]. With the

https://doi.org/10.1515/9783110654929-004

Table 1.4.1: Melting and boiling temperatures of selected rare earths and the respective method of reduction used to obtain the metal.

Element	Melting temperature (°C)	Boiling temperature (°C)	Method of reduction
La	920	3,457	Electrolysis in molten chloride salts
Ce	798	3,422	Electrolysis in molten chloride salts
Pr	931	3,506	Electrolysis in molten chloride salts
Nd	1,016	3,064	Electrolysis in molten chloride salts
Sm	1,072	1,788	Metallothermic reduction
Eu	817	1,524	Metallothermic reduction
Gd	1,312	3,262	Electrolysis in molten fluoride salts
Tb	1,357	3,216	Electrolysis in molten fluoride salts
Dy	1,409	2,558	Electrolysis in molten fluoride salts
Tm	1,545	1,944	Metallothermic reduction
Yb	824	1,196	Metallothermic reduction
Lu	1,663	3,391	Electrolysis in molten fluoride salts

methods presented, all 17 elements of rare earths can be converted into their respective metals. The reduction to the metal is particularly important for the elements Nd, Dy, Sm and Sc that are components of important alloys (cf. Table 1.2.1).

1.4.2 Metal refinement

The metals originating from smelt electrolysis or metallothermic reduction generally achieve a total purity of 98–99% [5]. Beside metallic impurities that originate from the starting material or the reactants utilized during the process, the missing 1–2% also include nonmetallic impurities, such as oxygen, nitrogen, carbon and hydrogen. The impurities originate from a large variety of sources. They come from the ore (e.g., Th), the commercial chemicals employed in the process (Ca, Pb), the reactor material (Fe, Ni, Co, W, Ta), the insufficient separation (other rare earths), the starting material of the reduction step (oxygen, fluorine and chlorine), the reductants (La, Ce, Li and Ca), the electrodes (carbon) and the melts (sodium, barium, chlorine, fluorine, etc.). Metallic impurities can be reduced by the use of refined starting materials and pure chemicals. This allows the production of rare earth metals with a purity of 99.99%, solely considering the metal atoms [5]. The prevention of nonmetallic impurities would require efforts that are, by far, more extensive. The task would be to operate the entire process under conditions that prevent any contact with atmospheric gases. The metallic purity achieved after the reduction step is usually sufficient for most of the applications. For achieving highest purity, the metals obtained additionally undergo a number of different process steps, namely vacuum melting, vacuum distillation, vacuum sublimation, zone refining, zone melting and/or solid-state electrotransport processing (Figure 1.4.1) [5]. The methods used and their order depend on the particular rare earth element, the required purity, the operational feasibility and the process economics.

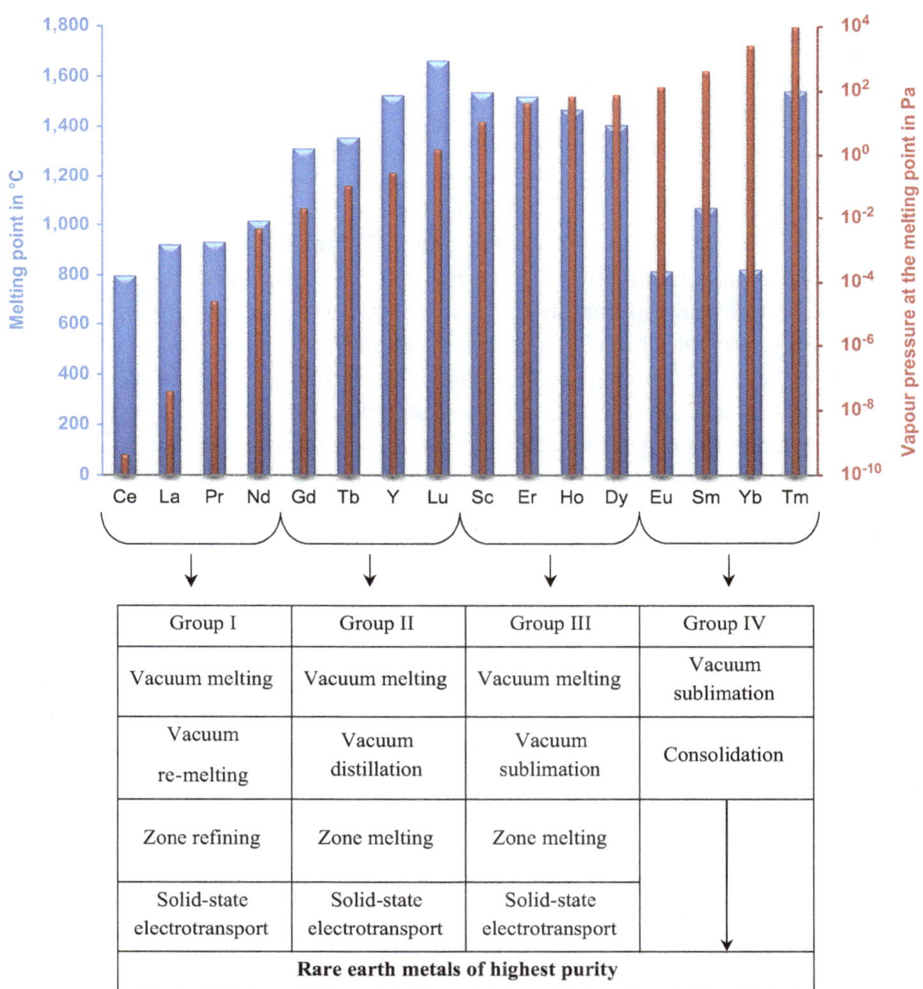

Group I	Group II	Group III	Group IV
Vacuum melting	Vacuum melting	Vacuum melting	Vacuum sublimation
Vacuum re-melting	Vacuum distillation	Vacuum sublimation	Consolidation
Zone refining	Zone melting	Zone melting	↓
Solid-state electrotransport	Solid-state electrotransport	Solid-state electrotransport	
Rare earth metals of highest purity			

Figure 1.4.1: Methods of metal refinement utilized to produce highly pure rare earths (modified from [5]).

Depending on the melting point and the vapor pressure at the melting point, rare earths can be divided into four groups, each group requiring a different set of refinement steps (Figure 1.4.1). Group I comprises all rare earth metals with low melting points (≤1,016 °C) and low vapor pressures. In group II, rare earths have melting points above 1,300 °C, but low vapor pressures. The more volatile rare earths Sc, Er, Ho and Dy belong to group III, in which elements have high melting points (>1,400 °C) and high vapor pressures. Last, but not the least, group IV comprises the most volatile rare earths combining low melting points with highest vapor pressures. In this regard, Tm is an exception. While Sm, Yb und Eu begin to melt at temperatures ≤1,072 °C, the melting point of Tm is the second highest among all rare earths (1,545 °C). However, Tm belongs to group IV, because of its vapor pressure achieving almost 10,000 Pa at the melting point.

In order to remove impurities, all four groups are initially treated employing methods summarized under the term "pyrovacuum treatment." *Vacuum melting* is conducted to evaporate metallic impurities that are more volatile than rare earths, which remain in the melt. The temperatures are adjusted to the particular melting points and vapor pressures (max. 1,850 °C). This method is applied to all rare earths except the four most volatile elements of group IV. Among all metallic impurities, Ca, Mg, Fe, silicon and aluminum can be removed by vacuum melting. The halides, foremost, chlorine and fluorine, as well as hydrogen can also be separated by this pyrovacuum treatment. The refractory metals, W and Ta, originating from the crucible used in the reduction step are less volatile than the rare earths and, therefore, remain in the melt. Consequently, *vacuum distillation* is conducted to separate rare earths from refractory metals. Now, like the volatile impurities before, the rare earth metals are evaporated while the less volatile impurities are left behind. Instead of distillation, the metals of group III and IV are evaporated from their solid state. Here, *vacuum sublimation* is possible due to high vapor pressures (Figure 1.4.1). For group IV, this procedure leads to small crystals that are vulnerable to contamination due to large surface areas. Therefore, the crystals are consolidated by remelting them under inert atmosphere and holding the temperature for up to 24 h. Nitrogen and oxygen cause a few problems when they need to be removed. Nitrogen readily forms nitrides with low decomposition pressures, rendering it impossible to use vacuum melting. The same applies to oxygen, which forms solid solutions with rare earth metals. However, pyrovacuum treatment reduces the oxygen content significantly by forming volatile rare earth sub-oxides [5].

For ultrapure metals, different types of *zone melting* and *solid-state electrotransport* are utilized to further reduce the concentration of all impurities to a total of <0.1% [5]. Zone melting is based on the mechanism that impurities accumulate in the molten metal. Therefore, a thin molten zone is passed slowly along a rare earth metal rod. The procedure is repeated multiple times, until the desired purity is achieved. The impurities accumulate at one end of the rod, which is cut off at the end of the refinement, and the rest of the rod consists of the purified metal. The solid-state transport makes use of the mobility of foreign ions in a metal structure, which occurs when direct current is passed through the metal [5]. In particular, nitrogen and oxygen can be removed this way. The rod has to be cooled, since the direct current generates much heat. The melting point of the metal, the vapor pressure and how effective the heat can be removed during the process, therefore, set limits for the method. In contrast to pyrovacuum treatment, zone melting and solid-state electrotransport do not remove impurities. Both methods redistribute the impurities so that one part of the metal rod or ingot is sacrificed for increasing purity. At the end of the ultrapurification process, rare earth metals of the highest purity available are obtained.

References

[1] Petr Vanýsek, "Electrochemical Series", can be found under http://sites.chem.colostate.edu/diverdi/all_courses/CRC%20reference%20data/electrochemical%20series.pdf.

[2] Ullmann's Encyclopedia of Industrial Chemistry, Wiley-VCH, Chichester, 2010. ISBN: 3527306730.

[3] R. E. Kirk, D. F. Othmer, J. I. Kroschwitz, M. Howe-Grant, Encyclopedia of Chemical Technology, Wiley, New York, Chichester, 1998. ISBN: 0471527041.

[4] R. Dittmeyer, W. Keim, G. Krevsa, A. Oberholz (Eds.) Winnacker-Küchler Chemische Technik. Prozesse und Produkte, Band 6: Metalle, Wiley-VCH, Weinheim, 2006. ISBN: 3527307710.

[5] C. K. Gupta, N. Krishnamurthy, Extractive Metallurgy of Rare Earths, CRC Press, Boca Raton, 2005. ISBN: 9780415333405.

Tom Lorenz, Martin Bertau

1.5 Recycling aspects – magnets

For the recycling of noble metals, there exists a series of established technologies that are used in industry. For rare earths, however, the situation is much more complex. Since they have negative electrochemical potentials between 1.991 V (Eu^{3+}/Eu^0) and 2.379 V (La^{3+}/La^0), all elements of the group belong to ignoble metals. Consequently, rare earth recycling involves the following steps: (i) collection, (ii) concentration, (iii) refining and (iv) reduction. These four steps have to be carried out in exactly the same order to recover rare earth metals from waste material. However, "rare earth recovery" and "waste material" are too generic in this context. Most waste materials presented in Section 1.2. are unsuitable for economic recycling due to their low rare earth contents, ineffective waste collection, low commodity prices and limited availability of waste streams. According to the availability of potential starting materials, (cf. Table 1.3.2) $Fe_{14}Nd_2B$ magnets are the most promising waste material. The problems and aspects associated with rare earth recycling are illustrated below using rare earth magnets as the most prominent example.

Primarily, it is important to decide if either of production waste or end-of-life magnets (EoL) serve as the starting material. For production waste, several recycling processes are already used to recover rare earths from powders, slurries, and chips accumulating during magnet production (Table 1.5.1).

Table 1.5.1: Selection of established processes for rare earth recycling from magnet production waste.

Company	Method and starting material	*RE* yield [%]	Ref.
Shin Etsu[(a)] (Japan)	Mixing and melting of magnet alloys with production waste (chips) at 1,550 °C	–	[1]
Showa Denko[(a)] (Japan)	Mixing and melting of magnet alloys with production waste (scraps) at 1,550 °C in an electric arc furnace	86.5	[2]
Pure Etch Company[(a,b)] (USA)	Hydrometallurgical method for recycling slurries, scraps, and chips	95	[3]
Santoku Metal Industries[(a,b)] (Japan/USA)	Acid digestion and rare earth precipitation utilizing oxalic acid	95	[4]
Panasonic[(a)] (Japan)	Dissolving the plastic mantle of magnets in organic solvents, followed by mixing and melting of magnet alloys with production waste at 1,550 °C	–	[5]
Hitachi[(a,b)] (Japan)	Only established postconsumer process; pyrometallurgical extraction of rare earths	95	[6]

[a]Suitable for production waste.
[b]Suitable for some kinds of EoL waste.

https://doi.org/10.1515/9783110654929-005

Pyrometallurgical methods are predominantly used to process production waste. These methods prove to be effective as long as the starting material is relatively pure. Impurities can hardly be removed. It is for this reason that pyrometallurgical methods are unsuitable for utilizing EoL magnets. Usually, they have either experienced corrosion or are covered by corrosion protective metallic layers of Ni or Zn. If impurities remaining in the material cause remanence losses of only 3%, the resulting magnets already fail to meet the strict requirements for new magnets [7].

As a consequence, there is a need for the development of new processes for the recycling of EoL. Rare earth magnets are used in a large variety of electric devices, rendering waste collection and separate sorting of magnet materials very difficult. Currently, recycling rates of rare earths from EoL waste are < 1%, regardless of which waste material is taken into consideration (magnets, phosphors, batteries, etc.) [8]. A large number of approaches are under consideration and an economic recycling process is still to be found. In fact, it is rather difficult to describe how an EoL recycling process has to be designed to be economic. Should EoL recycling rather focus on recovering whole magnets, magnet powder, a mixture of rare earths, or individual rare earths? If recycling aims at the latter, should they be recovered as oxides, metals or as a different compound? Most recycling approaches focus on individual rare earth oxides or mixtures thereof because rare earths are commonly traded as oxides or, to a smaller extent, as pure metals [9–12]. Rare earth-enriched solutions that originate from, for example, acid digestion steps are also considered a valid product. Intermediate products enable the processing of smaller, local waste streams that would not allow the entire process chain to be built up. Against the background of EoL waste amounting decentralized, the integration of recycling processes into existing process chains is the most likely way in which rare earth recycling will be carried out in future.

The possibilities of integrating such processes into industrial production are manifold. Since industrial processes from primary and secondary raw materials rely, to a certain extent, on similar methods and techniques, many combinations are conceivable. The basic combinations shown in Table 1.5.2 lead to the same products, regardless of what the starting material was. New processes related to primary production can make use of existing recycling facilities and vice versa. When processes are merged, different kinds of integrations are possible. Thus, recycling always aims to produce a new product in original quality from EoL waste. However, complete recovery of waste streams is possible, but rare to be achieved.

Recycling: The process recovers materials in primary product quality, enabling their unlimited reuse. For instance, aluminum recycling fulfils this requirement when single-origin waste is collected. The material is indistinguishable from primary product quality [13].

Transcycling: The process yields a product of comparative value but for a different use. As mentioned above, Umicore recovers Ni and Co from NiMH accumulators and introduces both metals to stainless steel production [14, 15].

Downcycling: The waste material is introduced to a process yielding a product of lower quality and functionality. For example, steel used in car chassis often contains Cu and Sn as impurities. Instead of removing both metals, vehicle steel scrap is remelted to construction steel that tolerates a higher impurities content [16].

Upcycling: Production waste, EoL and EoU products are transferred into new products of higher value and different functionality. This concept lacks a clear definition since all waste is considered to be of lower value. Consequently, any kind of treatment would inevitably lead to an increase in value. Instead, the original and the upcycled product should be compared. Here, precious metals may serve as an example, when they are recycled from EoL printed circuit boards for use in industrial catalysts.

Table 1.5.2: Integration of processes originating from different raw materials.

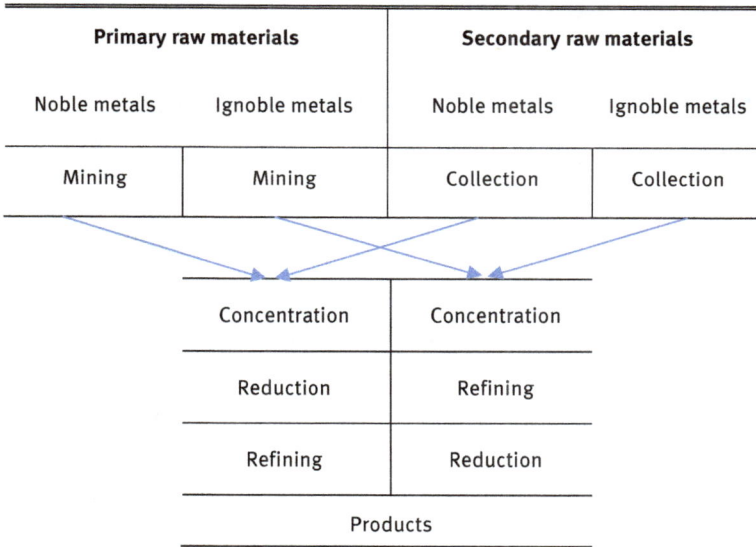

Primary raw materials		**Secondary raw materials**	
Noble metals	Ignoble metals	Noble metals	Ignoble metals
Mining	Mining	Collection	Collection

Concentration	Concentration
Reduction	Refining
Refining	Reduction
Products	

Besides recycling, process development can also lead to down, up and transcycling instead. Although recycling should always constitute the method of choice, downcycling is often inevitable as it allows for avoiding cost-intensive refining steps. Here, again, what is economically feasible dictates the type of process to be implemented. For one thing, recycling seems to be most favorable as it allows unlimited reuse, whereas down, up and transcycling only reduce the dependency on primary raw materials.

1.5.1 Strategies

For EoL magnets, there are three fundamental recycling approaches, each of which leads to different products and requires different steps of treatment. Bast et al. examined all three of them during the MORE project and their work is clearly something worth paying attention to [7]. Figure 1.5.1 shows a simplified scheme of how rare earth ores are processed to metals, magnets, and eventually electric engines. At the end of their life span, these engines become EoL waste which can be processed along three different routes: (i) mechanical recycling (ii) pyrometallurgical recycling and (iii) hydrometallurgical recycling.

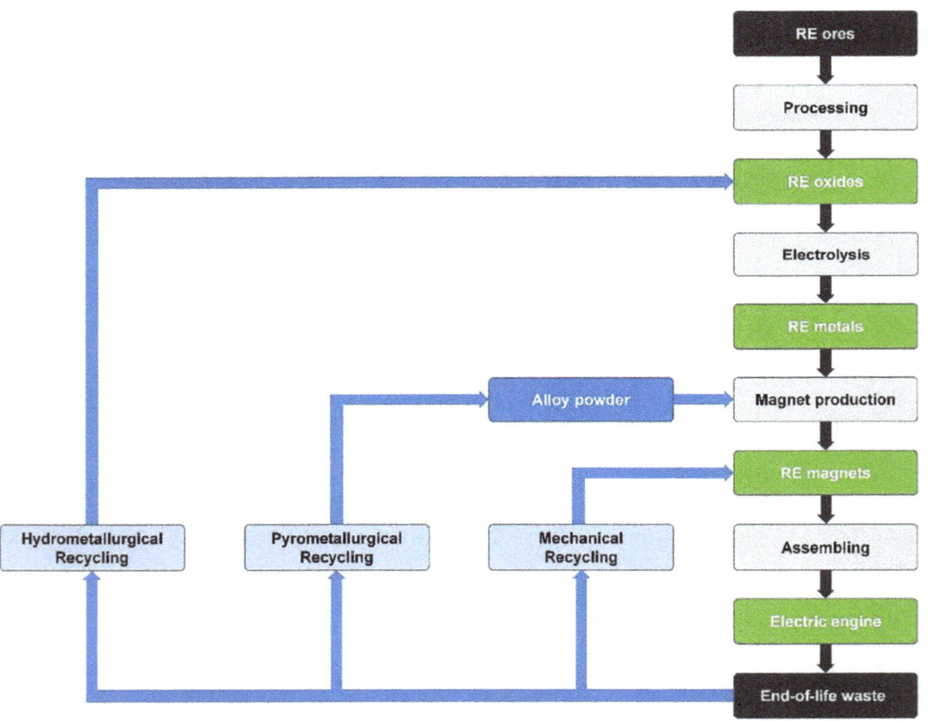

Figure 1.5.1: Strategies of recycling for rare earth magnets (modified from [7]).

Mechanical recycling constitutes the simplest kind of rare earth recovery considering process steps, energy consumption and demand of chemicals. It focusses on reusing whole magnets, for example, for assembling new electric engines. The method allows for recovering rare earths at very low costs but the advantage is realized at the cost of flexibility. When whole magnets are recycled, strict requirements have to be met. First, only undamaged magnets are suitable which makes separate sorting and disassembling of the waste material essential as key steps. Both requirements cause several problems. Even with the best collection imaginable, disassembling still has to be done manually or semimanually which inevitably leads to limited throughputs and considerable costs [17]. Furthermore, magnets differ in size, shape, composition and magnetic flux density. Their reuse is, therefore, limited to the products they initially came from. During the lifespan of a product, the original product is commonly improved with respect to design and performance. Spent magnets are often incapable of meeting the higher requirements resulting, thereof. If magnets are corroded or broken, there is no chance to apply mechanical recycling. The share of magnets complying with the stated requirements is too small and, consequently, electric devices are usually introduced to a different kind of recycling. Here, metal recovery focusses predominately on steel parts and copper.

Pyrometallurgical recycling does not depend on undamaged magnets but separate collection and disassembling are still necessary, yet for other reasons. Pyrometallurgical approaches are the methods of choice for recycling production waste and, as already mentioned, a variety of processes are applied in industrial production (Table 1.5.1). Initially, magnets are crushed and ground to a fine alloy powder. Later, the alloy powder is mixed with new starting material to produce new magnets. In general, this involves a melting step at 1,550 °C, giving the recycling its name. Compared to mechanical recycling, additional treatment steps are necessary but the requirements are less stringent. The main problem arises when impurities are present in the material, which is mostly the case for EoL waste. Pyrometallurgical approaches cannot remove all the impurities, rendering their application for EoL recycling impossible. To address the problem, additional pretreatment steps are introduced such as dissolving plastic parts in organic solvents [5]. The success of these countermeasures is rather limited. In this context, remanence and coercivity are usually measured as key parameters for determining whether magnet recycling was successful or not. Remanence constitutes the magnetic flux density provided by a magnet without interfering external magnetic fields. Roughly speaking, it is a measure of the strength of a magnet. Coercivity is a measure of the ability of magnets to withstand an external magnetic field without becoming demagnetized. Both parameters must meet high requirements. As mentioned above, remanence losses of 3% are already regarded as failures [7]. Therefore, pyrometallurgical recycling remains a method for processing production waste. As pyrometallurgical and mechanical recycling do not provide an answer to the treatment of all types of EoL waste, both strategies are off the list.

Hydrometallurgical recycling aims at recovering rare earths directly by employing techniques that are also applied in primary production. Although this kind of recycling always involves a plethora of process steps, hydrometallurgical approaches are suitable to treat all types of waste. This also comprises shredded electronic scrap that facilitates waste collection. The waste material is usually introduced to some kind of leaching to obtain a rare earth solution ready for further treatment. According to

Table 1.3.1, refining is conducted before reduction, meaning impurities have to be separated first. Without going too far into detail, this is accomplished by conducting several steps of purification such as solvent extraction, precipitation and ion exchange. Refining, usually yields rare earth oxides but other rare earth compounds, for example, chlorides, fluorides or nitrates, are also produced to some extent. Smelt electrolysis and metallothermic reduction (Sm, Yb, Eu, Tm) are used to obtain rare earth metals. For highest purity, these metals additionally undergo vacuum melting, vacuum distillation, vacuum sublimation, zone refining, zone melting and/or solid-state electrotransport processing (cf. Section 1.4) [18]. The production of rare earth metals considerably contribute to the costs of the new process and it is rather unlikely that the available magnet waste will justify these efforts. Since rare earths are mostly traded as oxides, they are better suited as end product. Table 1.5.3 gives a summary.

Table 1.5.3: Comparison of all three recycling approaches.

Recycling	Starting material	Suitable waste material	End Product
Mechanical	Whole magnets	–	Electric devices
Pyrometallurgical	Magnet powder	Production	Magnets
Hydrometallurgical	Magnet scraps	Production, EoL[1], EoU[2]	Oxides/compounds/metals

[1]EoL, end-of-life waste.
[2]EoU, end-of-use waste.

Chemistry or hydrometallurgy?

The development of new processes is a highly complex task in which different disciplines must work together. Within this field of application-oriented research, engineering sciences and chemistry are strongly connected. Against the background of magnet recycling, rare earth metals need to be dissolved in aqueous solutions, purified by solvent extraction, or precipitated. In this regard, engineers usually speak of hydrometallurgy, whereas chemists refer to chemical technology. In general, hydrometallurgy comprises all methods and processes of metal winning and refining, which, of course, includes all wet chemical methods mentioned above. However, the same applies to chemical technology, in which the very same methods are used to develop, optimise and scale-up processes. Although both disciplines use similar methods, process development may benefit from different approaches, as chemists focus on chemical reactions, for example, to effectively separate metals from impurities, while engineering sciences, for instance, consider reactor designs to enhance reactor throughputs. Modern process development therefore requires the expertise of engineers and chemists alike.

1.5.2 Obstacles

Any future EoL recycling process that can recycle all types of magnetic waste will most likely be a wet chemical/hydrometallurgical process. Fortunately, magnet recycling using hydrometallurgical approaches is a wide field of study, providing much information about processing and obstacles that require to be tackled. Although there are differences among the processes proposed so far, most of them are in accordance with the rough scheme depicted in Figure 1.5.2.

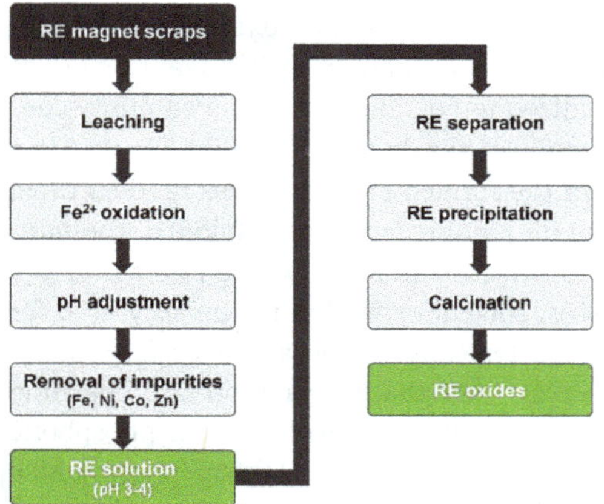

Figure 1.5.2: Scheme of the fundamental process steps of hydrometallurgical magnet recycling.

Overall, a minimum of seven steps are necessary to produce rare earth oxides from magnet scraps. Describing the functionality of each single step would go far beyond what is necessary to understand the obstacles that are usually encountered in hydrometallurgical processes. For detailed information, the reader is referred to the respective scientific literature [4, 9, 11, 12, 19]. Instead, this chapter deals solely with the obstacles arising from the scheme in Figure 1.5.2. As may be seen, the process chain has one intermediate and one end product, both colored in green. The three steps between them are identical with primary rare earth production (cf. Section 1.3). The actual challenge of recycling consists of producing the intermediate, which is a rare earth solution at pH 3–4. In most scientific approaches, the first step of treatment comprises some kind of acid leaching, where a mineral acid is used to dissolve the magnet material. It is this particular step that causes most problems and makes additional treatment steps necessary. Leaching always leads to a strongly acidic solution containing rare earth ions beside a mixture of $Fe^{2+/3+}$ ions, regardless of which mineral acid is used. The low pH value requires an adjustment. In general, a base like NaOH is added to the solution but any increase in pH values beyond 2.5 inevitably results in a precipitation of Fe hydroxide species. For instance, if hydrochloric acid was employed, akaganéite FeO(OH,Cl) precipitates while elevating the pH value [20], sulfuric acid would yield jarosite $KFe_3(OH)_6(SO_4)_2$

or schwertmannite $Fe_8O_8(OH)_6(SO_4) \cdot nH_2O$, whereas the use of nitric acid leads to goethite α-FeO(OH). All of these hydroxide species precipitate before the rare earth solution achieves the pH value of 4. As precipitation proceeds, RE^{3+} ions are partially removed from the solution. This co-precipitation can reduce rare earth yields by up to 10% [20]. Fe precipitation must, therefore, be carried out in a controlled manner. This procedure reduces rare earth losses to a minimum but further process steps become necessary. For Fe precipitation, an oxidation step is required to convert Fe^{2+} into Fe^{3+} ions. With the aid of an oxidizing agent (H_2O_2) or electric power, Fe^{2+} is quantitatively oxidized to Fe^{3+}, followed by a carefully conducted precipitation, where OH^- ions are added slowly to the solution [9, 11]. Even then, the small particle sizes (1–2 µm) of the formed Fe hydroxides renders them considerably difficult to filtrate [20]. With all requirements being fulfilled, the purified solution is subsequently subjected to solvent extraction where rare earths are separated from each other. Since the 1960s, solvent extraction is primarily used for rare earth separation. Nowadays, kerosene typically serves as an organic phase which is modified with e.g. 1-decanol to accelerate phase separation [21]. Chelating ligands, such as carboxylic acids like ethylenediaminetetraacetic acid or phosphoric acids and esters, are commonly used as extracting agents. Typical phosphorous based chelating agents are, for instance, tributyl phosphate, di-(2-ethylhexyl)phosphoric acid, 2-ethylhexyl hydrogen-2-ethylhexylphosphonate (PC-88A) or bis(2,4,4-trimetylpentyl)dithiophosphinic acid (Cyanex 302). The named extraction agents are, without exception, organic acids. Solvent extraction, therefore, strongly depends on the pH values of the aqueous phase, which is why pH values have to be elevated to 3–4 after acid leaching [9, 21, 22]. Solvent extraction and all steps that follow are state-of-the-art techniques. From here on, rare earth ores (primary raw materials) and rare earth waste (secondary raw materials) follow the same process chain. The process chain allows for merging their processes by introducing both streams to the same solvent extraction step. Before any EoL recycling can be integrated with an existing process chain, process development has to find a way to reduce process costs as well as chemical consumption of future recycling processes.

1.5.3 Alternatives: solid-state chlorination

In order to tackle obstacles related to acid digestion, alternative approaches are also taken into consideration. One of these approaches is called solid-state chlorination. This alternative method of digestion aims to replace the acid leaching that is usually conducted, so that chemical consumption, the amount of wastewater and the acidity of the rare earth solution can be decreased [15]. This enables a reduction of both chemical and disposal costs [23]. Instead of using mineral acids, solid-state chlorination converts the metal content of $Fe_{14}Nd_2B$-alloys into water-soluble metal chlorides by utilizing anhydrous hydrogen chloride (HCl) gas. The $HCl_{(g)}$, is, thereby, produced by the thermal decomposition of solid ammonium chloride (NH_4Cl). Gaseous NH_3 forms as co-product. It leaves the reactor along with unreacted $HCl_{(g)}$, $H_{2(g)}$ and $H_2O_{(g)}$ as exhaust gas. After chlorination, the reaction mixture is allowed to cool down to ambient temperature. An acetate buffer solution is then utilized to dissolve all formed metal chlorides at pH 3, thereby preventing the precipitation of Fe hydroxides. Both process steps eventually result in a pH-adjusted rare earth solution ready for further treatment, for example, solvent extraction. In contrast to acid leaching, the process splits chlorination and dissolution into two separate steps which provides several advantages compared to conventional leaching using mineral acids:

(i) **Consumption of chemicals**: NH_4Cl contains 68 wt.% of HCl, which is almost twice as much as hydrochloric acid (37 wt.%). This almost halves the demand for the digestion agent. Any $HCl_{(g)}$ that remained unreacted, recombines with $NH_{3(g)}$ in the exhaust gas, upon cooling. The recovered NH_4Cl is reused in the next solid-state chlorination run. This digestion step, consequently, generates no waste material.

(ii) **Costs of chemicals**: In addition to the higher HCl content, NH_4Cl is also cheaper than hydrochloric acid (37 wt.%). Due to the chlorination reaction withdrawing $HCl_{(g)}$ from the reaction atmosphere, there is always an equal stoichiometric amount of $NH_{3(g)}$ in the exhaust gas. When this $NH_{3(g)}$ is removed from the gas stream by a scrubber, an aqueous NH_3 is gained as a co-product, the revenues of which contribute to compensate for costs of chemicals.

(iii) **Wastewater**: To prevent the precipitation of Fe hydroxides, the buffer adjusts the pH of the rare earth solution. As a side effect, this also prevents the production of strongly acidic wastewater and circumvents the otherwise obligated neutralization step. The acetate buffer consists of a 3 wt% acetic acid solution and is biodegradable.

The solid-state chlorination comprises only a few simple reactions and its functionality is therefore easy to explain. At first, the EoL magnets that may come from, for example, wind turbines are ground to a powder which is then mixed with solid NH_4Cl. The mixture is introduced to a rotary kiln and heated to a temperature between 225 °C and 325 °C. With temperature rising, the following reactions take place:

$$\beta\text{-}NH_4Cl_{(s)} \overset{>200\,°C}{\longleftrightarrow} NH_{3(g)} + HCl_{(g)} \tag{1.5.1}$$

$$Fe_{(s)} + 2\,HCl_{(g)} \rightarrow FeCl_{2(s)} + H_{2(g)} \tag{1.5.2}$$

$$RE_{(s)} + 3\,HCl_{(g)} \rightarrow RECl_{3(s)} + \frac{3}{2}\,H_{2(g)} \tag{1.5.3}$$

When the temperature exceeds 200 °C, NH_4Cl starts to decompose into gaseous ammonia NH_3 and HCl (eq. (1.5.1)). The latter reacts with the magnet powder by converting all metals into their respective metal chlorides (eqs. (1.5.2) and (1.5.3)). In the case of $Fe_{14}Nd_2B$ magnets, chlorination produces at least three solid metal chlorides ($FeCl_2$, $NdCl_3$ and $DyCl_3$) with yields of up to 85%. Dy is usually added to $Fe_{14}Nd_2B$ magnet alloys to improve their temperature stability. These water-soluble chlorides can be readily dissolved.

The second step of the solid-state chlorination procedure is a leaching step at a pH value adjusted to 3, which is achieved by employing a dissolved acetate buffer. The acetate buffer was initially used to prevent the precipitation of Fe hydroxide species while dissolving the formed chlorides. Adjusting the pH value with an acetate buffer containing only 30 $g \cdot L^{-1}$ acetic acid already proved to be advantageous for rare earth yields [24]. Two kinds of reactions take place during buffer leaching: acetic acid-mediated metal dissolution (eqs. (1.5.4) and (1.5.5)) and cementation (eq. (1.5.6)) [15]:

$$Fe_{(s)} + 2\,CH_3COOH_{(aq)} \rightarrow Fe(CH_3COO)_{2(aq)} + H_{2(g)} \tag{1.5.4}$$

$$RE_{(s)} + 3\,CH_3COOH_{(aq)} \rightarrow RE(CH_3COO)_{3(aq)} + \frac{3}{2}H_{2(g)} \tag{1.5.5}$$

$$3\,Fe^{2+}{}_{(aq)} + 2\,RE_{(s)} \rightarrow 3\,Fe_{(s)} + 2\,RE^{3+}{}_{(aq)} \tag{1.5.6}$$

In eqs. (1.5.4) and (1.5.5), acetic acid dissolves Fe, Nd and Dy that remained unchlorinated, to yield their water-soluble metal acetates. At the same time, Fe^{2+} ions undergo a redox reaction with both rare earth metals. By this so-called cementation (eq. (1.5.6)), Fe^{2+} ions oxidize unchlorinated rare earth metals to Nd^{3+} and Dy^{3+} ions. Fe^{2+} ions are, thereby, reduced to their metallic state. The cementation, on its own, contributes to rare earth yields by up to 7% [15, 24]. Eventually, buffer leaching produces a rare earth-containing solution with the right pH value. An uncontrolled Fe hydroxide precipitation is not to be seen in the process. Therefore, two of the otherwise obligatory steps, neutralization and Fe^{2+} oxidation, become redundant. However, the Fe^{2+} ions should preferably be removed, for instance, as $FeS_{(s)}$ via sulfide precipitation. According to Figure 1.5.3, the replacement of acid leaching with solid-state chlorination simplifies the process significantly. The new recycling process that uses solid-state chlorination to recover rare earths from spent $Fe_{14}Nd_2B$- and SmCo-magnets is the MagnetoRec process.

Furthermore, solid-state chlorination produces a NH_3 solution, the purity of which exceeds 99.9998%, rendering it as a high-quality readily available-to-use solution for the fertilizer industry. Apart from nitrogen-based fertilizers, NH_3 is a platform chemical for the production of urea resins, nitric acid and many organic compounds (amines, amides, cyanides, nitrates and hydrazine). This way, the revenues from marketing the obtained NH_3 solution contribute to reduction of the process costs.

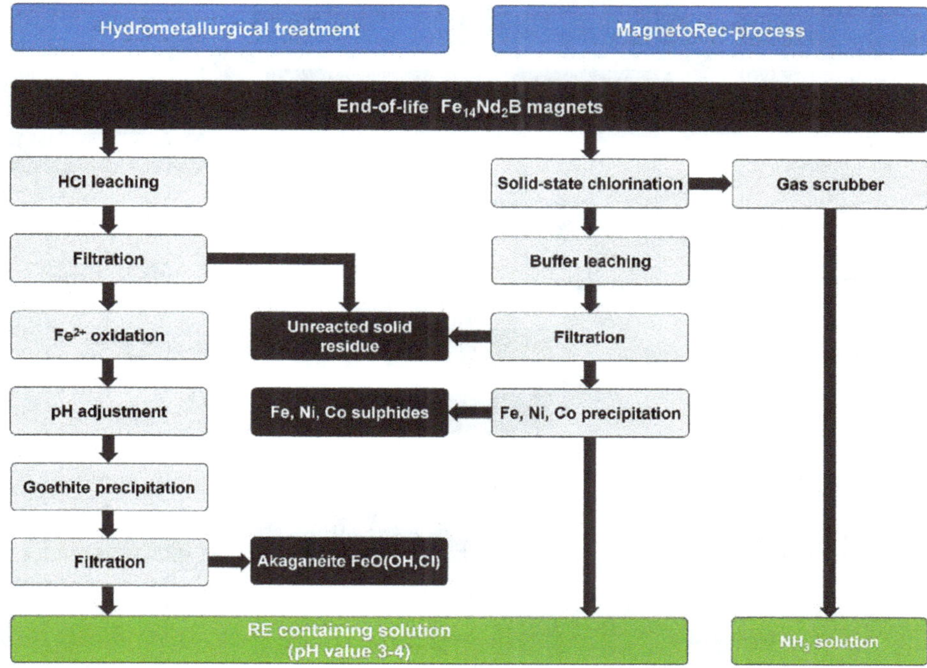

Figure 1.5.3: Comparison of the conventional HCl leaching and the MagnetoRec process.

As mentioned earlier, collectable waste streams are most likely too small for building up the entire process chain. Thus, magnet recycling has to be merged with one or more industrial processes to become economical. The MagnetoRec process is meant only to close the gap between EoL magnets and conventional rare earth production. In this respect, the process is to be understood more as a supplement to industrial production. Figure 1.5.4 presents a rough scheme of how processes can be combined to integrate magnet recycling. Beside EoL magnets, NH_4Cl is required as main reactant for the chlorination step. In principle, chemical costs could be reduced even further if NH_4Cl from waste streams could be used. In polysilane production, for example, hydrochloric acid is neutralized with an NH_3 solution during polymerization [16]. The formed NH_4Cl leaves the process as production waste for disposal and could instead be introduced to solid-state chlorination (upcycling).

The MagnetoRec process yields three different products: the rare earth solution (pH 3–4), the highly pure NH_3 solution and the sulfides FeS, CoS and NiS. In this regard, Ni and Co originate from a Ni layer covering $Fe_{14}Nd_2B$ magnets (corrosion protection) and missorted SmCo magnets. The sulfides can be roasted to give metal oxides if they are free of chlorine. The metal oxides might then be introduced to steel production (transcycling). When chlorine is present in the precipitate, sulfides

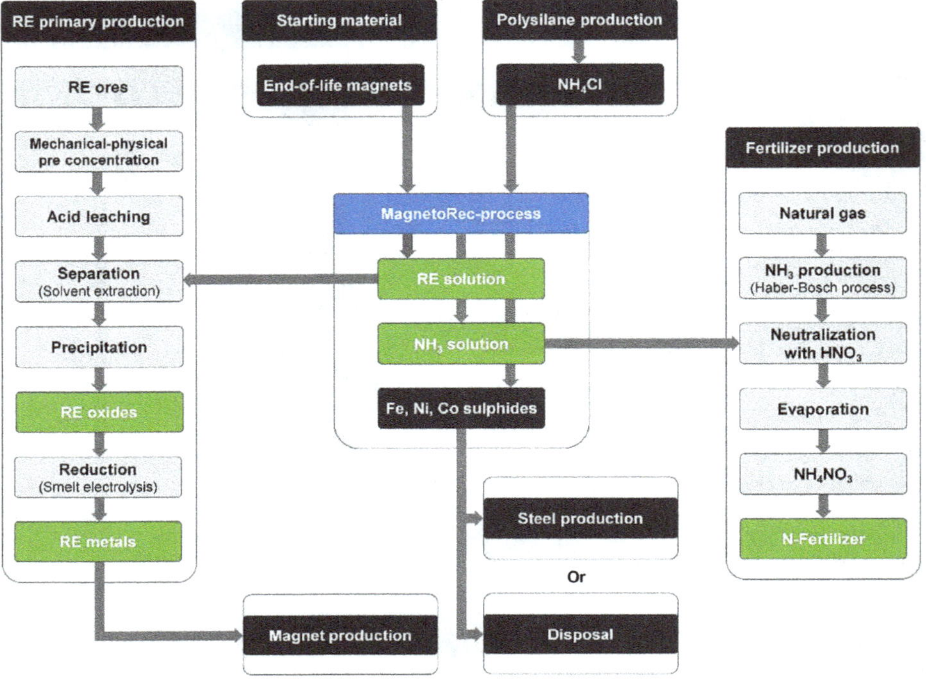

Figure 1.5.4: A rough scheme of how the MagnetoRec process can be combined with industrial processes.

have to be disposed. Although recycling should always be preferred, this option is usually not applied once disposal is more economical. When the coproduct of the process, the NH_3 solution, is added to nitric acid, neutralization yields NH_4NO_3, which is a N-based fertilizer (transcycling). Since NH_3 is a platform chemical, the solution can in principle be fed into a large number of different processes. A reuse in polysilane production is, therefore, possible too (recycling). The main product, the rare earth solution, is expected to be introduced to the refining part of rare earth primary production. At this point, magnet recycling merges with the established process chain. The resulting rare earth metals can be used to manufacture new permanent magnets (recycling). The MagnetoRec process, with all its products and reactants, is then integrated with four different process chains (polysilanes, fertilizer, rare earth metals and steel). Depending on the chemical compound under consideration, the MagnetoRec process is a recycling, transcycling, upcycling or disposal process. Even though solid-state chlorination has already proven to be an effective method for magnet recycling, scale-up will still take years before it can be operated at an industrial scale. Although large-scale EoL recycling of rare earths remains an issue to be solved in future, the MagnetoRec process constitutes the first step towards economic rare earth recycling. This process is already operated economically on a technical scale. The demonstration plant is located in Freiberg, Germany, and constitutes the world's first reactor, enabling solid-state chlorination to be operated continuously.

References

[1] K. Hirota, T. Minowa, US6960240 B2, 2002.
[2] Y. Hirose, T. Go, M. Renda, N. Kawamura, A. Otaki, WO2000039514 A1, 2000.
[3] B. Greenberg, WO1994026665 A1, 1994.
[4] A. Asada, US5728355 A, 1998.
[5] H. Onishi, T. Terada, Y. Yamagata, F. Yamashita, EP1096517 B1, 2005.
[6] Pressemitteilung, "Hitachi Develops Recycling Technologies for Rare Earth Metals", can be found under http://www.hitachi.com/New/cnews/101206.pdf, 2010.

[7] U. Bast, R. Blank, M. Buchert, T. Elwert, F. Finsterwalder, G. Hörnig, T. Klier, S. Langkau, F. Marscheider-Weidemann, J.-O. Müller et al., Recycling von Komponenten und strategischen Metallen aus elektrischen Fahrantrieben (MORE). Abschlussvorhaben zum Verbundvorhaben, 2015.
[8] T. E. Graedel, J. Allwood, J.-P. Birat, M. Buchert, C. Hagelüken, B. K. Reck, S. F. Sibley, G. Sonnemann, Recycling Rates of Metals. A Status Report, United Nations Environment Programme, Nairobi, Kenya, 2011. ISBN: 9789280731613
[9] T. Elwert, D. Goldmann, F. Schmidt, R. Stollmaier, World Metall.-Erzmet. 2013, 66, 209.
[10] T. Lorenz, K. Golon, P. Fröhlich, M. Bertau, Chem. Ing. Tech. 2015, 87, 1373.
[11] M. A. R. Önal, C. R. Borra, M. Guo, B. Blanpain, T. van Gerven, J. Rare Earths 2017, 35, 574.
[12] P. Venkatesan, Z. H. I. Sun, J. Sietsma, Y. Yang, Sep. Purif. Technol. 2018, 191, 384.
[13] T. Jüstel, "Seltene Erden – Vorkommen und Anwendung", can be found under https://www.fh-muenster.de/ciw/downloads/personal/juestel/juestel/Seltene_Erden-Vorkommen_und_Anwendungen-1.pdf, 2012.
[14] a) T. Huckenbeck, R. Otto, E. Haucke, WO2012143240 A2, 2012; b) J.-J. Braconnier, A. Rollat, US8501124 B2, 2013; c) J.-J. Braconnier, A. Rollat, US9102998 B2, 2015.
[15] T. Lorenz, M. Bertau, Phys. Sci. Rev. 2017, 20160067.
[16] C.-F. Hoppe, C. Götz, H. Rauleder, G. Uhlenbruck, WO2014023470 A1, 2014.
[17] M. Schluep, C. Hagelueken, R. Kuehr, F. Magalini, C. Maurer, C. Meskers, E. Mueller, F. Wang, Recycling from E-waste to Resources. Studie im Auftrag der UNEP und UNU, 2009.

[18] C. K. Gupta, N. Krishnamurthy, Extractive Metallurgy of Rare Earths, CRC Press, Boca Raton, 2005. ISBN: 9780415333405
[19] a) T. Itakura, R. Sasai, H. Itoh, J. Alloys Compd. 2006, 408–412, 1382; b) M. A. R. Önal, E. Aktan, C. R. Borra, B. Blanpain, T. van Gerven, M. Guo, Hydrometallurgy 2017, 167, 115.
[20] T. Lorenz, Dissertation, TU Bergakademie Freiberg, Freiberg, 2018.
[21] T. Elwert, Dissertation, TU Clausthal, Clausthal, 2015.
[22] a) R. Dittmeyer, W. Keim, G. Krevsa, A. Oberholz (Eds.) Winnacker-Küchler Chemische Technik. Prozesse und Produkte, Band 6: Metalle, Wiley-VCH, Weinheim, 2006. ISBN: 3527307710; b) R. E. Kirk, D. F. Othmer, J. I. Kroschwitz, M. Howe-Grant, Encyclopedia of Chemical Technology, Wiley, New York, Chichester, 1998. ISBN: 0471527041
[23] a) T. Lorenz, M. Bertau, J. Cleaner Prod. 2019, 215, 131; b) T. Lorenz, M. Bertau, J. Cleaner Prod. 2019, 118980.
[24] T. Lorenz, P. Fröhlich, M. Bertau, Chem. Ing. Tech. 2017, 66, 209.

Jörg Zimmermann

1.6 Recycling aspects – phosphors

Currently, rare earth elements (REE) are needed for modern technologies such as wind turbines, E-mobility or energy-efficient lighting. However, REE are crucial for many more different applications, as is depicted in Figure 1.6.1.

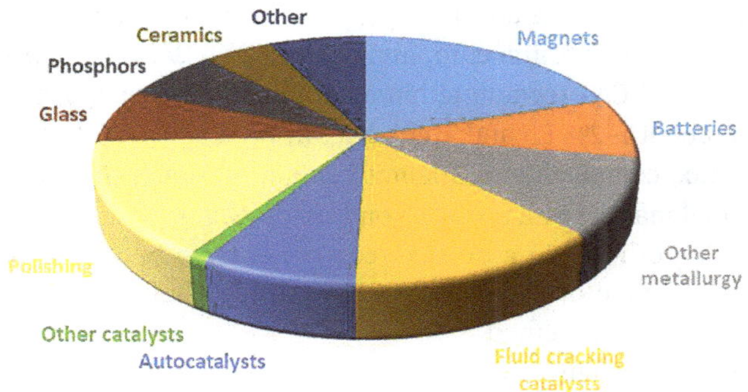

Application	Share in %
Magnets	20
Batteries	8
Other metallurgy	10
Fluid cracking catalysts	13
Autocatalysts	8
Other catalysts	1
Polishing	15
Glass	7
Phosphors	7
Ceramics	5
Other	7

Figure 1.6.1: Share of REE in current applications. Values are taken from the report on Critical Material for the EU [1].

Different REEs are used in different applications. For instance, for grinding and polishing of glasses in the optical industry, only Ce in the form of Ce(IV) oxide is used as basic material. In contrast, in the field of phosphors, at least about half of the REEs is used. Some of the main applications and the respective REEs are listed below:

Permanent magnets (Nd, Pr, Dy, Tb, Sm)
Phosphors (Eu, Tb, Y, Ce, Gd, La, Lu)
Batteries (La, Ce, Nd, Pr)

https://doi.org/10.1515/9783110654929-006

Polishing compounds (Ce)
Catalysts (La, Ce, Pr, Nd, Y)

The content of REE is extremely variable, depending on the application. Here we want to compare outstanding examples regarding REE that form the main part of bulk materials and are used as dopants in matrices.

In some materials or products, the REE concentrations are below the natural concentrations in ores [2]. For example, in a fluorescent lamp, the content of phosphor is around 1%, that is, for a weight of approximately 100 g, the phosphor content is 1 g with actually 10–500 µg REE, such as Eu, Ce and Tb. A typical mixture in a fluorescent lamp is a triband phosphor comprising yttrium europium oxide (YOX), BAM (Eu-doped barium magnesium aluminate) and CAT (cerium terbium aluminate) The composition is 60% YOX (i.e., 48% Y) and about 1–2% Eu and Tb from YOX, BAM and CAT, respectively. In terms of luminous flux, comparable LED lamps contain phosphor quantities of around 100 µg, and the actual mass of REE is correspondingly lower.

Although the concentration of REEs in a product/component can be below the concentration in ores, the advantage of the recycling compared to mining is that the material mix in the product is well defined.

In terms of REE content, the situation is completely different in materials for strong magnets, such as NdFeB. In $Nd_2Fe_{14}B$, the share of Nd is ca. 36 mass%, and in $SmCo_5$ the share of Sm is ca. 30 mass%. However, regardless of the quantity of the REEs contained in modern functional materials, the need for the recovery of those comes also from the fact that mining requires elaborate processes while causing groundwater pollution, and is inevitably coupled with exposure to the radioactive elements, uranium and thorium [3].

As in mining processes, where rare earth ores are just a small fraction mixed with gangue, in urban mining, the first step is a good presorting and a dismantling of products to concentrate the desired elements. In the case of lamps, the presorting is required to separate rare-earth containing lamps (fluorescent lamps and LED lamps) from others like incandescent lamps, halogen lamps or discharge lamps without phosphors, among others. The REEs are in the phosphors as dopants. Furthermore, the separation between fluorescent lamps and LED lamps is mandatory, since LED lamps (especially those with white LEDs based on fluorescent converters) contain around 1000 times less rare earths, when compared to fluorescent lamps. Additionally, the fluorescent lamps contain mercury, that is, the phosphors are contaminated.

After presorting, the next step in the recycling process for fluorescent lamps is the safe capture of the hazardous material (particularly, elemental Hg). In the case of tube-like lamps, for example, the "Kapp-Trenn-(trim separate)" [4] process is used. The metal parts are removed first, and then the mercury containing phosphor layer is blown out by compressed air. Thereafter, a thermal treatment and a distillation process follow to separate and recover mercury and phosphors. Subsequently, the phosphor mix can be recycled by a chemical treatment.

As in the case of lamp phosphors, more than half of the REE is used in the form of inorganic compounds, a lot of them as oxides, which can only be extracted by wet chemical leaching [2]. The first patent for the recovery of REE-containing phosphors was issued in 1969 [5]. The process is roughly described to consist of the following steps:

Heating up between around 540–590 °C to remove organic contaminants, such as polymers and photoresists → production of an aqueous slurry containing 1–25% solid material → heating up the slurry to about 80 °C and adjusting the pH value between 3 and 5 with acid → removal of residual solids by filtration → addition of oxalic acid to the resulting liquid phase → precipitation of the rare earth oxalates and recovery of the relatively pure solids by filtration → drying of the oxalates and conversion to oxides by a two-step heating to above 815 °C and then up to about 1,370 °C.

In the patent specification, an example is given for the recycling of a phosphor mix typical for this area: europium-activated gadolinium-yttrium oxide as the main part with smaller quantities of yttrium vanadate, zinc cadmium sulfide and zinc sulfide.

In Europe, recycling processes for phosphors have been developed during the rare earth crisis. Osram [6] uses a chemical leaching on a pilot scale; Solvay is the only European company with a plant in La Rochelle operating a REE recycling process on an industrial scale [7]. The Entsorgungsdienste Freiberg (FNE) operated a plant for the recycling of production waste developed in cooperation with the Technische Universität Bergakademie Freiberg [8], with a focus on the recovery of europium from the red phosphor Y_2O_3:Eu. In Figure 1.6.2 an overview of the three recycling routes is given.

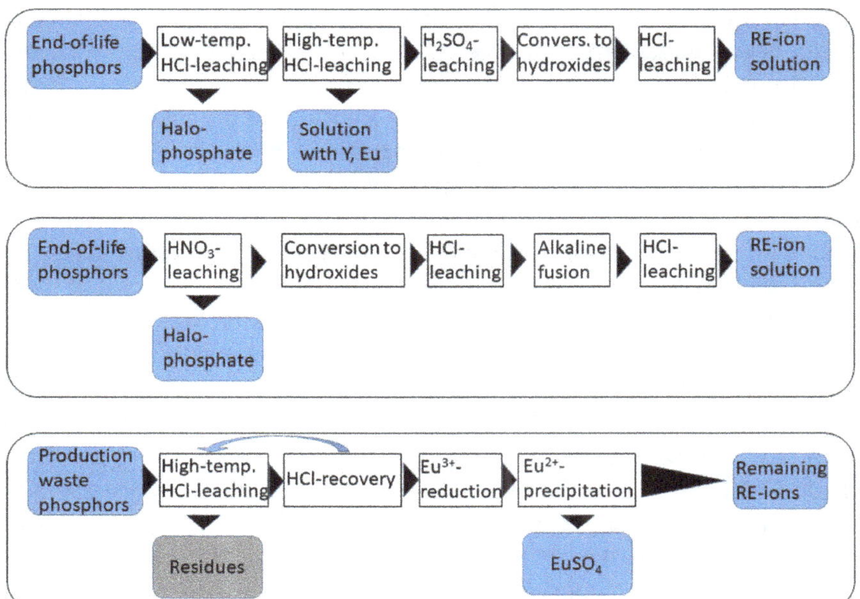

Figure 1.6.2: Overview of the recycling routes of OSRAM, Solvay and FNE.

Since the prices of *RE* compounds have drastically decreased after the REE crisis, all the mentioned plants cannot operate economically anymore.

For REE in LED lamps, the recycling process is even more challenging than in the case of fluorescent lamps. In the EU-project cycLED [9], it was found that the recovery of the REE is currently not economically viable. However, the biggest challenge is the preconcentration of the REE containing components. This approach requires a selective fragmentation of the LED lamps as described by Gassmann et al. [10]. An alternative route is the use of a recyclable phosphor layer as described by Hämmer et al., which can easily be dissolved to liberate the phosphor. The phosphor is filtered out, collected and reused in a new phosphor layer [11, 12].

In contrast, for NdFeB, the content of REEs is around 30%. Although neodymium can be substituted by the less-expensive praseodymium by up to 20% [13], the more expensive dysprosium is added in amounts of up to 10% [14] for the improvement of the coercive field strength at higher temperatures. The REEs are, however, matrix-elements, thus having a great share. For neodymium from NdFeB magnets, a variety of processes exist: wet chemical leaching with NaOH [15], with liquid metals through alloying [16, 17] or with acidic leaching and *RE* precipitation using oxalic acid [18].

Recycling can take place at different levels. In some cases, it is more efficient in terms of economy and recovery rate to process the functional material, that is, recovering of a compound instead of separating its elements. Shin Etsu mixes new FeNdB alloys with production waste and melts new materials [19]. Panasonic treats scrap magnets with organic agents to remove the plastic coating and then mix and melt it with new material [20]. Smart methods are hydrogen decrepitation [21] and hydrogen decrepitation-deabsorbation-recombination [22] processes, where scrap magnets are comminuted by hydrogen embrittlement [23] in the first step to get a starting material for new magnets. It has to be mentioned that for cost reduction, the development of functional materials aims (among other things) at minimizing expensive REEs. For instance, the reduction of expensive dysprosium in magnets is of economic interest for the producing industry. On the other hand, this makes recycling and the recovery of these metals, which may result in the disposal of those magnets followed by dissipation of REEs economically less attractive.

However, in many applications, the REEs exist only in small concentrations and the challenge is to preconcentrate and extract these elements from the matrix efficiently. The LED phosphors in lamps are an impressive example of the need of preconcentration, as depicted in Figure 1.6.3.

A typical LED lamp contains, semi-conductor (In, Ga) and electrical contacts (Ag, Au) besides REEs (Y, Lu, La, Ce, Eu, Tb). Together, they hold a mass share of about 0.003% in the lamp. By shredding and sorting the LED, chips could be separated and collected, thus holding a share of 0.4%. Pre-sorting, selective fragmentation and preconcentration are key in an efficient recycling process.

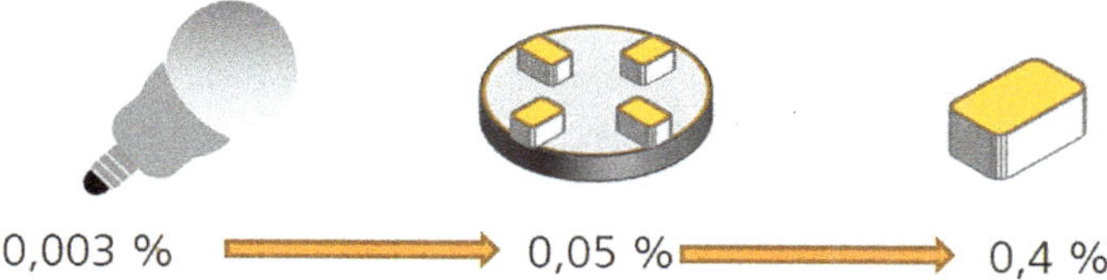

Figure 1.6.3: Preconcentration of critical materials in a typical LED lamp.

References

[1] Critical Raw Materials Profiles 2014, p. 153, based on data by Roskill Information Services and IMCOA.
[2] K. Kuchta, Recycling Seltener Erdmetalle, TU Hamburg-Harburg, TU International 69, 2012.
[3] K. Binnemans, P. T. Jones, B. Blanpain, T. Van Gerven, Y. Yang, A. Waltone, M. Buchertf, J. Cleaner Production 2013, 51, 1.
[4] P. Herborn, WO1997032332 A1.
[5] R. E. Dodds, US-Patent 3577 351 1969/1 respectively C.A. 75 S.8043 1971.
[6] R. Otto, W. Agnieszka, DE 10 2006 025 945 A1 2007.
[7] The Solvay "LOOP" Project: (© 2014 Solvay, Layman's Report, Projekt «LOOP»), can be found under https://www.solvay.com/sites/g/files/srpend221/files/2018-07/solvay-loop-project-de-en.pdf.
[8] M. Stelter, Hydrometallurgische Raffination Seltener Erden, can be found under http://www.acatech.de.
[9] O. Deubzer, R. Jordan, M. Marwede, P. Chancerel, Deliverable 2.1: Categorization of LED Products for the EU-project cycLED (Cycling Resources Embedded in Systems Containing Light Emitting Diodes), can be found under http://www.cyc-led.eu/Files/CycLED_D2_1_120507.pdf.
[10] A. Gassmann, J. Zimmermann, R. Gauß, R. Stauber, O. Gutfleisch, LED Professional Review (LpR) 2016, 56, 74.
[11] M. Hämmer, A. Gassmann, A. Reller, H. von Seggern, O. Gutfleisch, R. Stauber, J. Zimmermann, J. Electron. Mater. 2019, 48, 2249.
[12] M. Hämmer, A. Gassmann, A. Reller, H. von Seggern, O. Gutfleisch, R. Stauber, J. Zimmermann, Mater. Technol. 2019, 34, 178.
[13] G. Hatch, can be found under http://www.techmetalsresearch.com/seagate-rare-earths-and-the-wrong-end-of-the-stick/.
[14] S. Hoenderdaal, L. T. Espinoza, F. Marscheider-Weidemann, W. Graus, Energy 2013, 49, 344.
[15] B. Greenberg, WO 026 665, 1994.
[16] O. Takeda, T. H. Okabe, Y. Umetsu, J. Alloys Compd. 2006, 408–412, 387.
[17] O. Takeda T. H. Okabe, Y. Umetsu, J. Alloys Compd. 2004, 379, 305.
[18] A. Asada, US5728355 A, 1998.
[19] K. Hirota, T. Minowa, US6960240 B2, 2002.
[20] H. Onishi, T. Terada, Y. Yamagata, F. Yamashita, EP1096517 B1, 2005.
[21] O. Diehl, M. Schönfeldt, E. Brouwer, A. Dirks, K. Rachut, J. Gassmann, K. Güth, A. Buckow, R. Gauß, R. Stauber, O. Gutfleisch, J. Sustain. Metall. 2018, 4, 163.
[22] A. Lixandru, I. Poenaru, K. Güth, R. Gauß, O. Gutfleisch, J. Alloys Compd. 2017, 724, 51.
[23] A. Walton, A. Williams, Mater. World 2011, 19, 24.

Rainer Pöttgen

1.7 The structures of the elements

The rare earth metals crystallize with structures that are derived from the well-known simple face-centered cubic (*fcc*; Cu type), hexagonal closest packing (*hcp*; Mg type) and body-centered cubic (*bcc*; W type) arrangements [1, 2]. The complete structural data has already been summarized and reviewed competently in 1982 by *Donohue* in his basic textbook *The Structures of the Elements* [3]. Most of the diffraction data available at that time were room temperature data and they generally relied on Debye-Scherrer photographs with limited resolution. Meanwhile, these studies have significantly been extended, since diffraction setups for the very low- and high-temperature regimes are more or less routinely available. Another advantage is the use of powerful synchrotron X-ray sources along with high-resolution image plate detectors. Furthermore, diamond-anvil cells (DAC) allow for high-pressure studies up to 300 GPa. Yet, not all phases of the rare earth metals are fully characterized. High-pressure studies as a function of temperature are still an open field.

We start with the three basic structure types. The room-temperature structures of holmium (Pearson symbol *hP*2, Mg type, $P6_3/mmc$), cerium (Pearson symbol *cF*4, Cu type, $Fm\bar{3}m$) and europium (Pearson symbol *cI*2, W type, $Im\bar{3}m$) are presented in Figure 1.7.1. Before discussing these structures in detail, we briefly introduce the nomenclature. Consequently, we list the Pearson symbol, which includes the crystal system, the Bravais lattice and the number of atoms per unit cell.

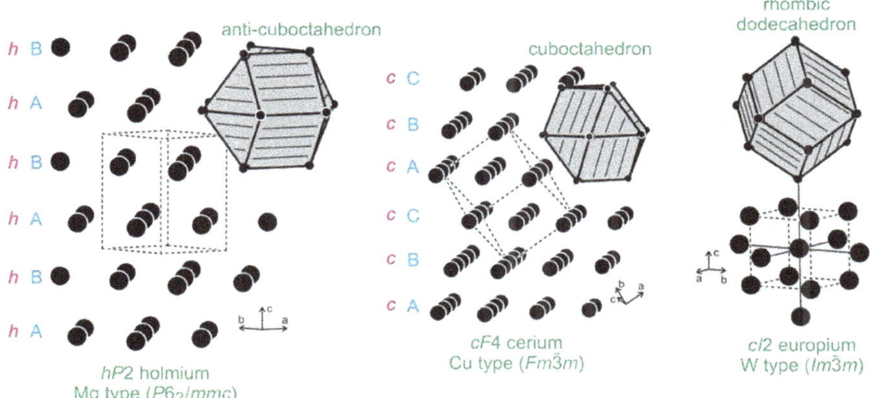

Figure 1.7.1: Crystal structures of *hP*2 holmium (Mg type), *cF*4 cerium (Cu type), and *cI*2 europium (W type). The coordination polyhedra, the stacking sequences (in AB/ABC and *hc* notation) and the space group symmetry are indicated.

https://doi.org/10.1515/9783110654929-007

The close-packed layers in the holmium and cerium structures show stacking sequences of ABAB and ABCABC, respectively, with an *anti*cuboctahedral coordination (CN 12) for the *hcp* and a cuboctahedral one (CN 12) for the cubic-closest packing (*ccp* or *fcc*), both with 74% space filling. For longer stacking sequences than ABC, the *Jagodzinski hc* notation [1, 4] is helpful. We have therefore marked the *h* and *c* layers in Figure 1.7.1 along with the ABC ones. A layer is entitled "*h*" if the one above and below are the same (*anti*cuboctahedral coordination) and "*c*" if they are different (cuboctahedral coordination). The structures of holmium and cerium are thus complete *h* and *c* stackings, respectively. Europium adopts the *bcc* structure with a space filling of 68%. Each europium atom has 8+6 europium neighbors (CN 14; rhombic dodecahedron).

Neodymium and samarium exhibit more complex stacking sequences at room temperature: ABAC for neodymium and ABACACBCB for samarium. The AB/ABC stacking sequences are transformed to the *hc* nomenclature in Figure 1.7.2. The advantage of the *hc* nomenclature is readily evident. The neodymium and samarium structures can be described by the short repeating sequences *hc* and *hhc*.

*hP*2 holmium → *h*
A B A B A B A B
h h h h h h h h

*cF*4 cerium → *c*
A B C A B C A B C
c c c c c c c c c

*hP*4 neodymium → *hc*
A B A C A B A C
c h c h c h c h

*hR*9 samarium → *hhc*
A B A C A C B C B
h h c h h c h h c

Figure 1.7.2: Stacking sequences of the *hP*2 holmium, *cF*4 cerium, *hP*4 neodymium and *hR*9 samarium structures in the AB/ABC and *hc* notation. For details, see text.

The mix-up of different stackings of A, B and C layers has drastic consequences on the symmetry of the resulting structure. The monomeric layers are all equal, but the different sequences lead to different space group symmetries. Only the ABC sequence is cubic, while all others are hexagonal, with either *P* or *R* lattices. A switch to another stacking sequence forces translation of one or more layers and is thus a reconstructive phase transition by breaking bonds and forming new ones. These structures are not related by group-subgroup schemes [5]. For these room temperature modifications, the *c*/*a* ratios are approximately integer multiples of 1.633, the ideal value of the *hcp* structure. Some of the rare earth elements show small anomalies in the temperature dependence of their *c*/*a* ratio because of the onset of magnetic ordering.

We now turn to the temperature dependence of the rare earth structures under ambient pressure conditions. The crystallographic data for these phases was taken from the Pearson database [6] and the compilations in [7] and [8]. The structure types adopted by the rare earth elements are graphically presented on a temperature scale in Figure 1.7.3.

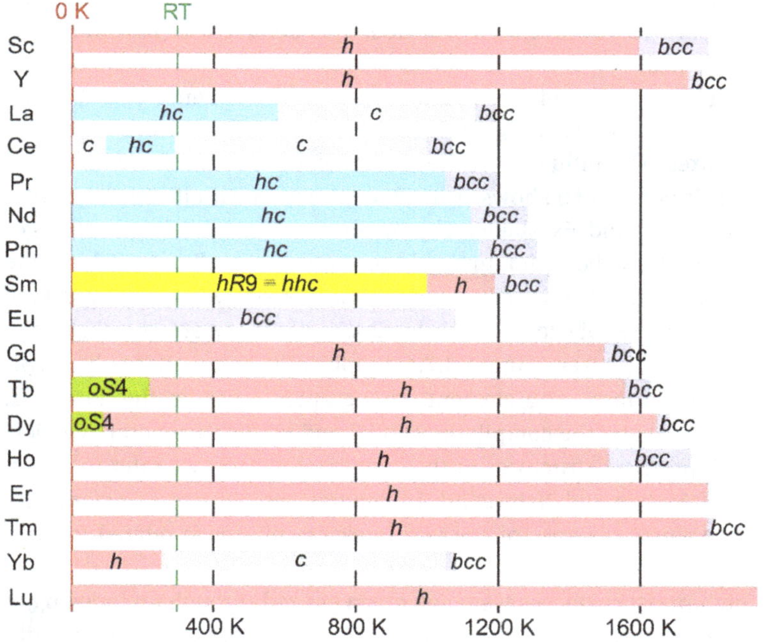

Figure 1.7.3: Structures of the rare earth elements as a function of temperature, drawn with the data listed in references [6–8]. For details see text.

Lutetium and erbium keep their simple *hcp* structure up to the melting point. Similar behavior is observed for *bcc* europium. All other rare earth metals show at least one temperature-induced structural phase transition. Scandium, yttrium, gadolinium, holmium and thulium display a reconstructive phase transition from the *hcp* to the *bcc* type at high temperature. In all cases, this *bcc* arrangement is only stable within a small temperature regime just before the melting point. This is, in principle, similar to terbium and dysprosium; however, these two elements show an additional low-temperature phase – terbium below 220 and dysprosium below 86 K. Here, the *hcp* structure shows an orthorhombic distortion (a *translationengleiche* transition of index 3 from $P6_3/mmc$ to *Cmcm*). In the low-temperature regime, terbium and dysprosium are isotypic with α-uranium.

Praseodymium, neodymium and promethium adopt the *hc* stacking sequence and transform to a *bcc* type at high temperatures. Again, the *bcc* regime extends only to about 150 K below the melting points. All other rare earth metals show their own temperature sequence. Lanthanum switches in the sequence $hc \rightarrow c \rightarrow bcc$ and cerium has an additional low-temperature phase, leading to the sequence $c \rightarrow hc \rightarrow c \rightarrow bcc$. Samarium is the only rare earth metal that shows the *hhc* sequence under ambient pressure conditions. It then transforms to a simple *h* sequence and finally to a *bcc* arrangement, approximately 150 K before its melting temperature. Finally, we turn to ytterbium, which does not behave like europium. Its sequence is $h \rightarrow c \rightarrow bcc$.

The temperature-dependent phase transitions are a consequence of the different influences of the *f* electrons. A higher *f* electron concentration leads to more *h* layers in the structures. The influence of the 4*f* shell decreases with increasing ordering number. The inner 4*f* shell shows stronger contraction with increasing ordering number than the 5*d* and 6*s* shells. The 4*f* electrons exhibit much stronger influence on the structural behavior under high-pressure conditions, where the outer 5*d* and 6*s* electrons show stronger compression. Generally, the high-pressure phases show a higher degree of *c* layers. This leads directly to the structural chemistry of the high-pressure phases. Figure 1.7.4 graphically summarizes the high-pressure phases for each rare earth element. This graphic relies on the data from the Pearson data base [6] and the compilations in [7] and [8]. For a general overview on the various high-pressure structures of the elemental metals, we refer to an excellent review by *McMahon* and *Nelmes* [9]. In general, we observe low-symmetry structures and the volume collapse of the high-pressure phases is related to pressure-induced 4*f* delocalization.

Most of the high-pressure phases were just characterized through an indexing of the diffraction patterns, along with a refinement of the lattice parameters and verification of the structural model by comparison with calculated powder patterns. Many of the low-symmetry structures were observed during energy-dispersive X-ray diffraction studies, and this technique does not allow for refinements of the crystal structures. More detailed studies use in situ synchrotron radiation and diamond anvil cells (nitrogen or argon are used as pressure medium) under angle-dispersive X-ray diffraction conditions. A typical example concerns the structural study of the high-pressure phases of gadolinium, which also led to the construction of the pressure–temperature phase diagram [10].

Between two and five high-pressure phases per rare earth element are known. The highest transition pressure of 240 GPa was observed for the *hP*6 phase of

Sc	20.5	*tP4*	104	?	140	?	240	*hP6*		
Y	12	*hR9*	25	*hP4*	50	*hR24*	99	*mS4*		
La	2.0	*c*	7.0	*hR24*	60	*c*				
Ce	0.76	*c*	5.1	*mS4*	12.2	*tI2*				
Pr	4.0	*c*	7.4	*hR24*	13.7	*oI16*	20.5	*oS4*	147	*oP4*
Nd	5	*c*	17	*hR24*	35	*hP3*	75	*mS4*	113	*oS4*
Pm	10	*c*	18	*hR24*						
Sm	6	*hP4*	11	*hR24*	37.4	*hP3*	105	*mS4*		
Eu	12.5	*h*	31.5	*mS4*	37	?				
Gd	1.5	*hR9*	6.5	*hP4*	26	*c*	33	*hR24*	60.5	*mS4*
Tb	3	*hR9*	10	*hP4*	30	*hR24*	51	*mS4*		
Dy	7	*hR9*	17	*hP4*	42	*hR24*	82	*mS4*		
Ho	7	*hR9*	19.5	*hP4*	54	*c*	58	*hR24*	102	*mS4*
Er	12.4	*hR9*	24	*hP4*	58	*hR24*	118	*mS4*		
Tm	9	*hR9*	32	*hP4*	61	*hR24*	124	*mS4*		
Yb	3.5	*bcc*	26	*h*	53	*c*	98	*hP3*		
Lu	25	*hR9*	45	*hP4*	88	*hR24*				

⟶ Pressure [GPa]

Figure 1.7.4: The structures of the rare earth elements as a function of pressure, drawn with the data listed in references [6–8]. The transition pressures are given in units of GPa. For details, see text.

scandium [11]. Determination of the correct structure type for a given high-pressure phase is not an easy task. The rare earth elements show significant compressibility and the strong changes in the lattice parameters lead to substantial shifts in the X-ray powder patterns. In many cases, comparison with simulated powder patterns of known simple distortion variants is a meaningful way to understand the high-pressure crystal chemistry.

Volume–pressure data of the rare earth elements were evaluated with the *Birch-Murnaghan* equation, resulting in values for the bulk moduli. In the low symmetry structures, at the transition pressures, one observes volume collapses that can be small (e.g., 1.5% at 120 GPa for thulium) or large (e.g., 16% at 0.7 GPa for cerium) [12], signaling distinct changes in chemical bonding. In the case of cerium, a transition from Ce^{3+} to $Ce^{4+} + e^-$ takes place. Also, the compressibility over the whole studied pressure range can be remarkable. A striking example is gadolinium: the volume per atom is ca. 31 Å^3 under ambient pressure conditions and decreases to ca. 15.5 Å^3 at 93 GPa, a reduction by 50%! [10].

Inspection of Figure 1.7.4 readily reveals that we do not observe a simple trend of high-pressure phases in the series of the rare earth metals; however, we can make out some smaller groups. At least for the first two pressure-induced phase transitions, we observe some similarities. This is the case with lanthanum, praseodymium, neodymium and promethium in the group of the light rare earth metals and for gadolinium, terbium, dysprosium, holmium, erbium, thulium and lutetium for the heavy rare earth metals. It is interesting to note that yttrium too fits into this group – a consequence of its atomic radius. The other rare earth metals and the highest-pressure phases manifest the individuality of the rare earth elements and the remarkable influence of the external pressure on the *4f* electrons, with a clear trend of pressure-induced delocalization. Of the many studies on the rare earth high-pressure phases (which are all stacking or distortion variants of the three basic structures discussed above), we can only highlight few data. For a compilation of crystallographic data up to 2005, we refer to a textbook [13].

We start with the smallest rare earth metal, scandium, which crystallizes with an *hcp* ambient pressure/ambient temperature structure (AB stacking of close-packed layers). Each scandium atom has coordination number 12 (*anti*cuboctahedron). Due to a slight deviation of the c/a ratio from the ideal value of 1.633 (compression to 1.594), each scandium atom shows 6×325 and 6×331 pm Sc–Sc distances. The first high pressure phase transition leads to a tetragonal structure, with four atoms per unit cell (β-neptunium structure) [14]. The high-pressure phase stable at >240 GPa [11] shows a substantial shift of the close-packed scandium layers. The stacking of these layers, however, differs from the ABC variants discussed earlier. The translation along the stacking axis covers six layers and they are symmetry related via a 6_1 screw axis. This leads to drastic differences for the scandium coordination. Now each scandium atom has 14 nearest neighbors and drastically shortened (!) Sc–Sc distances: 2×205, 2×216, 6×236 and 4×268 pm. The compression of scandium is remarkable. The volume per scandium atom is 25 $Å^3$ at ambient pressure, 17.8 $Å^3$ for the β-Np modification at 35 GPa and 8.4 $Å^3$ at 240 GPa.

Next, we turn to the high-pressure structures of cerium [15, 16]. The *fcc* ambient pressure modification shows coordination number 12 with 12×365 pm Ce–Ce and transforms to the *mS4* phase with 298 to 336 pm Ce–Ce at a transition pressure of 5.1 GPa. The highest-pressure phase (>12.2 GPa) adopts the indium type structure (tetragonally distorted *fcc*) with 4×283 and 8×312 pm Ce–Ce. The cerium atoms in both high-pressure phases keep coordination number 12, with different distance ranges, though, which account for the distortions. The unit cell relations are shown in [16].

Praseodymium is an example of a rare earth element with five high-pressure phases [17, 18], which are all distortion variants of classical closest packed layers (the relations of these unit cells with the *fcc* subcell are summarized in [18]). The *oI*16 phase (space group *Ibam*) exhibits two crystallographically independent praseodymium sites, both with coordination number 12, but with different distance ranges: 295–343 pm Pr–Pr for Pr1 and 298–332 pm Pr–Pr for Pr2 [18]. The *oI*16 phase then transforms to the *oS*4 phase (α-uranium type) which is stable between 20.5 and 147 GPa. Studies of praseodymium under ultrahigh pressure conditions revealed an additional structural distortion. *oP*4 praseodymium forms above 147 GPa and is stable at least up to 313 GPa. Ingoing from the α-uranium type phase to *oP*4 praseodymium, the space group symmetry is reduced from *Cmcm* to $P2_12_12_1$ in two steps [17].

Besides temperature-dependent high-pressure studies (*vide ultra*), magnetic field-dependent X-ray powder diffraction is also still a hardly investigated field. Studies with an applied external magnetic field help for a better understanding of magneto-structural phase transitions. A recent example is dysprosium, where the decoupling of a first-order magneto-structural transition was studied in fields up to 40 kOe [19].

The high-pressure structures of Gd, Tb, Dy, Ho, Er and Tm have recently been reinvestigated using synchrotron radiation [20]. These diffraction studies revealed the correct *oF*16 structure and additionally, an *oF*8 structure for neodymium. The unit cell relations are schematically presented in Figure 1.7.5.

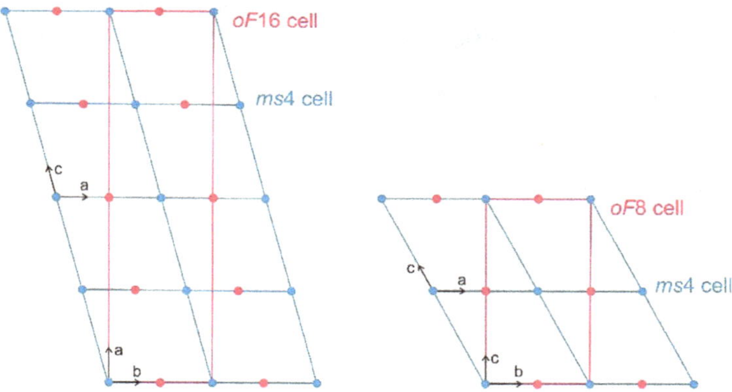

Figure 1.7.5: The relation of the pseudo-orthorhombic *C*-centered monoclinic lattices of the *mS*4 structures and the correct orthorhombic *F*-centered lattices for the *oF*16 (left) and *oF*8 (right) structures, redrawn from [20].

References

[1] U. Müller, Inorganic Structural Chemistry, 2nd Edition, Wiley-VCH, Weinheim, 2007.
[2] R. Pöttgen, D. Johrendt, Intermetallics, De Gruyter, Berlin, 2014.
[3] J. Donohue, The Structures of the Elements, Wiley, New York, 1974.
[4] H. Jagodzinski, Acta Crystallogr. 1949, 2, 201.

[5] a) U. Müller, Relating Crystal Structures by Group-subgroup Relations, in: International Tables for Crystallography, Vol. A1, Symmetry Relations Between Space Groups (ed. H. Wondratschek, U. Müller), 2nd Edition, John Wiley & Sons, Ltd., Chichester, 2010, 44–56; b) U. Müller, Symmetriebeziehungen zwischen verwandten Kristallstrukturen, Vieweg + Teubner Verlag, Wiesbaden, 2012.
[6] P. Villars, K. Cenzual, Pearson's Crystal Data: Crystal Structure Database for Inorganic Compounds (release 2019/20), ASM International®, Materials Park, 2019.
[7] R. Ferro, A. Saccone, Intermetallic Chemistry, Elsevier, Amsterdam, 2008.
[8] W. Steurer, J. Dshemuchadse, Intermetallics: Structures, Properties, and Statistics, IUCr Monographs on Crystallography, Volume 26, Oxford University Press, New York, 2016. ISBN-10: 0198714556

[9] M. I. McMahon, R. J. Nelmes, Chem. Soc. Rev. 2006, 35, 943.
[10] D. Errandonea, R. Boehler, B. Schwager, M. Mezouar, Phys. Rev. B 2007, 75, 014103.
[11] Y. Akahama, H. Fujihisa, H. Kawamura, Phys. Rev. Lett. 2005, 94, 195503.
[12] G. Fabbris, T. Matsuoka, J. Lim, J. R. L. Mardegan, K. Shimizu, D. Haskel, J. S. Schilling, Phys. Rev. B 2013, 88, 245103.

[13] E. Yu. Tonkov, E. G. Ponyatovsky, Phase Transformations of Elements under High Pressure, CRC Press, Boca Raton, 2005.
[14] Y. K. Vohra, W. Grosshans, W. B. Holzapfel, Phys. Rev. B 1982, 25, 6019.
[15] W. H. Zachariasen, F. H. Ellinger, Acta Crystallogr. A 1977, 33, 155.
[16] M. I. McMahon, R. J. Nelmes, Phys. Rev. Lett. 1997, 78, 3884.
[17] N. Velisavljevic, Y. K. Vohra, High Press. Res. 2004, 24, 295.
[18] S. R. Evans, I. Loa, L. F. Lundegaard, M. I. McMahon, Phys. Rev. B 2009, 80, 134105.
[19] A. S. Chernyshov, Ya. Mudryk, V. K. Pecharsky, K. A. Gschneidner Jr., Phys. Rev. B 2008, 77, 094132.

[20] M. I. McMahon, S. Finnegan, R. J. Husband, K. A. Munro, E. Plekhanov, N. Bonini, C. Weber, M. Hanfland, U. Schwarz, S. G. Macleod, Phys. Rev. B 2019, 100, 024107.

Mikhail G. Brik, Alok M. Srivastava

1.8 Electronic properties of the lanthanide ions

The lanthanides form a group of chemical elements consisting of 15 members, from lanthanum La (atomic number $Z = 57$) to lutetium Lu ($Z = 71$). They have a peculiar, at the first glance, order of the electron shell filling. Their inner 4*f*-electron shell is filled after the outer $5s^2 5p^6$ shells are completed. This is explained by the *Aufbau principle*, according to which, the states with the smallest $n + l$ sum (n and l stand for the principal and orbital quantum numbers of the electrons in a given electron shell, respectively) have a lower energy. If the $n + l$ values are the same for different n and l combinations, the states with smaller n have a lower energy. The unfilled 4*f* shell of these elements, which is screened by the filled outer shells, is not actively involved into the chemical bonding. Moreover, the interaction of the 4*f* electronic states with the nearest environment in solids and solutions, which manifests as very sharp absorption and emission lines corresponding to the intrashell *f*–*f* electronic transitions, vary weakly from one host to another. The most stable oxidation state for the lanthanides is +3 and with such an electric charge, the number of the 4*f* electrons varies smoothly across the whole series, from 0 (La^{3+}) to 14 (Lu^{3+}).

Since, for the *f*-electrons, the orbital quantum number is $l = 3$, there are $2l + 1 = 7$ *f*-orbitals, which differ by the magnetic quantum number m, which takes all integer values from −3 to 3. Each of these orbitals can accommodate not more than two electrons with opposite spins, which sets a limit for the maximal number of the *f*-electrons as 14. There are many possibilities of placing n 4*f* electrons into these 14 orbitals, which gives rise to a large number of allowed states (from the point of view of the Pauli principle). Many of those states are highly degenerated. Since the electrons are indistinguishable, the number of states N for a particular $4f^n$ configuration can be determined as a number of permutations of n electrons through 14 single-electron states:

$$N = \frac{14!}{n!(14-n)!} \tag{1.8.1}$$

It is not difficult then to check that the greatest number of states can be as high as 3,432 for the ion of Gd^{3+} with the $4f^7$ electron configuration (Table 1.8.1). It is also easy to see that the number of the states in the so-called *conjugate* electron configurations $4f^n$ and $4f^{14-n}$ is the same, because the latter can be considered as a filled 4*f* shell containing n positively charged holes.

Since the 4*f* orbitals exhibit well-pronounced difference in their shape (see, e.g., [1]), an average distance between the electrons would be different, if they are placed in different orbitals and so will be the Coulomb interaction between these

https://doi.org/10.1515/9783110654929-008

Table 1.8.1: Number of states and *LS* terms of configurations of 4*f* electrons. The subscript in parenthesis denotes the number of different terms with the same values of *L* and *S* (those terms with the same *L* and *S* values arise from different terms of the preceding "parent" f^{n-1} configuration and can be distinguished by additional quantum numbers). For the sake of brevity, the terms multiplicity is given only once as a superscript-to-the-left for the term listed in the first position of the considered group, for example, ^{1}SDGI denotes four terms ^{1}S, ^{1}D, ^{1}G, ^{1}I and so on.

Configurations of triply charged lanthanides	*LS* terms	Number of *LS* terms (number of *J*-multiplets is given in parentheses)	Number of states
f^1 (Ce^{3+}) and f^{13} (Yb^{3+})	^{2}F	1 (2)	14
f^2 (Pr^{3+}) and f^{12} (Tm^{3+})	^{1}SDGI, ^{3}PFH	7 (13)	91
f^3 (Nd^{3+}) and f^{11} (Er^{3+})	$^2PD_{(2)}F_{(2)}G_{(2)}H_{(2)}IKL$, ^{4}SDFGI	17 (41)	364
f^4 (Pm^{3+}) and f^{10} (Ho^{3+})	$^1S_{(2)}D_{(4)}FG_{(4)}H_{(2)}I_{(3)}KL_{(2)}N$, $^3P_{(3)}D_{(2)}F_{(4)}G_{(3)}H_{(4)}I_{(2)}K_{(2)}LM$, ^{5}SDFGI	47 (107)	1,001
f^5 (Sm^{3+}) and f^9 (Dy^{3+})	$^2P_{(4)}D_{(5)}F_{(7)}G_{(6)}H_{(7)}I_{(5)}K_{(5)}L_{(3)}M_{(2)}NO$, $^4SP_{(2)}D_{(3)}F_{(4)}G_{(4)}H_{(3)}I_{(3)}K_{(2)}LM$, ^{6}PFH	73 (198)	2,002
f^6 (Eu^{3+}) and f^8 (Tb^{3+})	$^1S_{(4)}PD_{(6)}F_{(4)}G_{(8)}H_{(4)}I_{(7)}K_{(3)}L_{(4)}M_{(2)}N_{(2)}Q$, $^3P_{(6)}D_{(5)}F_{(9)}G_{(7)}H_{(9)}I_{(6)}K_{(6)}L_{(3)}M_{(3)}NO$, $^5SPD_{(3)}F_{(2)}G_{(3)}H_{(2)}I_{(2)}KL$, ^{7}F	119 (295)	3,003
f^7 (Gd^{3+})	$^2S_{(2)}P_{(5)}D_{(7)}F_{(10)}G_{(10)}H_{(9)}I_{(9)}K_{(7)}L_{(5)}M_{(4)}N_{(2)}OQ$, $^4S_{(2)}P_{(2)}D_{(6)}F_{(5)}G_{(7)}H_{(5)}I_{(5)}K_{(3)}L_{(3)}MN$, ^{6}PDFGHI, ^{8}S	119 (327)	3,432

electrons. To calculate the energy of the multielectron electron states, one would have to first build proper wave functions of those multielectron configurations, which would be expressed as linear combinations of the products of the single-electron wave functions. The quantum mechanical theory of the angular momenta addition comes into play at this stage; the interested reader is advised to look into references [2–4, etc.]. The results of the addition of angular momenta are the so-called *LS* terms, also listed in Table 1.8.1. The standard notation is $^{\kappa}L$, where $\kappa = 2S + 1$ is called the term *multiplicity*. This is equal to the number of the total spin *S* projections and *L* denotes the total orbital momentum of the considered configuration. The following convention is used between the Latin alphabet letters and the *L* values:

L	0	1	2	3	4	5	6	7	8	9	10	11	12
	S	P	D	F	G	H	I	K	L	M	N	O	Q

For example, the ^{6}H term corresponds to the state with the total spin $S = 5/2$ and total orbital momentum $L = 5$, and so on. Since for each value of *S* and *L* momenta, there are $2S+1$ and $2L+1$ projections, respectively, each $^{\kappa}L$ term is highly degenerated with the degree of degeneracy equal to the product $(2S+1)\times(2L+1)$. The only nondegenerated term is ^{1}S and it cannot be split further. It is instructive to note that summing the $(2S+1)\times(2L+1)$ values over all terms in a considered configuration would yield the number of states given by eq. (1.8.1).

There is an empirical Hund's rule which allows to identify the ground term of any electron configuration. One has to first look at the terms with the highest multiplicity and then select the term with the highest L value. The ground terms of each configuration are given as the last entries in the corresponding lists of the LS terms in Table 1.8.1.

To get a complete picture of the relative positions of all energy levels in a considered electron configuration, it is necessary to calculate the matrix elements of the two-electron Coulomb interaction operator on the corresponding wave functions. Since there are repeating terms with the same L and S values, the Coulomb interaction would mix them up, which would involve calculations of non-diagonal matrix elements. Without going into further details (which can be found in the above-cited references), we mention that the energy of each LS term would be expressed in terms of the so-called Slater integrals F^k ($k = 0, 2, 4, 6$ for the f electrons):

$$F^k = e^2 \int\int \frac{r_<{}^{k}}{r_>{}^{k+1}} R_1(r_1)^2\, R_2(r_2)^2 r_1^2\, r_2^2 dr_1 dr_2 \tag{1.8.2}$$

which can be calculated numerically, since the radial parts of the corresponding wave functions $R_1(r_1)$, $R_2(r_2)$ can be easily found in the literature. Integration is performed in the spherical coordinates, the electrons radial coordinates are denoted by r_1, r_2, and $r_<$ and $r_>$ are the smallest and the greatest values of those two coordinates respectively. The energies of the LS terms and the matrix elements of the Coulomb interaction for all electronic configurations of the p, d and f electrons are collected in ref. [5]. The Coulomb interaction results in the separation of about 10^4 cm^{-1} or so between the individual LS states (although some states can come closer or even an accidental degeneracy of the energies of the different LS terms can take place).

However, considering only the Coulomb interaction is a very crude approximation for the lanthanides. The reason is that they are the elements from the second half of the Periodic Table with relatively large values of atomic numbers Z. It is well known that the spin–orbit (SO) interaction grows with Z as Z^4 and its effect on the overall appearance of the energy-level scheme is extremely important. It produces additional splitting of the LS terms into the so-called J-multiplets or manifolds, denoted as $^{\kappa}L_J$, with J being the total angular momentum varying with the step of 1 between $|L-S|$ to $L+S$. Since the values of S can be the half-integer for the configurations with odd number of electrons (see Table 1.8.1), the values of J are also half-integer for those configurations and they are integer for the configurations with the even number of the electron. The splitting of the LS terms into the J-multiplets is called the *terms fine structure*.

If $L \geq S$, then $2S+1$ values of J are possible (the LS term splits into $2S+1$ J-multiplets). If $L < S$, then $2L+1$ J values appear. Sometimes, the group of J states (they all have the same energy in the absence of any external field) arising from a given LS term is called a J-multiplet. The states with different J coming from the same parent LS term have different energies. Another useful empirical rule that describes the energy separation between the states with different J is the Lande's interval rule: the energy gap between the J and $J-1$ states is proportional to J:

$$E_J - E_{J-1} = A(LS)J \tag{1.8.3}$$

where A is the multiplet splitting constant. It depends on the values of L and S and can have different signs. If $A > 0$, the multiplet with the smallest J value for a given LS term has the lowest energy and the increased J values correspond to the states with higher energies. These multiplets are called the normal multiplets. Schematically, formation of the normal multiplet for the ^{5}D term (taken as an example, when the allowed J values are 0, 1, 2, 3, 4) is shown in Figure 1.8.1a. The fine structure of the electron configurations with less-than-half-filled shells is described by the normal multiplets; this is the case of the Ce^{3+} ($4f^1$), Pr^{3+} ($4f^2$), Nd^{3+} ($4f^3$), Pm^{3+} ($4f^4$), Sm^{3+} ($4f^5$), Eu^{3+} ($4f^6$) ions. It can be easily seen from eq. (1.8.3) and Figure 1.8.1 that the energy intervals between the neighboring levels increase with increased values of J.

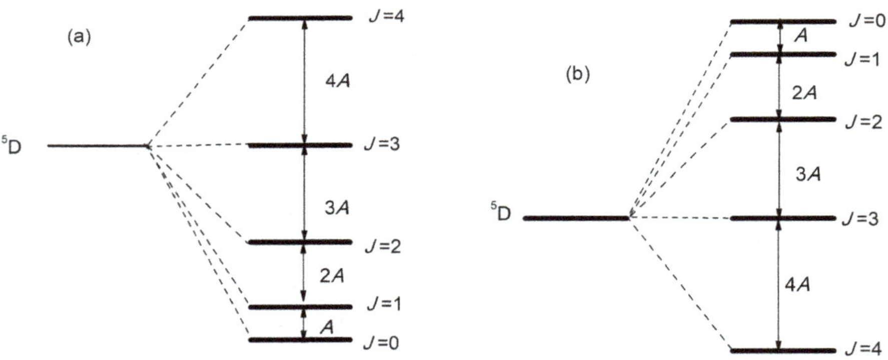

Figure 1.8.1: Formation of the normal (a) and inverted (b) multiplets from the ^{5}D term with J = 0, 1, 2, 3, 4. The barycenter of the group of J-multiplets corresponds to the LS term energetical position.

When $A < 0$, the situation becomes the opposite and the multiplet with the smallest J value for a given LS term has the highest energy, whereas the multiplet with the greatest J value has the lowest energy. This is the case of the so-called inverted multiplets (Figure 1.8.1b), which are characteristic of the more-than-half-filled configurations, that is, Tb^{3+} ($4f^8$), Dy^{3+} ($4f^9$), Ho^{3+} ($4f^{10}$), Er^{3+} ($4f^{11}$), Tm^{3+} ($4f^{12}$), Yb^{3+} ($4f^{13}$) ions.

The electron configuration with exactly half-filled shell (which is the case of the Gd^{3+} ion with $4f^7$ configuration) is a special case when $A = 0$, which means that there is no fine structure in the first order of the perturbation theory.

The numerical value of the A constant can be found from the experimental spectra. Alternatively, it can be calculated from the following equation:

$$A = \xi_{nl}\sqrt{\frac{l(l+1)(2l+1)}{S(S+1)(2S+1)L(L+1)(2L+1)}}(l^n\gamma SL\|V^{11}\|l^n\gamma SL) \tag{1.8.4}$$

where ξ_{nl} is the SO interaction constant, and the matrix elements $(l^n\gamma SL\|V^{11}\|l^n\gamma SL)$ can be found in the book by Nielsen and Koster [5].

Theoretical modeling of the lanthanide energy levels became possible with the development of the semi-empirical effective Hamiltonian [6, 7]. In this model, all effective interactions among the $4f$ electrons are represented as the product of the radial fitting parameters (depending on the f electrons radial coordinates only) and the angular operators (acting on the angular coordinates of the f-electrons wave functions only). The matrix elements of all operators entering the free ion Hamiltonian acting between the basis wave functions $|f^N\eta SLJM>$ can be derived from the Racah–Wigner algebra. Here S, L, J and M stand for the quantum numbers of the spin, orbital momentum, total angular momentum and its projection, respectively; η denotes all additional quantum numbers needed to distinguish between the states with identical S, L, J, M sets.

A usual way to obtain the radial fitting parameters is to minimize the root-mean-square deviation between the theoretically calculated and experimentally observed energy levels. It was successfully done to explain the spectra of the trivalent lanthanides and actinides in $LaCl_3$ and LaF_3, by Carnall and his coworkers, and the trend of the obtained fitting parameter values across the whole f-element series was analyzed and discussed (see, e.g., the review article [8]).

The standard way to write the free-ion Hamiltonian H_{FI} of a f^N (N = 1,. . .,13) electron configuration is as follows [6, 7, 9]:

$$\begin{aligned}\mathbf{H}_{FI}(f^N) &= E_{\text{avg}} + \sum_{k=2,4,6} F^k\mathbf{f_k} + \xi_f\sum_{i=1}^{N}\mathbf{l}_i\cdot\mathbf{s}_i + \alpha\mathbf{L}(\mathbf{L}+\mathbf{1}) + \beta\mathbf{G}(\mathbf{G_2})\\ &+ \gamma\mathbf{G}(\mathbf{R_7}) + \sum_{i=2,3,4,6,7,8} T^i\mathbf{t_i} + \sum_{h=0,2,4} M^h\mathbf{m_h} + \sum_{k=2,4,6} P^k\mathbf{p_k}\end{aligned} \tag{1.8.5}$$

The italic and bold letters denote the radial fitting parameters and the angular operators, respectively. The first term E_{avg} is the f^N configuration barycenter energy which does not cause any energy splitting. The second term is the Coulomb interaction between the f electrons, where F^k (k = 2, 4 and 6) are the Slater parameters defined in equation (1.8.2). The $\mathbf{f_k}$ values are the numerical constants [10]. The third term stands for the f electron's SO interaction, where ξ_f is the SO coupling constant and $l_i S_i$ is the SO interaction operator, with the first and second terms in the scalar product being the operators of the angular and spin momenta of an individual f electron with index i. The second and the third terms are the most important in the Hamiltonian (5), their contribution to the formation of the energy-level scheme is key. The next four terms are the two- and three-electron Coulomb correlation contributions from the higher configurations with the same parity. The last two terms represent the magnetically and electrostatically correlated interactions. All these six last terms produce very small corrections to the energy levels, determined by the second and third terms of Hamiltonian (5). However, taking them into account significantly improves agreement between the calculated and observed energy-level schemes of the trivalent lanthanides.

The Slater parameters F^k and the SO constant ξ_f can be calculated theoretically, see for example, references [11, 12]; the remaining Hamiltonian (5) parameters are better found from the fitting procedure. Systematic studies of the variation of the Hamiltonian parameters across the $4f$ ions series revealed certain trends in the ways the parameters depend on the atomic numbers [11–16]. Most of the parameters increase linearly with Z. The initial set of the free ion Hamiltonian parameters for all trivalent lanthanides can be taken, for example, from ref. [6]; for the reader's convenience, these parameters are given in Table 1.8.2.

Finding the eigenvalues of Hamiltonian (5) with all relevant parameters from Table 1.8.2 allows us to plot the energy-level diagram for all trivalent lanthanides in a free state in the energy range up to 40,000 cm^{-1} (Figure 1.8.2). It is analogous to the famous Dieke's diagram [17, 18] which was obtained experimentally by careful measurements of the absorption spectra of trivalent lanthanides in aqueous solutions. The densities of energy levels in some spectral regions are quite high and their wave functions are mixed up strongly by the SO coupling (the so-called J-mixture effects).

It is suggested to compare the energy-level schemes of the ions with conjugate electron configurations in Figure 1.8.2. We begin with the simplest case of $4f^1$ and $4f^{13}$ configurations (Ce^{3+} and Yb^{3+} ions, respectively). They have only one term ^2F, which is split by the SO coupling into the $^2F_{5/2}$ (ground) and $^2F_{7/2}$ (excited) states for the Ce^{3+} ions, separated by about 2200 cm^{-1}. This order of energy levels becomes inverted for the Yb^{3+} ions, with a simultaneous increase of the energy interval between them to about 10,000 cm^{-1} due to increased atomic number Z. The Yb^{3+} ions are widely used for the up-conversion in the codoped systems since the energy of the excited Yb pair (roughly at about 20,000 cm^{-1}) is favorably close to the many energy levels of the Pr^{3+}, Nd^{3+}, Ho^{3+}, Tm^{3+} ions. This facilitates the energy transfer from the pair of the Yb^{3+} ions to codopants.

The next pair of the conjugate configurations is $4f^2$ (Pr^{3+}) and $4f^{12}$ (Tm^{3+}). The ground LS term is ^3H with allowed J values of 4, 5 and 6. According to the order of the direct and inverted multiplets, the ground state of the Pr^{3+} ions in $^3H_4$ and that of Tm^{3+} ions is $^3H_6$. For the Pr^{3+} ions, the three states $^3H_{4,5,6}$ and three states $^3F_{2,3,4}$

Table 1.8.2: Parameters of the free ion Hamiltonian eq. (1.8.5) (in cm^{-1}) [6, 9].

	Ce^{3+}	Pr^{3+}	Nd^{3+}	Pm^{3+}	Sm^{3+}	Eu^{3+}	Gd^{3+}	Tb^{3+}	Dy^{3+}	Ho^{3+}	Er^{3+}	Tm^{3+}	Yb^{3+}
F^2		68878	73018	76400	79805	83125	85669	88995	91903	94564	97483	1,00134	
F^4		50347	52789	54900	57175	59268	60825	62919	64372	66397	67904	6,9613	
F^6		32901	35757	37700	40250	42560	44776	47252	49386	52022	54010	5,5975	
ξ_f	647	751.7	885.3	1025	1176	1338	1508	1707	1913	2145	2376	2,636	2928
α		16.23	21.34	20.50	20.16	20.16	18.92	18.40	18.02	17.15	17.79	17.26	
β		−567	−593	−560	−567	−567	−600	−591	−633	−608	−582	−625	
γ		1371	1445	1475	1500	1500	1575	1650	1790	1800	1800	1,820	
T^2			298	300	300	300	300	320	329	400	400	400	
T^3			35	35	36	40	42	40	36	37	43		
T^4			59	58	56	60	62	50	127	107	73		
T^6			−285	−310	−347	−300	−295	−395	−314	−264	−271		
T^7			332	350	373	370	350	303	404	316	308		
T^8			305	320	348	320	310	317	315	336	299		
M^0		2.08	2.11	2.4	2.6	2.1	3.22	2.39	3.39	2.54	3.86	3.81	
P^2		−88.6	192	275	357	360	676	373	719	605	594	695	

$M^2 = 0.56M^0, M^4 = 0.31M^0, P^4 = 0.5, P^2, P^6 = 0.1P^2$

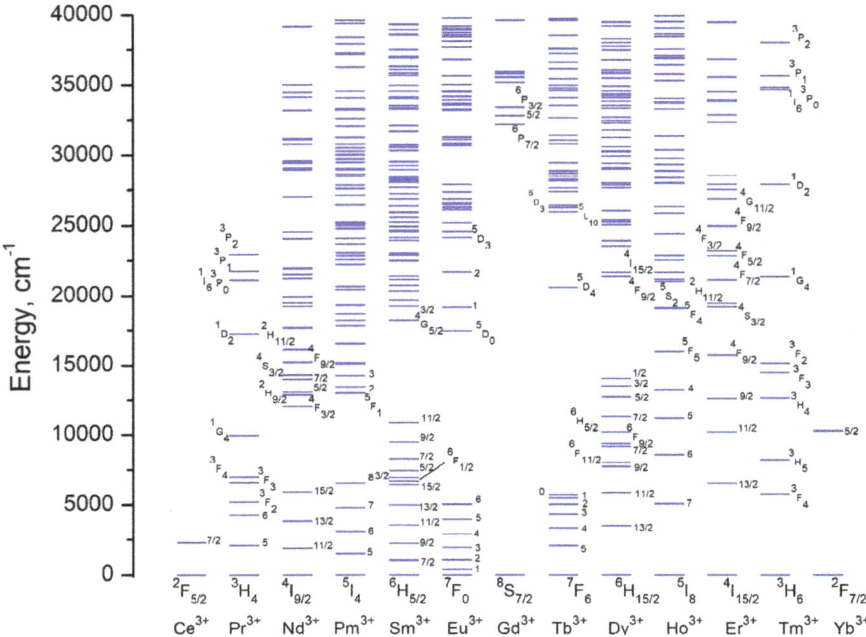

Figure 1.8.2: Energy-level scheme of free trivalent lanthanides in the spectral range until 40,000 cm^{-1} as obtained after diagonalization of the Hamiltonian (5) with its parameters from Table 1.8.2.

are clearly separated from each other. For the Tm^{3+} ions, the stronger SO coupling pushes the $^3F_4$ state further to make it the first excited state located between the $^3H_6$ and $^3H_{5,4}$ states. The energy levels such as $^1G_4$, $^1D_2$ and $^3P_0$, which are well separated from the lower states, are suitable emitting levels (a gap of at least several thousand cm^{-1} makes nonradiative relaxation from these states to the lower levels almost negligible).

Very important ions, from the view of laser applications, are Nd^{3+} ($4f^3$ configuration) and Er^{3+} ($4f^{11}$ configuration). The ground states are $^4I_{9/2}$($^4I_{15/2}$) for the Nd^{3+} (Er^{3+}) ions. The $^4F_{3/2}$ state is a lasing level of the Nd^{3+} ion in the infrared region (1064 nm, which can be converted into green light at 532 nm by means of the second harmonic generation). The $^4I_{13/2}$ level is a laser level of the Er^{3+} ion producing emission, at about 1,500 nm.

Radioactive Pm^{3+} ions with $4f^4$ configuration ($^5I_4$ ground state) are not used in optical applications. However, the Ho^{3+} ions with $4f^{10}$ configuration ($^5I_8$ ground state) are used for phosphor applications. The groups of the ^5I_J and ^5F_J multiplets, in the case of the Ho^{3+} ions, are well separated from each other.

The ions with the $4f^5$ (Sm^{3+}) and $4f^9$ (Dy^{3+}) configurations are widely used for solid state lighting, as emitting light, in the visible spectral range. The ground state is $^6H_{5/2}$ ($^6H_{15/2}$) for the f^5 (f^9) configurations; six J states (with J varying from 5/2 to 15/2 with a step of unity) arising from the ^{6}H term, are grouped together and are followed by another group of six levels coming from the ^{6}F term with J varying from 1/2 to 11/2. The next excited state separated from those levels is $^4G_{5/2}$, which is responsible for visible emission of the Sm^{3+} ions. For this ion, all energy levels located higher are very close to each other. Hence, the nonradiative relaxation from those levels would easily populate the $^4G_{5/2}$ state, leading to efficient emission. In the case of the Dy^{3+} ions, the $^4F_{9/2}$ level plays an analogous role to the Sm^{3+} $^4G_{5/2}$ state: the gap from it to the lower states is very wide and efficient emission can be observed.

Another pair of the conjugate electron configurations is that of $4f^6$ (Eu^{3+}) and $4f^8$ (Tb^{3+}). Their ground states are $^7F_0$ and $^7F_6$, respectively. The emitting level $^5D_0$ is well above the group of seven states, arising from the ground ^{7}F term (with J varying from 0 to 6) for the Eu^{3+} ions. For the Tb^{3+} ions, the emitting level is the $^5D_4$ state.

Finally, a very special case is the Gd^{3+} ion with a $4f^7$ configuration. It is a half-filled configuration and its ground state corresponds to the complete occupation of all seven f orbitals by the electrons with the aligned spins. The ground state is $^8S_{7/2}$ and the first excited state $^6P_{7/2}$ is at about 32,000 cm^{-1}. Such high energy implies realization of the spin-flip transitions. It is in the UV range, which does not permit for the visible range applications of these ions.

Quite a number of papers on the extension of the Dieke's diagram to the UV and vacuum ultraviolet (VUV) ranges were published in the past. We refer to [19–22] for more details.

When the trivalent lanthanides are placed in a crystal, their energy levels are split because of the crystal field. The total number of split levels depends on the symmetry of the nearest environment and cannot be greater than $2J + 1$ for the ions with an even number of electrons and $J + 1/2$ for the ions with an odd number of electrons. In the latter case of the so-called Kramers' ions, each energy level is double degenerated and only an external magnetic field can lift this degeneracy.

However, since the unfilled $4f$ shell is screened by the completed outer $5s$ and $5p$ shells, the crystal field splitting of the J-multiplet is very weak, compared to the SO coupling. It is just about few hundreds cm^{-1} for each multiplet, which allows to treat the crystal field as a small perturbation in the case of the lanthanides. This is the reason, the free ion energy-level diagram can still be successfully used to assign and interpret the optical absorption spectra of trivalent lanthanides, in crystals and glasses.

Mathematically, modeling of the $4f$ electron energy levels splitting in a crystal field can be realized by adding the following term to the Hamiltonian (5) (here, the so-called Wybourne notation [7] is used):

$$H_{CF} = \sum_{k,q,i} B_q^k C_q^k(i) \tag{1.8.6}$$

where C_q^k denote the normalized spherical tensor operators and B_q^k represents the crystal field parameters (CFPs), which depend on the geometrical arrangement of the crystal lattice ion around a $4f$ ion. They can be determined either by fitting the eigenvalues of the crystal field Hamiltonian to the experimental values or by direct calculations from the known crystal structure. The values of k can be 2, 4 and 6, $-k \le q \le k$ and $-i$ enumerates the electrons in the unfilled shell of an impurity ion. The number of CFPs can be as large as 27 and depends on the site symmetry of the impurity ions [9, 23].

Examples of calculations of the $4f$ states splittings in crystals are numerous. A few representative references are given as follows [24–32].

We present here the results of the calculations of the Nd^{3+} energy levels in the YZnPO crystal, with the experimental data from ref. [33]. The Nd^{3+} replaces the Y^{3+} ions, and the local symmetry of that site is C_{3v}. In this symmetry, the structure of the crystal field Hamiltonian is such that only six CFPs are nonzero. The least-square-fit of the calculated energy levels to those observed experimentally was performed and the obtained Hamiltonian parameters and Nd^{3+} energy levels are shown in Tables 1.8.3 and 1.8.4, respectively. A good agreement between the experimental and theoretical energy-level schemes confirms the reliability of the performed crystal field analysis.

Table 1.8.3: Free-ion and CFPs (C_{3v} symmetry) for Nd^{3+} in YZnPO (cm^{-1}).

Parameters	**Nd^{3+}:YZnPO**	**Parameters**	**Nd^{3+}:YZnPO**
E_{av}	25,123.38	M^0	0.4072
F^2	79,590.73	P^2	75.226
F^4	40,446.31	B_{20}	318.77
F^6	55,192.15	B_{40}	−261.32
ζ	879.24	B_{43}	1,476.53
α	22.818	B_{60}	325.01
β	−595.14	B_{63}	−457.83
γ	1,718.68	B_{66}	148.4
T^2	1,322.16	S_2	142.5583
T^3	41.664	S_4	701.4722
T^4	80.96	S_6	209.1913
T^6	−804.49	CF strength	857.6732
T^7	448.81	N (number of levels)	37
T^8	1,636.49	σ (rms deviation)	**17.6234**

Table 1.8.4 shows that the crystal field splittings of the J-multiplets do not exceed a few hundreds of cm^{-1}, a firm evidence of a weakened crystal field effects on the $4f$ states due to the completely filled outer $5s$ and $5p$ orbitals.

To sum up, the rich electronic energy-level scheme of trivalent lanthanides arises from the following interaction (in the order of their decrease): Coulomb interaction → SO interaction → crystal field interaction (Figure 1.8.3).

Table 1.8.4: Observed ($E_{obs.}$) [33] and calculated ($E_{cal.}$, this work) energy levels of Nd^{3+} in YZnPO ($\Delta E = E_{obs.} - E_{cal.}$). IR stands for the irreducible representations of the C_{3v} double group.

$^{2S+1}L_J$	$E_{obs.}$ (cm^{-1})	$E_{cal.}$ (cm^{-1})	ΔE (cm^{-1})	IR
$^4I_{9/2}$	0	−24	24	1/2
	63	69	−6	3/2
	79	80	−1	1/2
	118	138	−20	3/2
	363	360	3	1/2
$^4I_{11/2}$	1,880	1,895	−15	1/2
	1,919	1,904	15	3/2
	1,956	1,928	28	1/2
	2,002	1,996	6	1/2
	2,066	2,089	−23	3/2
	2,115	2,128	−13	1/2
$^4I_{13/2}$	3,868	3,864	4	1/2
	3,885	3,873	12	3/2
	3,900	3,894	6	1/2
	3,924	3,948	−24	1/2
	4,053	4,040	13	1/2
	4,093	4,102	−9	3/2
	–	4,138	–	1/2
$^4I_{15/2}$	–	5,866	–	1/2
	–	5,869	–	3/2
	–	5,919	–	1/2
	–	5,988	–	3/2
	–	6,099	–	1/2
	–	6,161	–	1/2
	–	6,244	–	3/2
	–	6,294	–	1/2
$^4F_{3/2}$	–	11,099	–	1/2
	11,153	11,148	5	3/2
$^4F_{5/2}$	12,042	12,036	6	1/2
	12,060	12,093	−33	3/2
	12,133	12,159	−26	1/2
$^2H_{9/2}(2)$	[12,216]*	12,486	[−269.6]	1/2
	–	12,543	–	3/2
	–	12,575	–	1/2
	–	12,612	–	1/2
	–	12,640	–	3/2
$^4F_{7/2}$	13,016	13,004	12	1/2
and	13,084	13,045	39	3/2
$^4S_{3/2}$	13,108	13,089	19	1/2
	13,133	13,134	−1	1/2
	13,199	13,209	−10	1/2
	13,226	13,211	15	3/2

Table 1.8.4 (continued)

$^{2S+1}L_J$	$E_{obs.}$ (cm^{-1})	$E_{cal.}$ (cm^{-1})	ΔE (cm^{-1})	IR
$^4F_{9/2}$	14,422	14,459	−37	1/2
		14,498	–	3/2
	–	14,551	–	1/2
	–	14,670	–	1/2
	–	14,675	–	3/2
$^2H_{11/2}(2)$	–	16,083	–	1/2
	–	16,096	–	3/2
	–	16,102	–	1/2
	–	16,148	–	1/2
	16,162	16,168	−6	3/2
	16,198	16,178	20	1/2
$^4G_{5/2}$ and $^2G_{7/2}(1)$	16,402	16,410	−8	1/2
	16,445	16,468	−23	1/2
	16,525	16,525	−0	3/2
	16,571	16,551	20	1/2
	16,593	16,607	−14	1/2
	16,746	16,728	18	3/2
	16,888	16,885	3	1/2

*The energy level is not included in the fitting procedure.

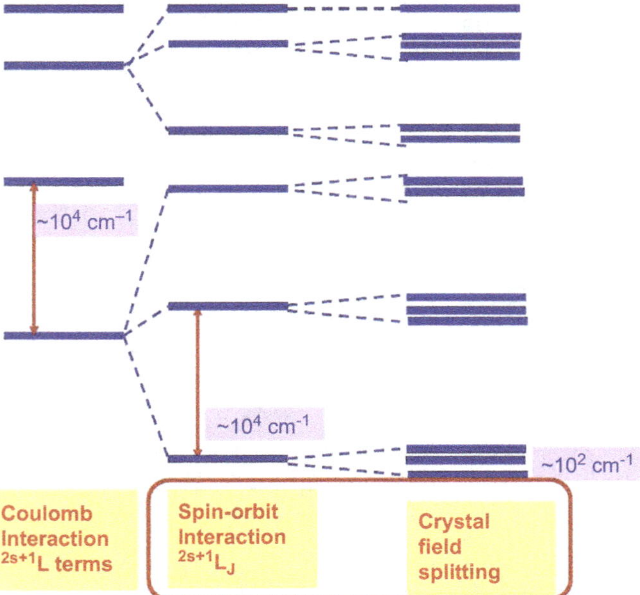

Figure 1.8.3: Formation of energy levels of lanthanide ions. The sequence of physical interactions (from left to right) and their approximate magnitude are also shown.

References

[1] http://science.widener.edu/svb/at_orbital/orbitals.html, https://chem.libretexts.org/Bookshelves/Physical_and_Theoretical_Chemistry_Textbook_Maps/Supplemental_Modules_(Physical_and_Theoretical_Chemistry)/Quantum_Mechanics/09._The_Hydrogen_Atom/Atomic_Theory/Electrons_in_Atoms/Electronic_Orbitals.

[2] A. R. Edmonds, Angular Momentum in Quantum Mechanics, Princeton University Press, 1957.

[3] I. I. Sobelman, Atomic Spectra and Radiative Transitions, Springer, 1992.

[4] R. D. Cowan, The Theory of Atomic Structure and Spectra, University of California Press, Berkeley, 1981.

[5] C. W. Nielsen, G. F. Koster, Spectroscopic Coefficients for the p^n, d^n, and f^n Configurations, MIT Press, Cambridge, Massachusetts, 1963.

[6] W. T. Carnall, G. L. Goodman, K. Rajnak, R. S. Rana, J. Chem. Phys. 1989, 90, 3443.

[7] B. G. Wybourne, Spectroscopic Properties of Rare Earths, Interscience Publishers, John Wiley & Sons, Inc., New York, 1965.

[8] G.-K. Liu, J. Solid State Chem. 2005, 178, 489.

[9] G.-K. Liu, Electronic Energy Level Structure, in Spectroscopic Properties of Rare Earths in Optical Materials (ed. G. K. Liu, B. Jacquier), Tsinghua University Press & Springer-Verlag, Berlin, Heidelberg, 2005, 1–84.

[10] C. A. Morrison, Crystal Fields for Transition-Metal Ions in Laser Host Materials, Springer-Verlag, 1992.

[11] C.-G. Ma, M. G. Brik, Q. X. Li, Y. Tian, J. Alloys Compd. 2014, 599, 93.

[12] C.-G. Ma, M. G. Brik, Y. Tian, Q. X. Li, J. Alloys Compd. 2014, 603, 255.

[13] Y. Y. Yeung, P. A. Tanner, J. Alloys Compd. 2013, 575, 54.

[14] Y. Y. Yeung, P. A. Tanner, Chem. Phys. Lett. 2013, 590, 46.

[15] P. A. Tanner, Y. Y. Yeung, L. X. Ning, J. Phys. Chem. A 2014, 118, 8745.

[16] Y. Y. Yeung, P. A. Tanner, J. Phys. Chem. A 2015, 119, 6309.

[17] G. H. Dieke, H. M. Crosswhite, Appl. Opt. 1963, 2, 675.

[18] G. H. Dieke, Spectra and Energy Levels of Rare Earth Ions in Crystals, Wiley Interscience, New York, 1968.

[19] K. Ogasawara, S. Watanabe, Y. Sakai, H. Toyoshima, T. Ishii, M. G. Brik, I. Tanaka, Jpn. J. Appl. Phys. 2004, 43, L611.

[20] K. Ogasawara, S. Watanabe, T. Ishii, M. G. Brik, Jpn. J. Appl. Phys. 2005, 44, 7488.

[21] P. S. Peijzel, A. Meijerink, R. T. Wegh, M. F. Reid, G. W. Burdick, J. Solid State Chem. 2005, 178, 448.

[22] C.-G. Ma, M. G. Brik, D.-X. Liu, B. Feng, Y. Tian, A. Suchocki, J. Lumin. 2016, 170, 369.

[23] P. Su, C.-G. Ma, M. G. Brik, A. M. Srivastava, Opt. Mater. 2018, 79, 129.

[24] F. S. Richardson, M. F. Reid, J. J. Dallara, R. D. Smith, J. Chem. Phys. 1985, 83, 3813.

[25] M. F. Reid, F. S. Richardson, J. Chem. Phys. 1985, 83, 3831.

[26] Y. Y. Yeung, P. A. Tanner, J. Alloys Compd. 2013, 575, 54.

[27] M. N. Popova, P. Chukalina, K. N. Boldyrev, T. N. Stanislavchuk, B. Z. Malkin, I. A. Gudim, Phys. Rev. B 2017, 95, 125131.

[28] M. C. Tan, G. A. Kumar, R. E. Riman, M. G. Brik, E. Brown, U. Hommerich, J. Appl. Phys. 2009, 106, 063118.

[29] C.-G. Ma, M. G. Brik, V. Kiisk, T. Kangur, I. Sildos, J. Alloys Compd. 2011, 509, 3441.

[30] C.-G. Ma, M. G. Brik, W. Ryba-Romanowski, H. C. Swart, M. A. Gusowski, J. Phys. Chem. A 2012, 116, 9158.

[31] M. Stachowicz, A. Kozanecki, C.-G. Ma, M. G. Brik, J. Y. Lin, H. Jiang, J. M. Zavada, Opt. Mater. 2014, 37, 165.
[32] H. Przybylińska, C.-G. Ma, M. G. Brik, A. Kamińska, J. Szczepkowski, P. Sybilski, A. Wittlin, M. Berkowski, W. Jastrzębski, A. Suchocki, Phys. Rev. B 2013, 87, 045114.
[33] K. Lemański, M. Babij, M. Ptak, P. J. Dereń, J. Lumin. 2017, 184, 130.

2 Reactivity and compounds

Mathias Wickleder

2.1 Halides of trivalent lanthanides

Halides are an important class of compounds for the rare earth elements. For the most typical oxidation state, that is, +III, all of the binary halides *RE*X_3 (X = F, Cl, Br, I) are known, with the nonexisting EuI_3 being the only exception. Meanwhile, low-valent halides are also known for various lanthanides (see also Chapter 2.3). In addition, a large number of ternary compounds, especially those containing alkali metals as further component, has been described. It is beyond the scope of this chapter to discuss all compounds in depth. Thus, emphasis will be laid on the binary species and on selected examples of ternary phases. Compilations that are more comprehensive can be found in the *Handbook on the Physics and Chemistry of Rare Earths*, for example, in [1].

2.1.1 Synthesis

The binary fluorides, *RE*F_3, can be precipitated from aqueous solutions of any soluble salt, for example, chlorides or nitrates. In a subsequent step, the crude material needs to be dried to remove any remaining water, preferably in an HF stream at 600 °C in order to prevent oxide or hydroxide contamination. These steps must be carried out in inert containers, at ambient and slightly elevated temperatures in polyethylene or teflon containers, and at higher temperatures in containers manufactured of nickel and its alloys, for example Monel, an alloy of composition $Cu_{32}Ni_{68}$. Alternatively, the reaction of the respective sesquioxides with $(NH_4)HF_2$ can be used for the preparation of fluorides. In principle, this is a two-step procedure, with the formation of a ternary fluoride first and, secondly, its thermal decomposition (Figure 2.1.1). Finally, the reaction of elemental fluorine with rare earth metals or any of their salts is also a possibility in preparing fluorides. Ternary fluorides are obtained by a solid state reaction of the respective binary components in sealed gold or platinum tubes.

A common way to prepare rare-earth chlorides and bromides, even in large quantities, is the so-called ammonium halide route [3]. In the first step, the respective rare earth sesquioxide, *RE*$_2O_3$, is converted into a ternary ammonium halide, $(NH_4)_3$*RE*X_6 or $(NH_4)_2$*RE*X_5 (X = Cl, Br). The latter are thermally decomposed in a second step, yielding the respective rare-earth trihalides, *RE*X_3 and the ammonium halide NH_4X. The first step may be carried out by dissolving the oxide and ammonium halide in hydrochloric or hydrobromic acid followed by evaporation to dryness ("wet" route); alternatively, the solids NH_4X and *RE*$_2O_3$ may be reacted directly ("dry" route). This procedure might be extended to the production of ternary compounds. Then, an alkali-metal halide is added to yield a quaternary halide, for example $(NH_4)_2LiGdCl_6$, that is subsequently decomposed to the ternary salt, $LiGdCl_4$ in the present example [4].

https://doi.org/10.1515/9783110654929-009

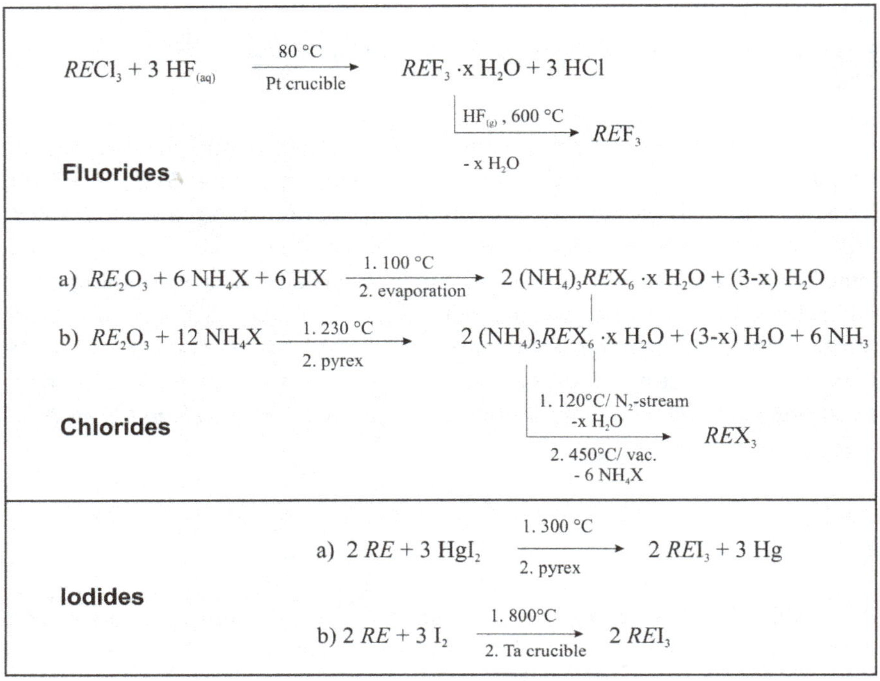

Fluorides

$$RECl_3 + 3\,HF_{(aq)} \xrightarrow[\text{Pt crucible}]{80\,^\circ C} REF_3 \cdot x\,H_2O + 3\,HCl$$

$$REF_3 \cdot x\,H_2O \xrightarrow[-x\,H_2O]{HF_{(g)},\ 600\,^\circ C} REF_3$$

Chlorides

a) $$RE_2O_3 + 6\,NH_4X + 6\,HX \xrightarrow[\text{2. evaporation}]{\text{1. }100\,^\circ C} 2\,(NH_4)_3REX_6 \cdot x\,H_2O + (3\text{-}x)\,H_2O$$

b) $$RE_2O_3 + 12\,NH_4X \xrightarrow[\text{2. pyrex}]{\text{1. }230\,^\circ C} 2\,(NH_4)_3REX_6 \cdot x\,H_2O + (3\text{-}x)\,H_2O + 6\,NH_3$$

$$(NH_4)_3REX_6 \cdot x\,H_2O \xrightarrow[\text{2. }450^\circ C/\text{ vac.},\ -6\,NH_4X]{\text{1. }120^\circ C/\ N_2\text{-stream},\ -x\,H_2O} REX_3$$

Iodides

a) $$2\,RE + 3\,HgI_2 \xrightarrow[\text{2. pyrex}]{\text{1. }300\,^\circ C} 2\,REI_3 + 3\,Hg$$

b) $$2\,RE + 3\,I_2 \xrightarrow[\text{2. Ta crucible}]{\text{1. }800^\circ C} 2\,REI_3$$

Figure 2.1.1: Synthesis strategies of binary rare earth halides (modified after [1]).

For the preparation of ternary (and quaternary) chlorides and bromides it is normally sufficient to dissolve the respective rare earth and alkali metal halides or carbonates in hydrohalic acid solution, evaporate to dryness and heat this intermediate product in a stream of the respective hydrogen halide gas at temperatures between 300 and 500 °C. Special care has to be taken if single crystals of the compounds are needed. These might be grown from stoichiometric amounts of the respective binary halide components in sealed containers of silica, tantalum, or even glass (Pyrex). However, knowledge of the respective phase diagrams of the AX/REX_3 systems is mandatory for the success of this route.

The preparation of rare-earth triiodides, REI_3, is more difficult, apparently due to the special stability of the oxyiodides, *RE*OI. The best route to pure triiodides is synthesis from the elements in sealed Pyrex containers [5]. The reaction of rare-earth metals with HgI_2 also leads to pure triiodides. The formation of elemental mercury is the major drawback of this procedure and it is, therefore, not favorable, especially for large batches. Annealing of stoichiometric amounts of the binary iodides or, alternatively, a one-batch synthesis from AI, *RE* and I_2 yields satisfyingly pure ternary rare-earth iodides.

2.1.2 Binary halides REX_3

The crystal structures of binary trihalides, REX_3, change within the lanthanide series as a result of the lanthanide contraction. High coordination numbers are found for the lighter lanthanides, especially with small ligands (fluoride) and vice versa. Figure 2.1.2 gives an overview of the crystal structures of the trihalides at ambient conditions.

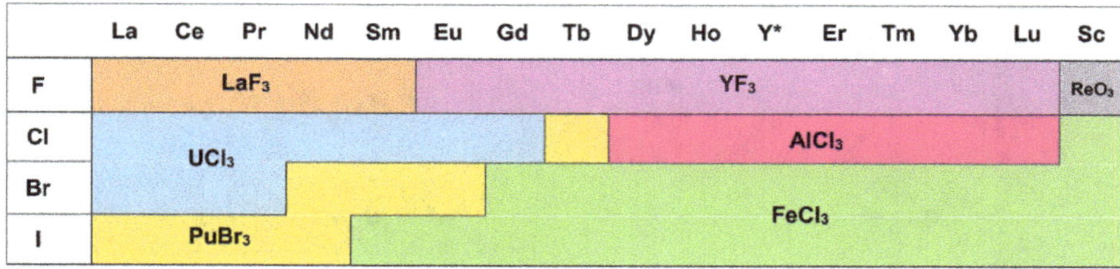

Figure 2.1.2: Structure types of the trihalides REX_3 of the rare-earth elements.

The trifluorides of lanthanum and cerium through samarium adopt the tysonite type of structure (LaF_3) with 11-fold RE^{3+} coordination [6]. The [REF_{11}] unit, the so-called *Edshammar* polyhedron may be described as a trigonal prism with all faces capped by additional ligands [7]. With the smaller lanthanide ions (Eu–Lu), the YF_3 type of structure occurs for the trifluorides, bearing a nine-fold (strictly speaking 8+1) RE^{3+} coordination. The coordination polyhedron is a distorted, tricapped trigonal prism. With the smallest of the rare-earth cations, Sc^{3+}, a third structure type is found. ScF_3 crystallizes with the ReO_3 (AlF_3) type of structure with octahedral coordination of the cations and all [ScF_6] octahedra share common corners [8].

The chlorides, bromides and iodides of the trivalent rare-earth cations occur in four different structure types: UCl_3, $PuBr_3$, $AlCl_3$ (YCl_3) and $FeCl_3$ (BiI_3) with coordination numbers of 9, 8 and 6 for the last two, respectively (Figure 2.1.3).

The UCl_3 type of structure consists of [$RECl_9$] polyhedra, trigonal prisms capped on each rectangular face. These are connected via the triangular faces to form columns running along the *c*-axis. The columns are shifted against each other by 1/2 *c*, so that ligands defining one prism serve as caps on adjacent prisms. The arrangement of the columns leads to channels in the [001] direction providing empty sites (octahedral and trigonal prismatic) for additional cations, an important feature in view of derivative structures, as discussed below. The $PuBr_3$ type of structure is closely related to that of UCl_3. The coordination number of the RE^{3+} ion is, however, only eight. The coordination polyhedron is a bicapped trigonal prism. The $AlCl_3$ type of structure is best derived from the cubic closest packing of anions. Octahedral interstices between alternating layers are either empty or are filled to two thirds with cations (Figure 2.1.3). In the closely related structure of RhF_3, octahedral holes

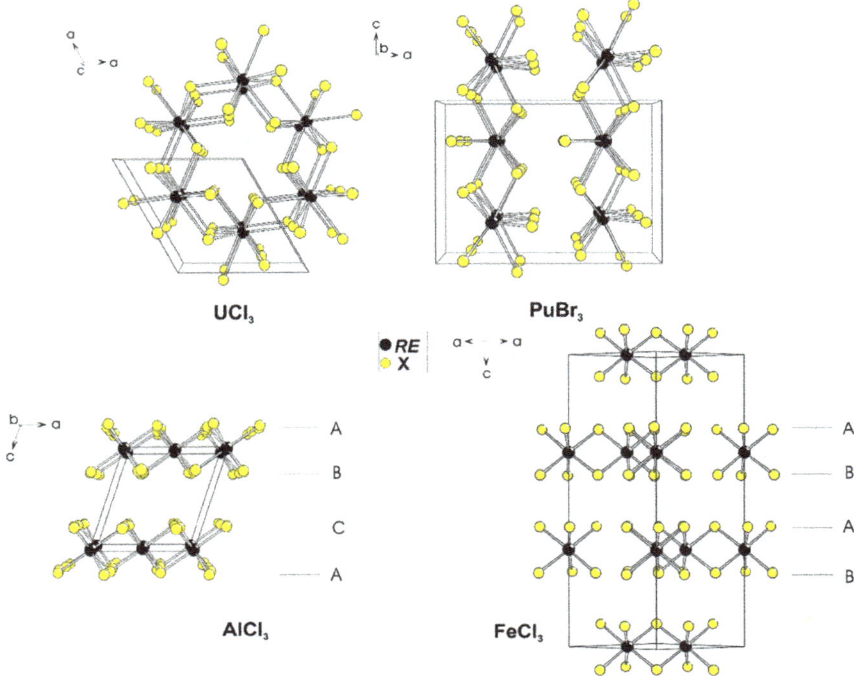

Figure 2.1.3: Perspective view of the typical REX_3-type structures (X = Cl, Br, I).

between each layer are filled to one third. In fact, this variety was found as the high-pressure modification for most of the $AlCl_3$-type trihalides [9]. Finally, the $FeCl_3$ type of structure is the hexagonally closest packed analogue to the $AlCl_3$ type of structure.

2.1.3 Binary halides REX_2

Halides with the rare-earth element in an oxidation state of less than +3 may be obtained by reduction of the respective (tri-)halides with their rare-earth metals or other reducing agents such as hydrogen or alkali metals [2], or even electrochemically. Two groups of dihalides, REX_2, of the rare-earth elements may be distinguished. The first group (**I**) are salt-like halides according to the formulation $(RE^{2+})(X^-)_2$ with typically ionic crystal structures, and the second group (**II**) are metallic or semi-metallic halides with delocalized electrons according to $(RE^{3+})(X^-)_2(e^-)$ which crystallize in more covalent/intermetallic structure types like those of MoS_2 or $MoSi_2$. Compounds of type **I** are found with the *RE* = Nd, Sm, Eu, Dy, Tm and Yb. With *RE* = Sm, Eu and Yb, even the difluorides exist, crystallizing with the CaF_2 type of structure. The other rare-earth elements form dihalides only with the “stronger reducing” halides I^- and (to a lesser extent) Br^- [10]. With *RE* = Ho, Er, Lu and Y, dihalides seem not to exist. Figure 2.1.4 summarizes the rare-earth dihalides known so far.

	La	Ce	Pr	Nd	Sm	Eu	Gd	Tb
F	—	—	—	—	CaF_2	CaF_2	—	—
Cl	—	—	—	$PbCl_2$	$PbCl_2$ CaF_2	$PbCl_2$ CaF_2	—	—
Br	$2H_2$-MoS_2	—	—	$PbCl_2$	$SrBr_2$	$SrBr_2$	—	SrI_2
I	$MoSi_2$	$MoSi_2$	$MoSi_2$ $2H_1$-MoS_2 3R-MoS_2 $CdCl_2$ PrI_2-V	$SrBr_2$ $MoSi_2$# CaF_2#	EuI_2	EuI_2 SrI_2	$MoSi_2$ 2H-MoS_2	—

	Dy	Ho	Y*	Er	Tm	Yb	Lu	Sc
F	—	—	—	—	—	—	—	—
Cl	$SrBr_2$	—	—	—	SrI_2	SrI_2	—	—
Br	SrI_2	—	—	—	SrI_2	SrI_2 α-PbO_2 $CaCl_2$	—	—
I	$CdCl_2$	—	—	—	CdI_2	CdI_2	—	CdI_2

Figure 2.1.4: Structure types of the dihalides of the rare-earth elements.
* Y is placed between Ho and Er.

The salt-like dihalides crystallize with well-known structures, thereby establishing the close relationship between the rare-earth elements in their divalent states and the respective alkaline earth cations of similar ionic radii. Only one modification of EuI_2 and SmI_2 adopt an otherwise unknown structure type. It consists of monocapped trigonal prisms [REI_7] (*RE* = Eu, Sm) that are connected in a rather complicated manner to a three-dimensional network. The $(RE^{3+})(e^-)(X^-)_2$-type dihalides often show interesting magnetic and electronic properties. For example, magnetic ordering occurs in GdI_2 below 313 K in a way that within the layers of the $MoSi_2$-type structure, the Gd^{3+} ions are coupled ferromagnetically, while the interlayer coupling is antiferromagnetic [11].

Resistivity and susceptibility measurements have been carried out for LaI_2 and the electronic band structure has been calculated [12]. Extended Hückel band structure calculations were also undertaken for the remarkable cubic modification of PrI_2 (PrI_2-V, see Chapter 2.3) [13]. It resembles the spinel structure ($MgAl_2O_4$) with all of the Mg^{2+} sites empty. Considerable metal–metal interactions occur (see Chapter 2.3.). Another interesting compound among the dihalides is neodymium diiodide, NdI_2. To date, three modifications are known and their occurrence depends on pressure and temperature. Under ambient conditions, the ionic $SrBr_2$ type of structure appears. At higher pressures, the metallic $MoSi_2$ type of structure is adopted and at even higher pressures, the CaF_2 type occurs [14]. The transformation from the $SrBr_2$ to the $MoSi_2$ type is accompanied by the electronic configuration crossover $5d^04f^4 \rightarrow 5d^14f^3$.

2.1.4 Ternary lanthanide halides

The crystal chemistry of the ternary rare-earth halides, with monovalent cations A^+ of the general composition $A_w RE_y X_{3y+w,}$ is very rich and a plethora of compounds has been reported [15–17], especially with A = K, Rb, Cs. The lanthanide halide systems with $A = NH_4{}^+$ have been investigated in more detail in order to understand the synthetic uniqueness of the ammonium ion [18], and the systems with Li^+, Na^+, Cu^+ and Ag^+ were explored aiming at new fast ionic conductors. The relative stability of the ternary halides in the AX/REX_3 systems and the diversity of compound types increases with increasing A^+/RE^{3+} ratio. For example, the system $CsCl/ErCl_3$ contains four compounds ($CsEr_2Cl_7$, $Cs_3Er_2Cl_9$, Cs_2ErCl_5 and Cs_3ErCl_6) [19], while the system $NaCl/ErCl_3$ has only two ($NaErCl_4$ and Na_3ErCl_6) [20].

2.1.4.1 Derivatives of the UCl_3 type of structure

The binary rare-earth chlorides $RECl_3$ with *RE* = La–Gd and the binary bromides $REBr_3$ with *RE* = La-Pr crystallize with the UCl_3 type of structure (Figure 2.1.5). In the direction of the crystallographic *c*-axis, this structure type provides empty sites (compressed octahedral and trigonal prismatic) that may accommodate additional cations. These are usually relatively small monovalent cations such as Na^+ (and K^+), but it is also possible to incorporate ions with higher charges. This is most prominently documented by the chloride $PrCl_{2.31}$ which is, in fact, a UCl_3 derivative according to $Pr_{0.3}[PrCl_3]$ [21, 22]. Due to the close proximity of the polyhedral centers, complete occupation is not possible.

There are two possibilities to compensate for the charge of the cations added. Firstly, RE^{3+} cations may be substituted by A^+ leading to the general formula $A_{2x}(A_x RE_{2-x})X_6$ (**I**); secondly, partial reduction of RE^{3+} to RE^{2+} might occur, leading to mixed-valent halides according to $A_x(RE^{2+})_x(RE^{3+})_{2-x}X_6$ (**II**). Type **I** compounds are

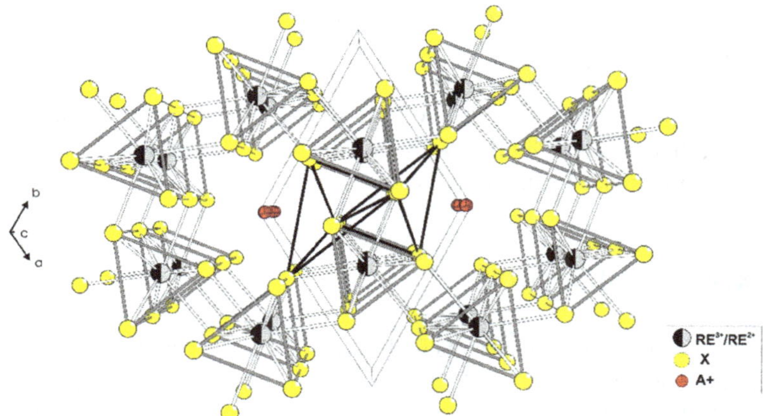

Figure 2.1.5: UCl_3 type of structure with possible sites for monovalent cations (A^+, in red). Charge compensation might occur by reduction of a part of the RE^{3+} cations to RE^{2+} cations, or by substitution of RE^{3+} by A^+.

substitution/addition derivatives of the UCl_3 type of structure and are found for chlorides with A = Na, K and *RE* = La-Sm [23]. They are sometimes referred to as "$A_3RE_5X_{18}$"-type compounds, but due to the more (La) or less (Sm) large phase widths, this can only be an approximate formula. For the sodium compounds (which are the best characterized ones), there is a strong relation between the maximum Na^+ content and the RE^{3+} ionic radius. The larger the ionic radius of the RE^{3+} cation, the more Na^+ can be incorporated. With $x = 0.5$ we arrive, according to $Na_{1.0}(Na_{0.5}RE_{1.5})Cl_6$ = $NaRECl_4$, at $AREX_4$-type compounds which are, in fact, known for a number of rare-earth fluorides and chlorides. For fluorides, the UCl_3-type derivate is indeed found, and many $NaREF_4$-type compounds are known even if none of the binary fluorides adopts the UCl_3 type of structure. The chlorides $ARECl_4$, however, (even though the chlorides with *RE* = La-Gd are isotypic with UCl_3), adopt other structure types.

Type **II** compounds are pure *addition* derivatives of the UCl_3 type and occur with *RE* = Pr-Eu and A = Na. They have the ideal composition $NaRE_2Cl_6$ (=$Na_{0.5}RECl_3$), but deviations of this formula are also known, for example, $Na_{0.75}Eu_2Cl_6$. The crystals show blue ($NaEu_2Cl_6$), green ($Na_{0.75}Eu_2Cl_6$) and dark red ($NaSm_2Cl_6$) colors, or are black ($NaPr_2Cl_6$, $NaNd_2Cl_6$). In $NaPr_2Cl_6$, XANES measurements proved that there is only trivalent praseodymium present. Thus, the compound has to be written as $Na(Pr^{3+})_2Cl_6(e^-)_{1.0}$.

In the mixed-valent $A(RE^{2+})(RE^{3+})Cl_6$-type compounds, the RE^{2+} ion may be substituted by an alkaline-earth ion, B^{2+}, of preferably the same size. For example, $NaEu_2Cl_6$ and $NaSm_2Cl_6$ are paralleled by $NaSrEuCl_6$ and $NaSrSmCl_6$ without structural changes [24]. As these compounds are no longer mixed-valent and with the reducing Eu^{2+} and Sm^{2+} cations, Ag^+ can now be built in instead of Na^+, yielding $AgSrEuCl_6$ and $AgSrSmCl_6$. Even bromides like $NaBaLaBr_6$ are available, although $NaLa_2Br_6$ is not known. Slight deviations from the exact stoichiometry seem to be tolerable as is indicated by elemental analysis. Iodides of this class of compounds are not known. On the other hand, a fluoride with the idealized composition $NaCaYF_6$ and the same structure even occurs in nature, as the mineral *gagarinite* [25].

2.1.4.2 ARE_2X_7-type halides

ARE_2X_7-type halides occur mainly in systems with large alkali cations (K-Cs) [26–29] and pseudoalkali cations like In^+, Tl^+ and NH_4^+ [30]. They have not been observed with Li^+ and Ag^+. With Na^+, there is evidence from thermoanalytical investigations for $NaRE_2Cl_7$-type compounds with *RE* = Gd-Ho [31, 32]. For ARE_2F_7-type fluorides, four structure types have been determined for KHo_2F_7, KEr_2F_7, KYb_2F_7 and $RbEr_2F_7$. The monoclinic structure of KHo_2F_7 shows the Ho^{3+} ions in eightfold coordination with a cube, a square antiprism and a dodecahedron as the coordination polyhedra [33]. The same is true for KEr_2F_7 although the symmetry is orthorhombic [34]. The coordination polyhedron of Yb^{3+} in KYb_2F_7 is a pentagonal bipyramid [34].

Single crystals of a size up to 10 mm of $CsGd_2F_7$ and CsY_2F_7 doped with Ce^{3+} have been grown under hydrothermal conditions. Scintillation properties (Chapter 4.6) have been measured. However, the light yield after high-energy irradiation (662 keV) is low [35].

ARE_2X_7-type chlorides and bromides crystallize with at least three crystallographically distinguishable structures, two of which have been determined for KDy_2Cl_7 and $RbDy_2Cl_7$ [36]. The principal features of the crystal structures of KDy_2Cl_7 and $RbDy_2Cl_7$ (Figure 2.1.6) are the same, although they exhibit different symmetry (monoclinic and orthorhombic, respectively). The structures consists of dimeric units built up from two monocapped-trigonal prisms sharing a common triangular face. These are connected via four edges to form infinite layers stacked in the [100] direction and held together by A^+ cations. In accordance with the face connection of the $[DyCl_7]$ polyhedra, the Dy^{3+}–Dy^{3+} distance is quite short (385 pm), a situation that is also found in $Cs_3RE_2X_9$-type compounds containing face-sharing octahedra (see below).

The close proximity of the RE^{3+} ions in the face-sharing $[Dy_2Cl_{11}]$ groups appears to have great influence on the spectroscopic properties – upconversion luminescence has been investigated for orthorhombic $RbGd_2Cl_7$ and $RbGd_2Br_7$ single crystals doped with various amounts of Er^{3+} and grown via the Bridgman technique [37]. The interpretation of the spectra shows that energy transfer processes supported by the closely neighbored ions are involved.

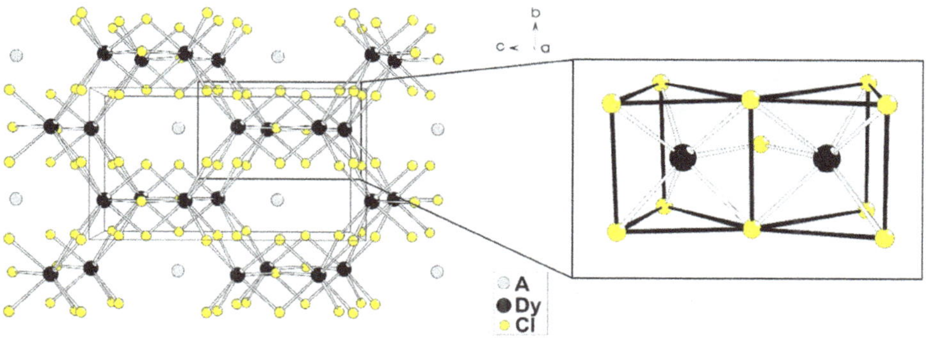

Figure 2.1.6: Crystal structure of ADy_2Cl_7 (A = K, Rb) (modified after [1]).

2.1.4.3 $AREX_4$-type halides

Ternary fluorides of the $AREF_4$ type with A = Li-Cs, Ag, In and Tl are long known and pretty well characterized [38–40]. Five structure types are known for $AREF_4$-type fluorides: the *anti*-scheelite type and the structure types that were first determined for $NaNdF_4$, $KErF_4$, $TlTmF_4$ and β-$KCeF_4$. In addition, the CaF_2 type occurs with high-temperature modifications for most of the $AREF_4$-type fluorides, with statistical distribution of the cations A^+ and RE^{3+} on the Ca^{2+} sites. The hexagonal $NaNdF_4$ type of structure is the aristotype for the α-$NaREF_4$ phases. $NaNdF_4$ is an addition/substitution variant of the UCl_3 type of structure (refer above) [39]. One fourth of the Nd^{3+} ions are replaced by Na^+ in a way that chains of pure [$NdCl_9$] and mixed [$Na_{0.5}Nd_{0.5}Cl_9$] polyhedra occur (Figure 2.1.7).

$KErF_4$ (trigonal) and $TlTmF_4$ (hexagonal)-type compounds crystallize with CaF_2-type superstructures, showing complex ordering schemes with 18 and 24 formula units in the unit cell, respectively [40]. The ß-$KCeF_4$ type of structure which occurs with the larger lanthanide and alkali metal ions might be understood as a "mixture" of the UCl_3 and $PbCl_2$ types of structure. Both ions are nine-coordinate in the form of tricapped trigonal prisms. The [CeF_9] polyhedra are linked just like the [UCl_9] polyhedra in UCl_3, and the linkage of the [KCl_9] polyhedra represent a part of the $PbCl_2$ type of structure.

$ARECl_4$-type chlorides occur for A = Li with *RE* = Sm, Eu, Gd only. Systems with larger RE^{3+} ions seem to be simply eutectic; systems with smaller RE^{3+} ions are dominated by Li_3RECl_6-type chlorides. $LiGdCl_4$ has an *anti*-scheelite type of structure ($Gd^{[8]}Li^{[4]}Cl_4 = Ca^{[8]}W^{[4]}O_4$) [41]. The $NaRECl_4$ compounds were obtained with *RE* = Gd-Lu, Y, Sc [42–45]. With A^+ ions other than Li^+ and Na^+, no formation of $AREX_4$-type compounds with X = Cl, Br, I has been observed for ternary rare-earth chlorides, bromides and iodides. With A = Na, three structure types are now known. For *RE* = Gd-Tb the triclinic structure of $NaGdCl_4$ is found [42]. $NaGdCl_4$ is an ordered variety of EuI_2, although the symmetry is only triclinic for $NaGdCl_4$ rather than

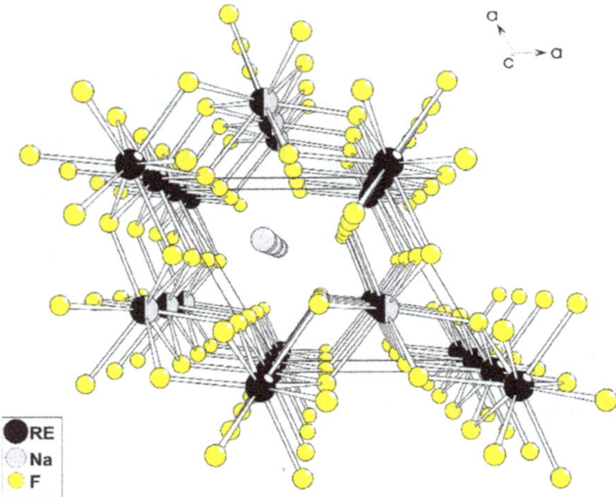

Figure 2.1.7: Crystal structure of α-NaREF_4-type phases. Every second row of face-sharing [$RECl_9$] polyhedra contains Na^+ and RE^{3+} (50% each) (modified after [1]).

monoclinic, as for EuI_2. The structure of $NaErCl_4$ exhibits a hexagonal closest packing of chloride ions, with the octahedral holes between alternating layers being filled with Na^+ and Er^{3+}, respectively.

2.1.4.4 A_2REX_5-type halides

Examples for ternary rare-earth halides of the A_2REX_5 type are known for all of the halides and most A^+ ions, although there are only few examples with A = Na, Cs. The sodium chlorides Na_2RECl_5 (*RE* = Sm-Gd) melt incongruently [46–48], and no compound formation at all could be detected in the NaCl/$RECl_3$ systems with the RE^{3+} ion larger than Sm^{3+}; these systems are simply eutectic. With RE^{3+} ions smaller than Gd^{3+}, chlorides with other compositions such as Na_3RECl_6 and $NaRECl_4$ become more stable and no Na_2RECl_5 formation is observed. No Na_2REX_5 halides other than chlorides are known and the same is true for A = Cs.

All of the A_2REX_5-type halides crystallize with one of two structure types, the K_2PrCl_5 and the Cs_2DyCl_5 type of structure [49, 50]. The fluorides A_2REF_5 adopt the K_2SmF_5 type of structure. Its structural arrangement is the same as in K_2PrCl_5 [51]. The K_2PrCl_5 type of structure (Figure 2.1.8) consists of monocapped trigonal prisms [$PrCl_7$], which are connected via common edges to form infinite chains running along the crystallographic *b*-axis. The chains are arranged as hexagonal packing of rods; these are held together by A^+ ions. The coordination number of the latter is between 8 and 9 and varies slightly, depending upon the ionic size.

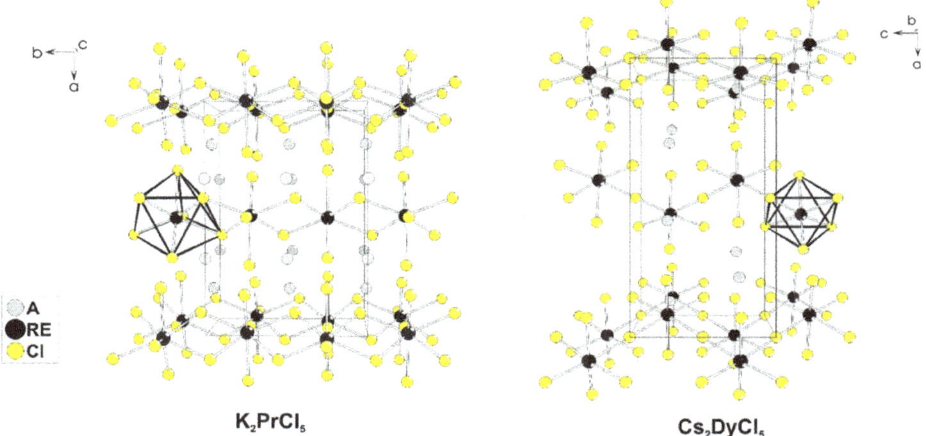

Figure 2.1.8: Crystal structures of K_2PrCl_5 and Cs_2DyCl_5 emphasizing the chains of connected [$PrCl_7$] and [$DyCl_6$] polyhedra (modified after [1]).

In $A_2\mathit{RE}X_5$ halides with smaller RE^{3+} ions and/or larger ligands, the Cs_2DyCl_5 type of structure occurs. Chains of cis-vertex connected [$\mathit{RE}Cl_6$] octahedra, which is also found in several oxide halides, for example, $K_2VO_2F_3$ are the characteristic feature of this structure type. The chains are oriented parallel to [001], and in analogy to K_2PrCl_5, packed hexagonally like rods. Single crystals of K_2LaCl_5 and K_2LaBr_5 doped with Er^{3+} were grown via the Bridgman technique and their IR to VIS up conversion properties have been investigated. Remarkable differences can be seen, subject to the two host structures in accordance with the lower phonon energy of the bromide structure [52].

2.1.4.5 $A_3\mathit{RE}_2X_9$-type halides

One of the best known classes of ternary rare-earth halides is that of the $A_3\mathit{RE}_2X_9$-type compounds. The A^+ cation takes part in closest packed layers of the composition AX_3 and, therefore, requires with respect to its coordination number of 12, large A^+ ions, namely Cs, Rb, and to a much lesser extent K. They also do not occur with X = F.

Two structure types are observed in rare earth containing $A_3\mathit{RE}_2X_9$-type compounds, namely those of $Cs_3Cr_2Cl_9$ and $Cs_3Tl_2Cl_9$. Both may be derived from closest packed layers of the composition [$CsCl_3$] (Figure 2.1.9). In principle, such layers may be stacked in an infinite number of ways. The two structures represent ABACBC and ABABAB stacking sequences, respectively.

If all possible sites are occupied, one arrives at the composition $A\mathit{RE}X_3$, known for example for $CsNiCl_3$ (and "$CsScCl_3$") in the case of the hexagonal closest packing and perovskite-type compounds for the cubic closest packing motif.

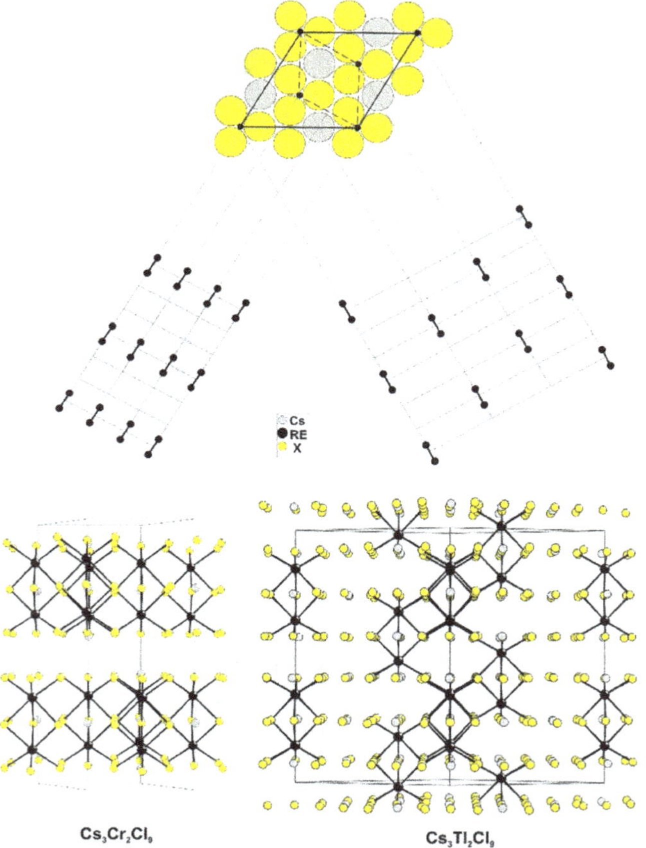

Figure 2.1.9: Crystal structures of $Cs_3Tl_2Cl_9$ and $Cs_3Cr_2Cl_9$ as derived from hexagonally closest packed $[CsCl_3]$ layers (modified after [1]).

The common feature of the $Cs_3Cr_2Cl_9$ and the $Cs_3Tl_2Cl_9$ types of structure are bi-octahedral $[RE_2X_9]$ dimers, resulting from the occupation of neighboring octahedra. This arrangement allows for the shortest distances between the cations RE^{3+} in a closest-packed array of anions. Therefore, exchange phenomena between neighboring centers including the IR to VIS up conversion properties of the respective erbium halides [53–55] have been studied extensively.

2.1.4.6 A_3REX_6-type halides and derivatives

A_3REX_6-type compounds occur quite frequently in the chemistry of rare earth halides. At least seven structure types have been identified solely for the A_3REF_6-type fluorides. These structures are all strongly related in a way that they can be derived from the elpasolite type of structure (K_2NaAlF_6), by tilting the [$\mathit{RE}F_6$] octahedra. This has been shown extensively [38] and will not be outlined here. For Rb_2KScF_6, temperature-dependent neutron diffraction studies have been carried out, resulting in the determination of three modifications of the structures. The low temperature form has the structure of cryolite (monoclinic), at ambient conditions the β-$(NH_4)_3ScF_6$ type (tetragonal) and at high temperatures, the elpasolite type (cubic) occurs [56]. The cryolite and the elpasolite types of structure are also known for the higher halides, along with a number of further structure types, namely Cs_3BiCl_6, K_3MoCl_6, Na_3CrCl_6, Na_3GdCl_6 (stuffed $LiSbF_6$), Na_3GdI_6, Li_3YCl_6 and Li_3YbCl_6. The structures of these halides depend strongly on the size of the A cation: If A is large enough (e.g., Cs) closest packings of A and X ions are found (e.g., elpasolite); small A ions are, on the other hand, mostly located in octahedral sites of pure anion packings (e.g., Na_3CrCl_6). There is a common feature in all of these structures: The rare-earth ion RE^{3+} is always octahedrally coordinated, and the [$\mathit{RE}X_6$] octahedra are isolated from each other. This is the tribute to the high A content of $A_3\mathit{RE}X_6$-type halides which makes the connection of [$\mathit{RE}X_6$] octahedra rather unlikely.

The survey of the $A_3\mathit{RE}X_6$ structure types starts with the elpasolite type of structure (K_2NaAlF_6). Although this is not strictly a ternary compound, many of the ternary $A_3\mathit{RE}X_6$ rare-earth halides adopt this structure, at least at elevated temperatures. There is a plethora of $A_2B\mathit{RE}X_6$ halides with the cubic elpasolite-type structure, its tetragonally (or otherwise) distorted varieties or its hexagonal stacking variants, depending also upon temperature and pressure. Many of the features have been reviewed previously and shall be left aside here [36]. The cubic face-centered elpasolite type of structure (K_2NaAlF_6) is strongly related to the K_2PtCl_6 type of structure. Both, with their general formulae $A_2B\mathit{RE}X_6$ and $A_3\mathit{RE}X_6$, consist of [AX_3] layers stacked in the ABC (cubic closest packed) fashion in the [111] direction of the cubic face-centered Bravais lattice. Half of the available octahedral sites are occupied by Pt^{4+} cations in K_2PtCl_6 and all are occupied by Na^+ and Al^{3+} in K_2NaAlF_6, in an ordered way. Note that, starting formally from $A_2B\mathit{RE}X_6$-type compounds, with B = *RE* and with B = A, we will get to the classes of perovskite ($A_2\mathit{RE}_2X_6$ = $A\mathit{RE}X_3$) and $A_3\mathit{RE}X_6$-type compounds with all their structural varieties.

It is obvious that the elpasolite type of structure is preferred with A^+ and B^+/RE^{3+} cations of strongly different sizes, because A is part of the closest packing and B and RE^{3+} occupy the octahedral sites within the packing. Therefore, structural difficulties must arise when B becomes chemically equal to A as is the case in $A_3\mathit{RE}X_6$ (= $A_2A\mathit{RE}X_6$)-type halides. These difficulties are dealt with in two ways:

a) With large ions like Cs^+ and Rb^+, too big for octahedral coordination, structures are formed with these ions more or less incorporated in the layers that are then stacked. Examples are the structures of K_3MoCl_6 and Cs_3BiCl_6. As the packing is not ideal, rather low symmetries are normally observed, monoclinic for

K_3MoCl_6 and Cs_3BiCl_6, although higher symmetries are possible, as can be seen from the low-temperature modification of the $Cs_3\mathit{RE}Cl_6$-type chlorides [57].

b) With all A^+ ions in $A_3\mathit{RE}X_6$ structures equal and small, the elpasolite type of structure cannot be retained. The most prominent example is the cryolite type of structure, Na_3AlF_6, where according to $Na_2(NaAl)F_6$, the parent elpasolite is resembled in that the coordination number of the two Na^+ of formerly 12 is reduced to 8, through tilts of the $[\mathit{RE}X_6]$ octahedra. Many of the $Na_3\mathit{RE}X_6$-type chlorides and bromides also crystallize with the cryolite type of structure.

If the A^+ ions are small enough for octahedral coordination, they may all be incorporated in the octahedral holes of a closest packing of anions. The so-called stuffed $LiSbF_6$ type is a prominent example which is found in numerous $A_3\mathit{RE}X_6$-type compounds with A = Na, Ag [58–61]. There is a strong structural similarity between the cryolite and the stuffed $LiSbF_6$ type. The latter may be derived from a hexagonal closest packing of anions (Cl^- in Na_3GdCl_6) with two of the six octahedral holes filled regularly with Gd^{3+} and $Na1^+$, and the remaining two Na^+ ions distributed statistically over four empty interstices, according to $Na_{4/2}(NaGd)Cl_6$. The stuffed $LiSbF_6$ type of structure might also be understood as a derivative of the RhF_3 type of structure that occurs as the high-pressure form of various binary rare-earth halides of the $FeCl_3$ (BiI_3) type. According to $2 \times \mathit{RE}X_3 = \mathit{RE}_2X_6 \rightarrow (A\mathit{RE})X_6 \rightarrow A_2(A\mathit{RE})X_6$, the stuffed $LiSbF_6$ type is an addition/substitution variant of the RhF_3 type (Figure 2.1.10).

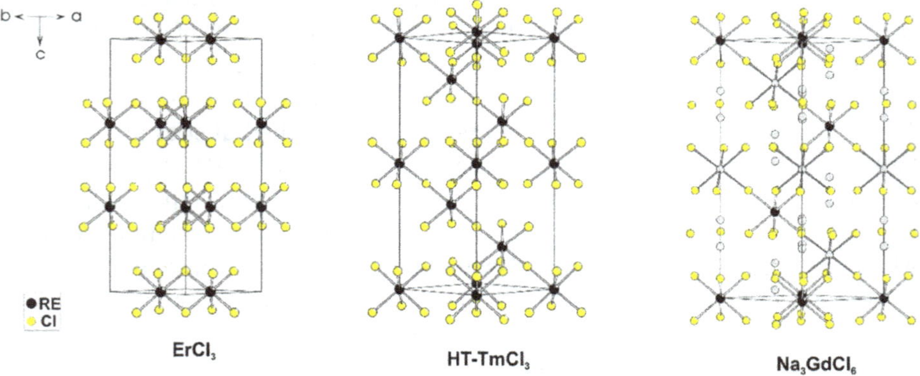

Figure 2.1.10: A comparison of the crystal structures of $ErCl_3$ ($FeCl_3$), HT-$TmCl_3$ (RhF_3) and Na_3GdCl_6 (stuffed $LiSbF_6$ type) (modified after [1]).

An analogous addition/substitution variant now derived from the cubic closest packing of halide anions is found in Na_3GdI_6, Li_3ErBr_6 and Li_3ScCl_6, to name only three [62–64]. The parent structure type is that of $AlCl_3$, which appears in numerous binary rare-earth halides at ambient conditions (see above). Two thirds of the Na^+ ions of Na_3GdI_6 are statistically distributed over the four possible sites.

The chlorides $Li_3\mathit{RE}Cl_6$ appear in three structure types of which Li_3ScCl_6 is isostructural with, for example, Li_3ErBr_6 and Na_3GdI_6. The other two structure types, first determined for Li_3YCl_6 and Li_3YbCl_6, are very similar and connected via a phase transition. They have the hexagonal closest packing of chloride ions in common, in which the cations occupy octahedral sites. The structure of Na_3ScBr_6 (Na_3CrCl_6 type) represents another variant of occupation of four out of the six octahedral sites in the hexagonal closest packing of anions [64]. In the $A_3\mathit{RE}X_6$-type compounds with A = Li, Na, Ag ionic conductivity has been found. The best conductivity is found for structures with a hexagonal closest packing of anions, polarizing and/or very small cations (Li^+, Ag^+) and a polarizable anionic matrix (Br^-, I^-). The highest lithium ionic conductivity is observed in Li_3YCl_6. Very good ionic conduction is also found in Ag_3YCl_6 (stuffed $LiSbF_6$ type) [65–67].

2.1.4.7 Complex halides in $AX_2/\mathit{RE}X_3$ systems

The systems $AX_2/\mathit{RE}X_3$ with A now representing an alkaline-earth element have been investigated to a much lesser extent than the respective alkali metal halide systems. Phase diagrams indicate the formation of $A_4\mathit{RE}X_{11}$-, $A_3\mathit{RE}X_9$-, $A_2\mathit{RE}X_7$- and $A\mathit{RE}X_5$-type compounds and their occurrence is obviously related to the ionic radii of RE^{3+} and A^{2+} [68, 69]. The existence of $Ba_2\mathit{RE}Cl_7$-type chlorides with *RE* = Gd-Yb and Y has been proven for the examples of Ba_2ErCl_7 and Ba_2GdCl_7 [70]. They crystallize with a three-dimensional network of egde- and face-connected [$BaCl_9$] polyhedra and RE^{3+} ions in seven-coordinate interstices. The [$\mathit{RE}Cl_7$] polyhedra may be described as monocapped trigonal prisms and are isolated from each other. Another variant of $Ba_2\mathit{RE}Cl_7$-type compounds is found in Ba_2ScCl_7 [71], which shows the Sc^{3+} ions octahedral coordination of six Cl^- ligands, while the Ba^{2+} ions have coordination numbers of 9 and 10, respectively.

The approximate composition "$A_2\mathit{RE}X_7$" is also observed with *RE* = La–Sm and A = Sr. These phases show, as well as most of the $A\mathit{RE}X_5$-type compounds, a certain phase width which might be understood with respect to their crystal structures. These represent superstructures of the CaF_2 type of structure in the following way: Tilting of cubes of the anionic substructure (Figure 2.1.11) generates space for four additional anions leading to a "$(A,\mathit{RE})_6X_{36}$" building unit. The cuboctahedral site in the center of this unit can be filled with an additional anion. This is quite well known and described for fluorides. Because the occupation of the void need not necessarily be complete, slight deviations from the exact stoichiometry may occur. Instead of additional halide anions, oxide ions may be incorporated.

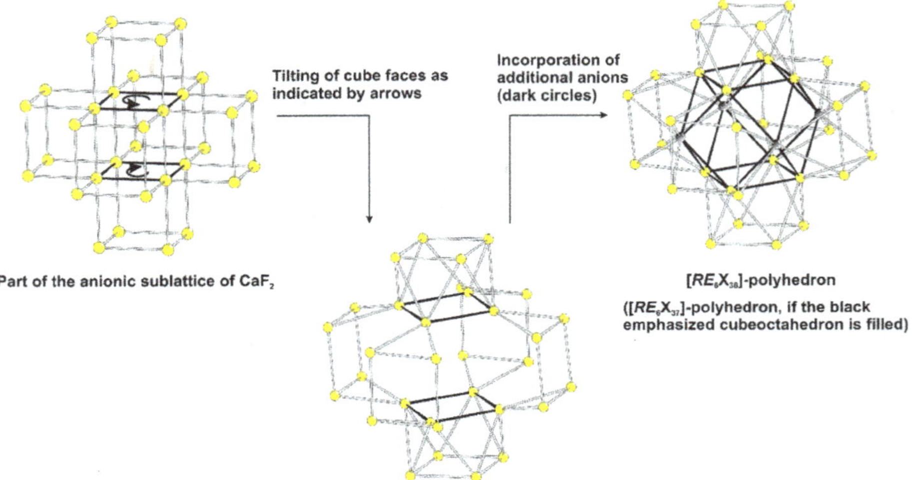

Figure 2.1.11: Generation of the $(A,RE)_6X_{36}$ polyhedral "cluster" from a section of the parent CaF_2 type of structure (modified after [1]).

The $(A,RE)_6Cl_{36+x}$ polyhedral "clusters" may be linked in two different ways, leading to two different crystal structures, one with tetragonal and the other with trigonal rhombohedric symmetry, with the approximate compositions $(A,RE)_{15}X_{34+x}$ ($\cong A_{10}RE_5X_{35}$=A_2REX_7) and $(A,RE)_{14}X_{32+x}$, respectively. Based on X-ray investigations alone, it cannot be judged often, whether an additional anion occupies the cuboctahedron or not. As A^{2+} and RE^{3+} ions occupy, at least partly, equal crystallographic sites, electroneutrality can be achieved in any case. Very thorough research by Bärnighausen et al. on isotypic mixed-valent rare-earth halides (A = *RE*) gave evidence of partial occupation of the cuboctahedron with oxide or chloride in some cases [72].

A second type of a CaF_2 superstructure occurs in the crystal chemistry of rare-earth/alkaline-earth halides, the so-called *Vernier* phases [73–75]. These are one-dimensional derivatives of the fluorite type in a sense that part of the cubic closest packed 4^4 anion layers are transformed to the more dense 6^3 layers. As these layers cannot be superimposed exactly, Vernier type structures are often incommensurate.

As mentioned earlier, $AREX_5$-type compounds crystallize with *RE* = La, Ce-Sm with a three-dimensional CaF_2 superstructure. In $BaGdCl_5$, however, another structure has been found [76]. The compound crystallizes monoclinically with strictly ordered cations. Both are in eightfold coordination, Ba^{2+} in form of a slightly distorted dodecahedron, and Gd^{3+} nearly in form of a square antiprism.

References

[1] G. Meyer, M. S. Wickleder, Simple and Complex Rare Earth Halides, in Handbook on the Physics and Chemistry of Rare Earths (ed. L. Eyring, K. Gschneidner), 28, Elsevier Science Publishers, New York, 2000, 53–129.

[2] G. Meyer, L. R. Morss (Eds.) Synthesis of Lanthanide and Actinide Compounds, Kluwer, Dordrecht, 1991.

[3] G. Meyer, Inorg. Synth. 1989, 25, 146.

[4] G. Meyer, Z. Anorg. Allg. Chem. 1984, 511, 193.

[5] J. D. Corbett, Inorg. Synth. 1983, 22, 31.

[6] A. F. Wells, Structural Inorganic Chemistry, 4th Edition, Clarendon Press, Oxford, 1975.

[7] B. G. Hyde, S. Andersson, Inorganic Crystal Structures, Wiley, New York, 1989.

[8] R. Lösch, C. Hebecker, Z. Ranft, Z. Anorg. Allg. Chem. 1982, 491, 199.

[9] H. P. Beck, E. Gladrow, Z. Anorg. Allg. Chem. 1983, 498, 75.

[10] M. Krings, M. Wessel, R. Dronskowski, Acta Crystallogr. 2009. C65, i66.

[11] A. Kasten, P. H. Müller, M. Schienle, Solid State Commun. 1984, 51, 919.

[12] J. H. Burrow, C. H. Maule, P. Strange, J. N. Tothill, J. A. Wilson, J. Phys. C: Solid State Phys. 1987, 20, 4115.

[13] E. Warkentin, H. Bärnighausen, Z. Anorg. Allg. Chem. 1979, 459, 187.

[14] H. P. Beck, M. Schuster, J. Solid State Chem. 1992, 100, 301.

[15] R. Blachnik, D. Selle, Z. Anorg. Allg. Chem. 1979, 454, 90.

[16] R. Blachnik, A. Jäger-Kasper, Z. Anorg. Allg. Chem. 1980, 461, 74.

[17] H. J. Seifert, G. Thiel, J. Chem. Thermodyn. 1982, 14, 1159.

[18] G. Meyer, The Ammonium Ion for Synthesis, in: Advances in the Synthesis and Reactivity of Solids (ed. T. E. Mallouk), 2, JAI Press, Greenwich, 1994, 1.

[19] R. Blachnik, D. Selle, Z. Anorg. Allg. Chem. 1979, 454, 90.

[20] H. J. Seifert, J. Sandrock, Z. Anorg. Allg. Chem. 1990, 587, 110.

[21] J. D. Corbett, Rev. Chim. Minér. 1973, 10, 239.

[22] G. Meyer, T. Schleid, K. Krämer, J. Less-Common Met. 1989, 149, 67.

[23] F. Lissner, K. Krämer, T. Schleid, G. Meyer, Z. Hu, G. Kaindl, Z. Anorg. Allg. Chem. 1994, 620, 444.

[24] M. S. Wickleder, G. Meyer, Z. Anorg. Allg. Chem. 1998, 624, 1577.

[25] J. M. Hughes, J. W. Drexler, Can. Min. 1994, 32, 563.

[26] H. J. Seifert, H. Fink, G. Thiel, J. Less-Common Met. 1985, 110, 139.

[27] H. J. Seifert, J. Sandrock, G. Thiel, J. Therm. Anal. 1986, 31, 1309.

[28] H. J. Seifert, J. Sandrock, J. Uebach, Z. Anorg. Allg. Chem. 1987, 555, 143.

[29] H. J. Seifert, H. Fink, J. Uebach, J. Therm. Anal. 1988, 33, 625.

[30] R. Blachnik, E. Enninga, Z. Anorg. Allg. Chem. 1983, 503, 133.

[31] H. J. Seifert, J. Sandrock, J. Uebach, Acta Chem. Scand. 1995, 49, 653.

[32] H. J. Seifert, J. Sandrock, Z. Anorg. Allg. Chem. 1997, 623, 1525.

[33] Y. Le Fur, S. Aleonard, M. F. Gorius, M. T. Roux, Acta Crystallogr. 1982, 38, 1431.

[34] S. Aléonard, Y. Le Fur, M. F. Gorius, M. T. Roux, J. Solid State Chem. 1980, 34, 79.

[35] D. R. Schaart, P. Dorenbos, C. W. E. van Eijk, R. Visser, C. Pedrini, B. Moine, N. M. Khaidukov, J. Phys.: Condens. Matter 1995, 7, 3063.

[36] G. Meyer, Prog. Solid State Chem. 1982, 14, 141.

[37] T. Riedener, K. Krämer, H. U. Güdel, Inorg. Chem. 1995, 34, 2745.

[38] O. Greis, J. M. Haschke, Handbook on the Physics and Chemistry of Rare Earths, 5, North Holland Publishing Company, 1982, 387.

[39] J. H. Burns, Inorg. Chem. 1965, 4, 881.

[40] S. Aléonard, Y. Le Fur, L. Pontonnier, M. F. Gorius, M. T. Roux, Ann. Chim. Sci. 1978, 3, 417.
[41] G. Meyer, Z. Anorg. Allg. Chem. 1984, 511, 193.
[42] T. Schleid, G. Meyer, Z. Anorg. Allg. Chem. 1990, 622, 173.
[43] M. S. Wickleder, G. Meyer, Z. Anorg. Allg. Chem. 1995, 621, 546.
[44] M. S. Wickleder, H. U. Güdel, T. Armbruster, G. Meyer, Z. Anorg. Allg. Chem. 1996, 622, 785.
[45] M. S. Wickleder, A. Bohnsack, G. Meyer, Z. Anorg. Allg. Chem. 1996, 622, 675.
[46] G. Thiel, H. J. Seifert, Thermochim. Acta 1988, 133, 275.
[47] H. J. Seifert, J. Sandrock, Z. Anorg. Allg. Chem. 1990, 587, 110.
[48] H. J. Seifert, J. Sandrock, G. Thiel, Z. Anorg. Allg. Chem. 1991, 598/599, 307.
[49] G. Meyer, E. Hüttl, Z. Anorg. Allg. Chem. 1983, 497, 191.
[50] G. Meyer, Z. Anorg. Allg. Chem. 1980, 469, 149.
[51] R. I. Bochkova, Y. N. Saf´yanov, E. A. Kuz´min, N. v. Belov, Dokl. Akad. Nauk. SSSR 1973, 211, 357.
[52] K. Krämer, H. U. Güdel, J. Alloys Compd. 1995, 207/208, 128.
[53] M. Hehlen, H. U. Güdel, Q. Shu, J. Rai, S. C. Rand, J. Chem. Phys. 1996, 104, 1232.
[54] P. Allenspach, A. Furrer, H. U. Güdel, H. Büttner, Physica B 1997, 234, 744.
[55] M. Pollnau, W. Lüthy, H. P. Weber, K. Krämer, H. U. Güdel, R. A. McFarlane, J. Appl. Phys. B 1996, 62, 1339.
[56] H. Faget, J. Grannec, A. Tressaud, V. Rodriguez, T. Roisnel, I. N. Flerov, M. V. Gorev, Eur. J. Solid State Inorg. Chem. 1996, 33, 893.
[57] G. Reuter, J. Sebastian, M. Roffe, H. J. Seifert, Thermochim. Acta 1997, 296, 47.
[58] G. Meyer, Z. Anorg. Allg. Chem. 1984, 517, 191.
[59] M. S. Wickleder, G. Meyer, Z. Anorg. Allg. Chem. 1995, 621, 457.
[60] G. Meyer, P. Ax, T. Schleid, M. Irmler, Z. Anorg. Allg. Chem. 1987, 554, 25.
[61] T. Staffel, G. Meyer, Z. Anorg. Allg. Chem. 1988, 557, 40.
[62] A. Bohnsack, G. Meyer, Z. Anorg. Allg. Chem. 1997, 623, 837.
[63] A. Bohnsack, F. Stenzel, A. Zajonc, G. Balzer, M. S. Wickleder, G. Meyer, Z. Anorg. Allg. Chem. 1997, 623, 1067.
[64] A. Bohnsack, G. Balzer, M. S. Wickleder, H. U. Güdel, G. Meyer, Z. Anorg. Allg. Chem. 1997, 623, 1352.
[65] A. Bohnsack, G. Meyer, Z. Anorg. Allg. Chem. 1997, 623, 837.
[66] K. Lerch, W. Laqua, Z. Anorg. Allg. Chem. 1990, 591, 47.
[67] F. Stenzel, G. Meyer, Z. Anorg. Allg. Chem. 1993, 619, 652.
[68] R. Blachnik, J. E. Alberts, Z. Anorg. Allg. Chem. 1982, 490, 235.
[69] R. Blachnik, G. Alberts, E. Enninga, Z. Anorg. Allg. Chem. 1985, 522, 207.
[70] M. S. Wickleder, P. Egger, T. Riedener, N. Furer, H. U. Güdel, J. Hulliger, Chem. Mater. 1996, 8, 2828.
[71] S. Masselmann, G. Meyer, Z. Anorg. Allg. Chem. 1998, 624, 551.
[72] H. Bärnighausen, T. Lange, see: Th. Lange, PhD thesis, Universität Karlsruhe, 1992.
[73] B. G. Hyde, A. N. Bagshaw, S. Andersson, M. O´Keeffee, Ann. Rev. Mater. Sci. 1974, 4, 43.
[74] E. Makovicky, B. G. Hyde, Struct. Bonding 1981, 46, 101.
[75] H. Bärnighausen, A. Lumpp, see: A. Lumpp, PhD thesis, Universität Karlsruhe, 1988.
[76] S. Masselmann, G. Meyer, Z. Anorg. Allg. Chem. 1998, 624, 357.

Mathias Wickleder

2.2 Lanthanide compounds with tetrahedral oxoanions

Many important natural sources of the lanthanides are oxoanionic minerals: *Gadolinite* is a silicate, *xenotime* and *monazite* are phosphates, *bastnaesite* is a carbonate. In the early days of rare-earth chemistry, double nitrates and double sulfates were used to separate the elements by fractional crystallization [1]. Even if more sophisticated separation procedures such as chromatographic methods and solvent extraction are used currently for lanthanide separation [2], oxoanionic compounds are still of interest as a fundamental class of compounds. Our knowledge of these compounds roughly correlates with their thermal stability, at least with respect to the binary solvent-free species. For example, there is a plethora of well-known silicates and phosphates, while nitrates, carbonates, and perchlorates are still not fully investigated. This chapter will focus on the *structures* of oxoanionic rare-earth compounds. In some cases, the preparative aspects and selected properties will also be mentioned. To stay within the scope of this chapter, we will restrict ourselves to tetrahedral oxoanions. Furthermore, we will focus on "binary" compounds and compounds containing additional metal ions are more or less completely neglected. However, mixed-anionic compounds will be occasionally discussed. Other oxoanions are treated in several reviews, for example, rare-earth borates [3, 4], nitrates [5, 6], selenates and selenites [7]. A more comprehensive review of rare-earth compounds with complex anions was presented some years ago [8]. Also, *Gmelin´s* handbook provides a good overview of the early literature [9].

Figure 2.2.1 gives an outline for this chapter. It also summarizes some generalities concerning tetrahedral oxoanions. First, the thermal lability decreases in the row $[ClO_4]^-$ - $[SO_4]^{2-}$ - $[PO_4]^{3-}$ - $[SiO_4]^{4-}$. Second, the tendency to condensate to large oligo- or even polyanions increases in the same row. Thus, extended anionic structures are often found in silicates and phosphates. Furthermore, the figure depicts that the sulfate derivatives $[HSO_4]^-$, $[NH_2SO_3]^-$, and $[CH_3SO_3]^-$, which will be also discussed in this chapter, are isoelectronic with the perchlorate anion.

2.2.1 Silicates

2.2.1.1 Orthosilicates

Silicates belong to the most important rare-earth minerals, being a natural reservoir of the rare-earth elements. Examples are the minerals *thortveitite* ($Sc_2Si_2O_7$) and *gadolinite* ($Be_2FeY_2O_2[SiO_4]_2$), with the former being a nearly pure mineral in scandium. Due to the above mentioned tendency of the $[SiO_4]$ tetrahedra to condense to

https://doi.org/10.1515/9783110654929-010

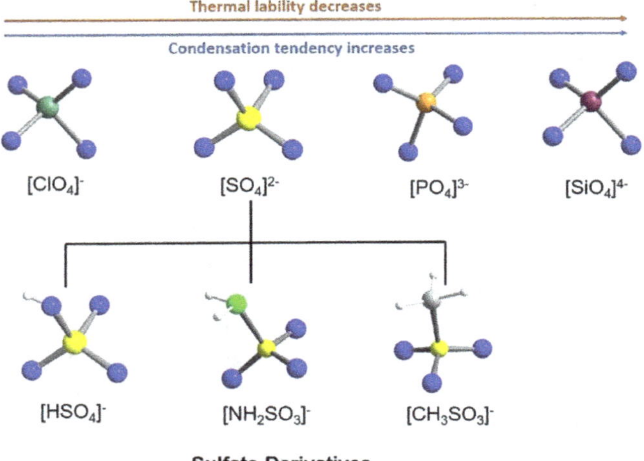

Figure 2.2.1: Tetrahedral oxoanions that are treated in this chapter. Note that with increasing charge of the anions, the tendency to form larger polyanions increases significantly.

disilicates, trisilicates or even more extended polyanionic arrays, the number of different compositions is very large. Extensive systematic studies were presented early by *Felsche* in an excellent review [10].

Interestingly, even currently, there is no simple orthosilicate with the composition "$RE_4[SiO_4]_3$," known for the trivalent rare-earth elements. On the other hand, with divalent europium, two modifications of Eu_2SiO_4 are known which adopt the monoclinic β-Ca_2SiO_4 and the orthorhombic α-K_2SO_4 type of structure, respectively [11, 12]. With the trivalent rare-earth ions oxide silicates, $RE_2O[SiO_4]$, frequently occur. They adopt two different structures, depending on the RE^{3+} radius [10, 13, 14]. Both types of structures are similar, in the sense that they contain oxide centered [ORE_4] tetrahedra as the characteristic structural feature (Figure 2.2.2).

The substitution of the oxide ions in $RE_2O[SiO_4]$ by the heavier chalcogenides Se and Te is possible and leads to the selenide and telluride orthosilicates $M_2X[SiO_4]$ (X = Se [15–17],Te [18]), which also adopt two different structure types. Halide orthosilicates of the rare-earth elements are also known, essentially with Cl^- as the halide component [19–25]. The only structurally investigated bromide orthosilicate known so far is $Gd_3Br[SiO_4]_2$ [26]. With $La_3Cl_5[SiO_4]$ and $Ce_3Cl_5[SiO_4]$, two examples of chlorine-rich chloride orthosilicates are known [27, 28]. They are isotypic and show a certain similarity to the trichlorides $LaCl_3$ and $CeCl_3$ (UCl_3 type of structure, see chapter 2.1) when four Cl^- ions of the tripled formula are substituted for $[SiO_4]^{4-}$. A chloride orthosilicate is also known for divalent europium with the composition $Eu_5Cl_6[SiO_4]$. It contains three crystallographically

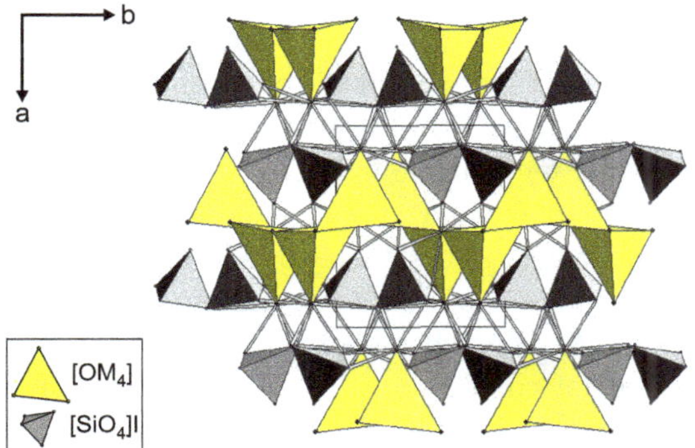

Figure 2.2.2: The oxide-silicates $RE_2O(SiO_4)$ are composed of oxide-centered $[ORE_4]$ tetrahedra and tetrahedral $[SiO_4]^{4-}$ ions (modified after [8]).

different Eu^{2+} ions which are in eight- and ninefold coordination of oxygen atoms and chloride ligands [29]. The two modifications of $La_2I_2[SiO_4]$ are the only iodide orthosilicates that have been described so far [30]. Interestingly, the number of fluoride orthosilicates is very limited, examples are the mixed disilicate-orthosilicate $Er_4F_2[Si_2O_7][SiO_4]$, the mixed-valent apatite $Eu_5[SiO_4]_3F$ and the oxide fluoride $La_7OF_7[SiO_4]_3$ [31].

A huge number of rare-earth orthosilicates can be derived from the structure of the mineral *fluoroapatite*, $Ca_5[PO_4]_3F$ [8]. As shown in the scheme in Figure 2.2.3, the structure is retained when the $[SiO_4]^{4-}$ ion replaces $[PO_4]^{3-}$ and F^- is substituted for O^{2-}. However, to maintain the charge balance, the Ca^{2+} ions cannot be replaced completely

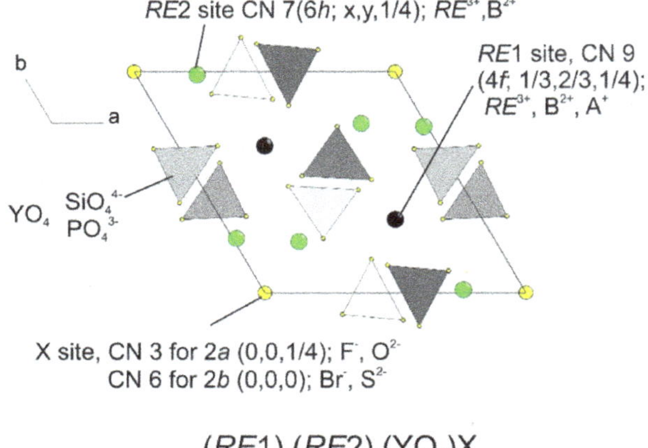

Figure 2.2.3: Possible variations for the hexagonal apatite type of structure, $Ca_5[PO_4]_3(OH)$. The crystallographic labeling of different sites refers to space group $P6_3/m$ (modified after [8]).

by RE^{3+} ions. In the so-called rare-earth *oxyapatites*, the 4 *f*-site is only partly occupied with RE^{3+} ions, leading to cation defect compounds $RE_{4.667}[SiO_4]_3O$. Another possibility to achieve electroneutrality is when there is a mixed occupancy of cationic sites with rare-earth ions and lower valent ions like Ca^{2+}, K^+ or Na^+. This may occur in both of the crystallographically different sites, as found for example in $CaLa_4[SiO_4]_3O$ or in one specific site as in $NaSm_9[SiO_4]_6S_2$. The latter compound shows another frequently observed variety of the apatite structure: The smaller F^- or O^{2-} may be substituted for the larger Br^- or S^{2-} ions because the structure offers not only a site with trigonal planar coordination (Wyckoff notation *2a*; 0, 0, 1/4), which is preferred by F^- and O^{2-}, but also by a site (*2b*; 0, 0, 0) with octahedral coordination, necessary for the large Br^- and S^{2-} ions. With respect to the naturally abundant minerals, these compounds are called *bromoapatites*. Finally, it should be emphasized that even mixed valent rare-earth orthosilicates with an apatite structure have been prepared [32]. In the ruby red $Eu_5[SiO_4]_3F$, the larger Eu^{2+} ions occupy the *4f* site leading to a ninefold coordination while the Eu^{3+} ions are in sevenfold coordination on the *6h* position. Thus, one can formulate the compound as $(Eu^{III})_3(Eu^{II})_2[SiO_4]_3F$. In the black ytterbium apatite $Yb_5[SiO_4]_3S$ (*bromoapatite* structure), there must be a certain mixture of Yb^{2+} and Yb^{3+} that may be expressed by the formula $(Yb^{III})_3(Yb^{III/II})_2[SiO_4]_3S$.

2.2.1.2 Disilicates

For the rare-earth disilicates of the type $RE_2[Si_2O_7]$, a large number of different structures are known [8]. They are labeled with the capital letters A to I. The large number of structure types arise from the variable packing patterns of the $[Si_2O_7]^{6-}$ ions and can be attributed to the different coordination requirements of the RE^{3+} ions within the lanthanide series. Furthermore, there is a great flexibility in the shape of the disilicate anion, as can be seen from the broad range of the bridging angle Si–O–Si (128° to 180°). Special emphasis should be put on the B-type of the disilicates and the I-type, which are known only for lanthanum [33]. Both types are *no* real disilicates but mixed *catena*-trisilicate-orthosilicates of the type $RE_4[Si_3O_{10}][SiO_4]$ (*RE* = Eu–Er), in the case of the B-type and a mixed *catena*-tetrasilicate-orthosilicate according to $La_6[Si_4O_{13}][SiO_4]_2$ for I-$La_2Si_2O_7$ (Figure 2.2.4).

The first structurally characterized derivative of a disilicate, $Sm_4S_3[Si_2O_7]$, was obtained as a side-product during the chemical transport of SmS_2 with I_2 in silica ampoules [34]. Up to now, a number of isotypic compounds were obtained either with larger or smaller M^{3+} ions and with S^{2-} or Se^{2-} as chalcogenide anion [35, 36]. Derivatization of rare-earth disilicates is also possible by introducing chloride ions in the crystal structure, for example in the compounds $La_3Cl_3[Si_2O_7]$ and $Pr_3Cl_3[Si_2O_7]$ [37]. Pure disilicates containing additional F^- ions are not known, but with $Er_4F_2[Si_2O_7][SiO_4]$, an interesting crystal structure was described containing disilicate and orthosilicate ions [38].

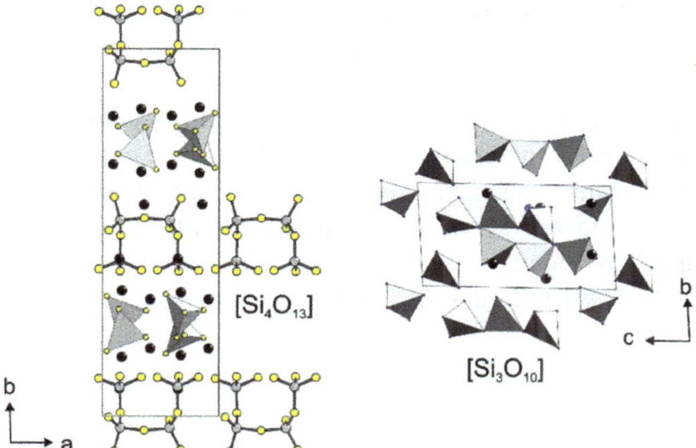

Figure 2.2.4: The "disilicates" $La_2[Si_2O_7]$ (left) and $Er_2[Si_2O_7]$ (right) do not contain disilicate anions but tetra- and trisilicate groups according to $La_6[Si_4O_{13}][SiO_4]_2$ and $Er_4[Si_3O_{10}][SiO_4]$ (modified after [8]).

2.2.1.3 Higher silicates

Lanthanide compounds with the *catena*-trisilicate anion $[Si_3O_{10}]^{8-}$ are the fluoride silicates of the form $RE_3F[Si_3O_{10}]$ (RE = Dy, Ho, Er, Y) [39–41]. They adopt the *thalenite* type of structure, named after a mineral that has nearly the composition of $Y_3F[Si_3O_{10}]$, but which usually contains various amounts of other rare-earth ions and OH^- instead of F^-. The *catena*-tetrasilicate anion $[Si_4O_{13}]^{10-}$ is known from two other lanthanide compounds, $Ba_2Nd_2[Si_4O_{13}]$ and $Na_4Sc_2[Si_4O_{13}]$. Tri- and tetrasilicates are also known as cyclosilicates and for the rare-earth elements, a limited number of crystal structures were determined. $La_3F_3[Si_3O_9]$ is a fluoride cyclotrisilicate that shows honeycomb shaped layers of La^{3+} and F^- which are alternatingly stacked with the silicate anions along the [001] direction of the hexagonal unit cell [42]. Structural data of cyclotetrasilicates have been provided for $RE_6Cl_{10}[Si_4O_{12}]$ (RE = Sm, Y) [43, 44] and the divalent europium compound $Eu_8Cl_8[Si_4O_{12}]$ [45]. Finally, further connection of the silicate tetrahedra lead to the *tecto*-silicates, of which $K_2Ce[Si_6O_{15}]$ is an example. The three-dimensional silicate network provides the space for the incorporation of K^+ and Ce^{4+} ions, the latter being an octahedral coordination of oxygen atoms.

2.2.2 Phosphates

Two of the natural abundant rare-earth phosphates, *monazite* and *xenotime*, are important sources for the production of rare-earth metals. Both are orthophosphates, $RE[PO_4]$, with *monazite* containing the larger lanthanide ions La–Nd (and Th) and *xenotime* containing mainly yttrium and various amounts of the smaller rare-earth ions. These minerals and their artificial analogues have been intensively studied. Similar to the silicates, phosphates also show the tendency to condense to larger aggregates, forming either ring-shaped (cyclophosphates) or chain-like (*catena*-phosphates) polyanions. Thus, the number of different rare-earth phosphates is large. For example, the phase diagram La_2O_3/P_2O_5 shows six compositions with the molar ratios 3:1, 7:3, 1:1, 1:2, 1:3 and 1:5 [46]. Structural investigations of the phosphates started in the 1940s and more than 100 structures have been documented till date. A number of them have been reviewed by Palkina in 1982 [47] and Niinistö in 1986/7 [5, 6].

2.2.2.1 Orthophosphates

The orthophosphates of the larger lanthanides *RE* = La–Gd crystallize with the monoclinic structure of *monazite* and contain the RE^{3+} ions in ninefold coordination of oxygen atoms, which belong to seven $[PO_4]^{3-}$ ions [8]. With the lighter rare-earth elements *RE* = Tb–Lu, Y, Sc, the tetragonal structure of $ZrSiO_4$ is adopted. The RE^{3+} ions are eightfold, coordinated by oxygen atoms from two chelating and four monodentate phosphate ions. A few Eu(II) phosphates are known, for example, $Eu_3[PO_4]_2$ and $Eu_5[(PO_4)_3](OH)$ [48].

The chloride phosphate $Ce_3Cl_6[PO_4]$ obtained by reacting $CeCl_3$, Ce, CeO_2 and P_2O_5 in silica tubes is isotypic with the respective vanadates and shows the Ce^{3+} ions in coordination of seven Cl^- ions and three oxygen ligands [49]. The structure can be seen as built up from $[Ce_3Cl_6]$ units in form of three vertex-connected $[CeCl_3]$ triangles. These units are arranged in a way that pseudohexagonal channels occur along [001], which incorporate the $[PO_4]^{3-}$ ions.

2.2.2.2 Polyphosphates

Diphosphates of the lanthanides are known to exist as the alkali metal containing species $AY[P_2O_7]$ (A = Na–Cs) and $CsYb[P_2O_7]$. They contain the $P_2O_7^{4-}$ ion in a staggered conformation with the angles within the P–O–P bridge typically measuring 125°. The cyclotriphosphate anion, $P_3O_9^{3-}$, is found in the crystal structure of $Ce[P_3O_9] \cdot 3H_2O$ [50]. There are a number of other compounds with the composition "$RE(P_3O_9)$," but *none* of them is a cyclotriphosphate. Most of the compounds are *catena*-polyphosphates with infinite chains of vertex sharing tetrahedra. They should be formulated as $RE(PO_3)_3$. Some are cyclotetraphosphates according to $RE_4[P_4O_{12}]_3$ containing the $P_4O_{12}^{4-}$ anion [51, 52]. The cyclohexaphosphates have the compositions $RE_2[P_6O_{18}] \cdot xH_2O$ and the content of the crystal water *x* may be 6, 10, 12 or 16. The $[P_6O_{18}]^{6-}$ anion can be either "boat"-shaped, like in $Nd_2[P_6O_{18}] \cdot 6H_2O$, or "chair"-shaped, as in $Nd_2[P_6O_{18}] \cdot 12H_2O$ [53, 54].

The highest condensation of phosphate groups is found in the so-called ultraphosphates. They can be prepared from the lanthanide oxides, either with phosphoric acid or with $(NH_4)_2HPO_4$ at 600 °C, if a large amount of the phosphorous component is used. As in the *catena*-phosphates part of the $[PO_4]$ tetrahedra share two common vertices according to $[PO_2O_{2/2}]$ but some of the tetrahedra are linked via three corners, as represented by $[PO_1O_{3/2}]$. These building units are in the ratio of 3:2, yielding an anionic network $[P_5O_{14}]^{3-}$, as found in the *RE*(P_5O_{14}) compounds (Figure 2.2.5) which show different structure types. The ultraphosphates were of interest as laser materials. Unfortunately, large single crystals are hard to grow due to the decomposition of the compounds at higher temperature.

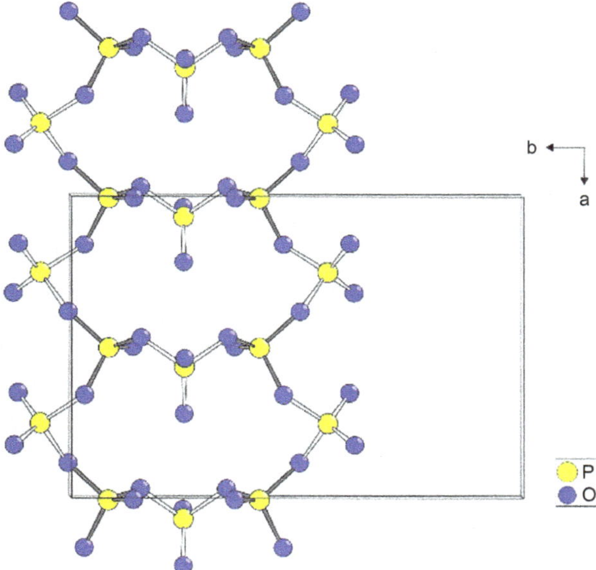

Figure 2.2.5: The structures of the so-called ultraphosphates *RE*(P_5O_{14}) originate from the connection of $[PO_4]^{3-}$ ions via two corners (show with white P–O bonds) and three corners, respectively (shown by gray P–O bonds) (modified after [8]).

2.2.3 Sulfates

Simple and complex sulfates are intensively investigated compounds and they were the subject of several reviews [8, 55]. These compounds are most often hydrates or multinary phases, containing various additional cations. Anhydrous sulfates have been described to a much lesser extent, especially for binary species. This limited information is due to the difficulties in growing single crystals for structure determinations.

Compared to silicates and phosphates, which can be prepared even by high temperature methods, sulfates decompose at elevated temperatures and hence, they cannot be obtained from their melt. On the other hand, the use of a solvent, usually water, does not lead to solvent-free compounds.

2.2.3.1 Sulfate hydrates

The most common binary hydrates are the octahydrates $RE_2[SO_4]_3 \cdot 8H_2O$, which have been characterized throughout the whole lanthanide series (La-Lu), including yttrium [8]. It turned out that the compounds are isotypic with each other and contain the RE^{3+} in eightfold coordination of oxygen atoms. This is one of the very rare examples where the lanthanide contraction does *not* lead to a structural change for a given compound. Sulfates, with other water contents, have been found mainly for the larger RE^{3+} ions La–Nd. The hydrates, $RE_2[SO_4]_3 \cdot 9H_2O$, are known for RE = La, Ce and contain the RE^{3+} ions in 12-fold and ninefold coordination of oxygen atoms. The crystal structures of lower hydrates are known for $RE_2[SO_4]_3 \cdot 5H_2O$ (RE = Ce, Nd) and $RE_2[SO_4]_3 \cdot 4H_2O$ (RE = La, Ce, Nd, Tb, Er). In the pentahydrates, the RE^{3+} ions are surrounded by six sulfate groups and two H_2O molecules. With $Sc_2[SO_4]_3 \cdot 5H_2O$, another pentahydrate is known. It is triclinic and shows the Sc^{3+} ions in an octahedral coordination of oxygen atoms. In the crystal structure of the tetrahydrates, $RE_2[SO_4]_3 \cdot 4H_2O$ with RE = La, Ce, Nd the CN 8 for the RE^{3+} ions is achieved by the coordination of two H_2O molecules and five SO_4^{2-} ions.

Lower hydrates of the lanthanide sulfates could also be prepared for the smaller lanthanides RE = Tb, Er and Lu. The isotypic tetrahydrates, $Tb_2[SO_4]_3 \cdot 4H_2O$ and $Er_2[SO_4]_3 \cdot 4H_2O$, were obtained from acidic solutions at elevated temperatures in single crystalline form. In the triclinic crystal structure, two crystallographically different RE^{3+} ions are present, which are in eight- and sevenfold coordination, respectively. The structure of the trihydrate $Lu_2[SO_4]_3 \cdot 3H_2O$ is very complex and contains three crystallographic Lu^{3+} ions and five different sulfate groups [56].

The thermal behavior of the sulfate hydrates has been studied several times by means of DTA/TG measurements, mainly for the octahydrates. According to these investigations, the second step after dehydration is the formation of the oxide-sulfate $RE_2O_2[SO_4]$, which finally decomposes to the respective oxide RE_2O_3.

2.2.3.2 Anhydrous sulfates

Compared to the respective hydrates, the knowledge of binary anhydrous sulfates is rather limited. For large lanthanides, the structure of $Nd_2[SO_4]_3$ is known [57]. The structure of the smaller ions is known for RE = Ho–Lu, Y and Sc [58]. The latter have been obtained as crystalline materials from alkali halide melts, noticeably LiF and

NaCl, respectively. In the crystal structure of $Nd_2[SO_4]_3$, Nd^{3+} is surrounded by seven sulfate groups. Two of the latter are chelating ligands, yielding a coordination number of nine for Nd^{3+}. According to the formulation $[Nd(SO_4)_{5/5}(SO_4)_{2/4}]_2$, the $[SO_4]^{2-}$ ions are coordinated by five or four neodymium ions. For the crystal structures of the sulfates $RE_2[SO_4]_3$ with RE = Ho–Lu, Y a coordination number of six, with octahedral coordination of the RE^{3+} ions is found. The sulfate groups are attached to four RE^{3+} ions. Thus, the $[REO_6]$ octahedra and the $[SO_4]^{2-}$ tetrahedra are linked by all the vertices. If the complex anions are regarded as spheres, the analogy of the crystal structure to the binary sulfide Rh_2S_3 is obvious, as can be seen from Figure 2.2.6. The same figure compares the crystal structures of $Sc_2[SO_4]_3$ and Lu_2S_3 (α-Al_2O_3 type of structure). The connectivity of the $[ScO_6]$ octahedra and the sulfate tetrahedra in $Sc_2[SO_4]_3$ is the same as just described. Another crystal structure of an anhydrous rare-earth sulfate is the one of tetravalent cerium, Ce $[SO_4]_2$. This is the sole sulfate of a tetravalent rare-earth element. It contains the Ce^{4+} ions in an eightfold coordination of eight monodentate $SO_4{}^{2-}$ groups, each of them being attached to four Ce^{4+} ions [59].

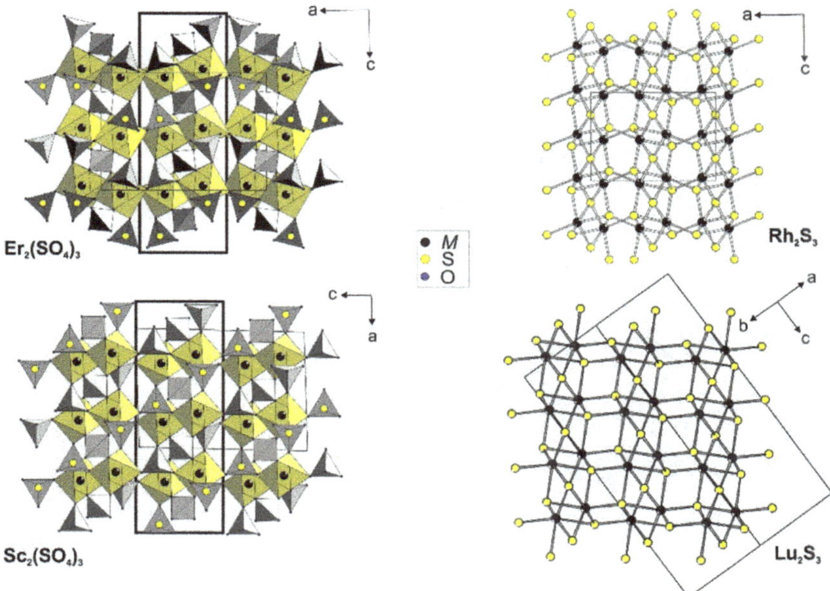

Figure 2.2.6: Crystal structures of the sulfates of the smaller rare-earth elements Ho–Lu (shown for RE = Er) and of $Sc_2(SO_4)_3$. The structures are oxoanionic representatives of the sulfides Rh_2S_3 and Lu_2S_3 (α-Al_2O_3 type) (modified after [8]).

2.2.3.3 Acidic sulfates

The simplest acidic sulfates are the hydrogensulfates $RE[HSO_4]_3$. They are known to adopt two different structures which are orthorhombic and hexagonal, respectively. The orthorhombic structure occurs for the smaller lanthanides *RE* = Gd–Lu and Y [8]. It is a typical layer structure with the RE^{3+} ions being in an eightfold coordination of oxygen atoms. The hexagonal hydrogensulfates of the larger lanthanides *RE* = La–Sm show the RE^{3+} ions as ninefold coordinated by oxygen atoms in the form of tricapped trigonal prism. The oxygen atoms belong to nine monodentate $[HSO_4]^-$ ions which are attached to three RE^{3+} ions. The OH groups of the ions remain uncoordinated and point towards channels which are formed in the hexagonal structure along [001]. If the hydrogensulfate tetrahedra are reduced to their centers, the structure corresponds to the one of UCl_3, which is well know for the trichlorides of the larger lanthanides. Interestingly, the structure of the hexagonal hydrogensulfates has also been found for anhydrous lanthanide perchlorates, amidosulfates and methanesulfonates (Figure 2.2.7).

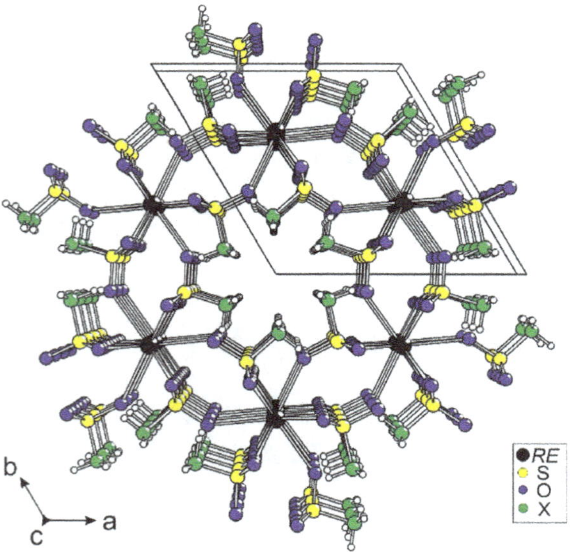

Figure 2.2.7: Crystal structure of the hexagonal hydrogensulfates $RE[HSO_4]_3$ (X = OH) for the larger rare-earth elements. Note, that the same structure occurs also for perchlorates $RE[ClO_4]_3$ (X = O), amidosulfates $RE[NH_2SO_3]_3$ ($X = NH_2$), and methanesulfonates $RE[CH_3SO_3]_3$ ($X = CH_3$) (modified after [8]).

The thermal decomposition of the rare-earth hydrogensulfate occurs in two steps, resulting in anhydrous sulfates. As an intermediate, the respective sulfate-hydrogensulfates $RE[HSO_4][SO_4]$ are found. This is in strong contrast to the finding of hydrogensulfates of alkali or alkaline earth metals, which usually decompose with the formation of disulfates ("pyrosulfates"). For $Er[HSO_4][SO_4]$, this intermediate could be gained [60]. The crystal structure contains Er^{3+} ions surrounded by four monodentate $[SO_4]^{2-}$ and three monodentate $[HSO_4]^-$ ions.

The hydrogensulfates of the rare-earth elements are very moisture sensitive. The reaction products are often compounds containing oxonium ions. For example, the sulfates $(H_5O_2)RE[SO_4]_2$ (RE = Gd, Er, Ho, Y) show H_3O^+ ions and the water molecules joined to $(H_5O_2)^+$ ions, with very short hydrogen bonds [61]. The hydrogensulfate-hydrates $Y[HSO_4]_3 \cdot H_2O$ and $Gd[HSO_4]_3 \cdot H_2O$ are rare examples where the hydrogen atom is not part of an oxonium ion but located on the sulfate group. Interestingly, the compounds are isotypic with the perchlorate $Yb[ClO_4]_3 \cdot H_2O$. There are even more complicated acidic sulfates known, for example $(H_3O)La[SO_4]_2 \cdot 3H_2O$ with strongly hydrogen bonded H_3O^+ ions and ninefold coordinated La^{3+} ions, $(H_3O)_2Nd[HSO_4]_3[SO_4]$ or $(H_5O_2)(H_3O)_2Nd[SO_4]_3$ [62]. The big variations in the composition and the crystal structures of acidic sulfates suggest that there might be further compounds to be discovered. More clarity is achieved if the compounds so far known are written as combinations of $RE_2[SO_4]_3$, H_2SO_4 and H_2O, respectively, as has been done in Figure 2.2.8.

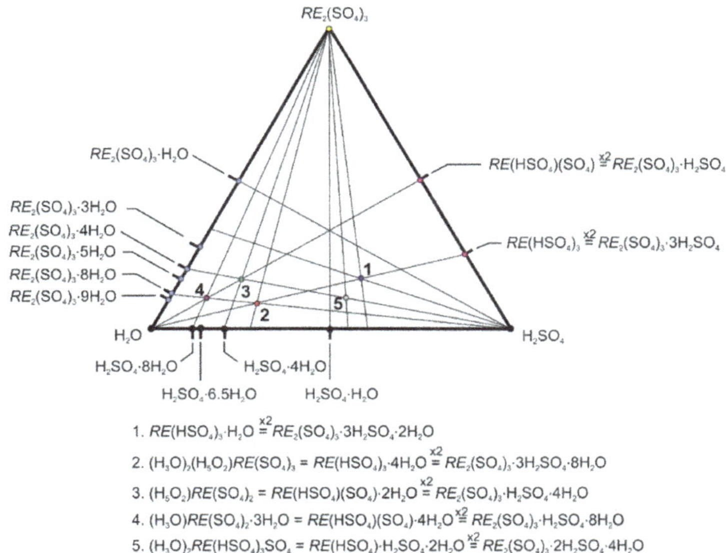

Figure 2.2.8: Acidic rare-earth sulfates represented in the triangle $RE_2[SO_4]_3/H_2SO_4/H_2O$ (modified after [8]).

In this way, the acidic compounds can be arranged in a triangle with these three compounds as the vertices.

2.2.3.4 Halide sulfates

The chloride sulfates, *RE*Cl[SO_4], occur in two different structure types. With the larger lanthanides *RE* = La–Pr, a structure is formed which contains RE^{3+} ions coordinated by three Cl^- and five $[SO_4]^{2-}$ ions [63]. The anions are arranged as layers which extend in the (100) plane and are surrounded by three and five RE^{3+} ions, respectively, so that one may formulate the linkage according to $[RECl_{3/3}(SO_4)_{5/5}]$. For the smaller lanthanides, Sm–Tb, another type of structure has been found which shows the lanthanide ions in ninefold coordination of three Cl^- ions and six oxygen atoms [64]. The main difference between the two structures is the arrangement of the anions.

Fluoride sulfates of the rare-earth elements can be obtained from the reaction of the anhydrous sulfates with LiF in sealed gold ampoules. For *RE* = Gd, Tb, Lu and Y, the crystal structures of *RE*F[SO_4] type compounds have been solved based on single crystal data. They are isotypic with each other and can be seen as a rod packing from cationic $[RE(F)_{2/2}]^{2+}$ and anionic $[SO_4]^{2-}$ columns (Figure 2.2.9, left). The coordination number of the RE^{3+} ions is eight. If analogous reactions are performed with the sulfates of the larger lanthanides, the lithium containing species $LiRE_2F_3(SO_4)_2$ (*RE* = La–Eu) form [65]. A characteristic feature of the monoclinic crystal structure is the dimeric building unit $[RE_2F_5(SO_4)_{10}]$ which arises from the connection of two $[LaF_4(SO_4)_5]$ polyhedra via three common F^- ions. The dimeric units are linked with

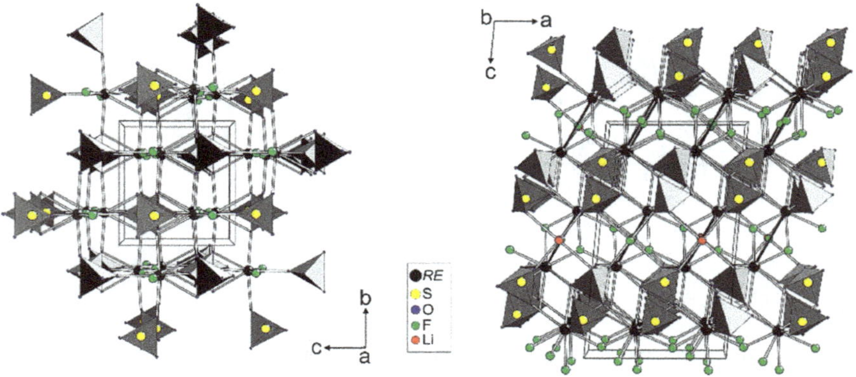

Figure 2.2.9: Crystal structures of the fluoride sulfates *RE*F[SO_4] (left, *RE* = Gd–Yb, Y) and $LiRE_2F_3[SO_4]_2$ (right, *RE* = La–Eu).

each other to chains by fluoride and sulfate anions (Figure 2.2.9, right). It is possible to replace Li^+ by Na^+ and the respective fluoride sulfates as has been shown for $NaPr_2F_3(SO_4)_2$ and $NaEu_2F_3(SO_4)_2$.

2.2.4 Sulfate derivatives

2.2.4.1 Amidosulfates

The investigation of rare-earth amidosulfates (sulfamates) started quite late although sulfamic acid is a long known derivative of sulfuric acid [8]. Some years ago, the crystal structures of $La[NH_2SO_3]_3 \cdot 2.5H_2O$ and $RE[NH_2SO_3]_3 \cdot 2H_2O$ (RE = Pr–Lu, Y) and their thermal behavior were investigated [66, 67]. Accordingly, in $La[NH_2SO_3]_3 \cdot 2.5H_2O$, La^{3+} is coordinated by nine oxygen atoms in the form of a monocapped square antiprism. The connection can be written as $La(H_2O)_2[NH_2SO_3]_{3/3}[NH_2SO_3]_{4/2}$ and the linkage occurs exclusively in the (001) plane, yielding infinite sheets that are held together by hydrogen bonds. These bonds also involve the water molecules located between the sheets. The dihydrates $RE[NH_2SO_3]_3 \cdot 2H_2O$ are isotypic throughout the series Ce-Lu, including Y. The coordination polyhedron of the RE^{3+} ions is a distorted square antiprism of oxygen atoms. The connectivity of the anions and the rare-earth ions is given by $[RE(H_2O)_{2/1}(NH_2SO_3)_{6/2}]$. Just as in $La[NH_2SO_3]_3 \cdot 2.5H_2O$, the linkage occurs only in two dimensions, leading to sheets which are again connected only via hydrogen bonding, but in contrast to the lanthanum compound, there are no additional water molecules between the sheets. The thermal decomposition of amidosulfate-hydrates leads via monohydrates as intermediates to the anhydrous compounds. The structure of the monohydrates is known for $Sm[NH_2SO_3]_3 \cdot H_2O$ and $Tb[NH_2SO_3]_3 \cdot H_2O$. They show coordination number eight for the RE^{3+} ions and a linkage according to $[RE(H_2O)(NH_2SO_3)_{3/3}(NH_2SO_3)_{4/2}]$. The anhydrous amidosulfates, obtained after complete dehydration, were shown to be isotypic for RE = La, Ce–Nd, Sm and adopt the structure which was previously described for the respective hydrogensulfates (cf. Figure 2.2.7). The anhydrous amidosulfates of the rare-earth elements decompose in two steps, yielding the respective sulfates.

2.2.4.2 Methanesulfonates

The methanesulfonates of the rare-earth elements are known as hydrates and adducts with organic molecules like TMSO (tetramethylenesulfoxide) and others [68–70]. In a few cases, crystal structures of the anhydrous compounds are known. Hydrates were obtained from the solutions of the rare-earth sesquioxides in aqueous solutions of methane sulfonic acid. They crystallize with two or three molecules of water. For the dihydrates, two different crystal structures occur. The first modification which has been found for La and Nd contains the RE^{3+} ions in a ninefold coordination of oxygen atoms with slightly distorted tricapped trigonal prisms as coordination polyhedra [71]. The oxygen atoms belong to six $[CH_3SO_3]^-$ ions and three water molecules. Four of the methane sulfonate groups and two of the water molecules connect the polyhedra to infinite chains along [100]. The remaining two $[CH_3SO_3]^-$ ions link two of these chains to double chains. The connectivity may be written as $[RE(H_2O)(H_2O)_{2/2}(CH_3SO_3)_{6/2}]$. The second modification for the dihydrates which has been found with RE = Ce–Nd and Sm–Tb also shows a layer structure. The crystal structure contains the RE^{3+} ions in an eightfold coordination of oxygen. The oxygen atoms belong to six $[CH_3SO_3]^-$ groups and two H_2O molecules. Each methanesulfonate ion is connected to a further RE^{3+} ion according to $[RE(H_2O)_{2/1}(CH_3SO_3)_{6/2}]$. The linkage of the $[REO_8]$ polyhedra via the $[CH_3SO_3]^-$ groups occurs only in the (100) plane and leads to puckered layers that are held together by hydrogen bonds.

In the crystal structure of the trihydrates $RE[CH_3SO_3]_3 \cdot 3H_2O$ (RE = Sm, Gd, Dy), three H_2O molecules and five $[CH_3SO_3]^-$ ions are attached to a RE^{3+} ion. According to $[RE(H_2O)_{3/1}(CH_3SO_4)_{4/2}(SO_4)_{1/1}]$, four of the latter act as bidentate bridging ligands, while one of them is monodentate. The chains are held together by hydrogen bonds. The thermal decomposition of the hydrates leads in the first step to the anhydrous compounds. In $La[CH_3SO_3]_3$, the structure can be derived from its powder pattern. This can be indexed based on the single crystal data of $La[NH_2SO_3]_3$ [71]. Thus, the methanesulfonate crystallizes with the hexagonal structure that can be seen as a derivative of the UCl_3 type of structure and which has been discussed for the hydrogensulfates of the larger lanthanides (cf. Figure 2.2.7). This structure obviously occurs only for lanthanum. With the slightly smaller Nd^{3+} ion, another structure has been determined which consists of layers built up from $[NdO_8]$-polyhedra and $[CH_3SO_3]^-$ ions. Nd^{3+} is surrounded by eight methanesulfonate groups. Six of the latter are attached to three Nd^{3+} ions, while the remaining two are coordinated to only two Nd^{3+} ions. The noncoordinating oxygen atoms of these $[CH_3SO_3]^-$ ions act as acceptors in hydrogen bonds with the CH_3 groups of adjacent layers as donors. These hydrogen bonds are known to be much weaker than those with oxygen atoms of water molecules as donors, for instance. This modification of $Nd[CH_3SO_3]_3$ undergoes a phase transition at higher temperature, leading to a crystal structure that is known in Er $[CH_3SO_3]_3$ and $Yb[CH_3SO_3]_3$ [72]. Thus, the structure changes in a way that the layers found in $Nd[CH_3SO_3]_3$-I are cut into double chains running along [010] which are again connected via hydrogen bonds. If the two structures are viewed in the [100] direction (I) and [010] direction (II), respectively, they turn out to be quite similar (Figure 2.2.10). Upon heating, the anhydrous methanesulfonates decompose in a complicated way to give the oxide disulfides $RE_2O_2S_2$ and finally, RE_2O_2S.

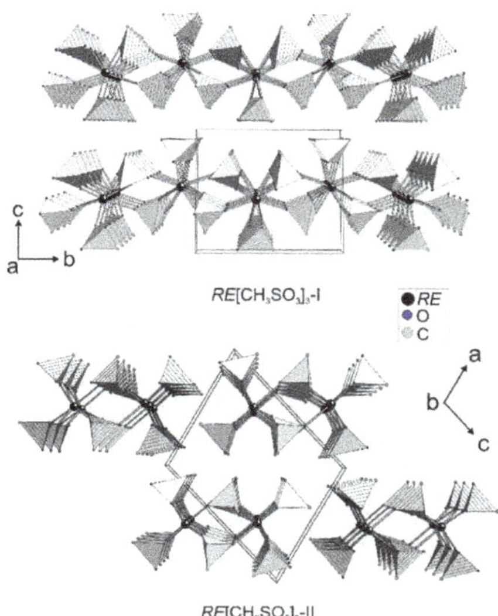

Figure 2.2.10: Crystal structures of the anhydrous rare-earth methanesulfonates. Modification I has a layers structure while modification II show a structure composed of double chains (modified after [8]).

2.2.5 Perchlorates

2.2.5.1 Perchlorate hydrates

The perchlorate ion, $[ClO_4]^-$, is known to be a very weak ligand. Thus, it shows only a little tendency to enter the inner coordination sphere of a cation. This is especially true for the lanthanide ions. From aqueous solutions containing RE^{3+} and $[ClO_4]^-$ ions, the hydrates $RE[ClO_4]_3 \cdot xH_2O$ are obtained which show only water molecules coordinated to the RE^{3+} ions. The number of water molecules in the crystal structures are reported to be 9, 8 or 7. The respective hexahydrates are the best investigated ones and their structure is in line with the formula $[RE(H_2O)_6][ClO_4]_3$ [8]. Upon heating, the hexahydrates melt around 100 °C and the melt loses water molecules over a wide temperature, ranging from 160 to 300 °C, followed by a complete decomposition of the perchlorate, yielding the oxide-chlorides *RE*OCl. It is possible to obtain lower hydrates by heating the hexahydrates to a certain temperature. For example, the heating of $RE[ClO_4]_3 \cdot 6H_2O$ to 140 °C under an argon stream led to the perchlorates $RE[ClO_4]_3 \cdot 3H_2O$, $RE[ClO_4]_3 \cdot 2H_2O$ and $RE[ClO_4]_3 \cdot H_2O$ [73, 74]. The decreasing water content necessarily leads to a coordination of the perchlorate ions with the RE^{3+} ions. For example, in the crystal structure of $Yb[ClO_4]_3 \cdot 2H_2O$, the Yb^{3+} ions are in an eightfold coordination of the oxygen atoms and the latter belong to two water molecules and five ClO_4^- groups. The perchlorate ions connect the Yb^{3+} ions to chains according to $[Yb(H_2O)_2(Cl(1)O_4)_{1/1}(Cl(2)O_4)_{4/2}]$. Further removal of water molecules leads a an even stronger connection and in the crystal structure of $Yb[ClO_4]_3 \cdot H_2O$, seven monodentate $[ClO_4]^-$ groups create a three-dimensional network according to $[RE(H_2O)_{1/1}$

$(ClO_4)_{3/3}(ClO_4)_{4/2}]$ (Figure 2.2.11). The trihydrate $Lu[ClO_4]_3 \cdot 3H_2O$ shows also infinite chains running along *c* in the crystal structure. The chain is connected with other ones only via hydrogen bridges.

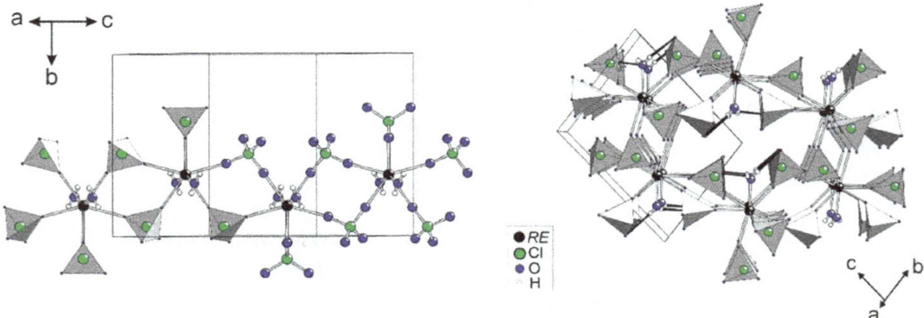

Figure 2.2.11: The perchlorates $RE[ClO_4]_3 \cdot 2H_2O$ (left) and $RE[ClO_4]_3 \cdot H_2O$ (right) for the example of RE = Yb. With decreasing water content, the connectivity increases, that is, a chain structure is found for the dihydrate, while the monohydrate shows already a three-dimensional network (modified after [8]).

2.2.5.2 Anhydrous perchlorates

The anhydrous perchlorates $RE[ClO_4]$ can be obtained by a complete dehydration of the respective hydrates. Usually, this procedure leads only to powder samples so that no single crystal structure can be determined. Nevertheless, a successful elucidation of the crystal structures of anhydrous-rare-earth perchlorates was possible using synchrotron or neutron powder data [75, 76]. Furthermore, it has been shown that the dehydration can be carried out in a way that even single crystals can be grown so that single crystal data become available [77]. Two structure types for anhydrous rare-earth perchlorates are known. Most of the lanthanides, namely La–Nd and Sm–Tm and Y, crystallize with a hexagonal structure. This has already been found for the hydrogensulfates and amidosulfates of the lighter rare-earth elements (see above). Thus, the RE^{3+} ions are coordinated by nine perchlorate groups which are attached to three RE^{3+} ions, leading to a three-dimensional structure, according to $[RE(ClO_4)_{9/3}]$. If the tetrahedra are regarded as spheres, the well-known UCl_3 type of structure arises. The anhydrous perchlorates of the smallest lanthanides, Yb and Lu, crystallizes with trigonal symmetry and the acentric space group *R*3*c*. The RE^{3+} ions are surrounded by six perchlorate groups with the chlorine atoms forming an octahedron around the cations. Three of the $[ClO_4]^-$ groups act as chelating ligands, so that a coordination number of nine results for RE^{3+} with a coordination polyhedron which may be viewed as a distorted tricapped trigonal prism. Because there is crystallographically only one $[ClO_4]^-$ group present, each chelating perchlorate ion to one RE^{3+} is a monodentate ligand to a second one and vice versa. Thus, the connectivity may be written as $[RE(ClO_4)_{6/2}]$ (Figure 2.2.12).

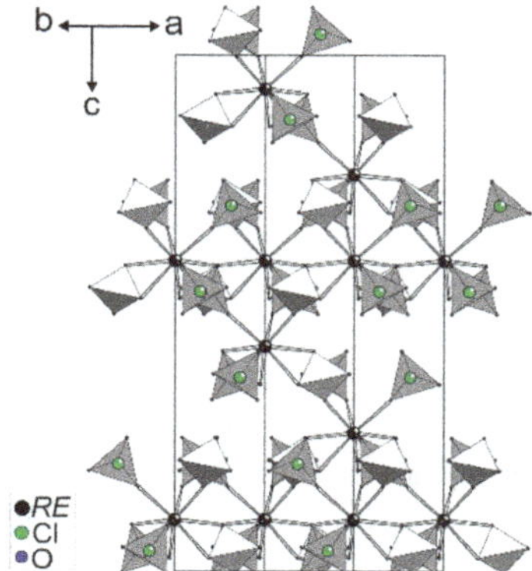

Figure 2.2.12: Crystal structure of the anhydrous perchlorates of the smaller lanthanides (*RE* = Yb, Lu) (modified after [8]).

The thermal decomposition of the perchlorates occurs around 300 °C and usually leads to the oxide chlorides, *RE*OCl. In $Yb[ClO_4]_3$ also, the oxide chloride Yb_3O_4Cl was observed as a decomposition product and $Ce[ClO_4]_3$ yielded CeO_2 exclusively [75].

2.2.5.3 Basic perchlorates

The anhydrous scandium perchlorate cannot be prepared by dehydration of the respective hydrate. Instead, decomposition to the basic compound $Sc(OH)[ClO_4]_2 \cdot H_2O$ is always observed. Recently, it was possible to obtain single crystals of the compound and to determine the crystal structure [74]. According to this investigation, the crystal structure contains the Sc^{3+} ions octahedrally coordinated by three $[ClO_4]^-$ groups, two OH^- ions and one H_2O molecule. Two of these octahedra share one common edge, which is built from OH^- groups, to form pairs according to $[Sc_2(OH)_2(H_2O)_2(ClO_4)_6]^{2-}$ with a Sc^{3+}–Sc^{3+} distance of 3.20 Å. The pairs are connected to infinite layers according to the formulation $[Sc_2(OH)_2(H_2O)_2(ClO_4)_{4/2}(ClO_4)_{2/1}]$. The layers are connected via hydrogen bridges with the water molecules as donors and noncoordinating oxygen atoms O of the ClO_4^- ions as acceptors.

Another kind of basic perchlorate is obtained when an aqueous solution of lanthanide perchlorate is treated with an alkali hydroxide. The compounds have compositions such as, for example, $[RE_6(\mu\text{-}O)(\mu_3\text{-}OH)_8(H_2O)_{24}](ClO_4)_8 \cdot 8H_2O$ and $[RE_6(\mu\text{-}O)(\mu_3\text{-}OH)_8(\eta^2\text{-}ClO_4)_2(H_2O)_{20}](ClO_4)_6 \cdot 4H_2O$ [78]. The common features of all compounds are oxygen-centered $[ORE_6]$ clusters which are capped with OH^- ions over each triangular

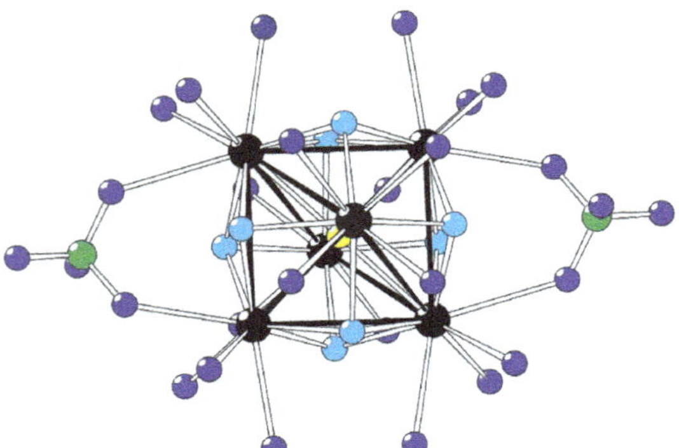

Figure 2.2.13: The oxide (yellow) centered $[RE_6(\mu\text{-O})(\mu_3\text{-OH})_8(\eta^2\text{-ClO}_4)_2(H_2O)_{20}]^{6+}$ cluster as it is found in several basic perchlorates. Note, that the octahedral $[ORE_6]$ cluster is capped by hydroxide ions over each of the triangular faces (light blue spheres) (modified after [8]).

face (Figure 2.2.13). The crystal structures contain further perchlorate groups for charge compensation and nonbonded H_2O molecules. The special arrangement of the metal ions leads to short distances of RE^{3+}–RE^{3+} of about 3.7–3.8 Å, but magnetic measurements show that there are no magnetic interactions, at least under ambient conditions.

References

[1] W. Prandtl, Z. Anorg. Allg. Chem. 1938, 238, 321.

[2] R. Bock, Angew. Chem. 1950, 62, 375.

[3] P. Becker, Adv. Mater. 1998, 10, 979.

[4] P. Becker, Z. Kristallogr. 2001, 216, 523.

[5] M. Leskelä, L. Niinistö, Handbook on the Physics and the Chemistry of Rare Earths (ed. L. Eyring, H. Gschneider), 16, Elsevier Science Publishers, New York, 1986, 203.

[6] L. Niinistö, M. Leskelä, Handbook on the Physics and the Chemistry of Rare Earths (ed. L. Eyring, H. Gschneider), 18, Elsevier Science Publishers, New York, 1987, 91.

[7] M. S. Wickleder, Oxo-Selenates of Rare Earth Elements, in Handbook on the Physics and Chemistry of Rare Earths (ed. K. Gschneidner Jr., J.-C. G. Bünzli, V. K. Pecharsky), 35, Elsevier Science Publishers, New York, 2005, 45–106.

[8] M. S. Wickleder, Chem. Rev. 2002, 102, 2011–2087.

[9] Gmelin Handbook of Inorganic Chemistry, 3, D2, D3, Springer Verlag, Berlin, 1980–84.

[10] J. Felsche, Struct. Bonding [Berlin] 1973, 13, 99, and references therein.

[11] R. L. Marchand, P. Haridon, P. Y. Laurent, J. Solid State Chem. 1978, 24, 71.

[12] J. Felsche, Naturwissenschaften 1971, 58, 218.

[13] U. Kolitsch, V. Ijevskii, H. J. Seifert, I. Wiedmann, F. Aldinger, J. Mater. Sci. 1997, 32, 6135.

[14] H. Müller-Bunz, T. Schleid, Z. Anorg. Allg. Chem. 1999, 625, 613.

[15] M. Grupe, W. Urland, Z. Naturforsch. 1990, 45b, 465.

[16] T. D. Brennan, J. A. Ibers, Acta Crystallogr. 1991, C47, 1062.

[17] K. Stöwe, Z. Naturforsch. 1994, 49b, 733.

[18] F. A. Weber, T. Schleid, Z. Anorg. Allg. Chem. 1999, 625, 2071.
[19] P. Gravereau, B. Es-Sakhi, C. Fouassier, Acta Crystallogr. 1988, C44, 1884.
[20] G.-C. Guo, Y.-G. Wang, J.-N. Zhuang, J.-T. Chen, J.-S. Huang, Q.-E. Zhang, Acta Crystallogr. 1995, C51, 2471.
[21] C. Sieke, I. Hartenbach, T. Schleid, Z. Anorg. Allg. Chem. 2000, 626, 2235.
[22] C. Sieke, T. Schleid. Z. Anorg. Allg. Chem. 1999, 625, 377.
[23] L.-S. Chi, L.-F. Zhou, H.-Y. Chen, H.-H. Zhang, J.-S. Huang, Jiegon Huaxue 1997, 16, 219.
[24] I. Hartenbach, T. Schleid, Z. Anorg. Allg. Chem. 2001, 627, 2493.
[25] C. Ayasse, H. A. Eick, Inorg. Chem. 1973, 12, 1140.
[26] J.-G. Mao, H.-H. Zhuang, J.-S. Huang, Jiegon Huaxue 1996, 15, 280.
[27] C. Sieke, T. Schleid, Z. Anorg. Allg. Chem. 2001, 627, 761.
[28] P. Gravereau, B. Es-Sakhi, C. Fouassier, Acta Crystallogr. 1989, C45, 1677.
[29] H. Jacobsen, G. Meyer, W. Schipper, G. Blasse, Z. Anorg. Allg. Chem. 1994, 620, 451.
[30] H. P. Beck, M. Schuster, L. Grell, J. Solid State Chem. 1993, 103, 433.
[31] H. Müller-Bunz, T. Schleid, Z. Kristallogr. 2002, 19, 115.
[32] C. Wickleder, P. Lauxmann, I. Hartenbach, T. Schleid, Z. Anorg. Allg. Chem. 2002, 628, 1602.
[33] H. Müller-Bunz, T. Schleid, Z. Anorg. Allg. Chem. 2002, 628, 564.
[34] M. Grupe, W. Urland, Naturwissenschaften 1989, 76, 327.
[35] M. Grupe, F. Lissner, T. Schleid, W. Urland, Z. Anorg. Allg. Chem. 1992, 616, 53.
[36] C. Sieke, T. Schleid, Z. Anorg. Allg. Chem. 2000, 626, 196.
[37] C. Sieke, T. Schleid, Z. Anorg. Allg. Chem. 1999, 625, 131.
[38] H. Müller-Bunz, T. Schleid, Z. Anorg. Allg. Chem. 2001, 627, 218.
[39] T. Schleid, H. Müller-Bunz, Z. Anorg. Allg. Chem. 1998, 624, 1082.
[40] O. V. Yakubovich, A. V. Ya, A. Pakhomovskii, M. A. Simonov, Kristallografiya 1988, 33, 605.
[41] H. Müller-Bunz, T. Schleid, Z. Anorg. Allg. Chem. 2000, 626, 845.
[42] H. Müller-Bunz, T. Schleid, Z. Anorg. Allg. Chem. 1999, 625, 1377.
[43] C. Sieke, I. Hartenbach, T. Schleid, Z. Kristallogr. 1999, Suppl. 16, 62.
[44] I. Hartenbach, T. Schleid, Z. Anorg. Allg. Chem. 2001, 627, 2493.
[45] H. Jacobsen, G. Meyer, W. Schipper, G. Blasse, Z. Anorg. Allg. Chem. 1994, 620, 451.
[46] H. D. Park, E. R. Kreider, J. Am. Ceram. Soc. 1984, 67, 23.

[47] K. K. Palkina, Inorg. Mater. 1982, 18, 1199.
[48] I. Mayer, Israel J. Chem. 1969, 7, 717.
[49] C. Sieke, T. Schleid, Z. Anorg. Allg. Chem. 2001, 627, 761.
[50] M. Bagieu-Beucher, I. Tordjman, A. Durif, Rev. Chim. Miner. 1971, 8, 753.
[51] Y. I. Smolin, Y. F. Shepelev, A. I. Domanskii, N. V. Belov, Kristallografiya 1978, 23, 187.
[52] M. Bagieu-Beucher, J. C. Guitel, Acta Crystallogr. 1978, 34, 1439.
[53] O. S. M. Elmokhtar, M. Rzaigui, Bull. Soc. Chim. Belg. 1996, 105, 307.
[54] V. K. Trunov, N. N. Chudinova, L. A. Borodina, Dokl. Akad. Nauk SSSR 1988, 300, 1375.
[55] L. Niinistö, Systematics and the Properties of the Lanthanides (ed. S. P. Sinha), Reidel, Dordrecht, 1983, 125.
[56] P. C. Junk, C. J. Kepert, B. W. Skelton, A. H. White, Austr. J. Chem. 1999, 52, 601.
[57] S. P. Sirotinkin, V. A. Efremov, L. M. Kovba, A. N. Prokovskii, Kristallografiya 1977, 22, 1272.
[58] M. S. Wickleder, Z. Anorg. Allg. Chem. 2000, 626, 1468.
[59] L. D. Iskhakova, N. P. Kozlova, V. V. Marugin, Kristallografiya 1990, 35, 1089.
[60] M. S. Wickleder, Z. Anorg. Allg. Chem. 1998, 624, 1347.
[61] M. S. Wickleder, Chem. Mater. 1998, 10, 3212.
[62] M. S. Wickleder, Z. Anorg. Allg. Chem. 1999, 625, 474.
[63] M. S. Wickleder, Z. Anorg. Allg. Chem. 1999, 625, 93.
[64] M. S. Wickleder, Z. Anorg. Allg. Chem. 1999, 625, 725.

[65] M. S. Wickleder, Z. Anorg. Allg. Chem. 1999, 625, 302.
[66] M. S. Wickleder, Z. Anorg. Allg. Chem. 1999, 625, 1794.
[67] M. S. Wickleder, J. Alloys Compd. 2000, 303–304, 445.
[68] G. Vicentini, G. Chiericato Jr., An. Acad. Brasil. Ciênc. 1979, 51, 217.
[69] Zinner, G. Vicentini, J. Inorg. Nucl. Chem. 1980, 42, 1349.
[70] J. E. X. de Matos, L. Niinistö, J. R. Matos, G. Vicentini, L. B. Zinner, Acta Chem. Scand. 1988, A42, 111.
[71] M. S. Wickleder, Z. Anorg. Allg. Chem. 2001, 627, 1675.

[72] E. M. Aricó, L. B. Zinner, C. Apostolidis, E. Dornberger, B. Kanellakopulos, J. Rebizant, J. Alloys Compd. 1997, 249, 111.
[73] C. Belin, F. Favier, J. L. Pascal, M. Thillard-Charbonnel, Acta Crystallogr. 1996, C52, 1872.
[74] M. S. Wickleder, Z. Anorg. Allg. Chem. 1999, 625, 1556.

[75] F. Favier, J. L. Pascal, F. Cunin, A. N. Fitch, G. Vaughan, Inorg. Chem. 1998, 37, 1776.
[76] M. S. Wickleder, W. Schäfer, Z. Anorg. Allg. Chem. 1999, 625, 309.
[77] M. S. Wickleder, Z. Anorg. Allg. Chem. 1999, 625, 11.
[78] R. Wang, M. D. Carducci, Z. Zheng, Inorg. Chem. 2000, 39, 1836.

Arndt Simon

2.3 Lanthanide compounds with low valence

Low valence of the lanthanides comprises normal, intermediate and mixed valence compounds in binary and multinary systems of hydrides, halides, carbides, borides and d-transition metals. Their crystal structures frequently exhibit close analogy with those of refractory metal compounds in low oxidation states, characterized by the occurrence of discrete and condensed clusters that show localized or extended metal–metal bonding. With only a few exceptions, these structures, in the case of lanthanides, need stabilization by interstitial atoms due to their small number of valence electrons. Disorder phenomena are discussed based on their relevance for a detailed understanding of the respective physical properties.

2.3.1 Mono- and dihalides of lanthanides

The general scenario of low valence of the lanthanides concerns the bonding situation between the extremes of salts and metals, and includes phases with localized metal–metal bonding as well as possible contaminants introduced accidentally. After decades of research on low-valence lanthanide halides, the only existing monohalide, LaI, was discovered only recently [1]. Strange synthesis conditions could have raised doubts in the well-defined character of LaI; however, the metallothermic reaction of LaI_2 and Na corroborates the existence of LaI as the first binary lanthanide monohalide [2, 3]. Polycrystalline LaI can be prepared within a few days by reacting LaI_2 and Na at 550 °C. Subsequent extraction with diglyme removes NaI without a visible attack of LaI. This unique compound deserves a more detailed description.

LaI crystallizes in the hexagonal NiAs-type structure (Figure 2.3.1), composed of interpenetrating lattices of hexagonal close-packed anions and trigonal prismatic packing of cations. In contrast to the normal c/a ratio of 1.63 for a columnar type, LaI exhibits a layered structure with a c/a ratio of 2.47 and La–La distances within the triangular layers of 393 pm, much shorter than the distances of 486 pm for two-dimensional metallic behavior. Powder samples exhibit an electrical resistivity in the order of 100 mΩcm, although low conductivity can be due to grain boundary effects that resemble anisotropic conductors like graphite where the in-plane and out-of-plane resistivities differ by many orders of magnitude. A critical point concerns phase purity of LaI toward incorporation of hydrogen; however, in spite of different routes of preparation, the only rare-earth monohalide with an atomic ratio exists as $X/Ln = 1$.

Traces of hydrogen introduced accidentally via moisture stabilized the other "monohalides," for example, $GdClH_x$ [4]. Enclosed in a tantalum capsule, at high

https://doi.org/10.1515/9783110654929-011

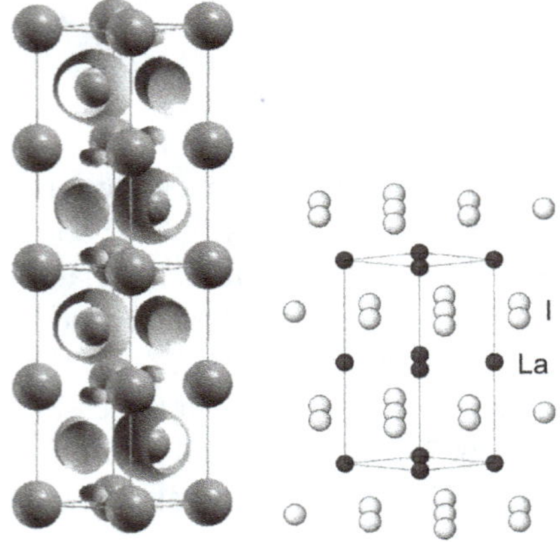

Figure 2.3.1: The crystal structure of LaI (right) and the corresponding electron localization function (left). Reproduced from [3] with permission of Elsevier.

enough temperature this compound decomposes into Gd and $GdCl_3$. The monohalide hydrides $LnXH_x$ all contain Ln^{3+} cores; the excess of metal valence electrons are either used for metal–metal bonding in the metallic phases at $0.7<x<1$ [5] or transferred to additional H atoms in the $LnXH_2$ phases which are ionic insulators bonded as hydride anion, and numerous dihalides of salt-like chlorides and bromides are structurally well characterized [6].

LaI_2 and CeI_2 crystallize in layered tetragonal structures, stacked by I–*Ln*–I square packages that connect via van der Waals bonding between cubes of I atoms (Figure 2.3.2). Interest arose from a speculation that LaI_2 might get superconducting. LaI_2 and CeI_2 are strictly stoichiometric compounds; in particular, they do not incorporate ubiquitous hydrogen because the pattern of σ bonds between the metal atoms does not offer bonding capability [7]. Detailed hydrogenation experiments show that the open square nets collapse into close-packed triangular nets stabilized by 2-electron 4-center bonds between H and *Ln*, and the trigonal LnI_2H_x phases

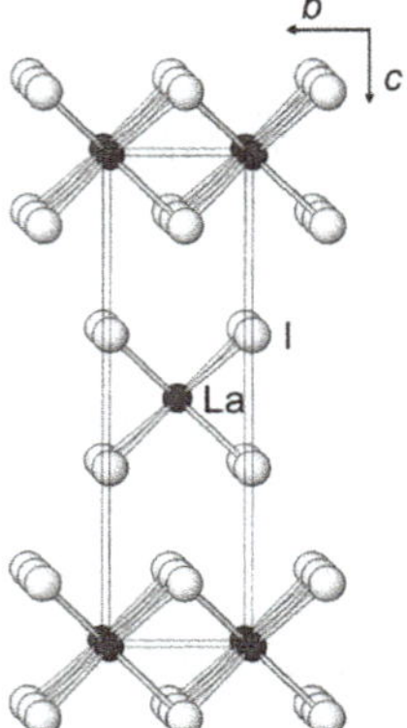

Figure 2.3.2: LaI_2 composed of square planar nets of La atoms each surrounded by a cube of I atoms. Reproduced from [3] with permission of Elsevier.

continuously incorporate hydrogen starting from $x = 0.5$ in the case of La, and $x = 0.35$ in case of Ce, up to the ionic limit $x = 1$. PrI_2 exhibits an amazingly complicated structural chemistry including cubane-like Pr_4I_4 clusters [8] in its position between the well-understood antipodes LaI_2, CeI_2 on one hand and GdI_2 on the other.

The crystallographic characteristic of GdI_2 is quite different as the compound adopts the (2H)-NbS_2-type structure, based on a triangular net of Gd atoms surrounded by prisms of I atoms. In contrast to the miscibility gaps with LaI_2 and CeI_2, GdI_2 absorbs hydrogen, and the range of homogeneity stretches from GdI_2 to GdI_2H. The compound GdI_2 is a two-dimensional ferromagnetic metal at room temperature with interesting physical properties [9].

2.3.2 Discrete clusters

Rare earth metal atoms in low-valence compounds frequently mimic valence electron-rich halides of Mo and W, with octahedral units of M_6X_8 and M_6X_{12} clusters with an inner M_6 core surrounded by X atoms (Figure 2.3.3). The monovalent halides [9] leave a large number of electrons for metal–metal bonding, one with eight nonmetal atoms above the faces M_6X_8, and one with 12 above the edges of the octahedron cluster M_6X_{12}.

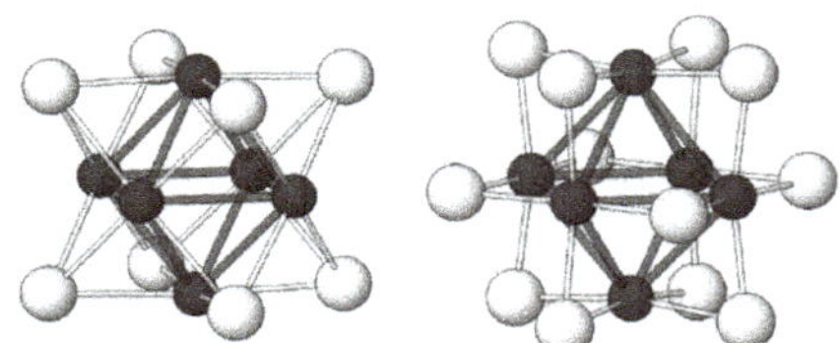

Figure 2.3.3: M_6X_8 and M_6X_{12} clusters. Reproduced from [3] with permission of Elsevier.

Decreasing the number of valence electrons by exchanging the M atoms with the group 5 metals Nb and Ta only allows the stabilization of the M_6X_{12} cluster in halides, with just one exception, Nb_6I_{11}. Moving on to a group 4 element, for Zr, only M_6X_{12} clusters exist, filled with all sorts of interstitial atoms; an extreme case is known from Th chemistry where the compound $Th_6Br_{15}H_7$ is stuffed with 7H atoms distributed on the faces of the octahedron. The still further reduced number of valence electrons in group 3 metals of the metal-rich lanthanide compounds mimic the feature in the structural chemistry of the refractory metals with additional interstitial atoms. Reducing the relative number of nonmetal atoms around a single cluster core leads to condensation of M–M bonding, and besides the aforementioned clusters, oligomeric species do exist up to the infinite chain.

Experiments in the LaI/Al system produced sharp Bragg reflections in agreement with metal-deficient NaCl-type $La_{1-x}I$, $x \approx 0.1$, by partial substitution of La and

15 at% Al [10, 11]. Diffuse Bragg spots at a level of three orders of magnitude less, however, evidence a wrong description where, actually, Al atoms replace I atoms in the NaCl-type substructure of Al-centered $La_6I_{12}Al$ clusters. Comparing observed and calculated intensities pixel-wise results in a disorder model shown with no random disorder but a uniform distribution with preference for certain connections of the clusters bridged by I atoms (Figure 2.3.4).

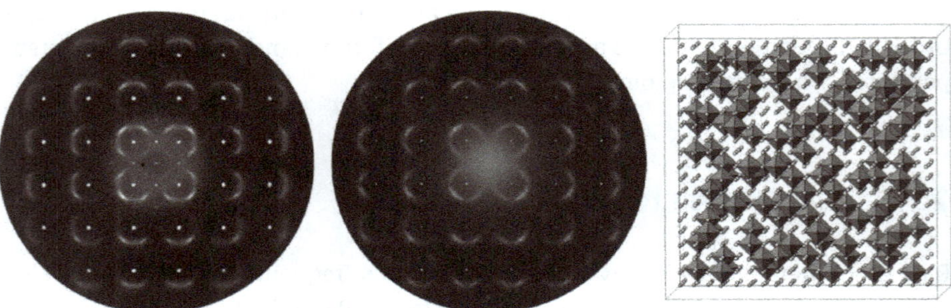

Figure 2.3.4: Experimental (left) and simulated (middle) X-ray diffractograms of $La_{0.7}I_{0.86}Al_{0.14}$ along with the disordered structure model (right). Reproduced from [3] with permission of Elsevier.

A large family of compounds is described as being built from discrete clusters with general formula $Ln_7X_{12}Z$, known for chlorides, bromides and iodides [12]. Interstitials Z range from main group elements to transition metals in an *fcc* packing of the X^- ions substituted by, for example, C^{4-} ions, a single cation occupying octahedral voids in discrete cations according to $Ln_6X_{12}CLn$ (Figure 2.3.5).

To combine the Ln_6Z unit and its surrounding X_{12} cuboctahedron to the $Ln_6X_{12}Z$ cluster, a close-packing constitutes the structure where notorious disorder appears as the main problem arising with the discrete single Ln atoms. Their electron distribution is strangely elongated in all cases where Z is a main group element, but quite spherical when Z is a transition metal and octahedral Ln_6 units dominate when carbon and transition metals act as interstitials. An unusual modification of the ligand shell is found in the structure of $La_6Br_{10}Fe$, where the La_6 octahedron is both coordinated above the edges and some of the faces [13]. Besides single atoms in the center of the octahedron (Figure 2.3.6), quasimolecular groups occur, as found in $Ln_6I_{10}(C_2)_2$, with Ln = La, Ce [14] as well as C_2 pairs in trigonal bipyramidal units (Figure 2.3.7).

The first step of condensation in the carbide halides $Gd_{10}Cl_{18}(C_2)_2$, $Gd_{10}Cl_{17}(C_2)_2$, $Gd_{10}Br_{17}(C_2)_2$ [15] and $Gd_{10}I_{16}(C_2)_2$ [16] joins common edges of the C_2 centered units (Figure 2.3.8), similar to $Gd_{10}Br_{15}B_2/Tb_{10}Br_{15}B_2$ [17]. Above all free edges, the Ln_{10} units are coordinated by X atoms. The loss of X atoms marks the crossing of a critical borderline in the series of halide carbides and borides from normal valence compounds to low-valence phases.

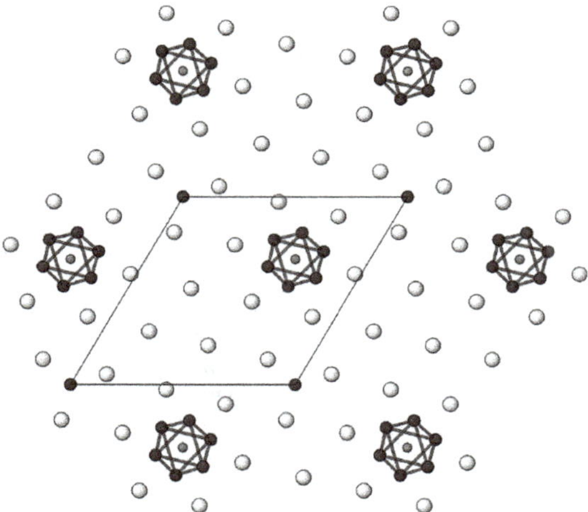

Figure 2.3.5: Close-packed layer of halogen atoms *X*, substituted by Ln_6Z units in the structure of $Ln_7X_{12}Z$. Reproduced from [3] with permission of Elsevier.

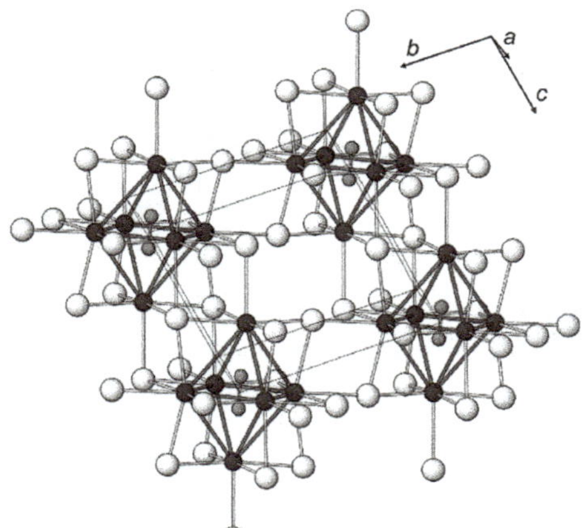

Figure 2.3.6: Section of the crystal structure of $La_6I_{10}(C_2)$ drawn with decreasing sizes for I, La and C. Reproduced from [3] with permission of Elsevier.

The C–C distance in the compounds $Gd_{10}X_x(C_2)_2$ comes close to 145 pm and thus corresponds to a single bond in the ethanide ion (C_2^{6-}). A hydrolysis reaction of $Gd_{10}Cl_{18}(C_2)_2$ leads to the evolution of nearly pure C_2H_6. Reducing the halogen content in going from $Gd_{10}Cl_{18}(C_2)_2$ to $Gd_{10}Cl_{17}(C_2)_2$ leads toward a low valence compound. A central longer metal–metal bond in $Gd_{10}I_{16}(C_2)_2$ with even lower halogen content is due to the larger iodine atoms in spite of two electrons being involved in metal–metal bonding.

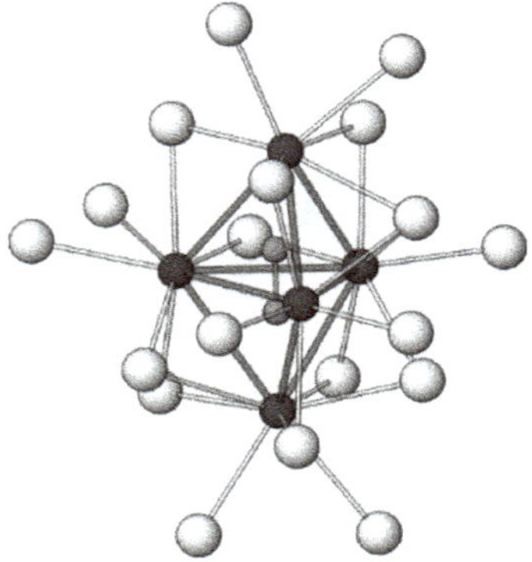

Figure 2.3.7: Cluster core Ln_5 known from the ternary compounds $La_5Cl_9(C_2)$ and $La_5I_9(C_2)$. Reproduced from [3] with permission of Elsevier.

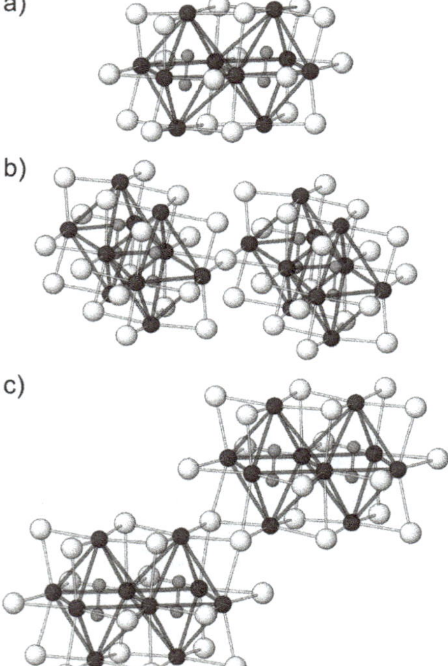

Figure 2.3.8: Bioctahedral $Gd_{10}X_{18}(C_2)_2$ unit in (a) $Gd_{10}Cl_{18}(C_2)_2$, (b) $Gd_{10}Cl_{17}(C_2)_2$ and (c) $Gd_{10}I_{16}(C_2)_2$, interconnected by X atoms. Reproduced from [3] with permission of Elsevier.

One way to increase the oligomer size is to add a next octahedron to a bioctahedron in a linear fashion, forming straight octahedron triples identified in the compounds $Ln_{14}I_{20}(C_2)_3$ [18] of La and Ce (Figure 2.3.9).

Disorder phenomena frequently cause problems that are investigated and interpreted by electron microscopy.

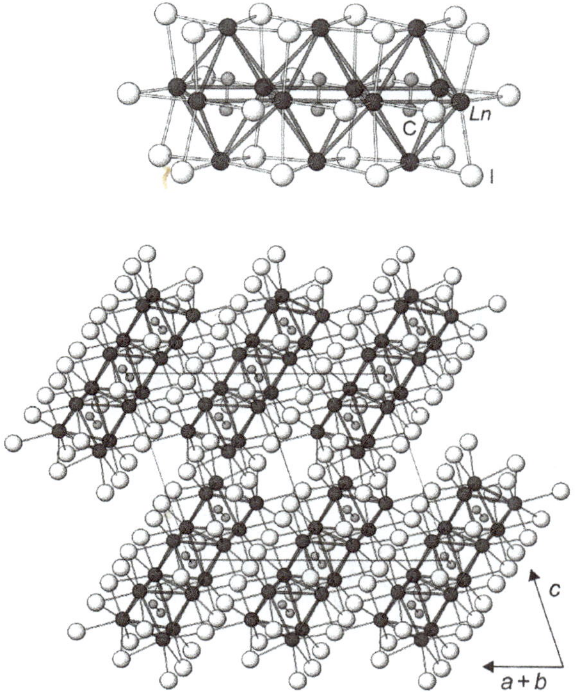

Figure 2.3.9: $Ln_{14}I_{24}(C_2)_3$ cluster formed from three $Ln_6(C)_2$ units via condensation and connected by I atoms. Reproduced from [3] with permission of Elsevier.

2.3.3 Chain structures

The discovery of Gd_2Cl_3 was a landmark, and numerous new metal-rich halides of the rare earth metals based on discrete and condensed clusters followed [18–21]. Ironically, intended purification actually added sufficient amounts of C or H, for example, to the metals from traces of impurity phases introduced by arc-welding of tantalum tubes, that paved the way to the field of reduced phases of lanthanides. In particular, the discovery of (C_2) units in Ln_6 octahedra of clusters and condensed cluster compounds of the lanthanides except Gd_2Cl_3 and their isotypic phases have been discarded as binary compounds. The remarkable compound Gd_2Cl_3 presented is the first example of a lanthanide metal to occur in an oxidation state below 1.5 that contains linear chains of *trans*-edge sharing Gd_6 octahedra (Figure 2.3.10). Two features are unique: octahedra do not contain interstitial atoms and the chains are formally derived from M_6X_8- instead of M_6X_{12}-type clusters as the X atoms coordinate the octahedra above faces. The octahedral basis is considerably elongated in the chain direction, 390 pm compared to 337 in the joined edges.

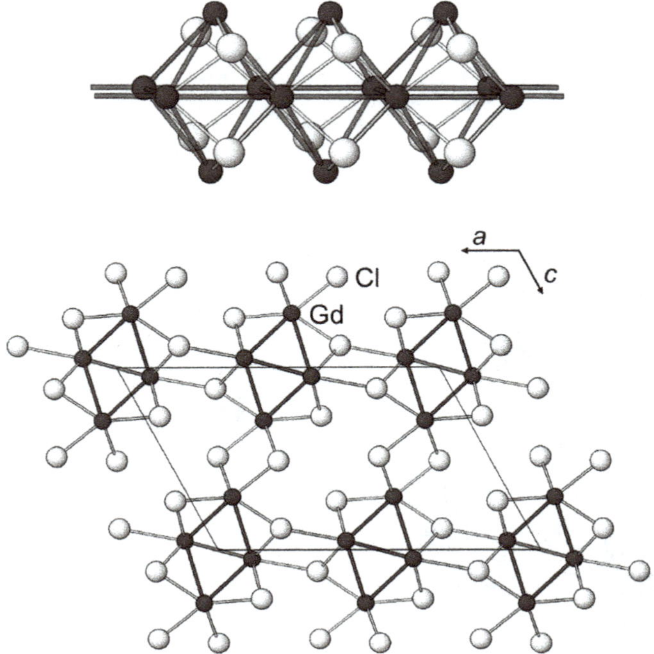

Figure 2.3.10: Gd_4Cl_6 chain (top) in the crystal structure of Gd_2Cl_3, projected along *b*-axis (bottom). Reproduced from [3] with permission of Elsevier.

The isotypic phases of Y, Gd and Tb can be prepared in quantitative yield and are well characterized; however, reactions of the metals with their corresponding trihalides suffer from slow reaction rates. A special strategy worked well in the case of Gd_2Cl_3. The compound was synthesized at the anode of Gd in a melt of $GdCl_3$ that produced crystals of several mm in size within a few hours [22].

The chemical bonding in Gd_2Cl_3 was characterized by photoelectron spectra of Gd and Gd_2Cl_3. They reveal no multiplet splitting giving evidence of a $4f^7$ (Gd^{3+}) core state, and only a small portion of the density of states around the Fermi level, mainly of *d* character, is left. To describe how 1.5 electrons per Gd atom distribute in metal–metal bonding, the *d* orbitals have to be involved in σ and π bonding along the joined edges of the octahedron and σ and δ bonding along the edges in chain direction. This avoids crossing of bands, which opens the gap, and electrons completely occupy the three lowest bands. Gd_2Cl_3 is only weakly stable against disproportionation, and the stability of chlorides and bromides with other rare earth metals is even less [23].

Salt-like colorless Gd_2Cl_3N [24] exhibits quite a different structure, containing linear chains of Gd_4N tetrahedra connected via their edges. The compound is the final member of a series derived by condensation of Ln_4N tetrahedra, the first step being realized in Gd_3Cl_6N, representing normal valence compounds.

In particular, carbon and transition metals in the octahedral Ln_6Z unit are dominant constituents of the chain structures. Clusters belong to series $Ln_{4n+2}X_{6n+4}Z_n$ and $Ln_{4n+2}X_{5n+5}Z_n$, with final members Ln_4X_6Z and Ln_4X_5Z form the compounds with infinite linear chains. Numerous phases are known to exist with this composition, for example, Gd_4I_5C, Gd_4I_5Si, Pr_4I_5Z with Z = Co, Ru, Os and La_4I_5Ru [25]. In La_4I_5C, the larger Ln_6 cage allows substitution of C^{4-} by isoelectronic species (C_2^{4-}). In the isotypic bromide and iodide carbides of La, Ce and Pr, a description as ethenides of composition $Ln_4X_5C_x$, $1 \le x \le 2$ gives evidence for a range of homogeneity between methanides and ethenides [26].

The remarkably open structure of Tb_4Br_6Z (Z = C, Si) [27] (Figure 2.3.11), isotypic with Sc_4Cl_6C, connects chains that create channels in a close similarity to the refractory oxomolybdate $NaMo_4O_6$ [28]. In $Gd_4Cl_3I_3(C_2)$ [29] the chain exhibits yet another alternating pattern (Figure 2.3.12). Bioctahedral clusters alternate with single octahedra, which are stretched in chain direction. The C–C distances in these clusters with 144 pm indicate the presence of ethanide ions (C_2^{6-}) whereas the other, 134 pm, rather corresponds to an ethenide ion (C_2^{4-}). The alternating substitution of the (C_2^{6-}) ion by two N^{3-} ions in the compound Gd_4CNI_6 [30] results in an even more pronounced stretching of the octahedron around the two N^{3-} ions, actually forming edge-shared Gd_4N tetrahedra.

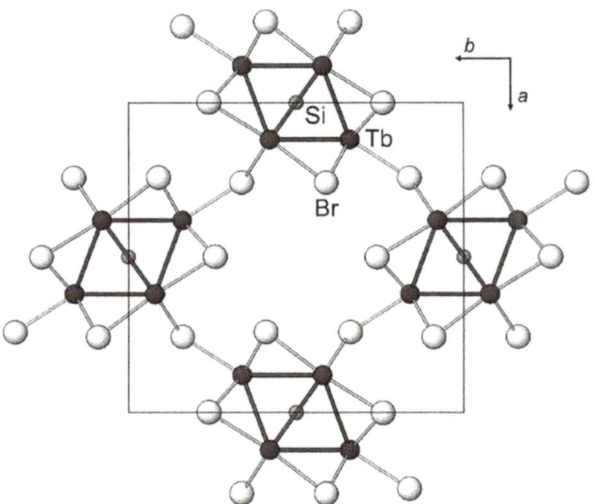

Figure 2.3.11: Projection of the tetragonal crystal structure of Tb_4Br_6Si along the c-axis. Reproduced from [3] with permission of Elsevier.

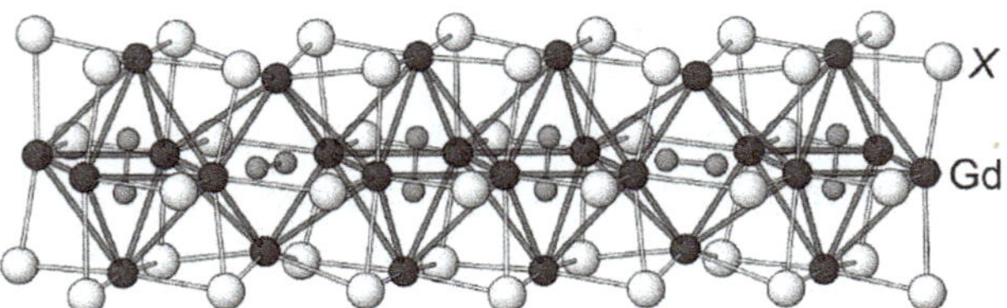

Figure 2.3.12: Chains with varying orientations of (C_2) units in the structure of $Gd_4Cl_3I_3(C_2)$. Reproduced from [3] with permission of Elsevier.

Other examples of chain types are realized for the compounds with composition $Ln_{12}X_{17}(C_2)_3$ [31], found in the iodides of the early lanthanides from La to Pr, Gd and Dy, where the chain can be decomposed into the familiar bioctahedral clusters connected via an intermediate $Ln_6(C_2)$ unit (Figure 2.3.13).

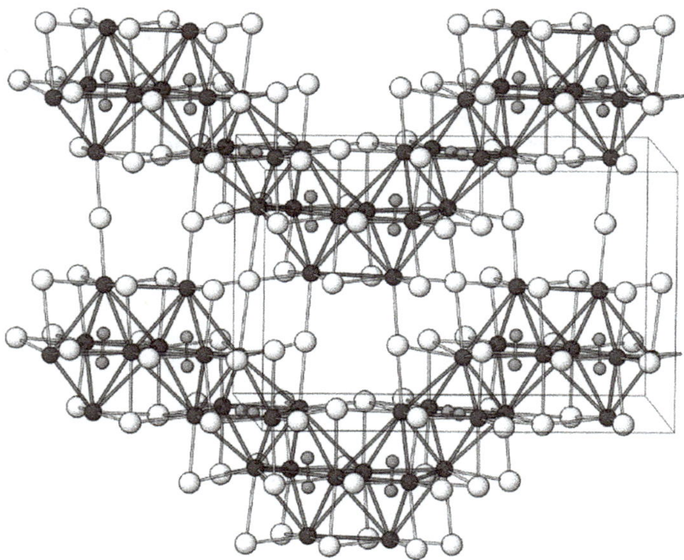

Figure 2.3.13: Nonlinear chain of condensed $Ln_6(C_2)$ units in the phases of composition $Ln_{12}X_{17}(C_2)_3$. Reproduced from [3] with permission of Elsevier.

Linear twin chains occur with lanthanides in quite different crystal structures. In particular, lanthanides with twin chain compounds have the general formulae $Ln_6X_6Z_2$ and $Ln_6X_7Z_2$, respectively, represented by $Gd_6I_6C_2$ [32] (Figure 2.3.14) and also $Pr_6Br_6Ru_2$ [33]. The structural relationship with $Gd_6I_7C_2$ and the parent compound Gd_4I_5C is shown in Figure 2.3.15, also emphasizing the different orientation and dimensionality of the cluster units.

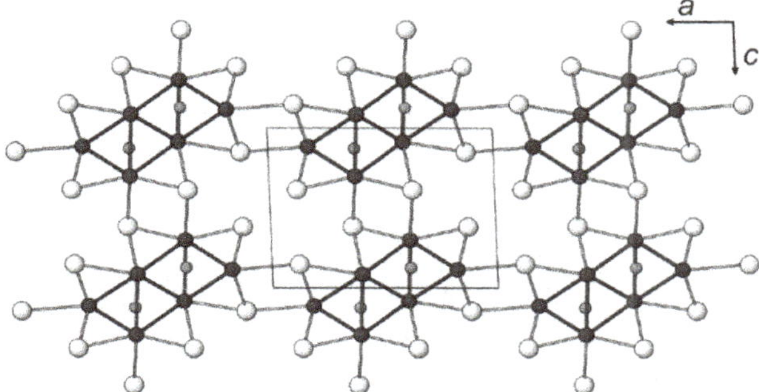

Figure 2.3.14: Projection of the crystal structure of $Gd_6I_6C_2$ along the *b*-axis. Reproduced from [3] with permission of Elsevier.

2.3.4 Layer structures

Two-dimensional metal–metal bonded systems of condensed clusters exist for hydrides, with all lanthanides in twin layers of *Ln* atoms sandwiched by *X* atoms above and below the octahedral voids of M_6X_8-type clusters. The H atoms fill the tetrahedral voids and thus, the distances between H^- and X^- are maximized. With all tetrahedral voids filled, the layer package has the composition Ln_2X_2H, and the hydrogen content can be continuously decreased before the compounds become instable at an approximate composition of $LnXH_{0.7}$. The increase of the *a* lattice parameter is proportional to the hydrogen content, which reflects the in-plane repulsion between the H^- ions. In contrast, the *c* parameter contracts, resulting in a volume change that is negligibly small.

All phases $LnXH_x$ ($x \leq 1$) are metallic conductors. Heated in hydrogen atmosphere under ambient pressure, they transform to transparent nonmetallic salt-like compounds, $LnXH_2$. The reaction proceeds topochemically and is reversible, accompanied by the usual volume increase of approximately 2 cm^3 mol^{-1}. H and D atoms enter opposite faces of the trigonal antiprisms ("octahedra") located in the centers of Ln_3 triangles (Figure 2.3.16). In the case of Tb, the Tb–Tb distances increase from 358 to 396 pm, and the *X* atoms move to positions above the tetrahedral voids, changing M_6X_8 to M_6X_{12} type condensed clusters. No miscibility between the phases *LnX*H and $LnXH_2$ can be detected, which finds an explanation in electronic band structure calculations [34].

What has been discussed with the layered halide hydrides also holds true for the corresponding halide carbides Ln_2X_2C and $Ln_2X_2(C_2)$, as well as Ln_2XC, where interstitial C atoms or (C_2) units occupy octahedral voids in the twin layers. Phases of composition $Ln_2X_2(C_2)$ [35] crystallize with the same kinds of stacking variants as the phases Ln_2X_2C; however, they have monoclinic symmetry due to the inclined orientation of the (C_2) unit (Figure 2.3.17). The electron-precise descriptions $Ln^{3+}{}_2X^-{}_2C^{4-}$ and

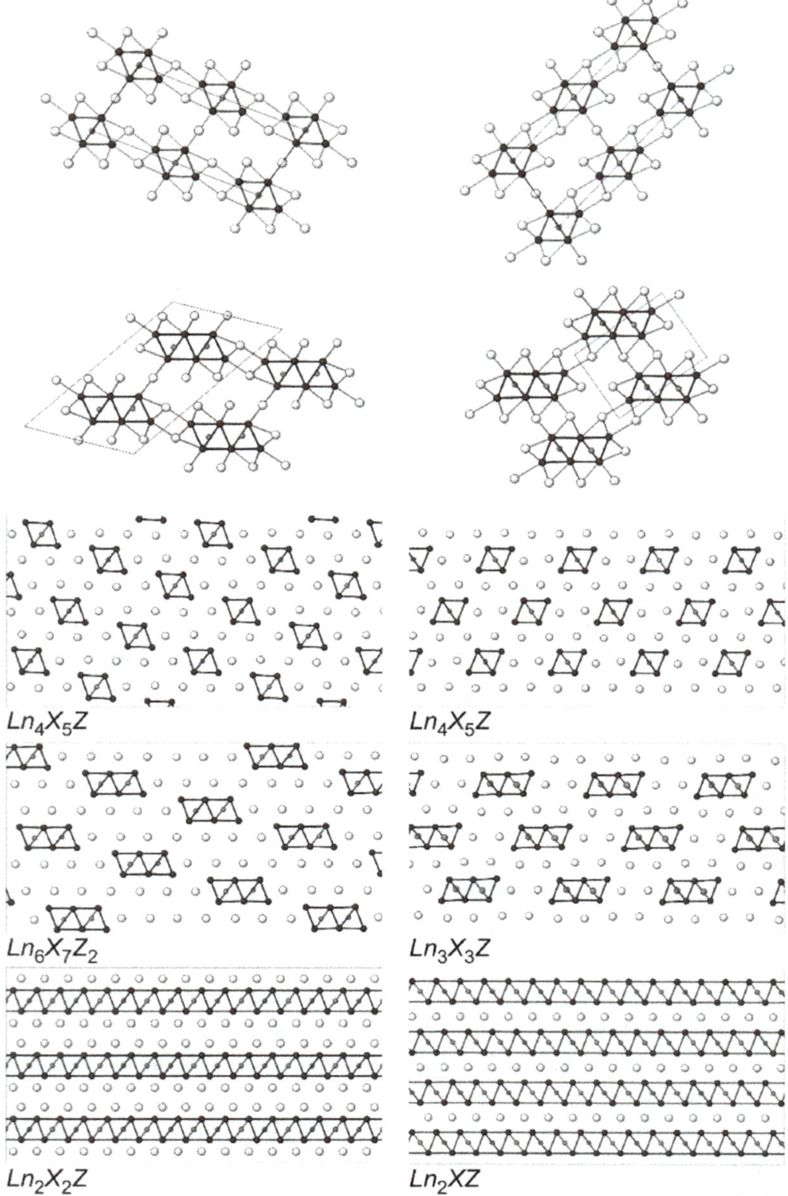

Figure 2.3.15: Series of differently oriented planes with increasing condensation. Reproduced from [3] with permission of Elsevier.

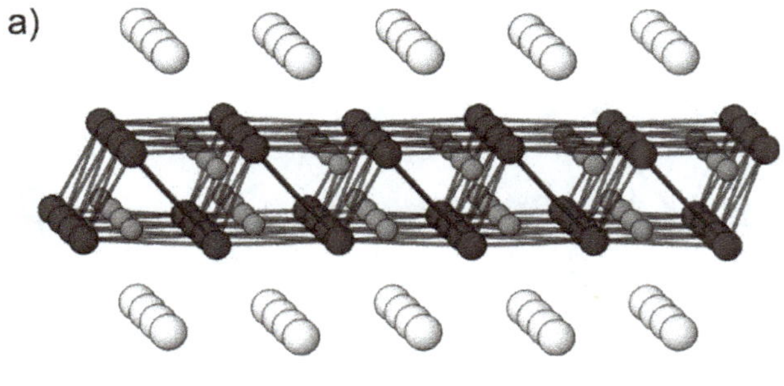

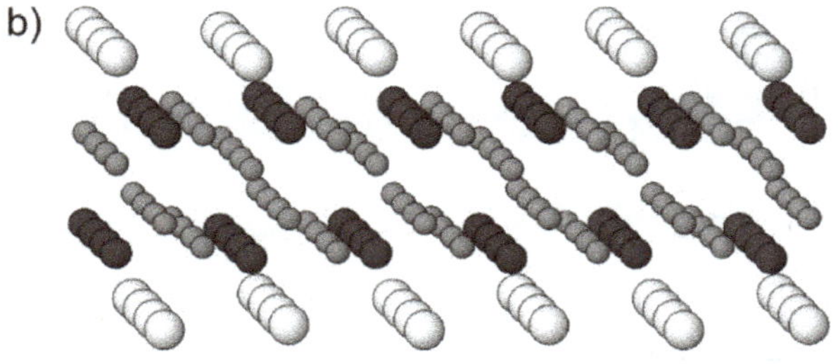

Figure 2.3.16: Layer packages Cl–Tb–Tb–Cl in chloride hydrides $TbClH_x$, the H atoms occupying tetrahedral voids inside the package for $x \leq 1$ (top), and voids in centers of triangular faces for $x = 2$ (bottom), loss of metal–metal bonding indicated. Reproduced from [3] with permission of Elsevier.

$Ln^{3+}{}_2X^-{}_2(C_2{}^{4-})$ lack metal–metal bonding. The black compounds Ln_2X_2C are semiconductors; however, the phases $Ln_2X_2(C_2)$ are metallic and, in certain cases, become superconductors at low temperature.

Reduction of the halogen content in the phases from Ln_2X_2C *to* Ln_2XC closes the van der Waals gap between the *X* atom layers and leaves one electron per formula unit for metal–metal bonding. Depending on the kind of *X* atom, the connection between the twin-layers is different. In Gd_2ClC, the Cl atom is coordinated by a trigonal antiprism of Gd atoms, whereas the coordination around Br and I is trigonal prismatic [36]. The Cl compound shows antiferromagnetic order at low temperature, whereas the Br and I compounds become ferromagnets (Figure 2.3.18).

The nonlinear chain structure of $Ln_{12}I_{17}(C_2)_3$ extends from a one-dimensional to a two-dimensional system in the structures of $Gd_6Cl_5C_3$ [37] and $Gd_4Br_3C_2$ [38] derived by twin chains via intermediate rows of Gd_6C octahedra in nonplanar layer structures, offering flexibility toward local distortions. In the case of $Gd_6Cl_5C_3$, it allows for a partial substitution of C^{4-} by $(C_2{}^{6-})$ ions (Figure 2.3.19).

Last, but not the least, systematic changes from planar to undulated sheets are observed in a remarkable series of compounds [39]. Chains join into ribbons connected into undulating sheets linked via halogen atoms between them, where the

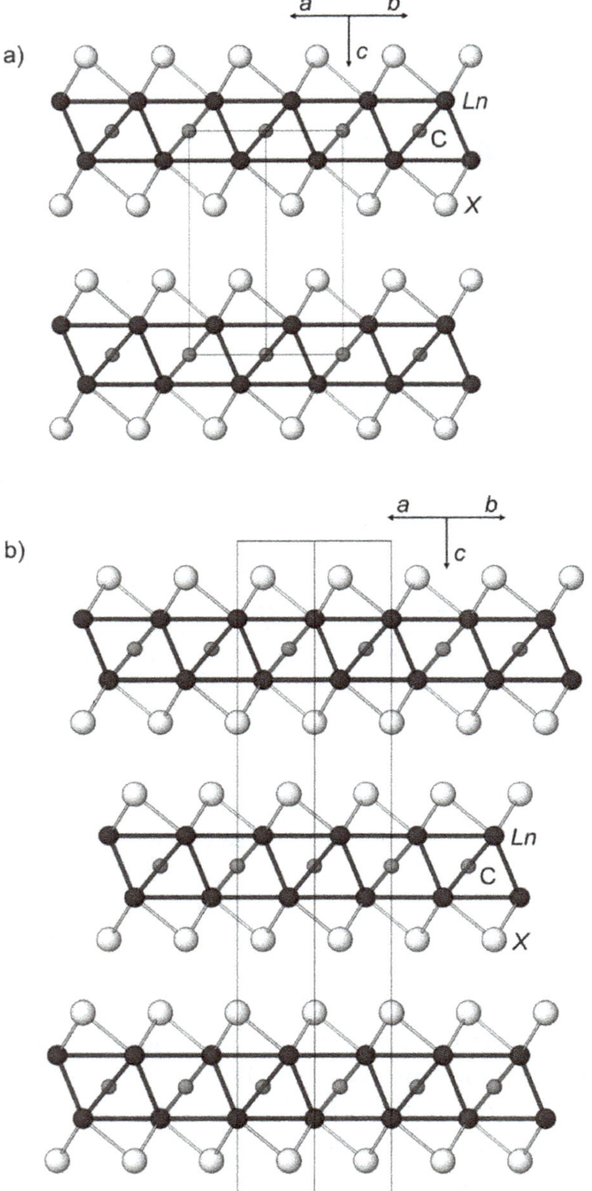

Figure 2.3.17: Crystal structures of (a) 1 T- and (b) 3 R-Ln_2X_2C. Reproduced from [3] with permission of Elsevier.

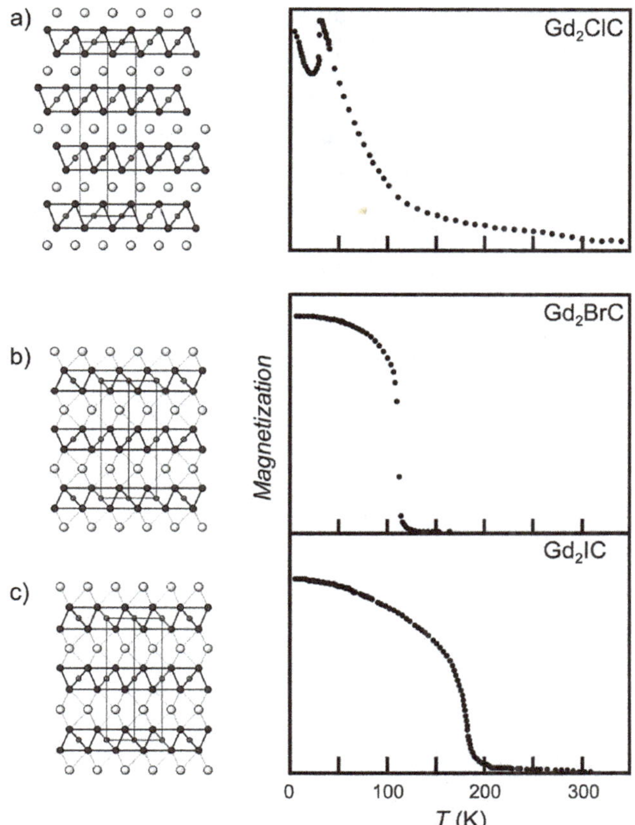

Figure 2.3.18: Crystal structures and magnetization behavior of (a) Gd_2ClC, (b) Gd_2BrC and (c) Gd_2IC. Reproduced from [3] with permission of Elsevier.

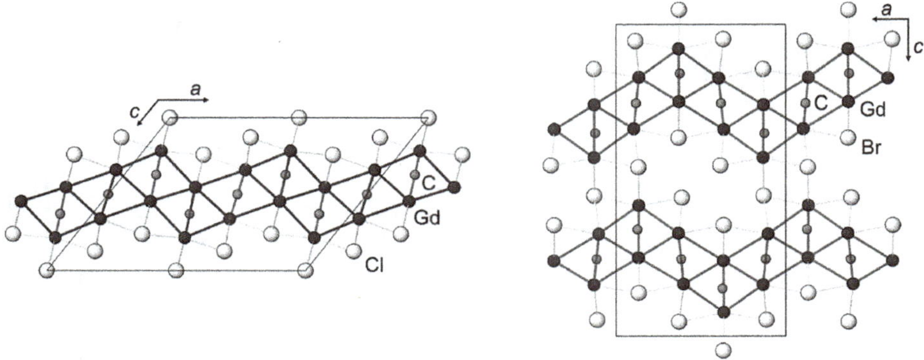

Figure 2.3.19: Corrugated layers of condensed Gd_6C octahedra in the structures of $Gd_6Cl_5C_3$ and $Gd_4Br_3C_2$. Reproduced from [3] with permission of Elsevier.

structural principle allows an enormous variety of different variable ribbon sizes leading to increasingly complex structures (Figure 2.3.20).

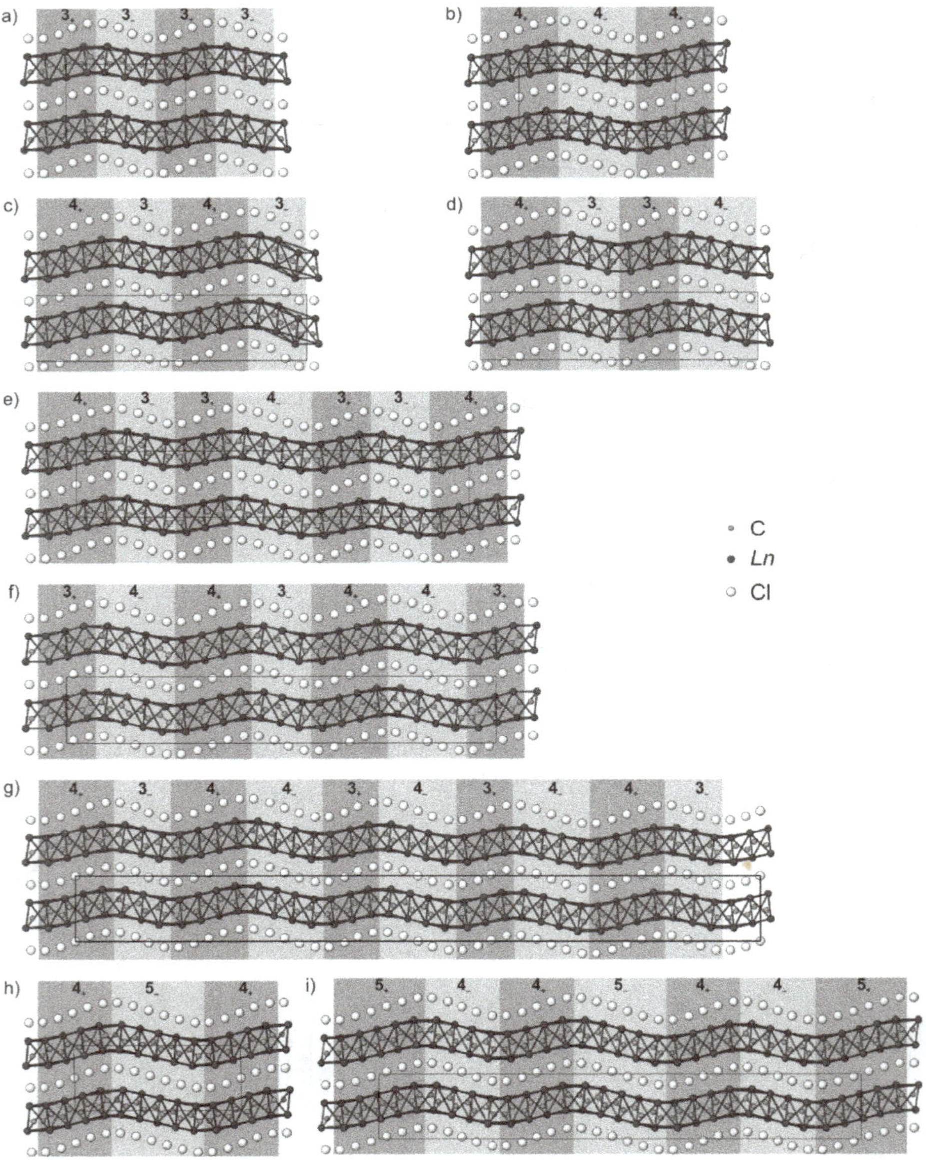

Figure 2.3.20: Undulating layers in the crystal structures of (a) $La_6Cl_4(C_2)_3$, (b) $La_8Cl_5(C_2)_4$, (c, d) $La_{14}Cl_9(C_2)_7$, (e) $La_{20}Cl_{13}(C_2)_{10}$, (f) $La_{22}Cl_{14}(C_2)_{11}$, (g) $La_{36}Cl_{23}(C_2)_{18}$, (h) $Ce_{18}(C_2)_9Cl_{11}$ and (i) $Ce_{26}(C_2)_{13}Cl_{16}$. Reproduced from [3] with permission of Elsevier.

2.3.5 Three-dimensional frameworks

Octahedral Ln_6Z units build up a very versatile structure family. A striking example is Gd_3Cl_3C [40], which forms black cubic crystals, a low-valence compound with metallic behavior (Figure 2.3.21). According to $Gd^{3+}{}_3Cl^{-}{}_3C^{4-}\cdot 2e^{-}$ significant metal–metal bonding is present in the condensed cluster phase. $Gd_6Cl_{12}C$ clusters share three of the octahedron edges, with others capped by the Cl atoms. A particularly interesting phase is the auride La_3I_3Au, where nearly half of the Au is replaced by iodine, due to the similarity between gold and iodine. The low-valence character of La_3I_3Au is readily evident through a vigorous hydrogen evolution when exposed to water, producing a colloidal product of deep purple color similar to the famous Cassius's purple [41].

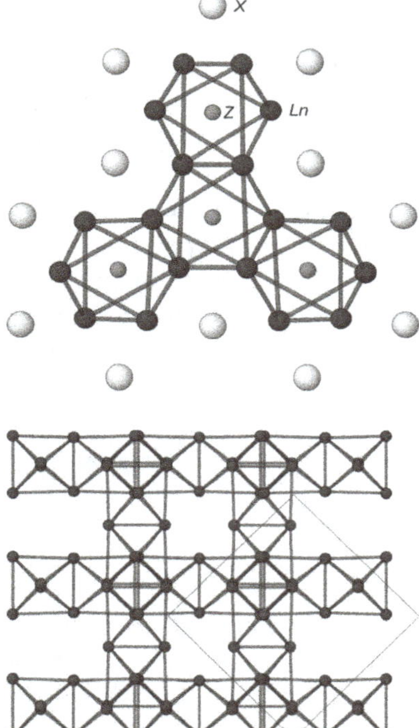

Figure 2.3.21: One layer of the crystal structure of Gd_3Cl_3C projected along [111] of the cubic cell and along [100]. The Cl atoms are omitting for clarity. Reproduced from [3] with permission of Elsevier.

Gold acts like a main group element, which participates in bonding with its *s* and *p* orbitals above the closed d^{10} shell, as frequently found in other heavy transition metals. The platinum metals behave quite similarly in compounds Pr_3Br_3Z with $Z = Rh$, Ir and Pt in their cubic structures [42]. When the single atom interstitial is substituted by a (C_2) unit, for example, in $La_3Br_3(C_2)$, the interconnection pattern remains the same as in the cubic phases, only the elongation of the octahedron by the (C_2) unit forces a change to orthorhombic symmetry. Hence, with the substitution of C^{4-} by (C_2^{6-}), one switches from a metal–metal bonded condensed cluster phase to a saturated valence compound [43].

2.3.6 A paracrystalline system

Great effort has gone into the characterization of the golden colored cubic compounds Ln_2C_3 that form with all lanthanides except Eu, and particular interest arose from the observation of superconductivity in phases of the diamagnetic lanthanides. The structures contain C_2 units, which are coordinated by tetragonal bisphenoids (Figure 2.3.22) of *Ln* atoms [44]. The puzzling feature of the orientation of the displacement ellipsoids in the carbon atoms, which is found in single crystal X-ray as well as neutron powder diffraction work, seems to be typical for all representatives of this structure type. The C–C distances must be taken with caution, possibly being too short due to an artefact. A description as "ethenide ions" therefore seems feasible, the metallic conduction resulting from back bonding effects from occupied C_2 π^* states. Therefore, the Ln_2C_3 phases have to be placed into the category of normal valence compounds, according to $Ln^{3+}{}_4(C_2{}^{4-})_3$.

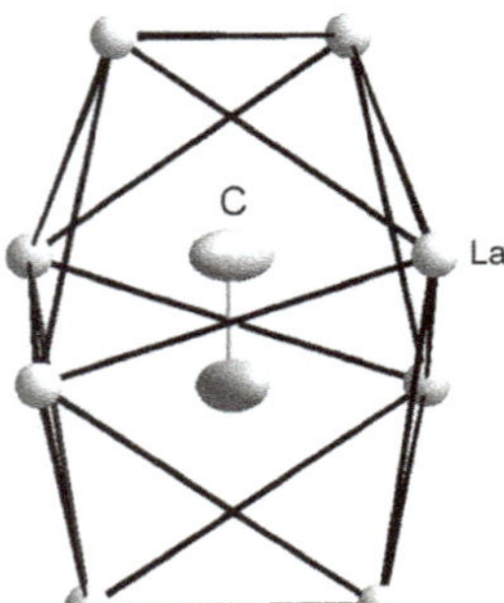

Figure 2.3.22: The bisdishenoidal lanthanum coordination of the (C_2) units in La_2C_3. Reproduced from [3] with permission of Elsevier.

At a surprisingly low temperature of approximately 170 °C, La_2C_3 begins to react with hydrogen, under ambient pressure [45]. The product becomes amorphous to X-rays and at temperatures up to 300 °C, it reaches its final composition $La_2C_{3-x}H_{1.5+x}$, keeping its X-ray amorphous state. Both the composition of the final product and its formation at such a low temperature show a remarkable topochemical reaction. By heating in vacuum, crystalline La_2C_{3-x} is recovered. From inelastic neutron scattering experiments, the hydrogen can be located as H^- in tetrahedral voids. Starting from the stoichiometric composition La_2C_3, all tetrahedral voids are filled in $La_4(C_2)_3H_3$. In order to form three H^- ions per formula unit, the electrons need to be recovered from the C_2 units, leading to a statistical distribution of (C_2^{2-}) and (C_2^{4-}) ions in a 2:1 ratio in $La_4(C_2)_3H_3$. The carbide hydride then represents a paracrystalline system [46] related to the glassy state. Whereas in a normal glass, units of constant shape are connected with no periodic order, in $La_4(C_2)_3H_3$, the periodic order is preserved, however, with building units of statistically changing shape.

References

[1] J. D. Martin, J. D. Corbett, Angew. Chem. Int. Ed. Engl. 1995, 34, 233–235.
[2] M. Ryazanov, L. Kienle, A. Simon, H. Mattausch, Inorg. Chem. 2006, 45, 2068–2074.

[3] A. Simon, H. Mattausch, Compr. Inorg. Chem. II: From Elem. Appl. 2013, 2, pp. 535–581.
[4] H. Mattausch, W. Schramm, R. Eger, A. Simon, Z. Anorg. Allg. Chem. 1985, 530, 43–59.
[5] A. Simon, H. Mattausch, R. Eger, Z. Anorg. Allg. Chem. 1987, 550, 50–56.
[6] O. Jepsen, O. K. Andersen, Z. Phys. 1995, B97, 35–47.
[7] C. Felser, K. Ahn, R. K. Kremer, R. Seshadri, A. Simon, J. Solid State Chem. 1999, 147, 19–25.
[8] E. Warkentin, H. Bärnighausen, Z. Anorg. Allg. Chem. 1979, 459, 187–200.

[9] H. Schäfer, H. G. von Schnering, Angew. Chem. 1964, 76, 833–849.
[10] O. Oeckler, T. Weber, L. Kienle, H. Mattausch, A. Simon, Angew. Chem. Int. Ed. Engl. 2005, 44, 3917–3921.
[11] T. Weber, A. Simon, H. Mattausch, L. Kienle, O. Oeckler, Acta Crystallogr. 2008, A64, 641–653.
[12] T. Hughbanks, J. D. Corbett, Inorg. Chem. 1988, 27, 2022–2026.
[13] C. Zheng, H. Mattausch, C. Hoch, A. Simon, Inorg. Chem. 2008, 47, 2356–2361.
[14] H. Mattausch, C. Hoch, A. Simon, Z. Anorg. Allg. Chem. 2005, 631, 1423–1429.
[15] E. Warkentin, R. Masse, A. Simon, Z. Anorg. Allg. Chem. 1982, 491, 323–336.
[16] H. Mattausch, E. Warkentin, O. Oeckler, A. Simon, Z. Anorg. Allg. Chem. 2000, 626, 2117–2124.
[17] H. Mattausch, A. Simon, L. Kienle, C. Hoch, C. Zheng, R. K. Kremer, Z. Anorg. Allg. Chem. 2006, 632, 1661–1670.
[18] J. E. Mee, J. D. Corbett, Inorg. Chem. 1965, 4, 88–93.
[19] J. D. Corbett, Rev. Chem. Miner. 1973, t10, 233–257.
[20] D. A. Lokken, J. D. Corbett, Inorg. Chem. 1973, 12, 556–559.
[21] A. Simon, N. Holzer, H. Mattausch, Z. Anorg. Allg. Chem. 1979, 456, 207–216.
[22] N. B. Mikheev, A. N. Kamenskaya, I. A. Rumer, V. L. Novitschenko, A. Simon, H. Mattausch, Z. Naturforsch. 1992, 47b, 992–994.
[23] A. Simon, H. Mattausch, G. J. Miller, W. Bauhofer, R. K. Kremer, Handbook on the Physics and Chemistry of Rare Earths, Vol. 15, K. A. Gschneidner Jr., L. Eyring (eds); North Holland, Amsterdam-London-New York-Tokyo, 1991, 191–285.
[24] U. Schwanitz-Schüller, A. Simon, Z. Naturforsch. 1985, 40b, 705–709.

[25] D. Nagaki, A. Simon, H. Borrmann, J. Less-Common Met. 1989, 156, 193–205.
[26] H. Mattausch, O. Oeckler, A. Simon, Z. Kristallogr. NCS 2003, 218, 282.
[27] H. Mattausch, M. C. Schaloske, C. Hoch, A. Simon, Z. Anorg. Allg. Chem. 2008, 634, 498–502.
[28] C. C. Torardi, R. E. McCarley, J. Am. Chem. Soc. 1979, 101, 3963–3964.
[29] H. Ließ, H.-J. Meyer, G. Meyer, Z. Anorg. Allg. Chem. 1996, 622, 494–500.
[30] H. Mattausch, H. Borrmann, R. Eger, R. K Kremer, A. Simon, Z. Anorg. Allg. Chem. 1994, 620, 1889–1897.
[31] M. Ryazanov, H. Mattausch, A. Simon, J. Solid State Chem. 2007, 180, 1372–1380.
[32] H. Mattausch, R. K. Kremer, A. Simon, W. Bauhofer, Z. Anorg. Allg. Chem. 1993, 619, 741–747.
[33] R. Llusar, J. D. Corbett, Inorg. Chem. 1994, 33, 849–853.
[34] J. K. Burdett, G. J. Miller, J. Am. Chem. Soc. 1987, 109, 4092–4104.
[35] A. Simon, J. Solid State Chem. 1985, 57, 2–16.
[36] C. Bauhofer, H. Mattausch, G. J. Miller, W. Bauhofer, R. K. Kremer, A. Simon, J. Less-Common Met. 1990, 167, 65–79.
[37] A. Simon, C. Schwarz, W. Bauhofer, J. Less-Common Met. 1988, 137, 343–351.
[38] C. Bauhofer, H. Mattausch, R. K. Kremer, A. Simon, Z. Anorg. Allg. Chem. 1995, 621, 1501–1507.
[39] H. Mattausch, A. Simon, Z. Anorg. Allg. Chem. 2011, 637, 1093–1100.
[40] E. Warkentin, A. Simon, Rev. Chem. Miner. 1983, 20, 488–495.
[41] H. Mattausch, C. Zheng, L. Kienle, A. Simon, Z. Anorg. Allg. Chem. 2004, 630, 2367–2372.
[42] P. K. Dorhout, M. W. Payne, J. D. Corbett, Inorg. Chem. 1991, 30, 4960–4962.
[43] C. Zheng, O. Oeckler, H. Mattausch, A. Simon, Z. Anorg. Allg. Chem. 2001, 627, 2151–2162.
[44] G. Auffermann, A. Simon, T. Gulden, G. J. Kearley, A. Ivanov, Z. Anorg. Allg. Chem. 2001, 627, 307–311.
[45] A. Simon, T. Gulden, Z. Anorg. Allg. Chem. 2004, 630, 2191–2198
[46] R. Hosemann, F.J. Balta-Calleja, Ber. Bunsenges. Phys. Chem. 1980, 84, 91–95.

Oliver Janka, Rainer Pöttgen

2.4 Intermetallic compounds

The rare-earth (*RE*) elements are typical ignoble metals. This is well underpinned by their low standard electrode potentials, low electronegativity (*EN*) as well as low electron affinity (*EA*) [1]. The structures of *RE* metals have been discussed in Chapter 1.7. They crystallize with the classical metal structures *fcc*, *hcp* and *bcc* or with close packed stacking variants. *Bcc* europium already shows its individualism with a clear preference for a divalent ground state, that is, a half-filled 4*f* shell (we draw back to this peculiar behavior later).

The little electronegative character defines the reactivity and compound formation of the *RE* elements. They form no binary compounds with the likewise low electronegative alkali and alkaline earth metals, Li, Na, K, Rb, Cs, Ca, Sr and Ba. However, they react with beryllium and magnesium. In the case of intermetallic compounds, beryllium and magnesium often do not play the role of a typical alkaline earth element [2, 3]. Both have sufficient *EN* and *EA*, allowing electron density acceptance and forming negatively charged substructures in binary compounds or covalent bonds in ternaries [3]. Typical binaries are the cage compounds $RE\mathrm{Be}_{13}$ with NaZn_{13}-type structure or the CsCl-type series *RE*Mg [4].

The present chapter gives a general overview of the characteristics of *RE* intermetallics. The basic crystallographic data of the many known phases is summarized in the Pearson database which contains several tens of thousand phases [4]. It is simply impossible to address all details of *RE* intermetallics in such a short chapter. We summarize the basic structural/crystal chemical principles. For further reading, we refer to relevant textbooks [3, 5] and review articles. For the many *RE*-based intermetallics, the *Handbook on the Physics and Chemistry of Rare Earths* [6] is an indispensable source with respect to crystal chemistry and structure-property relations.

2.4.1 Synthesis conditions

We start our discussion with a short overview of the synthesis conditions. Since the *RE* elements are all highly oxophilic, synthesis of intermetallics requires inert conditions, that is, argon as a protective atmosphere or vacuum. Mostly, one deals with typical high-temperature reactions since at least one of the reaction partners needs to be melted. The widely used techniques are arc-melting and induction melting, also on an industrial scale. In the case of reaction partners with comparatively high-vapor pressures (this is especially important for samarium, europium and ytterbium), reactions need to be conducted in sealed high-melting metal ampoules (mostly niobium, tantalum or molybdenum) to avoid partial evaporation of one of

https://doi.org/10.1515/9783110654929-012

the reaction partners. Typical crucible materials for the syntheses are ceramics (Al_2O_3, CeO_2, ZrO_2, MgO), metals (Nb, Ta, Mo, W), *h*-BN, graphite or glassy carbon. Single crystals of *RE* intermetallics can be grown on a centimeter scale with the Czochralski or Bridgman technique. Smaller crystals (µm to mm scale) can be obtained from metal or salt fluxes. All these techniques are summarized in a recent textbook [3].

Before starting with a discussion on representative compounds, we need to focus on the terms alloying/alloy versus intermetallic compound. Unfortunately, the terms alloying/alloy are often used in literature in a wrong context. Alloys (e.g., $Cu_{1-x}Au_x$ with $x = 0–1$) and solid solutions (e.g., $Ce(Rh_{1-x}Ru_x)Sn$ with $x = 0–1$) always show disorder (at least partial), a consequence of substitution on a given crystallographic Wyckoff position. Often, broader homogeneity ranges are observed. An intermetallic compound, on the other hand, has a defined composition. For a broader discussion of this topic, we refer to [3].

2.4.2 Reactivity and compound formation

When looking at the periodic table, one realizes immediately that ~80% of the elements are metals, which explains the huge diversity found in intermetallics. However, not all combinations of metals form compounds. Like with all chemical reactions, a certain energy gain is needed in order to form a new compound. In the case of solids, this energy is usually the lattice energy. In ionic solids, usually the Coulomb interactions increase due to changes in coordination number and interatomic distances. In intermetallics too, similar interactions are observed. In binary compounds, the electrostatic interactions between the different atom types along with a certain electron transfer stabilize these materials, while in ternary compounds, often covalent heteroatomic bonds yield an additional stabilizing effect. A common concept to explain the electron transfer in compounds and bonding characteristics is *EN*. While this concept works for ionic and covalent compounds, the *EN* differences in (inter)metallic compounds are sometimes quite small. A concept that plays along the *EN* is the *EA* which also quantifies the tendency to accept electrons within a given scenario. Since these values are at least in principle directly measureable, the use of the *EA* is, in some case, more intriguing than the comparison of *EN* values. Figure 2.4.1 depicts the electronegativities (blue) and first electron affinities (magenta) of some selected main group and transition elements (*T*) along with the data of the *RE* elements. When looking at the numbers, elements like Hf or Ta exhibit low electronegativities and electron affinities. Therefore, the reactivity with *RE* elements is hampered. Hence, the use of these elements as crucible materials can be explained. On the other hand, noble metals exhibit, in comparison with the *RE* metals, rather large *EN* and *EA* values, explaining the structural diversity (see Figure 2.4.3, phase diagram Er–Pd). For the *RE*

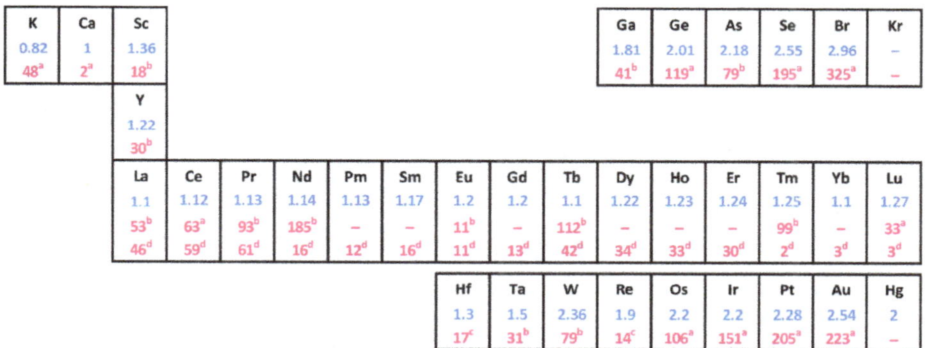

K 0.82 48^{a}
Ca 1 2^{a}
Sc 1.36 18^{b}
Y 1.22 30^{b}
Ga 1.81 41^{b}
Ge 2.01 119^{a}
As 2.18 79^{b}
Se 2.55 195^{a}
Br 2.96 325^{a}
Kr – –
La 1.1 53^{b} 46^{d}
Ce 1.12 63^{a} 59^{d}
Pr 1.13 93^{b} 61^{d}
Nd 1.14 185^{b} 16^{d}
Pm 1.13 – 12^{d}
Sm 1.17 – 16^{d}
Eu 1.2 11^{b} 11^{d}
Gd 1.2 – 13^{d}
Tb 1.1 112^{b} 42^{d}
Dy 1.22 – 34^{d}
Ho 1.23 – 33^{d}
Er 1.24 – 30^{d}
Tm 1.25 99^{b} 2^{d}
Yb 1.1 – 3^{d}
Lu 1.27 33^{a} 3^{d}
Hf 1.3 17^{c}
Ta 1.5 31^{b}
W 2.36 79^{b}
Re 1.9 14^{c}
Os 2.2 106^{a}
Ir 2.2 151^{a}
Pt 2.28 205^{a}
Au 2.54 223^{a}
Hg 2 –

Figure 2.4.1: Electronegativity (*EN*) (blue) and first electron affinity (*EA*) (magenta) data exemplarily shown for some main group and transition elements along with the rare-earth metals. The *EA* data was calculated or measured by different methods (^{a}tunable laser photodetachment threshold; ^{b}laser photodetachment electron spectrometry; ^{c}slow-electron velocity-map imaging spectrometry; ^{d}calculations). The numerical values were taken from [8].

elements, the numbers for the *EA* found in the literature are scarce and sometimes differ largely. In Eu, for example, experimental investigations indicated an *EA* of 83 kJ mol^{-1} [7], while more recent studies reported only 11 kJ mol^{-1} [8e]. Therefore, we differentiate experimental from calculated numbers. Regarding the calculated numbers, two studies have been published [8d, 9] which are in good agreement, however, the one used in Figure 2.4.1 contains comprehensive data.

Besides these basic parameters, several other empirical models were applied to predict the phase stability of intermetallics. For a brief summary, we refer to [5c, 5e]. One of the famous models is the one by *Miedema* [10] which was used in the Philips Research Laboratories. This semiempirical approach is *inter alia* based on the *EN* differences and estimates the formation of an alloy/a compound through its enthalpy of formation. The *Miedema* rules are still used in many current modelling studies, designing multinary alloys and intermetallic compounds [11].

2.4.3 Volume/Iandelli plots

Before discussing select examples for binary and ternary intermetallics, we briefly focus on the existence ranges of compound series. With an increasing ordering number, *RE* elements show the well-known lanthanide contraction (Chapter 1.7). The decreasing size of the *RE* atoms is also reflected in the binary or ternary intermetallic compounds. As an example, we have presented the course of the cell volumes (often called *Iandelli* plot) of the *RE*Pd_3 (Cu_3Au type) and *RE*PdZn (ZrNiAl or TiNiSi type) series in Figure 2.4.2.

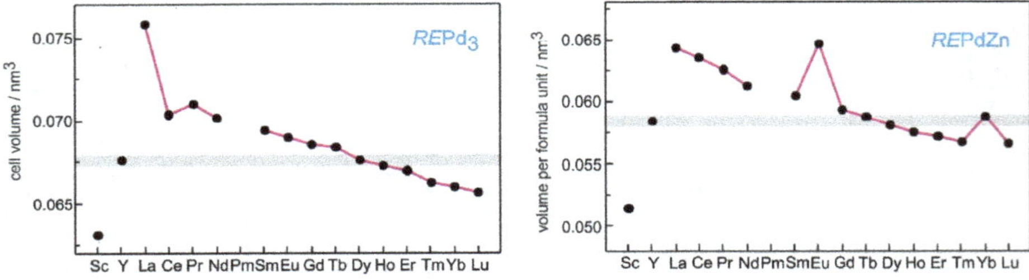

Figure 2.4.2: Plot of the cell volume (volume per formula unit) for the series of *RE*Pd_3 (left) and *RE*PdZn (right) compounds as a function of the rare-earth element (Iandelli plot). The volume ranges for the yttrium compounds are marked by a gray bar. The cell volumes were taken from [4].

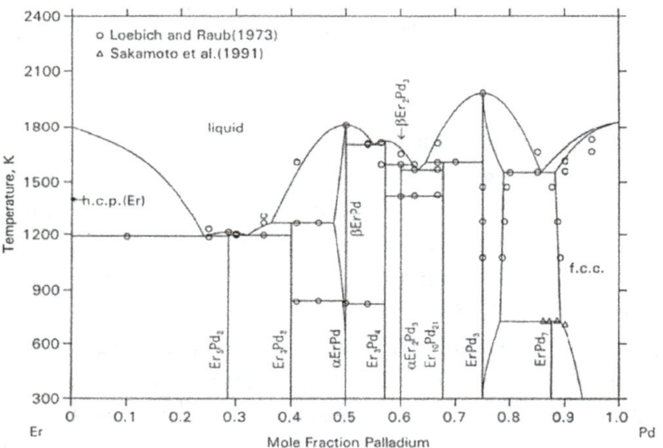

Figure 2.4.3: The Er–Pd phase diagram. Figure reproduced from reference [18] with permission from Elsevier.

Such *Iandelli* plots nicely illustrate many peculiarities of *RE* intermetallics. As it is readily evident for the *RE*Pd_3 series (Figure 2.4.2, left), we observe no strictly linear decrease of the cell volume. This is mostly a clear sign for small homogeneity ranges and/or structural disorder. The cell volumes of the yttrium representatives of both series fit in between the volumes of the gadolinium and terbium ones. This is generally the case but the precise volume of the yttrium compound also depends on the structure type. There even exist series where the yttrium compound is slightly smaller than the terbium one.

Now we turn to the anomalies in both plots. If all *RE* elements are trivalent in a given series of compounds, one observes a smooth curve. In the case of $CePd_3$, we observe a negative anomaly, indicating intermediate-valent cerium (partial formation of tetravalent cerium with smaller radius). Especially in combination with ruthenium, intermediate-valent or almost tetravalent cerium intermetallics have been observed. Typical compounds are CeRuSn or Ce_2RuZn_4. An overview is given in [12].

On the other side, we observe a positive anomaly for EuPdZn and YbPdZn. These two ternaries contain divalent europium and divalent ytterbium. Given the close radii of Ca^{2+} (100 pm)/Yb^{2+} (102 pm) and Sr^{2+} (118 pm)/Eu^{2+} (117 pm) [13], both europium and ytterbium intermetallics are often isotypic with alkaline earth compounds, due to a match of size and valence. Typical pairs of compounds are Ca_2Sn/Yb_2Sn and Sr_2Sn/Eu_2Sn for binaries and CaZnSn/YbZnSn or SrZnSn/EuZnSn for ternaries [4]. $EuPd_3$ in the *RE*Pd_3 series is one of the rare examples of trivalent intermetallic europium compounds [14].

Another issue concerns the role of the lanthanum compounds. In the *RE*PdZn series, LaPdZn perfectly fits into the volume plot while $LaPd_3$ shows a cell volume that is much larger than expected. We need to keep in mind that La^{3+} is a diamagnetic cation with empty 4*f* shell and these positive deviations for the volumes of lanthanum compounds are frequently observed. Strictly speaking, lanthanum does not belong to the *f* elements. This is also the case for yttrium and of course scandium, the by far smallest element in these series of compounds. Although it is much smaller than the smallest *RE* element lutetium, it shows crystal chemical behavior similar to that of the *RE* elements and one observes the formation of $ScPd_3$ and ScPdZn. Nevertheless, scandium plays its own role in the crystal chemistry of *RE* intermetallics [15]. In some cases, scandium is too small to form the same binary or ternary compounds as lutetium does. Then superstructure formation (small structural distortions) is frequently observed for the scandium representative and the resulting compound then has a structure comparable to the lutetium representative. If the superstructure formation is not a suitable solution, scandium forms ternary compounds with their own peculiar structure types. An excellent tool for the structural elucidation of intermetallic scandium compounds is ^{45}Sc solid state NMR spectroscopy (Chapter 3.6), especially as complementary tool to X-ray diffraction [16].

The two series presented in Figure 2.4.2 do not represent the usual case for *RE* intermetallics. Due to the lanthanide contraction, we deal with differently large RE^{3+} ions (La–Lu: 116–98 pm; CN = 8) [13]. The more complex binary and especially ternary structure types usually do not show the large geometric flexibility and the formation of compounds is restricted to a kind of a "size window." Very often, one observes formation of a structure type only for the larger (typically La–Sm) ones or only with the smaller (typically Gd–Lu) ones, also called the early and late *RE*s, or in special cases only for a few neighboring *RE* elements.

2.4.4 Some structural principles – binaries

We now try to summarize some of the structural principles of *RE* intermetallics. The presented compounds are only a small cutout of the enormous number of representatives.

Most of the binary phases are summarized in the phase diagram compilations [17]. As an example, we present the Er–Pd phase diagram [18] in Figure 2.4.3. Such phase diagrams show phase formation as a function of temperature under ambient pressure conditions and they exhibit the typical features: homogeneity ranges, congruently and incongruently melting phases and the thermal stability ranges of compounds. Of the many binary phase diagrams, only those with technologically highly important phases have deeply and reliably been characterized. Yearly, several of the binary diagrams are at least partially revised, since compositions or equilibrium lines have been wrong or crystal structures were not determined in the early days of X-ray diffraction.

Amongst the binary intermetallics consisting of a *RE* element with an element of groups 13 to 15 (for chalcogenides and halides see Chapters 2.1 and 2.2), two possible scenarios can be observed: the formation of the valence-precise Zintl phases and the classical intermetallics.

Zintl phases can be positioned between intermetallic compounds and insulating valence compounds since they combine covalent bonding and ionic interactions. "Classical" Zintl phases contain an electropositive alkali or alkaline-earth metal, which formally reduces a more electronegative element, which can be a triele, tetrele or a pnictogen. Besides the classical binary Zintl phases, numerous ternary compounds also containing *RE* and transition metals have been reported. For the theory of the Zintl concept, we refer to the literature [19]. Eu_5Ge_3 and $NdAs_2$ are two representative examples we want to discuss in detail. As stated before, Zintl phases are valence-precise compounds and therefore, our examples can be separated with respect to their ionic contributions. The Eu atoms are divalent and therefore, 10 e^- have to be distributed over three Ge atoms in the best possible way to achieve electronic situations similar to the ones of neighboring elements. In the present case, a Ge^{4-} anion (noble gas electron configuration) and a Ge_2^{6-} (isoelectronic to Br_2) are formed. In the case of $NdAs_2$ with trivalent neodymium, a four-membered As_4^{6-} chain fragment ($As^{2-}As^{1-}As^{1-}As^{2-}$) can be found which is a combination of chalcogen and halogen chemistry. These Zintl ions and their respective *RE* coordination environments are shown in Figure 2.4.4.

Figure 2.4.5 depicts some common and important structure types found in the field of binary *RE* intermetallics. On the top, representatives of the cubic (left) and hexagonal (right) Laves phases are depicted. At the bottom are examples for the AlB_2 (left) and the $CaCu_5$ (right)-type structure. The Laves phases [3, 22] are AB_2 line compounds with close packed structures. They form with radius ratios around the ideal value of 1.225. The basic structural building units are empty tetrahedral entities formed by the B atoms. Condensation of the B_4 tetrahedra via common corners or faces leads to two basic structure types, cubic $MgCu_2$ and hexagonal $MgZn_2$. In the cubic variant, the tetrahedra are connected via all corners. In the hexagonal structure, they share common corners and faces to form strands. Within the tetrahedral

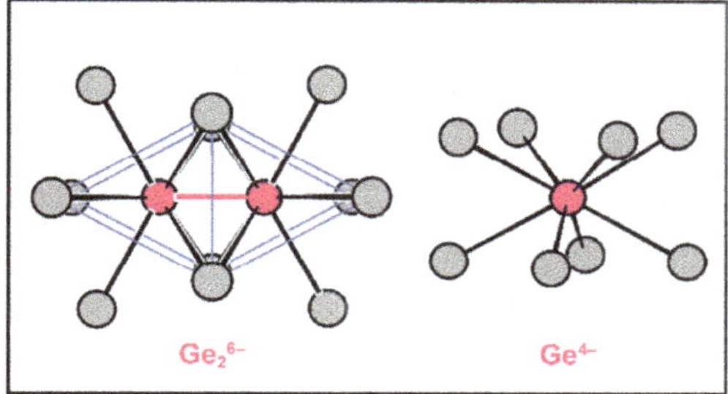

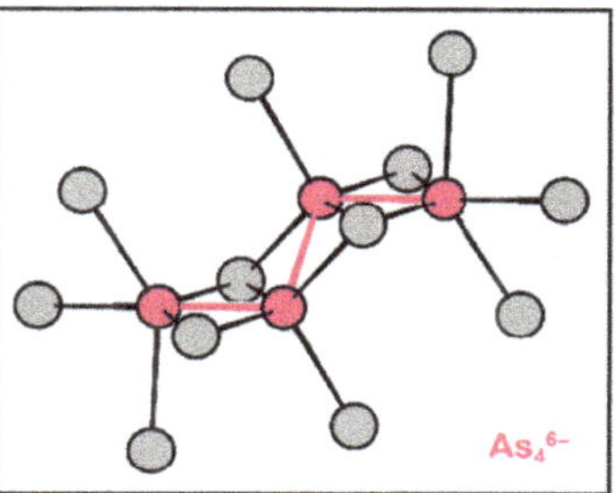

Figure 2.4.4: The Zintl anions of Eu_5Ge_3 [20] and $NdAs_2$ [21]. Rare-earth atoms are shown in gray, and main group atoms in magenta.

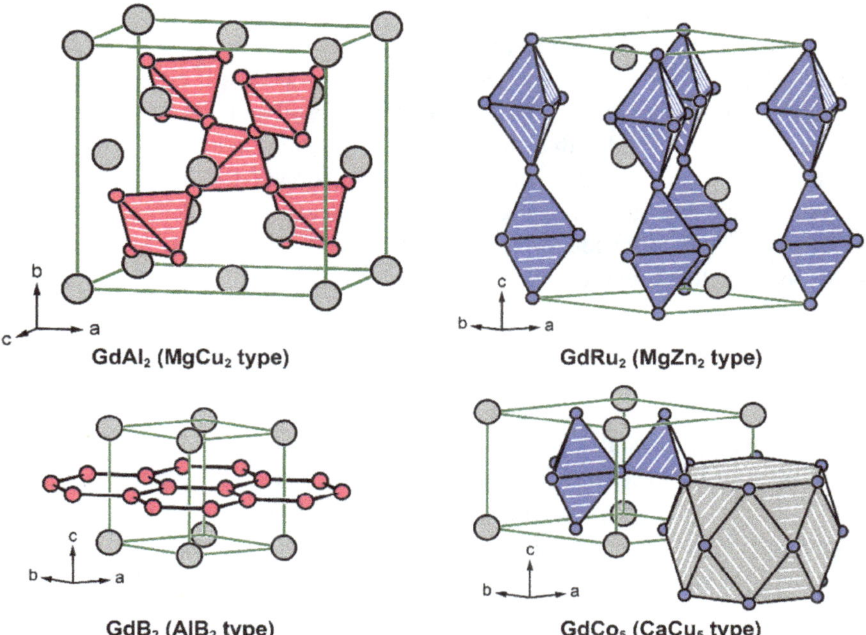

Figure 2.4.5: The crystal structures of $GdAl_2$ (top left, cubic Laves phase, $MgCu_2$ type), $GdRu_2$ (top right, hexagonal Laves phase, $MgZn_2$ type), GdB_2 (bottom left, AlB_2 type) and $GdCo_5$ (bottom right, $CaCu_5$ type). Data for the drawing were taken from the Pearson data base [4]. Rare-earth atoms are shown in grey, transition metals in blue and main group atoms in magenta.

units, rather strong covalent homoatomic interactions dominate, while the heteroatomic (*RE–T* and *RE–X*) distances are usually above the sum of the covalent radii.

In the $CaCu_5$-type structures ($GdCo_5$ is shown as an example), a similar scenario can be observed; however, with a different interconnection of the tetrahedral units.

The large *RE* atoms have the coordination number 18 by the transition metal atoms in the form of a hexacapped hexagonal prism. Another basic AB_2 structure type is aluminum boride, AlB_2. GdB_2 is depicted as an example in Figure 2.4.5. The gadolinium atoms show hexagonal-prismatic boron coordination between the planar honeycomb layers. The latter can be formed by a *p* element (the present case) or by a transition metal, for example, in $LaCu_2$.

Some statistics to underpin the importance of the discussed structures: several thousand binary and ternary *RE*-based intermetallic compounds crystallize with the four structure types presented in Figure 2.4.5 [4]. Such materials are of high importance in fundamental and applied research with respect to their broadly varying magnetic and transport properties.

In contrast to the electron-precise Zintl phases discussed above, we observe large electronic flexibility for the Laves phases and the AlB_2/$CaCu_5$-type variants discussed in the present paragraph. The valence electron count (VEC) of such phases can cover broad ranges. To give some examples, for the AlB_2-type phases, GdB_2 is isotypic with $GdSi_2$ with a higher VEC and for the Laves phases, we observe the complete series $GdMn_2$, $GdFe_2$, $GdCo_2$ and $GdNi_2$. These Laves phases are all metallic and the increase or decrease of the VEC is else than filling or depopulating bands. The same holds true for the $CaCu_5$-type phases where $GdFe_5$, $GdCo_5$, $GdNi_5$, $GdCu_5$ and $GdZn_5$ are known. For the main group-metal networks, we observe isotypic $GdMg_2$ and $GdAl_2$ with a lower electron count for the tetrahedral magnesium network. Many of such series can be extracted from the Pearson database [4].

2.4.5 Some structural principles – ternaries

In the introduction part, we discussed that no binaries form between the *RE* elements and the alkali metals. However, when combing a *RE* metal with an alkali metal and a *p* element, structurally, highly interesting Zintl phases with unusual Zintl anions form. Some representative examples are $La_2Li_2Ge_3 \equiv 2La^{3+}2Li^{+}Ge^{4-}2Ge^{2-}$ [23a] with isolated Ge^{4-} Zintl anions and germanium zig-zag chains (two-bonded Ge^{2-}), Na_8EuSn_6 with cyclopentadienyl-like Sn_5^{6-} anions (besides isolated Sn^{4-}) [23b] or $Ce_5Li_2Sn_7$ with an open heptane-like Sn_7^{16-} chain [23c]. The combination of a larger *RE* cation with Li^+ or Na^+ realizes the perfect cationic matrix for embedding the Zintl anions.

In going from binary to ternary systems, we significantly enlarge the combinatorial diversity. So far, only a small number of the possible combinations has been explored. The same holds true for the systematic phase analytical work in ternary systems. Some statistic overviews are summarized in [5d]. For ternary *RE*-based systems of the type *RE*-*T*-*X* (*X* = *p* element), a simple empirical rule is mostly fulfilled. If the underlying binary phase diagrams are already rich in compounds, one also observes formation of many ternaries and *vice versa*. Usually for ternary systems, one studies the so-called isothermal sections. The samples are equilibrated at a given temperature and quenched. The Ce–Pd–Ge system is presented as an example for such an isothermal section (870 K data) in Figure 2.4.6. This is one of the *rich* phase diagrams with 17 ternary germanides.

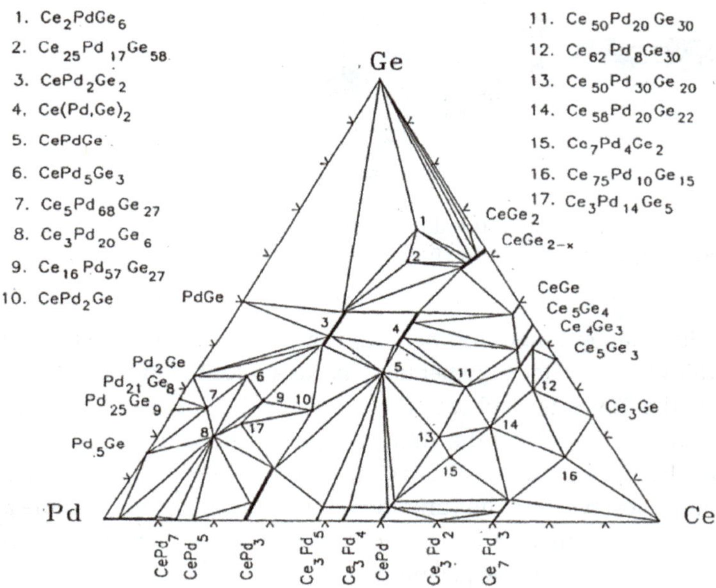

Figure 2.4.6: Isothermal section of the Ce–Pd–Ge diagram at 870 K. Figure reproduced from reference [24] with permission from Elsevier.

Such ternary *RE*-based intermetallics can roughly be divided into two groups: (i) complex structures which crystallize with their own type or (ii) so-called coloring variants that are superstructures or ordering variants of a binary structure. Figure 2.4.7 depicts two examples for such coloring variants. On the left, CeCuCd, a representative of the SrPtSb-type structure is shown. A comparison with the lower left-hand part of Figure 2.4.5 immediately illustrates the structural relationship. The honeycomb layers of the AlB_2-type structure now exhibit alternatively arranged Cu and Cd atoms, forming heteroatomic bonds. The Ce atoms still reside in the cavities of the polyanionic framework. Besides this coloring, further structural distortions (e.g. shifting, tilting and puckering) can occur. This is the case for EuZnSn (Figure 2.4.7, right) which crystallizes in the TiNiSi-type structure. The latter also derives from the aristotype AlB_2 by coloring and puckering. Interestingly, the close shell d^{10}-metals Zn and Cd act like main group elements, as also illustrated earlier in the case of the *RE*PdZn series (Figure 2.4.2). These examples are only two out of many more that can be frequently found in the crystal chemistry of intermetallics. Very often, the coloring and distortion/puckering variants are strictly related by group-subgroup theory. Striking examples are summarized in the literature [27].

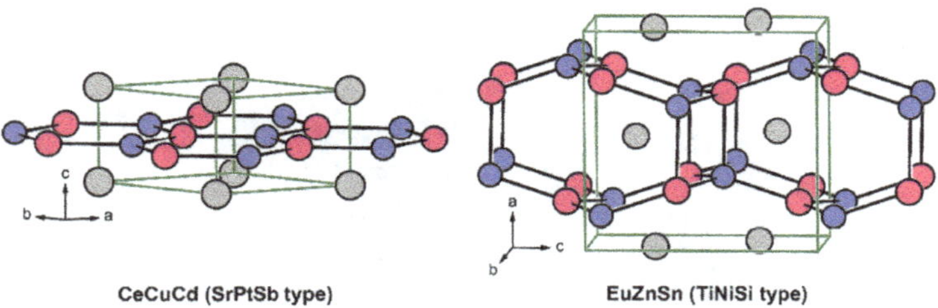

Figure 2.4.7: Colored and puckered variants of the AlB_2-type structure. The crystal structures of (left) CeCuCd (SrPtSb type) [25] and (right) EuZnSn (TiNiSi type) [26].

The electronic flexibility (larger ranges for the VEC) is also observed for the ternary *RE*-based intermetallics for the same arguments as discussed for the binaries. Some typical examples are the TiNiSi-type phases (LaIrAl [VEC = 15], LaPtAl [VEC = 16], LaIrSb [VEC = 17], LaNiSb [VEC = 18]) or the $ThCr_2Si_2$-type representatives ($CeMn_2Ge_2$ [VEC = 25], $CeFe_2P_2$ [VEC = 29], $CeCo_2As_2$ [VEC = 31], $CeAg_2Si_2$ [VEC = 33]) [4]. For a given structure type, we observe both the variation of the transition metal and the *p* element site. In these simple structure types, nature optimizes through changes in the lattice parameters and/or the atomic positions.

2.4.6 Complex intermetallics

Finally, we turn to intermetallic *RE* compounds with very complex crystal structures. It is difficult to make a clear cut between simple and difficult since we deal with the full range, from simple CsCl-type phases such as CeCd with just two atoms and structures with several thousand atoms per unit cell. The fingerprint for a structurally complex phase can be manifold: (i) a large unit cell (long translation period), (ii) defect formation or mixed site occupancies, (iii) breaking of symmetry, (iv) structural hierarchy, (v) complex ordering of structural motifs or (vi) modulations.

Such structurally complex phases are frequently called CMAs (complex metallic alloys) in the current literature [28]. This is a misleading term, since most of the discussed phases indeed exhibit large atomic ordering and (as discussed above) belong to inorganic compounds. The term CIM (complex intermetallics) is more adequate [5d].

As an example, for such phases, we present the cubic structures of YCd_6 (168 atoms per unit cell, $a = 1548.2$ pm) and $Dy_{117}Co_{57}Sn_{112}$ (1144 atoms per unit cell,

$a = 2983.1$ pm) in Figure 2.4.8. If one simply looks at the filled unit cell, it looks like a large box with many different spheres. A closer look at the individual atom coordination leads to a simpler picture. Such structures can often be described by simpler packing motifs of a small number of different polyhedra, for example, empty Cd_8 cubes and seven-fold capped pentagonal $Y@Cd_{17}$ prisms in the YCd_6 structure. Filled $Co@Sn_8$ cubes are one of the building units in the $Dy_{117}Co_{57}Sn_{112}$ structure besides icosahedra, bicapped cubes, Frank-Kasper polyhedra or again the seven-fold capped pentagonal prisms. Sometimes, their structures can also be described by clusters of a "Russian Doll"/onion type. A small inner cluster is surrounded by different "layers" which form a regular polyhedron. Depending on the shape of the respective polyhedron, one can distinguish between Bergman, Mackay and Tsai clusters. Often, these structures can be regarded as the so called quasicrystal approximants (QCA) [5c, d].

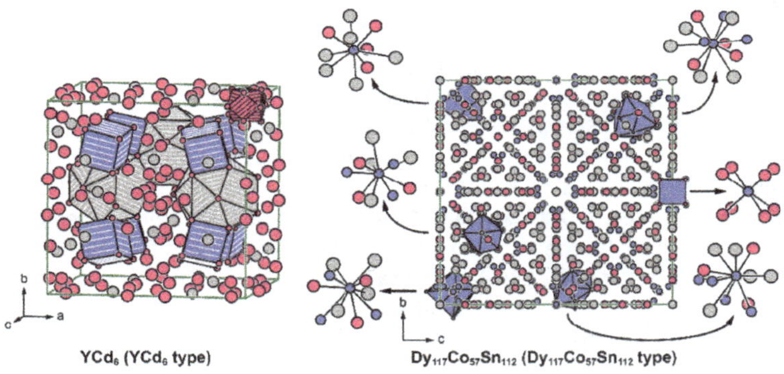

Figure 2.4.8: The crystal structures of two complex intermetallic compounds. (left) Binary YCd_6 (YCd_6 type [29]) and (right) ternary $Dy_{117}Co_{57}Sn_{112}$ ($Dy_{117}Co_{57}Sn_{112}$ type [30]).

The structurally complex intermetallic *RE* phases and QCAs build the bridge to the quasicrystals (QCs); metastable and stable ones, icosahedral (*i*-QCs), octagonal, decagonal (*d*-QCs) and dodecagonal quasicrystals [5d, 31, 32]. This class of materials has been intensively studied in recent years with respect to their demanding crystallography (e.g., icosahedral diffraction symmetry) as well as some promising properties (surface coatings, optical and magnetic properties, catalysis). Typical compositions in the field of *RE*-based quasicrystals are *i*-$YbCd_{5.7}$ or *i*-$Yb_{16}Al_{42}In_{42}$. Current trends in quasicrystalline research and crystallography are reviewed in [5d, 32].

2.4.7 Applications and function

RE-based intermetallics are not only studied with respect to their structural behavior but most intensively with respect to their *RE*-centered functionalities. Typical examples are (i) $LaNi_5$ and its substituted variants as hydrogen storage materials (the so-called metal-hydride batteries), (ii) $SmCo_5$, Sm_2Co_7 or $Nd_2Fe_{14}B$ as permanent magnets, (iii) $Gd_5Ge_2Si_2$ as magnetocaloric materials (see Chapter 4.8), (iv) *RE*Pd_3 phases for catalytic hydrogenation, (v) *RE* metal additions for precipitation strengthening of light weight alloys on the basis of Al, Mg or Zn, or (vi) *RE*-based skutterudites as thermoelectrics.

References

[1] J. Emsley, The Elements, Oxford University Press, Oxford, 1999.
[2] U. Ch. Rodewald, B. Chevalier, R. Pöttgen, J. Solid State Chem. 2007, 180, 1720.
[3] R. Pöttgen, D. Johrendt, Intermetallics, 2nd Ed., De Gruyter, Berlin, 2019.
[4] P. Villars, K. Cenzual, Pearson's Crystal Data: Crystal Structure Database for Inorganic Compounds (release 2019/20), ASM International®, Materials Park, Ohio (USA) 2019.
[5] a) E. Parthé, L. Gelato, B. Chabot, M. Penzo, K. Cenzual and R. Gladyshevskii, TYPIX–Standardized Data and Crystal Chemical Characterization of Inorganic Structure Types, Gmelin Handbook of Inorganic and Organometallic Chemistry, 8th edition, Springer, Berlin (Germany), 1993; b) A. Szytuła, J. Leciejewicz, Handbook of Crystal Structures and Magnetic Properties of Rare Earth Intermetallics, CRC Press, Boca Raton, 1994; c) R. Ferro, A. Saccone, Intermetallic Chemistry, Elsevier, Amsterdam, 2008; d) W. Steurer, J. Dshemuchadse, Intermetallics: Structures, Properties, and Statistics, IUCr Monographs on Crystallography, Volume 26, Oxford University Press, New York, 2016. ISBN-10: 0198714556; e) G. S. Rohrer, Structure and Bonding in Crystalline Materials, Cambridge University Press, Cambridge, UK, 2001.
[6] J.-C. G. Bünzli, V. K. Pecharsky (Eds.), Handbook on the Physics and Chemistry of Rare Earths, Volumes 1–54, North-Holland, Elsevier, Amsterdam, 1978–2020.
[7] V. T. Davis, J. S. Thompson, J. Phys. B: At. Mol. Opt. Phys. 2004, 37, 1961.
[8] a) H. Hotop, W. C. Lineberger, J. Phys. Chem. Ref. Data 1985, 14, 731; b) T. Andersen, H. K. Haugen, H. Hotop, J. Phys. Chem. Ref. Data 1999, 28, 1511; c) J. Rumble, CRC Handbook of Chemistry and Physics, 99th Edition, CRC Press, Boca Raton, 2018; d) S.-B. Cheng, A. W. Castleman Jr, Sci. Rep. 2015, 5, 12414; e) Z. Felfli, A. Z. Msezane, D. Sokolovski, Phys. Rev. A 2009, 79, 012714.
[9] S. M. O'Malley, D. R. Beck, Phys. Rev. A 2008, 78, 012510.
[10] a) A. R. Miedema, Philips Techn. Rev. 1973, 33, 149; b) A. R. Miedema, Philips Techn. Rev. 1973, 33, 196.
[11] R. F. Zhang, X. F. Kong, H. T. Wang, S. H. Zhang, D. Legut, S. H. Sheng, S. Srinivasan, K. Rajan, T. C. Germann, Sci. Rep. 2017, 7, 9577. DOI:10.1038/s41598-017-09704-1.
[12] a) W. Hermes, S. F. Matar, R. Pöttgen, Z. Naturforsch. 2009, 64b, 901; b) T. Mishra, R.-D. Hoffmann, C. Schwickert, R. Pöttgen, Z. Naturforsch. 2011, 66b, 771.
[13] R. D. Shannon, Acta Crystallogr. A 1976, 32, 751.
[14] F. Stegemann, T. Block, S. Klenner, O. Janka, Chem. Eur. J. 2019, 25, 3505.

[15] a) K. A. Gschneidner Jr., in: Scandium. Its Occurrence, Chemistry, Physics, Metallurgy, Biology and Technology, Ed. C. T. Horowitz, Academic Press, London, pp. 152–322, 1975; b) B. Ya. Kotur, Croatica Chem. Acta 1998, 71, 635; c) B. Ya. Kotur, E. Gratz, Scandium Alloy Systems and Intermetallics, in: K. A. Gschneidner Jr., L. Eyring (eds.), Handbook on the Physics and Chemistry of Rare Earths, Elsevier, Amsterdam, Volume 27, Chapter 175 (pp. 339–533) 1999.
[16] H. Eckert, R. Pöttgen, Z. Anorg. Allg. Chem. 2010, 636, 2232.
[17] a) W. G. Moffat (Ed.), The Handbook of Binary Phase Diagrams, Genium Publishing Corporation, New York, 1984; b) T. B. Massalski, Binary Alloy Phase Diagrams, Vols. 1 and 2, American Society for Metals, Ohio, 1986; c) G. Petzow, G. Effenberg (Eds.), Ternary Alloys – A Comprehensive Compendium of Evaluated Constitutional Data and Phase Diagrams, VCH, 1988.
[18] Z. Du, H. Yang, J. Alloys Compd. 2000, 299, 199.
[19] a) H. Schäfer, B. Eisenmann, W. Müller, Angew. Chem. 1973, 85, 742; b) S. M. Kauzlarich, Chemistry, Structure, and Bonding of Zintl Phases and Ions, VCH Publishers, 1996; c) T. F. Fässler, Zintl Phases: Principles and Recent Developments, Springer-Verlag, Berlin, 2011; d) O. Janka, S. M. Kauzlarich, Zintl Compounds, in: Encyclopedia of Inorganic and Bioinorganic Chemistry, John Wiley & Sons, Ltd, 2013.
[20] R. Pöttgen, A. Simon, Z. Anorg. Allg. Chem. 1996, 622, 779.
[21] Y. Wang, R. D. Heyding, E. J. Gabe, L. D. Calvert, J. B. Taylor, Acta Crystallogr. B 1978, 34, 1959.
[22] K. A. Gschneidner, Jr., V. K. Pecharsky, Z. Kristallogr. 2006, 221, 375.
[23] a) S.-P. Guo, T.-S. You, S. Bobev, Inorg. Chem. 2012, 51, 3119; b) I. Todorov, S. C. Sevov, Inorg. Chem. 2004, 43, 6490; c) I. Todorov, S. C. Sevov, Inorg. Chem. 2007, 46, 4044.
[24] Yu. D. Seropegin, A. V. Gribanov, O. I. Bodak, J. Alloys Compd. 1998, 269, 157.
[25] A. I. Horechyy, V. V. Pavlyuk, O. I. Bodak, Polish J. Chem. 1999, 73, 1681.
[26] R. Pöttgen, Z. Kristallogr. 1996, 211, 884.
[27] a) U. Müller, Z. Anorg. Allg. Chem. 2004, 630, 1519; b) U. Müller, Symmetry Relationships between Crystal Structures, Oxford University Press, 2013; c) R. Pöttgen, Z. Anorg. Allg. Chem. 2014, 640, 869.
[28] J.-M. Dubois, E. Belin-Ferré, Complex Metallic Alloys, Wiley-VCH, Weinheim, 2011.
[29] A. C. Larson, D. T. Cromer, Acta Crystallogr. 1971, 27, 1875.
[30] P. Salamakha, O. Sologub, G. Bocelli, S. Otani, T. Takabatake, J. Alloys Compd. 2001, 314, 177.
[31] a) W. Steurer, Chem. Soc. Rev. 2012, 41, 6719; b) J. Dolinšek, Chem. Soc. Rev. 2012, 41, 6730; c) J.-M. Dubois, Chem. Soc. Rev. 2012, 41, 6760.
[32] W. Steurer, Acta Crystallogr. 2018, A74, 1.

Bassem S. Bassil, Ulrich Kortz

2.5 Polyoxometalates

Polyoxometalates (POMs), or polyoxoanions, are discrete polynuclear, anionic oxo-bridged clusters of early-transition metal ions in high oxidation states, and they exhibit a unique structural and compositional variety. Although the first POM was prepared by *Berzelius* already in 1826 [1], it was not until about 150 years later that the structural chemistry of polyoxomolybdates and -tungstates was studied and reported systematically. The use of powder X-ray diffraction in 1933 allowed *Keggin* to determine the first POM structure as $H_3[(PO_4)W_{12}O_{36}] \cdot 6H_2O$, but subsequent progress on the structural chemistry of POMs during the following decades was slow, and this only changed after eventually single-crystal X-ray diffraction became a routine tool [2]. POMs are usually built up of edge- and/or corner-shared $\{MO_6\}$ octahedra, and these metal atoms M, also known as addenda, are usually of groups V or VI in high oxidation states, such as V^V for polyoxovanadates, Mo^{VI} for polyoxomolybdates and W^{VI} for polyoxotungstates [3]. Oxygen atoms are the common ligands in POMs, hence the terminology "polyoxo", and are usually present as bridging and terminal oxo ligands. In some cases, protonation of oxygen atoms, in particular the bridging ones, can also be observed. Terminal oxo ligands in closed-shell (plenary) POMs are coordinated to the addenda atoms M with a significant degree of π-bonding (leading to short bonds), hence rendering such POMs rather inert and not prone to further condensation reactions.

The general area of POM synthesis and characterization has seen a fast growth over the years, and every year, a good number of novel polyanion structures are being reported. However, it must also be mentioned that many of these are solid-state structures, with little to no solution investigation. Ideally, a POM represents a discrete molecular metal-oxide with an independent existence in solution, rather than being just a structural fragment in the solid state. Currently, the number of the latter structures dominates the former. Only if the POM is discrete, soluble and solution stable, a systematic study of properties and reactivity in solution is possible. The physiochemical properties and structural robustness of POMs have rendered them attractive candidates in different areas, ranging from analytical chemistry to magnetic devices, nanoscience, industrial catalysis and even medicine. Numerous reviews and special issues on POM chemistry have been published over the years, reflecting the fast growth of this interesting class of inorganic compounds that merges the classical areas of metal oxides and coordination complexes, all the way to nanochemistry and supramolecular chemistry [4].

The formation of POMs from simple, tetrahedral oxometalate $\{MO_4\}$ ions (e.g., tungstate, WO_4^{2-}) usually takes place in aqueous, acidic medium, and is driven by the formation of $\{MO_6\}$ octahedra followed by dimerization and further condensation to the kinetic and eventually thermodynamic products in equilibria (Figure 2.5.1),

https://doi.org/10.1515/9783110654929-013

which may differ significantly (e.g., polytungstates usually equilibrate much more slowly than molybdates). A systematic variation of synthetic parameters such as pH, temperature, ionic strength, concentration and ratio of reagents allows for the preparation of a large variety of polyanions. On the other hand, POMs are usually subject to decomposition in basic media, although the pH required for complete degradation may, at times, be rather high (in particular for niobates and tantalates). POMs can be classified into two main categories depending on their constituents: isopolyanions and heteropolyanions. *Isopolyanions* are composed exclusively of the same kind of addenda atoms, whereas *heteropolyanions* incorporate one or more *heteroatoms*, metallic or nonmetallic. As many elements in the periodic table can function as heteroatoms, the number of known heteropolyanions is vast and continually increasing.

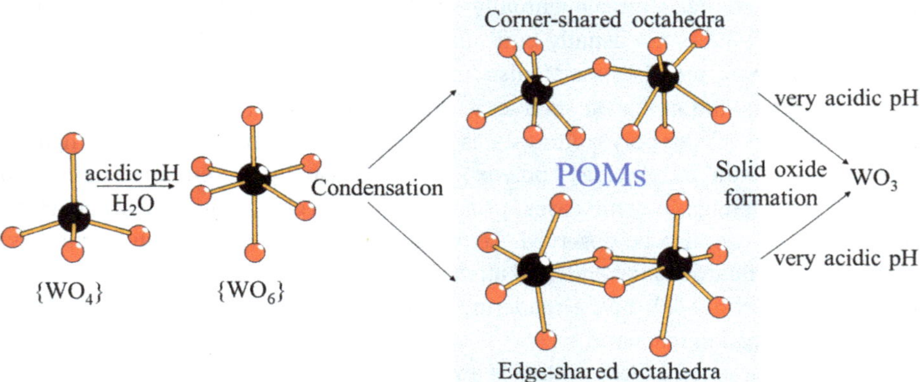

Figure 2.5.1: General scheme for POM (e.g. polyoxotungstate) formation *via* condensation of simple oxoanions in aqueous acidic media. Color code: addenda atoms (black) and oxygen atoms (red).

2.5.1 Isopolyanions

In particular, V^V, Nb^V, Ta^V, Mo^{VI} and W^{VI} (Cr^{VI} is too small to form octahedral $\{MO_6\}$ units) can form isopolyanions, due to an expansion of the coordination number of the starting oxometalate ion from 4 to 5, 6 and occasionally 7. The overall condensation reaction of mononuclear, tetrahedral oxoanions to polyoxoanions can be represented as follows:

$$aMO_4^{n-} + 2bH^+ \leftrightarrow \left[M_aO_{(4a-b)}\right]^{m-} + bH_2O$$

This equilibrium is strongly dependent on the reaction conditions such as pH, concentration, ionic strength, countercation type and temperature. Of the group 5 elements, polyoxovanadates, in particular, form a significant range of structures. In alkaline medium, the species present are mainly based on tetrahedral coordination of vanadium, whereas in acidic medium (pH ~ 3–6) the orange decavanadate $[H_xV_{10}O_{28}]^{(6-x)-}$ [5] with six-coordinate vanadium(V) dominates. On the other hand, the aqueous solution chemistry of polyoxoniobates and -tantalates is mainly based on the hexanuclear Lindqvist-type anions, $[M_6O_{19}]^{n-}$, which is also known for Mo^{VI} and W^{VI} (but only in organic medium) [6].

In group 6, Mo^{VI} and W^{VI} form several isopolyanions, most of which incorporate six-coordinate metal centers. In aqueous solution, the initial stable polyanions formed in moderately acidic medium are the heptametalates $[M_7O_{24}]^{6-}$ [7]. Further condensation in more acidic media can lead to large molybdates such as $\{Mo_{36}\}$ which also contains hepta-coordinated Mo centers [8]. Under such conditions, tungstates convert eventually to different species such as metatungstate $[H_2W_{12}O_{40}]^{6-}$ and paratungstate $[H_2W_{12}O_{42}]^{10-}$ [6]. Figure 2.5.2 shows a few representative isopolyanion structures.

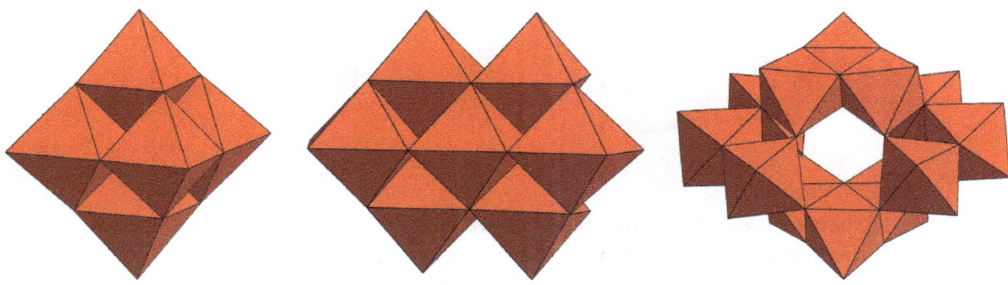

Figure 2.5.2: Polyhedral representation of the Lindqvist-type anion $[M_6O_{19}]^{n-}$ (left), heptametalate ion $[M_7O_{24}]^{6-}$ (center) and paratungstate ion $[H_2W_{12}O_{42}]^{10-}$. The red octahedra represent $\{MO_6\}$ units.

2.5.2 Lanthanide-containing isopolyanions

Not all reported isopolyanions containing lanthanide ions will be discussed herein, but only some prominent structures. Lanthanide-containing isopolyanions can be classified mainly into two categories: (i) sandwich-type and (ii) lanthanide-stabilized structures.

Sandwich-type isopolyanion structures correspond to a central, octacoordinated lanthanide ion (usually with square-antiprismatic geometry), which is coordinated by two lacunary POM units, each via four oxo ligands, resulting in coordination number eight for the lanthanide ion. Lacunary POMs are defect plenary structures that have lost one or more tungsten addenda. A large structural family is based on a lanthanide ion sandwiched by two monolacunary Lindquist-type units $[M_5O_{18}]^{6-}$, resulting in the general formula $[Ln(M_5O_{18})_2]^{n-}$, which is known for M = tungsten and molybdenum and Ln = lanthanide ions. Such sandwich-type structure has D_{4d}

symmetry and is also known as the *Weakley–Peacock* structure type [9]. Several lacunary POMs are also known for heteropolyanions, and these will be discussed in the following section. As shown in Figure 2.5.3, lacunary POMs are more nucleophilic than their plenary analogues, and hence can act as inorganic ligands for coordination to electrophiles such as lanthanide ions.

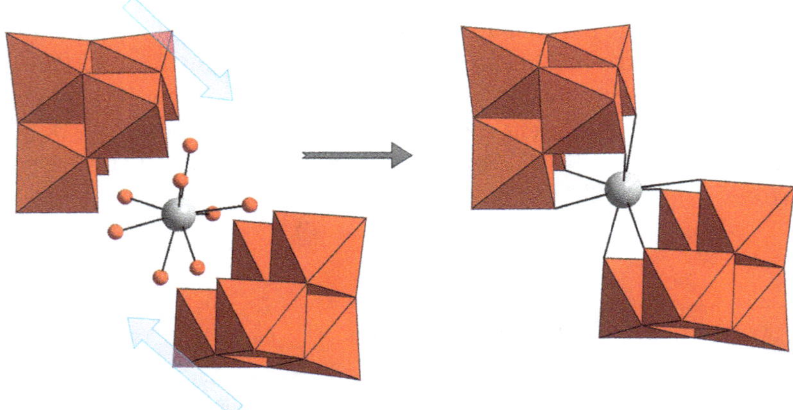

Figure 2.5.3: Scheme showing the coordination of two monolacunary Lindqvist units to the lanthanide center (light gray ball) to form a sandwich-type Weakley–Peacock polyanion. The red octahedra represent $\{MO_6\}$ units and the red balls represent oxygen atoms.

Lanthanide-stabilized structures result from coordination of lanthanide ions to isopolyanion units, leading to assemblies which are otherwise difficult to isolate. For example, lanthanides can stabilize macrocyclic isopolyanion assemblies, such as $[H_6Ce_2(H_2O)Cl(W_5O_{18})_3]^{7-}$, where two cerium(III) ions cap the cyclic assembly from opposite sides [10]. Similar examples also exist for vanadates in $[LaV_{10}O_{30}]^{7-}$ and $[LnV_9O_{27}]^{6-}$ ($Ln = Ce^{3+}$, Pr^{3+}), where the lanthanide ions act as templates for the POM macrocycle formation [11]. An example for niobates, $[Eu_3(H_2O)_9\{Nb_{24}O_{69}(H_2O)_3\}_2]^{27-}$, is also known, where three europium(III) ions bridge two $\{Nb_{24}\}$ trianglar units [12]. Discrete isopolyanions can also be stabilized via grafting of lanthanide ions, as seen in $[Ln_2(H_2O)_{10}W_{22}O_{71}(OH)_2]^{8-}$ and $[Ln_2(H_2O)_{10}W_{28}O_{93}(OH)_2]^{14-}$ [13], where the lanthanide ions stabilize $\{W_{22}\}$ and $\{W_{28}\}$ assemblies derived from the known $[H_4W_{11}O_{38}]^{6-}$ isopolytungstate (Figure 2.5.4) [14].

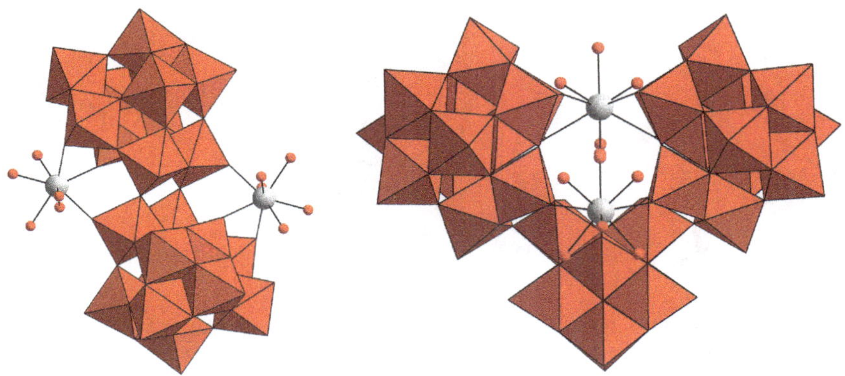

Figure 2.5.4: Combined polyhedra/ball-and-stick presentation of $[Ln_2(H_2O)_{10}W_{22}O_{71}(OH)_2]^{8-}$ (left) and $[Ln_2(H_2O)_2W_{28}O_{93}(OH)_2]^{14-}$ (right). Color code same as in Figure 2.5.3.

2.5.3 Heteropolyanions

Heteropolyanions form a much wider and more versatile subclass than isopolyanions, due to their higher stability and structural diversity, and are quite abundant for polytungstates, and to a lesser extent for polymolybdates [6]. These species contain an additional, usually central element called *heteroatom,* forming with its coordinated oxygen atoms, the so-called *heterogroup*. The nomenclature of heteropolyanions is based on the type of addenda element and the respective heteroelement; hence, polytungstates with the heteroelement phosphorus are called *tungstophosphates*, those with silicon as heteroelement, *tungstosilicates*, and so on. As stated above, polyanions are formed via self-assembly, and for heteropolyanions, the heterogroup plays an important templating role in the final POM structure and is usually bound weakly to the surrounding MO_6 octahedra by long (X)O-M bonds. The coordination geometry of the heterogroup defines the class of the heteropolyanion. For example, octahedral $\{XO_6\}$ heterogroups are observed in the *Anderson–Evans* structure $[XM_6O_{24}]^{n-}$ ($X = I^{VII}$, Te^{VI}, etc.) [15]. Tetrahedral $\{XO_4\}$ heterogroups, on the other hand, are known to form the *Keggin* structure $[XM_{12}O_{40}]^{n-}$ ($X = P^{V}$, Si^{IV}, Al^{III}, etc.) with one heterogroup [2], the *Wells–Dawson* structure $[X_2M_{18}O_{62}]^{n-}$ ($X = P^{V}$, As^{V}, etc.) with two heterogroups [16] and the *Pope–Jeannin–Preysssler* structure $[P_5W_{30}O_{110}]^{15-}$ with five heterogroups [6]. Some of these structures can be seen in Figure 2.5.5. A representative equation for the synthesis of heteropolyanions is as follows:

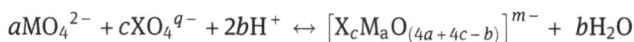

$$a\mathrm{MO_4}^{2-} + c\mathrm{XO_4}^{q-} + 2b\mathrm{H}^+ \leftrightarrow \left[\mathrm{X}_c\mathrm{M_a}\mathrm{O}_{(4a+4c-b)}\right]^{m-} + b\mathrm{H_2O}$$

Due to the low basicity of the terminal oxygen atoms, heteropolyanions with closed-shell structures (also known as *plenary* structures) are chemically rather unreactive. Therefore, the interaction of *f*-block metal ions with plenary POMs is usually restricted to ion pairing, resulting in solid-state arrangements with the metal ions

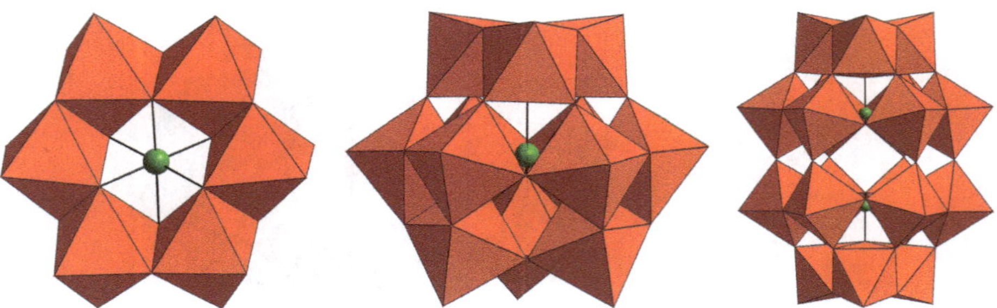

Figure 2.5.5: Polyhedral representation of the Anderson–Evans structure (left) with an octahedrally-coordinated heteroatom (green ball), the Keggin structure (center) with a tetrahedrally-coordinated heteroatom, and the Wells–Dawson structure (right) with two tetrahedrally-coordinated heteroatoms. The red octahedra represent $\{MO_6\}$ units.

acting mainly as counter cations for the POM. The reactivity of plenary POMs can be significantly increased by the formation of *lacunary* derivatives, which can be prepared by controlled hydrolysis in basic aqueous medium. For heteropolyanions, stable lacunary derivatives are known for polytungstates, which will be mainly discussed in this and the following sections. In contrast to their isopolyanionic counterparts, heteropolytungstates of the *Keggin* and *Wells–Dawson* type can lose more than one tungsten center (and its terminal oxo-ligand), leading to various mono-, di-, tri- or even hexavacant derivatives (e.g., $[PW_{11}O_{39}]^{7-}$, $[SiW_{10}O_{36}]^{8-}$, $[P_2W_{15}O_{56}]^{12-}$ and $[H_2P_2W_{12}O_{48}]^{12-}$; see Figure 2.5.6) [3, 6]. These derivatives are quite reactive and nucleophilic, and can hence act as inorganic multidentate ligands for cationic centers, in particular lanthanide ions. Due to their larger size compared to transition metal ions, lanthanide ions cannot be fully incorporated into the vacant sites of lacunary heteropolyanion precursors. The lanthanide ions usually have coordination numbers >6 and hence, a tendency to coordinate to two or more lacunary POM units. The variety of lacunary heteropolyanions coupled with the flexibility of *Ln*-POM linkages allows for a rather extensive library of lanthanide-containing heteropolyanions.

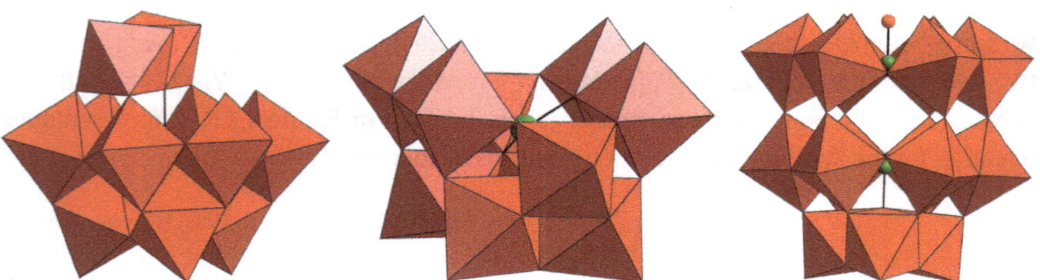

Figure 2.5.6: Polyhedral representation of the monolacunary Keggin ion (left), the dilacunary Keggin ion (center) and the trilacunary Wells–Dawson ion (right). The green balls represent heteroatoms and the red octahedra represent $\{MO_6\}$ units.

2.5.4 Lanthanide-containing heteropolyanions

The structural library of lanthanide-containing heteropolytungstates includes a multitude of reports in the scientific literature, exhibiting a variety of shapes, sizes and number of incorporated lanthanide centers within the POM. This section will focus on the most prominent structural features of lanthanide-heteropolytungstate chemistry.

Sandwich-type structures are also known for heteropolytungstates, and with more variations than for isopolyanions. As such, assemblies with one lanthanide center are also known for monolacunary *Keggin* and *Wells–Dawson* POM units. In fact, *Peacock* and *Weakley* were the first to also report such structures, although the structural characterization by single-crystal XRD was performed later by other groups [9, 17]. The *Weakley–Peacock* dimers can be represented by the annotation $\{LnL_2\}$, "L" being a monolacunary *Keggin* $\{XW_{11}\}$ or *Wells–Dawson* $\{X_2W_{17}\}$ unit (see Figure 2.5.7). However, other sandwich dimers with the annotation $\{(LnL)_2\}$ are also known. As mentioned above, the *Weakley–Peacock* dimer has one lanthanide center sandwiched between two monolacunary POM units, each coordinating four oxo atoms of the lacunary site to the lanthanide ion. However, in the $\{(LnL)_2\}$ dimer, each of the two monolacunary POM units has a lanthanide center coordinated to the lacunary site by four oxo-bonds to form a monomer $\{LnL\}$, which then dimerizes with an identical unit, resulting in $\{(LnL)_2\}$ (see Figure 2.5.7). Dimerization occurs *via* *Ln*-O-W bridges, where the coordinatively unsaturated lanthanide center of each $\{LnL\}$ monomer bridges to a terminal addenda oxygen atom of the other $\{LnL\}$ monomer. This bridging mode is

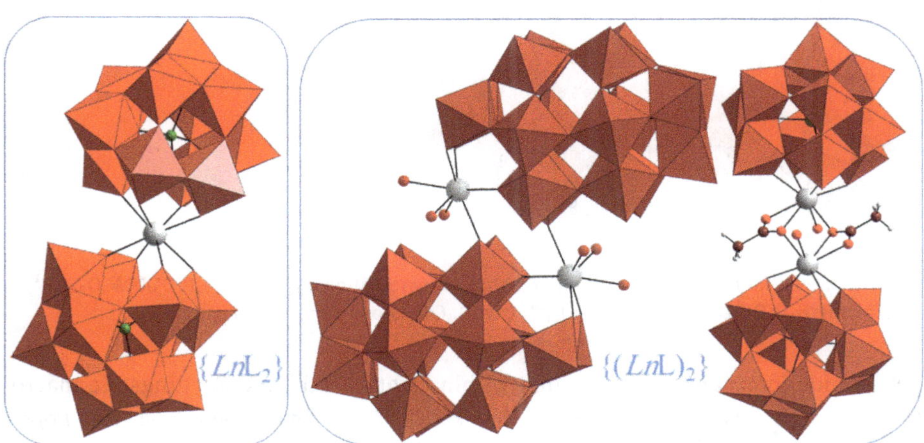

Figure 2.5.7: Combined polyhedral/ball-and-stick representations of $\{LnL_2\}$ (left) and $\{(LnL)_2\}$ structures (right). In $\{LnL_2\}$ the coordination sphere of the lanthanide ion is saturated by the two lacunary POM units, which is not the case for the $\{(LnL)_2\}$ structures. Color code: heteroatom (green balls), oxygen (red balls), carbon (brown balls) and hydrogen (small white balls). The red octahedra represent $\{MO_6\}$ units and the terminal atoms on the lanthanide ions are water molecules (hydrogens not shown).

seen in the solid-state for *Wells–Dawson*-type tungstophosphates and -arsenates of general formula $[\{Ln(H_2O)_n(X_2W_{17}O_{61})\}_2]^{14-}$ (X = P, As) and is likely to dissociate to two {*Ln*L} monomers when in solution, since the Ln-(O_{term}-W) bonds are rather weak [18]. Another dimerization mode involves *Ln*-O-*Ln* linkages via acetate bridges and this is known for monolacunary *Keggin* tungstosilicates and -germanates and *Wells–Dawson* tungstophosphates with overall formulas $[\{Ln(CH_3COO)(H_2O)_n(XW_{11}O_{39})\}_2]^{12-}$ (X = Si, Ge) [19a], and $[\{Ln(CH_3COO)(H_2O)_n(P_2W_{17}O_{61})\}_2]^{16-}$ [19b], respectively.

Lanthanide-containing sandwich-type POM structures are also known for trilacunary heteropolytungstates, which can coordinate to multiple lanthanide centers, such as the *Keggin*-based, trilanthanide-carbonato-encapsulating tungstophosphate $[\{Ln(H_2O)\}_3(CO_3)(PW_9O_{34})_2]^{11-}$ [20], and the *Wells–Dawson*-based, hexaytterbium-containing tungstophosphate $[\{Yb_6O(OH)_6(H_2O)_6\}(P_2W_{15}O_{56})_2]^{14-}$ [21] (see Figure 2.5.8).

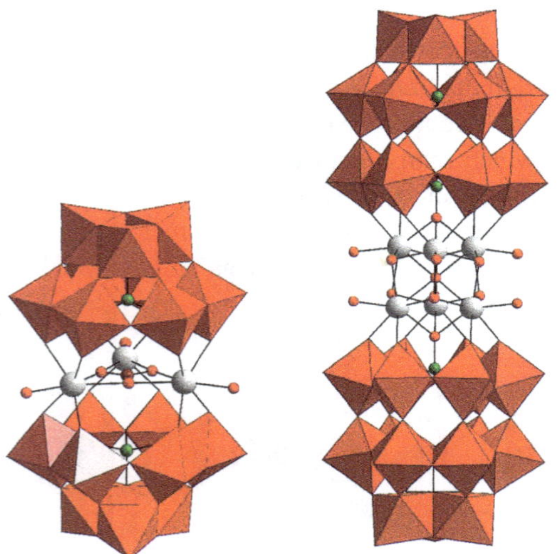

Figure 2.5.8: Combined polyhedral/ball-and-stick representation of $[\{Ln(H_2O)\}_3(CO_3)(PW_9O_{34})_2]^{11-}$ (left) and $[\{Yb_6O(OH)_6(H_2O)_6\}(P_2W_{15}O_{56})_2]^{14-}$ (right). Color code same as in Figure 2.5.7.

Lanthanide ions can also be grafted within cavities of heteropolytungstate macrocycles which possess nucleophilic "pockets". The prime example is the *Pope–Jeannin–Preyssler* ion, which usually has a sodium ion in the central cavity. However, Pope's group has demonstrated that this guest cation can be replaced by a multitude of lanthanide ions, resulting in $[LnP_5W_{30}O_{110}]^{12-}$ [22]. The cyclic, multilacunary tungstophosphate $[P_8W_{48}O_{184}]^{40-}$ can graft two lanthanide centers within its cavity [23], and up to four lanthanide ions can be grafted within the nucleophilic pockets of the tetrameric tungstoarsenate(III) $[As_4W_{40}O_{140}]^{28-}$ [24].

As stated earlier, the large size and coordination number of lanthanide ions allows linking several lacunary POM units to form very large oligomeric aggregates with nanometer dimensions. Fascinating structures are known for heteropolytungstates, with tungsten and lanthanide nuclearities surpassing 100 atoms, such as the two tungstoarsenates(III) $[Ce_{16}(H_2O)_{36}As_{12}W_{148}O_{524}]^{76-}$ [25a] (Figure 2.5.9), and $[Gd_8(H_2O)_{22}As_{12}W_{124}O_{432}]^{60-}$ [25b]. The former is one of the very few gigantic *Ln*-POMs which are solution stable.

Figure 2.5.9: Combined polyhedral/ball-and-stick representation of $[Ce_{16}(H_2O)_{36}As_{12}W_{148}O_{524}]^{76-}$. Color code same as in Figure 2.5.3.

For many large *Ln*-POMs only the solid-state structure is reported, as the salt is not soluble or not solution-stable. Also, many such materials exhibit lacunary Keggin isomers which are formed in situ (often through hydrothermal synthesis) and hence the reaction mechanism is hard to comprehend. A nice exception is a structural family involving the $\{Ln_2XW_{10}\}$ building block, which appears in various *Ln*-POMs. As explained above, lanthanide ions are too large to be fully incorporated in the lacunary sites of the POM precursors. As a result, the lanthanide centers can link several $\{Ln_2XW_{10}\}$ units in order to form oligomeric (tetrameric, pentameric, hexameric) arrangements (Figure 2.5.10). Tetrameric assemblies were obtained for the tungstophosphate $[(Ln_2PW_{10}O_{38})_4(W_3O_{14})]^{30-}$ [26], whereas pentameric assemblies were formed for the tungstogermanate $[Ce_{10}(H_2O)_{15}Ge_5W_{50}O_{188}(OH)_2]^{28-}$, which dimerizes in the solid state to an assembly containing 20 Ce and 100 W atoms [27]. Hexameric assemblies are seen in the tungstogermanate $[Ln_{12}(H_2O)_nGe_6W_{60}O_{228}]^{36-}$ with two different structural arrangements (see Figure 2.5.10) [28, 29], and the largest member of this family being a dodecameric tungstogermanate, $[Ln_{24}(H_2O)_{64}Ge_{12}W_{120}O_{456}(OH)_{12}]^{61-}$ [29, 30]. Many more lanthanide-containing heteropolytungstates are known. In terms of solid-state structural chemistry, there appear to be virtually endless possibilities of combining *Ln* ions with the various types of lacunary POM precursors (including rotational isomers).

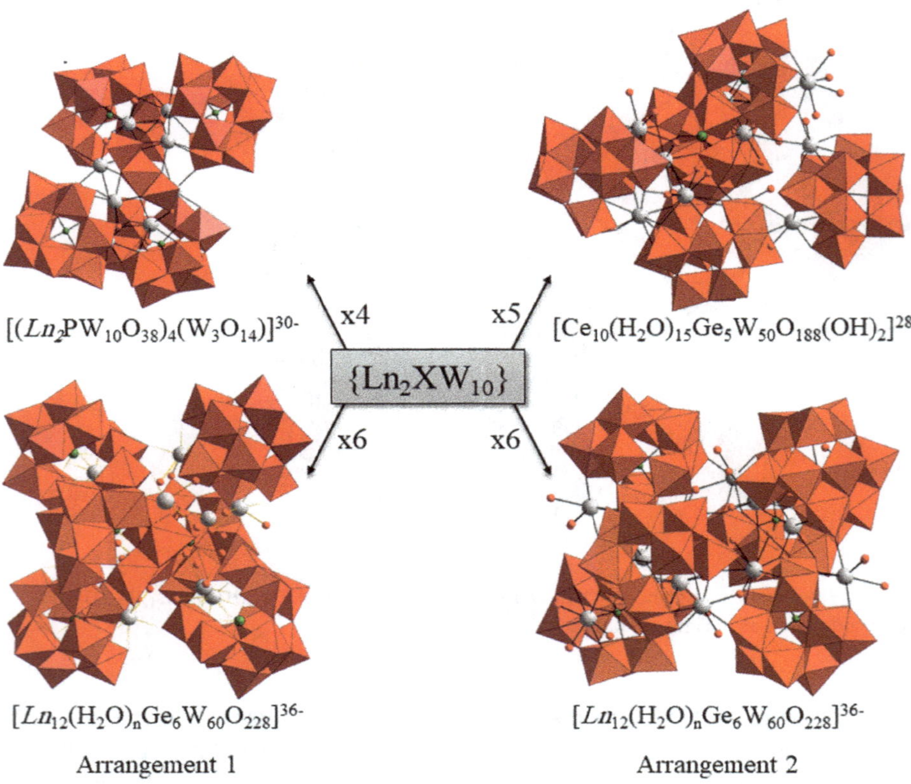

Figure 2.5.10: Structural representation of different assemblies based on the $\{Ln_2XW_{10}\}$ unit. Color code same as in Figure 2.5.3.

2.5.5 General properties of lanthanide-containing POMs

In analogy to lanthanide-containing coordination complexes, *Ln*-POMs possess a multitude of physiochemical properties, resulting from the insertion of lanthanide centers in the POM framework. Three main properties are known for *Ln*-POMs: (i) *luminescence*, (ii) *magnetism* and (iii) *catalysis*.

(i) Luminescence in *Ln*-POMs is well known, as charge transfer transitions allow for an excitation of the coordinated trivalent lanthanide centers, in particular Eu^{3+} and Tb^{3+}, leading to luminescent relaxation [31]. For example, the europium(III)-containing polyoxoniobate $[Eu_3(H_2O)_9\{Nb_{24}O_{69}(H_2O)_3\}_2]^{27-}$ exhibits an intense emission in the red region ($\lambda = 500–750$ nm) [12], and the terbium(III)-picolinate-containing tungstoarsenate(III) $[Tb_8(pic)_6(H_2O)_{22}(AsW_8O_{30})_4(WO_2(pic))_6]^{12-}$ exhibits an intense emission in the green region at around $\lambda = 490$ nm [32].

(ii) Interesting magnetic properties of *Ln*-POMs can arise when the lanthanide ion is in a square-antiprismatic coordination environment, such as *single molecule magnet* (SMM) behavior. This has been seen for classical sandwich-type POMs, in particular for the *Weakley*-type isopolytungstate $[Ln(W_5O_{18})_2]^{9-}$, which displays SMM behavior for various encapsulated magnetic lanthanide ions [33]. SMM behavior for heteropolytungstates is also known, for example in the monolanthanide-encapsulated 30-tungsto-5-phosphate $[LnP_5W_{30}O_{110}]^{12-}$ (*Pope-Jeannin-Preyssler* ion) [34], and the monolanthanide-sandwiched silico- and germanotungstates $[Ln(XW_{11}O_{39})_2]^{13-}$ (X = Si, Ge) [33b, 35].

(iii) Being Lewis acid centers, lanthanide ions can act as efficient catalytic sites for various chemical conversions, especially when combined with lacunary POM ligands, which display chemical robustness and thermal stability. As a result, some remarkable catalytic processes have been reported, in particular for lanthanide-containing heteropolytungstates. For example, the organic-soluble salt of the tungstosilicate $[\{Ln(H_2O)_2(acetone)\}_2\{\gamma\text{-}SiW_{10}O_{36}\}_2]^{10-}$ effectively promotes cyanosilylation of carbonyl compounds with trimethylsilyl cyanide [36], and the tungstophosphate $[Ln(H_2O)_4(P_2W_{17}O_{61})]^{7-}$ showed stereoselective catalytic behavior in Mannich-type and Diels–Alder reactions [37]. Apart from Lewis acid catalysis, lanthanide-containing heteropolytungstates may show reactivity in reduction and oxidation catalysis. For instance, the 3,4-pyridinedicarboxylate-capped tungstosilicate $[(pdc)_2Ln(H_2O)_2(SiW_{11}O_{39})]^{9-}$ displays hydrogen evolution reactivity under homogenous photochemical as well as heterogenous electrochemical conditions, depending on the encapsulated lanthanide center [38]. Very recently, a new class of cerium(IV)-peroxo-containing POMs was discovered, $[Ce^{IV}_6(O_2)_9(GeW_{10}O_{37})_3]^{24-}$, which displays selective oxidation catalysis of the model substrate methionine [39].

References

[1] J. Berzelius, Poggendorff's Ann. Phys. 1826, 6, 369.
[2] J. F. Keggin, Proc. Roy. Soc. A 1934, 144, 75.
[3] a) M. T. Pope, Handbook on the Physics and Chemistry of Rare Earths, 2008, 38, 337; b) B.S. Bassil, U. Kortz, Z. Anorg. Allg. Chem. 2010, 636, 2222; c) M. T. Pope, U. Kortz, Encyclopedia of Inorganic and Bioinorganic Chemistry, John Wiley and Sons, Ltd: Hoboken, NJ, 2012.
[4] L. Cronin, A. Müller, Eds. Chem. Soc. Rev.; Royal Society of Chemistry, 2012, 41, (special issue on polyoxometalates).
[5] a) H. T. Evans, Jr, Inorg. Chem. 1966, 5, 967; b) P. A. Durif, M. T. Averbuch-Pouchot, Acta Crystallogr. B 1980, 36, 680.
[6] M. T. Pope, Heteropoly and Isopoly Oxometalates, Springer, Berlin, 1983.
[7] H. T. Evans, Jr, J. Chem. Soc., Dalton Trans. 1975, 505.
[8] B. Krebs, I. Paulat-Böschen, Acta Crystallogr. 1982, B38, 1710.
[9] a) R. D. Peacock, T. J. R. J. Weakley, Chem. Soc. A 1971, 1836; b) J. Iball, J. N. Low, T. J. R. Weakley, J. Chem. Soc., Dalton Trans. 1974, 2021.
[10] T. Li, F. Li, J. Lu, Z. Guo, S. Gao, R. Cao, Inorg. Chem. 2008, 47, 5612.
[11] M. Nishio, S. Inami, Y. Hayashi, Eur. J. Inorg. Chem. 2013, 1876.
[12] S. Chen, P. Ma, H. Luo, Y. Wang, J. Niu, J. Wang, Chem. Commun. 2017, 53, 3709.
[13] a) A. H. Ismail, M. H. Dickman, U. Kortz, Inorg. Chem. 2009, 48, 1559; b) A. H. Ismail, B. S. Bassil, A. Suchopar, U. Kortz, Eur. J. Inorg. Chem. 2009, 5247.
[14] T. Lehmann, J. Fuchs, Z. Naturforsch. 1988, 43b, 89.
[15] H. T. Evans, Jr, J. Am. Chem. Soc. 1948, 70, 1291.
[16] B. Dawson, Acta Crystallogr. 1953, 6, 113.
[17] T. Ozeki, T. Yamase, Acta Crystallogr. 1994, C50, 327.
[18] Q. Luo, R. C. Howell, J. Bartis, M. Dankova, W. D. Horrocks, Jr., A. L. Rheingold, L. C. Francesconi, Inorg. Chem. 2002, 41, 6112.
[19] a) P. Mialane, A. Dolbecq, E. Riviere, J. Marrot, F. Sécheresse, Eur. J. Inorg. Chem. 2004, 33; b) U. Kortz, J. Cluster Sci. 2003, 14, 205.
[20] M. J. Giansiracusa, M. Vonci, W. van den Heuvel, R. W. Gable, B. Moubaraki, K. S. Murray, D. Yu, R. A. Mole, A. Soncini, C. Boskovic, Inorg. Chem. 2016, 55, 5201.
[21] X. Fang, T. M. Anderson, C. Benelli, C. L. Hill, Chem. Eur. J. 2005, 11, 712.
[22] I. Creaser, M. C. Heckel, R. J. Neitz, M. T. Pope, Inorg. Chem. 1993, 32, 1573.
[23] M. Zimmermann, N. Belai, R. J. Butcher, M. T. Pope, E. V. Chubarova, M. H. Dickman, U. Kortz, Inorg. Chem. 2007, 46, 1737.
[24] a) G. L. Xue, J. Vaissermann, P. Gouzerh, J. Cluster Sci. 2002, 13, 409; b) K. Wassermann, M. T. Pope, Inorg. Chem. 2001, 40, 2763.
[25] a) K. Wassermann, M. H. Dickman, M. T. Pope, Angew. Chem. Int. Ed. 1997, 36, 14458; b) F. Hussain, F. Conrad, G. R. Patzke, Angew. Chem. Int. Ed. 2009, 48, 9088.
[26] R. C. Howell, F. G. Perez, S. Jain, W. D. Horrocks, Jr., A. L. Rheingold, L. C. Francesconi, Angew. Chem. Int. Ed. 2001, 40, 4031.
[27] B. S. Bassil, M. H. Dickman, I. Römer, B. von der Kammer, U. Kortz, Angew. Chem. Int. Ed. 2007, 46, 6192.
[28] K.-Y. Wang, B. S. Bassil, Z. Lin, I. Römer, S. Vanhaecht, T. N. Parac-Vogt, C. Sáenz de Pipaón, J. R. Galán-Mascarós, L. Fan, J. Cao, U. Kortz, Chem. Eur. J. 2015, 21, 18168.
[29] B. Artetxe, S. Reinoso, L. San Felices, J. M. Gutiérrez-Zorrilla, J. A. García, F. Haso, T. Liu, C. Vicent, Chem. Eur. J. 2015, 21, 7736.
[30] S. Reinoso, M. Giménez-Marqués, J. R. Galán-Mascarós, P. Vitoria, J. M. Gutiérrez-Zorrilla, Angew. Chem. Int. Ed. 2010, 49, 8384.

[31] T. Yamase, Handbook on the Physics and Chemistry of Rare Earths, 2009, 39, 297.

[32] C. Ritchie, E. G. Moore, M. Speldrich, P. Kögerler, C, Boskovic, Angew. Chem. Int. Ed. 2010, 49, 7702.

[33] a) M. A. AlDamen, J. M. Clemente-Juan, E. Coronado, C. Martí-Gastaldo, A. Gaita-Ariño, J. Am. Chem. Soc. 2008, 130, 8874; b) M. A. AlDamen, S. Cardona-Serra, J. M. Clemente-Juan, E. Coronado, A. Gaita-Ariño, C. Martí-Gastaldo, F. Luis, O. Montero, Inorg. Chem. 2009, 48, 3467; c) M. Vonci, M. J. Giansiracusa, W. van den Heuvel, R. W. Gable, B. Moubaraki, K. S. Murray, D. Yu, R. A. Mole, A. Soncini, C. Boskovic, Inorg. Chem. 2017, 56, 378.

[34] S. Cardona-Serra, J. M. Clemente-Juan, E. Coronado, A. Gaita-Ariño, A. Camón, M. Evangelisti, F. Luis, M. J. Martínez-Pérez, J. Sesé, J. Am. Chem. Soc. 2012, 134, 14982.

[35] a) B.S. Bassil, M.H. Dickman, B. von der Kammer, U. Kortz, Inorg. Chem., 2007, 46, 2452. b) A. S. Mougharbel, S. Bhattacharya, B. S. Bassil, A. Rubab, J. van Leusen, P. Kögerler, J. Wojciechowski, U. Kortz, Inorg. Chem. 2020, 59, 4340.

[36] K. Suzuki, M. Sugawa, Y. Kikukawa, K. Kamata, K. Yamaguchi, N. Mizuno, Inorg. Chem. 2012, 51, 6953.

[37] C. Boglio, G. Lemière, B. Hasenknopf, S. Thorimbert, E. Lacôte, M. Malacria, Angew. Chem. Int. Ed. 2006, 45, 3324.

[38] M. Arab Fashapoyeh, M. Mirzaei, H. Eshtiagh-Hosseini, A. Rajagopal, M. Lechner, R. Liu, C. Streb, Chem. Commun. 2018, 54, 10427.

[39] H. M. Qasim, W. W. Ayass, P. Donfack, A.S. Mougharbel, S. Bhattacharya, T. Nisar, T. Balster, A. Solé-Daura, I. Römer, J. Goura, A. Materny, V. Wagner, J. M. Poblet, B. S. Bassil, U. Kortz, Inorg. Chem. 2019, 58, 11300.

Jean-Claude G. Bünzli

2.6 Coordination chemistry

2.6.1 Basic concepts and definitions

At the end of the nineteenth century, Alfred Werner (1866–1919) developed a theory to explain the chemical bonding in metal-containing compounds. He postulated that a metal ion surrounds itself with several *ligands*, ions or neutral molecules, and that the physical and chemical properties of the resulting assembly are determined by the nature of the cation-to-ligand chemical bonds and by the geometrical arrangement of the ligands around the central metal ion. This concept, which earned Werner a Nobel Prize in 1913, is valid for both solid-state compounds and solvated ions. It is now commonly expanded to describe polynuclear compounds, clusters containing metal–metal bonds and organometallic molecules in which the metal is linked to ligands via metal-to-carbon bonds. Here are some definitions associated with this concept:

- A metal complex or *coordination compound* is a chemical assembly consisting of a central metal ion and its surrounding ligands. These entities are written within square brackets, for example, $[Er(hfa)_3(H_2O)_2]$ (Figure 2.6.1, top).
- The group of ligands, neutral or ionic, directly bonded to the central metal cation is called the *inner* or *first coordination sphere* (first CS, Figure 2.6.1, bottom).
- The group of ligands polarized by the electric field generated by the positive charge of the cation and connecting with particles in the first CS via weak interactions is called the *outer* or *second coordination sphere* (second CS, Figure 2.6.1, bottom).
- The *coordination number* (CN) is the number of atoms directly bonded to the cation.
- The group of atoms directly coordinated to the metal ion and their corresponding 3D geometry is called the *coordination polyhedron*.
- The *nuclearity* (nu) of a complex is the number of metal centers it is made of.
- The *denticity* (de) of a ligand is the number of donor atoms able to bind the cation.
- A *chelate* is a complex with a multidentate ligand (de $\geq$ 2).
- The *solvation* of a cation is the complex system arising from its interaction with the solvent and the anions ensuring electric neutrality of the solution. Molecules and ions are involved in exchange reactions between the first and second CS and between the second CS and the bulk (Figure 2.6.1, bottom).

The coordination chemistry of d-transition metal ions is essentially governed by steric requirements, dictated by the shapes of the valence d-orbitals. CNs usually

https://doi.org/10.1515/9783110654929-014

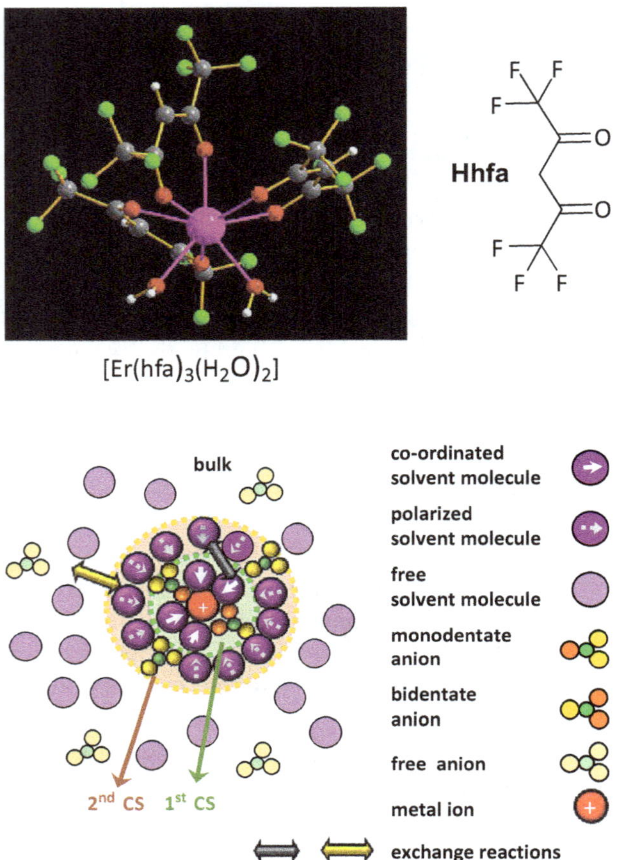

Figure 2.6.1: Top: Example of an Er^{III} coordination compound (from CIF data in CCDC 694,403); bottom: schematic of the solvation of a metal ion.

range between 4 and 6. For rare earths, the situation is quite different. While ligand-field splitting in d-transition metal complexes commonly reaches 15,000 to 25,000 cm^{-1} (180–300 kJ mol^{-1}), it is much smaller in 4f-element complexes, approximately 500 cm^{-1} (6 kJ mol^{-1}). This is due to the shielding of the f-orbitals by filled $5s^2$ $5p^6$ subshells.

The coordination chemistry of 4f-elements is, therefore, subtle with variable and often large CNs ranging between 3 and 12, and with geometrical arrangements essentially determined by the steric repulsion between ligands. Since *RE* ions are strong Lewis acids, their bonding with ligands is mainly ionic or ion-dipolar. Covalent contributions are frequently smaller than 5%. As a consequence, *RE* ions preferentially bind ligands with hard donor atoms, typically oxygen or the somewhat more polarizable nitrogen, which generates a quite interesting coordination chemistry. *RE* coordination compounds play a crucial role in catalysis, lighting, electronics, single-molecule magnets, biology and medicine [1].

2.6.2 Electronic properties of 4f-element ions

Rare earths have a preferred valence state of +3. For a long time, only a few RE^{II} compounds were known to be stable but organometallic complexes with the entire RE^{II} series have recently been isolated [2] (Figure 2.6.2). On the other hand, very few RE^{IV} coordination compounds are known, most of them with Ce^{IV} and, presently, only two with Tb^{IV} [3]. The other tetravalent ions, Pr^{IV}, Nd^{IV}, Dy^{IV}, have essentially been detected in solid inorganic compounds, double fluorides, oxides, perovskites or in solutions of heteropolyanions. In this chapter, we concentrate on RE^{III} coordination chemistry.

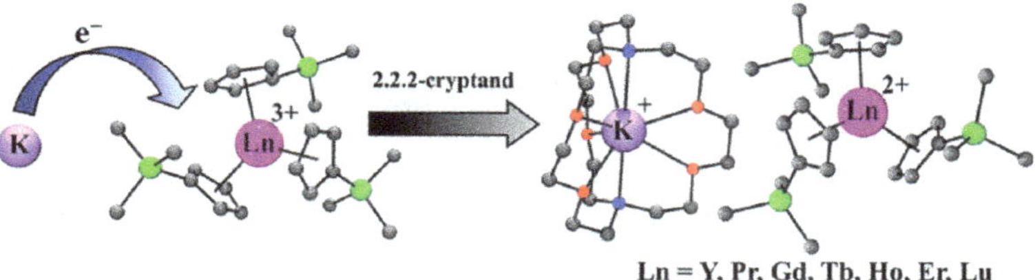

Figure 2.6.2: Strategy used for synthesizing some RE^{II} organometallic complexes. Reproduced, with permission from ref. [2], © 2013, American Chemical Society.

The stability of a coordination compound depends on the electronic structure, charge density and polarizability of both the cation and the ligands. Electronic configurations of the rare earths and their ions are reported in Table 2.6.1. Most of them are well separated from the first excited configuration, except when 5d orbitals lie at relatively low energy, for instance in Ce^{III}, Pr^{III}, Tb^{III}, or in the presence of low-lying ligand-to-metal charge-transfer states, for instance for easily reducible Sm^{III}, Eu^{III}, Tm^{III} and Yb^{III} ions.

2.6.3 Complexation reactions

The complexation reaction between a cation and a ligand can be viewed as a two-step procedure. The first step implies (partial) desolvation of the cation and ligand (eq. (2.6.1)) and the second step is the binding between the partially desolvated

Table 2.6.1: Electronic structure of *RE* atoms and their divalent and trivalent ions[a].

RE	*RE*^{0}	*RE*^{II} [b]	*RE*^{III}
Sc	[Ar] $3d^1\ 4s^2$	[Ar] $3d^1$	[Ar]
Y	[Kr] $4d^1\ 5s^2$	[Kr] $4d^1$	[Kr]
La	[Xe] $5d^1\ 6s^2$	[Xe] $5d^1$	[Xe] $4f^0$
Ce	[Xe] $4f^1\ 5d^1\ 6s^2$	[Xe] $4f^2$	[Xe] $4f^1$
Pr–Eu	[Xe] $4f^n\ 6s^2$, $n = 3–7$	[Xe] $4f^n$ $n = 3–7$	[Xe] $4f^n$ $n = 2–6$
Gd	[Xe] $4f^7\ 5d^1\ 6s^2$	[Xe] $4f^7\ 5d^1$	[Xe] $4f^7$
Tb–Yb	[Xe] $4f^n\ 6s^2$ ($n = 9–14$)	[Xe] $4f^n$ $n = 9–14$	[Xe] $4f^n$ $n = 8–13$
Lu	[Xe] $4f^{14}\ 5d^1\ 6s^2$	[Xe] $4f^{14}\ 6s^1$	[Xe] $4f^{14}$

[a]The electronic structure of all *RE*^{IV} is inferred to be [Xe] $4f^n$ ($n = 0–13$, Ce–Lu)
[b]For organometallic compounds $[RE^{II}(Cp')_3]^-$, Cp' = $C_5H_4SiMe_3$:
[Xe]$4f^n$ for Sm, Eu, Tm, Yb
[Xe]$4f^{n-1}5d^1$ for La, Ce, Pr, Nd, Gd, Tb, Dy, Ho, Er, Lu
Nd and Yb are trans configurational (their electronic configuration changes with the nature of the surrounding ligands) [4].

species (eq. (2.6.2)), resulting in the overall reaction (2.6.3) characterized by an equilibrium (stability) constant K_1:

$$\left[RE(\text{solv})_x\right]^{n+} + \left[\text{L}(\text{solv})_y\right]^{m-} \leftrightharpoons \left[RE(\text{solv})_{x'}\right]^{n+} + \left[\text{L}(\text{solv})_{y'}\right]^{m-} + (x - x' + y - y')\text{solv} \tag{2.6.1}$$

$$\left[RE(\text{solv})_{x'}\right]^{n+} + \left[\text{L}(\text{solv})_{y'}\right]^{m-} \leftrightharpoons \left[RE(\text{L})(\text{solv})_{(x'+y')}\right]^{(n-m)+} \tag{2.6.2}$$

$$\left[RE(\text{solv})_x\right]^{n+} + \left[\text{L}(\text{solv})_y\right]^{m-} \leftrightharpoons \left[RE(\text{L})(\text{solv})_{(x'+y')}\right]^{(n-m)+} + (x - x' + y - y')\text{solv}$$

$$\text{equilibrium constant } K_1 \tag{2.6.3}$$

The decrease in ion solvation results in a positive entropy change, reflecting the increase in randomness of the solution and a positive (endothermic) enthalpy change due to the breaking of stable cation–solvent bonds. The formation of the cation–ligand bonds corresponds to negative enthalpy and entropy changes. The stability of the resultant complex depends on the enthalpy and entropy of the total reaction (eq. (2.6.3)) that can be split into desolvation (d) and complexation (c) contributions:

$$\Delta G_r = \Delta H_r - T\Delta S_r = -RT\ln K_1 \tag{2.6.4}$$

$$\Delta G_r = \Delta G_d + \Delta G_c = \Delta H_d + \Delta H_c - T(\Delta S_d + \Delta S_c) \tag{2.6.5}$$

The difference between ΔH_d and ΔH_c reflects the difference in bond strength between *RE*-solv and *RE*-L. Therefore, if the reaction is conducted in a solvent that strongly coordinates *RE* ions such as water, dimethylsulfoxide (DMSO), dimethylformamide (DMF), or to a lesser extent, alcohols, the resulting ΔH_r is often positive.

As a consequence, the complex has to be entropically stabilized ($\Delta S_r > 0$) which can be realized by using multidentate ligands: each ligand displaces several solvent molecules and renders $\Delta S_r \gg 0$ (the so-called chelate effect). This is particularly important when water is the solvent because hydration enthalpies are large and increase with atomic numbers (Table 2.6.2, 5th col.). Hard bases are therefore preferred as ligands, particularly those bearing a negative charge such as carboxylates.

Table 2.6.2: Selected properties of rare earths: Pauling electronegativity, reduction potentials (V),[a] as well as experimental hydration enthalpies (kJ mol^{-1})[b,c], hydrolysis constants,[d] hydroxide precipitation pH[e] and ionic radii (pm)[f] for RE^{III} ions.

Ln	χ_P	$E^0_{r,3-0}$	$E^0_{r,3-2}$	ΔH^0_h	$\log {}^*\beta_{11}$	pH	$r_i(6)$	$r_i(9)$	$r_i(12)$
Sc	1.36	−2.077		−3,897	−5.15	n.a.	75	n.a.	n.a.
Y	1.22	−2.372	n.a.	−3,640	−8.36	n.a.	90	108	n.a.
La	1.10	−2.379	−3.1	−3,326	−9.01	7.47	103	122	136
Ce	1.12	−2.336	−2.92	−3,380	−10.6	7.10	101	120	134
Pr	1.13	−2.353	−2.84	−3,421	−8.55	6.96	99	118	132
Nd	1.14	−2.323	−2.62	−3,454	−8.43	6.78	98	116	130
Pm	1.13	−2.30	−2.44	−3,482	n.a.	n.a.	97	114	128
Sm	1.17	−2.304	−1.50	−3,512	−8.34	6.65	96	113	127
Eu	1.2	−1.991	−0.34	−3,538	−8.31	6.61	95	112	125
Gd	1.20	−2.279	−2.85	−3,567	−8.35	6.58	94	111	124
Tb	1.1	−2.28	−2.83	−3,600	−8.16	6.47	92	110	123
Dy	1.22	−2.295	−2.56	−3,634	−8.10	6.24	91	108	122
Ho	1.23	−2.33	−2.79	−3,663	−8.04	6.20	90	107	121
Er	1.24	−2.331	−2.87	−3,692	−7.99	6.14	89	106	119
Tm	1.25	−2.319	−2.22	−3,717	−7.95	5.98	88	105	118
Yb	1.1	−2.19	−1.18	−3,740	−7.92	5.87	87	104	117
Lu	1.27	−2.28	n.a.	−3,759	−7.90	5.74	86	103	116

[a] N. B. Mikheev, *Inorg. Chim. Acta* **1984**, *94*, 241. [b] D. W. Smith, *J. Chem. Ed.* **1967**, *54*, 540.
[c] E. N. Rizkalla *et al.*, *Handb. Phys. Chem. Rare Earths*, 1991, Vol. 15, Ch. 103, 393. [d] V. Haase *et al.*, *Gmelin's Handb. Inorg. Chem.* "Sc, Y, La-Lu" 8th Ed., 1979, 1,1. [e] Y. Suzuki et al., *J. Less-Common Met.* **1986**, *126*, 351. [f] R. D. Shannon, *Acta Crystallogr. A* **1976**, *32*, 751.

At pH levels higher than 5.5–6, rare-earth ions sustain noticeable hydrolysis because of their large Lewis acidity. The reaction is complex, leading to different mononuclear and polynuclear hydroxo species; it is characterized by the hydrolysis constants ${}^*\beta_{z,y}$:

$$y[RE(H_2O)_x]^{3+} \leftrightarrows [RE_y(OH)_z(H_2O)_{(yx-z)}]^{(3y-z)+} + z\,H^+$$

$$^{*}\beta_{z,y} = \frac{\left[RE_y(\text{OH})_z^{(3y-z)+}(\text{H}_2\text{O})_{(yx-z)}\right][\text{H}^+]^z}{\left[RE(\text{H}_2\text{O})_x{}^{3+}\right]^y} \tag{2.6.6}$$

Log$^*\beta_{1,1}$ values for RE^{III} are given in Table 2.6.2 (6th col.). Care should be exercised since literature values sometimes correspond to $K_1 = {}^*\beta_{11} \cdot K_{\text{w}}$ (K_{w} being the ionic product of water). With the exception of scandium and cerium, they are in the range −8 to −9 and increase slightly with increasing atomic number. The effect of hydrolysis may be large and must not be disregarded, even in the presence of a complexing agent. In this respect, a useful indication is the pH value at which hydroxide precipitation starts (Table 2.6.2, 7th col.). Solvolysis also happens in other solvents, mainly alcohols and particularly methanol, in which methanolates form readily.

2.6.4 Ionic radii, coordination numbers and coordination polyhedra

Trivalent lanthanide ions have large ionic radii and accommodate variable CNs. Ionic radii of trivalent ions display a progressive and rather smooth decrease (disregarding second order effects) with increasing atomic number, that is with increasing charge density, resulting in the well-known "lanthanide contraction." However, differences between two consecutive ions are very small, in the order of 1 pm only. Yttrium has an ionic radius similar to holmium, while scandium has a much smaller radius, more similar to those of d-transition metal ions. Ionic radii strongly depend upon CNs (Table 2.6.2, col. 8–10): for instance, differences between CNs 6 and 12 amount to about 30 pm, which makes Ln^{III} ions highly adaptable to many coordination environments.

The geometry of the coordination polyhedron (Table 2.6.3) is governed by the steric properties of the ligands, so that suitable design of the ligating molecules leads to easy tuning of their CNs. As a matter of fact, in crystals, CNs between 3 and 12 are documented, the former with bulky ligands such as bis(trimethylsilyl)amine and the latter, either with small bidentate ligands or macrocyclic ligands. However, a study carried out on 1,389 structurally characterized Ln^{III} and Y^{III} complexes reported between 1935 and 1995 revealed that CNs 8 and 9 are, by far, the most frequent (about two-thirds of the complexes) [5]. CN 6 represents a special historical case. The crystal structure of $[\text{Nd(H}_2\text{O)}_9](\text{BrO}_3)_3$ had been determined in 1939 and clearly pointed to 9-coordination but a commonly expressed opinion among inorganic chemists was that rare-earth ions formed 6-coordinate, octahedral complexes, similar to 3d-transition metals. Work on polyaminocarboxylate complexes of yttrium and cerium in the early 1960s, started casting doubt on this thinking. The 6-coordinate theory finally faded out in 1965 when

Table 2.6.3: Examples of observed coordination numbers of RE^{III} ions in the solid state.

CN	Idealized geometrical arrangement of REX_n	Idealized sym.	Examples[a]
3	Pyramidal	C_{3v}	$[Eu\{N(SiMe_3)_2\}_3]$, $[RE\{N(i\text{-}prop)_2\}_3]$
4	Tetrahedral	T_d	$[RE\{N(i\text{-}prop)_2\}_3]^-$
5	Trigonal bipyramid	D_{3h}	$[Tm\{P(SiMe_3)_2\}_3(THF)_2]$
6	Octahedral	O_h	$[Er(NCS)_6]^{3-}$
	Trigonal prism	D_{3h}	$[Er(dpm)_3]$
7	Pentagonal bipyramid	D_{5h}	$[Eu(dpm)_3(DMSO)]$
	Capped octahedron	C_{3v}	$[Ho(dbm)_3(H_2O)]$
	Capped trigonal prism	C_{2v}	$[LaI_3(i\text{-}propOH)_4]$
8	Square antiprism	D_{4d}	$[Lu(H_2O)_8]^{3+}$
	Bicapped trigonal prism	C_{2v}	$[LaCl_3(15C5)]$
	Dodecahedron	C_{2v}	$[Tb(teta)_4]^-$
9	Tricapped trigonal prism	D_{3h}	$[Eu(H_2O)_9]^{3+}$, $[Ln(MeCN)_9]^{3+}$
	Capped square antiprism	C_{2v}	$[Pr\ Cl_3(terpy)(H_2O)_3]$
10	Bicapped square antiprism	D_{4d}	$[La(edta)(H_2O)_4]^-$
	C_{2v} dodecahedron	C_{2v}	$[Eu(NO_3)_5]^{2-}$
	Tetracapped hexagon	D_{2h}	$[Nd(NO_3)_2(18C6)]^+$
11	(Pentacapped trigonal prism)[b)]	$(D_{3h})^b$	$[La(NO_3)_3(H_2O)_5]$
	Heptadecahedron	C_S	$[Eu(NO_3)_3(15C5)]$
12	Icosahedron	I_h	$[Ce(NO_3)_6]^{3-}$
	Distorted icosahedron	C_1	$[Nd(NO_3)_3(18C6)]$

[a]i-prop = isopropyl; THF = tetrahydrofuran; dpm = dipivaloylmethane; DMSO = dimethylsulfoxide; dbm = dibenzolymethane; 15C5 = 15-crown-5 ether; tetra = 1,4,8,11-tetraazacyclotetradecane-1,4,8,11-tetraacetate; terpy = terpyridine; edta = ethylenediamine-tetraacetate; 18C6 = 18-crown-6 ether.

[b]Known 11-coordinated lanthanide complexes deviate strongly from this idealized geometry.

crystal structures of $NH_4[La(edta)(H_2O)_3]\cdot 5H_2O$ and $[La(Hedta)(H_2O)_4]\cdot 3H_2O$ (H_4edta = ethylenediaminetetraacetic acid) pointed to La^{III} being 9- and 10-coordinate, respectively [6]. The most representative examples of 6-coordination are elpasolites containing $[REX_6]^{3-}$ hexahalide anions. Finally, Sc^{III} is mainly hexacoordinate.

Coordination environments are usually assigned to an ideal polyhedron, using the shape measure metrics $S(\delta,\theta)$ which compares dihedral angles between the normals to adjacent binding faces of the polyhedron in the crystal structure (δ_i) with those of the ideal polyhedron θ_i for the m edges of the polyhedron [7]:

$$S(\delta,\theta) = \min\sqrt{\frac{1}{m}\sum_{i=1}^{m}(\delta_i - \theta_i)^2} \qquad (2.6.7)$$

Alternatively, the continuous symmetry method (CSM) compares distances between vertices of the experimental and idealized polyhedra. In addition to determining the polyhedron closest to the experimental one, the CSM concept enables defining the symmetry content of a structure. This is useful in interpreting spectroscopic data [8]:

$$S(P,Q) = \min \frac{\sum_{i=1}^{N} (Q_i - P_i)^2}{\sum_{i=1}^{N} (Q_i - Q_0)^2} \qquad (2.6.8)$$

where Q_i and P_i are the coordinates of the vectors defining the N vertices of the experimental and idealized polyhedra, respectively, while Q_0 is the coordinate vector of the center of mass of the investigated structure; therefore $S(P,Q)$ lies between 0 and 1.

2.6.5 Solvation and solvent exchange

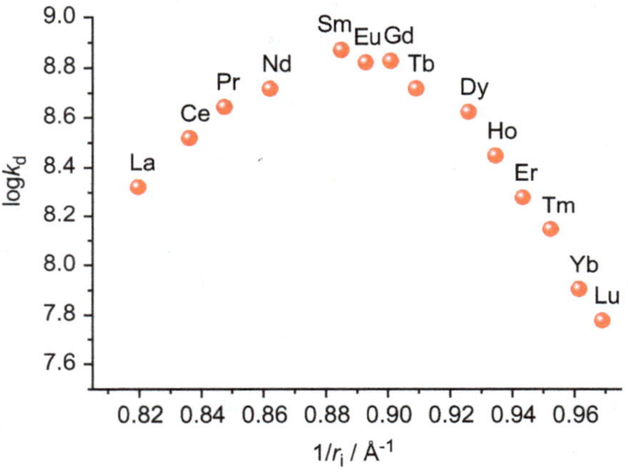

Figure 2.6.3: Water dissociation rates at 298 K versus the inverse of the ionic radii for CN = 9. Redrawn from data in D. P. Fay *et al.*, *J. Phys. Chem.* 1969, *73*, 544.

RE^{III} ions are highly labile, as shown by water dissociation rates depicted on Figure 2.6.3 for the following reaction:

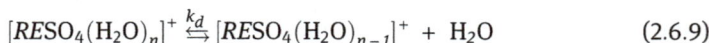

$$[RESO_4(H_2O)_n]^+ \overset{k_d}{\leftrightarrows} [RESO_4(H_2O)_{n-1}]^+ + H_2O \qquad (2.6.9)$$

The lability of RE^{III} complexes spans a large range, ligand exchange being fast for monodentate ligands but increasingly slow for polydentate ones, for example, $k_{ex} = 7.7 \times 10^{-2}\ s^{-1}$ for $[Lu(dtpa)]^{2-}$ ($dtpa^{5-}$ is diethylenetriaminepentaacetate). This is also true for nonaqueous solutions. In some specific cases, complexes may become even more inert, for instance, complexes with macrocyclic ligands such as $dota^{4-}$ (1,4,7,10-tetraazacyclododecane-1,4,7,10-tetraacetate): $k_{ex} \sim 10^{-4}\ s^{-1}$ for $[Lu(dota)]^-$ [9].

Studying solvation is difficult in view of the high lability of the RE^{III} ions, the presence of inner and outer sphere species, and of the interaction with the counter ions. For instance, F^-, SO_4^{2-}, CO_3^{2-}, NO_3^- form inner-sphere complexes in water with $\log K_1$(Eu) ~ 3.2, 1.4, 5.9 and 0.2, respectively. In nonaqueous solvents, interactions with anions are much stronger and are detected even with "noncoordinating" anions. For instance, the main species in 0.05 M solutions in acetonitrile are $[RE(NO_3)_3(MeCN)_x]$, $[RE(SO_3CF_3)_2(MeCN)_x]^+$, and $[RE(ClO_4)(MeCN)_x]^{2+}$ [10]. Although all details are not yet clarified, the main trends in CNs are as follows. In water, CN is 9 at the beginning of the series and 8 at the end, with equilibria between the two species, in the middle of the series [11]. A similar trend holds true in nonaqueous solvents but depending on its nature and on the nature of the counter ion, CN can vary more, typically from 10 to 8 [10].

2.6.6 Complexes with classical ligands

Carboxylates and aminocarboxylates represent a large class of ligands, commonly used for tailoring RE^{III} coordination compounds. In principle, stability constants have the tendency to increase along the lanthanide series in view of the increasing charge density of the cation but this effect may be counteracted by the ligand steric hindrance, when the cation becomes smaller (Figure 2.6.4). Another feature is the large increase in stability offered by multidentate chelating agents such as edta^{4-} (denticity de = 6) or dtpa^{5-} (de = 8).

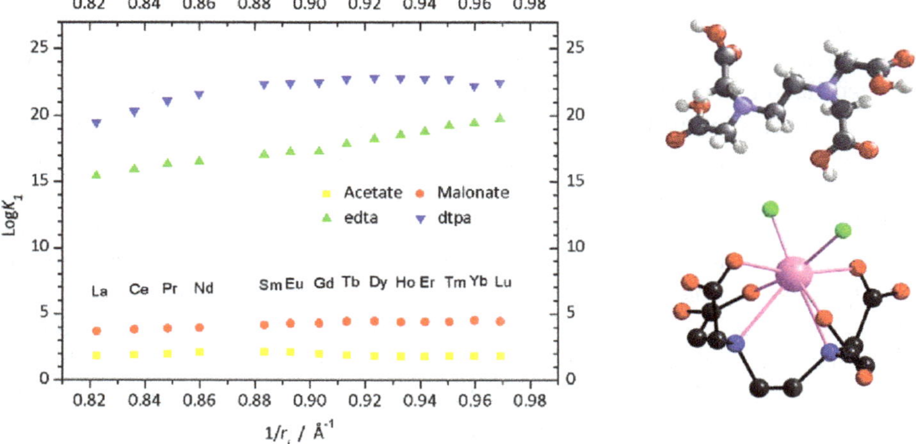

Figure 2.6.4: Left: Stability constants for the formation of 1:1 complexes in water at 298 K versus reciprocal ionic radii for CN = 9. Redrawn from data in A.E. Martell, R.M. Smith, Critical Stability Constants, Plenum Press, New York, (1974). Right: Structures of H_4edta and $[Y(edta)F_2]^4$.

Another important class of lanthanide complexes features β-diketonates. Several types of complexes are found: (i) anhydrous *tris* complexes, (ii) hydrated or ternary *tris* complexes in which the coordination sphere is saturated by an ancillary ligand which allows one to fine-tune luminescence properties, (iii) anionic *tetrakis* complexes, (iv) *bis*(diketonates) used in bioanalyses, and, more recently, (v) *tetrakis* (β-diketone) podates, which have the advantage of generating a large chelate effect. Diketonates are not particularly stable in water, but are ubiquitous in luminescence applications, rare-earth separation by gas chromatography, or as Diels–Alder catalysts [12].

Approximately 75% of RE^{III} complexes for which structural data are available contain at least one *RE*–O bond, but bonding to nitrogen donors is quite important with 25% of the lanthanide complexes featuring at least one *RE*–N bond [5]. Therefore, aliphatic amides, silylamides, pyridines, benzimidazoles, and Schiff base ligands also represent suitable building blocks for the design of complexes and associated materials with specific magnetic and/or optical properties.

2.6.7 Macrocyclic complexes

When macrocyclic chemistry based on host-guest (supramolecular) interactions developed in the 1980s, there was hope that sizeable selectivity could be induced within the *RE* series. Several classes of ligands, crown ethers, aza-crown ethers, cryptands, calixarenes, cyclic Schiff bases, porphyrins, phthalocyanines, and metallacrowns were investigated [13]. However, contrary to the selectivity found between, for instance, sodium and potassium with crown ethers and cryptands, little discrimination was found within the *RE* series, because tuning of the preorganized ligand cavity is too coarse with respect to the minute differences in *RE* ionic radii. Turning to more flexible macrocycles or to small macrocycles fitted with pendant coordinating arms resulted in much more stable complexes, though, again, to the detriment of selectivity. Figure 2.6.5 illustrates this point: (i) while stability constants of 1:1 complexes with 15C5 and 18C6 ethers display some size discrimination, it remains small; (ii) replacing two ether functions with amine groups to yield the more flexible (2.1) and (2.2) coronands increases the stability by 7–8 orders of magnitude due to better fit of the ligand around the metal ion (Figure 2.6.6b, c). Introducing an additional dioxyethylene chain to yield cryptand (2,2,1) further increases the thermodynamic stability but not the selectivity. The influence of denticity is seen in the stability increase in going from (2,1) (de = 5) to (2,2) (de = 6) and (2,2,1) (de = 7).

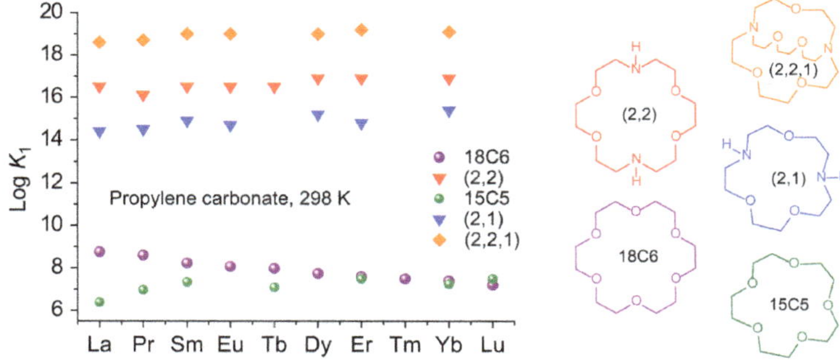

Figure 2.6.5: Stability constants for 1:1 complexes in propylene carbonate at 298 K and $I = 0.1$ M Et_4NClO_4. From data reported in ref. [14].

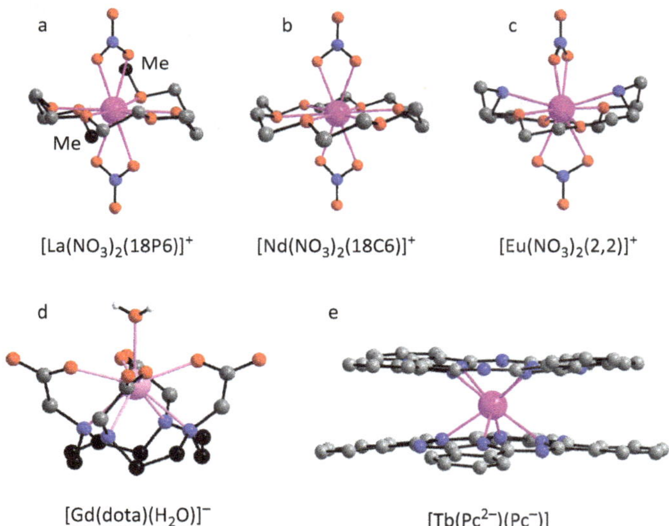

Figure 2.6.6: Structures of (a) an 18P6 podate, (b) 18C6 and (c) (2,2) coronates, (d) a Gd^{III} contrast agent, and (e) a Tb^{III} single-molecule magnet. Data from J.-C. G. Bünzli *et al.* (a) *Helv. Chim. Acta* 1984, *67*, 1121; (b) *Inorg. Chim. Acta* 1984, *54*, L43; (c) *Helv. Chim. Acta* 1986, *69*, 288. Data from (d) CCDC 1,188,959, (e) CCDC 748,937.

When compared with their noncyclic analogues called podands, macrocycles have the tendency to lead to more stable complexes. The origin of this macrocyclic effect is diverse but one important contribution is entropic: the free podand has a larger degree of freedom compared to the macrocycle. This freedom is lost upon complexation, since podates usually have a structure very similar to macrocyclic complexes (Figure 2.6.6a,b). For crown ethers versus linear Me-$(OCH_2CH_2)_n$-OMe podands, the $\Delta \log K_1$ improvement depends on the ring size and ionic radius; it is small for $n = 3$ and 6 (≈ 1) but reaches ≈ 3 for $n = 5$.

One of the richest classes of macrocyclic ligands is the molecules derived from cyclen (1,4,7,10-tetraaza-cyclododecane), the amine functions of which can easily be derivatized, particularly with carboxylic acid or amide functions. An archetype is H_4dota (cyclen bearing four -CH_2CO_2H *N*-substituents) forming among the most thermodynamically stable and kinetically inert *RE* complexes; its Gd^{III} complex is a universally used contrast agent for magnetic resonance imaging in medicine (Figure 2.6.6d). Other derivatives find applications as optical sensors or probes in bioanalysis and bioimaging.

Macrocyclic Schiff bases and calixarenes can be easily modulated with pendant arms. Their *RE* complexes, mono- or polymetallic, enable taking full advantage of the chemical (catalysts, extraction/separation agents) and spectroscopic (luminescent probes) properties of *RE*s. Tetrapyrrole derivatives such as porphyrins and phthalocyanines feature large delocalized π-systems and their coupling with *RE*s results in a variety of mono- and polynuclear complexes with various stoichiometries, 1:1, 1:2 (double deckers), 2:2, or 2:3 (triple-decker), for example. These compounds are instrumental in a wide range of applications, from electrochromic materials to organic field-effect transistors, optical-limiting materials, NIR luminescent complexes and sensors, probes for photodynamic therapy of cancer, activation of small molecules, single-molecule magnets (Figure 2.6.6) and quantum bits (qubits) for information processing [15]. Metallacrowns, in which the framework of the macrocycle ligand itself contains metal ions, represent an extension of Werner's concept. They offer a large degree of structural control, giving access to programmed magnetic and luminescent properties [16].

2.6.8 Self-assembly processes

Self-assembly is a tool of supramolecular chemistry, expanding the field of chemistry to molecular recognition, bringing it closer to natural phenomena observed in biology. In the process, small molecules or ions aggregate spontaneously, thanks to weak interactions (e.g., H-bonds, π-stacking interactions) to form an ordered superstructure. Programming specific functions in the components to be self-assembled enables recognizing a given molecule (guest). Distinction between coordination chemistry and supramolecular chemistry is not always clear, in that the arrangement of ligands around a central metal ion to form a complex is often assimilated to self-assembly. This should not, unless weak interactions between the ligands lead to the formation of a "coordinative cage," enabling recognition of a specific cation (or molecule). Inspired by the secondary helical structure of DNA, chemists have used metal-ligand bonds for inducing the wrapping of polytopic helical strands around a central axis, leading to so-called polynuclear helicates. Incorporation of at least two metal ions into the structure opens the way to controlling their mechanical, optical or magnetic interaction, particularly if mixed d–f helicates are elaborated [17].

The self-assembly process is exemplified in Figure 2.6.7 (top left), where three benzimidazole–pyridine ligands wrap around a RE^{III} ion, building a tight cage around it. The figure shows the ligands being held together by strong π-π interactions between aromatic rings. The figure also shows two potential applications of polynuclear helicates: the design of a Cr–Er–Cr trinuclear bimetallic helicate, the first example of molecular upconversion and the recognition of a pair of different RE^{III} ions by making the two coordinative moieties of a ditopic ligand slightly different.

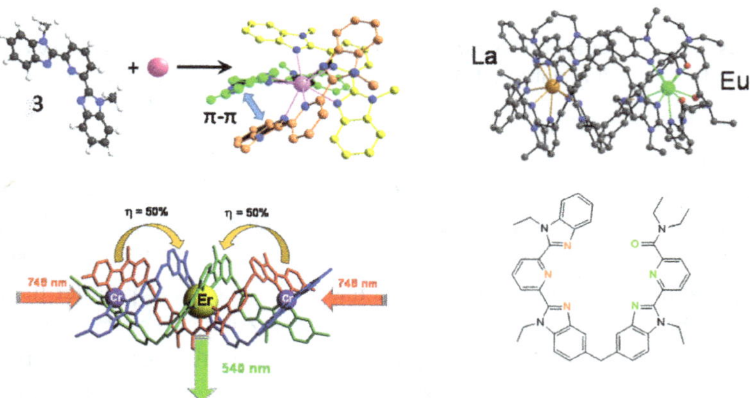

Figure 2.6.7: Top left: Schematics of a self-assembly process (redrawn from C. Piguet et al., Inorg. Chem. 1993, 32, 4139) and two examples of potential applications: design of the first molecular upconversion compound (bottom left, redrawn after L. Aboshyan-Sorgho *et al.*, Angew. Chem. Int. Ed. 2011, 50, 4108) and recognition of a pair of different RE^{III} ions (right, redrawn from data in ref. [17]).

A more complex structure is shown in Figure 2.6.8, representing self-assembly of a chiral tetrahedral cage. This system discriminates between RE^{III} ions with large separation factors, for example, ~14 between Pr and Sm.

2.6.9 Coordination polymers

A coordination polymer is a coordination compound repeating itself in 1, 2 or 3 dimensions, thanks to polydentate bridging inorganic or organic ligands and/or cross-

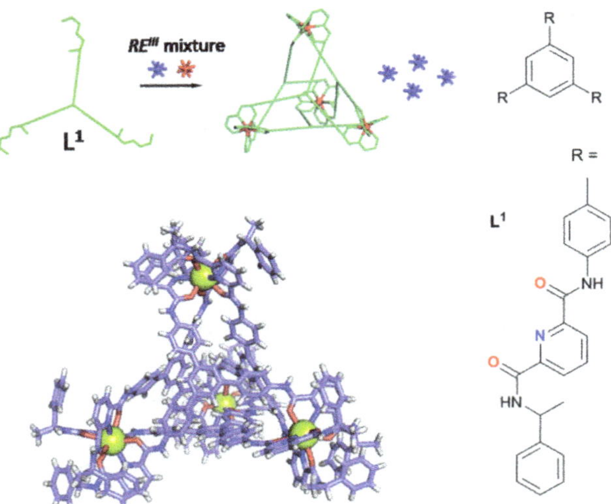

Figure 2.6.8: Self-assembly of a tetrahedral tetranuclear chiral RE^{III} cage enabling supramolecular recognition of rare-earth ions; redrawn after L. L. Yan *et al.*, *J. Am. Chem. Soc.* 2015, *137*, 8550.

linking units. When the structure is porous, the polymer is named metal-organic framework (MOF). Another variant of coordination polymers are the so-called polyoxometalates (POMs), in which a central metal ion, in its highest oxidation state (e.g., V, Mo, W), is surrounded by oxygen ions and serves as a ligand for other "guest" metal ions. The present surge of interest for these materials, particularly under their nanoform, is due to their ability for size-dependent gas separation and storage, catalysis, as analytical sensors, single-particle magnets, nonlinear optical properties and wavelength-conversion capability for lighting or photovoltaic materials. This chapter does not provide enough space for a detailed discussion but to give an idea of the complex structures that can be achieved, two examples are shown in Figure 2.6.9.

2.6.10 Conclusion

This short description of trivalent rare-earth coordination chemistry did not include purely inorganic complexes, clusters, extended structures, nanoparticles or organometallic complexes, nor did it deal with RE^{II} and RE^{IV} coordinative properties. Nevertheless, it shows how rich this field is. The absence of metal-induced directionality in *RE*-ligand bonds due to the shielding of 4f orbitals is more than compensated by the versatility of *RE* ions to adapt to vastly different chemical environments and to build materials with unmatched chemical, optical, and magnetic properties that are taken advantage of in numerous applications, ranging from catalysis to cancer therapy.

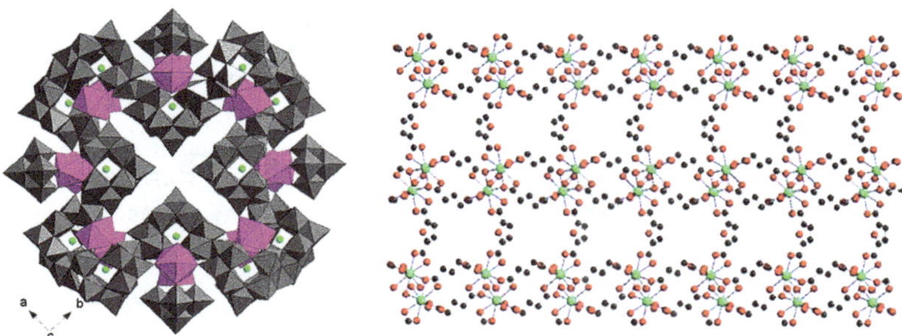

Figure 2.6.9: Left: Top view along the $C_2(z)$ axis of the $[(As^{III})_{12}(Ce^{III})_{16}(H_2O)_{36}W_{148}O_{524}]^{76-}$ structure with D_{2d} symmetry. WO_6 polyhedra are drawn in gray, CeO_8 (square antiprism) and CeO_9 (monocapped square antiprism) polyhedra are purple while As^{III} ions are green. Drawn from the data in ICSD-406,326. Right: Packing along the *c*-axis of the luminescent MOF $\{[Eu(FDA)_{0.5}(glu)(H_2O)_2]\}_n$ (FDA = 2,5-furane-dicarboxylate, glu = glutarate); color code: red, O, grey, C, green Eu; drawn from data in CCDC-1,868,175.

References

[1] J.-C. G. Bünzli, J. Coord. Chem. 2014, 67, 3706.

[2] M. R. MacDonald, J. E. Bates, J. W. Ziller, F. Furche, W. J. Evans, J. Am. Chem. Soc. 2013, 135, 9857.

[3] N. T. Rice, I. A. Popov, D. R. Russo, J. Bacsa, E. R. Batista, P. Yang, J. Telser, H. S. La Pierre, J. Am. Chem. Soc. 2019, 141, 13222.

[4] D. H. Woen, W. J. Evans, Expanding the +2 Oxidation State of the Rare-Earth Metals, Uranium, and Thorium in Molecular Complexes, in: J.-C. G. Bünzli and V. K. Pecharsky, Handbook on the Physics and Chemistry of Rare Earths, Elsevier Science, B.V., Amsterdam, 2016, Vol. 50, Ch. 293, pp. 337–394.

[5] C. Huang, Rare Earth Coordination Chemistry, Fundamentals and Applications, John Wiley & Sons (Asia), Singapore, 2010.

[6] L. C. Thompson, Complexes, in: K. A. Gschneidner Jr. and L. Eyring, Handbook on the Physics and Chemistry of Rare Earths, North Holland Publ. Co., Amsterdam, 1979, Vol. 3, Ch.25, pp. 209–297.

[7] J. D. Xu, E. Radkov, M. Ziegler, K. N. Raymond, Inorg. Chem. 2000, 39, 4156.

[8] S. Alvarez, P. Alemany, D. Casanova, J. Cirera, M. Llunell, D. Avnir, Coord. Chem. Rev., 2005, 249, 1693.

[9] M. Kodama, T. Koike, A. B. Mahatma, E. Kimura, Inorg. Chem. 1991, 30, 1270.

[10] J.-C. G. Bünzli, A. Milicic-Tang, Solvation and anion interaction in organic solvents, in: K. A. Gschneidner Jr. and L. Eyring, Handbook on the Physics and Chemistry of Rare Earths, Elsevier Science Publishers B.V., Amsterdam, 1995, Vol. 21. Ch.145, pp. 305–366.

[11] E. N. Rizkalla, G. R. Choppin, Hydration and hydrolysis of lanthanides, in: K. A. Gschneidner Jr. and L. Eyring, Handbook on the Physics and Chemistry of Rare Earths, Elsevier Science Publ. B.V., Amsterdam, 1991, Vol. 15, Ch.103, pp. 393–442.

[12] K. Binnemans, Rare earth β-diketonate complexes: functionalities and applications, in: K. A. Gschneidner Jr., J.-C. G. Bünzli and V. K. Pecharsky, Handbook on the Physics and Chemistry of Rare Earths, Elsevier Science B.V., Amsterdam, 2005, Vol. 34, Ch. 225, pp. 107–272.
[13] V. S. Sastri, J.-C. G. Bünzli, V. R. Rao, G. V. S. Rayudu, J. R. Perumareddi, Modern Aspects of Rare Earths and Complexes, Elsevier Science B.V., Amsterdam, 2003, pp. Ch. 4B, 303–375.
[14] J.-C. G. Bünzli, Complexes with Synthetic Ionophores, in: K. A. Gschneidner Jr. and L. Eyring, Handbook on the Physics and Chemistry of Rare Earths, Elsevier Science Publ. B.V., Amsterdam, 1987, Vol. 9, Ch. 60, pp. 321–394.
[15] E. Moreno-Pineda, C. Godfrin, F. Balestro, W. Wernsdorfer, M. Ruben, Chem. Soc. Rev. 2018, 47, 501.
[16] C. Y. Chow, E. R. Trivedi, V. Pecoraro, C. M. Zaleski, Comments Inorganic Chem. 2014, 35, 214.
[17] C. Piguet, J.-C. G. Bünzli, Self-assembled Lanthanide Helicates: from Basic Thermodynamics to Applications, in: K. A. Gschneidner Jr., J.-C. G. Bünzli and V. K.Pecharsky, Handbook on the Physics and Chemistry of Rare Earths, Elsevier Science, B.V., Amsterdam, 2010, Vol. 40, Ch. 247, pp. 301–553.

Marc D. Walter, Grégory Nocton

2.7 Organometallic rare-earth chemistry

2.7.1 Historic perspective

Soon after the discovery of ferrocene [(η^5-C_5H_5)$_2$ Fe] in 1951 [1], the groups of Wilkinson and Fischer started with their investigation on cyclopentadienyl (Cp) containing rare-earth metal compounds, isolating the tris(cyclopentadienyl) derivatives [(C_5H_5)$_3$Ln] (Ln = Sc, Y, La, Ce, Pr, Nd, Sm, Gd, Tb, Dy, Ho, Er, Tm, Yb and Lu) [2]. These complexes are sublimable, air and moisture sensitive solids with poor solubility in apolar solvents. The bonding in these compounds was assumed to be predominantly ionic, as exemplified by their reaction with $FeCl_2$ in tetrahydrofuran to form [(η^5-C_5H_5)$_2$Fe]. The intrinsic property of rare-earth metals to act as strong Lewis acids and therefore to achieve high coordination numbers (C.N.), results in different structural motifs in the solid state for [(C_5H_5)$_3$Ln] complexes (Figure 2.7.1). Within this series, the molecular structures and, therefore, the total C.N. vary systematically depending on the ionic radii of the lanthanide ion (Figure 2.7.2) [3]. Note that a η^5-coordinate Cp ligand acts as a tridentate ligand with a C.N. = 3, occupying one coordination side per electron pair.

$$LnX_3 + 3\ MCp \longrightarrow Cp_3Ln + 3\ MX$$

MCp = NaCp, KCp, Cp_2Mg and Cp_2Be

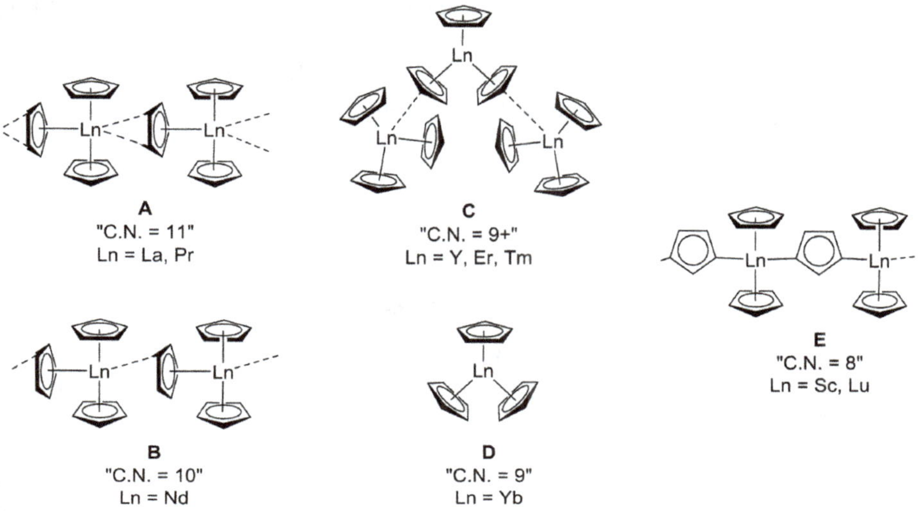

Figure 2.7.1: Synthesis of [(C_5H_5)$_3$Ln] and structural motifs observed in solid state.

https://doi.org/10.1515/9783110654929-015

Sc														
0.88														
Y 1.04 (-2.8)														
La 1.17 (-3.1)	Ce 1.15 (-3.2)	Pr 1.12 (-2.9)	Nd 1.12 (-2.6)	Pm 1.11	Sm 1.10 (-1.55)	Eu 1.09 (-0.35)	Gd 1.08 (-3.9)	Tb 1.06 (-3.7)	Dy 1.05 (-2.5)	Ho 1.04 (-2.9)	Er 1.03 (-3.1)	Tm 1.02 (-2.3)	Yb 1.01 (-1.15)	Lu 1.00 (-2.7)

Figure 2.7.2: Lanthanoids: their ionic radii (r(Ln^{3+}) [4] (in Å; values derived from the metal oxides, C.N. = 6). Values given in parenthesis correspond to the estimated standard reduction potentials $E^0(Ln^{3+}/Ln^{2+})$ (±0.2 V) based on experimental, spectroscopic and thermodynamic data [5].

Only for $[(C_5H_5)_3Yb]$, a simple molecular structure is obtained with C.N. = 9, while for the slightly larger derivatives $[(C_5H_5)_3Ln]$ (Ln = Y, Er, Tm), associates tied together by weak van der Waals' forces are formed, resulting in a C.N. marginally larger than 9. The tendency to form associates and polymeric structures in solid state also rationalizes their poor solubility in apolar solvents. Therefore, initial studies mainly focused on the formation of Lewis acid–base adducts between the lanthanide atom (acting as Lewis acid) and coordinating ligands such as isonitriles and ethers (acting as Lewis bases).

The first divalent complexes such as $[(\eta^5\text{-}C_5H_5)_2Eu]$ and $[(\eta^5\text{-}C_5H_5)_2Ln(thf)_n]$ (Ln = Eu, Sm, Yb) were prepared by either direct synthesis of Eu+HC_5H_5 in liquid ammonia [2c] or by transmetallation between Ln + $Hg(C_5H_5)_2$ (Figure 2.7.3). Similar to their Ln(III) counterparts, they are strong Lewis acids and readily bind donor ligands. However, they are also strong reducing reagents participating in single-electron transfer (SET) processes (also see Section 2.7.3) and therefore serve as ideal starting materials for the synthesis of $[Cp'_2LnX]$ and $[Cp'LnX_2]$ complexes (Cp' = Cp and substituted Cp derivatives) (Figure 2.7.3) [6].

$$Eu + 3\ C_5H_6 \xrightarrow{NH_3(l)} [(\eta^5\text{-}C_5H_5)_2Eu] + C_5H_8$$

$$Ln + Hg(C_5H_5)_2 \xrightarrow{THF} [(\eta^5\text{-}C_5H_5)_2Ln(thf)_n] + Hg \qquad (Ln = Eu, Sm, Yb)$$

$$[(\eta^5\text{-}C_5Me_5)_2Yb] + 0.5\ Ph_2Hg \longrightarrow [(\eta^5\text{-}C_5Me_5)_2YbPh] + 0.5\ Hg$$

$$[(\eta^5\text{-}C_5Me_5)_2Yb] + 2\ MeCu \longrightarrow [(\eta^5\text{-}C_5Me_5)_2Yb_2(Me)(\mu\text{-}Me)] + 2\ Cu$$

$$1.0\ [(\eta^5\text{-}C_5Me_5)_2Yb(OEt_2)] + (1.0 + a)\ RX \longrightarrow (1.0 - a)\ [(\eta^5\text{-}C_5Me_5)_2YbX] + a\ [(\eta^5\text{-}C_5Me_5)YbX_2] + a\ C_5Me_5R + 1.0\ (total)\ [R\text{-}R, R\text{-}H]$$

Figure 2.7.3: Preparation of divalent rare-earth metal metallocenes and their application in SET reactions.

After some years of stagnation, the renaissance of organolanthanide chemistry was again connected with the discovery of a new class of metallocenes, that is, the preparation of uranocene, $[(\eta^8\text{-}C_8H_8)_2U]$ in 1968 by Müller-Westerhoff and Streitwieser [7]. Similar to ferrocene, the uranium atom in $[(\eta^8\text{-}C_8H_8)_2U]$ is "sandwiched" between two cyclooctatetraenyl (COT, $[C_8H_8]^{2-}$) ligands, being 10π-Hückel aromatic systems. The COT ligand was also applied to prepare neutral half-sandwich compounds of the types $[(\eta^8\text{-}C_8H_8)LnX]_2$ or $[(\eta^8\text{-}C_8H_8)LnX(L)_n]$ (X = Cl, L = THF; $n = 1$ (Ln = Sc, Er, Lu), $n = 2$ (Ln = La, Ce, Nd, Pr, Sm)) or anionic sandwich complexes such as $K[(\eta^8\text{-}C_8H_8)_2Ln]$ (Ln = Sc, Y, La, Ce, Pr, Nd, Sm, Gd, Tb), whose molecular structures were determined, and their magnetic properties were evaluated (Figure 2.7.4) [3].

One notable exception within this series of COT-containing molecules represents the neutral cerocene $[(\eta^8\text{-}C_8H_8)_2Ce]$, since it constitutes the first example of what was considered to be a well-characterized organometallic Ce(IV) compound, from which the triple-decker complex $[(\eta^8\text{-}C_8H_8)_3Ce_2]$ could be prepared (Figure 2.7.4) [8]. However, the electronic structure of $[(\eta^8\text{-}C_8H_8)_2Ce]$ is more complicated than a simple ionic bonding model would actually predict (see Section 2.7.2).

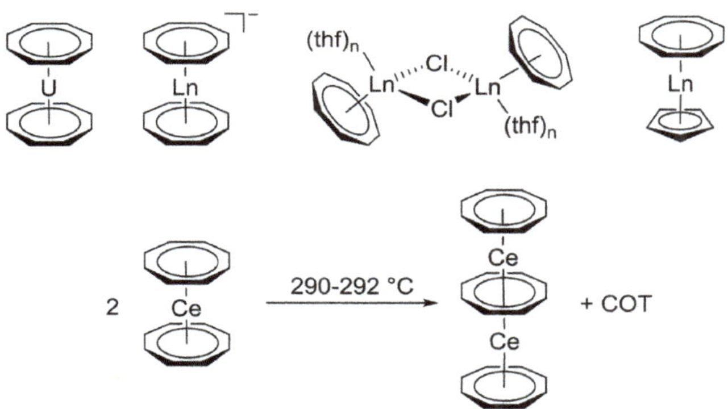

Figure 2.7.4: Examples of COT-containing rare-earth metal compounds.

Concomitant with investigations concerning the influence of substituents on the COT ligand, the development of a high yield synthesis of pentamethylcyclopentadienyl ($Cp^* = \eta^5\text{-}C_5Me_5$) constituted a milestone in organometallic rare-earth metal chemistry. Starting in the 1980s, the groups of Watson [9], Andersen [10] and Evans [11] have prepared various Cp*-containing rare-earth metal complexes. The introduction of Cp* improves the solubility of the resulting metal compounds and the permethylation of the Cp-ring also increases the reducing power of the divalent lanthanoidocenes. These complexes are readily accessible via salt-metathesis reaction between LnI_2 and NaCp* in diethyl ether, and subsequent removal of the coordinated diethyl ether yields the base-free decamethylmetallocenes (Figure 2.7.5) [12].

Their bent sandwich structures in gas phase and in solid-state resemble those found in the heavy-alkaline earth metallocenes [Cp*$_2$M] (Cp* = η^5-C_5Me_5; M = Ca, Sr, Ba), which is not surprising, considering that Eu(II), Sm(II) and Yb(II) have similar ionic radii to those of Ca(II) and Sr(II) [13]. To account for the observed bending, several explanations have been proposed, including attractive interannular dispersion (van der Waals) interactions between the methyl substituents on both Cp-rings [14].

Nevertheless, besides divalent Cp* containing species, several trivalent species have also been prepared; some prominent representatives include [Cp*$_3$Sm] [15] and [Cp*$_2$Lu(CH_3)(μ-CH_3)]$_2$ (Figure 2.7.5) [16]. Over the years, the steric demand on the Cp ligand has systematically been varied and more sterically hindered Cp-ligands were introduced. This allowed divalent metallocenes of lanthanide metals, which are difficult to reduce such as those of Tm, Dy and Nd to be isolated (see Figures 2.7.3 and 2.7.6) and their reaction chemistry was explored:

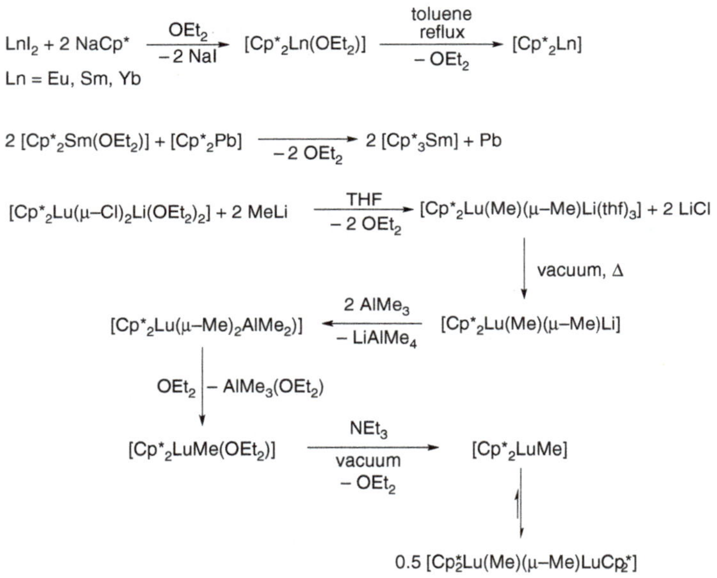

$$\text{LnI}_2 + 2\,\text{NaCp}^* \xrightarrow[-2\,\text{NaI}]{\text{OEt}_2} [\text{Cp}^*_2\text{Ln(OEt}_2)] \xrightarrow[-\,\text{OEt}_2]{\text{toluene, reflux}} [\text{Cp}^*_2\text{Ln}]$$

Ln = Eu, Sm, Yb

$$2\,[\text{Cp}^*_2\text{Sm(OEt}_2)] + [\text{Cp}^*_2\text{Pb}] \xrightarrow[-2\,\text{OEt}_2]{} 2\,[\text{Cp}^*_3\text{Sm}] + \text{Pb}$$

$$[\text{Cp}^*_2\text{Lu}(\mu\text{-Cl})_2\text{Li(OEt}_2)_2] + 2\,\text{MeLi} \xrightarrow[-2\,\text{OEt}_2]{\text{THF}} [\text{Cp}^*_2\text{Lu(Me)}(\mu\text{-Me})\text{Li(thf)}_3] + 2\,\text{LiCl}$$

$$[\text{Cp}^*_2\text{Lu(Me)}(\mu\text{-Me})\text{Li(thf)}_3] \xrightarrow{\text{vacuum, }\Delta} [\text{Cp}^*_2\text{Lu(Me)}(\mu\text{-Me})\text{Li}]$$

$$[\text{Cp}^*_2\text{Lu(Me)}(\mu\text{-Me})\text{Li}] \xrightarrow[-\,\text{LiAlMe}_4]{2\,\text{AlMe}_3} [\text{Cp}^*_2\text{Lu}(\mu\text{-Me})_2\text{AlMe}_2)]$$

$$[\text{Cp}^*_2\text{Lu}(\mu\text{-Me})_2\text{AlMe}_2)] \xrightarrow[-\,\text{AlMe}_3(\text{OEt}_2)]{\text{OEt}_2} [\text{Cp}^*_2\text{LuMe(OEt}_2)]$$

$$[\text{Cp}^*_2\text{LuMe(OEt}_2)] \xrightarrow[\text{vacuum}, -\,\text{OEt}_2]{\text{NEt}_3} [\text{Cp}^*_2\text{LuMe}] \rightleftharpoons 0.5\,[\text{Cp}^*_2\text{Lu(Me)}(\mu\text{-Me})\text{LuCp}^*_2]$$

Figure 2.7.5: Preparation of selected divalent and trivalent Cp*-containing rare-earth metal compounds.

Besides Cp ligands, other Hückel aromatic systems such as neutral arenes (6π-electrons) [17] and the anionic cyclononatetraenyl (10π-electrons) [18] have also been isolated (Figure 2.7.7).

Neutral [(η^6-1,3,5-(tBu)$_3C_6H_3$)$_2$Ln] (Ln = Sc, Y, La, Pr, Sm, Gd, Tb, Dy, Ho, Er, Lu) complexes constitute the f-element analogues to bis(benzene)chromium [(η^6-C_6H_6)$_2$Cr], but their synthesis and isolation require metal-atom-ligand-vapor cocondensation techniques and sterically encumbered arene ligands. Furthermore, the stability of

Ln = Nd, Dy

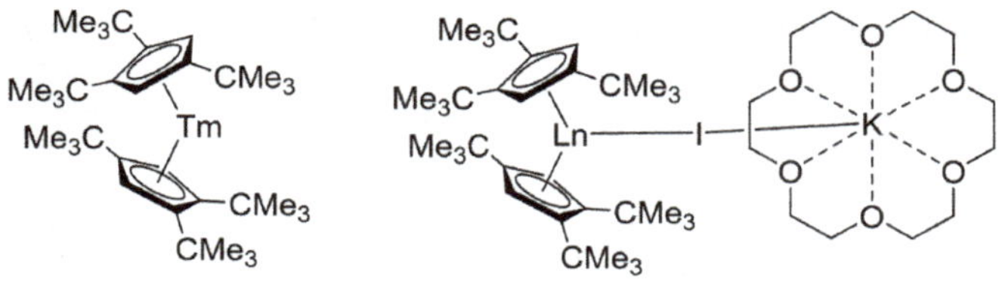

Figure 2.7.6: Synthesis of divalent metallocenes of Tm, Dy and Nd.

Ln = Eu, Sm, Tm, Yb

Ln = Sc, Y, La, Pr, Sm, Gd, Tb, Dy, Ho, Er, Lu

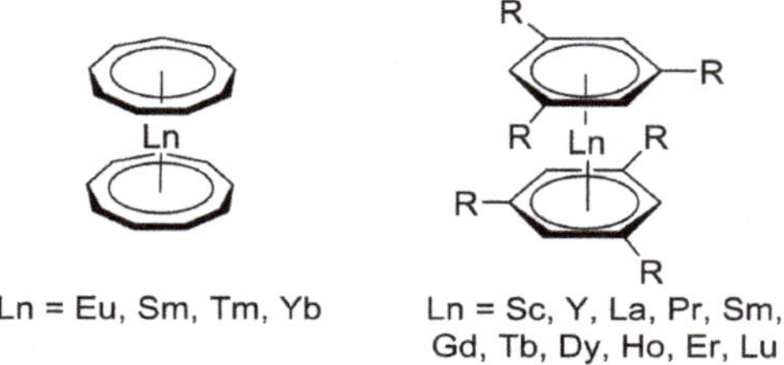

Figure 2.7.7: Cyclononatetraenyl and arene containing rare-earth metal compounds.

these compounds varies significantly within the series of lanthanoid metals, which is based on the general bonding mechanism in these molecules. The 15 valence electron species (Ln = Sc, Y, La) with their d^1s^2 electron configuration display backbonding between the Ln metal atom and the arene ring. However, when this picture is extended to the lanthanide metals, a promotion from the electronic ground state configuration of f^ns^2 to the electronic excited state $f^{n-1}d^1s^2$ must occur. This proposition explains the trend that for lanthanide atoms, for which the promotion energy is particularly large, only unstable complexes are obtained. Nevertheless, the inherent donor–acceptor synergy between the lanthanide metal and the arene ligand results in a considerable covalent Ln-arene bonding with short Ln–C bonds, expansion of the C–C bond distances and strongly deviating effective magnetic moments for $[(\eta^6\text{-}1,3,5\text{-}(t\text{Bu})_3C_6H_3)_2\text{Ln}]$ complexes from those of the free Ln^0 atoms. The bond dissociation energy BDE(Ln-arene) = 285 kJ mol^{-1} observed for $[(\eta^6\text{-}1,3,5\text{-}(t\text{Bu})_3C_6H_3)_2\text{Ln}]$ is also significantly larger than that found for $[(\eta^6\text{-}C_6H_6)_2\text{Cr}]$ of 170 kJ mol^{-1}. These compounds represented one of the first examples that question the purely ionic bonding model in organo rare-earth metal chemistry.

In contrast to complexes with carbocyclic ligands, the numbers of homoleptic pentadienyl, allyl and alkyl compounds are rather limited [19]. Although pentadienyls are generally considered to be open-Cp variants, only a small number of pentadienyl rare-earth metal compounds were prepared since the early 1980s [20]. Only recently, some new structural motifs originating from pentadienyl ligands have been discovered (Figure 2.7.8) [21], and novel applications as precatalysts in diene polymerization have been established [22].

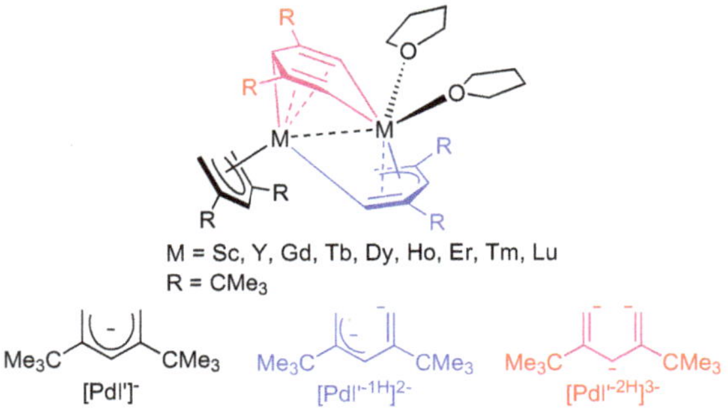

Figure 2.7.8: Di- and trianionic ligand scaffolds derived from a pentadienyl ligand.

Allyl complexes of the lanthanide metals are also known, but their molecular structures strongly depend on the ionic radius of the lanthanide atom as well as the substitution patterns on the allyl ligand itself [23]. In general, there is a strong tendency to yield ate-complexes of the general type $[Li(solv)]^+[Ln(\eta^3\text{-}C_3H_5)_4]^-$. However, given sufficient steric bulk of the allyl ligand, either THF adducts $[(\eta^3\text{-}1,3\text{-}(Me_3Si)_2C_3H_3)_3Ln(thf)]$ (Ln = Ce, Nd, Tb) or base-free species $[(\eta^3\text{-}1,3\text{-}(Me_3Si)_2C_3H_3)_3Ln]$ (Ln = Dy, Ho, Er, Tm, Lu) are isolated.

Even less explored are neutral, homoleptic σ-organyl compounds of the general type $[LnR_n]$, since their stability is rather limited and they suffer from β-H elimination and bimolecular degradation [19b]. However, with sufficient steric protection, a few stable ate-derivatives have successfully been prepared, in which the lanthanoid ion is coordinated in a tetrahedral fashion (Figure 2.7.9). Alternatively, smaller alkyl groups such as methyl groups can also be employed to satisfy the coordination sphere at the metal ion, which, however, results in an octahedral ligand arrangement at the lanthanide atom (Figure 2.7.9).

Unlike $SiMe_3$-substituted alkyl derivatives such as $R = CH_2SiMe_3$ and $CH(SiMe_3)_2$, neutral, homoleptic alkyl complexes are accessible (Figure 2.7.10). While a tetrahedral structure is found for $[Ln(CH_2SiMe_3)(thf)_2]$ (Ln = Sc, Y, Sm, Tb, Er, Tm, Yb, Lu), the more sterically hindered derivatives $[Ln(CH(SiMe_3)_3]$ (Ln = Y, La, Ce, Sm, Lu) feature a pyramidal structure. The origin of this pyramidalization has been subject to some debate. Originally, agostic γ-C-H···Ln interactions have been invoked to account for this experimental structure, but more recent studies unambiguously demonstrate that this structural feature is caused by a 3-center-2-electron β-Si-γ-C···Ln interaction [24]. This conclusion is also supported by NMR spectroscopic studies performed in the solid state and in solution. Furthermore, computational studies establish that the

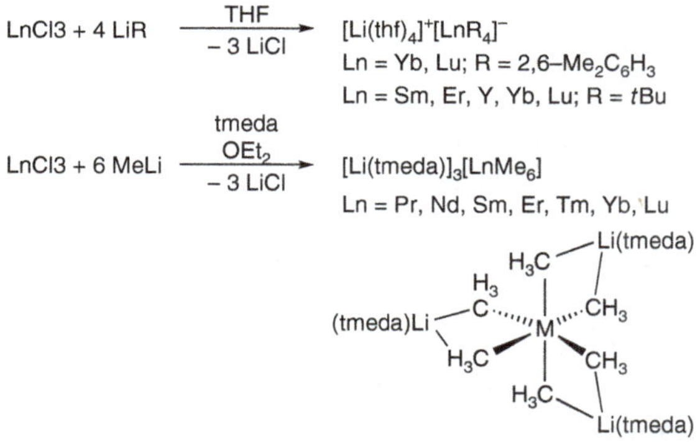

Figure 2.7.9: Homoleptic σ-organyl ate-complexes of the rare-earth metals.

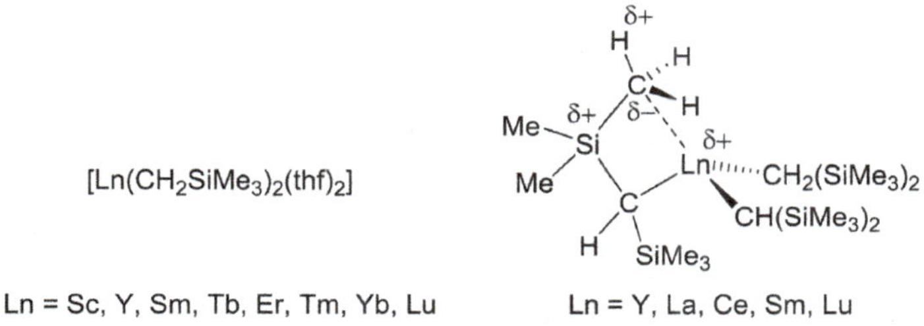

Figure 2.7.10: Neutral alkyl compounds of the rare-earth metals.

γ-C atom carries a negative charge, whereas Ln, γ-H and β-Si carry a positive charge, ruling out an agostic γ-C-H···Ln interaction.

2.7.2 Bonding considerations of rare-earth organometallics

Lanthanides exist principally in their most stable trivalent oxidation state, which is caused by the high energy required for the fourth ionization. Only cerium and terbium can actually reach this fourth oxidation state. However, while it is a relatively common oxidation state for coordination compounds of Ce [25], it is more rare for terbium [26]. In organometallic chemistry, the most prominent example of a formally tetravalent complex is $[(\eta^8\text{-}C_8H_8)_2Ce]$, also known as cerocene (see previous paragraph). On the other hand, divalent lanthanides have been well known for decades and numerous studies have been reported with so-called low-valent lanthanides. Since they rapidly oxidize to their preferred trivalent oxidation state, their reductive chemistry is well-studied and several examples will be discussed later. It is convenient to classify the divalent lanthanide compounds into two classes: the classical ones, based on Eu(II), Sm(II) and Yb(II), and the nonclassical, based on the other rare-earth metals [27]. Originally, the classic representatives have been the most widely used ones. They are the easiest to reduce (Figure 2.7.2); and their cyclopentadienyl derivatives are relatively stable and rather soluble in hydrocarbon solvents, explaining their great popularity [10, 28]. Until the end of the twentieth century, the occurrence of other examples remained scarce and their relative instability limited their application. However, after the development of feasible synthetic strategies for divalent lanthanide halide precursors, several rare-earth metal organometallic compounds in nonclassic oxidation states emerged, generally in the context of small molecule activation [27]. Additionally, inspired by the seminal work of Lappert on the divalent lanthanum complex, $[Cp''_3La^{II}]^-$ ($Cp'' = \eta^5$-1,3-$Me_3Si_2C_5H_3$), recent studies now include divalent complexes of the entire lanthanide series [5]. However, the formal oxidation state in organolanthanides needs to be taken cautiously, and the spectroscopic oxidation state of low- and high-valent compounds can be tendentious (see further). From the point of view of their f-electron count, the divalent complexes of Sm, Eu, Yb and Tm are viewed as $4f^{n+1}$ species [29], that is, the additional electron is placed in the f-shell of the corresponding trivalent state ($4f^n$). In contrast, the divalent complexes of the remaining rare-earth metals are viewed as $4f^n5d^1$ systems, which implies that the supplementary electron does not fill the 4f-shell but the low-lying empty 5d-shell [30].

The overall bonding of organolanthanides is primarily ionic. This means that the charges are well localized and separated, and their electrostatic interactions form the principal component of the bond. As noted in the previous paragraph, $[(C_5H_5)_3Yb]$ and $[(\eta^8\text{-}C_8H_8)_2Ce^{III}]^-$ complexes react rapidly with iron and uranium salts to form ferrocene and uranocene, respectively, indicating that ligand exchange processes are fast. Similar arguments were also given to explain the bent structure of the divalent $[(C_5Me_5)_2Yb^{II}]$ complex, since no rational explanation could be provided based on molecular orbital models derived from computations and photoelectron spectroscopy [31]. This archetypical ionic bonding model in rare-earth metal complexes is based on the limited radial extension of the 4f orbitals, namely 4f orbital manifold remains largely unperturbed by the ligand field. As such, most applications developed from organolanthanide compounds are based on the steric influence of the ligand, which modulates the reactivity and/or the physical properties of the complexes.

However, a few complexes challenge this traditional bonding picture. The problem began with one of the fist organolanthanide complexes prepared. For example, the reported magnetic susceptibility of $[(C_5H_5)_3Yb]$ [2b] is lower than expected; the effective magnetic moment $\mu_{eff} \approx 4\ \mu_B$ instead of the 4.5 μ_B that is expected for an ionic trivalent ytterbium ion, in which the ligand has little influence. Furthermore, the Mößbauer spectrum of [Cp_3Eu] is ambiguous, since it is neither in agreement with a trivalent nor a divalent oxidation state [32]. [$Cp^*_2Yb(OEt_2)$] exhibits a prominent emission band at $\lambda_{max} = 780$ nm [33], despite the fact that divalent ytterbium is a closed-shell system with no plausible f-shell excited state. Finally, the magnetic susceptibility for [$(\eta^8\text{-}C_8H_8)_2Ce$], a f^0 complex, is not negative, as expected for a diamagnetic molecule, but it is positive and temperature independent [8b]. Thus, more than 50 years after their first report, an accurate and general bonding scheme for all organolanthanide compounds was still missing. The next paragraphs will focus on three representative examples of this dichotomy that are formally trivalent [$(C_5H_5)_3Yb$] [2a], formally divalent [$Cp^*_2Yb(bipy)$] [34] and formally tetravalent [$(\eta^8\text{-}C_8H_8)_2Ce$] complexes [8a].

The formal charges of these three molecules represented in Figure 2.7.11, [$(C_5H_5)_3Yb$], [$Cp^*_2Yb(bipy)$] and [$\eta^8\text{-}(C_8H_8)_2Ce$], are given as follows: The first one contains a trivalent, Yb(III), f^{13}, atom coordinated to three monoanionic cyclopentadienyl ligands, Cp^-; the second one features a divalent, Yb(II), f^{14}, atom with two cyclopentadienyl ligands, Cp^- and a neutral bipyridine ligand; and in the third, a tetravalent Ce(IV), f^0, center is coordinated by two cyclooctatetraenide ligands, $[C_8H_8]^{2-}$.

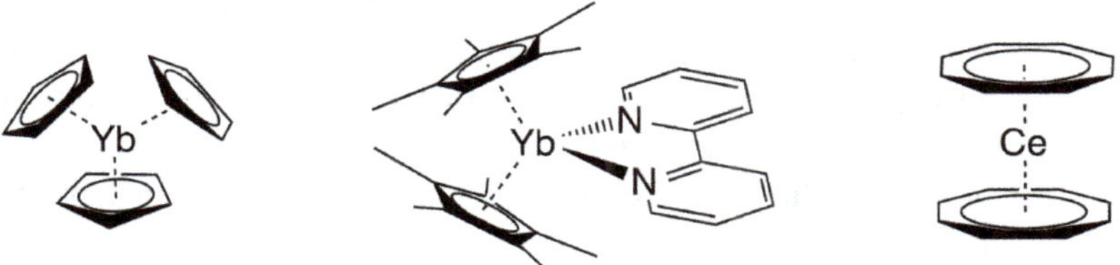

Figure 2.7.11: Representation of the molecular structures of the complexes [Cp_3Yb] (left), [Cp^*_2Yb(bipy)] (center) and [$(\eta^8\text{-}C_8H_8)_2Ce$] (right).

In [$Cp^*_2Yb(bipy)$], given the strong reducing nature of the divalent lanthanides, it is plausible to assume that an intramolecular electron transfer occurs and thus the Yb(II) center is oxidized to Yb(III), f^{13}, while the bipyridine ligand accepts this electron in its LUMO to form a bipy radical anion. These two resonance forms Yb(III), f^{13}-bipy$^{\bullet-}$ and Yb(II), f^{14}-bipy with two different charge separations are shown in Figure 2.7.12. In the case of the limiting Yb(II), f^{14}-bipy resonance form, the complex is diamagnetic, whereas in the other resonance structure assuming uncorrelated electrons, the complex should be paramagnetic with a theoretical value of the effective moment of the order of $\mu_{eff} = 4.8\ \mu_B$ (based on a trivalent ytterbium (^2F) and radical anion (^2S): $\mu_{eff} = \sqrt{(4.5)^2 + (1.72)^2}\ \mu_B = 4.8\ \mu_B$). However, the effective moment at

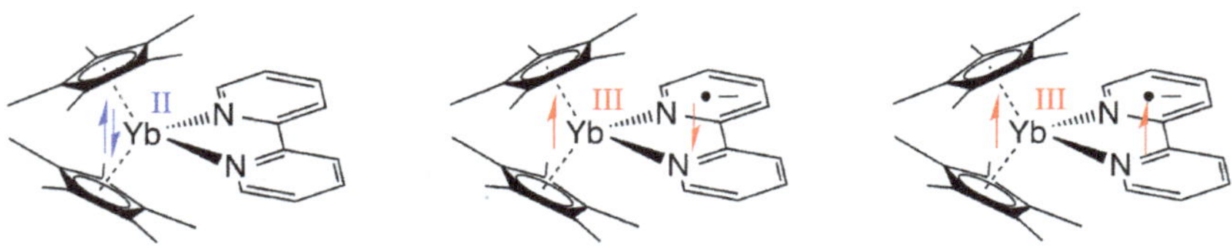

Figure 2.7.12: Representation of the resonance structures of the [Cp*$_2$Yb(bipy)] complex.

room temperature of [Cp*$_2$Yb(bipy)] is reduced to $\mu_{eff} = 2.3\ \mu_B$ [34]. Furthermore, the plot of magnetic susceptibility (χ) as a function of temperature (T) clearly points to a nonmagnetic ground state with a strong temperature-independent paramagnetism (TIP or VanVleck paramagnetism) [35].

From these observations, it is possible to envisage a nonmagnetic ground state for which the bipyridine electron is antiferromagnetically coupled to that of the Yb atom, so that the ground state is an open-shell singlet with a low-lying triplet excited state, rationalizing the presence of the VanVleck paramagnetism at low temperature and the effective moment of $\mu_{eff} = 2.3\ \mu_B$ at room temperature. This situation is, however, in conflict with the idea that no interaction occurs between the f-electrons and the electrons located at the ligand. Moreover, this situation is not sufficient to model the magnetism curve of [Cp*$_2$Yb(bipy)] [36]. A key spectroscopic method to unravel formal vs. spectroscopic oxidation states in lanthanide chemistry constitutes X-ray absorption at the near edge spectroscopy (XANES). Although it requires the use of synchrotron radiation, it gives a precise insight into the oxidation state, since the white line energy is characteristic for each metal oxidation state. Measurements of L_{III}-edge XANES on [Cp*$_2$Yb(bipy)] confirmed the peculiarity of this molecule: Both divalent and the trivalent signatures are present, and the molecule has, therefore, a multiconfigurational ground-state composed of both configurations, that is, Yb(III), f^{13}-bipy$^{\bullet-}$ and Yb(II), f^{14}-bipy, with different weights [37]. Adapted multireference theoretical calculation (CASSCF) confirmed this proposition and rationalized the magnetic data [36].

Similar spectroscopic studies (magnetism, XANES) [8b] coupled with adapted theoretical computations demonstrated a similar bonding situation for cerocene [η^8-$(C_8H_8)_2$Ce]: with two resonances structures, that is a Ce(IV), f^0 atom coordinated to two dianionic $[C_8H_8]^{2-}$ ligands or alternatively a resonance form, in which a Ce(III), f^1 center binds to two intermediate charged $[C_8H_8]^{1.5-}$ ligands (Figure 2.7.13) [38]. Again, only the model of a multiconfigurational ground state is consistent with the experimentally observed behavior of cerocene.

Finally, this model accounts for unusual properties in [$(C_5H_5)_3$Yb] (EPR, solid-state magnetism and visible absorption spectra). The bonding in this molecule can also be rationalized by the formation of a multiconfigurational ground state composed of a reduced ytterbium, Yb(II) and one oxidized Cp radical, Cp$^\bullet$, along with traditional Lewis form, Yb(III) and three anionic Cp$^-$ ligands [31b].

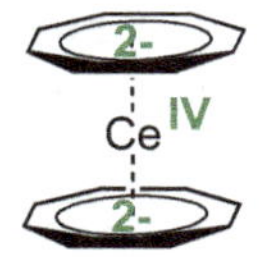

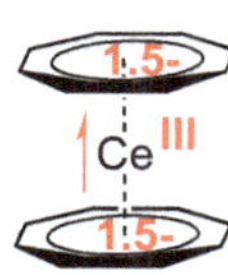

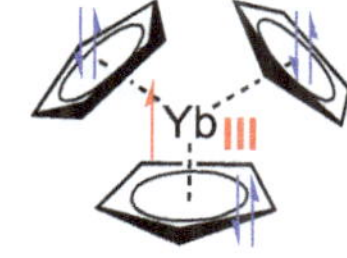

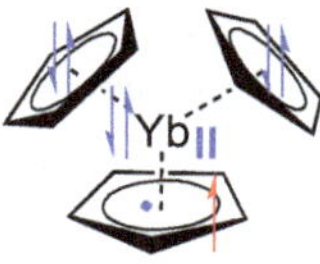

Figure 2.7.13: Representation of the resonance structures of the $[(\eta^8\text{-}C_8H_8)_2Ce]$ (left) and $[Cp_3Yb]$ (right) complexes.

These observations are important. From a fundamental point of view, this implies that lanthanide organometallics can mimic covalency with the surrounding ligands, without a strong electronic overlap between metal and ligand orbitals. Although the bonding within the individual resonance structures of a multiconfigurational species is predominately ionic, the overall averaged electronic density reflects ligand participation in the bonding. To realize a multiconfigurational ground state, the energy gap between the metal and ligand orbitals [39] and the respective orbital symmetry need to be considered [40], because they influence the relative contribution of the different resonance forms to the multiconfigurational ground state. It is important to note, in this context, that the concept of redox noninnocence, which is very well developed for transition metals [41], reemerges in organolanthanide complexes in the form of multiconfigurational ground states. Furthermore, it may also extend to the excited states, and therefore the emission properties [33]. Here, the 5d-shell may be involved in the description of the individual resonances forms.

In summary, while the bonding of lanthanide compounds is principally ionic, multiconfigurational ground states allow them to accommodate the Pauling electroneutrality principle, by mixing different configurations. As a consequence, the ligand symmetry and its relative orbital energies (compared to the f-orbital energies of all possible oxidation states) are equally important as the steric bulk. Considering the relative low-lying empty π^* orbitals and high-energy π system of typical organometallic ligands combined with the high symmetry of these compounds, organolanthanides are excellent candidates for multiconfigurational ground and excited states that influence their physical properties and/or reactivity.

2.7.3 Reaction mechanism in rare-earth metal chemistry and small molecule activation

The traditional mechanisms of oxidative addition and reductive elimination that dominate d-transition chemistry and require a (formal) oxidation state change of ±2, are not available for lanthanide metals. Instead, the reactivity of organometallic lanthanide complexes is determined by σ-bond metathesis, insertion and SET processes. The first two processes require an initial adduct formation, which naturally depends on the steric congestion at the metal atom. Therefore, the rates for σ-bond metathesis and insertion reactions vary significantly with the substitution patterns at the ligand as well as the ionic radii of the rare-earth metal atom.

2.7.3.1 σ-Bond metathesis

The σ-bond metathesis mechanism is a concerted process with a relatively nonpolar transition state (Figure 2.7.14) [42]. Furthermore, the Ln(III) metal atom is highly electrophilic, and the group R is strongly nucleophilic, as a consequence of the ionic bonding within the rare-earth metal complexes, so that this reaction can also be regarded as a concerted proton transfer mechanism, and it shows large and temperature-dependent kinetic isotope effects. Furthermore, the reaction rate decreases with decreasing s character of the participating σ-bond. The first example for this reactivity was provided by Patricia L. Watson at the beginning of the 1980s. She demonstrated that poorly C–H acidic methane ($pK_a(CH_4) = 49$) can readily be activated under mild conditions using [$Cp^*_2LnCH_3$] (Ln = Sc, Y, Lu) (Figure 2.7.15) [16, 42a].

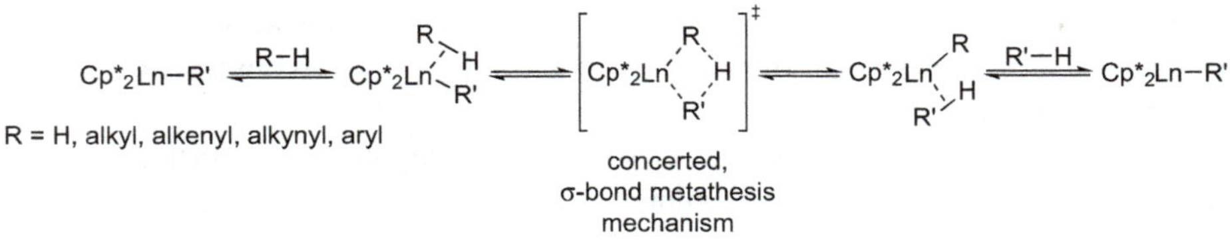

$$Cp^*_2Ln{-}R' \xrightleftharpoons{R-H} Cp^*_2Ln(R{-}H)(R') \rightleftharpoons [Cp^*_2Ln(R)(H)(R')]^{\ddagger} \rightleftharpoons Cp^*_2Ln(R)(H{-}R') \xrightleftharpoons{R'-H} Cp^*_2Ln{-}R'$$

R = H, alkyl, alkenyl, alkynyl, aryl

concerted, σ-bond metathesis mechanism

Figure 2.7.14: General mechanistic scheme for σ-bond metathesis reactions.

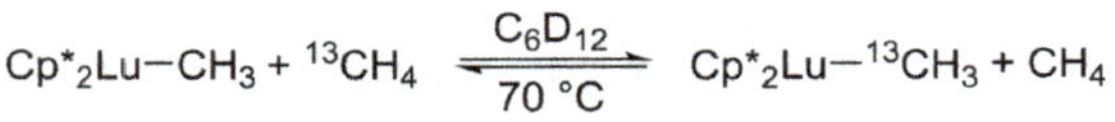

$$Cp^*_2Lu{-}CH_3 + {}^{13}CH_4 \xrightleftharpoons[70\ °C]{C_6D_{12}} Cp^*_2Lu{-}{}^{13}CH_3 + CH_4$$

Figure 2.7.15: Methane activation accomplished by [$Cp^*_2LuCH_3$].

Furthermore, [Cp'_2LnR] ($Cp' = \eta^5\text{-}C_5H_5$ or substituted derivatives, R = organyl) complexes are valuable starting materials – for example, in the preparation of the corresponding rare-earth metal hydrides, [Cp'_2LnH], which are either monomeric or dimeric depending on the steric demand of the Cp ligand and the ionic radius of the rare-earth metal. These Ln-H species are very reactive and may even activate strong C–F bond such as those found in CH_3F or CH_2F_2 (Figure 2.7.16) [43].

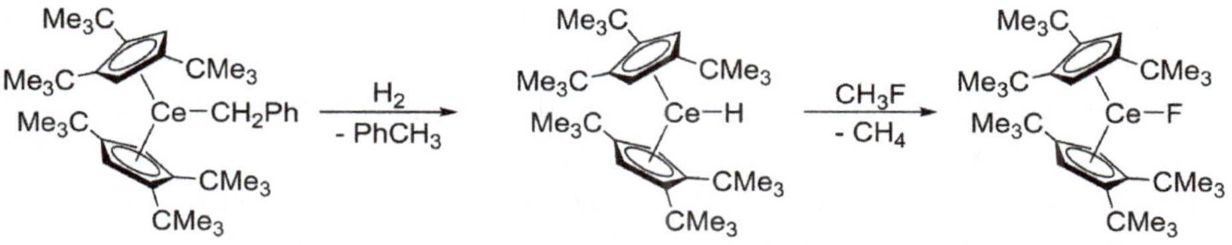

Figure 2.7.16: CF bond activation by a Ln–H species.

2.7.3.2 Insertion reactions

Early on, P. L. Watson also investigated the reaction of [$Cp^*{}_2LuCH_3$] towards unsaturated substrates such as ethylene and α-olefins such as 2-butene, and different reactivity patterns emerged depending on the substitution pattern of the olefin [16]. When the unsaturated π-system can approach the metal center closely, insertion occurs resulting in polymerization of ethylene or oligomerization of propene (Figure 2.7.17). Chain termination for lanthanide complexes may either occur by an β-H or β-alkyl elimination. The reaction of [$Cp^*{}_2LuCH_3$] with ethylene and propene also represented one of the first model systems for early-transition metal Ziegler–Natta catalysts. Nevertheless, when steric crowding at the metal atom prevents a close approach, σ-bond metathesis may occur instead (Figure 2.7.17).

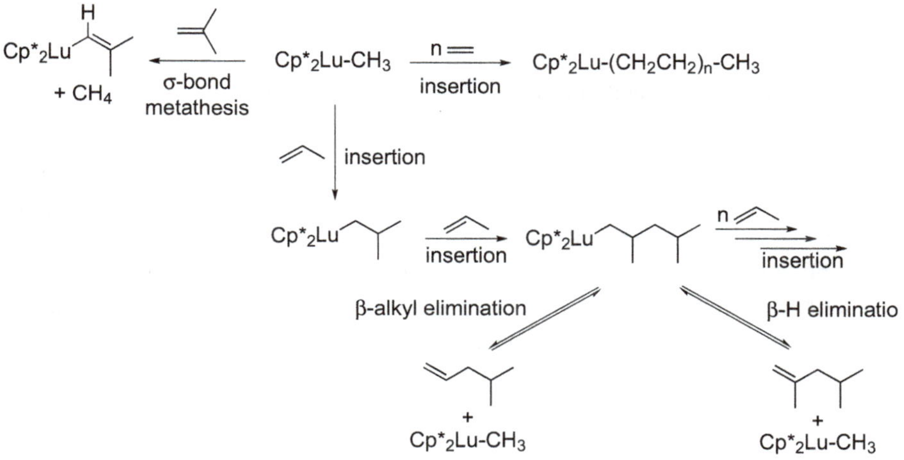

Figure 2.7.17: Reactivity of [$Cp^*{}_2LuCH_3$] toward olefins.

Another important application of organolanthanide complexes includes catalytic intramolecular hydroamination of olefins, which combines σ-bond metathesis with insertion reactions. Very detailed studies in this area were performed by Tobin Marks, and the generally accepted reaction mechanism is shown in Figure 2.7.18 [44]. In an initial σ-bond metathesis reaction, the precatalyst [$Cp^{*}_2LnCH(SiMe_3)_2$] reacts with the substrate to an amido species, which undergoes intramolecular olefin insertion into the Ln–N bond to form the alkyl intermediate, which reacts with free amine to release the hydroamination product to close the catalytic cycle.

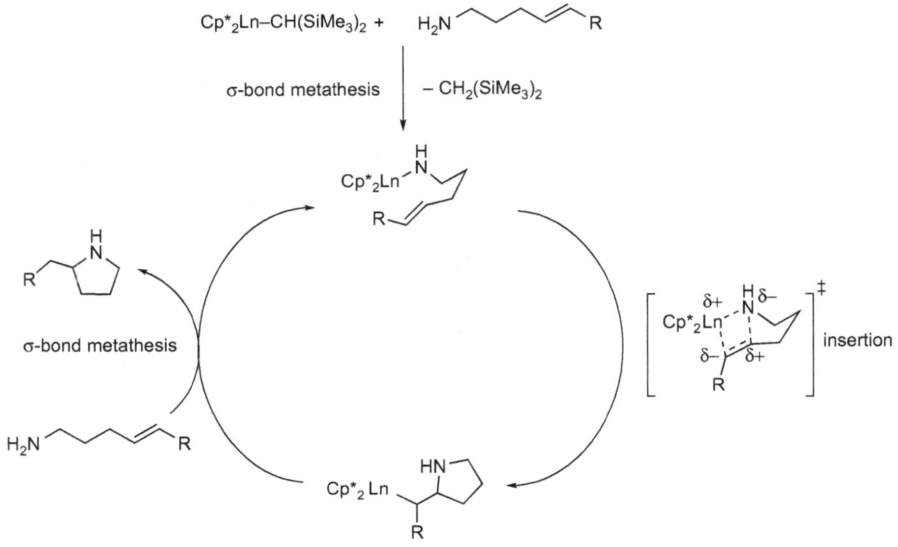

Figure 2.7.18: Catalytic cycle for the rare-earth metal catalyzed hydroamination of olefin.

Besides olefins, CO can also insert, for instance, into a Ln–alkyl or Ln–H bond to form acyl or formyl species (Figure 2.7.19). Acyl complexes are readily isolated, but may also react with an excess of CO to bimetallic species, featuring a dianionic enedione diolate scaffold [45]. In general, formyl derivatives are significantly more reactive and can, for example, react with another Ln–H species to yield dimeric compounds in which an oxomethylene group (or the formaldehyde dianion) bridges two metallocene fragments [46].

2.7.3.3 Single-electron transfer reactions

As already alluded to in the introduction, organolanthanide complexes in the formal oxidation state +2 are highly reducing species that readily engage in SET processes, transferring one electron from lanthanide atom into the σ*- or π*-orbital of a

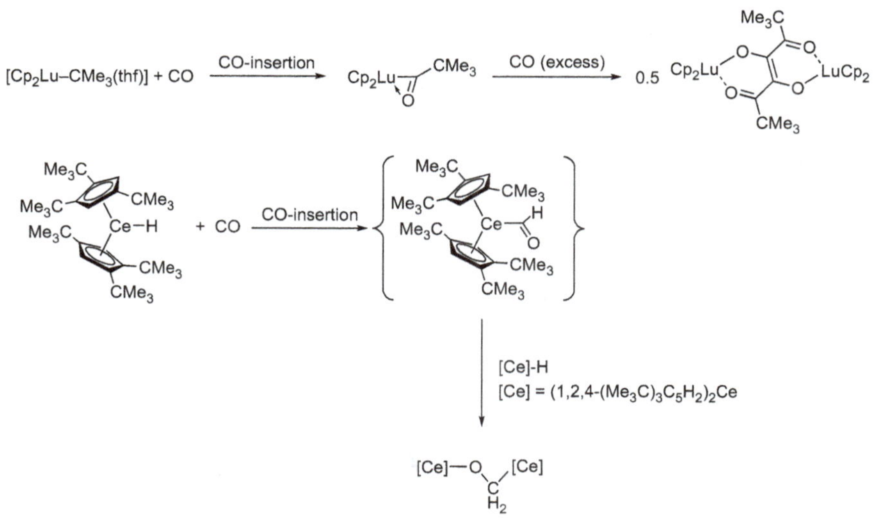

Figure 2.7.19: CO insertion into Ln–alkyl and Ln–H bonds.

substrate. Some of these reactions have found synthetic applications: for example, in the synthesis of Ln-σ-organyl complexes or the preparation of $[Cp^*_2LnX]$ species (Figures 2.7.3 and 2.7.20).

Furthermore, this behavior has also been exploited in the activation of small molecules such as N_2O, CO_2, O_2 and N_2 [27, 47]. Small molecule activation occurs through substrate coordination followed by a SET. After this step, two pathways are plausible: (i) the direct radical–radical coupling to form symmetrically coupled products; or (ii) by coordination to a second organolanthanide fragment, forming a dimeric species bridged by a doubly reduced substrate. The steric bulk of the ligand has a significant influence on the relative energy of the transition states for both pathways and thereby influences the outcome of the reactions. For example, the reaction between organolanthanide fragments and CO_2 may form either oxalates [48], carbonates [49], or a mixture of both [50], while the reaction with N_2O yields the oxido-bridged compounds [51]. Similarly, the reaction with R_3PSe leads to Se^{2-} bridged species [52]. In contrast, the reaction with O_2 is more difficult, since it is fast for divalent organolanthanides, but only few adducts with O_2 (superoxo and peroxo) are known from coordination chemistry [53] and only one organometallic side-on coordinate peroxo dimer has been isolated from the reaction of $[Cp^{ttt}_2Sm]$ (Cp^{ttt}=η^5-$(Me_3C)_3C_5H_3$) with O_2 [49b].

More importantly, considering the very low redox potential of divalent lanthanides, especially those of nonclassical divalent ones, they may even react with inert and apolar molecules such as N_2. This has been achieved for most rare-earths coordination compounds, using reducing conditions that yield side-on coordinate N_2^{2-} bridged dimers [54]. Divalent molecular organometallic species such as $[Cp^{ttt}_2Nd(\mu\text{-}I)$

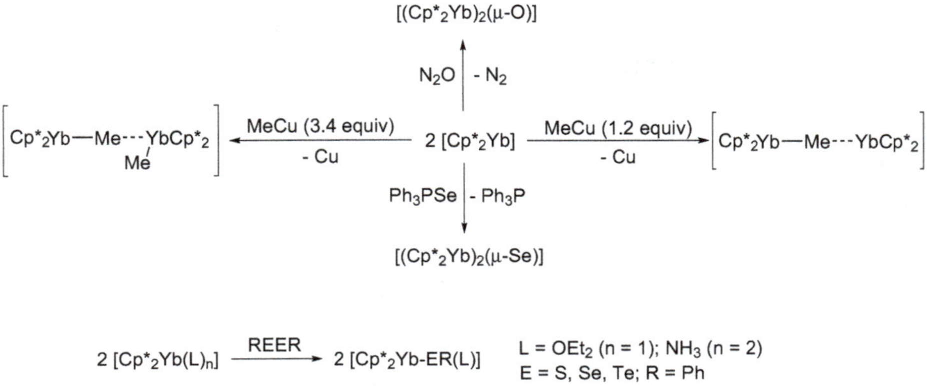

$[(Cp^*_2Yb)_2(\mu\text{-}O)]$

N_2O | $- N_2$

$[Cp^*_2Yb\text{—}Me\cdots YbCp^*_2(Me)]$ ← MeCu (3.4 equiv), − Cu — $2\ [Cp^*_2Yb]$ — MeCu (1.2 equiv), − Cu → $[Cp^*_2Yb\text{—}Me\cdots YbCp^*_2]$

Ph_3PSe | $- Ph_3P$

$[(Cp^*_2Yb)_2(\mu\text{-}Se)]$

$2\ [Cp^*_2Yb(L)_n] \xrightarrow{REER} 2\ [Cp^*_2Yb\text{-}ER(L)]$ $L = OEt_2$ (n = 1); NH_3 (n = 2); E = S, Se, Te; R = Ph

Figure 2.7.20: Single electron transfer (SET) process originating from $[Cp^*_2Yb]$ or $[Cp^*_2Yb(L)_n]$ ($L = OEt_2$, n = 1; $L = NH_3$, n = 2).

K(18-c-6)] ($Cp^{ttt} = \eta^5\text{-}(Me_3C)_3C_5H_3$) show clean N_2 reduction [55]. Small molecule activation accomplished by organolanthanide metals are important contributions, but further functionalization steps and (formally reductive) elimination are also required in order to achieve catalytic chemical transformations [56]. Substrate elimination from sterically encumbered Sm and Tm complexes results in spontaneous reduction of the lanthanide metal ion, with a concomitant elimination of an oxidized ligand [57]. This reactivity has been named sterically induced reduction, since it requires sterically encumbered metal atoms to proceed. Formally, this reduction, occurring at the metal atom, allows for further small molecules activation.

2.7.3.4 Application for single molecule magnets

Single molecule magnets (SMMs) are molecules that can behave as permanent magnets in the absence of an external field at a given temperature, namely the blocking temperature. Such properties lead to a molecular magnetic hysteresis (not a bulk property), which is an important concept in the development of miniaturized technologies used in spintronics and quantum computing devices [58]. For a good SMM, a large magnetic susceptibility and high anisotropy are required. Thus, since the discovery of this physical behavior in the early 1990s [59], most reports focused on transition metal compounds with the goal of increasing the maximum spin value. The first SMM containing a single rare-earth metal ion (terbium) was published in 2003 [60]. This report sheds some light on the important intrinsic physical properties of the f-elements in which the spin-orbit coupling is large. Thus, a description of these properties in terms of the quantum number J is appropriate [61]; and the ligand field splits the latter into m_J states. Using a qualitative electrostatic model, the lowest m_J state can be traced to the geometry around the lanthanide ion [62]: an oblate metal-ion (Dy) was predicted to require a strong axial coordination, while a prolate metal-ion (Er, Tm, Yb) requires a strong equatorial coordination to maximize the m_J microstate and maximize the anisotropy. Thus, from 2010 on, the first organolanthanide complexes were designed to fulfill the required coordination environment.

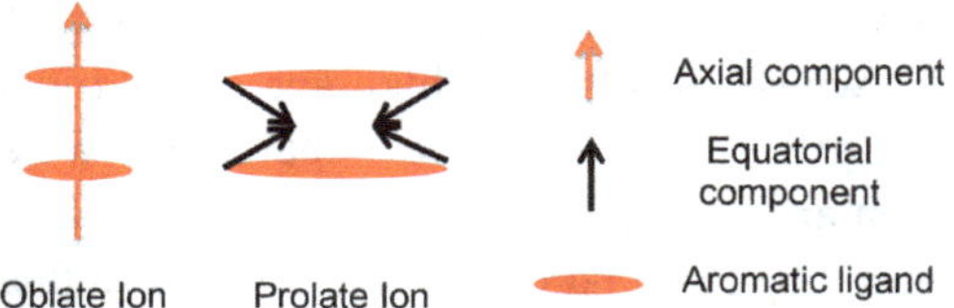

Figure 2.7.21: Qualitative scheme of the geometric requirement for designing SMMs with oblate and prolate ions.

The first organolanthanide complexes carried a $\{Cp_2Dy\}^+$ unit bridged by benzotriazolide- and amidobridged ligands (Figure 2.7.21) [63]. This was the starting point of an elegant magnetostructural study, consequently manipulating the $\{Cp'Dy\}^+$ (Cp′ = Cp, Cp*, MeCp) units in these dimers by steric and electronic modulation of the bridging unit [64]. Initial studies focused on the design of dimers, in which the two metals would couple or align their spins, so that the magnetization is maximized. However, as discussed in the bonding paragraph, the extent of the coupling of f-shell is very small, and this strategy was not very successful until the synthesis of $\{[((Me_3Si)_2N)_2Ln_2](\mu\text{-}N_2)\}^-$ (Ln = Dy, Gd, Tb), containing a bridging $(N_2)^{3-}$ radical anion was realized [65]. Strong antiferromagnetic coupling between the radical anion and the metal is possible (see bonding paragraph and Figure 2.7.22) and the spins at the metal atom align ferromagnetically, leading to increased magnetization and larger blocking temperatures up to 14 K [65b]. Building on this observation, related organometallic molecules such as $[(Cp^*{}_2Dy)_2(bipym)]^-$ (Figure 2.7.22) [66] and the organometallic derivative of the $N_2{}^{3-}$ bridged terbium complex discussed above, $\{[(C_5Me_4H)_2Tb(thf)_2](N_2)\}^-$ [67] were prepared. A similar approach led to the report of trimers with the radical

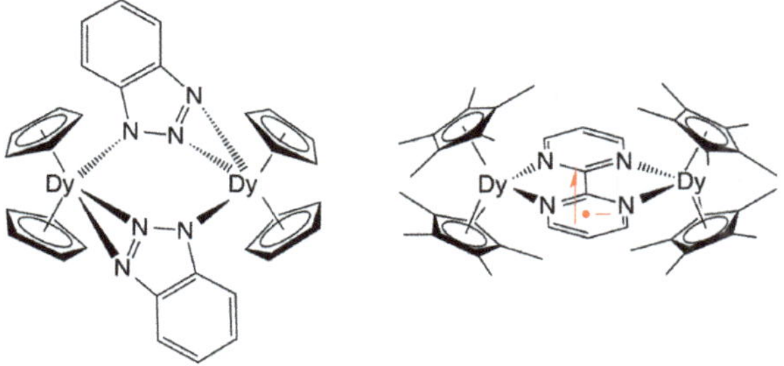

Figure 2.7.22: Representation of the molecular structures of the Dy dimers behaving as SMMs.

hexaazatrinaphthalene ligand [68]. However, although these compounds are SMMs with a high coercive field (much higher than commercial rare-earth magnets), their blocking temperatures are still in the He temperature regime. Indeed, although substituted Cp ligands provide a good axial field, the equatorial coordination of the bridges decreased the potential of these interesting compounds.

In 2011, the first single ion organolanthanide compound behaving as a SMM was reported [69]. The complex is based on erbium and is a heteroleptic complex of COT (η^8-(C_8H_8)) and Cp*, [(η^8-C_8H_8)ErCp*], which results in high-energy barriers (Figure 2.7.23). Several complexes containing the COT ligand were prepared, but were mostly based on Er, since it is a prolate ion and requires equatorial contributions to the ligand field. The large COT ligand with its equatorial charge distribution fulfills this requirement, whereas the smaller Cp ligand with its axial point charge distribution is more suitable for oblate lanthanide ions. A judicious metal–ligand combination is key to designing SMMs and is referred to as metal–ligand pair anisotropy [70]. Thus, the [(η^8-C_8H_8)$_2$Er]$^-$ motif leads to a high-blocking temperature [71] of 10 K, while the sandwich motif containing COT or substituted COT derivatives has led to several other SMMs with less prominent lanthanides, such as Ce [72] and Tm [73].

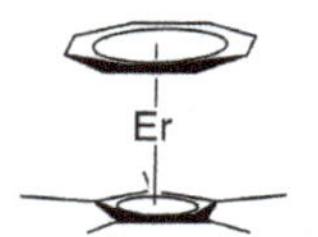

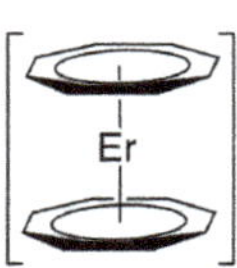

Figure 2.7.23: Representation of the molecular structures of the Er organometallic complexes with SMM behavior.

Based on the metal–ligand pair anisotropy approach, an axial symmetry with strongly focused point charges would be ideal for Dy. Thus a [Cp'$_2$Dy]$^+$ motif with no ligands in the equatorial plane would be the ideal target structure for SMMs. This was finally realized by the preparation of [Cp$^{ttt}{}_2$Dy]$^+$[B(C_6F_5)$_4$]$^-$ (Cp^{ttt}=η^5-(Me_3C)$_3C_5H_2$) in which the steric bulk at the Dy atom is large enough to prevent any solvent coordination, and [B(C_6F_5)$_4$]$^-$ anion results in well-separated anion–cation pairs (Figure 2.7.24) [74]. This strategy dramatically increased the blocking temperature to 60 K [75].

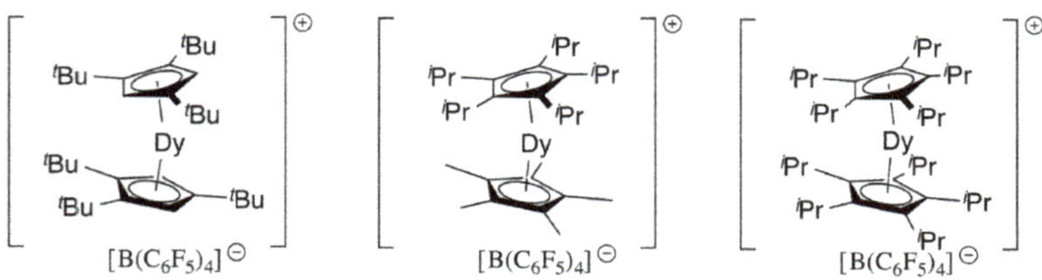

Figure 2.7.24: Representation of the molecular structures of the Dy organometallic complexes with record blocking temperature.

Modification of the steric bulk of the Cp ligand to pentaisopropyl ^{iPr5}Cp [76] increases the Cp-Dy-Cp angle slightly, getting closer to 180° and pushing again the blocking temperature of the liquid nitrogen up to more than 77 K, and offering potential applications for these molecules in spintronics and quantum computing.

References

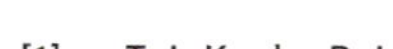

[1] T. J. Kealy, P. L. Pauson, Nature 1951, 168, 1039.

[2] a) G. Wilkinson, J. M. Birmingham, J. Am. Chem. Soc. 1954, 76, 6210; b) J. M. Birmingham, G. Wilkinson, J. Am. Chem. Soc. 1956, 78, 42–44; c) E. O. Fischer, H. Fischer, J. Organomet. Chem. 1965, 3, 181.

[3] H. Schumann, J. A. Meese-Marktscheffel, L. Esser, Chem. Rev. 1995, 95, 865, and references cited therein.

[4] R. Shannon, Acta Crystallogr. 1976, A32, 751.

[5] W. J. Evans, Organometallics 2016, 35, 3088.

[6] a) R. G. Finke, S. R. Keenan, P. L. Watson, Organometallics 1989, 8, 263; b) M. D. Walter, P. T. Matsunaga, C. J. Burns, L. Maron, R. A. Andersen, Organometallics 2017, 36, 4564.

[7] D. Seyferth, Organometallics 2004, 23, 3562.

[8] a) A. Greco, S. Cesca, W. Bertolini, J. Organomet. Chem. 1976, 113, 321; b) M. D. Walter, C. H. Booth, W. W. Lukens, R. A. Andersen, Organometallics 2009, 28, 698.

[9] P. L. Watson, J. Chem. Soc., Chem. Commun. 1980, 652.

[10] T. D. Tilley, R. A. Andersen, B. Spencer, H. Ruben, A. Zalkin, D. H. Templeton, Inorg. Chem. 1980, 19, 2999.

[11] A. L. Wayda, W. J. Evans, Inorg. Chem. 1980, 19, 2190.

[12] a) D. J. Berg, C. J. Burns, R. A. Andersen, A. Zalkin, Organometallics 1989, 8, 1865; b) M. Schultz, C. J. Burns, D. J. Schwartz, R. A. Andersen, Organometallics 2000, 19, 781.

[13] a) W. J. Evans, L. A. Hughes, T. P. Hanusa, J. Am. Chem. Soc. 1984, 106, 4270; b) W. J. Evans, L. A. Hughes, T. P. Hanusa, Organometallics 1986, 5, 1285; c) R. A. Andersen, J. M. Boncella, C. J. Burns, R. Blom, A. Haaland, H. V. Volden, J. Organomet. Chem. 1986, 312, C49.

[14] T. K. Hollis, J. K. Burdett, B. Bosnich, Organometallics 1993, 12, 3385.

[15] W. J. Evans, K. J. Forrestal, J. T. Leman, J. W. Ziller, Organometallics 1996, 15, 527.

[16] P. L. Watson, G. W. Parshall, Acc. Chem. Res. 1985, 18, 51.

[17] a) D. M. Anderson, F. G. N. Cloke, P. A. Cox, N. Edelstein, J. C. Green, T. Pang, A. A. Sameh, G. Shalimoff, J. Chem. Soc., Chem. Commun. 1989, 53; b) D. M. Anderson, F. G. N. Cloke, P. A. Cox, N. Edelstein, J. C. Green, T. Pang, A. A. Sameh, G. Shalimoff, J. Chem. Soc., Chem. Commun. 1990, 284.

[18] M. Xemard, S. Zimmer, M. Cordier, V. Goudy, L. Ricard, C. Clavaguera, G. Nocton, J. Am. Chem. Soc. 2018, 140, 14433.

[19] a) F. T. Edelmann, D. M. M. Freckmann, H. Schumann, Chem. Rev. 2002, 102, 1851; b) M. Zimmermann, R. Anwander, Chem. Rev. 2010, 110, 6194.

[20] R. D. Ernst, Chem. Rev. 1988, 88, 1255.

[21] J. Raeder, M. Reiners, R. Baumgarten, K. Münster, D. Baabe, M. Freytag, P. G. Jones, M. D. Walter, Dalton Trans. 2018, 47, 14468.

[22] a) D. Barisic, D. A. Buschmann, D. Schneider, C. Maichle-Moessmer, R. Anwander, Chem. Eur. J. 2019, 25, 4821; b) D. Barisic, J. Lebon, C. Maichle-Moessmer, R. Anwander, Chem. Commun. 2019, 55, 7089; c) D. Barisic, D. Schneider, C. Maichle-Moessmer, R. Anwander, Angew. Chem., Int. Ed. 2019, 58, 1515.

[23] J.-F. Carpentier, S. M. Guillaume, E. Kirillov, Y. Sarazin, C. R. Chim. 2010, 13, 608.
[24] M. P. Conley, G. Lapadula, K. Sanders, D. Gajan, A. Lesage, I. del Rosal, L. Maron, W. W. Lukens, C. Copéret, R. A. Andersen, J. Am. Chem. Soc. 2016, 138, 3831.
[25] a) L. A. Solola, A. V. Zabula, W. L. Dorfner, B. C. Manor, P. J. Carroll, E. J. Schelter, J. Am. Chem. Soc. 2016, 138, 6928; b) T. Cheisson, E. J. Schelter, Science 2019, 363, 489.
[26] a) C. T. Palumbo, I. Zivkovic, R. Scopelliti, M. Mazzanti, J. Am. Chem. Soc. 2019, 141, 9827; b) N. T. Rice, I. A. Popov, D. R. Russo, J. Bacsa, E. R. Batista, P. Yang, J. Telser, H. S. La Pierre, J. Am. Chem. Soc. 2019, 141, 13222.
[27] F. Nief, Dalton Trans. 2010, 39, 6589.
[28] W. J. Evans, I. Bloom, W. E. Hunter, J. L. Atwood, J. Am. Chem. Soc. 1981, 103, 6507.
[29] M. Xémard, A. Jaoul, M. Cordier, F. Molton, O. Cador, B. Le Guennic, C. Duboc, O. Maury, C. Clavaguéra, G. Nocton, Angew. Chem. Int. Ed. 2017, 56, 4266.
[30] M. E. Fieser, M. R. MacDonald, B. T. Krull, J. E. Bates, J. W. Ziller, F. Furche, W. J. Evans, J. Am. Chem. Soc. 2015, 137, 369.

[31] a) M. Coreno, M. de Simone, R. Coates, M. S. Denning, R. G. Denning, J. C. Green, C. Hunston, N. Kaltsoyannis, A. Sella, Organometallics 2010, 29, 4752; b) R. G. Denning, J. Harmer, J. C. Green, M. Irwin, J. Am. Chem. Soc. 2011, 133, 20644.
[32] G. Depaoli, U. Russo, G. Valle, F. Grandjean, A. F. Williams, G. J. Long, J. Am. Chem. Soc. 1994, 116, 5999.
[33] A. C. Thomas, A. B. Ellis, Organometallics 1985, 4, 2223.
[34] M. Schultz, J. M. Boncella, D. J. Berg, T. D. Tilley, R. A. Andersen, Organometallics 2002, 21, 460.
[35] Van Vleck J. H., The Theory of Electric and Magnetic Susceptibilities, Oxford University Press, London, 1932.
[36] W. W. Lukens, N. Magnani, C. H. Booth, Inorg. Chem. 2012, 51, 10105.

[37] C. H. Booth, M. D. Walter, D. Kazhdan, Y.-J. Hu, W. W. Lukens, E. D. Bauer, L. Maron, O. Eisenstein, R. A. Andersen, J. Am. Chem. Soc. 2009, 131, 6480.
[38] a) M. Dolg, P. Fulde, H. Stoll, H. Preuss, A. Chang, R. M. Pitzer, Chem. Phys. 1995, 195, 71; b) R. L. Halbach, G. Nocton, C. H. Booth, L. Maron, R. A. Andersen, Inorg. Chem. 2018, 57, 7290.
[39] G. Nocton, C. H. Booth, L. Maron, R. A. Andersen, Organometallics 2013, 32, 5305.
[40] G. Nocton, W. L. Lukens, C. H. Booth, S. S. Rozenel, S. A. Melding, L. Maron, R. A. Andersen, J. Am. Chem. Soc. 2014, 136, 8626.
[41] P. J. Chirik, K. Wieghardt, Science 2010, 327, 794.
[42] a) P. L. Watson, in Selective Hydrocarbon Activation (Eds.: J. A. Davies, P. L. Watson, J. F. Liebman, A. Greenberg), VCH, Weinheim, 1990, pp. 79–112; b) D. Balcells, E. Clot, O. Eisenstein, Chem. Rev. 2010, 110, 749.
[43] E. L. Werkema, E. Messines, L. Perrin, L. Maron, O. Eisenstein, R. A. Andersen, J. Am. Chem. Soc. 2005, 127, 7781.
[44] S. Hong, T. J. Marks, Acc. Chem. Res. 2004, 37, 673.
[45] W. J. Evans, A. L. Wayda, W. E. Hunter, J. L. Atwood, J. Chem. Soc., Chem. Commun. 1981, 706.
[46] E. L. Werkema, L. Maron, O. Eisenstein, R. A. Andersen, J. Am. Chem. Soc. 2007, 129, 2529.
[47] W. J. Evans, J. Alloys Compd. 2009, 488, 493.
[48] W. J. Evans, C. A. Seibel, J. W. Ziller, Inorg. Chem. 1998, 37, 770.
[49] a) N. W. Davies,A. S. P. Frey, M. G. Gardiner, J. Wang, Chem. Commun. 2006, 4853; bM. Xémard, V. Goudy, A. Braun, M. Tricoire, M. Cordier, L. Ricard, L. Castro, E. Louyriac, C. E. Kefalidis, C. Clavaguéra, L. Maron, G. Nocton, Organometallics 2017, 36, 4660.
[50] J. Andrez, J. Pécaut, P.-A. Bayle, M. Mazzanti, Angew. Chem. Int. Ed. 2014, 53, 10448.
[51] W. J. Evans, J. W. Grate, I. Bloom, W. E. Hunter, J. L. Atwood, J. Am. Chem. Soc. 1985, 107, 405.
[52] W. J. Evans, G. W. Rabe, J. W. Ziller, R. J. Doedens, Inorg. Chem. 1994, 33, 2719.

[53] a) X. Zhang, G. R. Loppnow, R. McDonald, J. Takats, J. Am. Chem. Soc. 1995, 117, 7828; b) B. Neumüller, F. Weller, T. Gröb, K. Dehnicke, Z. Anorg. Allg. Chem. 2002, 628, 2365.

[54] a) W. J. Evans, D. S. Lee, C. Lie, J. W. Ziller, Angew. Chem. Int. Ed. 2004, 43, 5517; b) W. J. Evans, D. S. Lee, D. B. Rego, J. M. Perotti, S. A. Kozimor, E. K. Moore, J. W. Ziller, J. Am. Chem. Soc. 2004, 126, 14574; c) W. J. Evans, D. S. Lee, J. W. Ziller, J. Am. Chem. Soc. 2004, 126, 454; d) W. J. Evans, D. S. Lee, Can. J. Chem. 2005, 83, 375.

[55] F. Jaroschik, A. Momin, F. Nief, X. F. Le Goff, G. B. Deacon, P. C. Junk, Angew. Chem. Int. Ed. 2009, 48, 1117.

[56] H.-M. Huang, J. J. W. McDouall, D. J. Procter, Nature Catal. 2019, 2, 211.

[57] a) W. J. Evans, J. Organomet. Chem. 2002, 652, 61; b) L. Jacquot, M. Xémard, C. Clavaguéra, G. Nocton, Organometallics 2014, 33, 4100; c) C. Ruspic, J. R. Moss, M. Schürmann, S. Harder, Angew. Chem. Int. Ed. 2008, 47, 2121.

[58] a) M. N. Leuenberger, D. Loss, Nature 2001, 410, 789; b) S. Sanvito, Chem. Soc. Rev. 2011, 40, 3336.

[59] a) A. Caneschi, D. Gatteschi, R. Sessoli, A. L. Barra, L. C. Brunel, M. Guillot, J. Am. Chem. Soc. 1991, 113, 5873; b) R. Sessoli, D. Gatteschi, A. Caneschi, M. A. Novak, Nature 1993, 365, 141.

[60] N. Ishikawa, M. Sugita, T. Ishikawa, S.-y. Koshihara, Y. Kaizu, J. Am. Chem. Soc. 2003, 125, 8694.

[61] J. H. V. Vleck, J. Phys. Chem. 1937, 41, 67.

[62] J. D. Rinehart, J. R. Long, Chem. Sci. 2011, 2, 2078.

[63] R. A. Layfield, J. J. W. McDouall, S. A. Sulway, F. Tuna, D. Collison, R. E. P. Winpenny, Chem. Eur. J. 2010, 16, 4442.

[64] B. M. Day, F.-S. Guo, R. A. Layfield, Acc. Chem. Res. 2018, 51, 1880.

[65] a) J. D. Rinehart, M. Fang, W. J. Evans, J. R. Long, Nature Chem. 2011, 3, 538; b) J. D. Rinehart, M. Fang, W. J. Evans, J. R. Long, J. Am. Chem. Soc. 2011, 133, 14236.

[66] S. Demir, J. M. Zadrozny, M. Nippe, J. R. Long, J. Am. Chem. Soc. 2012, 134, 18546.

[67] S. Demir, M. I. Gonzalez, L. E. Darago, W. J. Evans, J. R. Long, Nature Commun. 2017, 8, 2144.

[68] a) C. A. Gould, L. E. Darago, M. I. Gonzalez, S. Demir, J. R. Long, Angew. Chem. Int. Ed. 2017, 56, 10103; b) F.-S. Guo, A. K. Bar, R. A. Layfield, Chem. Rev. 2019, 119, 8479.

[69] S.-D. Jiang, B.-W. Wang, H.-L. Sun, Z.-M. Wang, S. Gao, J. Am. Chem. Soc. 2011, 133, 4730.

[70] a) J. D. Hilgar, M. G. Bernbeck, B. S. Flores, J. D. Rinehart, Chem. Rev. 2018, 9, 7204; b) J. D. Hilgar, M. G. Bernbeck, J. D. Rinehart, J. Am. Chem. Soc. 2019, 141, 1913.

[71] K. R. Meihaus, J. R. Long, J. Am. Chem. Soc. 2013, 135, 17952.

[72] J. J. Le Roy, I. Korobkov, J. E. Kim, E. J. Schelter, M. Murugesu, Dalton Trans. 2014, 43, 2737.

[73] M. Xémard, M. Cordier, F. Molton, C. Duboc, B. Le Guennic, O. Maury, O. Cador, G. Nocton, Inorg. Chem. 2019, 58, 2872.

[74] F. Jaroschik, F. Nief, X. F. Le Goff, L. Ricard, Organometallics 2007, 26, 1123.

[75] a) F.-S. Guo, B. M. Day, Y.-C. Chen, M.-L. Tong, A. Mansikkamäki, R. A. Layfield, Angew. Chem. Int. Ed. 2017, 56, 11445; b) C. A. P. Goodwin, F. Ortu, D. Reta, N. F. Chilton, D. P. Mills, Nature 2017, 548, 439.

[76] a) K. R. McClain, C. A. Gould, K. Chakarawet, S. Teat, T. J. Groshens, J. R. Long, B. G. Harvey, Chem. Sci. 2018, 9, 8492; b) F.-S. Guo, B. M. Day, Y.-C. Chen, M.-L. Tong, A. Mansikkamäki, R. A. Layfield, Science 2018, 362, 1400.

3 Characterization and properties

Michael Sperling, Maximilian von Bremen-Kühne, Sabrina Funke, Lukas Schlatt, Uwe Karst

3.1 Analysis of rare-earth metals and their species

Due to their important applications in many different fields including the materials sciences and medicine, the analysis of rare-earth elements (REE) is required in environmental as well as in biological samples [1]. Only recently, the role of the REEs within the metallome of organisms [2] and especially their role in the health of animals and humans has found increasing attention [3]. On the other hand, the physical properties of rare earth metals make them valuable analytical tools as markers and labels for bioanalysis [4]. In this chapter, the authors will focus on the most common analytical methods for the determination of rare earth metals in various environmental and biological matrices.

The traditional techniques for rare earth metals analysis involve the most important groups of atomic spectroscopy, including atomic absorption spectroscopy (AAS), total reflection X-ray fluorescence (TXRF), inductively coupled plasma-optical emission spectroscopy (ICP-OES, often also referred to as ICP-atomic emission spectrometry, ICP-AES) and inductively coupled plasma-mass spectrometry (ICP-MS) [5]. While these methods differ in the applied physical principles, all of them are commonly characterized by good to excellent limits of detection (LOD) for trace analysis of the total REE concentration in liquid samples. Solid samples are accessible by these methods after prior dissolution by mineral acids or melt digestion [6]. The choice of atomic spectroscopy methods depends on the required LOD, with AAS typically reaching the mg kg^{-1} (ppm), TXRF and ICP-OES the μg kg^{-1} (ppb) and ICP-MS the ng kg^{-1} (ppt) range. If only one REE has to be determined, AAS may be an attractive option due to the limited costs of ownership and operating costs, while the other methods exhibit capabilities for (quasi)simultaneous multielement analysis, covering a range of up to 60 elements in one analysis.

Conventional AAS is based on recording the absorbance of element-specific radiation, generated by a hollow cathode lamp (HCL) [7]. The more common and cheaper variation of AAS uses a flame composed of a burning gas (methane or acetylene) and an oxidizer (oxygen or dinitrogen monoxide), into which the solution of the analyte is introduced by pneumatic nebulization. The simple setup allows its use even in remote areas but the general analytical principle of absorption hampers the limits of detection. At low-analyte concentrations, the large element-specific signal of the HCL will only be diminished slightly by analyte absorption, thus leading to the measurement of the (small) difference between two large numbers. Graphite furnace (GF)-based AAS uses the same physical principle and thus suffers from the same limitations. However, it still shows significantly improved LODs due to longer dwell times

https://doi.org/10.1515/9783110654929-016

of the atom cloud in the GF compared to the flame and more efficient transfer of the atoms from solution to the gas phase by electrothermal heating of the graphite furnace.

As many analytical questions cover a broader range of REEs, the methods with multielemental capabilities have gained importance in recent years. In TXRF, the beam generated by an X-ray tube is directed to the sample at a very small angle, thus causing total reflection [8]. This way, only a thin sample film resulting from a dried solution of the liquid sample on top of a quartz target is excited and emits characteristic red-shifted X-rays, which are recorded. The lack of stray light and emitted X-rays by the target material, leads to excellent limits of detection, which are superior by several decades of concentration, compared to the earlier methods based on X-ray fluorescence. Simultaneous determination of most elements of the periodic table, down to the µg kg^{-1} concentration range, is possible using semiconductor-based and silicon drift detectors. Elements with higher atomic numbers, starting from sodium and including the REEs, can be determined this way. The ease of calibration using a known concentration of another element, which is not likely to be present in the sample as an internal standard for all other elements, is an important advantage of this method. However, TXRF is available only in a limited number of laboratories, while ICP-OES, which covers similar applications areas and concentration ranges, is more common in routine analytical laboratories.

In ICP-OES, an inductively coupled argon plasma with temperatures ranging between 6,000 and 10,000 °C is used to vaporize and atomize the sample aerosol introduced into the axial central channel of the plasma created within a plasma torch, with three different argon gas flows. The outer or "cooling gas" is meant to keep the hot plasma away from the quartz tube of the torch, the intermediate or "auxiliary flow" is meant to keep the plasma away from the sample injector, and the injector gas flow is transporting the sample into the center axis of the plasma. After atomization, the same plasma is used to excite the outer electrons of the atoms, which upon relaxation to the ground state lead to the emission of element-specific radiation in the UV/vis range. The element-specific radiation is viewed either radially above the torch or for higher sensitivity, axially. While most elements of the periodic table, including the REEs, can be determined using ICP-OES, the plasma is open to the surrounding atmosphere, leading to significant background signals for those elements, being major constituents of ambient air [9]. Additionally, resonance emission lines of strongly electronegative elements are located in the vacuum UV range, so that fluorine, at its main emission line, is not accessible with commercial instrumentation. Chlorine, bromine, phosphorus and sulfur require the use of an optical detection system, which is operating in the vacuum UV region by using an evacuated polychromator for simultaneous multielemental detection. ICP-OES is an established analytical tool which is available in most routine analytical labs. Limits of detection mostly are in the µg kg^{-1} range, which compares favorably to AAS and is similar to TXRF. Major limitation is the broadband black body radiation of the plasma, from which the analyte signals have to be discriminated.

In ICP-MS, the same plasma source is used but the ions formed in the plasma are transferred to the mass selective detector via a two-staged vacuum interface. This is designed to minimize transfer time and distance from the hot plasma at atmospheric pressure to high vacuum, with the goal of limiting discharge of the analyte ions by collisions [10]. Despite only a minor fraction of ions reaching the detector, limits of detection for REEs from aqueous solutions are in the ng kg^{-1} range due to characteristic mass to charge ratios (m/z) of the rare-earth ions and only a few interferences in this m/z range. With these limits of detection, ICP-MS offers the lowest limits of detection of any common, laboratory-based analytical method for elemental analysis. A reaction/collision cell allows to selectively address polyatomic and other interferences, prior to entering the quadrupole mass analyzer, which is most commonly used in ICP-MS. Due to the low natural concentrations of REEs in the environment and living organisms, ICP-MS is the most frequently used analytical method to solve respective analytical questions.

The instrumentation setup and the basic principles of (flame) AAS (a), TXRF (b), ICP-OES (c) and ICP-MS (d) are presented in Figure 3.1.1.

A major alternative to the determination of REEs by atomic spectroscopy is the use of ion exchange chromatography with conductivity detection. After dissolution of a solid sample by mineral acids or melt digestion, REEs are most frequently present in their respective trivalent form. Due to identical charge but subtle differences in ionic radii (lanthanide contraction) over the series of REEs, sulfonate-based cation exchange columns allow separating the cations and conductivity detection can be applied as a moderately selective and sensitive approach [11]. Alternatively, UV/vis absorption detection may also be employed in ion exchange chromatography after the complexation of the rare-earth cations with colored ligands.

While the REEs most frequently do occur as trivalent cations in environmental and biological samples, some particular situations require the analysis of the REEs in specific binding forms (chemical species) [12]. The field of speciation analysis is based on the fact that the effects of an element on living organisms or the environment are not dependent on the total concentration of this element per se but on the species responsible for this effect. This is very obvious and widely accepted for the element carbon, where some organic compounds ("carbon species") are known to be important for human health, while others are known to exhibit toxic effects. While no one would refer the observed effects to the total concentration of carbon but rather to those of the respective individual substances, this is not always the case in metals analysis, where scientists, politicians and industry continue to talk about the "toxic heavy metals" instead of "toxic metal species" [13]. However, a scientifically sound discussion of the effects caused by REEs does require consideration of the influence of their chemical species using the speciation analysis. More information on this topic is available under reference [14].

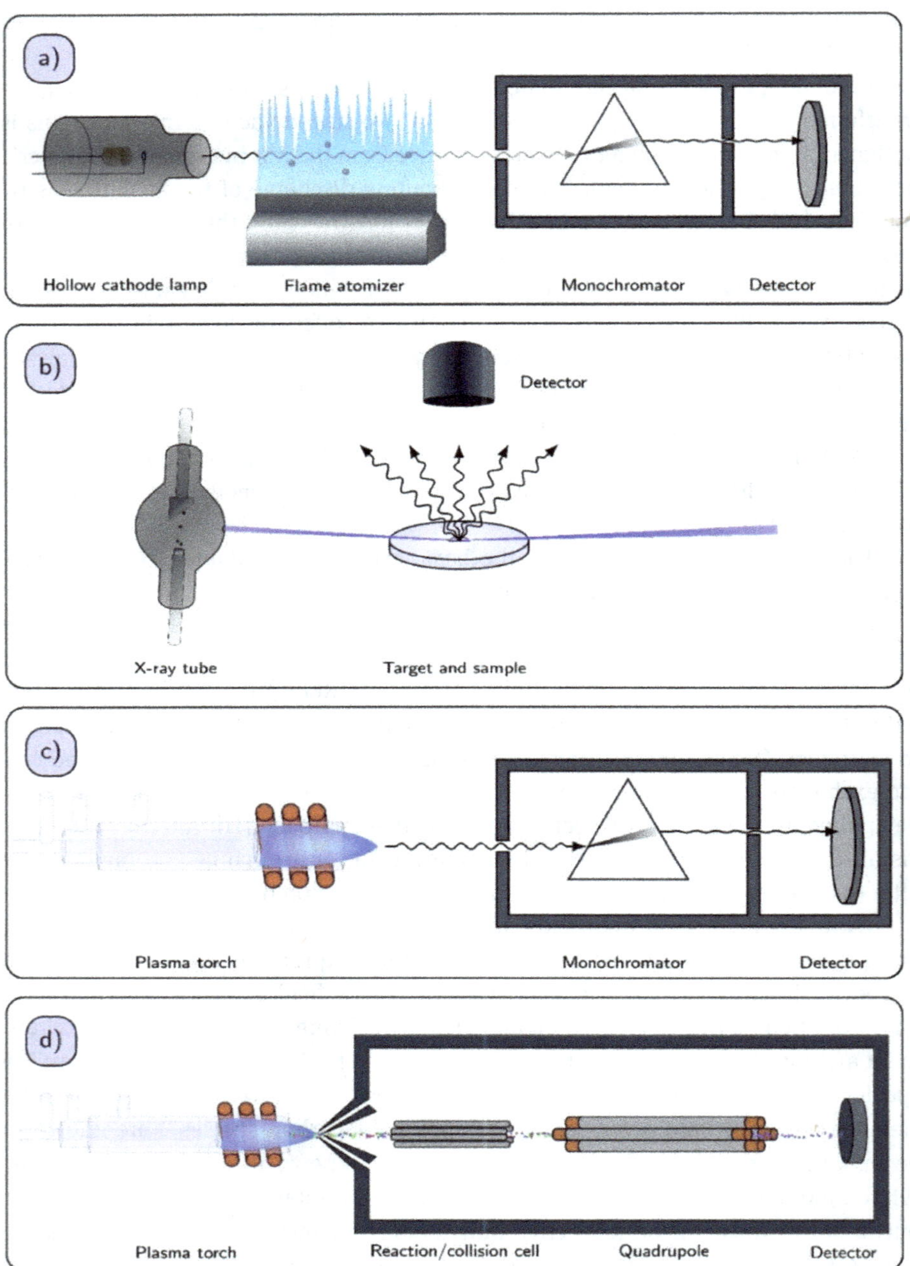

Figure 3.1.1: Basic analytical principles for AAS (a), TXRF (b), ICP-OES (c) and ICP-MS (d).

REEs play an important role in various medical applications. A prominent example is the use of Gd(III) complexes as contrast agents for magnetic resonance imaging. With its f^7 electron configuration, Gd(III) is paramagnetic and strongly influences the relaxation times of protons from coordinated water molecules, thus improving the contrast in the images. As Gd^{3+} in the bloodstream is highly toxic, its application in the patient depends on its conversion into stable complexes with multidentate ligands, to guarantee low acute toxicity. For almost three decades, several contrast agents are used that contain Gd(III) complexed to large polyaminocarboxylic acid ligands including DTPA and DOTA (see Figure 3.1.2).

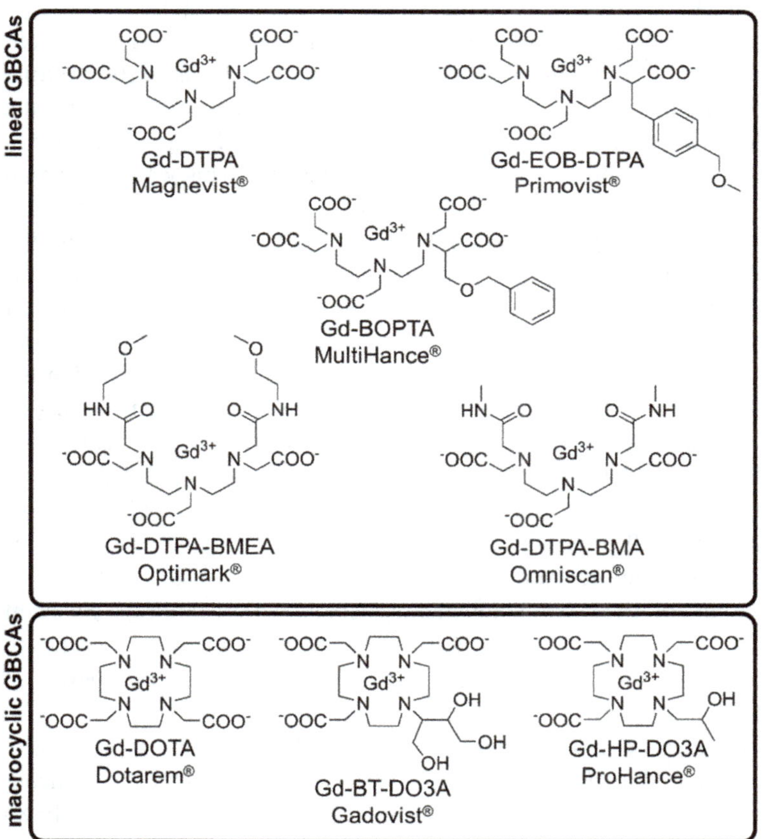

Figure 3.1.2: Structures of the most commonly used gadolinium-based complexes for magnetic resonance imaging.

Due to their excellent solubility in water, the half-life of the contrast agents is low as they are rapidly excreted via the kidneys and the urine. However, two important issues have become apparent in recent years which require strong consideration.

First, Bau et al. [15] discovered, using elemental analysis by ICP-MS, that rivers close to large medical institutions may contain strongly increased gadolinium concentrations compared to the other REEs. These findings have become known as "gadolinium anomaly." Considering the significant consumption of Gd-based contrast agents (GBCAs), which are injected in single doses of approximately 1 to 1.5 g of gadolinium for a person with 80 kg body weight, these were rapidly identified as the underlying cause. Patients who receive treatment with GBCAs in a hospital or a radiology practice, will excrete the GBCAs into the sewage system, which leads to a sewage treatment plant. As only a very small fraction of the GBCAs (<10%) appears to be removed there, it is important to know more about the chemical binding form of the gadolinium which is released into the aquatic environment [16].

While ICP-MS is known to provide excellent quantitative data on the REE concentration in the environment, the atomization in the plasma leads to the destruction of any information on the chemical species which were contained in the sample. Therefore, appropriate separation techniques for individual Gd species have to be developed and coupled to ICP-MS as the detector. This way, individual Gd species can be identified in chromatography by their respective retention times and quantified using various calibration strategies [17]. For separation of the GBCAs, hydrophilic interaction liquid chromatography (HILIC) has turned out to be particularly valuable [18]. Zwitterionic stationary phases based on sulfonates for cation exchange and quaternary ammonium salts for anion exchange in the stationary phase are often used for HILIC, as they allow separation via both the ion exchange and the HILIC effect. The latter is based on a hydrophilic partitioning of polar analytes from the stationary phase into a water film formed on the stationary phase which can be considered as the hydrate shell of the cations and anions. The combination of chromatography and atomic spectroscopy mostly requires the use of ICP-MS rather than ICP-OES, because the improved limits of detection are urgently required. This is due to both the dilution effects by diffusion in the chromatographic column and the inferior statistics associated with recording a transient chromatographic signal compared with the constant injection of an analyte solution over half a minute or even longer time. A typical HILIC-ICP-MS chromatogram of the most common GBCAs is presented in Figure 3.1.3.

If speciation analysis data are generated by LC-ICP-MS, it is important to carry out an additional total Gd analysis by ICP-MS to determine if the sum of the concentrations of the individual species equals the total Gd concentration [19]. This allows for investigating if there are losses of any species on the chromatographic column. Based on the respective analyses, it has been demonstrated that more than 90% of gadolinium discharged from a sewage treatment plant are present in the form of the intact GBCAs.

As the general mechanisms of drinking water purification resemble those of wastewater treatment, it is important to also monitor the kind and concentration of Gd species in drinking waters. This is particularly valid for areas in which surface

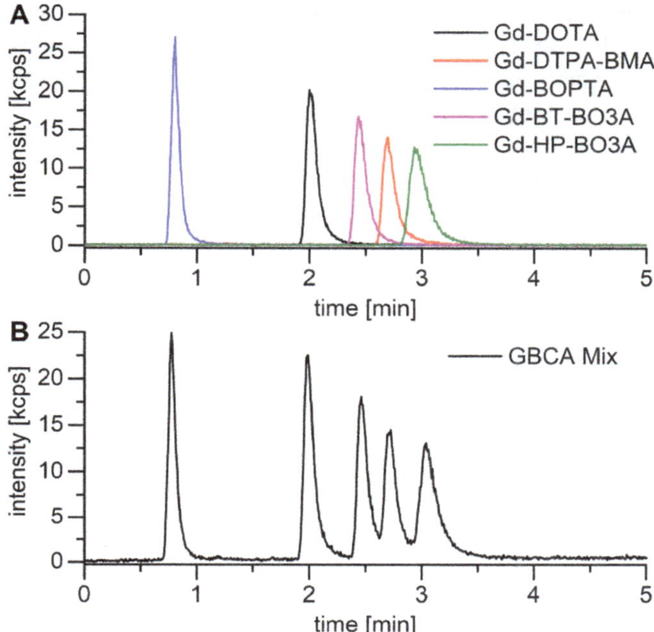

Figure 3.1.3: HILIC-ICP-MS chromatogram of common GBCAs in water. A: overlay of chromatograms of retention time standards; B: chromatogram of a mixture of the same GBCAs.

waters are cleared by bank filtration and less for areas with deep well water supplies. Using improved sample preparation and optimized sample introduction for the HILIC-ICP-MS system, it was found that intact GBCAs can even be detected in drinking water in major cities [20], although at significantly lower concentrations compared with those in surface waters. However, due to the similarities between the purification processes, an increasing concentration of Gd species can be expected in drinking waters in the future.

The second major issue, which became apparent with GBCAs in recent years, is the potential deposition of gadolinium in the human body. Due to the very high solubility of all major GBCAs in water, it was assumed and confirmed by various analyses that Gd is excreted quantitatively from the body within weeks [21]. In 2006, however, it was observed that patients with terminal renal insufficiency (dialysis patients) may develop a nephrogenic systemic fibrosis upon delivery of GBCAs with open-chain ligands, while patients with healthy kidneys or those who received GBCAs with macrocylic ligands are not affected [22]. These findings were explained by the limited kinetic stability of the open chain ligand-based GBCAs in combination with the massively prolonged dwell times of the GBCAs in dialysis patients compared to patients with healthy kidneys. Under these conditions, GBCAs have more time to exchange Gd versus other Pearson-hard cations. Gd is deposited, likely as phosphate, below the skin, the dermis. There, a complex biochemical response is triggered, leading to hardening of the affected parts of the skin, starting at the patients´ extremities and propagating to the torso and even to the membranes of the inner organs. This life-threatening disease is called nephrogenic systemic fibrosis as it is associated to kidney problems. It affects the complete body system and leads to hard fibers at the skin, which affects mobility and functionality of the organs.

To study Gd deposition in the body, thin tissue slices as frequently used by pathologists are investigated by elemental analysis. While microscale X-ray fluorescence (μXRF) can be used for quantification of gadolinium in a spatially resolved manner, its benchtop approach is limited by very poor limits of detection, while μXRF at a synchrotron facility provides excellent limits of detection but with very limited availability. Therefore, the combination of laser ablation (LA) with ICP-MS has evolved as the most important tool for the quantitative analysis of (rare earth) metals in biological samples. The general setup of the method is provided in Figure 3.1.4.

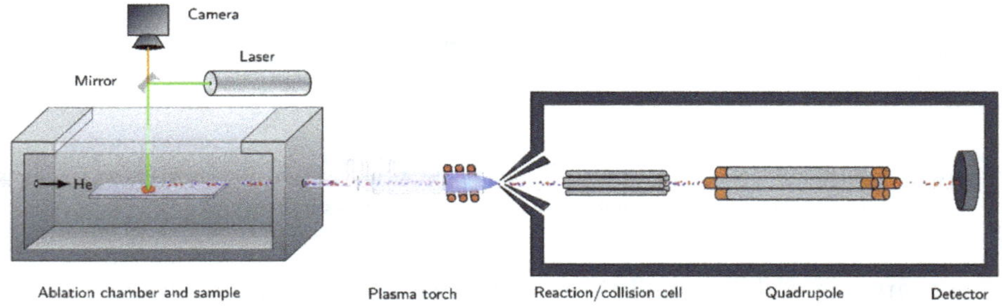

Figure 3.1.4: Set-up of an LA-ICP-MS instrument.

An ultraviolet laser beam, typically at wavelengths between 193 and 266 nm, is used to ablate the biological sample line-wise. Quantification is most frequently carried out by using external calibration based on matrix-matched standards, for example, tissue homogenates or gelatin containing defined concentrations of the analyte and their analysis under identical conditions as the sample. While it is not possible to identify individual Gd species in tissue slices, due to the destructive nature of the plasma, very low limits of detection can be achieved. Typical laser spot sizes at the tissue level range between 10 and 50 μm, but it is possible to use laser spot sizes down to 1 μm [17]. However, it has to be considered that the ICP-MS is a mass flow sensitive detector. Hence, a reduction of the spot size by a factor of ten in two dimensions reduces the amount of analyte transferred to the plasma by a factor of 100. Furthermore, a volume element of 10 μm × 10 μm × 10 μm corresponds to a volume of only 1 pL. Therefore, the balance between small spot size and therefore high resolution and a good limit of detection has to be balanced well. An LA-ICP-MS image of

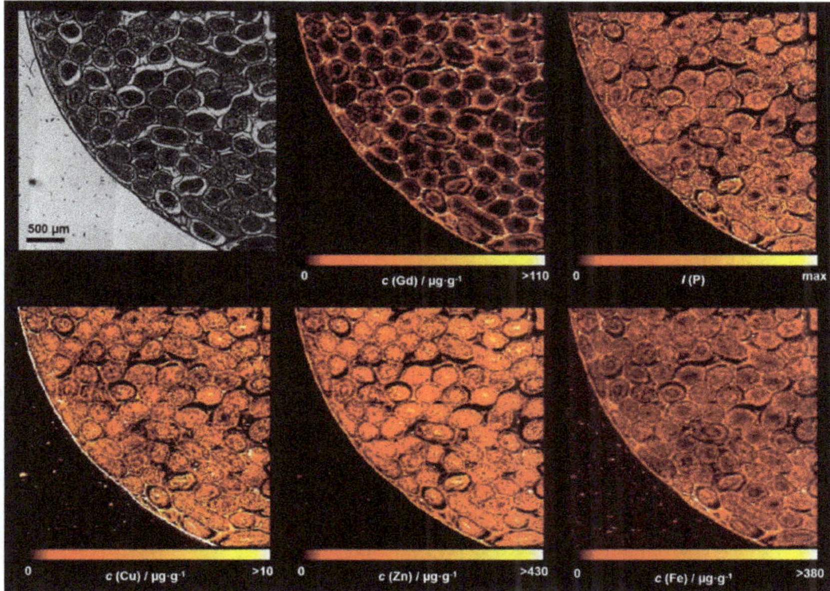

Figure 3.1.5: Distribution of Gd, P, Cu, Zn and Fe in the testicle of a rat after delivery of a GBCA as determined by LA-ICP-MS. The respective light microscopic image is provided in the upper left.

the distribution of gadolinium, phosphorus, copper, zinc and iron in the testicle of a rat after delivery of a GBCA is presented in Figure 3.1.5.

While LA-ICP-MS offers the best limits of detection for metals including REEs and transition metals, the analysis of moderately electronegative elements such as phosphorus or sulfur is also possible. This is important, as P and S as ubiquitous elements in biological systems can be used to depict the structure of the samples particularly well. This opens doors for "elemental pathology," which perfectly complements the current histological methods as well as methods of "molecular pathology," including matrix-assisted laser desorption/ionization mass spectrometry.

The last decade has provided excellent analytical tools for speciation analysis by LC-ICP-MS as well as for elemental bioimaging by LA-ICP-MS, but many challenges remain. Currently, there are only very few methods available for the spatially resolved, quantitative analysis of metal species in biological samples, and it is likely that several research groups will focus on this topic in the future.

References

[1] V. Balaram, Geosci. Front. 2019, 10, 1285.
[2] L. Daumann, Angew. Chem. 2019, 58, 12795.

[3] G. Pagano, F. Aliberti, M. Guida, R. Oral, A. Sicilano, M. Trifuoggi, F. Tommasi, Environ. Res. 2015, 142, 215.
[4] T. C. de Bang, S. Husted, Trends Anal. Chem. 2015, 72, 45.
[5] B. Zawisza, K. Pytlakowska, B. Feist, M. Polowniak, A. Kita, R. Sitko, J. Anal. At. Spectrom. 2011, 26, 2372.
[6] F. Garcia Pinto, R. Escalfoni Jr., T. Dillenburg St'Pierre, Anal. Lett. 2012, 45, 1537.
[7] B. Welz, M. Sperling, Atomic Absorption Spectrometry, 3rd Edition, Wiley-VCH, Weinheim, 1999.
[8] R. Klockenkämper, A. von Bohlen, Total-Reflection X-ray Fluorescence Analysis and Related Methods, 2nd Edition, Wiley-VCH, Weinheim, 2015.
[9] G. L. Moore, Introduction to Inductively Coupled Plasma Atomic Emission Spectroscopy, Elsevier, Amsterdam, 1989.
[10] J. G. Holland, D. R. Bandura, Plasma Source Mass Spectrometry, Royal Society of Chemistry, Cambridge, 2006.
[11] J. Weiss, Handbook of Ion Chromatography, 3rd Edition, Wiley-VCH, Weinheim, 2004.
[12] J. Wojcieszek, J. Szpunar, R. Lobinski, Trends Anal. Chem. 2018, 104, 42.
[13] J. H. Duffus, Arch. Environ. Health 2003, 58, 263.
[14] R. Cornelis, J. Caruso, H. Crews, K. Heumann, Handbook of Elemental Speciation: Techniques and Methodology, Wiley, Chichester, 2003.
[15] M. Bau, P. Dulski, Earth Planet. Sci. Lett. 1996, 143, 245.
[16] L. Telgmann, M. Sperling, U. Karst, Anal. Chim. Acta 2012, 764, 1.
[17] D. Clases, M. Sperling, U. Karst, Trends Anal. Chem. 2018, 104, 135.
[18] J. Künnemeyer, L. Terborg, B. Meermann, C. Brauckmann, I. Möller, A. Scheffer, U. Karst, Environ. Sci. Technol. 2009, 43, 2884.
[19] K. A. Francesconi, M. Sperling, Analyst 2005, 130, 998.
[20] M. Birka, C. A. Wehe, O. Hachmöller, M. Sperling, U. Karst, J. Chromatogr. A 2016, 1440, 105.
[21] E. Lancelot, Invest. Radiol. 2016, 51, 691.
[22] T. Grobner, Nephrol. Dial. Transplant. 2006, 21, 1104.

Walter Rambeck, Kerstin Redling, Cristian A. Strassert

3.2 Rare earth toxicology

3.2.1 Introduction and general considerations

Rare earth elements (*REE*s) have found applications in magnets, catalysts, metal alloys, superconductors, as glass additives, in luminescent materials, as contrast agents in magnetic resonance imaging (MRI), as phosphate binders in kidney disease patients in men [1] and in animals [2], as fertilizers, as feed additives in livestock, in lasers, in high-performance magnets and in ointments for burnt skin. Very recently, it was shown that these elements may play a catalytic role in certain enzymes [3] and that *REE* are even biologically essential [4]. In natural environments, *REE* only appear in low concentrations, but their use in technology has increased the exposure and provided new routes for human intake, including inhalation of dusts at production sites, intake through foodstuff, as well as administration of medicinal agents [1–9].

Inhalation of particles and the dust of lanthanides are the most frequent sources of accidental exposure to humans. Bioaccumulation and environmental aspects of *REE* have been documented at different distances from ore mining sites [10, 11]. Another source of air and soil *REE*-related pollutants is associated to the use of Ce_2O_3 nanoparticles as catalyst additives in diesel fuels which leads to airborne particles upon combustion [12–15]. Interestingly, enhanced concentrations of Ce, Pr and Sm related to the use of diesel fuel have been measured in the proximity of motorways [16]. Respiratory infections have been related to occupational *REE* exposure [17–20] and include some case reports [21], while epidemiological studies are missing. *REE* as feed additives for livestock have been studied for a long time [22–28], and lately, the European Food Safety Authority has declared the La–Ce citrate mixture (*Lancer*) given to pigs as a zootechnical additive as safe for the consumer and the environment [29]. In China, *REE* have also been used as performance enhancers for livestock and as fertilizers for crops [30, 31].

"Hormesis effects" (i.e., benefits vs. toxicity) have been reported for *REE* [32–36], explaining that low doses stimulate several biological processes that are inhibited at higher levels [37, 38].

REE are – as to the Hodge–Sterner toxicity classification – only slightly toxic. The oral toxicity with an LD50 beyond 1 g kg^{-1} body weight [39, 40] is ascribed to the poor gastrointestinal absorption. The toxicity varies strongly with the administration route: for subcutaneously or intramuscularly injected *REE,* it is, as expected a little higher. Inhalation, especially by chronic exposure, is more toxic but the most toxic effects are seen when *REE* are applied intravenously [31, 41]. *REE* play a role in the onset of cellular oxidative stress. Conversely, there are also

https://doi.org/10.1515/9783110654929-017

REE-associated antioxidative effects. Thus, redox imbalances in organisms related to *REE* have been discussed [21, 42–44], including the damage to sperm cells [45] and chromosomal aberrations in the bone marrow of mice [46]. Reports also exist about nephrogenic skin fibrosis caused by Gd as contrast agent in diagnostic medical imaging [47, 48].

Research efforts have been devoted to toxicological aspects of *REE* since 1960 [49], and several recent studies have been published regarding health issues [50, 51] and also highlighted by the European Agency for Safety and Health at Work in 2013 [51]. However, reports of *REE* toxicity have been focused on a few elements, particularly Ce, La and Gd. Thus, further studies on other elements are needed, especially beyond short- to medium-term intervals (i.e., longer than 3 months) [52] and particularly considering that long-term studies are completely absent. A series of review papers and books have been published addressing the impact of technological applications of *REE*, and bioaccumulation studies have also been discussed in recent review articles [9, 21, 23, 42, 43, 52, 53]. In the following sections, pathophysiological, metabolic and toxicological aspects will be discussed, as reviewed in further detail by Redling [31].

3.2.2 Pathophysiological mechanisms of action

REE and particularly lanthanides can substitute physiologically relevant cations, including Ca^{2+}, Mg^{2+}, Fe^{3+}, Fe^{2+} or Mn^{2+} [54]. Due to its relevance in living systems and its exchangeability by lanthanide cations, special attention is usually paid to Ca^{2+}. Even though rare earth (*RE*) ions are not able to penetrate cellular membranes under physiological conditions, Ca^{2+} can be replaced due to similar ionic radii, charge, hardness, ligand affinities and coordination geometries [5].

In some cases, the function of enzymes and biomolecules is not impeded but, often, inhibition is observed. On the other hand, some functional proteins show no affinity for lanthanide ions at all. For other biomolecules, a drop in the effect was observed with increasing atomic number.

Despite their similarities, substantial differences have been documented for Ca^{2+} and Ln^{3+}: *RE* ions display a higher charge/volume ratio, more stable coordination adducts with higher coordination numbers [55], stronger bonds and higher stability constants [56], which results in higher affinities for potential binding sites.

The interaction of *RE* ions with structures such as external plasmalemmae is key in their influence on cells, tissues and organs. These cations usually do not permeate across membranes and barriers of healthy cells but they can modulate transmembrane processes [5, 57]. However, dead and damaged cells with compromised integrity do show permeability for such charged species. In any case, binding of *RE* cations to membrane proteins, stabilizes phospholipidic bilayers while enhancing their rigidity as well as increasing their positive charge balance, their transmembrane potential and their resistance which ultimately compromises voltage-dependent Ca^{2+} currents. Also, receptor-operated Ca^{2+} channels and ion exchangers (e.g., Na^{+}–Ca^{2+}, and Ca^{2+}–Ca^{2+}) can be inhibited by *RE* ions [5]. In some cases (such as erythrocytes or adrenal cells), Ca^{2+} influx can be hampered, whereas the efflux is less susceptible. Thus, physiological processes depending on Ca^{2+} influx are antagonized, unlike those relying on intracellular release or mediated by insensitive deposits within the cells [58]. In this sense, diverse processes can be affected, including neurotransmission [59], blood clotting [55], contraction of muscles (smooth, skeletal and cardiac) [60], histamine release form mastocytes [61], reticuloendothelial function [62, 63] as well as several hormonal responses [64].

Investigations on murine neoblastoma cells suggested that *RE* ions bind to Ca^{2+} channels around acetylcholine receptors, inhibiting Ca^{2+} currents [65]. The *RE*-induced release of pancreatic insulin and amylase has also been documented as independent from Ca^{2+} influx depending on the intracellular release of Ca^{2+} [5], presumably by the depletion of extracellular Ca^{2+} reservoirs due to repeated activation of cellular processes. Since intracellular Ca^{2+} deposits are dependent on the extracellular availability of Ca^{2+}, further responses are avoided. In 3T3-L1 cells, adipogenesis and lipogenesis are stimulated by *REE*, presumably since the ions of La and Ce take over calcium binding sites [26].

It has been assumed that *RE* cations can also affect extracellular functions without inhibiting Ca^{2+} transport. For instance, Ghosh et al. [66] showed that La and Nd chloride reduced the enzymatic activity on erythrocytic membranes, upon intraperitoneal administration to chicken. Moreover, a dose-dependent in vitro enhancement of gastric acid production was documented for La^{3+} in isolated mice stomach [67]. In addition, A-type K^{+} channels of adrenal cortex were blocked by *RE* ions, upon binding to nonspecific Ca^{2+} sites [68]. On the other hand, γ-aminobutyric acid-activated chloride channels of dorsal root ganglion neurons of rats were enhanced [69]. *RE* ions have also been shown to coordinate with membrane structures, such as acetylcholine receptors [70], insulin [71], adenylate-cyclase [72] and Na^{+}–K^{+} ATPase [73]. They can also modulate the cell signaling pathways system by affecting intracellular Ca^{2+} ion concentrations or by hydrolysis of phosphatidyl-inositol [74].

Catalysis of RNA hydrolysis by *RE* ions has been reported by Eichhorn and Butzow [75] while having the ability to depolymerize RNA and nucleotides in vitro, whereas binding to DNA causes conformational changes and precipitation. However, their affinity for RNA is higher [5]. Yajima et al. [76] reported that *RE* cations form adenosine 3′,5′-cyclic monophosphate from adenosine triphosphate under physiological conditions.

For several Ca^{2+}-dependent G-protein coupled receptors, inhibition by lanthanide ions was suggested along with impaired hormone binding, as shown for ACTH [64] and other mediators [77–79]. Despite their general inhibitory activity, lanthanide ions can, if combined with intrinsic agonists, boost physiological processes, such as the release of serotonin (platelets), catecholamines (adrenal medulla) and acetylcholine or other neurotransmitters from neurons [5]. Depending on the concentration and ionic radii of the *RE* ions, this hormesis effect may vary (i.e., a dose-dependent reversal effect) and cause effects opposed to their intrinsic activity. Stimulation might be observed at low doses, whereas no (or inverse) effects can be associated to higher dosages [74].

A dose-dependent modulation of the immune response has also been documented. Secretion of histamine from mastocytes was inhibited at high concentrations whereas lower doses of lanthanide ions stimulated it [80]. Upon intravenous injection at rather low concentrations, they hampered phagocytosis by the reticuloendothelial system [63, 81]. A suppression of the immune system increases the susceptibility to infections caused by bacteria and yeast [62]. Splenic lymphocyte transformation in mice was stimulated upon enteral (oral) administration of $La(NO_3)_3$ at low concentrations, as reported by Wang et al. [82].

The activity of reactive oxygen species (ROS) is related to immune response and affected by lanthanide ions. Formation of ROS (including free radicals and peroxides), is also essential in the metabolism while showing serious cytotoxicity that compromises cellular function with potential cell death and mutagenesis. Lanthanide ions can enhance or suppress ROS-related processes, depending on their concentration: light *RE* have inhibitory effects with increasing concentrations whereas heavier ions activate them at lower doses. The mechanism has been related to hydroperoxide binding, thus precluding lipid and membrane-related protein peroxidation as well as to magnetic interaction with free radicals [23, 83].

Proliferation, differentiation and apoptosis are cellular features that are relevant in physiological and pathological conditions; the effect of lanthanides has been widely documented. Cell injury and proliferation as a function of concentration were correlated and mechanistically investigated. Preeta and Nair [84] showed proliferative effects at low doses of Ce in cardiac fibroblasts of rats, whereas Gd enhanced the proliferation of hepatocytes [85]. Gadolinium-based contrast agents are widely used as MRI contrast agents and many publications describe concentration-dependent deposition of Gd not only in the brain but also in the liver and bone. Potential side effects are discussed frequently [86]. Depending on the concentration, bovine chondrocytes were inhibited upon injecting gadodiamide (an MRI contrast agent) into the articulation [87]. Pretreatment with $GdCl_3$ inhibited the proliferation of oval cells in an artificially injured liver [88], whereas Ishiyama et al. [89] noticed that $GdCl_3$ applied after liver injury was followed by cell proliferation. Even though the details regarding this behavior are objects of ongoing research efforts, it is clear that *RE*-induced cell proliferation is related to an enhanced biosynthesis of the RNA and DNA proteins [74]. Sarkander and Brade [90] showed that variations in ionic radii of lanthanides affect RNA synthesis differently, depending on the species, cell types and experimental conditions. This could be assigned to an enhanced intracellular concentration of Ca^{2+}, which for instance, boosts the proliferative activity of rat fibroblasts [74]. Liu et al. [91] documented the inhibition of proliferation by $GdCl_3$, by preventing the influx of Ca^{2+}, whereas Rai et al. [85] and Preeta [84] agreed to the modulation of ROS by lanthanides, which in turn affects proliferative activity. Gd-related cytotoxicity was investigated in alveolar macrophages associated with phagocytosis in rats [92, 93], as well as in rat fibroblasts, HeLa and PC12 cells. Apoptosis in splenocytes and thymocytes [82] was documented upon feeding rats with a mixture of lanthanide nitrates (≥20 mg kg^{-1}, 14 days). In contrast, they also found that $LaCl_3$ (10 mg kg^{-1}) decreased the apoptotic response in mice of thymocytes to endotoxemia. Growth seems to be induced by low concentrations, whereas inhibition (up to cell death) is observed at higher concentrations, as suggested by the bulk of experimental data. This is in agreement with a hormesis effect.

3.2.3 Metabolic pathways

Administration, distribution, metabolism (i.e., detachment of ligands and binding to target structures without alteration of the metallic center) and excretion pathways determine the toxicity of *REE* [7]. Besides binding to bioorganic ligands, formation of insoluble hydroxides, carbonates and phosphates complicate related investigations with radionuclides [94]. Thus, precipitation facilitates uptake by phagocytes.

Absorption is usually poor in the gastrointestinal tract of adults [23, 25, 41, 95–101]. This is also true for animal models such as rats [96, 102], dogs [103], chicken [104] and quails [105]. Once blood vessels have been reached, the fate is similar to injected lanthanides, including liver, lungs, spleen, kidney, bone and teeth [5, 23, 41, 106–113]. Newborns, on the other hand, assimilate lanthanides to a larger extent upon oral administration, as observed in pigs [41, 114].

Intraperitoneally administered *REE* remain within the abdominal cavity while coating the surface of the organs, but a small fraction is transported to the bone, liver and teeth. They also remain in the pleural space upon intrapleural injection. However, transport seems to be enhanced with chelating agents. Intramuscular and subcutaneous injections lead to negligible uptake [5]. This is also true for wounds treated with $Ce(NO_3)_3$ ointment [115]. Excretion occurs via feces upon oral administration while urinary elimination was observed after intraarticular injection. No precipitation was observed after addition of lanthanides to blood or serum [116, 117], most likely due to the formation of soluble adducts with bioorganic ligands in the plasma. Clearance depends on the ligands they are bound to [5]. Beyond the saturation of the plasma complexation capacity, the clearance is slowed.

Due to variations in ionic radii and solubility, differences might be observed regarding biodistribution of *REE*: more basic lanthanides are found in the liver whereas acidic species prefer skeletal structures [94, 118], while in case of the latter, accumulation is enhanced in young animals [114]. As the liver releases them again, the skeleton appears to be the final fate [5]. Inhalation of insoluble species leads to retention for years [7].

Excretion via the gastrointestinal tract is usually preferred except for soluble complexes that favor the kidney-mediated urine production [6, 7, 41, 94, 95, 99, 103, 106, 118]. Elimination from the liver roughly occurs within a month, whereas up to 3 years are needed for the clearance of bone deposits [41, 96, 118].

In general, the following conclusions can be drawn:

- Soluble salts are most efficiently up-taken intravenously ≫ intraperitonally > intramuscularly > subcutaneously >> orally
- Accumulation occurs in the liver and in the bones, except upon inhalation of insoluble species being phagocyted by macrophages
- Efficient chelation enhances absorption and excretion rates while reducing dwell times in plasma; while oral administration of soluble chelates leads to fecal elimination, intravenous injections favors urinary excretion and diminishes uptake by the liver
- Weak chelating agents facilitate ligation to biogenic ligands and reduced excretion

Even though chelators hamper the uptake by liver and therefore favor muscular tissues and bones as targets for accumulation [41, 94] diethylenetriaminopentaacetate and related polydentate ligands may be used in case of intoxication to enhance excretion.

3.2.4 Toxicity of REE

According to the Hodge–Sterner system, *RE* are mostly considered as having low toxicity [41], despite depending on the administration route. High LD50 values for oral doses up to 1 g kg^{-1} body weight [23, 39, 40, 119] are related to a low bioavailability and have been confirmed for several animal models including rats, mice, guinea pigs and apes [23, 50, 97, 120, 121]. Medical studies reported innocuous oral doses of up to 3 g of $La_2(CO_3)_3$ per person and day for up to 4 years [107, 122–124]. In drinking water, 2 mg L^{-1} have been recommended [125].

Administration by injection is related to higher toxicities due to an enhanced bioavailability. Subcutaneous minimal lethal doses of 100–1000 mg kg^{-1} are reported. Intraperitoneal LD50 values are lower, ranging between 50–500 mg kg^{-1} body weight, depending on the susceptibility of the animal model.

Inhalation causes no immediate effects due to phagocytic capture of particles. However, emphysema has been observed upon chronic exposure as well as pneumonitis, bronchitis and pulmonary fibrosis.

The highest toxicity is observed for intravenous injection (LD50 in the range between 3 and 100 mg kg^{-1} body weight for rats and mice [5, 6, 41]). Liver as well as spleen degeneration with concomitant atrophy and central lobe necrosis have been observed [120]. Delayed lethality has been documented showing ataxia, labored respiration, walking on toes with arched back and sedation. Surviving individuals displayed generalized granulomatous peritonitis and focal hepatic necrosis. Intravenous application causes hypotension. Death is triggered by cardiovascular collapse with respiratory paralysis [41, 120]. Heavier *REE* seem to be more toxic while the counter-anion seems to play a role as well (for Nd, the following trend was observed: chloride < proprionate < acetate < 3-sulfoisonicotinate < sulfate < nitrate) [41].

Negligible genotoxicity and bone marrow aberration was observed in vitro and in vivo [127], upon tissue and plasma exposure to $La_2(CO_3)_3$, but chromosomal translocation increased in spermatocytes. Carcinogenic effects could not be detected [23, 128, 129], despite older studies that documented tumor formation in lung tissue upon intratracheal administration or inhalation, as well as in liver, stomach or intestinal tract [41]. However, chronical feeding showed growth depression in rats [5, 23, 41].

In general, no genotoxicity, carcinogenicity or teratogenicity is expected for oral administration, whereas application by injection is found to be threatening at very high doses. Further studies are needed to come up with the specific targets. *REE* may be regarded as low toxic elements, despite affecting different organs in an unspecific manner requiring further investigations to elucidate the mechanisms of action.

3.2.5 Toxicity on pulmonary tissues

Lung toxicity is mainly manifested as pneumoconiosis, a progressive pulmonary fibrosis depending on the type and properties of the inhaled materials that is enhanced by radioactive traces [5, 41, 130, 131]. Inhalation of *RE* dusts leads to accumulation of particles. They may cause interstitial disorder, emphysema as well as obstructive impairment. The deposits are traceable for long time periods [19, 130, 132, 133]. Compared to other fibrogenic dusts (e.g., silica and quartz), however, the potential displayed by *RE* is lower [39].

3.2.6 Toxicity on hepatic tissues

Light *REE* are mostly sequestered by the liver, leading to hepatotoxic manifestations. Magnusson [134] and Tuchweber [135] showed increased enzymatic values related to the liver that recovered after 10 days. Necrotic tissue was documented upon intravenous and subcutaneous administration [136, 137]. Fatty liver, however, is the most common phenomenon induced by the intravenous injection of light *RE* solutions, which was not observed after oral administration. A rise of free fatty acids in plasma was observed before fatty liver manifestation, with a concomitant drop in cholesterol and phospholipid levels [94]. This could be related to an altered lipidic metabolism [5] that might also be caused by an impaired adrenal function. Renaud [138] and Arvela [139] also reported a reduced microsomal activity. Fatty liver was observed in rats, mice and hamster, but not in guinea pigs, chicken or dogs. However, after 3 days, the effects had diminished [120].

3.2.7 Toxicity on spleen, kidney and gastrointestinal tract

Stineman et al. [140] noted hypertrophy, reticuloendothelial hyperplasia and hyperactive lymphoid follicles in mice spleen, upon subcutaneous as well as oral administration of cerium citrate, whereas focal gastric hemorrhages, necrosis of mucosa and neutrophil infiltration were observed only after oral intake.

With an oral LD50 > 1,000 mg kg^{-1} body (70,000 mg per person), previous studies showed that the risk of accidental intoxication is very low [141], as such high levels of ingestion are highly unlikely to occur. In fact, injection alone seems to be dangerous despite being highly improbable. Thus, the use of *REE* in feed additives for livestock seems to be safe for animals and humans [29]. The concentrations found in commercial edibles seem to be higher for vegetable feedstuff than in the tissue of animals raised with *RE*-based feed supplements.

3.2.8 Toxicity on bones

RE ions are incorporated into the organic network as well as into the inorganic mineral part of bones [5]. Despite constituting the second largest deposit in the body, no significant effects were observed in the bones. Mineralization defects in rats with chronical renal insufficiency are rather related to phosphate depletion upon administration of $La_2(CO_3)_3$ [142–144].

3.2.9 Toxicity on skin and eyes

Healthy skin is insensitive to *RE* salts, as opposed to damaged cutaneous tissues of rabbits [41]. Granulomas and local calcification with mild fibrosis was documented upon intradermal administration of *RE* [5, 41], presumably due to precipitation with pyrophosphate to form a crystallization nucleus on which calcium accumulates to form apatite [145].

Ocular irritation is observed after topical application of *RE* salts with concomitant ulceration in rabbits. Haley [120] and Swanson [146] found that lanthanides could play a role in cataractic diseases. Nevertheless, Ji et al. [23] documented only mild irritation on skin and eye mucosa of rabbits.

3.2.10 Toxicity on brain

RE ions affect Ca^{2+} currents and therefore can affect neurotransmission by interfering with neurotransmitter release and transport processes. However, the blood–brain barrier is not permeable to *RE* ions [5], thus preventing neurological effects upon systemic administration. Direct application of La^{3+} into the brain caused analgesia resembling opiates that were antagonized by naloxone and morphine tolerance [147]. Subarachnoidal administration of higher doses of $GdCl_3$, Gd-EDTA or Gd-DTPA can impair motoric function and cause epileptoid fits [148]. Oral administration of $La_2(CO_3)_3$ caused no adverse effects in the central nervous system, even after prolonged treatment [149].

References

[1] C. Zhang, J. Wen, Z. Li, J. Fan, BMC Nephrol. 2013, 14, 226.
[2] V. Bampidis, G. Azimonti, M. de L. Bastos, H. Christensen, B. Dusemund, M. Kouba, M. K. Durjava, M. López-Alonso, S. López Puente, F. Marcon, B. Mayo, A. Pechová, M. Petkova, F. Ramos, Y. Sanz, R. E. Villa, R. Woutersen, A. Chesson, J. Gropp, G. Martelli, D. Renshaw, G. López-Gálvez, A. Mantovani, EFSA J. 2019, 17, 5542. DOI:10.2903/j.efsa.2019.5542
[3] H. Lumpe, A. Pol, H. J. M. Op den Camp, L. J. Daumann, Dalton Trans. 2018, 47, 10463–10472.
[4] L. J. Daumann, Angew. Chem. Int. Ed. 2019, 58, 12795–12802.
[5] C. H. Evans, Biochemistry of the Lanthanides, Plenum Press, New York and London, 1990.
[6] S. Hirano, K. T. Suzuki, Environ. Health Perspect. 1996, 104, 85–95.
[7] R. A. Bulman, Metabolism and Toxicity of the Lanthanides, in The Lanthanides and Their Interrelations with Biosystems (eds. A. Sigel, H. Sigel), Vol. 40, Marcel Dekker, New York, 2003, 683–706.
[8] E. C. Giese, Clin. Med. Rep. 2018, 200.20.105.11.

[9] S. H. Shin, H. O. Kim, K. T. Rim, Saf. Health Work 2019, 10, 409–419.
[10] R. L. Peng, X. C. Pan, Q. Xie, Zhonghua Yu Fang Yi Xue Za Zhi 2003, 37, 20–22.
[11] S. L. Tong, W. Z. Zhu, Z. H. Gao, Y. X. Meng, R. L. Peng, G. C. Lu, J. Environ. Sci. Health A Tox. Hazard Subst. Environ. Eng. 2004, 39, 2517–2532.
[12] F. R. Cassee, A. Campbell, A. J. Boere, S. G. McLean, R. Duffin, P. Krystek, I. Gosens, M. R. Miller, Environ. Res. 2012, 115, 1–10.
[13] J. Y. Ma, S. H. Young, R. R. Mercer, M. Barger, D. Schwegler-Berry, J. K. Ma, V. Castranova, Toxicol. Appl. Pharmacol. 2014, 278, 135–147.

[14] S. J. Snow, J. McGee, D. B. Miller, V. Bass, M. C. Schladweiler, R. F. Thomas, T. Krantz, C. King, A. D. Ledbetter, J. Richards, J. P. Weinstein, T. Conner, R. Willis, W. P. Linak, D. Nash, C. E. Wood, S. A. Elmore, J. P. Morrison, C. L. Johnson, M. I. Gilmour, U. P. Kodavanti, Toxicol. Sci. 2014, 142, 403–417.
[15] J. G. Dale, S. S. Cox, M. E. Vance, L. C. Marr, M. F. Hochella Jr., Environ. Sci. Technol. 2017, 51, 1973–1980.
[16] N. I. Ward, Environmental Analysis Using ICP-MS, in Applications of ICP-MS (ed. A. R. Date, A. L. Grey), Chapman and Hall, New York, 1988, 189–219.
[17] H. Gong Jr., Curr. Opin. Pulm. Med. 1996, 2, 405–411.
[18] J. W. McDonald, A. J. Ghio, C. E. Sheehan, P. F. Bernhardt, V. L. Roggli, Mod. Pathol. 1995, 8, 859–865.
[19] E. Sabbioni, R. Pietra, P. Gaglione, G. Vocaturo, F. Colombo, M. Zanoni, F. Rodi, Sci. Total Environ. 1982, 26, 19–32.
[20] H. K. Yoon, H. S. Moon, S. H. Park, J. S. Song, Y. Lim, N. Kohyama, Thorax 2005, 60, 701–703.
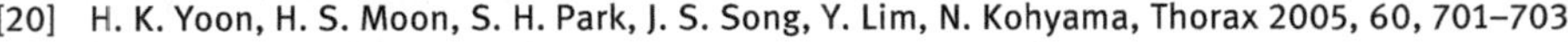
[21] G. Pagano, F. Aliberti, M. Guida, R. Oral, A. Siciliano, M. Trifuoggi, F. Tommasi, Environ. Res. 2015, 142, 215–220.
[22] M. L. He, D. Ranz, W. A. Rambeck, J. Anim. Physiol. Anim. Nutr. 2001, 85, 263–270.
[23] Y. Ji, M. Cui, Y. Wang, X. Zhang, Toxicological Study on Safety Evaluation of Rare Earth Elements Used in Agriculture, in New Frontiers in Rare Earth Science and Applications, Proceedings of the 1st International Conference on Rare Earth Development and Applications (ed. G. Xu, J. Xiao), Beijing, September 10–14, Science Press, Beijing, 1985, 700–704.
[24] W. A. Rambeck, U. Wehr, Pig News and Information 2005, 26, 41N–47N.
[25] S. von Rosenberg, W. Rambeck, U. Wehr, Proc. Soc. Nutr. Physiol. 2013, 22, 183.
[26] M. L. He, W. Z. Yang, H. Hidari, W. A. Rambeck, Asian-Australas. J. Anim. Sci. 2006, 19, 119–125.
[27] M. L. He, U. Wehr, W. A. Rambeck, J. Anim. Physiol. Anim. Nutr. 2010, 941, 86–92.
[28] S. A. Abdelnour, M. E. Abd El-Hack, A. F. Khafaga, A. E. Noreldin, M. Arif, M. T. Chaudhry, C. Losacco, A. Abdeen, M. M. Abdel-Daim, Sci. Total Environ. 2019, 672, 1021–1032.
[29] V. Bampidis, G. Azimonti, M. de L. Bastos, H. Christensen, B. Dusemund, M. Kouba, M. K. Durjava, M. López-Alonso, S. López Puente, F. Marcon, B. Mayo, A. Pechová, M. Petkova, F. Ramos, Y. Sanz, R. E. Villa, R. Woutersen, A. Finizio, A. Focks, K. Svensson, I. Teodorovic, L. Tosti, J. Tarrés-Call, P. Manini, F. Pizzo, EFSA J. 2019, 17, 5912. DOI:10.2903/j.efsa.2019.5912
[30] X. Pang, D. Li, A. Peng, Environ. Sci. Pollut. Res. Int. 2002, 9, 143–148.

[31] K. Redling, Rare Earth Elem. Agric. Emphasis on Anim. Husb. Deutsche Veterinärmedizinische Gesellschaft Giessen, München, 2006.
[32] F. Goecke, C. G. Jerez, V. Zachleder, F. L. Figueroa, K. Bišová, T. Řezanka, M. Vítová, Front. Microbiol. 2015, 6, 2.
[33] W. Jenkins, P. Perone, K. Walker, N. Bhagavathula, M. N. Aslam, M. DaSilva, M. K. Dame, J. Varani, Biol. Trace Elem. Res. 2011, 144, 621–635.
[34] D. Liu, J. Zhang, G. Wang, X. Liu, S. Wang, M. Yang, Biol. Trace Elem. Res. 2012, 150, 433–440.
[35] A. Pol, T. R. Barends, A. Dietl, A. F. Khadem, J. Eygensteyn, M. S. Jetten, H. J. Op den Camp, Environ. Microbiol. 2014, 16, 255–264.
[36] US Environmental Protection Agency, Rare Earth Elements: A Review of Production, Processing, Recycling, and Associated Environmental Issues, EPA 600/R-12/572, can be found under www.epa.gov/ord, 2012.
[37] E. J. Calabrese, Crit. Rev. Toxicol. 2013, 43, 580–586.
[38] A. R. Stebbing, Sci. Total Environ. 1982, 22, 213–234.
[39] H. Richter, Hinweise zur Toxikologie von Seltenen Erden, in XVI. Tage der Seltenen Erden, 4.-6. December 2003, Berlin, Germany, 2003, 18.

[40] H. Richter, K. Schermanz, Seltene Erden, in Chemische Technik – Prozesse und Produkte Band 6b Metalle (ed. R. Dittmayer, W. Keim, G. Kreysa, A. Oberholz), Vol. 5, Wiley-VCH Verlag, Weinheim, 2006, 147–208.
[41] T. J. Haley, Toxicity, in Handbook on the Physics and Chemistry of Rare Earths (ed. K. A. Gschneidner Jr., L. R. Eyring), Vol. 4, ElsevierNorth-Holland, Amsterdam, New York, Oxford, 1979, 553–579.

[42] G. Pagano, M. Guida, F. Tommasi, R. Oral, Ecotoxicol. Environ. Saf. 2015, 115C, 40–48.
[43] G. Pagano, M. Guida, A. Siciliano, R. Oral, F. Koçbaş, A. Palumbo, I. Castellano, O. Migliaccio, M. Trifuoggi, P. J. Thomas, Environ. Res. 2016, 147, 453–460.
[44] K. T. Rim, K. H. Koo, J. S. Park, Saf. Health Work 2013, 4, 12–26.
[45] R. Oral, P. Bustamante, M. Warnau, A. D'Ambra, M. Guida, G. Pagano, Chemosphere 2010, 81, 194–198.
[46] A. M. Jha, A. C. Singh, Mutat. Res. 1995, 341, 193–197.
[47] J. Ramalho, R. C. Semelka, M. Ramalho, R. H. Nunes, M. Al Obaidy, M. Castillo, AJNR Am. J. Neuroradiol. 2015 Dec 10 [PMID: 26659341].
[48] H. S. Thomsen, Eur. Radiol. 2006, 16, 2619–2621.
[49] T. J. Haley, K. Raymond, N. Komesu, H. C. Upham, Brit. J. Pharmacol. 1961, 17, 526–532.
[50] A. Izyumov, G. Plaksin, Cerium: Molecular Structure, Technological Applications and Health Effects, Nova Science Publishers, New York, 2013.
[51] a) U.S. E.P.A., Rare Earth Elements: A Review of Production, Processing, Recycling, and Associated Environmental Issues, National Service Center for Environmental Publications (NSCEP), EPA 600/R-12/572, can be found under https://nepis.epa.gov/Exe/ZyPDF.cgi/P100EUBC.PDF?Dockey=P100EUBC.PDF, 2012; b) EU-OSHA, Priorities for Occupational Safety and Health Research in Europe: 2013–2020 European Agency for Safety and Health at Work, 2013.

[52] G. Pagano, Rare Earth Elements in Human and Environmental Health: At the Crossroads between Toxicity and Safety, Pan Stanford Publishing Pte. Ltd. Penthouse Level, Suntec Tower 3 8 Temasek Boulevard Singapore 038988 Copyright © 2017 by Pan Stanford Publishing Pte. Ltd.
[53] K.-T. Rim, Toxicol. Environ. Health. Sci. 2016, 8, 189–200.

[54] C. H. Evans, Trends Biochem. Sci. 1983, 8, 445–449.
[55] M. A. Jakupec, P. Unfried, B. K. Keppler, Rev. Physiol. Biochem. Pharmacol. 2005, 153, 101–111.

[56] W. D. Horrocks. Lanthanide Ion Probes of Biomolecular Structure, in Advances in Inorganic Biochemistry (ed. G. L. Eichhorn, L. G. Marzilli), Vol. 4, Elsevier, New York, 1982.
[57] A. A. Korenevskii, V. V. Sorokon, G. I. Karavaiko, Mikrobiologiya 1997, 66, 198–205.
[58] D. A. Nachshen, J. Gen. Physiol. 1984, 83, 941–967.
[59] A. Vaccari, P. Saba, I. Mocci, S. Ruiu, Neurosci. Lett. 1999, 261, 49–52.
[60] C. R. Triggle, D. J. Triggle, J. Physiol. 1976, 254, 39–54.
[61] M. A. Beaven, J. Rogers, J. P. Moore, R. Hesketh, G. A. Smith, J. C. Metcalfe, J. Biol. Chem. 1984, 259, 7129–7136.
[62] B. Farkas, G. Karacsonyi, Mykosen 1985, 28, 338–341.
[63] G. Lazar, E. Husztik, S. Ribarsziki, Reticuloendothial Blockade Induced by Gadolinium Chloride. Effect on Humoral Immune Response and Anaphylaxis, in Macrophage Biology (ed. S. Reichard, M. Kojima), Alan Liss, New York, 1985, 571–582.
[64] J. J. Enyeart, L. Xu, J. A. Enyeart, Am. J. Physiol. Endocrinol. Metabol. 2002, 282, 1255–1266.
[65] E. El Fakahany, M. Pfenning, E. Richelson, J. Neurochem. 1984, 42, 863–869.
[66] N. Ghosh, D. Chattopadhayay, G. C. Chatterjee, Ind. J. Exp. Biol. 1991, 29, 226–229.
[67] X. Xu, H. Xia, G. Rui, C. Hu, F. Yuan, J. Rare Earths 2004, 22, 427.
[68] J. J. Enyeart, L. Xu, J. C. Gomora, J. A. Enyeart, J. Membrane Biol. 1998, 164, 139–153.

[69] T. Narahashi, J. Y. Ma, O. Arakawa, E. Reuveny, M. Nakahiro, Cell. Mol. Neurobiol. 1994, 14, 599–621.
[70] H. Rübsamen, G. P. Hess, A. T. Eldefrawi, M. E. Eldefrawi, Biochem. Biophys. Res. Commun. 1976, 68, 56–63.
[71] P. F. Williams, J. R. Turtle, Diabetes 1984, 33, 1106–1111.
[72] J. A. Nathanson, R. Freedman, B. J. Hoffer, Nature 1976, 261, 330–332.
[73] P. David, S. J. D. Karlish, J. Biol. Chem. 1991, 266, 14896–14902.
[74] K. Wang, Y. Cheng, X. Yang, R. Li, Cell Response to Lanthanides and Potential Pharmacological Actions of Lanthanides in Metal ions in Biological Systems, in Metal Ions in Biological Systems: Vol. 40: The Lanthanides and Their Interrelations with Biosystems (ed. A. Sigel, H. Sigel, S. Sigel), Marcel Dekker, Inc., New York, Basel, 2003.
[75] G. L. Eichhorn, J. J. Butzow, Biopolymers 1965, 3, 79–94.
[76] H. Yajima, J. Sumaoka, M. Sachiko, K. Makoto, J. Biochem. 1994, 115, 1038–1039.
[77] A. Haksar, D. V. Maudsley, F. G. Peron, E. Bedigian, J. Cell Biol. 1976, 68, 142–153.
[78] J. Segal, S. H. Ingbar, Endocrinology 1984, 115, 160–166.
[79] S. K. Lam, S. Harvey, T. R. Hall, Gen. Comp. Endocrinol. 1986, 63, 178–185.
[80] J. C. Foreman, J. L. Mongar, British J. Pharmacol. 1973, 48, 527–537.
[81] E. Husztik, G. Lazar, A. Parducz, Brit. J. Exp. Pathol. 1980, 61, 624–630.
[82] Y. Wang, F. Guo, K. Yuan, Y. Hu, Acta Acad. Medic. Jiangxi 2003, 43, 31.
[83] H. Shimada, M. Nagano, T. Funakoshi, S. Kojima, J. Toxicol. Environm. Health 1996, 48, 81–96.
[84] R. Preeta, R. R. Nair, J. Mol. Cell. Cardiol. 1999, 31, 1573–1580.
[85] R. M. Rai, S. Loffreda, C. L. Karp, S. Q. Yang, H. Z. Lin, A. M. Diehl, Hepatology 1997, 25, 889–895.
[86] B. J. Guo, Z. L. Yang, L. J. Zhang, Front. Mol. Neurosci. 2018, 20, 335.
[87] J. K. Greisberg, J. M. Wolf, J. Wyman, L. Zou, R. M. Terek, J. Orthopaedic Res. 2001, 19, 797–801.
[88] J. K. Olynyk, G. C. Yeoh, G. A. Ramm, S. L. Clarke, P. M. Hall, R. S. Britton, B. R. Bacon, T. F. Tracy, Am. J. Pathol. 1998, 152, 347–352.
[89] H. Ishiyama, M. Sato, K. Matsumura, M. Sento, K. Ogino, T. Hobara, Pharmacol. Toxicol. 1995, 77, 293–298.
[90] H. I. Sarkander, W. P. Brade, Arch. Toxicol. 1976, 36, 1–17.
[91] M. Liu, J. Xu, A. K. Tanswell, M. Post, J. Cellular Physiol. 1994, 161, 501–507.
[92] J. P. Mizgerd, R. M. Molina, R. C. Stearns, J. D. Brain, A. E. Warner, J. Leukocyte Biol. 1996, 59, 189–195.
[93] Y. Kubota, S. Takahashi, I. Takahashi, G. Patrick, Toxicol. in vitro 2000, 14, 309–319.
[94] P. Arvela, Progr. Pharmacol. 1977, 2, 69–73.
[95] G. Baehr, H. Wessler, Arch. Internal Med. 1909, 2, 517–531.
[96] J. G. Hamilton, New England J. Med. 1949, 240, 863–870.
[97] K. W. Cochran, J. Daull, M. Mazur, K. P. DuBois, Arch. Ind. Hyg. Occup. Med. 1950, 1, 637–650.
[98] G. Fiddler, T. Tanaka, I. Webster, 9th Asian Pacific Congress of Nephrology, 19.-20. February 2003, Pattaya, Thailand, 2003.
[99] P. C. D'Haese, G. B. Spasovski, A. Sikole, A. Hutchison, T. J. Freemont, S. Sulkova, C. Swanepoel, S. Pejanovic, L. Djukanovic, A. Balducci, G. Coen, W. Sulowicz, A. Ferreira, A. Torres, S. Curic, M. Popovic, N. Dimkovic, M. E. DeBroe, Kidney Int. 2003, 63, 73–78.
[100] W. A. Rambeck, M. L. He, U. Wehr, Proceedings of the British Society of Animal Science Pig and Poultry Meat Quality – Genetic and Non-genetic Factors, 14.-15. October 2004, Krakow, Poland, 2004.
[101] A. J. Hutchison, B. Maes, J. Vanwalleghem, G. Asmus, E. Mohamed, R. Schmieder, W. Backs, R. Jamar, A. Vosskuhler, Nephron. Clin. Pract. 2006, 102, 61–71.

[102] W. P. Norris, H. Lisco, A. Brues, The Radiotoxicity of Cerium and Yttrium, in Rare Earths in Biochemical and Medical Research (ed. G. C. Kyker, E. B. Anderson), U.S. Atomic Energy Commission, Report ORINS-12, 1956, 102–115.
[103] A. J. Hutchison, F. Albaaj, Expert Op. Pharmacother. 2005, 6, 319–328.
[104] F. R. Mraz, P. L.Wright, T. M. Ferguson, D. L. Anderson, Health Phys. 1964, 10, 777–782.
[105] G. A. Robinson, D. C. Wasnidge, F. Floto, Poultry Sci. 1978, 57, 190–196.
[106] T. S. Harrison, L. J. Scott, Drugs 2004, 64, 985–996.
[107] J. Fleckenstein, I. Halle, Z. Y. Hu, G. Flachowsky, E. Schnug, Analyse von Lanthaniden mittels ICP – MS in Futter- und Organproben im Broilermastversuch, 22. Arbeitstagung Mengen und Spurenelemente, Jena, Germany, 2004.
[108] F. Li, Y. Wang, Z. Zhang, J. Sun, H. Xiao, Z. Chai, J. Radioanal. Nuclear Chem. 2002, 251, 437–441.
[109] Y. Nakamura, Y. Tsumura Hasegawa, Y. Tonogai, M. Kanamoto, N. Tsuboi, K. Murakami, Y. Ito, Eisei Kagaku 1991, 37, 489–496.
[110] A. J. Wassermann, T. M. Monticello, R. S. Feldman, P. H. Gitlitz, S. K. Durham, Toxicol. Pathol. 1996, 24, 588–594.
[111] Y. Nakamura, Y. Tsumura, T. Shibata, Y. Ito, Fund. Appl. Toxicol. Off. J. Soc. Toxicol. 1997, 37, 106–116.
[112] H. Shimada, M. Nagano, T. Funakoshi, Y. Imamura, Toxicol. Mechan. Methods 2005, 15, 181–184.
[113] A. Shinohara, M. Chiba, Y. Inaba, Biomed. Environm. Sci. 1997, 10, 73–84.
[114] F. R. Mraz, G. R. Eisele, Health Phys. 1977, 33, 494–495.
[115] C. L. Fox, W. W. Monafo, V. H. Ayvazian, A. M. Skinner, S. Modak, J. Stanford, C. Condict, Surg. Gynecolo. Obstetrics 1977, 144, 668–672.

[116] G. C. Kyker, Rare Earths, in Mineral Metabolism (ed. C. L. Comar, F. Bronner), Vol. 2 B, Academic Press, New York, 1962, 499–541.
[117] G. M. Kanapilly, Health Phys. 1980, 39, 343–346.
[118] P. W. Durbin, M. H. Williams, M. Gee, R. H. Newman, J. G. Hamilton, Proc. Soc. Exp. Biol. Med. (New York, N. Y.) 1956, 91, 78–85.
[119] B. Venugopal, T. D. Luckey, Environ. Qual. Saf. 1975, Supplement 1, 4–73.
[120] T. J. Haley, J. Pharmaceut. Sci. 1965, 54, 663–670.
[121] D. P. Hutcheson, D. H. Gray, B. Venugopal, T. D. Luckey, J. Nutr. 1975, 105, 670–675.
[122] M. S. Joy, W. F. Finn, Am. J. Kidney Dis. 2003, 42, 96–107.
[123] F. Locatelli, M. D'Amico, G. Pontoriero, Drugs 2003, 6, 688–695.
[124] E. Ritz, Nephrol. Dialysis Transplant. 2004, 19, 1–3.
[125] J. L. M. de Boer, W. Verweij, T. van der Velde-Koerts, W. Mennes, Water Res. 1996, 30, 190–198.
[127] S. J. P. Damment, C. Beevers, D. G. Gatehouse, The 19th Congress of the European Renal Association and European Dialysis and Transplant Association 15.-18. May 2004, Lisbon, Portugal, 2004.
[128] H. A. Schroeder, M. Mitchener, J. Nutr. 1971, 101, 1431–1438.
[129] O. Strubelt, C. P. Siegers, M. Younes, Arzneim. Forsch. 1980, 30, 1690–1694.
[130] B. Nemery, Eur. Respir. J. 1990, 3, 202–219.
[131] P. J. Haley, Health Phys. 1991, 61, 809–820.
[132] P. Vogt, M. A. Spycher, J. R. Ruettner, Schweiz. Med. Wochenschr. 1986, 116, 1303–1308.
[133] S. Porru, D. Placidi, C. Quarta, E. Sabbioni, R. Pietra, S. Fortaner, J. Trace Elements Med. Biol. 2000, 14, 232–236.
[134] G. Magnusson, Acta Pharmacol. Toxicol. Suppl. 1963, 20, 1–95.
[135] B. Tuchweber, R. Trost, M. Salas, R. Sieck, Cana. J. Physiol. Pharmacol. 1976, 54, 898–906.

[136] G. A. Robinson, D. C. Wasnidge, F. Floto, Poultr. Sci. 1986, 65, 1178–1183.
[137] P. Salonpaa, M. Iscan, M. Pasanen, P. Arvela, O. Pelkonen, H. Raunio, Biochem. Pharmacol. 1992, 44, 1269–1274.
[138] G. Renaud, C. Soler Argilaga, C. Rey, R. Infante, Biochem. Biophys. Res. Commun. 1980, 92, 374–380.
[139] P. Arvela, N. T. Karki, Experientia 1971, 27, 1189–1190.
[140] C. H. Stineman, E. J. Massaro, B. A. Lown, J. B. Morganti, S. AlNakeeb, J. Environm. Pathol. Toxicol. 1978, 2, 553–570.
[141] P. H. Wald, Lawrence Livermore National Laboratory, UCID-21823 Rev. 1, 1990, 1–27.
[142] S. J. Damment, V. Shen, Clin. Nephrol. 2005, 63, 127–137.
[143] S. J. P. Damment, I. Webster, V. Shen, ERA – EDTA / World Congress of Nephrology 8.-12. June 2003, London, UK, 2003.
[144] A. Freemont, J. Denton, C. Jones, 19th Congress of the European Renal Association and European Dialysis and Transplant Association 15.-18. May 2004, Lisbon, Portugal, 2004.
[145] W. Boeckx, P. N. Blondeel, K. Vandersteen, C. DeWolf Peeters, A. Schmitz, Burns 1992, 18, 456–462.
[146] A. A. Swanson, A. W. Truesdale, Biochem. Biophys. Res. Commun. 1971, 45, 1488–1496.
[147] R. A. Harris, E. T. Iwamoto, H. H. Loh, E. L. Way, Brain Res. 1975, 100, 221–225.
[148] H. J. Weinemann, R. C. Brasch, W. R. Press, G. E. Wesbey, Am. J. Roentgenol. 1984, 142, 619–624.
[149] C. Jones, I. Webster, S. J. P. Damment, 19th Congress of the European Renal Association and European Dialysis and Transplant Association 15.-18. May 2004, Lisbon, Portugal, 2004.

Reinhard K. Kremer

3.3 Rare earths magnetism – condensed matter

3.3.1 Introduction

The series of chemical elements in the row from La to Lu, known as rare earths (*RE*) or lanthanides, form a group of elements which behave chemically very similarly. From La to Lu, the inner 4*f* electron shell gradually accepts electrons until complete filling is reached with $2(2l+1) = 14$ electrons, where $l = 3$ is the orbital angular moment quantum number for *f* electrons. Notwithstanding that the *RE* are chemically similar, the degree of filling of the 4*f* shell causes significant differences, for example of, the magnetic and optical properties of the *RE* elements and ions. The electrons in the partially occupied 4*f* shell have a magnetic moment composed of the spin and the orbital moment, the latter being partially very large, and consequently the magnetic properties of the free *RE* ions vary remarkably. These and the variety of the magnetic properties of the *RE* ions in solids, for instance, due to crystal electric field interactions that the *RE* ions experience in solids, or due to exchange interaction between the magnetic moments of *RE* ions in insulators or metals are the topic of this chapter.

The electron configuration of the *RE* elements with *n* electrons in the 4*f* shell is given by [Xe] $4f^n6s^2$ or [Xe] $4f^n6s^25d^1$ (see Table 3.3.1). [Xe] stands for the filled electron shells of the noble gas xenon with the electron configuration $1s^22s^22p^63s^23p^63d^{10}4s^24p^64d^{10}5s^25p^6$. In solids, the *RE* ions are mostly trivalent. However, if by giving away two or four electrons the *RE* ions can achieve a more stable electronic configuration, for example, an empty, a full or a half-filled 4*f* shell, they can also be divalent or fourvalent. The ions of the *RE* have rather large ionic radii of ~120 pm (1.2 Å) [1–3]. On the other hand, the 4*f* electrons are close to the nucleus and are thus well shielded by the completely filled 5*s* and 5*p* electron shells. The radii of the 4*f* electrons typically amount to less than one Bohr radius, that is, 52.9 pm (0.529 Å) (see Figure 3.3.1). Given the rather high electrical charge of the *RE* nuclei, spin–orbit interaction is rather large. The spin–orbit coupling constants ζ_{4f} for the *RE* are of the order of 10^3 cm^{-1} (~0.1 eV) (Table 3.3.2) leading to energy separation of the spin–orbit coupled multiplets of up to several electron volts.

3.3.2 Russel–Saunders coupling

The spin–orbit interaction couples the spin with the orbital motion of the electrons. In the *RE*, the Russel–Saunders coupling scheme is found to be the appropriate

https://doi.org/10.1515/9783110654929-018

Table 3.3.1: Electronic configuration of the $4f$ elements. [Xe] stands for the filled electron shells of the noble gas xenon, $1s^22s^22p^63s^23p^63d^{10}4s^2 4p^64d^{10}5s^25p^6$.

Element	Symbol	Electronic configuration [Xe]+
Lanthanum	**La**	$4f^0\ 6s^2\ 5d^1$
Cerium	**Ce**	$4f^1\ 6s^2\ 5d^1$
Praseodymium	**Pr**	$4f^3\ 6s^2$
Neodymium	**Nd**	$4f^4\ 6s^2$
Promethium	**Pm**	$4f^5\ 6s^2$
Samarium	**Sm**	$4f^6\ 6s^2$
Europium	**Eu**	$4f^7\ 6s^2$
Gadolinium	**Gd**	$4f^7\ 6s^2\ 5d^1$
Terbium	**Tb**	$4f^9\ 6s^2$
Dysprosium	**Dy**	$4f^{10}\ 6s^2$
Holmium	**Ho**	$4f^{11}\ 6s^2$
Erbium	**Er**	$4f^{12}\ 6s^2$
Thulium	**Tm**	$4f^{13}\ 6s^2$
Ytterbium	**Yb**	$4f^{14}\ 6s^2$
Lutetium	**Lu**	$4f^{14}\ 6s^2\ 5d^1$

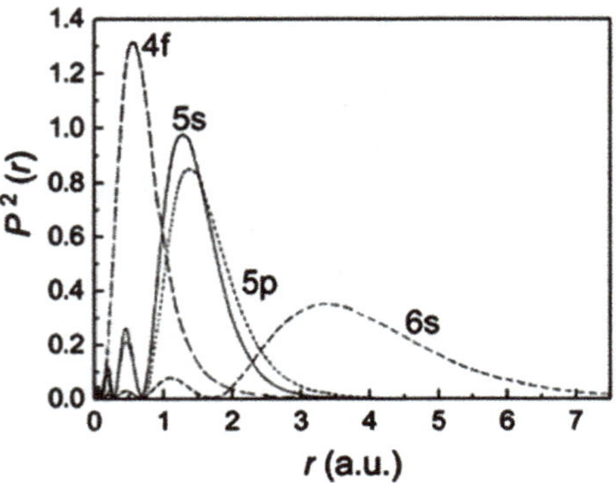

Figure 3.3.1: Hartree–Fock radial dependence of the charge density of the $4f$, $5s$,$5p$ and $6s$ orbitals for a Gd^+ cation versus distance from the nucleus in Bohr radii, a_0. The orbitals have been normalized according to $\int_0^\infty P^2(r)dr = 1$ [4]. Copyright (1962) by the American Physical Society.

Table 3.3.2: Theoretical and experimental spin–orbit coupling parameters ζ_{the} and ζ_{exp} for the trivalent *RE* ions. Except for Tm^{3+} and Gd^{3+} data were taken from [6]; (*) Tm^{3+}, see [7]; (**) Gd^{3+}, see [8].

Z	*RE*	Config.	ζ_{the} (cm^{-1})	ζ_{exp} (cm^{-1})
58	Ce^{3+}	$4f^1$	740	640
59	Pr^{3+}	$4f^2$	878	750
60	Nd^{3+}	$4f^3$	1,024	900
62	Sm^{3+}	$4f^5$	1,342	1,180
63	Eu^{3+}	$4f^6$	1,360	1,324
64	Gd^{3+}	$4f^7$	1,717	1,520(**)
65	Tb^{3+}	$4f^8$	1,915	1,620
66	Dy^{3+}	$4f^9$	2,182	1,820
67	Ho^{3+}	$4f^{10}$	2,360	2,018
68	Er^{3+}	$4f^{11}$	2,610	2,470
69	Tm^{3+}	$4f^{12}$	2,866	2,750(*)
70	Yb^{3+}	$4f^{13}$	3,161	2,950

approach how to couple the individual angular moment of the 4*f* electrons with their spin moment and to identify the ground state. Russel–Saunders coupling implies that the individual angular moments l_i and the individual spin moments s_i of the 4*f* electrons add up to the total orbital angular and spin moment vectors *L* and *S*, respectively, according to

$$L = \sum_{i=1}^{n} m_{l,i}$$

and

$$S = \sum_{i=1}^{n} s_i$$

where the summation is taken over all *n* electrons in the open 4*f* shell. Subsequently, *L* and *S* couple via spin–orbit coupling to the total angular moment vector $J = L + S$, to multiplets with the total angular moment quantum numbers spanning the range from $|L–S|$ to $L+S$, as follows:

$$|L-S|,\ |L-S+1|,\ |L-S+2|,\ \ldots, \le J \le, \ldots,\ (L+S-2),\ (L+S-1),\ (L+S)$$

Table 3.3.2 lists some theoretical and experimental values of the spin–orbit coupling parameters ζ_{4f} for trivalent *RE*. The theory values have been obtained from Hartree–Fock calculations [5, 6]. These data highlight the growth of the spin–orbit coupling with increasing atomic number by about a factor of 5. Consequently, the term splitting of Yb^{3+} amounts to approximately 10^4 cm^{-1}, whereas in Ce^{3+}, it was found to be of the order of 2,200 cm^{-1} [9]. The Russel–Saunders coupling scheme is adequate for the lighter elements in the periodic table for which the sum of Coulomb interaction between electrons in different orbits is much larger than the relativistic effect of spin–orbit coupling. For heavier elements, spin–orbit coupling grows, and the individual l_i and s_i couple to an individual j_i. The latter then combine to a total angular moment ("*jj* coupling"). In the *RE*, Coulomb interaction and spin–orbit coupling become of the same order of magnitude ("intermediate coupling"). Calculation of the exact energy levels and wave functions therefore requires a full quantum mechanical treatment of Coulomb and spin–orbit interaction, simultaneously. For example, the exact ground state wave function of Gd^{3+} (half-filled shell, $4f^7$ configuration) contains admixtures of states which lie about 3 eV above the ground state. Consequently, the g-factor of Gd^{3+} ions observed in electron spin resonance experiments in insulators is reduced from the spin-only value of 2 by about 0.4% [6].

3.3.3 Hund's rules

Let us consider the example of two electrons in the $4f$ shell corresponding to Pr^{3+} ions: obeying Pauli's exclusion principle Russel–Saunders coupling allows the states ^{1}I, ^{3}H, ^{1}G, ^{3}F, ^{1}D, ^{3}P and ^{1}S, where the capital letters S, P, D, F, G, H, I in the term symbol stand for total angular moment quantum number L = 0, 1, 2, 3, 4, 5, 6, respectively. The superscript is calculated according to $2S$+1 with S being the total spin quantum number, either zero or one for the chosen example of two electrons in the $4f$ shell. Which of these states becomes the ground state is not obvious per se. The Hund's rules provide a set of empirical guidelines how to select the ground state from the manifold of possible states:

Hund's rule 1
The ground state has the largest spin quantum number S consistent with the Pauli principle ('maximize S') will be chosen.

Hund's rule 2
The ground state is described by the maximum angular moment quantum number, L consistent with S, as required by the first Hund's rule ('maximize L').

Hund's rule 3
The total angular moment quantum number J amounts to either |L–S| if the 4f shell is less than half-filled, or L+S if the 4f electron shell is more than half-filled.

For our example of two electrons in the 4*f* shell with $L = \sum_{i=1}^{2} m_{l,i} = 3 + 2 = 5$ and $S = \sum_{i=1}^{2} s_i = 1/2 + 1/2 = 1$, the third Hund's rule gives us the ground state multiplet $^3H_4$, where the subscript in the term symbol represents the total angular momentum *J*. The energies of all other terms result from the competition of the Coulomb interaction between the two electrons and the spin–orbit coupling. They are shown in Figure 3.3.2 as measured for Pr^{3+} in $LaCl_3$ [10, 11]. For Ce^{3+} with one electron in the 4*f* shell and a total *L* of 3 (according to the second Hund's rule) only two multiplets, $^2F_{5/2}$ and $^2F_{7/2}$ can be combined, which have been found to be about 2,200 cm^{-1} apart. For the multiplet splitting of the other RE^{3+} ions, see [12].

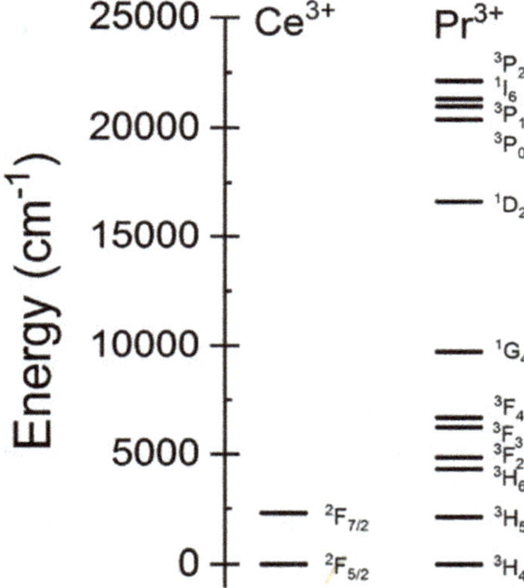

Figure 3.3.2: Energy levels of Ce^{3+} and Pr^{3+} as determined spectroscopically on the RE^{3+} ions substituted for La in $LaCl_3$. The terms are labeled by the Russel–Saunders classification [10, 11]. For other RE^{2+} and RE^{3+}, see [12] and also [9].

3.3.4 Magnetic susceptibility of free *RE* ions

The Zeeman energy of a magnetic moment, $\boldsymbol{\mu}_m$, in a magnetic field, $\boldsymbol{H}$, is given by $-\boldsymbol{\mu}_m\mathbf{H}$. The magnetic moment of the *RE* is combined of the orbital moment $\boldsymbol{\mu}_L = -\boldsymbol{\mu}_B \boldsymbol{L}$ and the magnetic moment of the electron spin $\boldsymbol{\mu}_S = -g_e\boldsymbol{\mu}_B\boldsymbol{S}$, where $\boldsymbol{\mu}_B$ is the Bohr magneton and g_e the *g*-factor of the free electron. g_e amounts to 2.00231930.... The difference to 2 arises from quantum electrodynamic corrections. For simplicity, we henceforth use $g_e = 2$, which is a very good approximation for all practical cases. If we merge the Zeeman energy parts of the orbital and the spin moments and rewrite it by introducing g_J as a new quantity

$$\mathcal{H} = \mu_0 \, \mu_B \, (\boldsymbol{L} + 2\boldsymbol{S})\boldsymbol{H} = g_J \mu_0 \, \mu_B \, \boldsymbol{JH}$$

g_J is commonly known as the Landé g-factor or sometimes also as the spectroscopic splitting factor. Assuming Russel–Saunders coupling holds that we can make use of the fact that $\boldsymbol{J}$, $\boldsymbol{L}$ and $\boldsymbol{S}$ commute and using the relation $\boldsymbol{J} = \boldsymbol{L} + \boldsymbol{S}$, g_J can be calculated as

$$g_J = \frac{3}{2} + \frac{S(S+1) - L(L+1)}{2\,J(J+1)}$$

where J, L and S are the respective quantum numbers. For an ensemble of noninteracting *RE* ions with a total angular moment $\boldsymbol{J}$, the magnetic susceptibility as a function of temperature is given by the "Curie law":

$$\chi(T) = \frac{N_\mathrm{A}}{V} \frac{\mu_0 \, \mu_\mathrm{B}^2 \, g_J^2 J(J+1)}{3k_\mathrm{B} T} = \frac{N_\mathrm{A}}{V} \frac{\mu_0 \, \mu_\mathrm{B}^2 \, p_\mathrm{eff}^2}{3k_\mathrm{B} T} \tag{3.3.1}$$

Here, we have implicitly assumed that the Zeeman energy is much smaller than the thermal agitation energy and a high-temperature approximation, $k_\mathrm{B}T >> \mu_\mathrm{B} \, g_J \, H$ can be applied. N_A and k_B are the Avogadro and Boltzmann constants, respectively, and V is the molar volume. The so-called "effective magnetic" moments are defined by $p_\mathrm{eff} = g_J \sqrt{J(J+1)}$. Typical experimentally observed effective moments, p_exp, for the RE^{3+} ions at room temperature are compiled in Table 3.3.3. The Curie law (eq. (3.3.1)) describes a hyperbolic dependence of the magnetic susceptibility on the temperature. Sometimes, especially in chemical literature, one finds plots of $\sqrt{\chi \cdot T}$ against temperature T to identify deviations from the Curie law. They may arise if spin exchange interactions become important or if the effective moment is temperature dependent. The latter is especially relevant for Eu^{3+} or Sm^{2+}, that is, for *RE* ions with $J = 0$ (see further).

Table 3.3.3 lists the spin and orbital moment quantum numbers, the quantum numbers of the total orbital angular moment and the effective moments of the *RE* ions in their most frequent oxidation states 2, 3 and 4. In the three rightmost columns, we have also given the Landé g-factor, g_J, the effective magnetic moments calculated from the Landé g-factor and the total angular moment quantum number J according to $g_J \sqrt{J(J+1)}$, and typical experimentally observed effective moments, p_exp. The values for p_exp for RE^{3+} can be found in standard textbooks and tables [13]. The data for Nd^{2+}, Pm^{3+}, Sm^{2+} and Tb^{4+} have been taken from [14–18], respectively. Figure 3.3.3 shows L, S and J highlighting the two "domes" with L and J vanishing at $n = 6$ and 7, respectively.

Here, we will discuss the case of Eu^{3+}. As shown in Figure 3.3.4, the Russel–Saunders ground state has $J = 0$ with a $^7F_0$ ground state. Consequently, for the ground state one has vanishing magnetization. However, at room temperature, experimentally, one observes an effective moment of $p_\mathrm{exp} \sim 3.3 - 3.5$, and one obtains a temperature dependence of the effective moment. This is due to the gradual thermal

Table 3.3.3: (from left to right) Typical oxidation states of the *RE* elements, electron configuration $4f^n$, total spin, total orbital and total angular moment quantum number according to Hund's rules, spectroscopic term symbol, Landé g-factor, g_J, effective magnetic moments, $p_{the} = g_J\sqrt{J(J+1)}$ and typical experimentally observed effective moments, p_{exp}, at room temperature (see, e.g., [13]). The values for Nd^{2+}, Pm^{3+}, Sm^{2+} and Tb^{4+} have been taken from [14–18], respectively.

RE		S	L	J	Term	g_J	$g_J J$	$g_J\sqrt{J(J+1)}$	p_{exp}
La^{3+}	$4f^0$	0	0	0	$^1S_0$	–	–	0	0
Ce^{3+}	$4f^1$	1/2	3	5/2	$^2F_{5/2}$	6/7	2.14	2.54	2.3–2.5
Pr^{3+}	$4f^2$	1	5	4	$^3H_4$	4/5	3.2	3.58	3.4–3.6
Nd^{3+}	$4f^3$	3/2	6	9/2	$^4I_{9/2}$	8/11	3.27	3.62	3.4–3.7
Nd^{2+}	$4f^4$	2	6	4	$^5I_4$	3/5	2.4	2.68	2.67–2.87
Pm^{3+}	$4f^4$	2	6	4	$^5I_4$	3/5	2.4	2.68	2.51
Sm^{3+}	$4f^5$	5/2	5	5/2	$^6H_{5/2}$	2/7	0.71	0.85	1.4–1.7
Sm^{2+}	$4f^6$	3	3	0	$^7F_0$	–	–	0	3.36
Eu^{3+}	$4f^6$	3	3	0	$^7F_0$	–	–	0	3.3–3.5
Eu^{2+}	$4f^7$	7/2	0	7/2	$^8S_{7/2}$	2	7	7.94	7.9
Gd^{3+}	$4f^7$	7/2	0	7/2	$^8S_{7/2}$	2	7	7.94	7.9–7.98
Tb^{4+}	$4f^7$	7/2	0	7/2	$^8S_{7/2}$	2	7	7.94	7.86
Tb^{3+}	$4f^8$	3	3	6	$^7F_6$	3/2	9	9.72	9.5–9.8
Dy^{3+}	$4f^9$	5/2	5	15/2	$^6H_{15/2}$	4/3	10	10.65	10.4–10.6
Ho^{3+}	$4f^{10}$	2	6	8	$^5I_8$	5/4	10	10.61	10.4–10.7
Er^{3+}	$4f^{11}$	3/2	6	15/2	$^4I_{15/2}$	6/5	9	9.58	9.4–9.6
Tm^{3+}	$4f^{12}$	1	5	6	$^3H_6$	7/6	7	7.56	7.1–7.6
Yb^{3+}	$4f^{13}$	1/2	3	7/2	$^2F_{7/2}$	8/7	4	4.54	4.3–4.9
Yb^{2+}	$4f^{14}$	0	0	0	$^1S_0$	–	–	0	0

population of the nearby multiplets, $^7F_1,\ldots,^7F_6$ distributed over an energy range of about 5×10^3 cm^{-1} [12]. Lueken ([19], page 153) has given the relationship to calculate the temperature dependence of the effective moment of Eu^{3+} as

$$
\begin{aligned}
& p_{\mathrm{eff}}^2(T) = \\
& \frac{1}{Z}\left\{144\frac{k_{\mathrm{B}}T}{\varsigma} + \left(\frac{27}{2} - 9\frac{k_{\mathrm{B}}T}{\varsigma}\right)\exp\left(-\frac{\varsigma}{6k_{\mathrm{B}}T}\right) + \left(\frac{135}{2} - 15\frac{k_{\mathrm{B}}T}{\varsigma}\right)\exp\left(-\frac{\varsigma}{2k_{\mathrm{B}}T}\right)\right. \\
& \quad + \left(189 - 21\frac{k_{\mathrm{B}}T}{\varsigma}\right)\exp\left(-\frac{\varsigma}{k_{\mathrm{B}}T}\right) + \left(405 - 27\frac{k_{\mathrm{B}}T}{\varsigma}\right)\exp\left(-\frac{5\varsigma}{3k_{\mathrm{B}}T}\right) \\
& \left. + \left(\frac{1,485}{2} - 33\frac{k_{\mathrm{B}}T}{\varsigma}\right)\exp\left(-\frac{5\varsigma}{2k_{\mathrm{B}}T}\right) + \left(\frac{2,457}{2} - 39\frac{k_{\mathrm{B}}T}{\varsigma}\right)\exp\left(-\frac{7\varsigma}{2k_{\mathrm{B}}T}\right)\right\}
\end{aligned}
\tag{3.3.2}
$$

where Z is the partition function given by

$$
\begin{aligned}
Z = & \left\{1 + 3\exp\left(-\frac{\varsigma}{6k_{\mathrm{B}}T}\right) + 5\exp\left(-\frac{\varsigma}{2k_{\mathrm{B}}T}\right) + 7\exp\left(-\frac{\varsigma}{k_{\mathrm{B}}T}\right) + 9\exp\left(-\frac{5\varsigma}{3k_{\mathrm{B}}T}\right)\right. \\
& \left. + 11\exp\left(-\frac{5\varsigma}{2k_{\mathrm{B}}T}\right) + 13\exp\left(-\frac{7\varsigma}{2k_{\mathrm{B}}T}\right)\right\}
\end{aligned}
$$

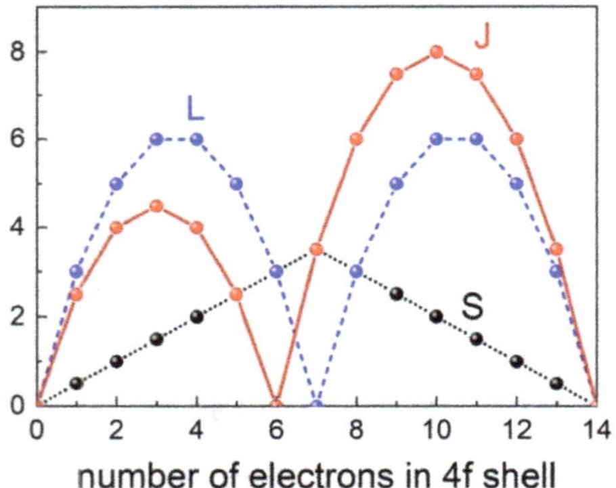

Figure 3.3.3: Orbital, spin and total angular moment quantum number against the number of electrons in the $4f$ shell.

Using the value of 1,360 cm^{-1} for the spin–orbit coupling constant ζ (see Table 3.3.2), the temperature dependence of p_{eff} for Eu^{3+} is displayed in Figure 3.3.5.

In Ce^{3+} where the excited $^2F_{7/2}$ multiplet is not too far away from the $^2F_{5/2}$ ground state (see Figure 3.3.2), the magnetic susceptibilities at sufficiently high temperatures are affected by the thermal population of the $^2F_{7/2}$ state and the effective magnetic moment also become moderately temperature dependent. By using the so-called van

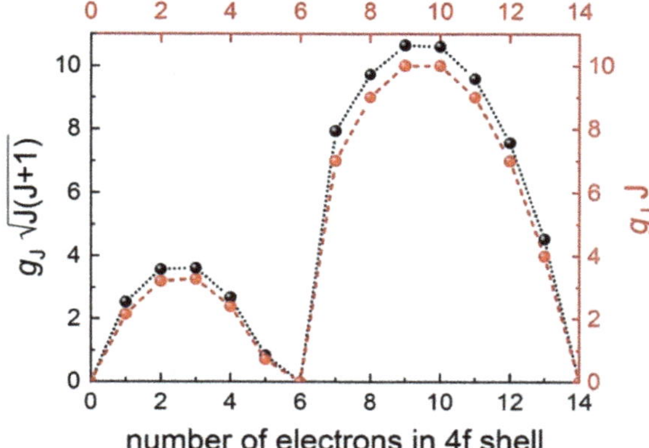

Figure 3.3.4: Effective moment $g_J\sqrt{J(J+1)}$ (black dots) and saturation moment $g_J J$ (red dots) versus number of electrons in the $4f$ shell.

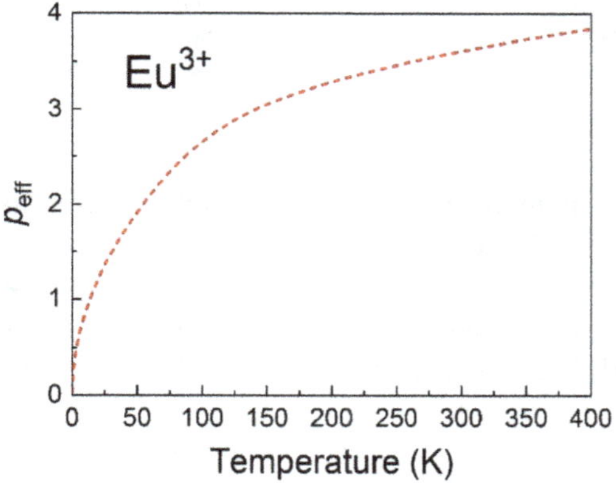

Figure 3.3.5: Effective moment p_{eff} of Eu^{+3} ions calculated according to eq. (2) by assuming a spin–orbit coupling parameter $\zeta = 1{,}360\ cm^{-1}$.

Vleck equation (see eq. (3.171) in [19]), the magnetic susceptibilities of Ce^{3+} ions were calculated as a function of the temperature (see Figure 3.3.6).

3.3.5 Magnetic saturation

The effective moment $p_{eff} = g_J\sqrt{J(J+1)}$ is a quantity that relates to the Curie law and must not be confused with the saturation moment. In the strong field case,

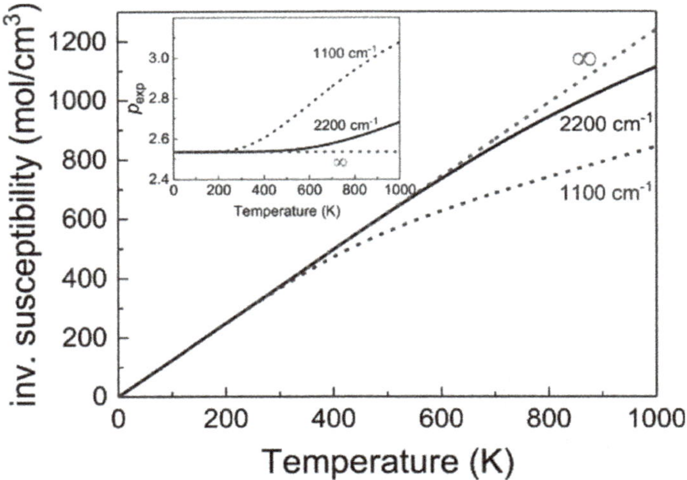

Figure 3.3.6: Inverse susceptibility of Ce^{3+} ions highlighting the deviations from the Curie law (eq. (1) at high temperatures when the $^2F_{7/2}$ excited state (at ~2,200 cm^{-1}) is thermally increasingly populated. For comparison, the susceptibilities obtained if the $^2F_{7/2}$ state is very far away (∞) or nearby at 1,100 cm^{-1} are also shown. The inset shows the temperature dependence of the effective moment.

$\mu_B\,\mu_0$H >> k_BT, the maximum magnetic moment that can be obtained for a *RE* ion characterized by g_J and J amounts to $g_J\mu_BJ$, a quantity that can be substantially different from the effective magnetic moment (see Figure 3.3.4). In the general case where thermal and Zeeman energies are comparable, the field and temperature dependence of the molar magnetization $M(T,H)$ is determined by the Brillouin function $B_J(x)$ according to

$$M(T,H) = N_A g_J \mu_B J\; B_J(x)$$

with the Brillouin function that depends only on the ratio of field and temperature, H/T,

$$B_J(x) = \left(\frac{2J+1}{2J}\right)\coth\left(\frac{2J+1}{2J}\,x\right) - \frac{1}{2J}\coth\left(\frac{x}{2J}\right)$$

with

$$x = g_J\mu_B\mu_0 HJ/k_BT.$$

As shown in Figure 3.3.7, at low temperatures and strong magnetic fields, saturation effects lead to deviation from the linear relationship $M \propto H$. Modern commercial SQUID magnetometers with superconducting magnets enable application of magnetic fields of several Tesla, and temperatures below the boiling point of liquid helium so that saturation can become important.

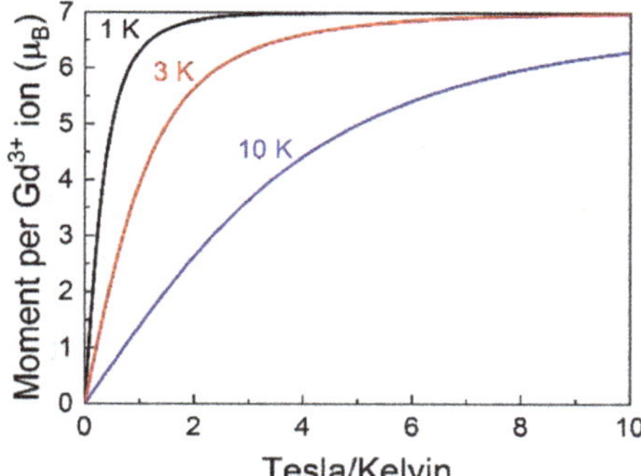

Figure 3.3.7: Saturation magnetization per atom for Gd^{3+} ($^8S_{7/2}$) with $J = S = 7/2$ and $g_J = 2$ against external magnetic field for various temperatures as indicated.

3.3.6 Crystal electric fields

A partially filled $4f$ shell in a *RE* ion has an aspherical charge distribution (see, e.g., Figure 3.3.8). A *RE* ion in a crystal experiences an inhomogeneous electrostatic field from its neighbors, which is commonly called "crystal electric field (CEF)" in

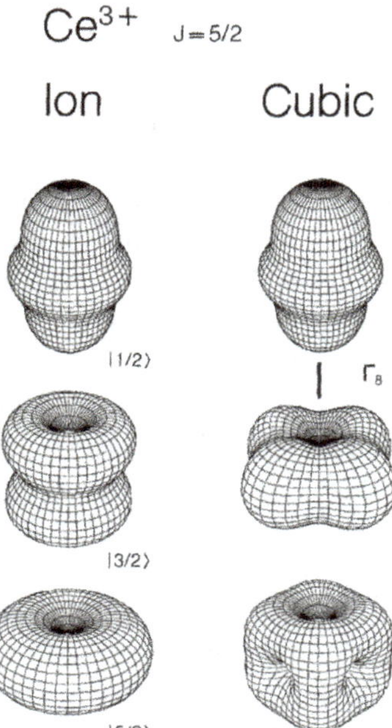

Figure 3.3.8: Spatial distribution of the $4f$ charge for Ce^{3+} with a $4f^1$ electronic configuration and a $^2F_{5/2}$ ground state. The right hand side displays the spatial distribution of the wave function in cubic symmetry where CEF states Γ_7 (doublet) and Γ_8 (quartet) are formed. Adapted from [21] by permission from Springer. Copyright 1972.

chemistry, often also called the "ligand field." The CEF splits the degeneracy of the ground state multiplet. The way the degeneracy is lifted has a symmetry and an energy aspect. The remaining degeneracy is determined by the site symmetry of the atom position of the *RE* in the crystal and can be obtained by group theory. Bethe in his groundbreaking work on the *Termaufspaltung in Kristallen* has conclusively solved the problem as to which degree the $2J$+1-fold degeneracy of a system with total angular moment J will be lifted, given a particular symmetry [20]. Two principally different situations have to be distinguished: for half-integer total moments (i.e., an odd number of electrons in the $4f$ shell) the electric field cannot lift the so-called Kramers degeneracy, which is a consequence of time-reversal symmetry ("which would be better called reversal of direction of motion" [6]). Kramers degeneracy implies that the states remain at least twofold degenerate ("Kramers doublets") and have a magnetic moment. For integer total angular momenta, degeneracy lifting can also result in singlets which are nonmagnetic. For the energy aspect of the splitting, that is, the magnitude of the splitting, several theoretical approaches have been put forward. The most simple is a point charge model wherein the surrounding ions are approximated by point charges, a model that almost never works for real systems. Other approaches that have been suggested to calculate the CEF are, for example, the *Superposition Model* by Newman or the *Angular Overlap Model (AOM)* by Urland [22, 23].

In practice, the magnitude of the CEF splitting is subject to experimental access. Optical spectroscopy and inelastic neutron scattering are the methods of choice to measure the CEF splitting.

Here, we want to briefly summarize the elementary theory of the crystal field. More extensive elaborations on this topic can be found in textbooks and more specialized monographs. An unsurpassable source in this respect is still the treatise by Abragam and Bleaney on *Electron Paramagnetic Resonance of Transition Ions*, which summarizes the state of the art before 1970 with an emphasis on spin resonance problems with many useful experimental references [6]. Lueken's textbook on *Magnetochemie* covers d and f systems with respect to CEF splitting on their magnetism [19]. Hüfner's monograph is focused on optical spectra of *RE* compounds [9]. Though a bit aged in its nomenclature, but still an extremely rich and very valuable source of CEF problems and implications on the optical, thermal and magnetic properties is Hellwege's textbook [24]. Blundell's more recent textbook is concise and modern with the emphasis on magnetism and a brief introduction to CEF theory [25].

Elementary CEF theory usually starts with an expansion of the CEF Coulomb potential around the position of the *RE* ion into a sum of spherical harmonics. Another more customary approach is to expand the CEF potential into homogeneous polynomials P_k^q. In order to calculate the matrix elements of the CEF potential, the operator equivalent technique is subsequently employed; it enables conversion of the spatial coordinates into vector operators with the components of the total angular momentum,

$\boldsymbol{J}_x, \boldsymbol{J}_y, \boldsymbol{J}_z$, or $\boldsymbol{J}_+, \boldsymbol{J}_-$ and $\boldsymbol{J}_z$ where the raising and lowering operators $\boldsymbol{J}_+$ and $\boldsymbol{J}_-$ are defined in the usual way as:

$$\boldsymbol{J}_+ = \boldsymbol{J}_x + i\boldsymbol{J}_y$$

$$\boldsymbol{J}_- = \boldsymbol{J}_x - i\boldsymbol{J}_y$$

Their action on a state $|J, m\rangle$ is given by

$$\boldsymbol{J}_+ \left|J, m_J\right\rangle = \sqrt{J(J+1-m(m+1)} \left|J, m_J+1\right\rangle$$

and

$$\boldsymbol{J}_- \left|J, m_J\right\rangle = \sqrt{J(J+1-m(m-1)} \left|J, m_J-1\right\rangle$$

Let us consider a regular octahedron whose vertices are occupied by six ions with negative charge $-Ze$. The expansions into polynomials up to the sixth order in the spatial coordinates are given by [6]

$$P_4 = 20\left(x^4 + y^4 + z^4 - \frac{3}{5}r^4\right)$$

$$P_6 = -224\left(x^6 + y^6 + z^6 + \right.$$

$$\left. + \frac{15}{4}\{x^4y^2 + y^4x^2 + x^4z^2 + z^4x^2 + y^4z^2 + z^4y^2\} - \frac{15}{14}r^6\right)$$

With P_4 and P_6, the CEF potential in the vicinity of the center of the octahedron where the *RE* ion resides is

$$V^{\text{cub}} = A_4P_4 + A_6P_6$$

The coefficients A_4 and A_6 depend on the actual charge distribution [6]. For an octahedron with six point charges $-Ze$ at the vertices A_4 and A_6 are given by

$$A_4^{\text{oct}} = \frac{7}{16}\frac{Ze^2}{R^5},\ A_6^{\text{oct}} = \frac{3}{64}\frac{Ze^2}{R^7}$$

and for a cube with eight point charges $-Ze$ at the vertices

$$A_4^{\text{cub}} = -\frac{7}{18}\frac{Ze^2}{R^5},\ A_6^{\text{cub}} = \frac{1}{9}\frac{Ze^2}{R^7}$$

and for a tetrahedron:

$$A_{4,6}^{\text{tet}} = \frac{1}{2}A_{4,6}^{\text{cub}}$$

The polynomials P_4 and P_6 can be rewritten in terms of homogeneous polynomials as $P_k^q(r)$; $(k=4,\ 6;\ q \le k)$, tabulated, for example, in Table 15 in [6].

Using the operators $O_k^q(J)$ (operator equivalent formalism by Stevens, [26]) which transform in the same way as $P_k^q(\vec{r})$, the matrix elements of $P_k^q(\vec{r})$ in the basis $|J, m\rangle$ can be calculated as

$$\left\langle J, m_J \middle| \sum_i P_k^q(\mathbf{r}_i) \middle| J, m_J' \right\rangle = a_k \langle r^k \rangle \; \langle J, m_J | \mathbf{O}_k^q(\mathbf{J}) | J, m_J' \rangle$$

where the summation is taken over the spatial coordinates $\mathbf{r}_i$ of all $4f$ electrons in the shell. It can be shown that matrix elements for terms with $k > 2l$ ($l = 3$ for f electrons) vanish, which reduces the number of expansion coefficients considerably.

Table 3.3.4: Expectation values $<r^k>$ ($k = 2, 4, 6$) for some *RE* ions. Data except for Tb^{3+} and Tm^{3+} (*) were taken from [4]. Data for Tb^{3+} and Tm^{3+} from [28]. $<r^k>$ are calculated according to $r^k = \int_0^\infty |f_l(r)|^2 r^{k+2} dr$, where $f_l(r)$ is the radial part of the Hartree–Fock wave functions. a_0 is the Bohr radius.

Z	Ion	Config.	$<r^2>$ (a_0^2)	$<r^4>$ (a_0^4)	$<r^6>$ (a_0^6)
58	Ce^{3+}	$4f^1$	1.200	3.455	21.226
59	Pr^{3+}	$4f^2$	1.086	2.822	15.726
60	Nd^{3+}	$4f^3$	1.001	2.401	12.396
61	Pm^{3+}	$4f^4$	–	–	–
62	Sm^{3+}	$4f^5$	0.883	1.897	8.775
62	Sm^{2+}	$4f^6$	–	–	–
63	Eu^{3+}	$4f^6$	–	–	–
63	Eu^{2+}	$4f^7$	0.938	2.273	11.670
64	Gd^{3+}	$4f^7$	0.785	1.515	6.281
65	Tb^{4+}	$4f^7$	–	–	–
65(*)	Tb^{3+}	$4f^8$	0.76	1.42	5.7
66	Dy^{3+}	$4f^9$	0.726	1.322	5.102
67	Ho^{3+}	$4f^{10}$	–	–	–
68	Er^{3+}	$4f^{11}$	0.666	1.126	3.978
69(*)	Tm^{3+}	$4f^{12}$	0.65	1.07	5.7
70	Yb^{3+}	$4f^{13}$	0.613	0.960	3.104
70	Yb^{2+}	$4f^{14}$	–	–	–

The constants a_k are the so-called reduced matrix elements and $\mathbf{O}_k^q(\mathbf{J})$, the Stevens equivalent operators. Tabulated values for the a_k and for the Stevens operators and their matrix elements for the various J relevant to the RE ions can be found in tables 16 and 17 in [6]. $\langle r^k \rangle$ are the expectation values of r^k averaged over the atomic wave functions [6]. Hartree–Fock results of the expectation values $\langle r^k \rangle$ $(k=2, 4, 6)$ are listed in Table 3.3.4 [4].

As for the interesting case of cubic symmetry of the CEF, Lea, Leask and Wolf [27] have calculated the energy eigenvalues and eigenvectors for J = 2 to 15/2 by numerical diagonalization methods and plotted the eigenvalues in compact diagrams. They also calculated the eigenvectors and classified them according to the irreducible representation of the cubic group, using the CEF Hamiltonian

$$\mathscr{H}_{CEF} = B_4\left(\boldsymbol{O}_4^0 + 5\boldsymbol{O}_4^4\right) + B_6\left(\boldsymbol{O}_6^0 - 21\boldsymbol{O}_6^4\right)$$

where the CEF parameters B_4 and B_6 relate to the A_4 and A_6 and the reduced matrix elements β_J and γ_J by

$$B_4 = A_4\,\beta_J\,\langle r^4 \rangle \text{ and } B_6 = A_6\,\gamma_J\,\langle r^6 \rangle$$

The Stevens equivalent operators $\boldsymbol{O}_4^0$ and $\boldsymbol{O}_4^4$ are (see, e.g., table 16 in [6])

$$\boldsymbol{O}_4^0 = 35\boldsymbol{J}_z^4 - 30J(J+1)\boldsymbol{J}_z^2 + 25\boldsymbol{J}_z^2 - 6J(J+1) + 3J^2(J+1)^2$$

and

$$\boldsymbol{O}_4^4 = \frac{1}{2}\left(\boldsymbol{J}_+^4 + \boldsymbol{J}_-^4\right)$$

For the equivalent operators O_{60} and O_{64} see e.g. [6]. With the definitions for $\boldsymbol{J}_+$ and $\boldsymbol{J}_-$ listed earlier, the matrix elements can be readily calculated. $\boldsymbol{O}_4^0$ has diagonal matrix elements only, whereas $\boldsymbol{O}_4^4$ generates matrix elements in secondary diagonals. Matrix elements for all RE-relevant J's can be found in table 17 in [6]. See also Rotter's tables [29].

Lately, Rudowicz and Chung have recalculated the Stevens operator and the matrix elements by using computer routines and spotted several misprints and typos in the standard textbook tables [30].

Lea, Leask and Wolf rewrote the cubic CEF Hamiltonian as

$$\mathscr{H}_{CEF}/W = x\,\boldsymbol{O}_4/F_4 + (1-|x|)\boldsymbol{O}_6/F_6$$

by merging fourth- and sixth-order terms to $\boldsymbol{O}_4 = \left(\boldsymbol{O}_4^0 + 5\boldsymbol{O}_4^4\right)$ and $\boldsymbol{O}_6 = \left(\boldsymbol{O}_6^0 - 21\boldsymbol{O}_6^4\right)$. The constants F_4 and F_6 depend on J, and they have been introduced to keep the eigenvalues in a similar numerical magnitude. x and W are defined as

$$Wx = B_4F_4, \text{with}(-1 \le x \le +1)$$

and

$$(1-|x|) = B_6F_6$$

With this transformation, the energy eigenvalues can be conveniently plotted in so-called Lea–Leask–Wolf diagrams as shown, for example, for $J = 7/2$ in Figure 3.3.9. For $J = 7/2$, the CEF lifts the eightfold degeneracy and creates two Kramers doublets (Γ_6 and Γ_7) and a quadruplet Γ_8. Depending on x and the sign of W, each of them can become the CEF ground state. The magnitude of the separation of the energy eigenvalues is determined by W. The eigenvectors are tabulated in [27].

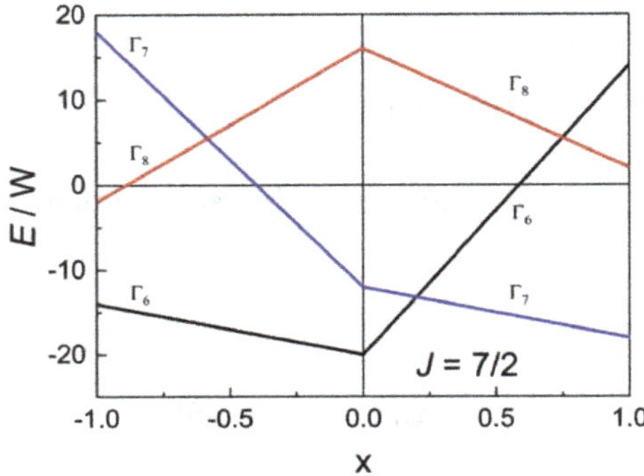

Figure 3.3.9: Energies of the Γ_6, Γ_7 and Γ_8 CEF states applicable to Yb^{3+} placed at a crystal site with cubic symmetry.

If the symmetry is lower than cubic, the quadruplet Γ_8 will further split into two Kramers doublets.

Since sixth-order components in the CEF potential have no matrix elements inside the $J = 5/2$ manifold of Ce^{3+}, the energy eigenvalues for Ce^{3+} are particularly simple and given by

$$\Gamma_7\text{: } E(x = \pm 1)/W = \mp 4$$

and

$$\Gamma_8\text{: } E(x = \pm 1)/W = \pm 2$$

The eigenvectors are:

$$\Gamma_7\text{: } 0.4083\,|\pm 5/2\rangle - 0.9129\,|\mp 3/2\rangle \quad \text{for} \quad x = \pm 1$$

and

$$\Gamma_8\text{: } \begin{matrix} 0.9129\,|\pm 5/2\rangle - 0.4083\,|\mp 3/2\rangle \\ \\ |\pm 1/2\rangle \end{matrix} \quad \text{for } x = \pm 1$$

The radial distribution of the CEF eigenstates Γ_7 and Γ_8 are displayed in Figure 3.3.8.

The CEF splitting of the J degeneracy has consequences on the magnetic and magnetothermal properties. Figure 3.3.10 displays the magnetic contribution to the heat capacity, $C_{mag}(T)$, of a system containing Ce^{3+} ions, with the Γ_7 doublet as ground state and the excited Γ_8 CEF state being located $k_B \times 100$ K above, and the reverse situation with Γ_8 high and Γ_7 low. It was calculated using the general equation

$$C_{mag}(T) = \frac{R}{(k_B T)^2} \frac{\left[\sum_n \exp(-E_n/k_B T)\right]\left[\sum_n E_n^2 \exp(-E_n/k_B T)\right] - \left[\sum_n E_n \exp(-E_n/k_B T)\right]^2}{\left[\sum_n \exp(-E_n/k_B T)\right]^2},$$

where the summation runs over $n = 1,\dots, 6$. Also shown in Figure 3.3.10 is the magnetic entropy obtained by integrating $C_{mag}(T)/T$ versus temperature. The magnetic heat capacity and the entropy versus temperature are markedly different for the two different CEF level splitting scenarios. The magnetic heat capacity and the entropy change can therefore be used to determine the CEF level sequence and the degeneracy of the CEF states. From a fit of the heat capacity, the energetic separation of the CEF levels can also be obtained.

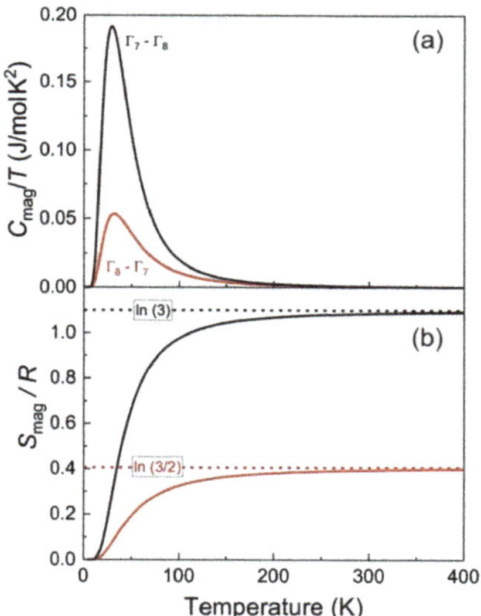

Figure 3.3.10: (a) Magnetic heat capacity and (b) magnetic entropy, S_{mag}, of a system with a doublet and a quadruplet state separated by 100 K, applicable to Ce^{3+} ions in a cubic CEF environment. The black curves represent the case with a doublet as ground and the Γ_8 quadruplet, the excited state; the red curve gives the magnetic contribution to the heat capacity with the reversed CEF state sequence. The magnetic entropies amount to R ln (3) and R ln (3/2) in the former and latter case, respectively. R is the molar gas constant.

If the CEF symmetry is lower than cubic, the Γ_8 quadruplet splits and the CEF scheme consists of three Kramers doublets. A magnetic field will further split the doublets. A CEF splitting also has consequences for the magnetic symmetry. Now, not only the level separation, but also the dependence of the magnetic moment of the CEF levels in the direction of the applied magnetic field with respect to the crystal axes matters.

With a CEF splitting of the ground state multiplet, the effective magnetic moment becomes temperature dependent. Depending on the magnitude of the CEF and the temperature range, the susceptibilities may deviate largely from the free ion values listed in Table 3.3.3. Figure 3.3.11 shows as an example the inverse magnetic susceptibility and the effective moment against temperature of a Ce^{3+} ion in a cubic crystal field, with the energy separation between the Γ_7 ground state and the Γ_8 excited CEF state amounting to $k_B \times 100$ K and $k_B \times 300$ K. Deviations from the linear temperature dependence of the free ion susceptibility and the constant effective moment become noticeable below approximately 100 and 300 K, respectively, that is, the CEF splitting energies. At low temperature, when essentially the Γ_7 ground state is thermally populated, the effective moment amounts to ~1.25 μ_B. This value can be obtained by calculating the effective g-factor from the Γ_7 eigenvector given above. The effective g-factor of the Γ_7 ground state is isotropic and calculated based on $g_J \times 2 \langle \Gamma_7 | \boldsymbol{J}_z | \Gamma_7 \rangle = 1.428$. With this g-factor, the effective moment for a doublet, that is, an effective spin $S = 1/2$ state amounts to $1.428 \sqrt{\frac{1}{2}\left(\frac{1}{2}+1\right)} = 1.237$. If in Figure 3.3.11(a) we apply a Curie law to the high-temperature inverse susceptibility, the straight line intersects with the temperature axis at a negative temperature. Often, such extrapolated intercepts are quoted as a "Curie–Weiss" temperature and taken as evidence for spin exchange interaction between the *RE* magnetic moments

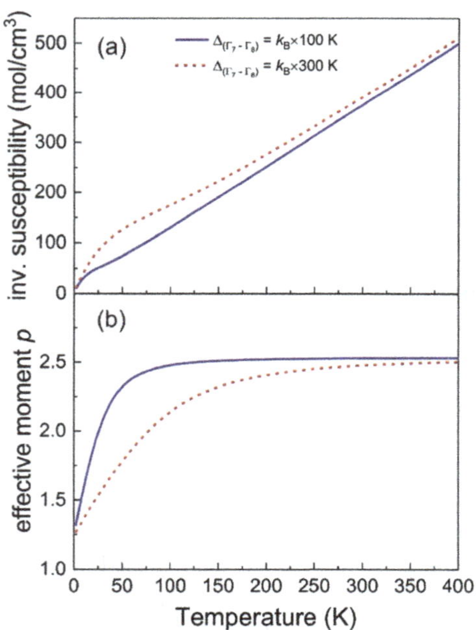

Figure 3.3.11: (a) Inverse magnetic susceptibility per Ce atom in a cubic crystal field (magnetic field || cubic axis) with separations between the Γ_7 ground state and the Γ_8 excited state of 100 K k_B and, as indicated. (b) Effective magnetic moments for the same CEF settings.

erroneously, even though we have been assuming a single-ion CEF setting with no spin exchange, when calculating the data plotted in Figure 3.3.11. However, if we take the low-temperature susceptibility data where essentially the ground state Γ_7 doublet is thermally populated, the extrapolated inverse susceptibility correctly intersects at $T = 0$ K, that is, vanishing exchange interaction.

Magnetic susceptibility and heat capacity measurements give an easy access to the CEF splitting in *RE*, though not always conclusive, since sometimes very high temperatures are necessary to populate CEF levels at high energies. Other frequently used experimental techniques to access the CEF level scheme are optical spectroscopy and neutron inelastic scattering methods. Both methods have their advantages and disadvantages. For example, optical spectroscopy experiments can be easily and relatively cheaply performed on insulating transparent materials in the laboratory. As a further advantage, optical spectroscopy also readily provides access to excited spin–orbit multiplets, also at higher energies (Figure 3.3.12). Neutron inelastic spectroscopy experiments require an expensive neutron source. On the other hand, neutron spectroscopy is well suited to study metals or intermetallic compounds.

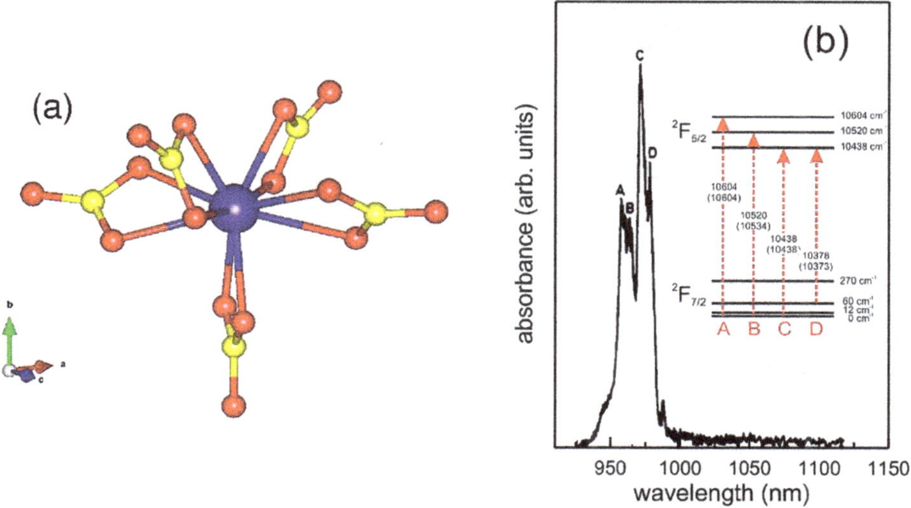

Figure 3.3.12: (a) Coordination of the Yb^{3+} cations by five NO_3^- groups in $[(C_6H_5)_4As]_2Yb(NO_3)_5$ [31]. (b) Optical absorbance spectrum of the compound tetraphenylarsonium pentakis(nitrato)ytterbate (III), $[(C_6H_5)_4As]_2Yb(NO_3)_5$, showing the optical transitions from the $^2F_{7/2}$ CEF ground state to the excited $^2F_{5/2}$ multiplet of the Yb^{3+} ions. The $^2F_{7/2}$–$^2F_{5/2}$ splitting amounts to about 10,500 cm^{-1}. The symmetry is low enough to split the $^2F_{5/2}$ and the $^2F_{7/2}$ states into three and four Kramers doublets, respectively. The structure in the spectrum allows to access the CEF splitting of the ground and the excited state. Adapted from [32] with permission. Copyright (1984) American Chemical Society).

Inelastic neutron scattering experiments are often limited by the available energy range of the neutrons (typically ≤ ~10^3 cm^{-1}). But in addition to CEF energies, they also allow measurement of dispersion of the CEF levels, if internal magnetic fields from exchange interactions are present. Here, we review exemplary optical and neutron spectra for two insulating compounds.

CEF splitting energies in *RE* ions are typically of the order of several hundred K × k_B, thus below 100 meV, matching well the energy of thermal neutrons obtained by moderating fission neutrons in a research nuclear reactor. Inelastic neutron spectroscopy, therefore, is a very powerful experiment in accessing CEF excitations. A typical instrument that measures CEF excitations is the triple-axis neutron spectrometer, a schematic sketch of which is shown in Figure 3.3.13. Neutrons from a source (e.g., a nuclear reactor) with a continuous energy distribution are monochromatized by a monochromator crystal, thus selecting the energy and the wave vector $\boldsymbol{k}$ of the neutrons impinging on and scattered by the sample. Wave vector and energy of the scattered neutrons are analyzed by the analyzer crystal reflecting the neutrons with wave vector $\boldsymbol{k}'$ finally into the detector.

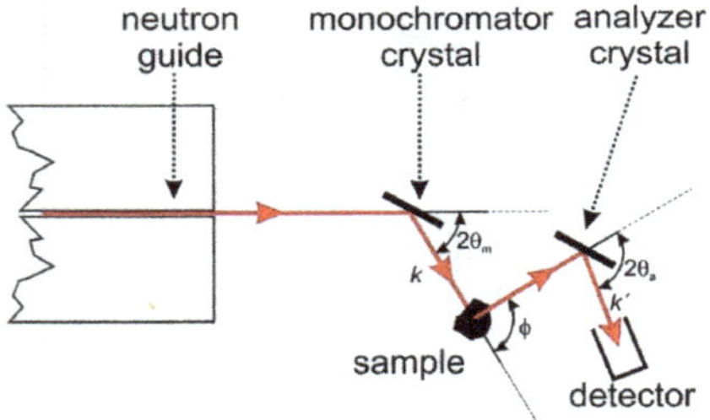

Figure 3.3.13: Schematic design of a neutron triple-axis spectrometer. After having left the neutron guide, the white neutron beam is scattered from the monochromator crystal, thus selecting neutrons with the wave vector $\boldsymbol{k}$. Scattering by the analyzer crystal allows to determine the wave vector $\boldsymbol{k}'$ of the scattered neutrons and to determine their energy.

A single crystalline sample can be aligned with respect to the direction of the incident neutron beam. The neutron is scattered about *three* parallel axes, explaining why such instruments have been named triple-axis spectrometers. Within the sample, the neutrons generate an excitation, E_{ex}, for example, between CEF levels, and the neutrons gain or lose energy. The scattering wave vector $\boldsymbol{q}$ is given by the difference of $\boldsymbol{k}$ and $\boldsymbol{k}'$ (up to a reciprocal lattice vector $\boldsymbol{G}$) as follows:

$$\boldsymbol{k} = \boldsymbol{k}' + \boldsymbol{q} + \boldsymbol{G}$$

The energy difference is calculated using

$$E_{ex} = E - E' = \frac{\hbar^2}{2m_N}\left(k^2 - k'^2\right)$$

where m_N is the mass of a neutron, and k and k' are the absolute values of the neutron wave vectors.

Typical inelastic neutron spectra measured with a triple-axis spectrometer on $PrCl_3$ and $PrBr_3$ are shown in Figure 3.3.14 [33]. In $PrCl_3$ and $PrBr_3$, the Pr^{3+} ions occupy sites with C_{3h} site symmetry and the resulting CEF Hamiltonian is

$$\mathcal{H} = B_2^0\mathbf{O}_2^0 + B_4^0\mathbf{O}_4^0 + B_6^0\mathbf{O}_6^0 + B_6^6\mathbf{O}_6^6$$

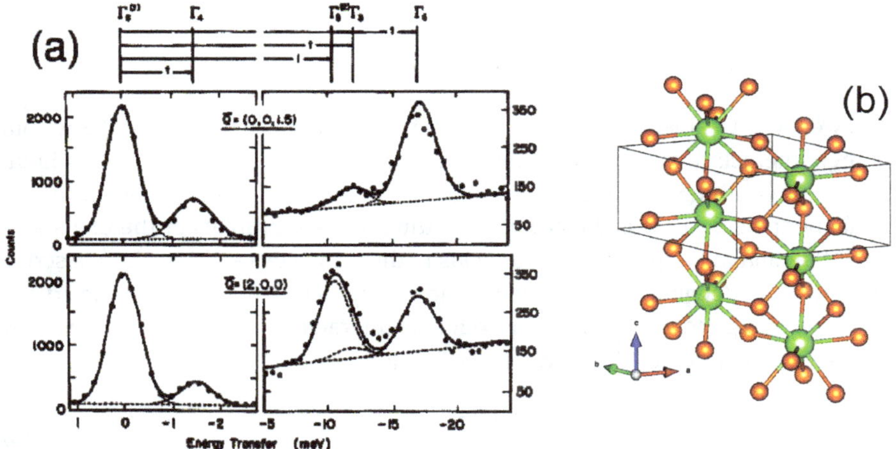

Figure 3.3.14: (a) Inelastic neutron spectra collected from single crystalline $PrBr_3$ at 1.5 K, with the scattering vector $\boldsymbol{q}$ oriented parallel and perpendicular to the hexagonal axis c. The solid lines correspond to a fit with a Gaussian plus a background to the measured spectra (dots). The derived CEF level scheme with the longitudinal (l) and transversal (t) transitions from the $\Gamma_5^{(1)}$ ground state to the excited states are indicated. Reprinted from [33], with the permission of AIP Publishing. (b) Crystal structure of $PrBr_3$ after Zachariasan [34]. Green spheres represent Pr^{3+} ions and orange spheres depict the Br^- anions. A hexagonal unit cell is outlined.

The ninefold-degenerate $^3H_4$ Russel–Saunders multiplet for Pr^{3+} splits into three doublets ($\Gamma_5^{(1)}$, $\Gamma_5^{(2)}$, Γ_6) and three singlets (Γ_1, Γ_3, Γ_4) (see, e.g., [6], page 312). Depending on the orientation of the scattering wave vector, different transitions between CEF levels (Table 3.3.5) can be excited, the energy loss or gain can be determined and the crystal field parameters B_n^m can finally be derived.

Table 3.3.5: CEF energy levels for $PrBr_3$ measured by inelastic neutron scattering and calculated by the angular overlap model (AOM) [33].

CEF state	E_{obs} (meV)	E_{calc} (meV)	$\|\Gamma_n$
$\Gamma_5^{(1)}$	0	0	$0.911\|\pm2\rangle - 0.412\|\mp4\rangle$
$\Gamma_5^{(1)}$	1.5±0.1	1.47	$0.707\|+3\rangle - 0.707\|-3\rangle$
$\Gamma_5^{(2)}$	10.6±0.3	1.61	$0.911\|\pm4\rangle + 0.412\|\mp2\rangle$
Γ_3	11.8±0.9	12.01	$0.707\|+3\rangle + 0.707\|-3\rangle$
Γ_6	17.0±0.3	16.83	$\|\pm1\rangle$
Γ_1	Not observed	26.77	$\|0\rangle$

3.3.7 Exchange interaction

So far, we have assumed that the *RE* ions are isolated from each other and any spin exchange interaction is prohibited, and thus, looked at single-ion properties in an ensemble of non-interaction *RE* ions embedded in a lattice, held at temperature T. In a magnetic field, there is the Zeeman splitting of the multiplets or the CEF states and the Boltzmann temperature factor leads to the Curie law (eq. (3.3.1)), to describe the temperature dependence of the magnetic susceptibility. If the ions in the crystal come close enough to each other that exchange interaction is present, the Curie law has to be modified to the "Curie–Weiss law" as follows:

$$\chi(T) = \frac{N_A}{V} \frac{\mu_0 \mu_B^2 \, g_J^2 \, J(J+1)}{3k_B(T - \Theta_{CW})} = \frac{N_A}{V} \frac{\mu_0 \mu_B^2 \, p_{eff}^2}{3k_B(T - \Theta_{CW})} \tag{3}$$

where Θ_{CW} is the so-called Curie–Weiss temperature. At sufficiently low temperatures, magnetic ordering phenomena like ferromagnetism, antiferromagnetism or ferrimagnetism may be observed. In simple systems, the magnetic ordering temperature is comparable to the absolute value of the Curie–Weiss temperature Θ_{CW}. The sign of the Curie–Weiss temperature allows a conclusion on the predominant exchange interaction. Negative Curie–Weiss temperatures indicate predominant antiferromagnetic exchange, that is, the exchange interaction favors antiparallel alignment of the magnetic moments, whereas positive Curie–Weiss temperatures indicate predominant ferromagnetic exchange, that is, the tendency to align the magnetic moments with a parallel arrangement. Generally, the Curie–Weiss temperature summarizes the exchange interaction of a magnetic moment with all its neighbor moments. If the exchange interaction to a particular neighbor z_{ij} is given by J_{ij}, where z_{ij} is the number of such neighbors in the jth coordination shell, Θ_{CW} is given by mean field approximation by

$$\Theta_{CW} = -\frac{1}{3}J(J+1)/k_B \sum_{ij} z_{ij}J_{ij}$$

Since, in many cases, the exchange interaction decreases very rapidly with the distance, it is often enough to sum over first and second nearest magnetic neighbors.

The origin of exchange interaction in *RE* materials can be manifold. In *RE* metals and intermetallic compounds (see further), spin polarization of the conduction electrons by the *RE* magnetic moment is the dominant source of exchange. In insulators, exchange interaction via intermediate anions is usually very small and mostly antiferromagnetic. In some cases, even ferromagnetism is found in the insulators. For example, $GdCl_3$, which is isotypic to $PrBr_3$ is a ferromagnet with a Curie temperature of 2.2 K [35]. $TbCl_3$ which crystallizes with the orthorhombic $PuBr_3$ structure type is a ferromagnet below 3.65 K [36, 37].

Figure 3.3.15 shows a schematic picture of inverse susceptibilities for a ferro- and an antiferromagnet. Equations to calculate the magnetic susceptibility of a ferrimagnet can be found, for example, in [19].

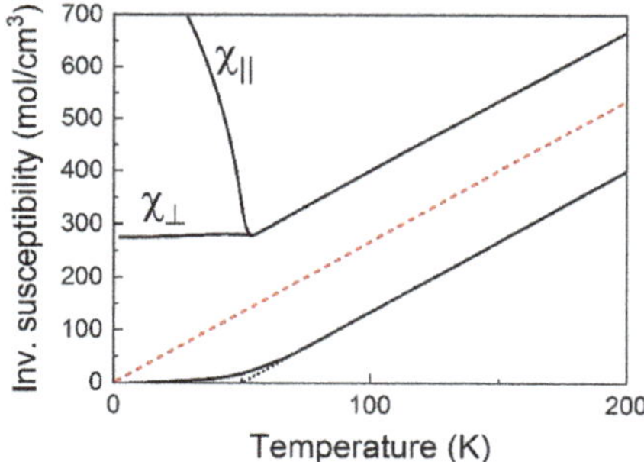

Figure 3.3.15: Inverse susceptibilities of a ferro-, an antiferro- and a paramagnet (red dashed line) with identical effective moments [38]. For the antiferromagnet, typical susceptibilities of an uniaxial system are shown. $\chi_{\|}$ and $\chi_{\perp}$ refer to the susceptibilities measured with the external field oriented parallel and perpendicular to the easy axes, respectively. The dotted line for the ferromagnet represents the extrapolation of the Curie–Weiss law from high temperatures. The curved susceptibility shortly above the critical temperature of ~30 K is due to critical fluctuations preceding long-range ordering.

3.3.8 Magnetic properties of the *RE* elements

Finally, we briefly review the magnetic properties of the *RE* elements themselves. All *RE* elements exhibit magnetic order with critical temperatures up to room temperature. Exchange coupling in the *RE* elements is an indirect mechanism and takes place via spin polarization of the conduction electrons by the so-called *s-f* exchange coupling. The sign of the exchange varies in an oscillatory manner, with the product of the distance from the *RE* atom and twice the Fermi wave vector of the conduction electrons determining the oscillation period, Ruderman-Kittel-Kasuya-Yosida (RKKY) exchange. The magnitude of the RKKY exchange falls off very rapidly with the third power of the distance to the *RE* atom.

Antiferromagnetic order prevails in the *RE*, but the magnetic structures are partially complex also depending very much on the crystal structure of the particular *RE* element. For the heavy *RE* with $n \geq 7$, *c*-axis modulated magnetic structures, helices and magnetic re-ordering processes are observed with the modulation wave vector being temperature dependent (Figure 3.3.16) [39]. Outstanding in its magnetic properties is Gd with a half-filled electron shell. Besides Fe, Co and Ni, Gd is one of the four elements in the periodic table which is ferromagnetic at room temperature with a Curie temperature close to ambient conditions, namely, 293 K. Neutron diffraction experiments on Gd single crystals showed neither satellite magnetic Bragg reflections nor any broadening of rocking curves limiting a possible spiral turn to under 2° [39]. According to these results, Gd behaves as a normal collinear ferromagnet. However, the moment direction is temperature dependent. The moments are aligned parallel to the *c*-axis between the Curie temperature and 232 K. Below 232 K they turn away and at ~180 K enclose an angle of 65°. Below 180 K, the moments rotate back and at lowest temperatures, they enclose an angle of ~32° with the *c*-axis [39].

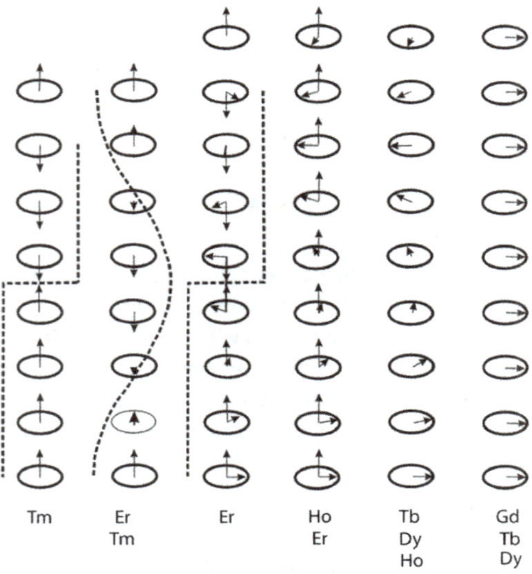

Figure 3.3.16: Schematic representations of the magnetic structures of the *RE* metals, Gd–Tm. The *c*-axis is vertical; the ellipsoids represent a hexagonal layer. Moments are parallel in a hexagonal layer. Adapted from [39] by permission from Springer. Copyright 1972.

The partially very large spin moments of the *RE* effectively polarize the spins of the conduction electrons, and the saturation moments are generally enhanced over their Russel–Saunders values of $g_J J \mu_B$ (see Table 3.3.3). In Gd, a saturation moment of 7.63 μ_B has been measured (see Figure 3.3.17). Table 3.3.6 lists the Néel and Curie temperatures of the *RE* elements as well, and the calculated and measured saturation magnetic moments μ_{sat}^{exp} and $g_J J \mu_B$.

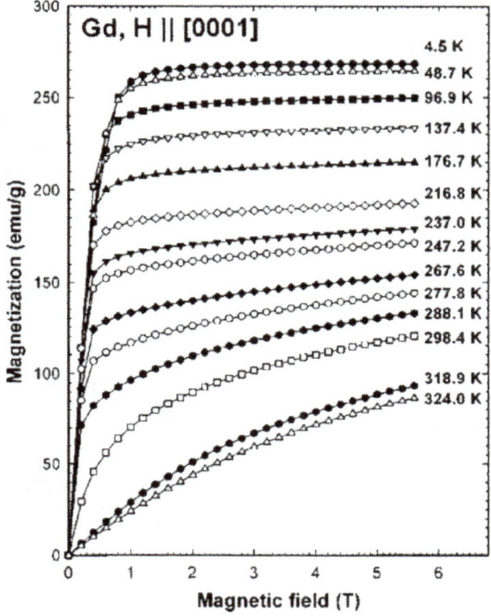

Figure 3.3.17: Magnetization of a Gd single crystal with the magnetic field aligned along [0001] [40]. Copyright (1998) by the American Physical Society. To convert the magnetization per gram in emu/g into μ_B per atom one multiplies by the molar mass and divides by the constant 5584.94.

Table 3.3.6: Theoretical and experimentally observed saturation moments and magnetic ordering temperatures of the *RE* elements (after [41]). ^{147}Pm containing about 2% ^{147}Sm has been reported to show magnetic order below ~98 K [42]. This appears to be the only investigation on Pm metal available, so far.

Element	gJ (μ_B)	μ_{sat}^{exp} (μ_B)	$\Theta_{\parallel}$ (K)	$\Theta_{\perp}$ (K)	T_N^{hex} (K)	T_N^{cub} (K)	T_C (K)
Ce	2.14	0.6			13.7	12.5	
Pr	3.20	2.7*			0.05		
Nd	3.27	2.2*			19.9	8.2	
Pm	2.40	<0.4(?)			98(?)		

Table 3.3.6 (continued)

Element	$g_J J$ (μ_B)	μ_{sat}^{exp} (μ_B)	$\Theta_{\parallel}$ (K)	$\Theta_{\perp}$ (K)	T_N^{hex} (K)	T_N^{cub} (K)	T_C (K)
Sm	0.71	0.13*			106	14.0	
Eu	7.0	5.1				90.4	
Gd	7.0	7.63	317	317			293
Tb	9.0	9.34	195	239	230		220
Dy	10.0	10.33	121	169	179		89
Ho	10.0	10.34	73	88	132		20
Er	9.0	9.1	62	33	85		20
Tm	7.61	7.14	41	−17	58		32

*Measured at 38 Tesla.

References

[1] R. D. Shannon, Acta Crystallogr. A 1976, 32, 751; see also [2] or [3].
[2] WebElements, can be found under https://www.webelements.com/cerium/atom_sizes.html
[3] Database of Ionic Radii, can be found under http://abulafia.mt.ic.ac.uk/shannon/ptable.php
[4] A. J. Freeman, R. E. Watson, Phys. Rev. 1962, 127, 2058.
[5] M. Blume, A. J. Freeman, R. E. Watson, Phys. Rev. 1964, 124, A320.
[6] A. Abragam, B. Bleaney, Electron Paramagnetic Resonance of Transition Ions, Clarendon Press, Oxford, 1970.
[7] M. F. Reid, F. S. Richardson, J. Chem. Phys. 1985, 83, 3831.
[8] A. Bronova, T. Bredow, R. Glaum, M. J. Riley, W. Urland, J. Comput. Chem. 2018, 39, 176.
[9] S. Hüfner, Optical Spectra of Transparent Rare Earth Compounds, Academic Press, New York, San Francisco, London, 1978.
[10] R. Sarup, M. H. Crozier, J. Chem. Phys. 1965, 42, 371.
[11] J. Sugar, Phys. Rev. Lett. 1965, 14, 731.
[12] G. H. Dieke, H. M. Crosswhite, Appl. Opt. 1963, 2, 675.
[13] Radiochemistry Soc., Richland, WA-99354, can be found under https://www.radiochemistry.org/periodictable/la_series/L8.html
[14] R. A. Sallach, J. D. Corbett, Inorg. Chem. 1964, 3, 993.
[15] H. Lueken, W. Bronger, U. Löchner, Rev. Chim. Miner. 1976, 13, 113.
[16] J. C. Sheppard, E. J. Wheelwright, F. P. Roberts, J. Phys. Chem. 1968, 67, 1568.
[17] M. Rudolph, W. Urland, Z. Anorg. Allg. Chem. 1997, 623, 1349.
[18] M. Guillot, M. El-Ghozzi, D. Avignant, G. Férey, J. Appl. Phys. 1993, 73, 5389.
[19] H. Lueken, Magnetochemie Eine Einführung, B. G. Teubner, Stuttgart, Leipzig 1999.
[20] H. Bethe, Ann. Phys. 1929, 5, 133.
[21] U. Walter, Z. Phys. B – Condensed Matter 1986, 62, 299.
[22] D. J. Newman, Betty Ng, Rep. Prog. Phys. 1989, 52, 699.
[23] W. Urland, Chem. Phys. 1976, 14, 393.

[24] K.-H. Hellwege, Einführung in die Festkörperphysik, Springer-Verlag, Berlin, Heidelberg, New York, 1981.
[25] S. J. Blundell, Magnetism in Condensed Matter, Oxford University Press, Oxford, 2001.
[26] K. W. H. Stevens, Proc. Phys. Soc. 1952, A65, 209.
[27] K. R. Lea, M. J. M. Leask, W. P. Wolf, J. Phys. Chem. Solids 1962, 23, 1381.
[28] T. O. Brun, G. H. Lander, G. P. Felcher, Bull. Am. Phys. Soc. 1971, 16, 325.
[29] https://www2.cpfs.mpg.de/~rotter/homepage_mcphase/manual/node132.html
[30] C. Rudowicz, C. Y. Chung, J. Phys.: Condens. Matter 2004, 16, 5825.
[31] W. Urland, E. Warkentin, Z. Naturforsch. 1983, 38b, 299.
[32] W. Urland, R. Kremer, Inorg. Chem. 1983, 23, 1550.
[33] B. Schmid, B. Hälg, A. Furrer, W. Urland, R. Kremer, J. Appl. Phys. 1987, 61, 3426.
[34] W. H. Zachariasen, Acta Crystallogr. 1948, 1, 265.
[35] W. P. Wolf, M. J. M. Leask, B. Mangum, A. F. G. Wyatt, J. Phys. Soc. Jpn. 1961, 17, Suppl. B-1, 487.
[36] A. Murasik, P. Fischer, A. Furrer, W. Szczepaniak, J. Less-Common Met. 1985, 111, 177.
[37] R. Kremer, E. Gmelin, A. Simon, J. Magn. Magn. Mater. 1987, 69, 53.
[38] J. M. D. Coey, Magnetism and Magnetic Materials, Cambridge University Press, Cambridge, 2010.
[39] W. C. Koehler, Magnetic Properties of Rare Earth Metals (ed. R. J. Elliott), Plenum Press, London, New York, 1972, 81.
[40] S. Y. Dan'kov, A. M. Tishin, V. K. Pecharsky, K. A. Gschneidner Jr., Phys. Rev. B 1998, 57, 3478.
[41] J. Jensen, A. R. Mackintosh, Rare Earth Magnetism Structures and Excitations, Clarendon Press, Oxford, 1991.
[42] W. C. Koehler, R. M. Moon, H. R. Child, AIP Conf. Proc. 1973, 10, 1319.

Jan van Leusen, Paul Kögerler

3.4 Rare earth magnetism – molecules

Over the past decade, the discovery of highly unusual magnetic features, frequently associated with the pronounced magnetic anisotropy of lanthanide ions in certain ligand environments, has considerably pushed forward the development of molecular coordination compounds of rare earth metals. As we will see, the characterization of the magnetic properties of molecular rare earth compounds usually starts with the experimental determination of thermodynamic parameters: the magnetization $\boldsymbol{M}$, that is, magnetic (dipole) moments per volume or the magnetic susceptibility χ [1, 2]. For small magnetic fields, both physical quantities and the magnetic field strength $\boldsymbol{H}$ obey the relation:

$$\boldsymbol{M} = \chi \boldsymbol{H} \tag{3.4.1}$$

Note that the susceptibility is, in general, a tensor reproducing the magnetic anisotropy of the material. The tensor components can be determined by measurement of oriented single crystals, while powder samples result in the mean values of the principal components. For magnetically isotropic compounds, the susceptibility can be simplified to a scalar, that is, a single value. Due to experimental considerations, the magnetic susceptibility often refers to the mass (χ_g) or the molar amount of substance (χ_m) instead of the volume of the sample:

$$\chi = \chi_g\, \rho = \chi_m \frac{\rho}{M_w} \tag{3.4.2}$$

where ρ is the volumetric mass density, and M_w the molar mass. Accordingly, eq. (3.4.1) has various implementations:

$$\boldsymbol{M} = \chi \boldsymbol{H} \text{ and } \boldsymbol{M}_g = \chi_g \boldsymbol{H} \text{ and } \boldsymbol{M}_m = \chi_m \boldsymbol{H} \tag{3.4.3}$$

For all materials, there is always a diamagnetic contribution in addition to potential paramagnetic contributions caused by unpaired electrons, eq. (3.4.4), while three-dimensional, long-range cooperative effects like ferromagnetism are typically absent in molecular compounds, at least. Exchange interactions restricted to small volumes or small, lower dimensional entities (layers, chains, etc.) are, however, possible.

$$\chi_{total} = \chi_{diam.} + \chi_{param.} \tag{3.4.4}$$

Since the intrinsic diamagnetic contributions are, to a very good approximation temperature independent, the analysis of the magnetic properties focuses on the other contributions and we, thus, will omit the index "param." in the following text.

Molecular compounds incorporating rare earth metals usually comprise one or a few rare earth metal ions that are coordinated by nonmetallic elements, small

https://doi.org/10.1515/9783110654929-019

inorganic or organic molecules and – on rare occasions – by other metals. These compounds form small and, therefore, spatially restricted units in contrast to the "infinite" extent of solid-state structures. Although molecular compounds can form crystalline solids, these solids are generally characterized by considerably larger lattice parameters and higher degrees of crystalline disorder, in comparison to crystals of solid-state compounds. Unlike the latter, the energetic structures of molecular compounds are, thus, composed of discrete energy levels (as for free atoms or ions) also in the solid phase instead of giving rise to a band structure. Each energy level is characterized by a specific magnetic moment. Therefore, the magnetization of the compound is usually temperature dependent according to Boltzmann statistics and is given by the fundamental equation for paramagnetic compounds [1, 2] (the direction of the magnetic field $\boldsymbol{B}$ is denoted by α):

$$M_{\mathrm{m},\alpha} = -N_{\mathrm{A}} \frac{\sum_n \frac{\partial E_{n,\alpha}}{\partial B} \exp\left(-\frac{E_{n,\alpha}}{k_{\mathrm{B}}T}\right)}{\sum_n \exp\left(-\frac{E_{n,\alpha}}{k_{\mathrm{B}}T}\right)} \quad \text{and} \quad \mu^{\mathrm{c}}_{n,\alpha} = \frac{\partial E_{n,\alpha}}{\partial B} \tag{3.4.5}$$

where E_n is the energy of the nth state, which additionally depends on the field direction for anisotropic centers, the denominator is the partition function and μ^{c} is the component of the magnetic moment parallel to the field.

Before discussing the magnetic characteristics of rare earth compounds, we start with the magnetism of the corresponding free ions since the ligand field usually causes a smaller perturbation for lanthanide centers [1–3]. In addition, we arbitrarily set the energy of the ground state to zero, thus treating the energies in terms of relative energies, which reflects the capability of the vast majority of all experimental data collection methods. Note that the determination of the total energies is a very sophisticated adventure without a guaranteed success. Unpaired valence electrons cause the paramagnetic behavior of free ions. The two main energetic contributions that determine the magnetic properties are the interelectronic repulsion (H_{ee}) and the spin–orbit coupling (H_{so}) of the valence electrons while interactions with the electrons of the other shells can be safely neglected under normal conditions. From a magnetochemical point of view, it is often sufficient for lanthanide ions (with the exception of $4f^4$, $4f^5$ and $4f^6$ electron configurations) to consider solely the $^{2S+1}L_J$ ground term, which is determined by the Russell–Saunders (or LS) coupling scheme. Note that this is an approximation and the energy states should be discussed only in terms of the quantum number J (total angular momentum), since L (total orbital angular momentum) and S (total spin angular momentum) are not good quantum numbers in this case. In addition, although H_{ee} is larger than H_{so} for rare earth ions, treating the spin–orbit coupling as a perturbation of the interelectronic repulsion also introduces errors of up to 5% (and even more) regarding the composition of the ground term. For example, the ground term of the Er^{3+} ion ($4f^{11}$) is $^4I_{15/2}$, according to the Russell–Saunders scheme, while the experimental data reveal the actual composition to be 96.8% $^4I_{15/2}$ + 3.1% $^2K_{15/2}$ + 0.04% $^2L_{15/2}$ (plus terms with even smaller contributions) [4].

Nevertheless, if we accept these approximations, the magnetism of the rare earth ions follows (at not too low temperatures and not too high magnetic fields) the Curie law [1, 2]:

$$\chi_m = \frac{C}{T} = \frac{\mu_0 N_A \mu_{eff}^2}{3k_B T} \quad \text{with}\ \mu_{eff} = g_J\sqrt{J(J+1)}\ \mu_B \tag{3.4.6}$$

The effective magnetic moment, μ_{eff}, is a characteristic magnetic property and is determined by the physical constant μ_B (Bohr magneton), the total angular momentum J and the Landé factor g_J, defined as

$$g_J = 1 + \frac{J(J+1) + S(S+1) - L(L+1)}{2J(J+1)} \tag{3.4.7}$$

An alternative representation is the product of the molar magnetic susceptibility and temperature $\chi_m T$:

$$\chi_m T = C = \frac{\mu_0 N_A g_J^2 J(J+1)\mu_B^2}{3k_B} \approx \frac{1}{8} g_J^2 J(J+1)\mathrm{cm^3\,K\,mol^{-1}} \tag{3.4.8}$$

This representation is very popular as it is linear in χ_m, that is, in the number of magnetic ions, and the physical constants combine to a factor of approximately 1/8 in the CGS system of units.

However, even if we only consider H_{ee} and H_{so}, in reality, μ_{eff} or $\chi_m T$ of the free ions do not match these values, but are slightly shifted. This is caused by the aforementioned mixing with excited states that is induced by spin–orbit coupling. Note that each energy state can be identified by a single quantum number m_J (secondary total angular momentum), though. Additionally, μ_{eff} or $\chi_m T$ are not constant values but are slightly temperature dependent, since the energy states of the ground term are not fully isolated from the other terms. Furthermore, we have to always apply a magnetic field to observe the paramagnetic (and diamagnetic) effects. The magnetic field, however, causes a splitting of the ground term. The corresponding effect is called the Zeeman effect (energetic contribution H_{mag}), named after its discoverer, the Dutch physicist Pieter Zeeman [1, 3, 5]. This effect causes a decrease of μ_{eff} or $\chi_m T$ at lower temperatures (e.g., below 3 K at 0.1 T, or below 60 K at 5.0 T), compared to the values at room temperature, and is more pronounced at stronger applied fields – even if the ground term is fully isolated.

Certain magnetic properties of a compound are revealed not only via the temperature dependence of the magnetic susceptibility at a static magnetic field, but also by the magnetization as a function of the applied magnetic field at a static temperature. In the latter case, most information can be gained by choosing the experimental conditions in a way that saturation or (pseudo-)saturation steps are observed. For paramagnetic molecular rare earth compounds, the temperature is adjusted to very low values, typically 2 K for applied fields up to 5–7 T. In the case of free rare earth ions, the molar magnetization (almost) reaches the saturation value at 2 K and 5 T, which is given by

$$M_{\mathrm{m,\,sat}} = g_J J\ N_{\mathrm{A}}\mu_{\mathrm{B}} \tag{3.4.9}$$

All characteristic values of the lanthanides are tabulated in Table 3.4.1. As mentioned above, the magnetic properties of $4f^4$, and notably $4f^5$ as well as $4f^6$ electron configurations deviate from the simple rules given in eqs. (3.4.6)–(3.4.9). This is due to the fact that several excited terms of these configurations are energetically very close to the ground term, which in turn leads to relevant contributions to the magnetization or magnetic susceptibility. Values close to the characteristic ones are found only in a small range at the lowest temperatures, in which, however, effects other than the Zeeman effect become relevant too.

Table 3.4.1: Characteristic values of rare earth centers [1] (+III is the most stable oxidation state); (a) Russell–Saunders ground term, (b) one-electron spin–orbit coupling parameter ζ [3], (c) experimental values at room temperature of the coordinated lanthanide center.

Ln^{3+}	$4f^N$	$^{2S+1}L_J$ (a)	ζ (b)/cm^{-1}	g_J	$g_J J$	$g_J[J(J+1)]^{1/2}$	$\mu_{\mathrm{eff,\,exp}}$ (c)/μ_B	$\chi_m T_{\mathrm{exp}}$ (c)/cm^3 K mol^{-1}
La^{3+}	$4f^0$	$^1S_0$				0		
Ce^{3+}	$4f^1$	$^2F_{5/2}$	625	6/7	15/7	2.535	2.3–2.5	0.66–0.78
Pr^{3+}	$4f^2$	$^3H_4$	758	5/4	16/5	3.578	3.4–3.6	1.4–1.6
Nd^{3+}	$4f^3$	$^4I_{9/2}$	884	8/11	36/11	3.618	3.4–3.5	1.4–1.5
Pm^{3+}	$4f^4$	$^5I_4$	1,000	3/5	12/5	2.683	2.9	1.1
Sm^{3+}	$4f^5$	$^6H_{5/2}$	1,157	2/7	5/7	0.845	1.6	0.32
Eu^{3+}	$4f^6$	$^7F_0$	1,326	0	0	0	3.5	1.5
Gd^{3+}	$4f^7$	$^8S_{7/2}$	1,450	2	7	7.937	7.8–7.9	7.6–7.8
Tb^{3+}	$4f^8$	$^7F_6$	1,709	3/2	9	9.721	9.7–9.8	11.8–12.0
Dy^{3+}	$4f^9$	$^6H_{15/2}$	1,932	4/3	10	10.646	10.2–10.6	13.0–14.1
Ho^{3+}	$4f^{10}$	$^5I_8$	2,141	5/4	10	10.607	10.3–10.5	13.3–13.8
Er^{3+}	$4f^{11}$	$^4I_{15/2}$	2,369	6/5	9	9.581	9.4–9.5	11.0–11.3
Tm^{3+}	$4f^{12}$	$^3H_6$	2,628	7/6	7	7.561	7.5	7.0
Yb^{3+}	$4f^{13}$	$^2F_{7/2}$	2,870	8/7	4	4.536	4.5	2.5
Lu^{3+}	$4f^{14}$	$^1S_0$				0		

So far, we have considered free (i.e., noncoordinated) rare earth ions, which usually represents a hypothetical scenario. If we continue towards actual molecular systems and consider a set of ligands coordinating the lanthanide ions, we arrive at a very good description of rare earth molecular compounds, in particular, of their magnetic properties. This is primarily a consequence of the fact that the effects of the ligands on 4f electrons are small in terms of energetic contributions, that is, the crystal or ligand field energy (H_{lf}) is, in general, a perturbation of H_{ee} and H_{so}. The ligand field is the electrostatic field imposed on the central lanthanide ion. Therefore, it combines the electric charge and geometrical information, and represents the spatial (partial) charge distribution of the ligands. In many instances, the ligand field can be described (at least approximately) by a point group higher than C_1, which allows for the simplification of the calculations by means of group theory. However, since paramagnetic effects are also rather small, the ligand field has a significant impact on the magnetic properties.

Now, what is the impact of the ligand field on the central rare earth ion? In the beginners' courses on coordination chemistry, we learn for example that the five 3d orbitals split into two (excited) e_g and three t_{2g} orbitals in an octahedral ligand field. This is usually oversimplified since interelectronic repulsion (for $3d^n$ configurations with $n > 1$) as well as the spin–orbit coupling are entirely ignored. In subsequent undergraduate courses, we then learn about the splitting of the microstates into ^{2S+1}L terms (disregarding the splitting with respect to J), according to the Russell–Saunders scheme based on H_{ee}. In the next step, the ligand field is introduced, splitting, for example, a ^4F term into A_{2g}, T_{1g} and T_{2g} states, when assuming octahedral ligand field symmetry. However, the spin–orbit coupling still remains ignored. In a realistic model, we have to account for all three contributions, H_{ee}, H_{so} and H_{lf}, besides further potential effects as the Zeeman Effect, particularly when discussing the magnetic properties of molecular compounds. In the case of rare earth compounds, the degeneration of the $^{2S+1}L_J$ ground term of the free ion at zero magnetic field is lifted (partially), if the ligand field is also considered. Furthermore, while the spin–orbit coupling generally enables the mixing of spin states, the ligand field and the spin–orbit coupling determine the degree of mixing in the energy states. Whereas, an energy state of the free lanthanide ion may be a mixture of states from different Russell–Saunders terms with the same quantum number m_J, the energy states of lanthanide coordination compounds may additionally be a mixture of states that are characterized by *different* quantum numbers m_J. As a consequence, the magnetically isotropic free ions become distinctly anisotropic centers, once embedded in certain ligand fields, for all $4f^N$ valence electron configurations, with the exception of $4f^7$ which is negligibly anisotropic, and the diamagnetic $4f^0$ and $4f^{14}$ configurations. Note that the definition of the quantum number m_J changes slightly for compounds. It was dubbed the crystal quantum number m_J by the German physicist Karl Heinz Hellwege [6] and represents the last remaining "good" quantum number for rare earth compounds. Also, note that the word "center" implies the respective ion and its ligand field.

We start the discussion of the implications on the magnetic properties with the effects on μ_{eff} or $\chi_m T$. At room temperature, the mean values (powder samples) of μ_{eff} or $\chi_m T$ (of all but the $4f^4$, $4f^5$ and $4f^6$ valence electron configurations) are in good agreement with the free ion values, although they are typically slightly lower and span a certain range of values (see Table 3.4.1). At lower temperatures, in particular below 100–150 K, they are distinctly temperature dependent with the exception of half-filled shell $4f^7$ centers as Gd^{3+} or Eu^{2+} (see Figure 3.4.1a,b). Due to the depopulation of excited states (see eq. (3.4.5)), the values of μ_{eff} or $\chi_m T$ typically decrease non-linearly with decreasing temperatures and at 2 K, for example, they reach values well below the corresponding room temperature values, which

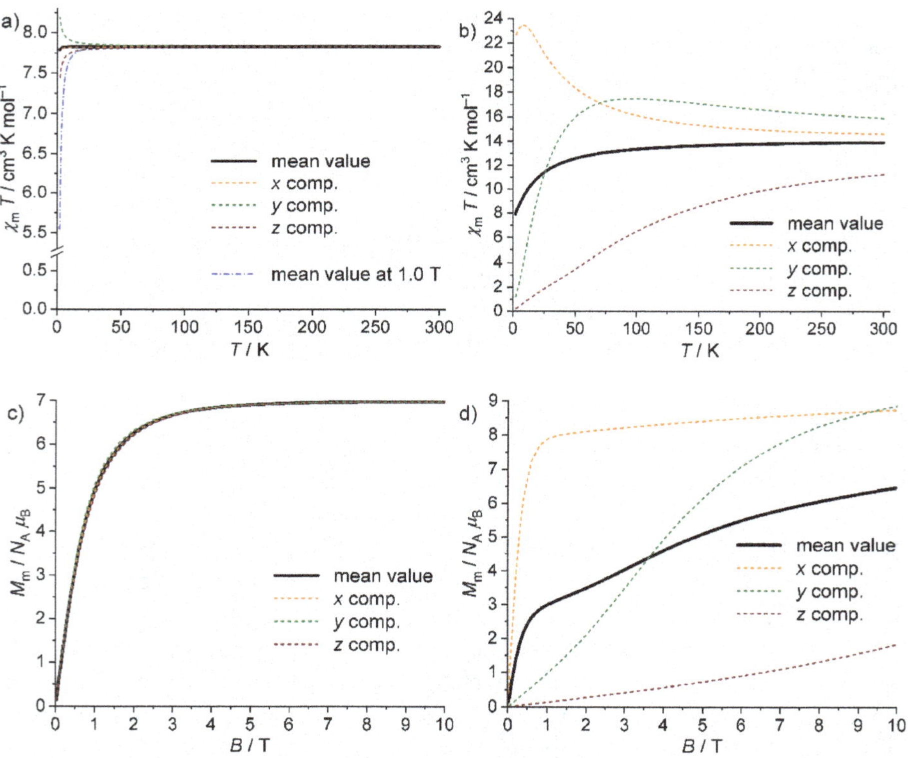

Figure 3.4.1: Examples for the temperature dependence of $\chi_m T$ in the range 2–300 K at 0.1 T of single, isolated, low symmetric Gd^{3+} (a, in detail) and Dy^{3+} (b) centers, and magnetic field dependence of M_m in the range 0–10 T at 2.0 K of the same Gd^{3+} (c) and Dy^{3+} (d) centers, respectively. Mean values (black solid line, powder) and components (colored dashed lines, single crystals); in (a) mean value at 1.0 T (blue dash-dot line, powder) illustrating the impact of the Zeeman effect.

essentially implies that the application of the Curie law (eq. (3.4.6)) to any lanthanide center (but $4f^7$) is physically meaningless. The decrease of μ_{eff} or $\chi_m T$ should not be confused with the Zeeman effect that can be additionally observed at higher applied magnetic fields (see Figure 3.4.1a). For single crystals, the magnetic susceptibility is additionally dependent on the relative orientation of the crystal and the magnetic field. For the plots in Figure 3.4.1a and b, we exemplarily choose a ligand field characterized by a low symmetry and the z-axis as the principal axis of the symmetry. For the (almost) magnetically isotropic Gd^{3+} center, the component values and the average of $\chi_m T$ are similar at most temperatures. Small differences are observed only at temperatures below 50 K. In contrast, the Dy^{3+} center, as a representative example for the usually anisotropic lanthanide centers, is characterized by three components that differ significantly from the mean $\chi_m T$ value. Note that most susceptibility measurements are performed on powder samples and thus yield the mean values.

The molar magnetization M_m of a Gd^{3+} center at 2.0 K as a function of the applied magnetic field is similar to that of the free Gd^{3+} ion (see Figure 3.4.1c). There is almost no difference between the mean value and the components of the magnetization and saturation is almost reached at about 5 T. In contrast, the magnetic anisotropy of the Dy^{3+} center results in a molar magnetization as a function of the applied magnetic field that distinctly differs from the respective plot of the free ion. As shown in Figure 3.4.1d, each component can exhibit its own characteristic behavior, reflecting the composition of the ground state with respect to m_J, and the gap(s) between the ground state and the first excited state(s). This yields a mean value of M_m that does not reach saturation at any field below 10 T. Even though the relatively small slope seen in the M_m versus B plot at 10 T seems to indicate saturation at fields not far above 10 T, the slope actually decreases only a little at higher fields. Therefore, reaching saturation is usually beyond the means of any standard experimental setup. However, as a rule of thumb, a value in the range of 40–60% of the saturation value is usually found at about 5 T at 2.0 K, for the magnetically anisotropic lanthanide centers.

The discussion so far has focused on isolated rare earth metal centers. In many compounds, the centers are, however, not isolated. If the distance between two or more paramagnetic centers is less than about 5–8 Å, they interact on a measurable scale either directly between the central ions or indirectly via bridging ligands that can act as exchange pathways. Rare earth centers, however, feature very weak Heisenberg exchange interactions (H_{ex}) [7] in comparison to transition metal centers, for example. Therefore, the interactions between lanthanide centers modify the energy structure of the relevant multiplets only to a small extent. Note that the discussion of further interactions is often a matter of the applied model, a classification issue that potentially causes confusion. For example, dipole–dipole interactions often have to be explicitly considered using effective theories, which reduce all magnetic moment components into a single main component (or few components), while such interactions can be also encoded in the spin operators using other more comprehensive models that make no such approximations.

As a consequence of the small coupling interactions, the room-temperature values of μ_{eff} or $\chi_m T$ and M_m are similar to the superposed values of the individual centers while characteristic deviations can be observed at lower temperatures, usually below 50 K and less. There are two kinds of interactions: ferromagnetic (magnetic moments strengthen each other) and antiferromagnetic exchange interactions (magnetic moments weaken each other). Note that these designations should not be confused with the long-range ordering of ferro- and antiferromagnetism typically observed for magnetically condensed (e.g., metallic) rare earth compounds (see Chapter 3.3). For example, the simplest system consists of two $S = 1/2$ centers; ferromagnetic implies a parallel, antiferromagnetic an antiparallel orientation of the spins. However, there are more possibilities for higher S, and for the lanthanides, only $4f^7$ centers are spin-like centers ($S = 7/2$). Regarding the ferromagnetic exchange interactions, the $\chi_m T$ versus T plot (mean values) shows a characteristic maximum at low temperatures (Figure 3.4.2a) and possibly a slightly larger value at room temperature. At 2.0 K, the molar magnetization (Figure 3.4.2b) increases with a slightly larger slope than without interactions. At fields $B > 2$ T, the M_m versus B curves representing interactions are almost congruent to the curve, showing magnetization in the absence of interactions. This example reveals another general property of the magnetization at low temperatures that we did not discuss so far since it is rather seldomly observed, usually at temperatures lower than 2.0 K. While the features hinting at steps are normally due to exchange interactions for 3d transition

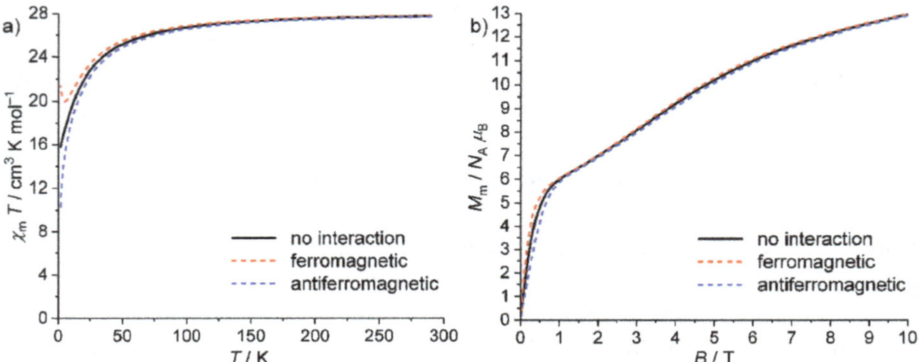

Figure 3.4.2: Examples of (a) the temperature dependence of $\chi_m T$ in the range 2–300 K at 0.1 T of two interacting Dy^{3+} centers, and (b) the corresponding magnetic field dependence of M_m in the range 0–10 T at 2.0 K of the two Dy^{3+} centers (mean values). No exchange interaction (black solid lines), ferromagnetic (red dashed lines, for an exchange energy of $-2J = -0.08$ cm^{-1}) and antiferromagnetic interactions (blue dashed lines, +0.08 cm^{-1}).

metals, they are often due to the properties of a single center for rare earth metals. For antiferromagnetic interactions, the increase of M_m below 2 T is slightly less steep than without interactions. In this case, the $\chi_m T$ versus T plot usually shows a slightly smaller value at room temperature (Figure 3.4.2a), and $\chi_m T$ decreases more rapidly upon cooling the compound compared to two isolated Dy^{3+} centers. Note that these comparingly small quantities of the exchange interaction effects are characteristic of rare earth metal compounds. In the case of interactions between lanthanide and transition metal centers, the effects may be larger by an order of magnitude, while interactions between transition metal centers are usually a few orders of magnitude larger.

The careful reader might have noticed that we have described the effects of the energy contributions (H_{ee}, H_{so}, H_{mag}, H_{lf} and H_{ex}) on the magnetic properties but avoided to specify the corresponding operators in detail, as there are various approaches to solve this multielectron problem, but none of them is (thus far) capable to fully include all desired aspects. Therefore, we limit ourselves to mention the most popular and/or promising models of molecular rare earth compounds at this point. We also add literature as a starting point for further self-studies. There are two main approaches to the problem: First, the data from spectroscopy and magnetometry are recorded and model parameters are fitted to reproduce the data. Representative models [8–11] are based on effective theories or semiempirical models as for example, extended ligand-field theory. While these approaches usually require little or average computing resources, they may offer multiple ways to reproduce the data mathematically, possibly yielding several physically divergent solutions. Thus, we must cautiously choose the right model for the respective system at hand to get *physically meaningful* results. If we are successful, the results can then allow for deep insights into the physical and electronic properties of the compound, since the fits usually agree very well with the experimental data. For the second approach, the expected spectroscopy and magnetometry data are calculated based on, for example, the structures of the compounds, and the calculated data are compared with the experimental data without further refinement. Commonly used ab initio methods include density functional theory or complete active space calculations [12]. In principle, this is the most desired way to go from a theoretical point of view since the data are calculated from first principles based on different data than from the magnetic data of a compound. However, the required computing resources can become limiting factors. Therefore, the analyst is forced to accept some approximations which may cause the calculated data to differ significantly from experimental data. Even if there are no issues arising due to approximations, the results (as of 2020) are more often than not incompatible with the experimental data. That is why the magnetic properties of compounds are, in general, still an interesting challenge for theorists, and much method development remains to be done.

The distinct magnetic anisotropy of a few of the rare earth metal centers allows for an interesting magnetic feature that may find applications in future information technologies as molecular spin qubits or qubits [13, 14]: single-molecule magnets or single-ion magnets (SIM) [15]. A single-ion magnet comprises a single magnetic center that exhibits magnetization hysteresis up to a certain temperature, the blocking temperature. This behavior is comparable to the Curie temperature of ferromagnets, although no long-range ordering and Weiss regions are involved. In lieu thereof, SIMs exhibit a preferred orientation of their magnetization in a magnetic field. To invert the orientation, an effective energy barrier U_{eff} must be overcome to enable thermal relaxation processes that are very slow. However, other magnetic (slow) relaxation processes, such as the Raman, quantum tunneling or direct relaxation [16, 17], are competing with the desired Orbach relaxation process. Therefore, the blocking temperatures are usually limited to very low temperatures (less than 10 K). As of mid-2020, the highest blocking temperature of all molecular compounds is 80 K for the SIM $[(\eta^5\text{-Cp}^*)\text{Dy}(\eta^5\text{-Cp}^{iPr5})][\text{B}(\text{C}_6\text{F}_5)_4]$ (see Figure 3.4.3) [18].

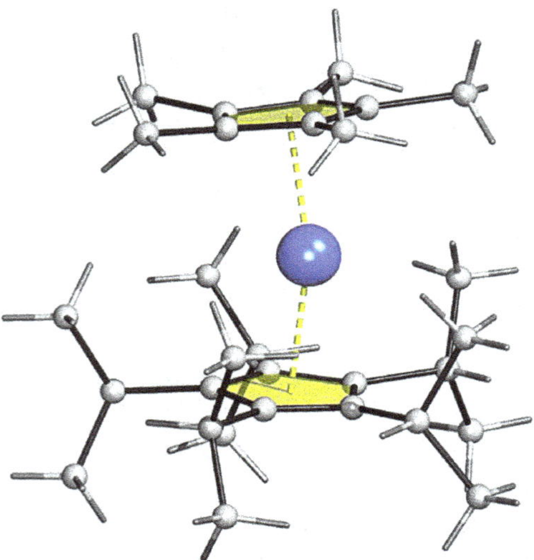

Figure 3.4.3: Molecular structure of $[(\eta^5\text{-Cp}^*)\text{Dy}(\eta^5\text{-Cp}^{iPr5})]^+$, as present in the $[\text{B}(\text{C}_6\text{F}_5)_4]^-$ salt, the molecular compound currently exhibiting the highest blocking temperature of 80 K [18]. Note the (almost) linear coordination of the central Dy^{3+} ion that is (almost) congruent with the magnetic anisotropy axis of the Dy^{3+} center.

Besides the observation of hysteresis, magnetic slow relaxation processes also cause a phase shift between source and detection signals in case of applying (very weak) alternating magnetic fields to the sample, yielding the in-phase (χ') and out-of-phase (χ'') components of the ac susceptibility. Typically, the corresponding data are analyzed in terms of a generalized Debye expression [19] in a first step, and subsequently in terms of effective expressions for the relaxation processes (Orbach,

Raman, quantum tunneling, etc. [16]). Note that this is a phenomenological description derived from concepts of solid-state physics, that is, specific phonon-induced contributions are formally included in the parameters, which are absent for molecular compounds. In this way, we obtain empirical fit parameters that can be used for a simple classification of various molecular compounds, while a comprehensive discussion based on quantum theory still remains elusive. A few models exist [15, 20], in particular, based on effective theories that try to explain the characteristic behavior. Unfortunately, they are not valid under all circumstances. There is experimental counterevidence that the suggested models fail in particular situations [21]. For example, the calculated values of the energy barrier U_{eff} considerably differ from the experimental values in most instances. Therefore, besides optimizing the syntheses to create SIMs based on experimental knowledge and routines, much effort of recent research is spent on developing theoretical models to connect molecular structure and SIM behavior to, ultimately, predict the structure of the perfect SIM.

References

[1] H. Lueken, Magnetochemie, Teubner Verlag, Stuttgart, 1999.
[2] J. H. Van Vleck, The Theory of Electric and Magnetic Susceptibilities, Oxford University Press, Oxford, 1932.
[3] E. U. Condon, G. H. Shortley, The Theory of Atomic Spectra, Cambridge University Press, Cambridge, 1970.
[4] S. Hüfner, Optical Spectra of Transparent Rare Earth Compounds, Academic Press, New York, 1978.
[5] P. Zeeman, Philos. Mag. 1897, 43, 226.
[6] H. K. Hellwege, Ann. Phys. (Berl.) 1948, 439, 95.
[7] W. Heisenberg, Z. Phys. 1928, 49, 619.

[8] O. Kahn, Molecular Magnetism, Wiley-VCH, Weinheim, 1993.
[9] A. Abragam, B. Bleaney, Electron Paramagnetic Resonance of Transition Ions, Clarendon Press, Oxford, 1970.

[10] J. S. Griffith, The Theory of Transition-Metal Ions, Cambridge University Press, Cambridge, 1980.
[11] J. van Leusen, M. Speldrich, H. Schilder, P. Kögerler, Coord. Chem. Rev. 2015, 289–290, 137.
[12] C. J. Cramer, Essentials of Computational Chemistry: Theories and Models, 2nd Edition, J. Wiley & Sons, West Sussex, 2004.
[13] Molecular Spintronics and Quantum Computing, E. Coronado, A. Epstein (Eds.) J. Mater. Chem. 2009, 19, 1661.
[14] E. Moreno-Pineda, C. Godfrin, F. Balestro, W. Wernsdorfer, M. Ruben, Chem. Soc. Rev. 2018, 47, 501.

[15] D. Gatteschi, R. Sessoli, J. Villain, Molecular Nanomagnets, Oxford University Press, New York, 2006.
[16] K. N. Shrivastava, Phys. Stat. Sol. b 1983, 117, 437.

[17] D. N. Woodruff, R. E. P. Winpenny, R. A. Layfield, Chem. Rev. 2013, 113, 5110.
[18] F.-S. Guo, B. M. Day, Y.-C. Chen, M.-L. Tong, A. Mansikkamäki, R. A. Layfield, Science 2018, 362, 1400.
[19] K. S. Cole, R. H. Cole, J. Chem. Phys. 1941, 9, 341.
[20] L. Bogani, W. Wernsdorfer, Nat. Mater. 2008, 7, 179.
[21] L. Escalera-Moreno, J. J. Baldoví, A. Gaita-Ariño, E. Coronado, Chem. Sci. 2018, 9, 3265.

Florian Baur, Thomas Jüstel

3.5 Optical characterization

3.5.1 Basics

This chapter covers the specifics concerning the optical characterization of rare earth elements (REE) and will not go into detail of basic fluorescence spectroscopy. It deals predominantly with the trivalent REE and divalent Sm^{2+}, Eu^{2+} and Yb^{2+}, as they possess electronic transitions that can result in luminescence: the intraconfigurational $[Xe]4f^n$–$[Xe]4f^n$ transitions (4f–4f transitions) and the interconfigurational $[Xe]4f^n$–$[Xe]4f^{n-1}5d$ transitions (4f–5d transitions) [1]. Furthermore, most REE show ligand-to-metal charge transfer (LMCT) transitions that can be used for excitation [2]. The 4f–4f transitions result in lines with a small full width at half maximum (FWHM); the 4f–5d and LMCT transitions result in broad bands. In this chapter, the term REE refers to trivalent cations and the three aforementioned divalent cations unless otherwise specified.

3.5.2 Reflection and absorption spectroscopy

Since absorption measurements require a transparent sample, they are usually performed on soluble REE complexes or REE crystals [3]. If the investigated compound is not soluble in any matrix without decomposition, transparent ceramics are an alternative. A problem associated with absorption spectroscopy of REE compounds is the low absorption cross section of the quantum-mechanically forbidden 4f–4f transitions. The extinction coefficient ε is typically of an order of magnitude of 10^0 L mol^{-1} cm^{-1} [4]. Furthermore, the narrow FWHM of these transitions requires a small step width of less than 1 nm for accurate determination of the peak values. The quantum-mechanically allowed 4f–5d or LMCT transitions exhibit a much larger extinction coefficient of around 10^3–10^4 L mol^{-1} cm^{-1} [5, 6]. Due to this vast difference in extinction, the required molar concentration or thickness of the sample differs strongly, and it is not possible to accurately measure both types of transitions in one recording.

Powder samples are typically investigated via reflection spectroscopy. To detect both diffuse and specular reflection, an integration sphere is used. The inner surface of the sphere has to offer high reflectance over the whole measurement range. To achieve this, the inside can either be coated with a highly reflective compound such as $BaSO_4$, or the whole sphere is made of optical-grade PTFE (e.g., Spectralon®). The reflection measurement is indirect, that is, measured against a standard. This causes the signal-to-noise ratio (SNR) to decrease further in regions with a low

https://doi.org/10.1515/9783110654929-020

intensity. Therefore, the weak 4f–4f transitions are mostly not easily observed in reflectance spectra of REE compounds. To determine the spectral position of these transitions, excitation or emission spectra are often the more suitable spectroscopic method.

3.5.3 Emission and excitation spectroscopy

When attempting to measure emission and excitation spectra, the narrow FWHM and low absorption cross-section, which the majority of the REE exhibit, has to be considered. An excitation source that closely matches the wavelength of the desired excitation transitions is required. In most cases, this is achieved by using a continuous emitting source such as a Xe high-pressure discharge lamp, in combination with a monochromator. Excitation with a laser is useful in realizing high excitation intensities. Additionally, Sm^{3+}, Eu^{3+} and Yb^{3+} can be excited via LMCT bands in the UV-C to UV-A range that show relatively strong absorption. The 4f–5d band of Tb^{3+} can also be used for excitation and is usually found between 200 and 300 nm. The REE cations Ce^{3+}, Sm^{2+}, Eu^{2+} and Yb^{2+} show broad 4f-5d absorption bands with a large absorption cross section in the UV-C to visible spectral range, allowing for relatively uncomplicated excitation [7].

Most trivalent REE exhibit luminescence spectra that cover a broad spectral range. For example, YPO_4:Nd^{3+} shows bands centered at 193 nm, arising from 4f–5d transitions and a variety of lines in the range from 500 to 1,100 nm, arising from 4f–4f transitions [8]. To record the full emission spectrum, an excitation source with a short wavelength, such as a D_2 lamp or an X-ray tube, is required. A further difficulty lies in the detector used, as the spectral response curve will strongly vary between the UV and NIR [9]. It is thus not uncommon that the emission spectrum is recorded by using two different detectors, for example, a photomultiplier tube (PMT) for the UV to visible region and a semiconductor-type detector for the NIR region.

3.5.4 Thermal quenching

A typical thermal quenching (TQ) curve of a phosphor with a low quenching temperature is shown in Figure 3.5.1. It can be fitted with a modified Arrhenius function to determine the activation energy of the thermal quenching process [10] as follows:

$$I(T) = \frac{I_0}{1 + B \cdot e^{-\frac{E}{kT}}} \tag{3.5.1}$$

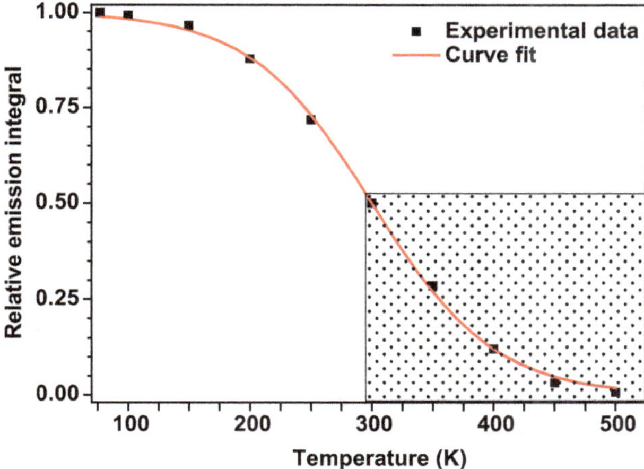

Figure 3.5.1: Plot of a typical thermal quenching curve showing the experimentally determined emission integrals (black squares) and the curve fit (red line). The box shows the part of the measurement results that would be available if no cooling equipment was used.

where $I(T)$ is the emission integral at temperature T, I_0 is the emission integral in absence of thermal quenching, B is a frequency factor that comprises the probability of the thermal quenching process, k is the Boltzmann constant and E is the activation energy. The physical unit of the activation energy is decided by choice of the unit of k and usually expressed in eV. The lower the activation energy and the higher the frequency factor, the lesser the temperature required to quench the luminescence.

The temperature at which the emission integral has declined to 50% of its maximum value is called $T_{1/2}$. It is a more tangible representation of the thermal quenching properties of a phosphor than the activation energy and frequency factor, and is often found in literature. It can be calculated by rearranging eq. (3.5.1) with $I/I_0 = 0.5$ as

$$T_{1/2} = -\frac{E}{k \cdot \ln(B^{-1})} \tag{3.5.2}$$

In general, many of the 4f–4f transitions of REE exhibit comparatively high TQ temperatures due to their small Stokes' shift, unless the LMCT is located at low energies [11], or high phonon frequencies quench the photoluminescence [12]. Consequently, sample holders for TQ measurements of REE have to be capable of reaching temperatures of 500 K and beyond. The lines of the 4f–4f transitions show pronounced broadening upon the temperature increase. Therefore, it is important to use emission integrals and not peak intensities for the plot of the TQ curve. To accurately determine the emission integral, the step size of the emission measurement must not be too large. Ten to 12 experimental data points per emission line should be recorded, that is, a step size of 0.1–0.2 nm is usually required. The peak of the transitions can shift with temperature, which means that over the course of the measurement, the chosen excitation wavelength might not match the respective transition of the REE, anymore. Broadening of the excitation line can influence the excitation intensity as well. To prevent this modification of the measurement, it is advisable to use an excitation peak that is broad enough. That means a continuous source for excitation spectra is preferred.

The recording of TQ curves of 4f–5d transitions is simpler to perform than that of the 4f–4f line emission, as the step size can be larger and temperature related shifts of the peak excitation wavelength do not strongly affect the measurement. On average, thermal quenching of 4f–5d transitions is stronger than that of 4f–4f transitions. For instance, Yb^{2+} is known to show strong quenching, even at room temperature [13]. In that case, a sample holder with cooling capability is mandatory to record the whole thermal quenching curve. Figure 3.5.1 shows the part of the TQ curve that lies within the range of 290–500 K. It is obvious that a curve fitting of this part solely will yield wrong results, in that the $T_{1/2}$ value would have been considerably overestimated.

3.5.5 Decay curves

For recording the decay curve, the respective sample is excited with a short pulse of radiation and the emission intensity is subsequently recorded over time. The faster the excited state relaxes to the ground state, the faster will the emission intensity decrease. The 4f–4f transitions of the REE are parity forbidden and – for the most part – spin forbidden, as well. Therefore, typical transition rates are of the order of 10^3 s^{-1}, and the decay times are consequently in the range of milliseconds. A Xe flash lamp with a pulse width of a few µs and flash frequency of around 25 Hz is sufficient to record decay curves of emission lines due to 4f–4f transitions. Pulsed lasers can be employed too; however, the narrow FWHM of the 4f–4f excitation lines can be a problem when matching a laser to a specific sample. The 4f–5d transitions of the REE are parity and mostly spin allowed and show high transition rates and short decay times in the range of 20 ns (Pr^{3+}) to 50 ns (Ce^{3+}) to 1 µs (Eu^{2+}). To record such decay curves, the so-called time-correlated single photon counting is often used [14], which requires special hardware in the computer. The excitation source needs to have a short pulse width in the picosecond to nanosecond range. Pulsed lasers, laser diodes or ns-Xe-flash lamps are commonly used sources. Additionally, the detector has to have a short decay time, which can be challenging when measuring emission in the NIR range.

In an idealized case, that is, single exponential decay, the obtained decay curve can be fitted by the following equation:

$$I(t) = A \cdot e^{-t/\tau} \tag{3.5.3}$$

where $I(t)$ is the intensity at time t, A is the initial intensity and τ is the decay time. The decay time is the average lifetime of the excited state whose depopulation is observed in the measurement. A constant can be added to account for background intensity.

Equation (3.5.3) can be logarithmized and simplified to yield $\ln(I) = -t/\tau$, which shows that the logarithmic intensity is linearly proportional to the time. However, decay curves do not necessarily follow this linear proportionality, as depicted in Figure 3.5.2. This is the case if two or more luminescent species are detected simultaneously, that is, when they both emit in the observed wavelength range. To fit such curves, eq. (3.5.3) has to be extended to

$$I(t) = A_1 \cdot e^{-\frac{t}{\tau_1}} + A_2 \cdot e^{-\frac{t}{\tau_2}} + \cdots \tag{3.5.4}$$

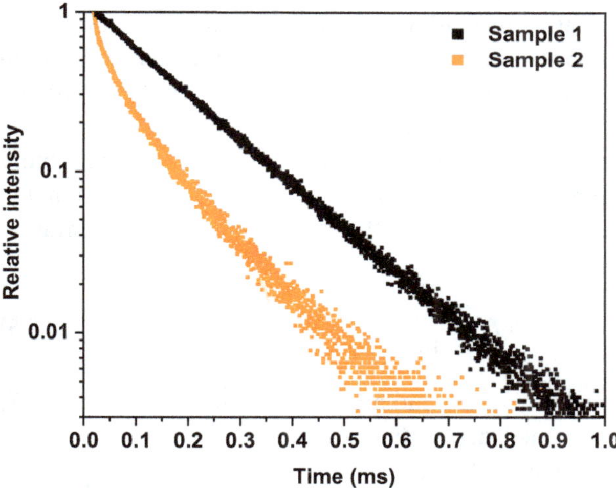

Figure 3.5.2: Plot of a typical decay curve with one component (black dots) and of a decay curve with two components (orange dots).

How many terms the decay fitting function should consist of, is often difficult to determine. The goodness of fit typically increases with each additional term due to mathematical reasons regardless of the actual number of luminescent species. Thus, the number of terms should be kept as small as possible, while maintaining a sufficiently high goodness of fit. Unfortunately, there is no simple rule to follow and decay curve fitting always encompasses some ambiguity.

The occurrence of more than one decay time can be caused by at least three factors. Firstly, several different REE might be excited, the same REE can occur on different crystallographic sites, or impurity phases might be present. For example, if two crystallographically distinct types of Eu^{3+} cations are present in the sample, any emission multiplet will yield two decay times. It is important to note that solely luminescent species will be observed. REE cations whose luminescence is quenched or very weak, for example, from a very minor impurity phase, are not detected.

Secondly, defects in the sample's crystal structure can also result in a decay curve following eq. (3.5.2). Those REE cations that are close enough to a defect to be influenced by its presence will show a different, typically shorter decay time than those that are not close to a defect.

Thirdly, energy transfer in a sample will also result in two decay times [12]. Those ions that are close to a potential acceptor will typically exhibit a shorter decay time, while those ions that are not in the vicinity of an acceptor will show the usual decay time. Thus, decay curves can serve as an indicator of defect density and of the occurrence of energy transfer.

Temperature-dependent decay measurements can be conducted to determine the ratio of radiative to nonradiative relaxation for a specific transition [15]. It can be assumed that at very low temperature[1] only radiative transitions occur and the measured decay time τ_{exp} corresponds to the decay time τ_0 of the radiative transition. The nonradiative relaxation has a very high rate and decay time close to zero, when compared to a radiative transition. Thus, with an increase of the probability of nonradiative relaxation, the measured decay time τ_{exp} decreases proportionally. The following equation can be used to calculate the ratio of radiative to nonradiative relaxation:

$$R(T) = \frac{\tau_{exp}}{\tau_0} \tag{3.5.5}$$

where $R(T)$ is the relative probability of the radiative relaxation at temperature T, τ_{exp} is the experimentally determined decay time at temperature T and τ_0 is the decay time measured at as low a temperature as possible.

3.5.6 Saturation measurements

Upon high excitation densities, such a large fraction of activator ions can be in the excited state that absorption will decrease significantly. In a phosphor converted LED, this will lead to a diminished light yield. A larger portion of the primary emission will be scattered instead of being converted to the desired emission wavelength. REE with a short decay time, such as Ce^{3+}, are less prone to saturation as

1 Ideally the temperature should be 4 K or lower, but if this is not possible in the measurement setup, 77 K will suffice.

the excited state quickly decays and the cation can be excited again [16]. Upon excitation via a 4f–4f transition, saturation is usually not observed despite their long decay time. This finding is a result of the low absorption cross section of these transitions which strongly decreases the effective excitation density as the majority of incident photons is scattered.

For the red emitting luminescent materials $Sr_2Si_5N_8:Eu^{2+}$ and $CaAlSiN_3:Eu^{2+}$ saturation sets in at around 200–300 W mm^{-2} [17]. To achieve such a high power density, a laser has to be used and additional focusing of the beam via a lens might be required. For accurate measurements, the laser has to feature stable output power over a long period of time. Furthermore, the excitation spot size has to be known with a high degree of certainty to calculate the excitation density. The sample will have to be cooled either passively or actively as otherwise thermal quenching might set in that could be confused with saturation quenching. The sample's temperature should be monitored and, if necessary, stabilized.

3.5.7 Site selective spectroscopy

The spectral position of the 4f–4f transitions does not strongly depend on the environment of the respective REE due to shielding from the 5s and 5p orbitals. The peak wavelength of the $^7F_0$–$^5D_0$ transition of Eu^{3+}, for example, can always be found close to around 580 nm. Since each crystallographically distinguishable Eu^{3+} species[2] will show solely a single $^7F_0$–$^5D_0$ line, this specific transition can be used for the so-called site selective spectroscopy. If more than one $^7F_0$–$^5D_0$ line is observed, at least this number of luminescent Eu^{3+} species has to be present in the sample [18].

If the FWHM of the individual transition is narrow enough, the lines might be fully separated and can be selectively excited. In most cases, however, the lines are not fully separated but overlap to form an asymmetric peak. Performing the measurement at low temperature helps decrease the FWHM and separate these lines further. Nevertheless, a very narrow excitation band is required, since if lines are not fully separated, only the edges of the combined peak can be used for excitation. Consequently, the recorded intensity is usually low. To obtain meaningful spectra with an acceptable SNR, an extended measurement period of several hours due to many repetitions is often required. Site-selective excitation spectra can be obtained in the same way, by monitoring the respective wavelengths of the $^5D_0 \rightarrow {}^7F_0$ transition.

---

2 This only applies to luminescent Eu^{3+} species; if luminescence from a specific site is quenched, it will not be accessible via this method. The number of $^7F_0$–$^5D_0$ lines indicates the minimum number of distinct Eu^{3+} species.

To significantly increase intensities, the obtained site-selective excitation spectra can be compared to find lines other than the $^7F_0 \rightarrow ^5D_0$ line that belong exclusively to one crystallographic site. In most cases, these lines, if they exist, will show a higher intensity than the $^7F_0 \rightarrow ^5D_0$ line. By using these lines for excitation, the SNR can be strongly increased and measurement times decreased. The same approach can be taken analogously by comparing site-selective emission spectra to find emission lines exclusive to one Eu^{3+} species, but with a higher intensity than that of the $^5D_0 \rightarrow ^7F_0$ line.

Site-selective spectroscopy is mostly undertaken on Eu^{3+} ions, but can be used for other REE as well, provided a suitable transition is available [19].

3.5.8 VUV spectroscopy

Most REE possess optical transitions in the vacuum-UV (VUV) spectral range between 100 and 200 nm [20]. However, measurements in the VUV range differ considerably from measurements in the UV-C to visible range. Most substances absorb VUV radiation, including CO_2, O_2 and water vapor. For that reason, the excitation source, the monochromators and detector of a VUV spectrometer are usually kept under vacuum. The sample chamber can also be kept under vacuum, but this requirement results in a time-consuming sample transfer due to the need of thorough pumping of the sample chamber. Alternatively, the sample chamber can be flushed with dry N_2 which yields slightly lower intensities but allows for comparatively fast sample changing.

VUV reflection measurements require a special setup as well. Materials that show strong absorption in the VUV range cannot be used as coating for the integration sphere. The fluorides LiF and MgF_2 can be used for reflection measurements down to 100–120 nm, but they are slightly hygroscopic and, therefore, to be handled with care, that is, they should not be in contact with humidity. CaF_2 is quite stable, but can only be used down to approximately 170 nm. For this reason, an indirect approach is more appropriate. The basic idea is to convert the reflected VUV radiation to UV or visible radiation, which is then detected for the measurement. This can be achieved by coating an integration sphere with a phosphor that shows strong and even absorption in the whole VUV with highly efficient conversion to UV or visible radiation, such as $LaPO_4:Ce^{3+}$ or $BaMgAl_{10}O_{17}:Eu^{2+}$. If no sample is present in the integration sphere, most of the incident VUV radiation is converted to UV or visible radiation and is subsequently detected. Once a sample absorbs a fraction of the incident VUV radiation, the intensity of the converted radiation decreases proportionally. By division of the respective intensities, a reflection spectrum is obtained. For a more detailed description of the measurement setup, please see reference [21].

References

[1] J.-C. G. Bünzli, S. V. Eliseeva, Basics of Lanthanide Photophysics, in Lanthanide Luminescence: Photophysical, Analytical and Biological Aspects (ed. P. Hänninen, H. Härmä), Springer-Verlag, Berlin, Heidelberg, 2011, 1.

[2] P. Dorenbos, J. Phys.: Condens. Matter 2003, 15, 8417.

[3] K. B. Yatsimirskii, N. K. Davidenko, Coord. Chem. Rev. 1979, 27, 223.

[4] K. Binnemans, C. Görller-Walrand, J. Phys.: Condens. Matter 1997, 9, 1637.

[5] A. Uehara, O. Shirai, T. Nagai, T. Fujii, H. Yamana, Z. Naturforsch. 2007, 62a, 191.

[6] A. Herrmann, H. A. Othman, A. A. Assadi, M. Tiegel, S. Kuhn, C. Rüssel, Opt. Mater. Express 2015, 5, 720.

[7] P. Dorenbos, J. Lumin. 2000, 91, 155.

[8] V. N. Makhov, N. Y. Kirikova, M. Kirm, J. C. Krupa, P. Liblik, C. Lushchik, E. Negodin, G. Zimmerer, Nucl. Instrum. Methods Phys. Res. A 2002, 486, 437.

[9] J. R. Lakowicz, Principles of Fluorescence Spectroscopy, 4th Edition, Springer, New York, 2010.

[10] C. W. Struck, W. H. Fonger, J. Lumin. 1975, 10, 1.

[11] W. M. Faustino, O. L. Malta, G. F. de Sá, Chem. Phys. Lett. 2006, 429, 595.

[12] G. Blasse, B. C. Grabmaier, Luminescent Materials, Springer, Berlin, 1994.

[13] S. Lizzo, E. P. Klein Nagelvoort, R. Erens, A. Meijerink, G. Blasse, J. Phys. Chem. Solids 1997, 58, 963.

[14] P. Kapusta, M. Wahl, A. Benda, M. Hof, J. Enderlein, J. Fluoresc. 2007, 17, 43.

[15] L.-J. Lyu, D. S. Hamilton, J. Lumin. 1991, 48-49, 251.

[16] J. Xu, A. Thorseth, C. Xu, A. Krasnoshchoka, M. Rosendal, C. Dam-Hansen, B. Du, Y. Gong, O. B. Jensen, J. Lumin. 2019, 212, 279.

[17] T. Jansen, D. Böhnisch, T. Jüstel, ECS J. Solid State Sci. Technol. 2016, 5, R91.

[18] R. Ternane, M. Ferid, G. Panczer, M. Trabelsi-Ayadi, G. Boulon, Opt. Mater. 2005, 27, 1832.

[19] M. B. Seelbinder, J. C. Wright, Phys. Rev. B 1979, 20, 4308.

[20] J. C. Krupa, M. Queffelec, J. Alloys Compd. 1997, 250, 287.

[21] T. Jüstel, J.-C. Krupa, D. U. Wiechert, J. Lumin. 2001, 93, 179.

Hellmut Eckert, Rainer Pöttgen

3.6 Solid-state NMR and Mößbauer spectroscopy

Both nuclear magnetic resonance (NMR) and Mößbauer spectroscopies are based on the interaction of nuclei with electromagnetic waves. NMR uses radio waves to effect transitions between nuclear ground state spin orientations within magnetic fields, whereas in Mößbauer spectroscopy, nuclear excited states with different spin are accessed by high-energy gamma rays. Although the rare earth elements feature many nuclear magnetic isotopes (both in the ground and excited states), only a few of them turn out to be suitable for structural investigations in the solid state. This chapter focuses on spectroscopic aspects and details that are directly related to one of the rare earth elements with respect to structure determination or for the understanding of structure–property relations.

We start with the *NMR spectroscopy*, beginning with a brief summary of the fundamental aspects. For more details, excellent textbooks are available [1–5]. In brief, NMR spectroscopy is based on the *spin* angular momentum of the atomic nuclei. This quantum mechanical property, described by a spin-quantum number I, implies the existence of $2I+1$ degenerate, orientation-quantized states, characterized by the quantum numbers m ε (I, I–1, . . ., –I). As predicted for any rotating charged particle, nuclear spin implies the existence of a *magnetic dipole moment*, which can be detected by the application of a magnetic field. Under this condition, the *Zeeman interaction* removes the degeneracy of the spin orientation states and results in energy splitting. Allowed transitions between adjacent states are stimulated by electromagnetic waves in the radio wave region, fulfilling the Bohr condition:

$$\omega_0 \sim \omega_p = \gamma B_{loc} \tag{3.6.1}$$

where ω_0, the *irradiation frequency*, is close or equal to the frequency ω_p with which the nuclei precess in the magnetic field. In NMR, we measure ω_p via electromagnetic detection of this spin precession. In eq. (3.6.1) γ, the *gyromagnetic ratio*, is a nucleus-specific constant proportional to the magnetic moment, while B_{loc} is the effective strength of the magnetic field that the nuclei experience along their quantization direction. The latter is provided in large part by the magnitude of the applied magnetic field B_0 (typically of the order of 4.7–23.6 Tesla), but is further influenced by internal magnetic fields, which arise from the interaction of the nuclei with their electronic and magnetic environments:

$$B_{loc} = B_0 + B_{int} \tag{3.6.2}$$

B_{int} is sample specific and contains information about the structure and bonding sought by experimentalists. It arises from the various distinct interaction mechanisms,

https://doi.org/10.1515/9783110654929-021

which must be quantitatively characterized and then related to the structural information. These mechanisms include (a) *magnetic shielding* caused by the electronic environment under the influence of B_0, (b) the *quadrupolar interaction* between the nonspherical nuclear charge distributions of spin > 1/2 nuclei and the local electric field gradients and (c) the *magnetic dipole–dipole coupling* between the nuclei under observation and their magnetic neighbors.

Magnetic shielding by the electronic environment comprises the cumulative effect of four distinct contributions,

$$\delta_{\mathrm{ms}} = \delta_{\mathrm{dia}} + \delta_{\mathrm{orb}} + \delta_{\mathrm{p}} + \delta_{\mathrm{K}} \tag{3.6.3}$$

where δ_{dia} is a diamagnetic effect caused by electron circulation within closed shells. The contribution δ_{orb} is the orbital angular momentum of excited electronic states admixed into the electronic ground state, while δ_{p} describes localized interactions with unpaired electrons. Furthermore, in metallic compounds, the *Knight shift* δ_{K} arises from the *Fermi contact* interaction of the nuclei with spin-polarized conduction electron density near the Fermi edge. While δ_{p} can be identified by its inverse temperature dependence, the remaining three contributions cannot be separated by experimental means. In nonmetallic compounds, the sum $\delta_{\mathrm{dia}} + \delta_{\mathrm{orb}}$ is known as the *chemical shift* (given in units of ppm) when the resonance frequency is compared to that of a standard reference material. For metallic materials, the Knight shift contribution dominates, while in closed-shell molecules and ionic/covalently bound solids, the de-shielding contribution δ_{orb} is generally larger than the diamagnetic effect.

Nuclei with $I > 1/2$ have nonspherical charge distributions, described by an *electric quadrupole moment eQ*. The interaction of this quadrupole moment with the local electric field gradients at the nucleus leads to shifts in the Zeeman energy levels. These levels depend on m^2, (m being the Zeeman orientation quantum number) and thus produce differences in the various $\Delta m = \pm 1$ Zeeman transition energies as depicted schematically in Figure 3.6.1 for a spin-7/2 nucleus. The effect is described quantitatively by standard perturbation theory [6]. The resonance frequency of the central $m = 1/2 \leftrightarrow m = -1/2$ transition remains unaffected to the first order, whereas those of the other six $\Delta m = \pm 1$ Zeeman transitions are orientation dependent, leading to broad *powder patterns* in polycrystalline materials (*satellite transitions*). Quadrupole couplings with energies exceeding 1/20 of the Zeeman energy need to be analyzed using second-order perturbation theory. In this regime, the central $m = 1/2 \leftrightarrow m = -1/2$ transition is anisotropically broadened as well. As the magnitude of the second-order effect depends on the ratio of the quadrupolar interaction to the Zeeman interaction energies, it decreases with increasing magnetic field strength. In addition, magnetic shielding anisotropies and dipolar interactions with the magnetic moments of proximal nuclei have an influence on the spectra. We can eliminate these effects by applying the technique of *magic angle spinning* (MAS, see Figure 3.6.2, left). Under this condition, the broad satellite transitions give rise to a

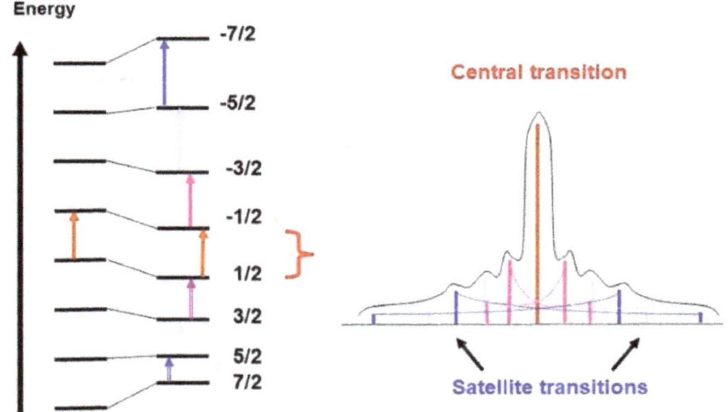

Figure 3.6.1: (Left) Effect of the nuclear electric quadrupolar interaction upon the Zeeman energy levels of an $I = 7/2$ nucleus (applicable to ^{45}Sc, ^{139}La and ^{175}Lu; see Table 3.6.1). Energy-level shifts predicted by first-order perturbation theory are schematically represented for one particular crystallite orientation. (right) Schematic representation of the resulting line shape in a polycrystalline material. While the energy of the central $m = 1/2 \leftrightarrow m = -1/2$ transition is independent of crystal orientation, all the other $\Delta m = \pm 1$ transitions are anisotropically broadened. Taken from [23] with permission from Wiley.

spinning sideband manifold, whose envelope mirrors their line shapes under static conditions. In the presence of second-order perturbations, the broadening of the central transition is not completely removed, resulting in characteristic line shapes from which the quadrupolar interaction parameters can be extracted with the help of simulations (Figure 3.6.2, right). The strength of the interaction is numerically described by the *quadrupolar coupling constant* C_Q (typically given in frequency units), while the deviation of the electric field gradient from cylindrical symmetry is characterized by the *asymmetry parameter* $\eta_Q = (V_{xx} - V_{yy}) / V_{zz}$, where V_{ii} denote the Cartesian components of the electric field gradient. The experimental results can be compared with quantum-chemically calculated values based on crystallographic input, using standard DFT software such as VASP, GAUSSIAN or WIEN2k. As discussed further, this comparison is an important method of validation for crystal structures proposed by single crystal or powder X-ray diffraction methods.

While all the rare-earth atoms feature at least one nuclear isotope, NMR experiments on those nuclei belonging to *RE* ions that feature *f*-electron paramagnetism, are not possible because the rapidly fluctuating hyperfine field associated with the electronic spin states produce extremely fast nuclear spin relaxation, that is, ultrashort lifetimes of the nuclear Zeeman states. This makes them undetectable according to Heisenberg's uncertainty principle (for the application of shift reagents we refer to Chapter 3.7). Thus, only the closed-shell atomic configurations of Sc^{3+}, Y^{3+}, La^{3+}, Yb^{2+} and Lu^{3+} are accessible to solid-state NMR spectroscopic experiments. Table 3.6.1 summarizes the properties of the relevant nuclear isotopes. More than 30 years ago, Thompson and Oldfield were the first to demonstrate the utility of high-resolution ^{45}Sc,

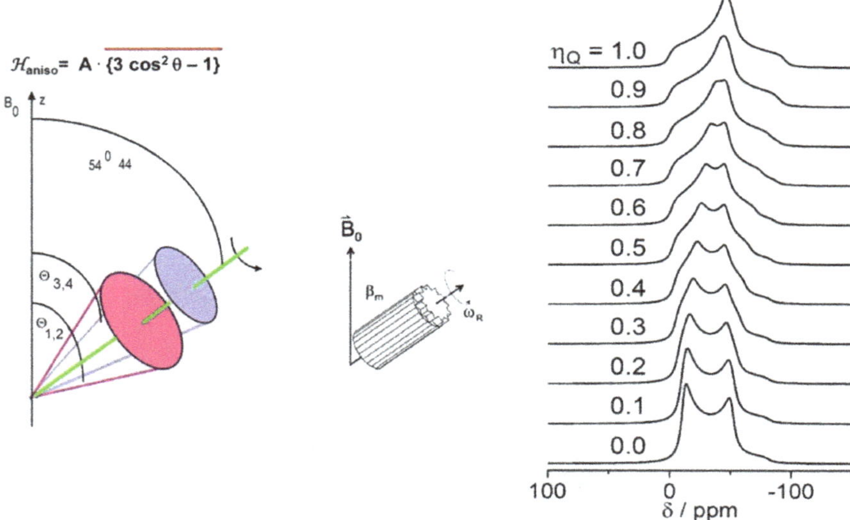

Figure 3.6.2: (Left) Orientational averaging of anisotropic interactions by the magic angle spinning technique. Each interaction is characterized by a principal direction whose orientation relative to the magnetic field is defined by an angle θ. As the precession frequency ω_p depends on θ via the term ($3\cos^2\theta$–1), the static NMR spectrum of a polycrystalline material presents a wide distribution of precession frequencies, depending on the magnitude A of the anisotropy. Fast sample rotation of the sample about an angle β_m produces the same average orientation for all crystallites, and for $\theta = 54.78°$, the term ($3\cos^2\theta$ –1) goes to zero, canceling the anisotropy. (Right) For nuclei with stronger quadrupolar interactions, the orientation dependence predicted by second-order perturbation theory is more complicated, leading to incomplete averaging. Characteristic MAS-NMR line shapes are observed, whose shapes reflect the value of the electric field gradient asymmetry parameter η_Q.

^{89}Y and ^{139}La NMR spectroscopy for structural investigations of inorganic solids [7]. Their study was followed by some empirical correlations of ^{89}Y chemical shifts with local environments in oxidic yttrium compounds [8]. In the last three decades,

Table 3.6.1: Available nuclear isotopes for NMR spectroscopic studies of rare-earth nuclei.

Isotope	Spin	Nat. abund. (%)	γ (10^7 rad s^{-1} T^{-1})	eQ (10^{-28} m^2)
^{45}Sc	7/2	100	6.509	−0.22
^{89}Y	1/2	100	−1.316	0
^{138}La	5	0.09	3.557	0.45
^{139}La	7/2	99.91	3.808	0.20
^{171}Yb	1/2	14.3	4.729	0
^{173}Yb	5/2	16.4	−1.303	2.8
^{175}Lu	7/2	97.4	3.055	3.49
^{176}Lu	7	2.6	2.168	4.97

hundreds of publications have appeared for these isotopes, addressing complex structural issues in solid solution systems [9], inorganic ceramics [10], glasses [11], endohedral fullerenes [12] and numerous functional materials, such as ceramic superconductors [13], lithium ion battery components [14] and zeolitic catalysts [15]. At the same time, sophisticated computational methods have become available for understanding NMR chemical shifts and nuclear electric quadrupole coupling parameters in terms of structural concepts [16]. In contrast, ^{175}Lu NMR finds itself still in its initial stages, while progress in ^{171}Yb NMR is limited by the availability of new materials featuring this element in its rare divalent state.

3.6.1 Scandium

Of all the nuclear isotopes listed in Table 3.6.1, ^{45}Sc is by far the most favorable one in terms of sensitivity and ease of detection, and a large number of compounds have been measured in the last two decades [17–26]. Owing to its moderately sized nuclear electric quadrupole moment, the MAS-NMR spectra are generally well resolved and can be analyzed in terms of magnetic shielding and quadrupolar interactions. As shown in Figure 3.6.3, the chemical shifts in oxidic inorganic compounds span more than 200 ppm. As is the case with other metal nuclei (e.g., ^{27}Al), separate ^{45}Sc chemical shift ranges are observed for Sc^{3+} in six- and eight-coordination, with the chemical shift decreasing with increasing coordination number.

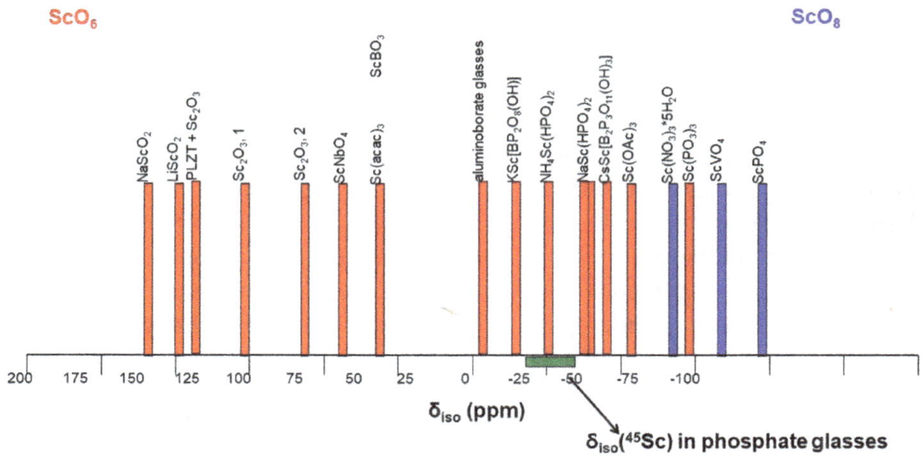

Figure 3.6.3: ^{45}Sc chemical shift range in inorganic scandium oxide compounds. Red color: six-coordinate Sc, blue color: eight-coordinate Sc [20].

A few ^{45}Sc MAS-NMR spectra of glassy systems have also been studied [20–22]. Based on the chemical shifts measured, scandium is six-coordinated in borate and phosphate glasses [20]. For scandium-containing aluminophosphate glasses, the second coordination sphere is dominated by phosphorus [21]. Scandium environments in fluoride phosphate laser glasses are characterized by a mixed coordination of fluoride and phosphate ligands, which can be controlled by the phosphate/fluoride ratio of the bulk composition. The detailed quantification of these environments by $^{45}Sc/^{19}F$ and $^{45}Sc/^{31}P$ double resonance experiments reveals a distinct preference for phosphate coordination [22]. For more details, the reader is referred to Chapter 4.13.

In the last two decades, we have further developed a large database for intermetallic compounds (see [23] for a review of the literature up to 2010 and [24–26] for additional examples). The C_Q values range from zero to about 20 MHz and Knight shifts between 0 and 2,000 ppm have been observed [23–26]. By comparing the experimental C_Q values with those computed from crystallographic input data, NMR can serve as a convenient method for crystal structure validation. If larger deviations between experimental and calculated field gradients are observed, either the proposed structure solution is invalid or there is a discrepancy between the single crystal mounted on the X-ray diffractometer and the bulk polycrystalline material used for the NMR measurement. Such structure validation becomes especially important for compounds based on elements with very similar scattering power, for materials presenting single crystals of low quality or twinning, and – of course – for those cases where no single crystals are available. Finally, the ability of NMR spectroscopy to resolve locally distinct environments allows for in-depth characterizations of order/disorder phenomena including: (1) occupancy deficiencies in non-stoichiometric compounds, (2) site multiplicities produced by positional disordering and (3) phase transitions/superstructure formation [23].

Figure 3.6.4 shows an interesting structure validation example concerning the intermetallic compound $Sc_5Pd_4Si_6$ [24]. This compound features three distinct scandium sites in a 2:2:1 ratio. The quadrupolar interaction parameters extracted from these spectra are close to those predicted theoretically, facilitating the assignment of the three NMR signals to the different scandium sites. As the electric field gradient at the Sc(2) site is close to zero, the corresponding central MAS-NMR peak includes contributions from all the $m \leftrightarrow m \pm 1$ Zeeman transitions, whereas for Sc(3) and Sc(1), the much stronger quadrupolar interaction spreads out the non-central Zeeman transitions over a wide frequency region. As a result, the intensity of the central peak is much higher for Sc(2) than for Sc(3) even though both sites have equal multiplicities.

Regarding the interpretation of the NMR interaction parameters deduced from the spectra, it is generally not possible to correlate nuclear electric quadrupolar coupling constants and/or asymmetry parameters with simple geometric distortion parameters, even within closely related series of isotypic compounds. To understand the sizes and symmetries of the electric field gradients in these compounds, quantum chemical calculations are essential. Similar conclusions hold for the isotropic shifts, $\delta_{ms}{}^{iso}$. For all intermetallic compounds measured thus far, the large

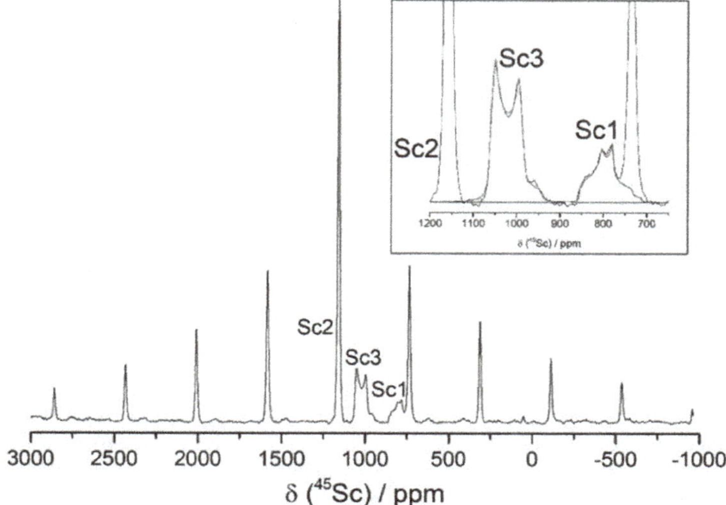

Figure 3.6.4: ^{45}Sc MAS-NMR spectrum of $Sc_5Pd_4Si_6$, featuring three sites Sc(1), Sc(2), and Sc(3) in a 1:2:2 ratio, which are spectroscopically differentiated by their Knight shifts and quadrupolar coupling parameters. Owing to the weak quadrupolar coupling of the Sc(2) site, its MAS central peak represents an overlap of the seven allowed Zeeman transitions, whereas for Sc(1) and Sc(3) the MAS central peak only arises from the central $m = 1/2 \leftrightarrow m = -1/2$ transitions [24]. This explains the large signal intensity differences of the MAS central peaks. All the unlabeled peaks are magic angle spinning sidebands. Taken from [24] with permission from RSC.

positive resonance displacements measured against an ionic reference standard (1 M Sc^{3+} in H_2O) suggests that the Knight shift contributions are dominant, revealing that the scandium atoms make significant contributions to the density of states in the vicinity of the Fermi level. Despite the general need of band structure calculations for understanding the Knight shifts on a theoretical basis, sometimes interesting trends can be observed for a series of isotypic Sc compounds. Figure 3.6.5 shows some results for compounds in the Sc*TX* series (*X* = Si, Ge and *T* = Co, Ni, Cu, Rh, Pd, Ag, Ru, Ir, Pt, Au) [25, 26]. For all those compounds which crystallize in the TiNiSi-type structure, the Knight shifts decrease monotonically with increasing atomic number of the transition metal atom. They further show a systematic dependence on the *T* group number (i.e., the valence electron concentration). In contrast, none of those compounds crystallizing in the ZrNiAl structure follow this trend.

3.6.2 Yttrium

As the NMR spectroscopy of the spin-1/2 isotope ^{89}Y isotope lacks the information content (or complications) from nuclear electric quadrupolar interactions, spectroscopic

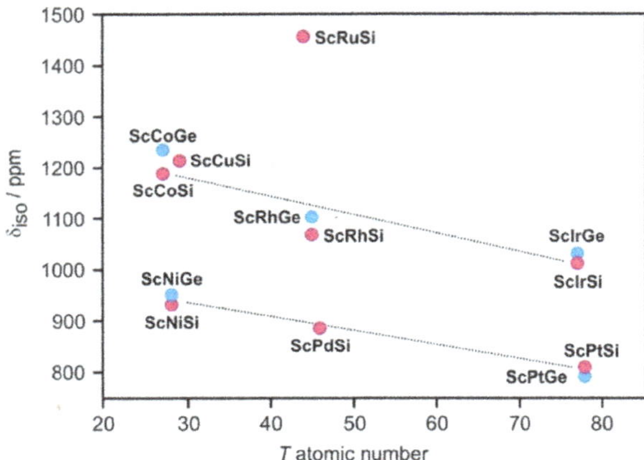

Figure 3.6.5: Correlation of the isotropic resonance shift versus 1 M $ScCl_3$ solution in Sc*T*Ge and Sc*T*Si compounds crystallizing in the TiNiSi-type structure with atomic number and valence electron concentration (group number). The dashed lines are linear least-squares fits to the data referring to compounds of the group 9 and the group 10 T atoms, respectively [25, 26]. Drawing adapted from [26].

investigations focus on the measurement of chemical shifts using MAS-NMR. Early work has been summarized in an excellent review by Sebald [27], covering topics from site populations in high-temperature superconductors [13] to paramagnetic dopant distributions in pyrochlores [28]. Although the isotope is 100% abundant, detection is hampered by the small magnetic moment and long spin-lattice relaxation times, requiring extended periods (typically ~24 h) for signal accumulation unless cross-polarization from ^{1}H nuclear spin reservoirs can be used [27]. At the low resonance frequencies encountered for ^{89}Y (e.g., 25 MHz at 11.7 T), probe ringing artifacts can be severe. In aprotic inorganic compounds, a combination of light paramagnetic doping (shortening spin-lattice relaxation times) and multiple-echo acquisition, using the *Carr–Purcell-Meiboom-Gill* (CPMG) technique (see Figure 3.6.6) [29] is helpful. As doping also diminishes spin-spin relaxation times (limiting the number of echoes that can be acquired), an optimum doping level has to be found empirically – typical values range from 0.1 to 1 mol%. The method works both under static and MAS conditions; in the latter case, the π pulses must be applied rotor synchronously, that is, at time intervals equal to the rotor period.

For insulating compounds ^{89}Y chemical shifts ($\delta_{dia} + \delta_{orb}$) differ between −400 ppm (organometallic compounds) and +314 ppm (Y_2O_3) versus 1 M YCl_3 solution. In oxidic compounds, systematics of the ^{89}Y chemical shifts have been explored by various authors [8–10, 16, 30, 31]. A quasilinear correlation between the mean Y–O distance and the ^{89}Y chemical shift is observed for most of the yttrium silicate compounds. There is a general trend of the isotropic chemical shift to decrease with increasing Y coordination number. However, the chemical shifts are also strongly

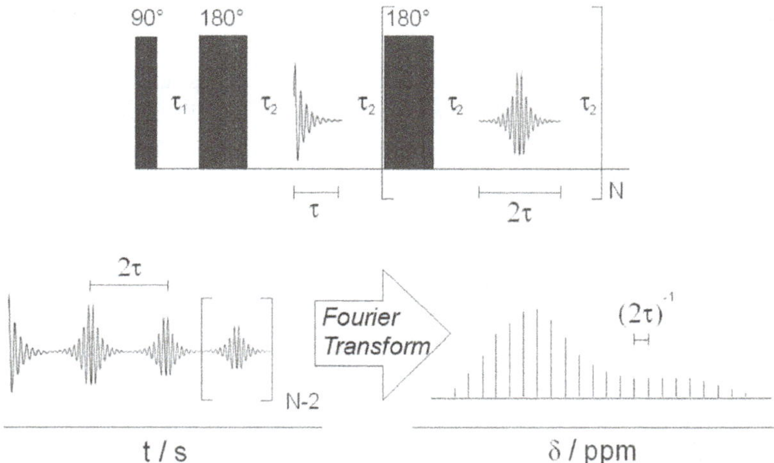

Figure 3.6.6: Carr-Purcell-Meiboom-Gill (CPMG) sequence with multiple echo acquisition for enhancing the signal to noise ratio of ^{89}Y NMR signals in compounds with long spin-lattice relaxation times [29]. The detector is closed when the pulses are applied and open in between the pulses during echo formation. Fourier transformation of the echo train acquired in this fashion leads to a *spikelet pattern* enveloping the static or MAS-NMR line shape.

influenced by next nearest neighbor effects, so that structural assignments merely on chemical shift basis need to be viewed with caution. In the first ^{89}Y NMR spectroscopic study of glasses, the chemical shifts suggest that the second coordination

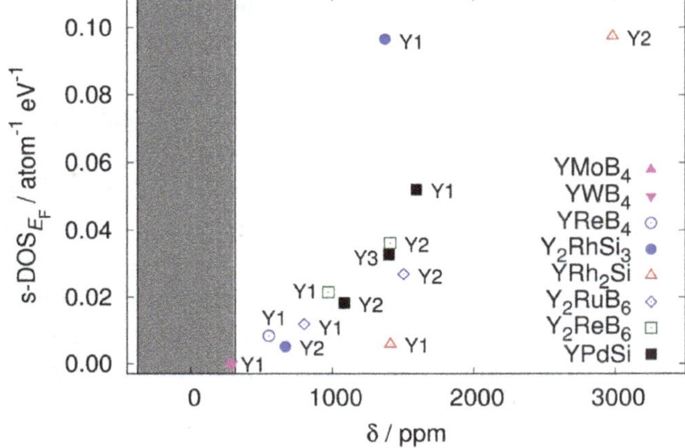

Figure 3.6.7: Correlation of isotropic ^{89}Y resonance shifts with calculated s-densities of states at the Fermi level in ternary yttrium borides [33]. The various Y-sites are indicated beside the symbols. The grey shaded area represents the typical range of chemical shifts relative to 1 M aqueous YCl_3 solution in non-metallic compounds. Taken from [33] with permission from RSC.

sphere of the Y^{3+} ions in the Y_2O_3-Al_2O_3-B_2O_3 system is dominated by meta- or pyroborate anions [31].

For intermetallic compounds, the spin-lattice relaxation times tend to be short (of the order of 1 s) and standard MAS-NMR spectroscopy is usually sufficient. The wide range of ^{89}Y resonance frequencies in metals (~3000 ppm) can be understood on the basis of band structure calculations [32]. For example, in a series of yttrium borides and silicides, one observes the expected linear correlation between ^{89}Y Knight shifts and the Y-centered calculated s-density of states in the conduction band near the Fermi level, see Figure 3.6.7 [33].

3.6.3 Lanthanum

Among the two available isotopes, ^{138}La ($I = 5$, 0.09%) and ^{139}La ($I = 7/2$, 99.91%), the choice is obvious and only a single historic reference is worth mentioning concerning the former isotope [34]. Although the quadrupole moment of ^{139}La is slightly smaller than that of ^{45}Sc, the crystal chemistry of lanthanum compounds features local La environments that produce larger electric field gradients at the rare earth site than in Sc compounds. In the majority of La compounds, the quadrupolar coupling constants are larger than 20 MHz, which precludes detection of the ^{139}La resonance by standard MAS-NMR methods. At magnetic field strengths of < 10 T, the $m = 1/2 \leftrightarrow m = -1/2$ Zeeman transitions are about 1 MHz wide under static conditions. Detection was initially done by plotting the amplitude of the spin echo, measured in steps of typically 100 kHz (*spin-echo mapping*) [35]. Alternatively, one can use the highest accessible field strengths to minimize second-order quadrupolar line-broadening effects. The first spectra of La complexes, having coordination numbers between 8 and 12, were measured in this manner [36]. At 17.6 T, reasonably undistorted spectra can be obtained for compounds with C_Q values up to 35 MHz. At these high magnetic field strengths, the chemical shift anisotropy may also influence the line shapes, which complicates the analysis [36].

In recent years, new and efficient wideband excitation methods have become available. They involve FID acquisition under linear variation (*sweep*), of either the magnetic field or, more commonly, the frequency of the applied radio wave. In particular, the latter method popularized under the acronym WURST (*wideband uniform-rate smooth truncation*) [37], which can also be applied in combination with CPMG multiple-echo acquisition [38], has made it possible to obtain high-fidelity spectra of metallocene complexes [39] and other structurally complex oxides [40], featuring C_Q values in the range of 50–100 MHz. Figure 3.6.8 shows an example from our laboratory for the semi-metal La_2AuP_2O [41]. Consistent with the crystal structure, there are two equally intense powder signals which are assigned to the sites, La(1) (9-fold coordinated by 3O, 4P, 2Au) and La(2) (11-fold coordinated by 1O,

7P, 3Au), based on electric field gradient calculations. While the latter is axially symmetric for La in the phosphorus-dominated environment, the η_Q-value for the La site in the oxygen-dominated environments is found at 0.53.

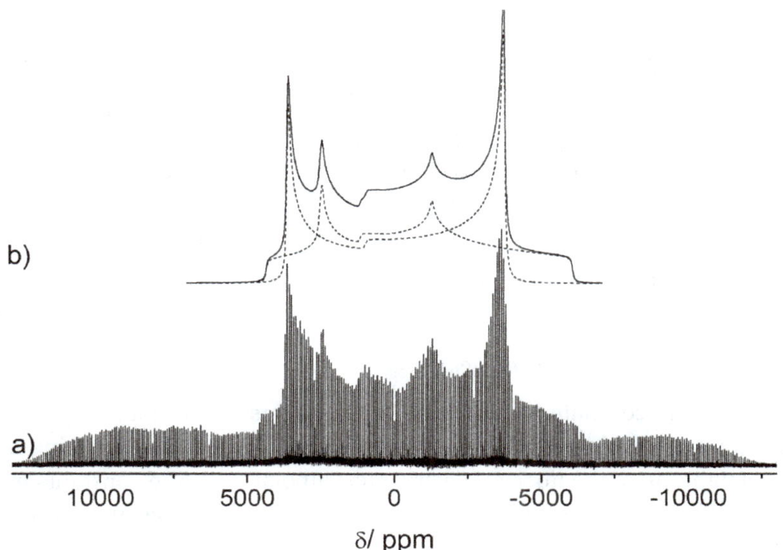

Figure 3.6.8: Static ^{139}La NMR spectrum of La_2AuP_2O obtained at 70.65 MHz [41] using the sequence shown in Figure 3.6.6. The spikelet pattern envelope comprises both the central transition and part of the satellite transitions. (b) Simulated spectrum of the central transition in terms of two distinct spectral contributions with an area ratio of 1:1. Spectral simulation parameters are: La(1) (9-fold coordinated by 3O, 4P, 2Au): δ_{iso} = 843 ppm, C_Q = 51.5 MHz, η_Q = 0.53; La(2) (11-fold coordinated by 1O, 7P, 3Au): δ_{iso} = 983 ppm, C_Q = 52.6 MHz, η_Q = 0. Taken from [41] with permission from ACS.

Regarding the correlation of the electric field gradients with local structural details, similar comments are applicable as in the case of scandium. An interesting approach has been suggested by the Zwanziger group [40], who have approximated the first La coordination spheres by ellipsoids and characterized the anisotropies of the latter by the parameters, *sphericity* $\sum$ and *ellipsoid span* ε (Figure 3.6.9) [42]. Sphericity is defined via the deviation of the local coordination environment from a sphere, according to

$$\Sigma = 1 - \sigma_S / r_S \tag{3.6.4}$$

where σ_s is the mean square deviation of the metal–ligand distance from the average distance r_s defining the centroid of the sphere (typically but not necessarily identical with the central metal atom). The ellipsoid span ε is a measure of the

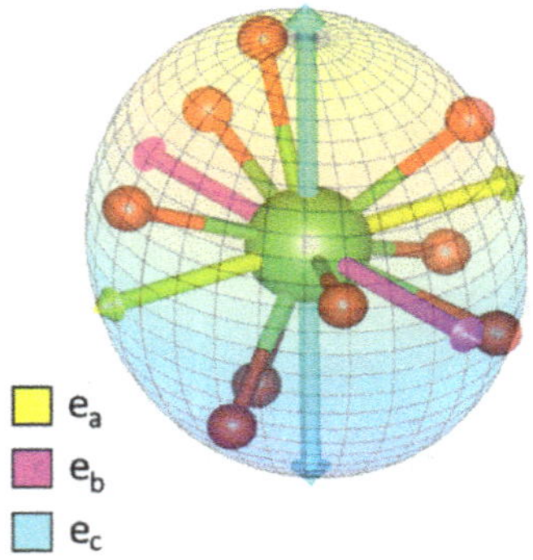

Figure 3.6.9: Approximation of the first coordination sphere by an ellipsoid and definition of the axes e_a, e_b and e_c for calculation of the ellipsoid span. Taken from [40] with permission from ACS.

difference between the lengths of the longest and the shortest ellipsoid axes, normalized by the average length $<e>$

$$\varepsilon = (e_a - e_b)/<e> \tag{3.6.5}$$

Indeed, C_Q was found to correlate well with both parameters, even though a number of exceptions were noted. Overall, these parameters are helpful in reaching a qualitative understanding of the nuclear electric quadrupolar coupling constants in lanthanum compounds and extensions to compounds with other quadrupolar nuclei should be explored.

3.6.4 Ytterbium

Ytterbium NMR spectroscopy is limited to the relatively rare cases of divalent ytterbium compounds having the $4f^{14}$ closed-shell configuration. Among the two ytterbium isotopes listed in Table 3.6.1, the spin-1/2 isotope ^{171}Yb is the preferred one owing to its significantly larger magnetic moment compared to ^{173}Yb and the absence of an electric quadrupole moment. Aside from the pioneering ytterbium magnetic moment measurements in Yb metal [43], not a single ^{173}Yb NMR work has been published. Successful detection of signals in YbS, $YbCl_2$, Yb metal, and the Laves phase $YbAl_2$ was later followed by measurements of yttrium dihydride (3,750 ppm) [44], the mixed-valent (Yb^{2+}/Yb^{3+}) compound $YbPtGe_2$ [45], where only the divalent species was detected and a small series of alkali metal ytterbium iodides [46].

The first-ever solution-state NMR spectroscopic study of this nuclide was published in 1989 on a series of Yb(II) amido complexes [47], proposing $Yb(cp^*)_2(thf)_2$ as the zero ppm reference standard, since aqueous solutions of Yb^{2+} salts are unstable. The overall chemical shift range observed in solutions and in the solid state spans about 1500 ppm [47–52]. For a series of η-cyclopentadienyl complexes, the magnetic shielding tensor components measured in the solid state were found to be sensitive to both the substituents at the cp* ligand and the nature of additional coordinating base molecules [52]. To date, there have been no reports of an understanding of the large ^{171}Yb chemical shift variations on a first-principles theoretical basis.

3.6.5 Lutetium

The nuclei of ^{175}Lu and ^{176}Lu may well be considered the most "exotic" ones. A recent update on the properties of the nuclear isotopes [53] reveals that they possess the largest nuclear electric quadrupole moments known in the group of stable isotopes. As a consequence, very few NMR spectroscopic structural studies have been reported, all of which utilize the much more highly abundant ^{175}Lu isotope [54–56]. We have recently obtained the first ^{175}Lu MAS-NMR spectra in metallic and semiconducting materials [57, 58]. Owing to the huge size of the quadrupole moment, such studies are limited to compounds with Lu atoms situated on sites with cubic point symmetry, and even for those, signal detection can be sabotaged by electric field gradients arising from defects. This is the case if there are mixed metal site occupancies, a defect frequently observed in intermetallic compounds, for example, in Heusler phases. In contrast, ^{175}Lu MAS-NMR spectra were observed for structures where covalent bonding renders such defects thermodynamically unfavorable. Examples include the binary pnictides Lu*X* (*X* = P, As, Sb) with NaCl-type structure and various ternary intermetallic compounds crystallizing in the MgAgAs type (Figure 3.6.10). Owing to its well-defined MAS-NMR spectrum which can be observed within a few minutes, we propose the compound LuP as the zero ppm reference standard.

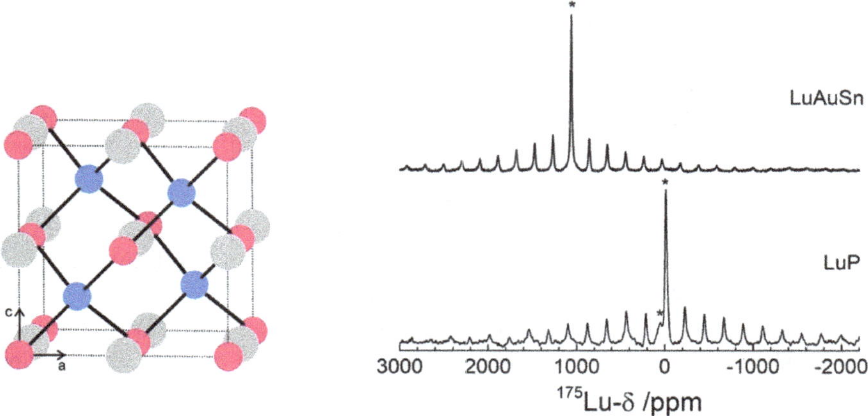

Figure 3.6.10: (left) Unit cell of the MgAgAs structure type adopted by some *RETX* compounds; rare-earth (*RE*), transition metal (*T*) and main group element (*X*) atoms are denoted by gray, blue and red colors. (right) ^{175}Lu MAS-NMR spectra of LuAuSn and LuP. Minor peaks denote spinning sidebands [58]. Taken from [58] with permission from Elsevier.

3.6.6 Basic Principles of Mößbauer Spectroscopy of Rare Earth Isotopes

We now turn to γ radiation absorption. The **Mößbauer effect**, originally denoted as *recoilless nuclear absorption and fluorescence* was discovered in 1958 by Rudolf L. Mößbauer at the Technical University of Munich. In his original experiment, he proved that recoilless absorption and emission of γ radiation is possible in ^{191}Ir and other nuclei embedded in a crystalline matrix [59]. Mößbauer received the Nobel Prize in Physics in 1961 and his wide-ranging discovery, the Mößbauer effect, has become an important and indispensable solid state spectroscopy in Chemistry, Physics, Biology and Geosciences. The most popular Mößbauer active isotope is ^{57}Fe, owing to the large importance of iron compounds in nearly all areas of chemistry. The basic physics of the Mößbauer effect and the technical details of instrumentation have been well summarized with many examples in review articles and textbooks [60–66].

In the present chapter, we solely focus on the application of the Mößbauer effect on rare earth nuclei. With the exception of cerium, all rare earth elements show the Mößbauer effect. Some basic data for the relevant nuclei are summarized in Table 3.6.2. The quantities listed in the table determine the suitability of the respective isotope for applications in chemical spectroscopy. A key parameter is the half-life of the Mößbauer source which poses limitations on their availability. Usually, the sources can be prepared in nuclear reactors via neutron or proton irradiation of suitable precursor materials. ^{145}Nd, ^{147}Pm, ^{151}Eu and ^{155}Gd sources have half-lives in the order of years and can be used for standard laboratory *in-house* experiments. For this reason, Eu- and Gd-based solids have been most widely studied. Mößbauer sources of the other rare-earth nuclei are short-lived and hence require close collaboration with a nuclear physics laboratory close to a reactor. The decay schemes for the diverse isotopes are all known and can be looked up in the relevant textbooks [67]. Another critical parameter is the energy difference E_γ, between the ground and the excited states, which controls the probability of a zero-phonon transition (*recoil-free fraction, Lamb Mößbauer factor*) f according to:

$$f \sim \exp - (E_\gamma{}^2 < x^2 >) \qquad (3.6.6)$$

where $<x^2>$ is the mean square deviation of the nucleus from its equilibrium position in the lattice due to the thermal motion of the parent atom. The Lamb Mößbauer factor depends on the vibrational characteristics of the lattice and decreases sharply with increasing temperatures. For source energies in excess of 100 keV, this fraction is very low, restricting experiments to low temperatures.

Depending on the nuclear spin quantum numbers of the ground state and the excited state (Table 3.6.2), the Mößbauer spectra are more or less complex. The extent to which individual transitions are resolved depends on the lifetime of the nuclear excited state, which controls the natural line width via the Heisenberg uncertainty principle. This quantity is also listed in Table 3.6.2. The important parameters that we extract from the Mößbauer spectra for addressing chemical issues are the isomer shift (δ), the experimental line width (Γ), the electric quadrupole splitting parameter (ΔE_Q) and in the case of magnetically ordered materials, the magnetic hyperfine field (B_{hf}).

The isomer shift (IS) is one of the key interaction parameters in Mößbauer spectroscopy. It is proportional to the product

$$\mathrm{IS} \sim [\Psi(0)^2{}_A - (\Psi(0)^2{}_S] \times (R_{ex}{}^2 - R_g{}^2) \tag{3.6.7}$$

where $[\Psi(0)^2{}_A - (\Psi(0)^2{}_S]$ is the difference between electronic charge densities at the nuclear site between the sample (absorber, A) and source (S) material, whereas $(R_{ex}{}^2 - R_g{}^2)$ is the difference in squared effective nuclear radii in the excited and ground states. For differentiating spectroscopically between two distinct nuclear environments (for example different valence states), the above product must be comparable to or larger than the natural linewidth. Favorable examples include

Table 3.6.2: Basic parameters for rare earth elements that show the Mößbauer effect: I_e, nuclear spin quantum number in the excited state; I_g, nuclear spin quantum number in the ground state. The data was taken from the compilations of *Shenoy* and *Wagner* [68], *Gubbens* [69] and some original literature [70]. Some of the rare earth elements show several Mößbauer active isotopes. Only the most relevant ones are summarized in this table.

Isotope	**^{139}La**	**^{141}Pr**	**^{145}Nd**	**^{147}Pm**	**^{149}Sm**	**^{151}Eu**	**^{155}Gd**
I_e and I_g	5/2 and 7/2	7/2 and 5/2	5/2 and 7/2	5/2 and 7/2	5/2 and 7/2	7/2 and 5/2	5/2 and 3/2
Energy (keV)	166	145.4	72.5	91	22.49	21.53	86.54
Half-life	140 d	32.5 d	340 d	17.7 y	106 d	87 y	1.81 y
Source material	$^{138}CeO_2$	CeF_3	Nd_2O_3:^{145}Pm	^{147}Pm	$^{150}Sm_2O_3$	$^{150}SmF_3$	$^{154}SmPd_3$
Line width (mm/s)	1.10	1.02	5.24	1.17	1.71	1.30	0.5
Isotope	**^{159}Tb**	**^{161}Dy**	**^{165}Ho**	**^{166}Er**	**^{169}Tm**	**^{170}Yb**	**^{175}Lu**
I_e and I_g	5/2 and 3/2	5/2 and 5/2	9/2 and 7/2	2 and 0	3/2 and 1/2	2 and 0	9/2 and 7/2
Energy (keV)	58	25.66	95	80.56	8.40	84.25	113.8
Half-life	144 d	6.9 d	2.4 h	26 h	9.4 d	130 d	101 h
Source material	$^{158}Dy_2O_3$	$^{160}Gd_{0.5}{}^{162}Dy_{0.5}F_3$	Dy_2O_3	$HoPd_3$	$^{168}ErAl_3$/Al	TmB_{12}	Lu_2O_3
Line width (mm/s)	44.9	0.38	160.5	1.89	8.33	2.03	26.7

^{151}Eu, ^{149}Sm and ^{155}Gd, for which the valence states can be unambiguously assigned based on isomer shift measurements. A second key parameter is the nuclear electric quadrupole splitting. As explained at the beginning, a nonspherical charge distribution around a nucleus creates an electric field gradient at the nuclear origin which interacts with the electric quadrupole moments of the nuclear ground and excited states. This interaction partially removes the degeneracy of the $2I + 1$ spin orientation states, leaving the states with quantum numbers $\pm$ m degenerate, such that $(2I + 1)/2$ distinct levels result. Allowed nuclear magnetic dipole transitions are those with $\Delta m = 0$ and ± 1. However, whether or not they can be resolved depends on the size of the quadrupole splitting in relation to the natural line width. The experimental line width parameter is often found larger than the natural line width, which may arise from (i) overlapping sub-spectra with close isomer shifts, (ii) domain distributions, (iii) unresolved quadrupole or Zeeman splitting or (iv) static distributions or fluctuations of interaction parameters in disordered or nanosized materials. For magnetically ordered materials, usually the magnetic hyperfine field increases from the ordering temperature to lower temperature, in the form of a Brillouin function and extrapolation leads to the saturation value. The schemes for level splitting in the magnetically ordered state are documented in the standard textbooks. The value of the hyperfine field strongly depends on the rare earth nucleus. While the hyperfine fields of europium and gadolinium compounds usually range from 20 to 30 T, thulium compounds show hyperfine fields up to 700 T. The Mößbauer line shapes are further influenced by the relative orientation of the direction of the magnetic hyperfine field and the principle axis of the electric field gradient.

3.6.7 Applications of Mößbauer Spectroscopy to Rare Earth Structural Chemistry

In the following section, some key structural applications of rare-earth Mößbauer spectroscopy are described. Work up to the late 1970s has been summarized in a review article by Barnes [71]. The most widely used Mößbauer isotope among the rare earth elements is ^{151}Eu. Its popularity arises from the long half-life time of 87 y. Another advantageous feature is the clear spectroscopic separation of the Eu(II) and Eu(III) isomer shifts [72], an excellent prerequisite for an unequivocal assignment of the europium valence. Depending on the degree of metallic, covalent or ionic bonding, the isomer shifts range from 0 to +5 mm/s for Eu(III) species and from −7 to −14 mm/s for Eu(II) species (Figure 3.6.11) [68]. The *s*-electron densities and thus the isomer shifts of Eu(II) and Eu(III) are susceptible to external pressure [73]. Through ^{151}Eu Mößbauer spectroscopy, one can also monitor pressure-induced valence transitions. A striking example is the series of EuT_2Ge_2 (T = Ni, Pd, Pt) germanides in which, $EuNi_2Ge_2$ and $EuPd_2Ge_2$ show a valence transition from Eu(II) to Eu (III) at pressures around 5 and 25 GPa along with changes in the isomer shift of −9.3

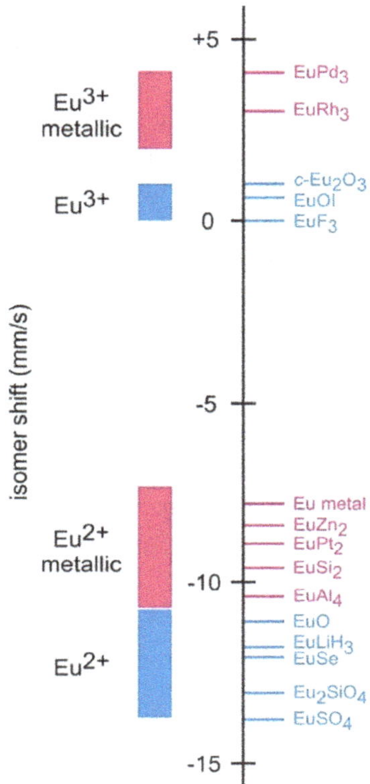

Figure 3.6.11: ^{151}Eu isomer shifts of different europium compounds (data taken from the compilation of Shenoy and Wagner [68]) with decreasing electron density from top to bottom.

and −10.1 mm/s, respectively [74]. Remarkable examples in the field of chalcogenides are Eu_3O_4 and Eu_3S_4. The oxide shows static mixed valence with distinct Eu (II) and Eu(III) sites from cryogenic temperatures to 300 K. The Eu(II) site with $4f^7$ configuration shows magnetic hyperfine field splitting with an internal field of 30.5 T at 1.2 K [75], whereas the $^7F_0$ state of Eu(III) is nonmagnetic. In contrast, Eu_3S_4 shows a thermally activated electron hopping process: the Eu(II) and Eu (III) signals are well separated at 83 K and show coalescence toward 325 K with a single signal at an intermediate isomer shift of −3 mm/s. The activation energy for the hopping process is 0.24 eV [76]. Further examples of ^{151}Eu Mößbauer spectroscopic studies of mixed valence materials are summarized [68, 77].

Although Eu(II) and Gd(III) are isoelectronic, the range of isomer shifts is much smaller in the case of ^{155}Gd Mößbauer spectroscopy [68, 78, 79]. While intermetallic gadolinium compounds cover the complete range from 0 to 0.7 mm/s, the isomer shift variations for ionic compounds are much smaller (0.5 to 0.7 mm/s). Published trends reveal a decrease in isomer shift with increasing electron density. In the case of metallic compounds, the shift range reflects a considerable variation of the character of the conduction electrons, larger 6s contributions leading to smaller isomer shift values and stronger 5d and/or 6p populations having the expected opposite effects.

Since natural gadolinium has a high absorption cross section, which strongly hampers neutron diffraction experiments, ^{155}Gd Mößbauer spectroscopy is a very helpful complementary tool for the determination of magnetic structures. For that reason, many magnetically ordered gadolinium phases have been characterized thoroughly by low-temperature ^{155}Gd spectroscopy. As an example, we present the 57 and 4.2 K ^{155}Gd spectra of the 49 K antiferromagnet Gd_2Ni_2Mg [80] in Figure 3.6.12. The resolution is not as good as for ^{151}Eu, but one can derive the basic parameters and, sometimes, it is possible to infer the orientation of the gadolinium magnetic moments from the ^{155}Gd spectra. Complex magnetic structures manifest themselves in ^{155}Gd spectra, characterized by hyperfine field distributions. A further interesting application concerns the experimental determination of the crystal field parameters relying on the proportionality of the V_{zz} value and the crystal electric field parameter A_2^0. Some examples are summarized in [79].

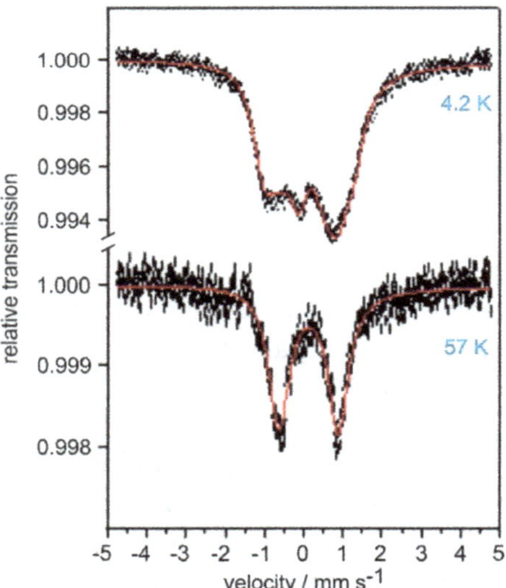

Figure 3.6.12: Experimental and simulated ^{155}Gd Mößbauer spectra of the 49 K antiferromagnet Gd_2Ni_2Mg at 4.2 and 57 K (adapted from [80]).

The following paragraphs briefly summarize Mößbauer spectroscopic work on the remaining isotopes. Since these have short half-life times or require special source preparation, they are less frequently used. Especially ^{159}Tb [70f], ^{165}Ho [70a] and ^{175}Lu [70 c] play only a subordinate role and only few compounds have been studied. ^{139}La Mößbauer spectra can be collected only at 4.2 K because the high transition energy results in very small recoil-free fractions. The variation of the isomer shifts for the trivalent compounds is small: −0.16 mm/s for LaC_2 and +0.09 mm/s for La_2O_3 [70e].

A small range of the isomer shifts is also observed for the ^{141}Pr [70d] Mößbauer spectra of trivalent praseodymium, for example, from −0.63 mm/s for $PrBr_3$ to −1.14 mm/s for $Pr_2(SO_4)_3$ [68]. The electron density increases towards $PrBr_3$ and one can clearly differentiate within these trivalent compounds. Tetravalent highly ionic praseodymium compounds show positive isomer shifts (+0.47 and +0.54 mm/s for $CsPrF_5$ and Cs_2PrF_6) and can be clearly distinguished from trivalent ones (−0.94 mm/s for PrF_3). Intermetallic praseodymium compounds show strong hyperfine fields in the magnetically ordered regimes. An interesting family of compounds are the silicides PrT_2Si_2 and the germanides PrT_2Ge_2, where the largest hyperfine field of 319 T was observed for the 30 K antiferromagnet $PrCo_2Si_2$ [69].

The situation for trivalent neodymium is similar to that for lanthanum. The ^{145}Nd isomer shifts show little variation [70b] but a clear differentiation is possible from divalent neodymium compounds which show negative δ values [68]. Similar trends are observed with ^{149}Sm [68], again with negative values for divalent samarium species (e.g., −0.90 mm/s for SmF_2). A striking application of ^{149}Sm Mößbauer spectroscopy is the study of the pressure-induced semiconductor-to-metal transition of SmS [81], which shows a clear increase of the isomer shift from −0.71 mm/s for the semiconducting to −0.21 mm/s for the metallic phase (11 kbar data).

Ionic and metallic trivalent dysprosium compounds show a range of isomer shifts from 0 to +4 mm/s and changes in the local electron density can be monitored through ^{161}Dy Mößbauer spectroscopy [68], allowing a clear differentiation between oxidation states. Divalent dysprosium (Dy^{2+} in a CaF_2 host) shows a distinctly negative isomer shift of −6.3 mm/s, while tetravalent dysprosium (Dy^{4+} in $Gd_{0.5}Dy_{0.5}F_3$) is characterized by a highly positive δ value of 7.0 mm/s.

The ^{166}Er and ^{169}Tm transitions belong to the difficult ones, especially due to the short half-life times of 26 h and 9.4 days of their Mößbauer sources. The basic transition data and the relevant expressions for the nuclear energy level splittings, including both the magnetic and electric quadrupole interactions, have been summarized in a resource letter by Cadogan and Ryan [82]. Typical examples are the ^{166}Er spectra of the pyrochlore representatives $Er_2Sn_2O_7$ and $Er_2Ti_2O_7$ [83]. Both oxides show no hyperfine field splitting down to 1.56 K. This is different in the spinel $CdEr_2S_4$, which shows a huge hyperfine field of 727.6 T at 5 K. $CdEr_2S_4$ shows slow paramagnetic relaxation behavior and a weak hyperfine field is still detectable at 45 K [83]. Similar behavior has been observed for the 5.8 K antiferromagnet $ErNiAl_4$ [84].

An interesting feature of ^{169}Tm Mößbauer spectroscopy is the low transition energy of 8.4 keV which enables the collection of data up to temperatures of 2,000 K on suitable materials having the necessary thermal stability. Thus, ^{169}Tm Mößbauer spectroscopy is an attractive tool for crystal field investigations. The majority of studies were performed on diverse intermetallic compounds and have been competently summarized by Gubbens [69]. Similar to ^{166}Er, ^{169}Tm spectra also show high magnetic hyperfine fields (up to ~700 T) in magnetically ordered states.

The ^{170}Yb isomer shift values for Yb(II) and Yb(III) vary only over a small range. Representative values are $\delta = -0.49$ mm/s for $Yb^{II}SO_4$ and +0.09 mm/s for $Yb^{III}Cl_3$ [68]. While this marginal separation of the Yb(II) and Yb(III) isomer shifts hampers studies of valence changes [77], the ^{170}Yb probes are very sensitive to magnetic ordering phenomena. The ytterbium substructure of pure *o*-$YbMnO_3$ shows no magnetic ordering down to 1.8 K, which contradicts an earlier report [85]. The second large field for ^{170}Yb Mößbauer spectroscopy concerns intermetallics. Due to the low magnetic ordering temperatures of Yb^{3+}-containing compounds, Mößbauer spectra have frequently been measured down to the mK range. An interesting result is the study of inhomogeneous hyperfine fields/hyperfine field distributions in the monopnictides YbP, YbAs and YbSb [86]. ^{170}Yb Mößbauer spectroscopy was employed to differentiate the ytterbium magnetic behavior of the isotypic compounds $Yb_2Co_3Al_9$ and $Yb_2Co_3Ga_9$ [87]. The aluminide shows a stable $^2F_{7/2}$ ground state and well resolved hyperfine field splitting (95 T) at 0.06 K, while $Yb_2Co_3Ga_9$ is a paramagnetic heavy-fermion material showing no hyperfine field splitting over the whole temperature range.

References

[1] C. P. Slichter, Principles of Magnetic Resonance, Springer Verlag, Berlin, 1990.
[2] M. H. Levitt, Spin Dynamics. Basics of Nuclear Magnetic Resonance, J. Wiley & Sons, 2nd Edition, 2008.

[3] M. J. Duer, Introduction into Solid State NMR Spectroscopy, Blackwell, London, 2005.
[4] R. K. Harris, Nuclear Magnetic Resonance Spectroscopy: A Physicochemical View, Halstead Press, Sydney, 1982.
[5] K. Schmidt-Rohr, H. W. Spiess, Multidimensional Solid State NMR and Polymers, Academic Press, London, 1994.
[6] D. Freude, J. Haase, NMR Basic Prin. Prog. 1993, 29, 1.
[7] A. R. Thompson, E. Oldfield, J. Chem. Soc. Chem. Commun. 1987, 27.
[8] R. Dupree, M. E. Smith, Chem. Phys. Lett. 1988, 148, 41.
[9] K. E. Johnston, M. R. Mitchell, F. Blanc, P. Lightfoot, S. E. Ashbrook, J. Phys. Chem. C 2013, 117, 2252.
[10] M. R. Mitchell, D. Carnevale, R. Orr, K. R. Whittle, S. E. Ashbrook, J. Phys. Chem. C 2012, 116, 4273.

[11] H. Eckert, Int. J. Appl. Glass Sci. 2018, 9, 167.
[12] H. Kato, A. Taninaka, T. Sugai, H. Shinohara, J. Am. Chem. Soc. 2003, 125, 7782.
[13] Z. P. Han, R. Dupree, D. M. Paul, A. P. Howes, L. W. J. Caves, Physica C 1991, 181, 355.
[14] L. Spencer, E. Coomes, E. Ye, V. Terskikh, A. Ramzy, V. Thangadurai, G. R. Goward, Can. J. Chem. 2011, 89, 1105.
[15] M. Hunger, G. Engelhardt, J. Weitkamp, Micropor. Mater. 1995, 3, 497.
[16] S. W. Reader, M. R. Mitchell, K. E. Johnston, C. J. Pickard, K. R. Whittle, S. E. Ashbrook, J. Phys. Chem. C 2009, 113, 18874.
[17] N. Kim, C. H. Hsieh, J. F. Stebbins, Chem. Mater. 2006, 18, 3855.
[18] M. D. Alba, P. Chain, P. Florian, D. Massiot, J. Phys. Chem. C 2010, 114, 12125.

[19] A. J. Rossini, R. W. Schurko, J. Am. Chem. Soc. 2006, 128, 10391.
[20] D. Mohr, Investigation of Scandium Coordination in Glasses and Crystalline Compounds – An NMR Study, PhD thesis, Universität Münster, 2010.
[21] D. Mohr, A. S. S. de Camargo, C. C. de Araujo, H. Eckert, J. Mater. Chem. 2007, 17, 3733.
[22] M. de Oliveira Jr., T. S. Gonçalves, C. Ferrari, C. J. Magon, P. S. Pizani, A. S. S. de Camargo, H. Eckert, J. Phys. Chem. C 2017, 121, 2968.
[23] H. Eckert, R. Pöttgen, Z. Anorg. Allg. Chem. 2010, 636, 2232.
[24] L. Schubert, C. Doerenkamp, S. Haverkamp, L. Heletta, H. Eckert, R. Pöttgen, Dalton Trans. 2018, 47, 13025.
[25] T. Harmening, H. Eckert, C. M. Fehse, C. P. Sebastian, R. Pöttgen, J. Solid State Chem. 2011, 184, 3303.
[26] B. Heying, S. Haverkamp, U. C. Rodewald, H. Eckert, C. P. Sebastian, R. Pöttgen, Solid State Sci. 2015, 39, 15.
[27] A. Sebald, NMR-Basic Prin. Prog. 1994, 31, 91.
[28] C. P. Grey, M. E. Smith, A. K. Cheetham, C. M. Dobson, R. Dupree, J. Am. Chem. Soc. 1990, 112, 4670.
[29] R. Siegel, T. T. Nakashima, R. E. Wasylishen, J. Phys. Chem. B 2004, 108, 2218.
[30] A. I. Becerro, A. Escudero, P. Florian, D. Massiot, M. D. Alba, J. Solid State Chem. 2004, 177, 2783.
[31] H. Deters, A. S. S. de Camargo, C. N. Santos, C. R. Ferrari, A. C. Hernandes, A. Ibanez, M. T. Rinke, H. Eckert, J. Phys. Chem. C 2009, 113, 16216.
[32] C. Höting, H. Eckert, F. Haarmann, F. Winter, R. Pöttgen, Dalton Trans. 2014, 43, 7860.
[33] C. Benndorf, M. de Oliveira, C. Doerenkamp, F. Haarmann, T. Fickenscher, J. Kösters, H. Eckert, R. Pöttgen, Dalton Trans. 2019, 48, 1118.
[34] O. Lutz, H. Oehler, J. Magn. Reson. 1980, 37, 261.
[35] T. J. Bastow, Solid State Nucl. Magn. Reson. 1994, 3, 17.
[36] M. J. Williams, K. W. Feindel, K. J. Ooms, R. E. Wasylishen, Chem. Eur. J. 2006, 12, 159.
[37] L. A. O`Dell, Solid State Nucl. Magn. Reson. 2013, 55–56, 28.
[38] L. A. O'Dell, A. J. Rossini, R. W. Schurko, Chem. Phys. Lett. 2009, 468, 330.
[39] H. Hamaed, A. Y. H. Lo, D. S. Lee, W. J. Evans, R. W. Schurko, J. Am. Chem. Soc. 2006, 128, 12638.
[40] A. L. Paterson, M. A. Hanson, U. Werner-Zwanziger, J. W. Zwanziger, J. Phys. Chem. C 2015, 119, 25508.
[41] T. Bartsch, T. Wiegand, J. Ren, H. Eckert, D. Johrendt, O. Niehaus, M. Eul, R. Pöttgen, Inorg. Chem. 2013, 52, 2094.
[42] T. Balić Žunić, E. Mackovicky, Acta Crystallogr. B 1996, 52, 78.
[43] A. C. Gossard, V. Jaccarino, J. H. Wernick, Phys. Rev. 1964, 133, A881.
[44] O. J. Zogal, B. Stalinski, Magn. Reson. Relat. Phenom., Proc. 20th Congr. Ampere (ed. E. Kundla, E. Lippma, T. Saluvere), Springer, Berlin, 1979, 433.
[45] X. Zhao, X.-A. Mao, S. Wang, C. Ye, J. Alloys Compd. 1997, 250, 409.
[46] R. Sarkar, R. Gumeniuk, A. Leithe-Jasper, W. Schnelle, Y. Grin, C. Geibel, M. Baenitz, Phys. Rev. B 2013, 88, 201101.
[47] A. G. Avent, M. A. Edelman, M. F. Lappert, G. A. Lawless, J. Am. Chem. Soc. 1989, 111, 3423.
[48] A. M. Dietel, O. Tok, R. Kempe, Eur. J. Inorg. Chem. 2007, 4583.
[49] S. Marks, J. G. Heck, M. H. Habicht, P. Ona-Burgos, C. Feldmann, P. W. Roesky, J. Am. Chem. Soc. 2012, 134, 16983.
[50] J. M. Keates, G. A. Lawless, Organometallics 1997, 16, 2842.
[51] G. W. Rabe, A. Sebald, Solid State Nucl. Magn. Reson. 1996, 6, 197.
[52] J. M. Keates, G. A. Lawless, M. Waugh, Chem. Commun. 1996, 1627.

[53] N. J. Stone, J. Phys. Chem. Ref. Data 2015, 44, 031215.
[54] O. Żogał, R. Wawryk, M. Matusiak, Z. Henkie, J. Alloys Compd. 2014, 587, 190.
[55] A. H. Reddoch, G. J. Ritter, Phys. Rev. 1962, 126, 1493.
[56] B. Nowak, D. Kaczorowski, Intermetallics 2013, 40, 28.
[57] C. Benndorf, Multinukleare Festkörper NMR spektroskopische Untersuchungen ausgewählter intermetallischer Verbindungen, PhD thesis, Universität Münster, 2017.
[58] C. Benndorf, M. de Oliveira Junior, H. Bradtmüller, F. Stegemann, R. Pöttgen, H. Eckert, Solid State Nucl. Magn. Reson. 2019, 101, 63.
[59] a) R. L. Mößbauer, Z. Phys. 1958, 151, 124; b) R. L. Mößbauer, Z. Naturforsch. 1959, 14a, 211; c) R. L. Mößbauer, Naturwissenschaften 1958, 22, 538.

[60] P. Gütlich, Chemie in unserer Zeit, 1970, 4, 133; ibid. 1971, 5, 131.
[61] D. Barb, Grundlagen und Anwendungen der Mößbauerspektroskopie, Akademie-Verlag Berlin und Editura Academiei Republicii Socialiste România, Berlin, Bucureşti, 1980.
[62] F. E. Wagner, Mößbauerspektroskopie, in Untersuchungsmethoden in der Chemie (ed. H. Naumer, W. Heller), 2. Auflage, Thieme Verlag, Stuttgart, 1990.
[63] G. J. Long, F. Grandjean (Eds.) Mößbauer Spectroscopy Applied to Magnetism and Materials Science, in Modern Inorganic Chemistry (ed. J. P. Fackler Jr.), Vol. 1, Plenum Press, New York, 1993.
[64] Y.-L. Chen, D.-P. Yang, Mößbauer Effect in Lattice Dynamics, Wiley-VCH, Weinheim, 2007.
[65] P. Gütlich, E. Bill, A. X. Trautwein, Mößbauer Spectroscopy and Transition Metal Chemistry, Springer, Berlin, Heidelberg, 2011.
[66] M. Kalvius, P. Kienle (Eds.) The Rudolf Mößbauer Story: His Scientific Work and Its Impact on Science and History, Springer-Verlag, Berlin, Heidelberg, 2012. DOI:10.1007/978-3-642-17952-5
[67] J. G. Stevens, V. E. Stevens, P. T. Deason Jr., A. H. Muir Jr., H. M. Coogan, R. W. Grant (Eds.) Mößbauer Effect Data Index, IFI/Plenum Data Company, Plenum Publishing, New York, 1975.
[68] G. K. Shenoy, F. E. Wagner (Eds.) Mößbauer Isomer Shifts, North Holland Publishing, Amsterdam, 1978.

[69] P. C. M. Gubbens, Rare Earth Mößbauer Spectroscopy Measurements on Lanthanide Intermetallics: A Survey, in Handbook of Magnetic Materials, Vol. 20, Ch. 4, 2012, 227–335. DOI:10.1016/B978-0-444-56371-2.00004-0
[70] a) T. Rousskov, T. Tomov, H. Popov, C. R. Acad. Bulg. Sci.: Sci. Mathém. Nat. 1966, 19, 701; b) G. Kaindl, R. L. Mößbauer, Phys. Lett. 1968, 26B, 386; c) K. A. Hardy, U. Atzmony, J. C. Walker, Nucl. Phys. 1970, A154, 497; d) W. H. Kapfhammer, W. Maurer, F. E. Wagner, P. Kienle, Z. Naturforsch. 1971, 26a, 357; e) F. E. Wagner, K. Thoma, M. Atoji, Z. Phys. 1973, 262, 265; f) B. R. Bullard, J. G. Mullen, Phys. Rev. B 1991, 43, 7416.

[71] R. G. Barnes, NMR, EPR and Mößbauer Effect: Metals, Alloys and Compounds, in Handbook on the Physics and Chemistry of Rare Earths (ed. K. A. Gschneidner, L. Eyring), Vol. 2, Ch. 18, North Holland Publishing, Amsterdam, 1979, 387–505.
[72] F. Grandjean, G. J. Long, Mößbauer Spectroscopy of Europium-Containing Compounds, in Mößbauer Spectroscopy Applied to Inorganic Chemistry (Ed. G. J. Long, F. Grandjean), Vol. 3, Ch. 11, Springer Science+Business Media, New York, 1989.
[73] U. F. Klein, G. Wortmann, G. M. Kalvius, Solid State Commun. 1975, 18, 291.
[74] H.-J. Hesse, G. Wortmann, Hyperfine Interactions 1994, 93, 1499.
[75] H. H. Wickman, E. Catalano, J. Appl. Phys. 1968, 39, 1248.
[76] O. Berkooz, M. Malamud, S. Shtrikman, Solid State Commun. 1968, 6, 185.
[77] H. Eckert, Mößbauer Spectroscopy of Mixed-Valent Compounds, in Mößbauer Spectroscopy Applied to Inorganic Chemistry (ed. G. J. Long), Vol. 2, Ch. 3, Plenum Press, New York, 1987.

[78] G. Czjzek, Mößbauer Spectroscopy of New Materials Containing Gadolinium, in Mößbauer Spectroscopy Applied to Magnetism and Materials Science (ed. G. J. Long, F. Grandjean), Vol. 1, Ch. 9, Plenum Press, New York, 1993.
[79] R. Pöttgen, K. Łątka, Z. Anorg. Allg. Chem. 2010, 636, 2244.
[80] K. Łątka, R. Kmieć, A. W. Pacyna, R. Mishra, R. Pöttgen, Solid State Sci. 2001, 3, 545.
[81] J. M. D. Coey, S. K. Ghatak, F. Holtzberg, J. Appl. Phys. Conf. Proc. 1975, 24, 38. DOI:10.1063/1.30149
[82] J. M. Cadogan, D. H. Ryan, Hyperfine Interact. 2004, 153, 25.
[83] A. Legros, D. H. Ryan, P. Dalmas de Réotier, A. Yaouanc, C. Marin, J. Appl. Phys. 2015, 117, 17C701.
[84] D. H. Ryan, N. Lee-Hone, G. A. Stewart, Solid State Phen. 2013, 194, 84.
[85] a) X. Fabrèges, I. Mirebeau, P. Bonville, S. Petit, G. Lebras-Jasmin, A. Forget, G. André, S. Pailhès, Phys. Rev. B 2008, 78, 214422; b) G. A. Stewart, H. A. Salama, C. J. Voyer, D. H. Ryan, D. Scott, H. StC. O'Neill, Hyperfine Interact. 2015, 230, 195.
[86] P. Bonville, J. A. Hodges, F. Hulliger, P. Imbert, G. Jéhanno, J. B. Marimon da Cunha, H. R. Ott, J. Magn. Magn. Mater. 1988, 76&77, 473.
[87] S. K. Dhar, C. Mitra, P. Bonville, M. Rams, K. Królas, C. Godart, E. Alleno, N. Suzuki, K. Miyake, N. Watanabe, Y. Onuki, P. Manfrinetti, A. Palenzona, Phys. Rev. B 2001, 64, 094423.

Hellmut Eckert

3.7 Rare-earth ion spin dynamics as a source of structural information

3.7.1 Introduction

Spin dynamics associated with the open 4f shell of rare-earth (*RE*) species produce unique physical properties that are opening up new materials science applications. Important examples discussed within various chapters in this book include (1) room temperature spintronics via single molecules or *RE* doped semiconductor nanoparticles (Chapter 4.10), (2) magnetic nanoparticles for magnetic resonance imaging contrast enhancement (Chapter 4.5) and (3) magneto-optic glassy and crystalline devices exhibiting the Faraday effect (Chapter 4.13). For the fundamental physics and device engineering aspects involved with these applications the reader is referred to those chapters and the references therein. The present chapter deals with the use of spin dynamics as a source of structural information. These include *RE* containing probes to study local environments via electron paramagnetic resonance (EPR) spectroscopy as well as studying their effects upon the NMR spectra in solution and in the solid state. The physical basis is the total angular momentum of the electronic ground state (and in a few cases (Eu^{3+}, Sm^{3+}) of low-energy excited electronic states), which arise from vectorial coupling of the orbital and spin angular momenta. Even though, strictly speaking, the quantum numbers denoting the total spin angular momentum, S, and the total orbital angular momentum, L, are not good quantum numbers for *RE* ions, the Russell–Saunders scheme ${}^{2S+1}\mathscr{L}_J$ is customarily used to label *RE* electron configurations. In this terminology the symbol $\mathscr{L}$ is represented by a capital letter S, P, D, F, G, H, or I for the possible orbital angular momentum quantum numbers $L = 0, 1, 2, 3, 4, 5$ and 6, respectively. $2S+1$ denotes the spin multiplicity, and J is the quantum number of the total angular momentum. Following Hund's rules, the electronic ground state configuration is the one with maximum spin multiplicity, and in cases of states with the same total spin quantum number, those having maximum orbital angular momentum. With regard to J, the electronic ground state has the minimum total angular momentum quantum number, $J = L-S$, for ions with less than half-filled shells, whereas it has the maximum total angular momentum quantum number, $L+S$, in case the shell is more than half-filled. From the values L, S and J, we can further calculate the Landé factors g, which determine the magnetic properties,

$$g = 1 + \frac{S(S+1) - L(L+1) + J(J+1)}{2J(J+1)} \tag{3.7.1}$$

https://doi.org/10.1515/9783110654929-022

Table 3.7.1 summarizes the free-ion ground state configurations, the *g*-factors and their predicted magnetic moments, $g_J\mu_B\{J(J+1)\}^{1/2}$ based on them. These values are contrasted with experimental magnetic moments, which in most cases are in good agreement with the predicted values. Two important exceptions are Eu^{3+} and Sm^{3+}, for which the experimentally observed magnetic moments are significantly influenced by low-energy excited electronic states, which become admixed to the ground state under the influence of an applied magnetic field.

Table 3.7.1: Free-ion ground state configurations, *g*-factors and predicted and experimental magnetic moments, $g_J\mu_B\{J(J+1)\}^{1/2}$, in units of the Bohr magneton and nuclear isotope characteristics of the rare-earth ions.

Ion	GS	g_J	$g_J\mu_B\{J(J+1)\}^{1/2}$		nuclear isotopes, (spin, natl. abundance)
			calc.	exp.	
La^{3+}	$^1S_0$	0	0	0	^{138}La (5, 0.1%), ^{139}La (7/2, 99.9%)
Ce^{3+}	$^2F_{5/2}$	6/7	2.54	(2.4)	none
Pr^{3+}	$^3H_4$	4/5	3.58	(3.5)	^{141}Pr (5/2, 100%)
Nd^{3+}	$^4I_{9/2}$	8/11	3.62	(3.5)	^{143}Nd (7/2, 12.2%), ^{145}Nd (7/2, 8.3%)
Sm^{3+}	$^6H_{5/2}$	2/7	0.84	(1.5)	^{147}Sm (7/2, 15%), ^{149}Sm (7/2, 13.8%)
Eu^{3+}	$^7F_0$	0	0	(3.4)	^{151}Eu (5/2, 47.8%), ^{153}Eu (5/2, 52.2%)
Eu^{2+}	$^8S_{7/2}$	2	7.94	(8.0)	^{151}Eu (5/2, 47.8%), ^{153}Eu (5/2, 52.2%)
Gd^{3+}	$^8S_{7/2}$	2	7.94	(8.0)	^{155}Gd (3/2, 14.8%), ^{157}Gd (3/2, 15.7%)
Tb^{3+}	$^7F_6$	3/2	9.72	(9.5)	^{159}Tb (3/2, 100%)
Dy^{3+}	$^6H_{15/2}$	4/3	10.63	(10.6)	^{161}Dy (5/2, 18.9%), ^{163}Dy (5/2, 24.9%)
Ho^{3+}	$^5I_8$	5/4	10.60	(10.4)	^{165}Ho (7/2, 100%)
Er^{3+}	$^4I_{15/2}$	6/5	9.59	(9.5)	^{167}Er (7/2, 22.9%)
Tm^{3+}	$^3H_6$	7/6	7.57	(7.3)	^{169}Tm (1/2, 100%)
Yb^{3+}	$^2F_{7/2}$	8/7	4.54	(4.5)	^{171}Yb (1/2, 14.3%), ^{173}Yb (5/2, 16.1%)
Lu^{3+}	$^1S_0$	0	0	0	^{175}Lu (7/2, 97.4%), ^{176}Lu (7, 2.6%)

The total angular moments associated with the *RE* ions' electronic ground states are listed in Table 3.7.1. Included are also the corresponding nuclear isotopes, their spin quantum numbers and natural abundances. The resulting electron and nuclear magnetic moments have important spectroscopic consequences that are relevant for structural characterization purposes, to be discussed in the present chapter: (1) they experience the Zeeman interaction in applied magnetic fields, resulting in energy splittings that form the basis of EPR spectroscopy, and (2) they influence the nuclear magnetic resonance (NMR) signals of nuclear spins in their vicinity, providing structure and bonding information. These two principal topics and their spectroscopic variants and opportunities, as well as some representative applications are the subject of the present chapter.

3.7.2 Rare-earth electron paramagnetic resonance

3.7.2.1 Basic concepts

General introductions into EPR spectroscopy with focus on transition metal and *RE* ions can be found in numerous comprehensive texts [1–6]. EPR is possible with any *RE* ion whose electronic ground states possess degeneracy. This degeneracy is removed by the *Zeeman interaction* between the magnetic moment associated with $\boldsymbol{J}$ and an applied magnetic field of magnetic flux density B_0. In the corresponding Hamiltonian

$$\mathcal{H}_z = -\mu_B J g_J B_0 \tag{3.7.2}$$

μ_B, the Bohr magneton, is a universal constant (9.24×10^{-24} J T^{-1}), and g_J is a second-rank tensor, defining the coupling between the electronic total angular momentum and the magnetic flux density. The energy levels defined by expression (3.7.2) are equally spaced and the transitions between them are governed by the magnetic dipole radiation selection rule $\Delta m_J = \pm 1$ (m_J being the orientational quantum number), leading to

$$h\nu = \mu_B g_J B_0 \tag{3.7.3}$$

For free ions, the *g*-factor is given by expression (3.7.1), whereas in the case of an ion incorporated in a solid, this is no longer the case. Customarily, however, one retains the form of eq. (3.7.1) by defining an *effective g tensor*, $\mathcal{g}_{eff}$, which describes the anisotropic interaction of an effective spin and the applied magnetic field. The effective spin approximation implies replacing the Hamiltonian of the ion in the lattice, with all its states and energies, by a simplified Hamiltonian that considers only the lowest-lying states. The relation of $\mathcal{g}_{eff}$ with the local atomic and electronic environment and symmetry is complex, and needs to be explored by quantum chemical calculations.

The energy of a paramagnetic ion in a solid is a complicated function of the coordinates and spins of the unpaired electrons in the ions and the coordinates and charges surrounding it in the lattice environment producing the crystalline field. The ground state term (defined by the spin-orbit coupling) is itself split by the *crystal field interaction*, yielding a lowest level whose wave functions depend on the point symmetry of the crystal field. The total Hamiltonian then is comprised of the Zeeman interaction, the crystal field interaction and the *nuclear hyperfine interaction* of the effective electron spin with nuclear magnetic moments associated with the atoms carrying the unpaired electrons and those of their directly bonded neighboring atoms.

In the effective spin formulation, we write the total spin Hamiltonian as

$$\mathcal{H}_{\mathrm{total}} = \mathcal{H}_{\mathrm{Z}} + \mathcal{H}_{\mathrm{CEF}} + \mathcal{H}_{\mathrm{hf}} \tag{3.7.4}$$

where

$$\mathcal{H}_{\mathrm{Z}} = -\mu_{\mathrm{B}}\, S\, g_{\mathrm{J}}\, B_0 \tag{3.7.5}$$

$$\mathcal{H}_{\mathrm{CEF}} = \Sigma_{n<2S}\, \Sigma_m\, B^m{}_n O^m{}_n \tag{3.7.6}$$

$$\mathcal{H}_{\mathrm{hf}} = S\mathcal{A}I \tag{3.7.7}$$

Within the Zeeman interaction term, g_{J} is a tensor whose components depend on the molecular structure and $\boldsymbol{S}$ is the effective spin of the lowest electronic level. Owing to the anisotropy of the spin Hamiltonian, the energy of a spectroscopic transition depends on the orientation of B_0 relative to the principal axes of the crystal field. The crystal field interaction term $\mathcal{H}_{\mathrm{CEF}}$ is compactly expressed in terms of numerical coefficients $\boldsymbol{B}^m{}_n$ and the so-called Stevens Operators $\boldsymbol{O}^m{}_n$ acting on the spin angular momentum wave functions. Only even n need to be considered, where the quantum number of orientation, m, can take the possible values $\{n, n-1, \ldots, -n\}$. The relevant coefficients and operators to be included in $\mathcal{H}_{\mathrm{CEF}}$ have been tabulated for various site symmetries [1, 2, 6]. Finally the electronic energy levels and hence the resonance condition are further modified by the interaction between the electronic angular momentum and the nuclear spins present in the atom under consideration, the so-called *nuclear hyperfine interaction*, where $\boldsymbol{I}$ represents the nuclear spin operator and $\mathcal{A}$ represents the hyperfine interaction tensor. The latter includes an isotropic part, reflecting the Fermi contact interaction between the nuclei and s-electron density at the nuclear sites and an anisotropic part, originating from through-space magnetic dipolar interactions. The latter part is responsible for inhomogeneous broadening contributions to the EPR line shape on polycrystalline and amorphous materials.

Table 3.7.1 includes the spin quantum numbers and natural abundances of the relevant RE nuclear isotopes. The majority of EPR experiments of the *RE* ions have been conducted with RE ions presenting an odd number of unpaired electrons, the so-called *Kramers ions*, which are Ce^{3+}, Nd^{3+}, Sm^{3+}, Eu^{2+}, Tb^{4+}, Gd^{3+}, Dy^{3+}, Er^{3+} and Yb^{3+}. Their electronic ground states possess $J + ½$ doubly degenerate energy levels and at sufficiently low temperatures only the doublet corresponding to the J - value with the lowest energy is thermally occupied. In external magnetic fields the Zeeman interaction produces energy level splittings equivalent to frequencies in the microwave region. In special situations, the non-Kramers ions (even number of f-electrons), Pr^{3+}, Tb^{3+}, Ho^{3+} and Tm^{3+} also can give rise to EPR spectra, even if their ground states are not doubly degenerate [7]. There is, however, no possibility for EPR on *RE* ions with diamagnetic electronic ground states, such as the closed shell ions La^{3+} and Yb^{2+}, Lu^{3+} and the $J = 0$ species Eu^{3+}.

3.7.2.2 Experimental aspects

Equation (3.7.3) predicts that at magnetic field strengths on the order of 1 T, resonance absorption will occur in the microwave region. The most commonly used microwave (MW) frequencies for EPR spectroscopy have been the X-band (9.5 GHz) and the Q-band (34.5 GHz), corresponding to magnetic field strengths of 0.34 and 1.25 T, respectively. To the present day, the majority of the EPR spectra are being measured in the continuous-wave, *field-sweep mode*, in which a microwave detector signal is monitored under continuous microwave irradiation and linear variation of the magnetic field (*continuous-wave* (cw-) EPR). Figure 3.7.1 shows a typical sketch of the hardware involved. The key components are the microwave source (a klystron or diode device), the microwave cavity containing the sample within the magnet, and the detector, which in standard equipment is a rectification device converting the microwaves into *dc* current. The construction of the cavity serves the purpose of maximizing the amplitude of the oscillating magnetic field component associated with the microwave at the sample locus. Source, detector and cavity are connected via waveguides (usually copper tubes) to the *microwave bridge*, which is balanced such that in the absence of absorption (off-resonance) no energy reaches the detector. If, at a

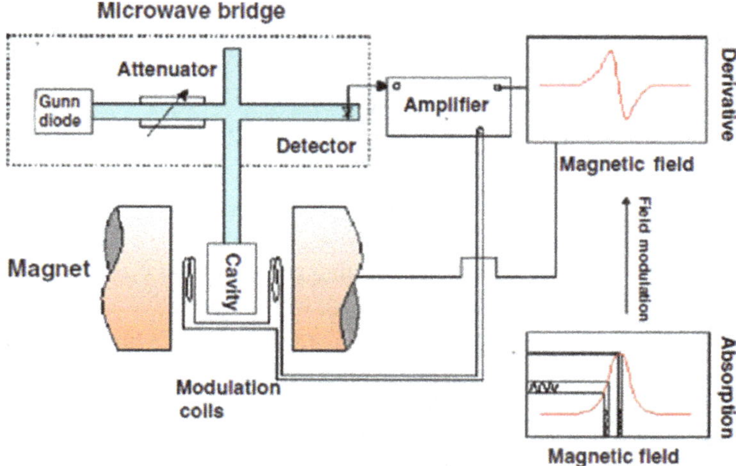

Figure 3.7.1: Schematic of a typical continuous-wave EPR spectrometer.

given value of the magnetic field resonance absorption occurs within the cavity, the bridge is thrown off-balance and a current is registered. Usually the field sweep is superimposed by a rapid (10-100 kHz) field modulation, which is synchronized with the detection (*lock-in detection*). In this manner the registered signal is the first derivative with respect to the magnetic field, which produces a large gain in the signal to noise ratio. The amplitude of the modulating field has to be chosen judiciously to avoid loss of resolution.

Starting in the 1970s and during the following decades time domain EPR spectroscopy was developed as a viable (and ultimately more powerful) alternative, following a parallel development in NMR spectroscopy. In this method, free induction decay signals are generated – usually in the form of spin echoes – by short, intense microwave pulses, and subsequently converted to EPR absorption spectra by Fourier-transformation. As the width of the overall line shape generally exceeds the excitation window of the microwave pulses this method is usually combined with linear field variation (*echo detected field sweep*, EDFS) [8]. During the 1980s and 1990s time-domain EPR spectroscopy also facilitated the extension of EPR to higher magnetic field strengths, and the emergence of versatile multi-pulse experiments creating a powerful spectroscopic toolbox for tailoring the relevant Hamiltonian and enhancing considerably the informational content of EPR spectroscopy.

RE EPR signals are most typically observed on samples containing the paramagnetic ion in a magnetically diluted state (concentrations between 0.01 to 1 mole %). Higher concentrations produce excessive line broadening due to strong magnetic dipole-dipole interactions, leading to a loss in resolution and short spin-spin relaxation times hampering signal detection. In addition, the coupling of their orbital angular momentum component to low-frequency vibrational modes via various distinct mechanisms produces extremely rapid spin-lattice relaxation and thus excessive Heisenberg line broadening at ambient temperature. To freeze out these relaxation mechanisms, rather low temperatures, typically $T < 20$ K, are required, necessitating a helium cryostat. An exception are the *S*-state ions Eu^{2+}, Gd^{3+}, and Tb^{4+} whose EPR spectra can be recorded at room temperature.

3.7.2.3 Continuous-wave measurements on single crystals

Once the spectrum has been recorded, it must be analyzed in terms of the various parameters characterizing the Hamiltonian, eq. (3.7.4). In the special case of the *S*-state ions Eu^{2+}, Gd^{3+}, and Tb^{4+}, the *g*- and magnetic hyperfine tensors are isotropic (owing to the absence of an orbital contribution), simplifying the numerical analysis of the spectra in terms of the crystal field parameters. For obtaining sufficient constraints for these multi-parameter fits, one needs to work with single crystals, whose signal positions are measured as a function of their orientation in the magnetic field. Along these lines, from the very beginnings of EPR up to the present date, spectra of *RE* dopants have been measured to characterize crystal fields in different point symmetries of a multitude of host structures [1, 2, 6, 9, 10].

EPR work on non-S-state ions has been frequently motivated by the desire to understand *RE* emission characteristics in important photonic devices such as luminescent phosphors [11–17] and laser materials [18–22]. The spin Hamiltonians are much more complex and simplified analysis approaches are chosen. The EPR spectra of such species are generally analyzed in terms of an effective g tensor, even for single-crystal specimens. Their relation with crystal field parameters has been explored, as for instance discussed in the case of Nd^{3+} ions in various garnet host structures [22]. In addition, magnetic hyperfine couplings to nuclei of spin quantum number I manifest themselves as multiplets of $2I + 1$ equally intense lines, reflecting the natural abundance of the isotope. If the element in question possesses different nuclear magnetic isotopes (Nd, Sm, Eu, Gd, Dy, Yb), the ratio of the hyperfine coupling constants can be predicted from the known nuclear gyromagnetic ratios. Figure 3.7.2 shows a typical result on an Yb-doped $ZrSiO_4$ crystal with the magnetic field oriented parallel to the tetragonal *c*-axis, highlighting the hyperfine splittings caused by the ^{171}Yb (I = ½) and ^{173}Yb (I = 5/2) isotopes [17]. For nuclei with spin > ½, the spectra may be additionally influenced by nuclear electric quadrupolar interactions, which must be included in the analysis.

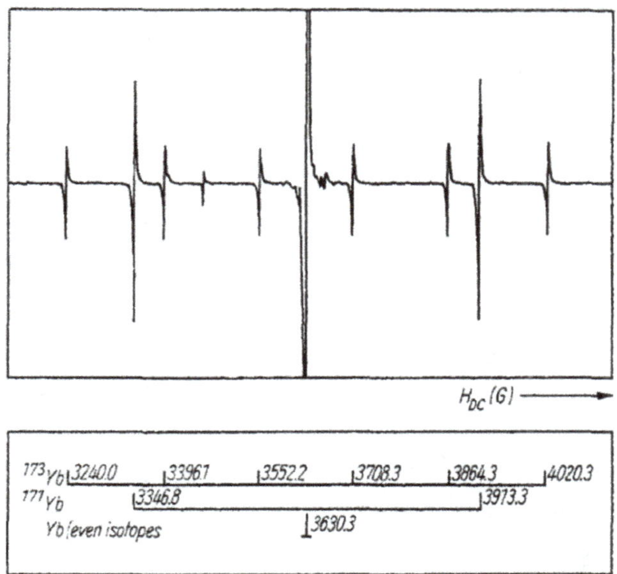

Figure 3.7.2: Q-band EPR spectrum of a Yb^{3+}-doped single crystal of tetragonal $ZrSiO_4$ at 15 K, with the magnetic field oriented parallel to the tetragonal *c*-axis and schematic illustration of the magnetic hyperfine splittings by the ^{171}Yb and ^{173}Yb isotopes. Reproduced from [17] with permission from Wiley.

3.7.2.4 EPR experiments on disordered materials

Despite the restrictions hampering a complete spin-Hamiltonian analysis, applications to poly-, nano-, semi- and non-crystalline solids have been reported in recent years [23–28], motivated by potential applications of these materials in photonics and lasers. For all non-S-state ions, excessive line broadening effects are observed, which is not only caused by *g*- and *A*-anisotropies, but also by a distribution of the tensor parameters which result from variations in the local environments produced by disorder. As very broad lines are difficult to detect via the continuous wave method one usually turns to pulsed excitation, such as the **E**cho **D**etected **F**ield **S**weep (EDFS) method [8], where the amplitude of an echo signal generated by a 90°-τ-180°-τ pulse sequence is recorded during linear field variation. In the absence of single crystals, the fundamental analysis of these powder line shapes in terms of the Hamiltonian parameters of eq. (3.7.4) would necessitate measurements at multiple magnetic field strengths. In practice such work, which would involve systematic comparisons of X-band, Q-band, W-band, etc. spectra has rarely been carried out. Rather, phenomenological analyses treating the spectra essentially as *fingerprint data* are more commonplace. For example, Figure 3.7.3 shows recent results on a series of Yb^{3+} doped *RE* fluoride phosphate glasses with different fluoride/phosphate ratios: with increasing F content of the glass, the signal measured at 9.5 GHz is shifted to successively higher magnetic field strengths, hence reflecting a systematic decrease in the effective *g*-value. Correlated changes are observed for the NMR parameters of diamagnetic *RE* mimics (^{45}Sc) and in the emission behavior of Eu^{3+} dopants, leading to the conclusion that the EPR spectra are monitoring systematic changes in the ligand distribution around the *RE* dopants [29] in this system.

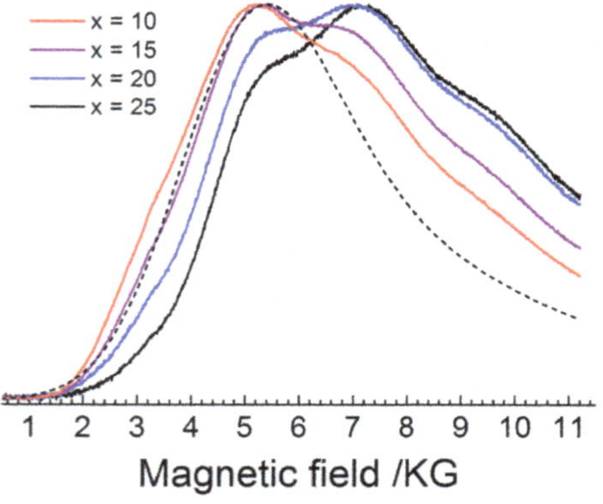

Figure 3.7.3: Pulsed EPR line shapes for fluorophosphate glasses with composition $25BaF_2$-$25SrF_2$-(30-*x*)$Al(PO_3)_3$-*x*AlF_3-$19.75YF_3$-$0.25YbF_3$.The black dashed curve is a simulated spectrum typically observed in pure phosphate glasses [26]. Note the systematic signal displacement towards higher magnetic field strength with increasing fluoride content.

3.7.2.5 Hyperfine interaction spectroscopy

Pulsed EPR methods offer the additional advantage of detecting nuclear hyperfine and magnetic dipole-dipole interactions via the electron spin echo envelope modulation (ESEEM) technique [8]. In this method, the amplitude of a stimulated echo, recorded by the three-pulse sequence shown in Figure 3.7.4 is recorded under systematic incrementation of the evolution time *T*. While it is expected that this signal decays exponentially as a function of *T* owing to spin-spin relaxation, these decays contain additional information if the electron spins experience magnetic hyperfine couplings or dipolar interactions with nearby nuclear magnetic moments. In such cases, the decay is modulated by characteristic frequencies reflecting a combination of the nuclear Zeeman frequencies and the electron-nuclear hyperfine coupling constants. By Fourier transformation of this time domain signal the modulation frequencies can be obtained. The method has been applied to various *RE* doped glass systems [29–38]. In the weak coupling limit (zero hyperfine coupling constant) the modulation behavior of the electron spin echo is dominated by the (known) nuclear Zeeman frequencies. In this way, nuclei found nearby the *RE* ion (typical distance range 300 to 800 pm) can be identified by their modulation frequencies in the ESEEM spectrum. Although the amplitude and shape of the ESEEM signal depends on many parameters (Zeeman frequency, distance, number of interacting nuclei,

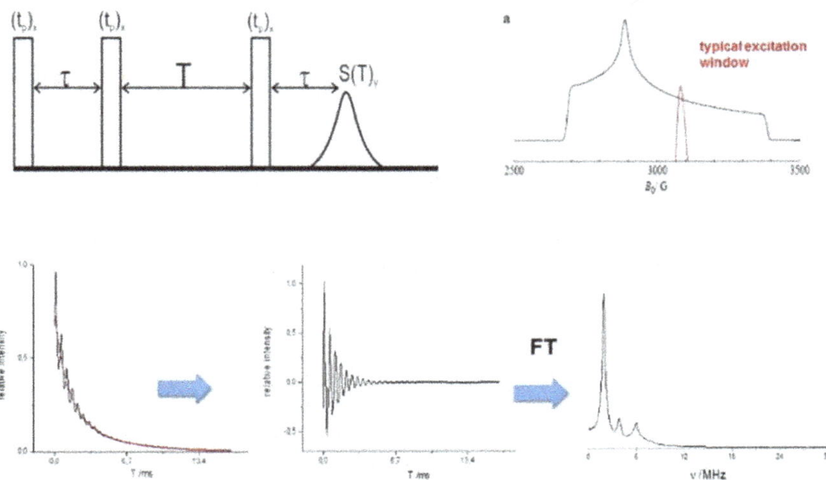

Figure 3.7.4: Principle of electron spin echo envelope modulation (ESEEM) spectroscopy: An isochronous set of spins resonating at a selected magnetic field strength is subjected to a stimulated spin echo sequence, and the echo amplitude is recorded as a function of the mixing time *T*. Further data processing involves removal of the exponential decay and Fourier transformation of the oscillating signal, resulting in the frequency components of the modulation.

quadrupolar interactions) in a complex way, measurements as a function of glass composition reveal that the peak intensity ratios produced by different nuclear species can be correlated with their relative concentration ratios. An example is given in Figure 3.7.5, showing ESEEM spectra of a series of Yb^{3+} doped yttrium fluoride phosphate glasses, where the systematic evolution of the $^{31}P/^{19}F$ peak intensity ratio as a function of composition is consistent with a homogeneous distribution of the Yb^{3+} dopant and a continuous structural evolution of the glass network structure as the P/F ratio in the glass composition is successively changed. In contrast, deviations from monotonic trends in intensities may hint at structural inhomogeneities such as clustering or phase separation.

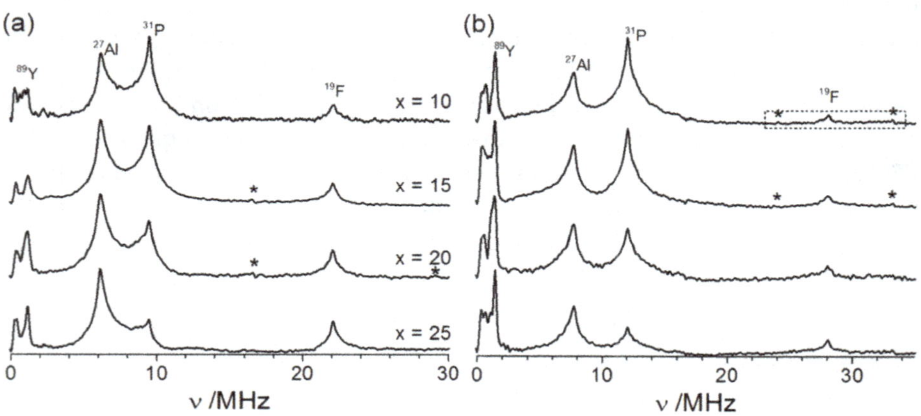

Figure 3.7.5: ESEEM spectra of glasses in the system $25BaF_2$-$25SrF_2$-(30-x)$Al(PO_3)_3$-xAlF_3-19.75YF_3-0.25YbF_3 at centerfield values of (a) 4.0 mT and (b) 7.0 mT. Zeeman frequencies of the various types of nuclear species detected are indicated. Yb-F covalent bonding producing magnetic hyperfine coupling is suggested by weak satellite peaks labeled with asterisks.

Additional information about the magnetic hyperfine coupling tensor is available using the two-dimensional hyperfine sublevel correlation (HYSCORE) method [8]. In this technique, two evolution times t_1 and t_2 are defined separated by a constant mixing period T. Fourier transformation with respect to these evolution times produces a two-dimensional spectrum, in which the diagonal peaks contain the same information as the regular ESEEM spectra. In the presence of non-zero magnetic hyperfine coupling, however, coherence transfer between spin populations takes place, such that some spins evolving with a frequency ω_1 during the evolution time t_1 now evolve with a different frequency ω_2 during the second evolution period t_2. Only those spins experiencing hyperfine coupling with the electrons are able to undergo this change, however, this coherence transfer then results in off-diagonal cross-peaks in the corresponding 2D-spectra (Figure 3.7.6). The shape of these cross-peaks can be further analyzed in terms of isotropic and anisotropic hyperfine coupling tensor components, using simulation programs such as the EASYspin software [39]. Their appearance constitutes direct evidence for hyperfine coupling interactions associated with covalent bonds between the *RE* atom carrying the unpaired electrons and the ligands whose nuclear spins are detected. In the structural analysis of fluoride phosphate glasses these off-diagonal signals prove the formation of direct Yb^{3+}–F bonds.

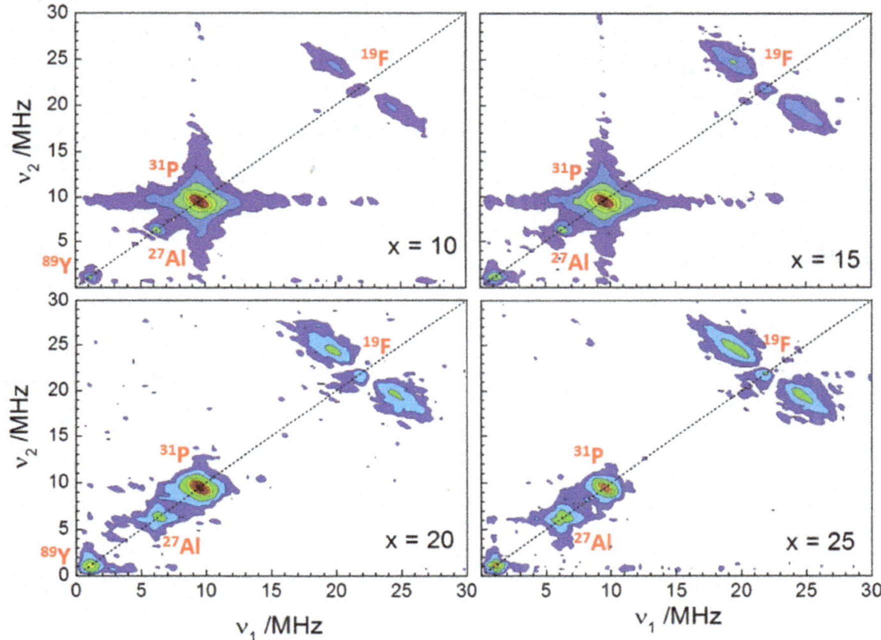

Figure 3.7.6: 2D-HYSCORE spectra of glasses in the system $25BaF_2$-$25SrF_2$-(30-*x*)$Al(PO_3)_3$-*x*AlF_3-$19.75YF_3$-$0.25YbF_3$. Diagonal signals belonging to the various types of nuclear species detected are indicated. Yb-F covalent bonding is proven by the off-diagonal signals near the ^{19}F Zeeman frequency.

3.7.2.6 Distance measurements by pulsed electron double resonance

Pulsed EPR methodology furnishes a variety of selective averaging experiments, allowing for a simplification of the spin Hamiltonian to address structural questions in a more selective fashion. For example, electron-electron dipole-dipole interactions can be selectively addressed by a method denoted **D**ouble-**E**lectron

Electron **R**esonance (DEER), sometimes also called **P**ulsed **El**ectron **Do**uble **R**esonance (PELDOR). As the strength of the magnetic dipole-dipole interaction is proportional to the inverse cubed inter-electronic separation, this method is extremely useful for distance measurements. In recent years, DEER has been used for determining the distances between strategic sites in biological macromolecules such as proteins, nucleic acids, and their assemblies [40–45]. Figure 3.7.7 illustrates the principle. A spin echo is generated on spins A by microwave pulses on resonance with the observation frequency ν_A, whereas spins B are selectively inverted by a π pulse on resonance with a different frequency $\nu_B = \nu_{pump}$. If the frequency separation $|\nu_A - \nu_B|$ is larger than the dipolar coupling constant in frequency units, the B spin inversion changes the resonance frequency of spins A. As a result the A-spin signal dephasing that has occurred up to this point can no longer be completely reversed by the final π pulse, and we see a diminution of the spin echo amplitude. The variable t, defining the timing of the pump pulse, is then systematically incremented and the A spin echo amplitude is monitored as a

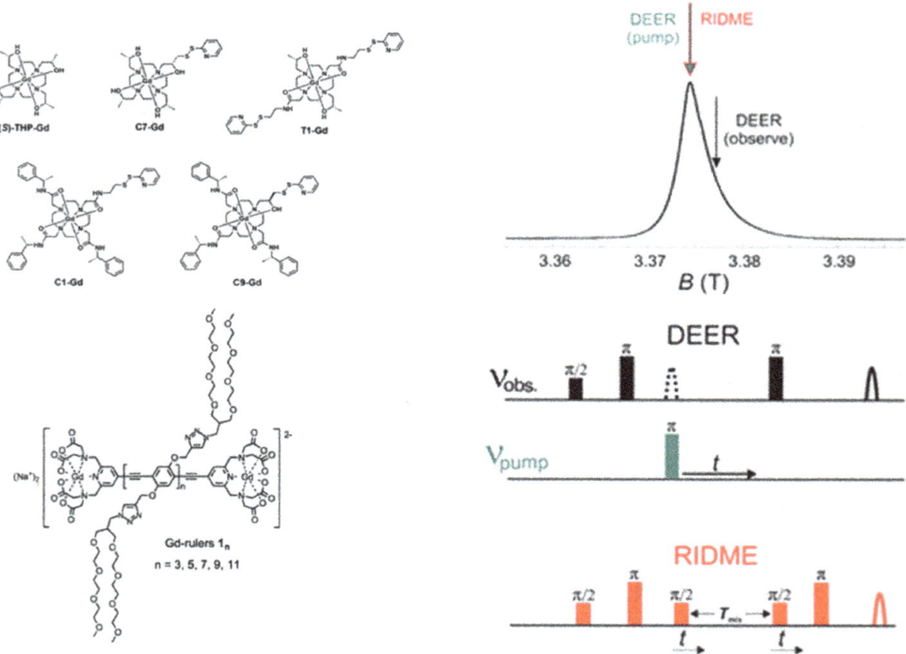

Figure 3.7.7: Left: Gd^{3+} spin markers used for DEER experiments. Right: Pulse sequences for the DEER and RIDME experiments. Delays between the pulse on the observe channel are kept fixed, while the echo intensity is measured as a function of the delay t between the (unobserved) first primary echo (dotted line) and the pump pulse. Reproduced from references [41–43] with permission from Elsevier and the American Chemical Society.

function of t. A two-spin system produces an oscillatory response in the echo intensity, from which the dipolar coupling constant can be extracted by simulation, which yields the inter-electronic distance.

At the heart of this methodology lies the availability of well defined paramagnetic probes between which the distances are measured. While earlier applications have mostly used nitroxide spin labels, Gd^{3+} complexes have become increasingly popular for such applications in recent years, owing to their facile handling, their simple spin dynamics and short relaxation times. Figure 3.7.7, left, shows the structures of typical Gd^{3+} spin tags and the pulse sequences used for these distance measurements. The principal drawback of the Gd^{3+} probe is the overall small modulation depth caused by the limited excitation window of the pump π pulse, compared to the overall width of the inhomogeneously broadened EPR spectrum. Thus, only a small fraction of the Gd^{3+} spins are being inverted by the pump pulse. In recent years, this difficulty has been overcome by the relaxation induced dipolar modulation enhancement (RIDME) sequence (Figure 3.7.7, bottom right), which avoids double irradiation, and replaces the pump pulse by a $\pi/2 - t_{mix} - \pi/2$ saturation recovery period. Both DEER and RIDME can measure Gd–Gd interatomic distances of up to 8 nm and have been widely used in dozens of insightful applications for structural studies in biological solids. Figure 3.7.8 shows a typical result obtained by

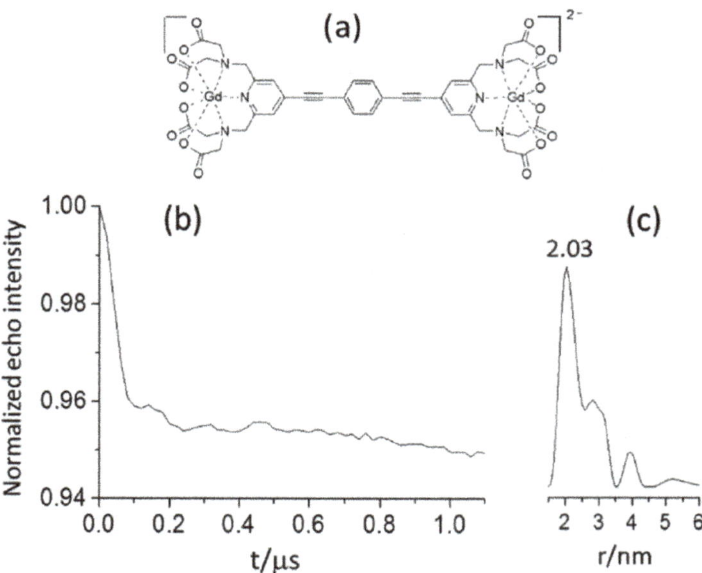

Figure 3.7.8: a) Structure of the dimeric gadolinium complex, (b) DEER time domain response, and (c) Gd-Gd distance distribution extracted from the DEER measurement. W-band measurement at 25 K. Reproduced from reference [45] with permission from the Royal Society.

the application of DEER to a dimeric gadolinium complex, in which the Gd-Gd distance is held fixed by a rigid spacer.

3.7.3 The effect of 4f paramagnetism upon NMR spectra of proximal nuclei

3.7.3.1 Fundamental concepts

The paramagnetism of unpaired f-electrons associated with *RE* ions also has a profound effect on liquid- and solid state NMR spectra of the nuclei in their proximity, including chemical shift displacements, line broadening effects and shortened spin-lattice and spin-spin relaxation times. These effects originate from the magnetic dipolar and hyperfine interactions between the unpaired electrons and the observed nuclei (already discussed in Section 3.7.2.1. with respect to its effect upon EPR spectra) and bulk susceptibility effects. These interactions potentially yield important structure and bonding information, the spatial atomic arrangement, the dynamics of the system, and, particularly relevant in the case of lanthanide ions, details about the crystal-field splitting and the resulting optical properties. Numerous theoretical descriptions and reviews have appeared in the literature [46–51]. Here we briefly summarize the key concepts from the excellent presentation of Pell et al. [46], focusing on the effect of unpaired electrons upon the NMR precession frequencies ω_P. A comprehensive treatment of relaxation phenomena can be found there as well and in other relevant literature. We start from the original Hamiltonian,

$$\widehat{H} = \hbar\omega_{0,I}\widehat{I}_z + \mu_B g_e B_0 \widehat{S}_z + \hat{S}\cdot A\cdot\hat{I} \tag{3.7.8}$$

where the first, second, and third terms describe the Zeeman interactions of the nuclear spins *I*, the electron spins *S* and the nuclear-electron hyperfine interactions and $\omega_{0,I} = \gamma B_0$ is the precession frequency of the bare nucleus. Here, the modification of the nuclear precession frequency ω_p by the third term is of interest. The effect of electronic environments upon ω_P is generally anisotropic, *i. e.* different in the three different directions of a molecular axis system. Owing to this anisotropy we must describe it by a tensor σ, the so-called *magnetic shielding tensor*, such that

$$\mathscr{H} = \gamma B_0(1 - \sigma)I_z \tag{3.7.9}$$

leading to

$$\omega_p = \gamma B_0 (1 - \sigma) \tag{3.7.10}$$

Note that this tensorial description implies that ω_p will in general depend on the orientation of the molecular axis system relative to the magnetic field direction, and hence produce inhomogeneous line broadening in polycrystalline samples. In diamagnetic compounds σ is largely temperature independent. It comprises a shielding effect from electron circulation in closed inner shells and a deshielding effect from the orbital angular momentum of excited electronic states that are admixed to the electronic ground state under the influence of the applied magnetic field. In paramagnetic compounds the tensor σ includes an additional term σ^S describing the effect of the electron-nuclear hyperfine interaction. For understanding the nature of σ^S it is important to keep in mind that electronic relaxation occurs on a timescale that is orders of magnitude shorter than nuclear relaxation, so that during the observation of the nuclear-spin transitions the electronic spins are effectively sampling all the electronic energy levels according to their equilibrium configuration. This means that, in effect, the hyperfine interaction couples the nuclear magnetic moment to the thermal average of the electronic magnetic moment. Thus, starting from the original Hamiltonian (3.7.8) the third term describing the nuclear-electron hyperfine interaction must be actually replaced by the expression,

$$\widehat{H}_{SI} = \left\langle \hat{S} \right\rangle \cdot A \cdot \hat{I} \tag{3.7.11}$$

where $\left\langle \hat{S} \right\rangle$ is the thermal (ensemble) average of the electron spin vector, given by the Boltzmann distribution. This effect produces a (generally anisotropic) magnetic shielding or de-shielding contribution influencing the resonance frequency of the nuclei that is inversely temperature dependent, which is given by

$$\sigma^S = \frac{\mu_B g_e S(S+1)}{3\hbar\gamma_I kT}\boldsymbol{A}. \tag{3.7.12}$$

in analogy to the well-known Curie law in the high-temperature approximation of the magnetic interaction energy. In expression (3.7.12) $\boldsymbol{A}$ is a tensor, consisting of an isotropic part (Fermi-contact interaction), $\boldsymbol{A}^{FC}$, and an anisotropic (spin-dipolar) part, $\boldsymbol{A}^{SD}$. These are given by the expressions,

$$\begin{aligned} A^{FC} &= \frac{\mu_0\mu_B g_e\hbar\gamma_I}{3S}\rho^{\alpha-\beta}(0), \\ A_{ij}^{SD} &= \frac{\mu_0\mu_B g_e\hbar\gamma_I}{8\pi S}\int\frac{3r_ir_j - \delta_{ij}r^2}{r^5}\rho^{\alpha-\beta}(r)d^3r \end{aligned} \tag{3.7.13a, b}$$

where i and j denote the cartesian components x, y, or z, and δ_{ij} is the Kronecker delta. The term $\rho^{\alpha\text{-}\beta}(r)$ is the total spin-unpaired electron density at position r, and the number of unpaired electrons is given by $2S$. The Fermi-contact coupling constant is proportional to the total unpaired electron spin density transferred to the nucleus, $\rho^{\alpha\text{-}\beta}(0)$, *i. e.* the sum of the contributions from the individual electrons. It is completely isotropic. In contrast, the spin-dipolar contribution A_{ij}^{SD} is a second rank tensor whose isotropic average is also known as the *„pseudo-contact shift"* measured in solution. For the special case of the lanthanide ions with 4f open shell paramagnetism and J being the relevant quantum number, the theory of σ^J developed by Bleaney [49] results in the expression,

$$\sigma^J \approx -\frac{\mu_B g_J J(J+1)}{3\hbar\gamma kT} A + \frac{\mu_B g_J J(J+1)(2J-1)(2J+3)}{30\hbar\gamma(kT)^2} B \cdot A \tag{3.7.14}$$

where the tensor $\boldsymbol{B}$ relates to the crystal field splitting parameters. As pointed out in reference [49], the contact and spin-dipolar contributions result in four temperature dependent contributions, listed in Table 3.7.2.

Table 3.7.2: The terms present in the paramagnetic shielding tensor in a lanthanide system, expressed in terms of both the molecular/atomic-level parameters σ^J, and the bulk magnetic susceptibility tensor σ^χ.

Type	Term	σ^J	σ^χ	Rank
Contact	1	$-\frac{\mu_B g_J J(J+1)}{3\hbar\gamma kT} A^{con}$	$-\chi^{(1)} C^{con}$	0
	2	$\frac{\mu_B g_J J(J+1)(2J-1)(2J+3)}{30\hbar\gamma(kT)^2} B \cdot A^{con}$	$-\chi^{(2)} C^{con}$	2
Dipolar	3	$-\frac{\mu_B g_J J(J+1)}{3\hbar\gamma kT} A^{dip}$	$-\chi^{(1)} C^{dip}$	2
	4	$\frac{\mu_B g_J J(J+1)(2J-1)(2J+3)}{30\hbar\gamma(kT)^2} B \cdot A^{dip}$	$-\chi^{(2)} \cdot C^{dip}$	0,1,2

Term (1), the so-called *Fermi-contact shift*, is frequently dominant, and manifests itself in significant positive or negative signal displacements. An additional source of signal shift arises from the isotropic component of term 4, the spin-dipolar contribution, known as *pseudo-contact shift*. The latter is calculated by Bleaney's theory [49] according to

$$\sigma_{iso}^{pcs} = \frac{\mu_0 \mu_B^2 g_J^2 J(J+1)(2J-1)(2J+3)}{360\pi(kT)^2 R^3} \times \left[\Delta B_{ax}(3\cos(\theta)-1) + \frac{3}{2}\Delta B_{rh}\sin^2(\theta)\cos(2\phi)\right] \tag{3.7.15}$$

where R is the electron-nuclear distance, and ΔB_{ax} and ΔB_{rh} define the axial and rhombic crystal field tensor components, while θ and ϕ are the polar and azimuthal angles describing the orientation of the susceptibility tensor relative to the dipolar coupling tensor. Note the bilinear inverse temperature dependence of this term, differentiating it from the Fermi-contact interaction.

3.7.3.2 Paramagnetic NMR in solution

3.7.3.2.1 Fermi contact shifts and pseudo-contact shifts

As mentioned above, in liquid solutions all anisotropic parts of the Hamiltonian listed in Table 3.7.2 are averaged to zero, leaving only the Fermi contact and the pseudo-contact shift contributions. The *Fermi contact shift* is transmitted through chemical bonding. In the simplest case the unpaired electron produces some unpaired spin density probability at the nucleus under consideration by polarizing the paired electrons in the molecular orbitals involved in metal-ligand bonding. Depending on the number and type of orbitals involved in this mechanism the sign of the spin density may change from one atom to the other. This often leads to alternating signs of signal displacements when passing from one atom to its neighbor. Significant *pseudo-contact shifts* are generally expected when there are energy levels close to the ground state. In this case, their orbital contributions to the ground state contribute to an anisotropy of the g-tensor. Detailed discussions on how to separate the Fermi contact and pseudo-contact contributions to the paramagnetic shifts are found in the literature [46, 47, 49, 50]. Among the numerous applications of paramagnetic NMR in solution the seminal work of the Bertini group on metalloproteins deserves to be highlighted specifically [50, 51].

3.7.3.2.2 NMR shift reagents

Paramagnetic shifts are frequently exploited to simplify liquid state NMR spectra (particularly of ^{1}H nuclei) of complex molecules, which are often complicated by signal crowding, particularly if the spectra are recorded at low magnetic field strengths. If the molecule contains nucleophilic groups (halogen, alcohol, carbonyl group, sulfur, nitrogen, etc.) capable of metal ion binding or complexation the presence of *RE*-containing *paramagnetic shift reagents* in the solution can produce large changes in the chemical shifts towards higher frequencies. Following the initial publication in 1969 [52], the effect was frequently used in subsequent decades to simplify spectra and make them more interpretable [53, 54]. Most commercially available shift reagents are paramagnetic chelates of europium (Eu) and ytterbium (Yb) with the diketones 2,2,6,6-tetramethyl-3,5-heptanedione and 1,1,1,2,2,3,3-heptafluoro-7,7-dimethyl-4,6-heptanedione, and camphor derivatives (see Figure 3.7.9), but Pr^{3+} and Er^{3+} complexes are also in use.

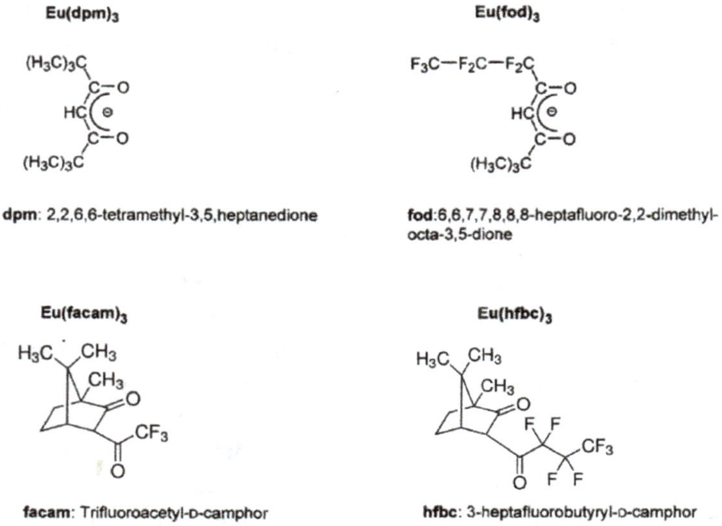

$Eu(dpm)_3$

dpm: 2,2,6,6-tetramethyl-3,5,heptanedione

$Eu(fod)_3$

fod:6,6,7,7,8,8,8-heptafluoro-2,2-dimethyl-octa-3,5-dione

$Eu(facam)_3$

facam: Trifluoroacetyl-D-camphor

$Eu(hfbc)_3$

hfbc: 3-heptafluorobutyryl-D-camphor

Figure 3.7.9: Molecular structures of some paramagnetic shift reagents. Modified from reference [53] with permission from Elsevier.

The *RE* ion is usually six-coordinated in these complexes and still possesses considerable electrophilic character. Thus, when a nucleophilic molecule is mixed with the solution of the *RE* complex such as $Eu(fod)_3$, a reversible complex can be formed, *e. g.*

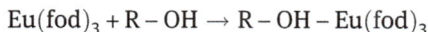

$$\mathrm{Eu(fod)_3 + R-OH \rightarrow R-OH-Eu(fod)_3}$$

The „on-off" equilibration is very fast on the NMR time scale such that a time-averaged spectrum is measured. As increased amounts of shift reagent are added, the above equilibrium is progressively shifted to the right. Consequently, the paramagnetic shifts become stronger, and it is generally useful to plot the chemical shifts for all the signals against the concentration of the added shift reagents. The dominant interaction mechanism is the *pseudo-contact shift*. The signal displacement caused by the dipolar mechanism for a particular nucleus of interest depends both on its distance *r* from the metal ion and the orientation of this vector relative to the molecular magnetic susceptibility tensor. In the simplest case the magnetic susceptibility tensor is axially symmetric with its principal axis parallel to the *RE*-nucleophile bond. In this case the paramagnetic shift is given by

$$\Delta\delta = K\frac{3\cos^2\Theta_a - 1}{r^3} \tag{3.7.16}$$

where K is proportional to the magnetic moment of the *RE* ion and Θ_a is the angle between the M–O bond and the electron-nucleus distance vector. More general situations of other magnetic axis orientations and asymmetric susceptibility tensors have been discussed as well [53, 54] and produce more complicated expressions. While the development of high-field, multidimensional NMR techniques has greatly reduced the need for shift reagents, there are still examples where the shift reagents can be successfully used. In particular, chiral shift reagents are useful to separately identify the resonances of the two enantiomeric forms of chiral molecules as the corresponding interaction complexes are diastereomeric. Thus, NMR can be conveniently employed to judge stereochemical purity of enantiomer [55].

3.7.3.3 Paramagnetic shifts in the solid state

The study of paramagnetic shifts in the solid state started in earnest with the seminal work of Grey et al., who investigated the effect in pyrochlores of composition $RE^{3+}{}_{y}Y_{2-y}M_2O_7$ (M = Ti, Sn, RE = La-Lu, $y \sim 0.2$), where the RE^{3+} ions substitute randomly on the Y^{3+} sites. The proximity of open-shell RE^{3+} ions to the nuclei on the Wyckoff sites of the constituent Y^{3+} and Sn^{4+} ions causes large paramagnetic shifts in the high-resolution ^{119}Sn and ^{89}Y MAS-NMR spectra, which are proportional to the magnetic moment of the *RE* ion. A strict additivity was found, i. e. the shift is proportional to the number of lanthanide ions substituted for yttrium in the coordination shell considered. In other words an ^{89}Y (or ^{119}Sn) nucleus next to two paramagnetic ions experiences twice the signal displacement as an ^{89}Y (or ^{119}Sn) nucleus next to only one. Thus, by counting signal intensities, the concentration of paramagnetic *RE* ions in the diamagnetic phase can be deduced and the mode of distribution (ordered or random) can be tested. The direction and magnitude of the shifts induced by the presence of paramagnetic ions suggest that there could be a significant contribution from the *pseudo-contact* mechanism in the case of the ^{89}Y NMR spectra, whereas for the ^{119}Sn MAS-NMR spectra of the stannates the data appear more consistent with a dominant *Fermi contact* interaction [56–58]. The temperature dependence of the paramagnetic shifts has been frequently exploited for temperature calibration in MAS-NMR experiments, as the actual temperature of the sample within the rotor may deviate substantially from that sensed by the thermocouple in the probe. In this respect the paramagnetic compounds in which the Fermi contact term dominates are preferred candidates as the latter produces a simple reciprocal temperature dependence of the chemical shift. Among the various „chemical shift thermometers“ proposed for the common temperature range between 100 and 400 K are samarium acetate [59], $Nd_2Sn_2O_7$ [60] $Sm_2Sn_2O_7$ [60, 61], $Pr_2Sn_2O_7$ [61] and $(VO)_2P_2O_7$ [62], and for ^{13}C, ^{119}Sn, and ^{31}P observe nuclei, respectively; detailed comparisons between them can be found in [60]. As reported in [63] for the ^{17}O chemical shifts of H_2O in aqueous solutions of lanthanide ions and for solid lanthanide oxides, the paramagnetic shifts of directly bonded oxygen atoms in trivalent *RE* oxides are positive for the lanthanides Ce-Sm and negative for the lanthanides Eu-Yb. In these cases, the shift originates from a polarization mechanism. The bonding interaction involves a lone pair of the oxide ion and the empty 6s orbital on the lanthanide, resulting in paramagnetic spin density on the ^{17}O nuclei. For the ions Ce^{3+} to Sm^{3+}, the orbital angular momentum exceeds the spin angular momentum and both of them are oppositely oriented (as the shell is less than half-filled). With the orbital angular momentum aligned parallel to the applied magnetic field, the spin angular momentum at the lanthanide center is antiparallel, leading to a parallel spin orientation at the neighboring oxygen atoms, i. e. a positive sign of $\langle S_z \rangle$, due to the polarization mechanism. For the ions Gd^{3+} to Yb^{3+}, the situation is opposite. Spin and orbital angular momenta are aligned parallel; thence the positive value of $\langle S_z \rangle$ at the lanthanide ion produces a negative sign of $\langle S_z \rangle$ at the neighboring oxide ion due to the polarization mechanism. Accordingly, strongly negative ^{17}O shifts are observed for the corresponding oxides. Consistent with a dominant Fermi contact mechanism, Figure 3.7.10 shows an excellent correlation between the theoretically calculated value of $\langle S_z \rangle$ (using the method detailed in ref. [48]) and the isotropic paramagnetic shifts measured in the solid state. An analogous correlation was found between the ^{17}O NMR signal of water in aqueous lanthanide ion solutions [48]. Assuming that there are no element-specific

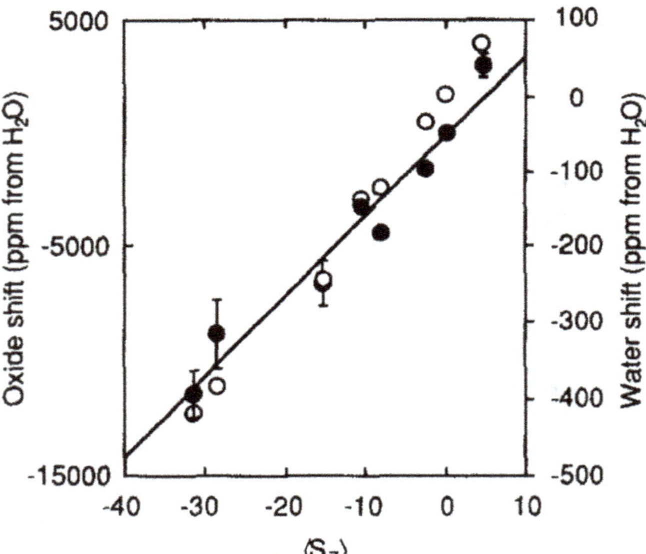

Figure 3.7.10: Plots of the ^{17}O NMR chemical shifts for solid lanthanide oxides (full circles, scale on the left), and of the ^{17}O NMR signal of water in aqueous lanthanide ion solutions (scale on the right, open circles), as a function of $\langle S_z \rangle$, the calculated spin moment according to ref. [48]. The straight line is a best fit to the lanthanide oxide data and was used to calculate the hyperfine coupling constant. Reproduced from reference [63] with permission from Elsevier.

differences in the mode of *RE*-oxygen bonding, the isotropic hyperfine coupling constant can be determined from the slope of the graph displayed in Figure 3.7.10, leading to a value of −2.7 MHz.

Figure 3.7.11 shows the ^{17}O MAS-NMR spectrum of monoclinic Sm_2O_3, a particular modification studied in reference [64]. This material features five distinct oxygen sites, whose resonances are well resolved at −44 °C and can be assigned based on intensity arguments and nutation characteristics [64]. The paramagnetic shifts detected here are rather weak because of a negative contribution arising from the van-Vleck paramagnetism of the energetically closest excited electronic state, which partially cancels the positive paramagnetic shift expected from the electronic ground state. ^{17}O shifts are significantly larger (strongly negative in the range −1000 to −3000 ppm) in the case of Eu_2O_3 [64]. Another instructive example is the case the yttrium aluminum garnet $Y_{3-x}Ce_xAl_5O_{12}$ (YAG, x = 0.09), where the large activator ion Ce^{3+} substitutes on the smaller Y^{3+} sites. The result is a yellow luminescing material that is nearly ideal for phosphor-converted solid-state white lighting. The YAG structure contains both four- and six-coordinated Al sites in a 3:2 ratio. As Figure 3.7.12 illustrates, the paramagnetic Ce^{3+} ion affects the resonance frequencies of the nearby ^{27}Al nuclei in their different coordination states in opposite directions. While the signals of the four-coordinate Al species move towards higher frequencies, those of the

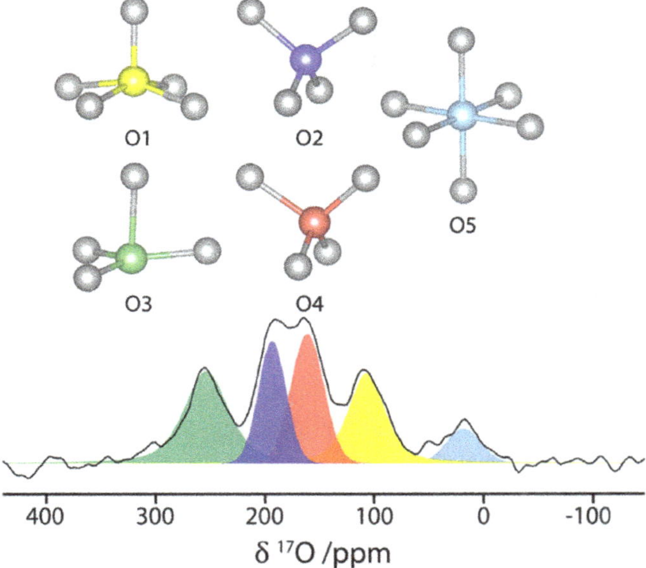

Figure 3.7.11: Local environments of the five crystallographically inequivalent oxygen sites in Sm_2O_3, ^{17}O solid state NMR spectrum, and suggested peak assignments. Reproduced from reference [64] with permission from Elsevier.

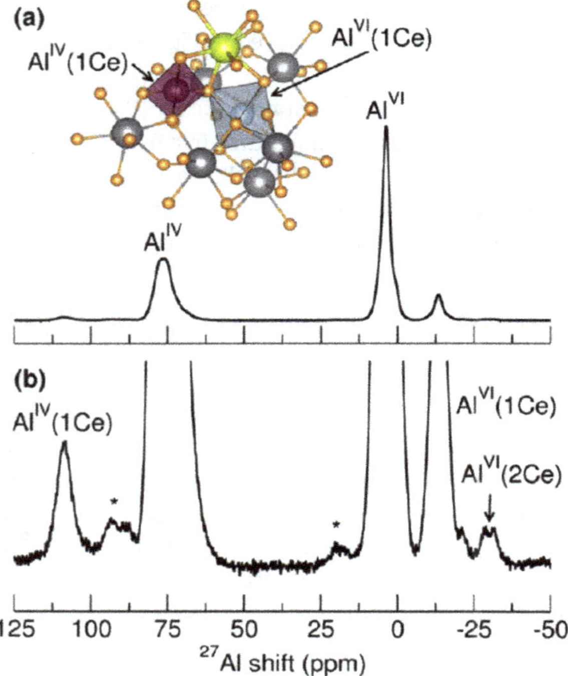

Figure 3.7.12: ^{27}Al MAS NMR spectrum of $Y_{2.91}Ce_{0.09}Al_5O_{12}$ acquired at room temperature, 23.5 T, and 60 kHz MAS. (a) full-scale spectrum, (b) with a 40× vertical Expansion of the intensity axis. Asterisks indicate small amounts of impurities. The inset depicts the local structures of $Al^{(4)}{}_{1Ce}$ and $Al^{(6)}{}_{1Ce}$ species near one Ce^{3+} substituent ion (yellow sphere). Reproduced with from reference [28] with permission from the American Chemical Society.

six-coordinate Al move to lower frequencies. In addition, a small fraction of Al species next to two Ce^{3+} ions is seen. This complex situation probably arises from contributions of Fermi contact shifts and pseudo contact shifts that are of comparable magnitude but different signs. Further insightful applications of paramagnetic NMR to polycrystalline Ce^{3+} doped phosphors include works on $CaSc_2O_4$ [27] and $La_{3-x}Ce_xSi_6N_{11}$ [65]. Augmented by parallel EPR work identifying the occupancies of crystallographically distinct sites, a reliable structural basis for understanding of the luminescent behavior was developed.

3.7.4 Conclusions

While the field of continuous-wave EPR on *RE* doped single crystals had become virtually extinct after the 1980s, simply because the number of manageable candidate materials became exhausted we are currently witnessing a renaissance of EPR and paramagnetic NMR spectroscopic applications to increasingly complex materials, which are usually polycrystalline, disordered or even glassy. This upsurge of activity has to be credited to the considerable advances in technique development, both on the NMR and the EPR side (most noteworthy pulsed EPR techniques), as well as the tremendous advances in computational methods that nowadays allow complex EPR Hamiltonians to be predicted from first principles and their effects on EPR and NMR spectra to be simulated. This latter part is absolutely essential in this research area, as EPR and paramagnetically affected NMR spectra can rarely be analyzed by inspection alone. Looking at the applications published so far on the side of paramagnetic NMR, one can say, that the most useful structural information is available from paramagnetic shifts in those cases in which the Fermi contact term dominates. Nevertheless, there are also useful applications of the spin-dipolar term. This term not only produces an isotropic pseudocontact shift contribution, but also is the source of wide spinning sideband manifolds. As shown in ref. [66], these side band manifolds can sometimes be analyzed in terms of distance information between the paramagnetic lanthanide and the nucleus under observation, producing surprisingly good agreement with diffraction results. To date this application has been under-utilized. Most recently, it was shown that small quantities of Gd^{3+} and Ho^{3+}, Tb^{3+}, and Dy^{3+} can also help to enhance the nuclear magnetic polarization effected by organic radicals in DNP (dynamic nuclear polarization) experiments, [67, 68]. Although the conditions for this enhancement have not yet been optimized, this discovery opens up new perspectives for the future use of *RE* element spin dynamics in obtaining structural information.

Acknowledgments: This work was supported by the Center of Research, Technology, and Education in Vitreous Materials (FAPESP 2013/07793-6). Support by the Deutsche Forschungsgemeinschaft, (DFG), Project 413550885 is also acknowledged.

References

[1] S. A. Altshuler, B. M. Kosyrev, Paramagnetische Resonanz, B. G. Teubner Verlagsgesellschaft, Leipzig, 1963.
[2] A. Abragam, B. Bleaney, Electron Paramagnetic Resonance of Transition Ions, Oxford University Press, Oxford, 1970.
[3] D. J. Ingram, Spectroscopy at Radio and Microwave Frequencies, Butterworths Scientific Publications, London, 1955.
[4] D. Goldfarb, S. Stoll (Eds.) EPR Spectroscopy, Fundamentals and Methods (eMagRes Books), Wiley-Interscience, 2018, 648 pp.
[5] J. A. Weil, J. R. Bolton, J. E. Wertz, Electron Paramagnetic Resonance: Elementary Theory and Applications, Wiley-Interscience, New York, 1994.

[6] R. G. Barnes, NMR, EPR and Mößbauer Effect: Metals, Alloys and Compounds, in Handbook on the Physics and Chemistry of Rare Earths (ed. K. A. Gschneidner Jr., L. Eyring), Vol. 2, Ch. 189, Elsevier, Amsterdam, 1979, 387–506.
[7] G. Schwab, W. Hillmer, Phys. Status Solidi B 1975, 70, 237.
[8] A. Schweiger, G. Jeschke. Principles of Pulsed Electron Spin Resonance, Oxford University Press, Oxford, 2001.
[9] J. M. Baker, B. Bleaney, W. Hayes, Proc. Roy. Soc. London 1958, 247, 141.
[10] S. K. Misra, P. Mikolaiczak, S. Korczak, J. Chem. Phys. 1981, 74, 922.
[11] S. Maron, G. Dantelle, T. Gacoin, F. Devreux, Phys. Chem. Chem. Phys. 2014, 16, 18788.
[12] O. Guillet-Noel, B. Viana, D. Vivien, D. Gourier, A. Antic-Fedancev, P. Porcher, J. Alloys Compd. 1998, 275–277, 177.
[13] O. Guillet-Noel, D. Simons, D. Gourier, J. Phys. Chem. Solids 1999, 60, 555.
[14] R. W. Reynolds, L. A. Boatner, C. B. Finch, A. Chatelain, M. M. Abraham, J. Chem. Phys. 1972, 56, 5607.
[15] U. Ranon, Phys. Lett. 1968, 28A, 228.
[16] J. Rosenthal, Phys. Rev. 1967, 164, 363.
[17] D. Ball, Phys. Status Solidi (b) 1982, 111, 311.
[18] J. A. Hodges, Phys. Status Solidi (a) 1975, 68, K-73.
[20] V. Nekvasil, Phys. Status Solidi (b) 2006, 87, 317.
[21] H. R. Asatryan, J. Rosa, J. Mares, Solid State Commun. 1997, 104, 5.
[22] G. R. Asatryan, A. P. Skvortsov, G. S. Shakurov, Phys. Solid State 2013, 55, 1039.
[23] G. Dantelle, M. Mortier, P. Goldner, D. Vivien, J. Phys.: Condens. Matter 2006, 18, 7905.
[24] S. Sen, S. B. Orlinskii, R. M. Rakhmatullin, J. Appl. Phys. 2001, 89, 2304.
[25] C. Carnevali, M. Mattoni, F. Morazzoni, R. Scotti, M. Casu, A. Musinu, R. Krsmanovic, S. Polizzi, A. Speghini, M. Betinelli, J. Am. Chem. Soc. 2005, 127, 14681.
[26] S. Sen, R. Rakhmatullin, R. Gubaidullin, A. Silakov, J. Non-Cryst. Solids 2004, 333, 22.
[27] N. C. George, J. Brgoch, A. J. Pell, C. Cozzan, A. Jaffe, G. Dantelle, A. Llobet, G. Pintacuda, R. Seshadri, B. F. Chmelka, Chem. Mater. 2017, 29, 3538.

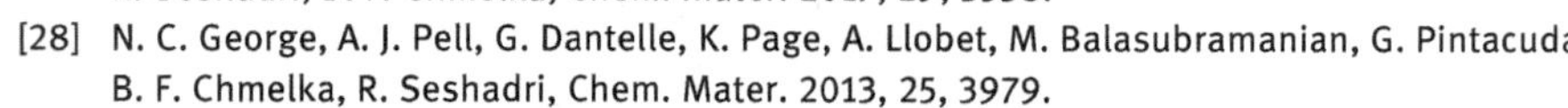

[28] N. C. George, A. J. Pell, G. Dantelle, K. Page, A. Llobet, M. Balasubramanian, G. Pintacuda, B. F. Chmelka, R. Seshadri, Chem. Mater. 2013, 25, 3979.

[29] M. de Oliveira Jr., T. Uesbeck, T. S. Gonçalves, C. J. Magon, P. S. Pizani, A. S. S. de Camargo, H. Eckert, J. Phys. Chem. C 2015, 119, 24574.
[30] T. Deschamps, H. Vezin, C. Gonnet, N. Ollier, Opt. Express 2013, 21, 8382.
[31] K. Arai, S. Yamasaki, J. Isoya, N. Namikawa, J. Non-Cryst. Solids 1996, 196, 216.
[32] R. R. Gubaidullin, S. B. Orlinskii, R. M. Rakhmatullin, S. Sen, J. Appl. Phys. 2007, 101, 063529.
[33] A. Saitoh, S. Matsuishi, M. Oto, T. Miura, M. Hirano, H. Hosono, Phys. Rev. B 2005, 72, 212101.
[33] T. Deschamps, N. Ollier, H. Vezin, C. Gonnet, J. Chem. Phys. 2012, 136, 014303.
[34] H. Deters, J. F. de Lima, C. J. Magon, A. S. S. de Camargo, H. Eckert, Phys. Chem. Chem. Phys. 2011, 13, 16071.
[35] R. Zhang, M. de Oliveira, Z. Wang, R. G. Fernandes, A. S. S. de Camargo, J. Ren, L. Zhang, H. Eckert, J. Phys. Chem. C 2017, 121, 741.
[36] M. de Oliveira, J. Amjad, A. S. S. de Camargo, H. Eckert, J. Phys. Chem. C. 2018, 122, 23698.
[37] S. Sen, R. Rakhmatullin, R. Gubaidullin, A. Pöppl, Phys. Rev. B 2006, 74, 100201.
[38] M. de Oliveira Jr., T. S. Gonçalves, C. Ferrari, C. J. Magon, P. S. Pizani, A. S. S. de Camargo, H. Eckert, J. Phys. Chem. C 2017, 121, 2968.
[39] S. Stoll, A. Schweiger, J. Magn. Reson. 2006, 178, 42.
[40] G. Jeschke, Y. Polyhach, Phys. Chem. Chem. Phys. 2007, 9, 1895.

[41] S. Razzaghi, M. Qi, A. I. Nalepa, A. Godt, G. Jeschke, A. Savitsky, M. Yulikov, J. Phys. Chem. Lett. 2014, 5, 3970.
[42] A. Doll, M. Qi, N. Wili, S. Pribitzer, S. A. Godt, G. Jeschke, J. Magn. Reson. 2015, 259, 153.
[43] G. Prokopiou, M. D. Lee, A. Collauto, E. H. Abdelkader, T. Bahrenberg, A. Feintuch, M. Ramirez-Cohen, J. Clayton, J. D. Swarbrick, B. Graham, G. Otting, D. Goldfarb, Inorg. Chem. 2018, 57, 5048.
[44] A. Potapov, H. Yagi, T. Huber, S. Jergic, N. E. Dixon, G. Otting, D. Goldfarb, J. Am. Chem. Soc. 2010, 132, 9040.
[45] D. Goldfarb, Phys. Chem. Chem. Phys. 2014, 16, 9685.
[46] H. Pell, G. Pintacuda, C. P. Grey, Prog. Nucl. Magn. Reson. Spectrosc. 2019, 111, 1.
[47] M. Enders, Assigning and Understanding NMR Shifts of Paramagnetic Metal Complexes, in Modeling of Molecular Properties (ed. P. Comba), Wiley VCH Verlag GmbH & Co., Weinheim, 2011, 49–63.
[48] R. Golding, M. A. Halton, Aust. J. Chem. 1972, 25, 2577.
[49] R. Bleaney, J. Magn. Reson. 1972, 8, 91.
[50] I. Bertini, C. Luchinat, G. Parigi, Solution NMR of Paramagnetic Molecules, Elsevier Science, Amsterdam, 2001, 1–384.
[51] I. Bertini, C. Luchinat, G. Parigi, R. Pieratelli, ChemBioChem 2005, 6, 1536.
[52] C. C. Hinckley, J. Am. Chem. Soc. 1969, 91, 5160.
[53] M. Balci, NMR Shift Reagents and Double Resonance Experiments: Simplification of NMR Spectra, in Basic ^{1}H- and ^{13}C-NMR Spectroscopy, Ch. 7, Elsevier, Amsterdam, 2005, 199–212.
[54] G. E. Hawkes, C. Marzin, D. Leibfritz, S. R. Johns, K. Herwig, R. A. Cooper, D. W. Roberts, J. D. Roberts, Nuclear Magnetic Resonance Shift Reagents, Acad. Press, New York, 1973, 129.
[55] M. Kainosho, K. Ajisaka, W. H. Pirkle, S. D. Beare, J. Am. Chem. Soc. 1972, 94, 5924.
[56] A. K. Cheetham, C. M. Dobson, C. P Grey, R. J. B. Jakeman, Nature (London) 1988, 328, 706.
[57] C. P. Grey, M. E. Smith, A. K. Cheetham, C. M. Dobson, R. Dupree, J. Am. Chem. Soc. 1990, 112, 4670.
[58] C. P. Grey, C. M. Dobson, A. K. Cheetham, R. J. B. Jakeman, J. Am. Chem. Soc. 1989, 111, 505.
[59] G. C. Campbell, R. C. Crosby, J. F. Haw, J. Magn. Reson. 1986, 69, 191.
[60] A. K. Cheetham, C. P. Grey, C. M. Dobson, J. Magn. Reson. A 1993, 101, 293.
[61] G. J. M. P. van Moorsel, E. R. H. Van Eck, C. P. Grey, J. Magn. Reson. A 1993, 113, 159.
[62] H. Pan, B. C. Gerstein, J. Magn. Reson. 1991, 92, 618.
[63] S. Yang, J. Shore, E. Oldfield, J. Magn. Reson. 1992, 88, 408.
[64] M. A. Hope, D. M. Halat, J. Lee, C. P. Grey, Solid State Nucl. Magn. Reson. 2019, 102, 21.
[65] N. C. George, A. Birkel, J. Brgoch, B. C. Hong, A. A. Mikhailovsky, K. Page, A. Llobet, R. Seshadri, Inorg. Chem. 2013, 52, 13730.
[66] A. R. Brough, C. P. Grey, C. M. Dobson, J. Am. Chem. Soc. 1993, 115, 7318.
[67] P. Niedbalski, C. Parish, A. Kiswandhi, L. Fidelino, C. Khemtong, Z. Hayati, L. Song, A. Martins, A. D. Sherry, L. Lumata, J. Chem. Phys. 2017, 146, 014303.
[68] L. Friesen-Waldner, A. Chen, W. Mander, T. J. Scholl, C. A. Mc Kenzie, J. Magn. Reson. 2012, 223, 85.

Michael Bredol

3.8 Electrical and dielectric properties

3.8.1 General aspects

Exposing a material to an electrical field will cause charge carriers to either be displaced locally (expressed as polarizability of the material) or flow stationary (expressed as conductivity of the material). Carriers involved in functional materials may be electrons, ions or *Cooper* pairs of electrons. Rare earth-based functional materials are found for all of these cases and are the subject of this chapter. Mobile (conducting) electrons are useful at ambient temperature in metals or semiconductors, whereas ionic conductivity in solids, in most cases, is a high-temperature effect. *Cooper* pairs, on the other hand, are a typical low-temperature case and have not been observed at room temperature yet. For electrons, the conventional classification of conductivity types using band relations is depicted in Figure 3.8.1. It excludes *Cooper* pairs (superconductivity), which is a purely quantum-mechanical phenomenon (pairing of electrons to bosons with zero spin). The mobility of ions can be best understood in the framework of defect chemistry.

The rare earth metals, in their elemental form, are only modest metallic conductors and, therefore, do not find much attention as such; their conductivities are in the order of 10^6 S m^{-1} (for comparison: copper and silver are in the order of 60×10^6 S m^{-1}). The compounds, LaB_6 and CeB_6, are exceptions; they are metals with extremely low work function (approximately 2.5 eV) and, therefore, find application in thermal electron emitters, for example, in the electron gun in electron microscopes. Other, often complex, rare earth compounds may show a rich variety of electrical and dielectric phenomena related to either their high atomic number (that is high polarizability of the electronic system), their large and variable ion size to adjust local crystalline environments, their specific redox chemistry or to the presence of the many very sharp optical absorption lines. Many rare earth pnictides are semimetals; some of the heavier ones (ErAs, ErSb, TbAs) find application in nanocomposites with III/V semiconductors for thermoelectrics. Since most rare earths do show complex magnetic behavior, some of the semiconducting compounds have triggered interest in spintronics. In this chapter, we will group the material along the basic electrical and dielectric phenomena. We will then discuss the contributions of the relevant rare earths in their respective context, in view of existing and potential applications, but we will not discuss the individual properties of each rare earth comprehensively.

Optical, magnetic and dielectric properties are closely related. These frequency-dependent quantities are all formulated as complex functions to represent elastic and inelastic components. The key quantity in our context is the dielectric function $\varepsilon(\omega)$, which, for the static case and elastic response, can be simplified to describe

https://doi.org/10.1515/9783110654929-023

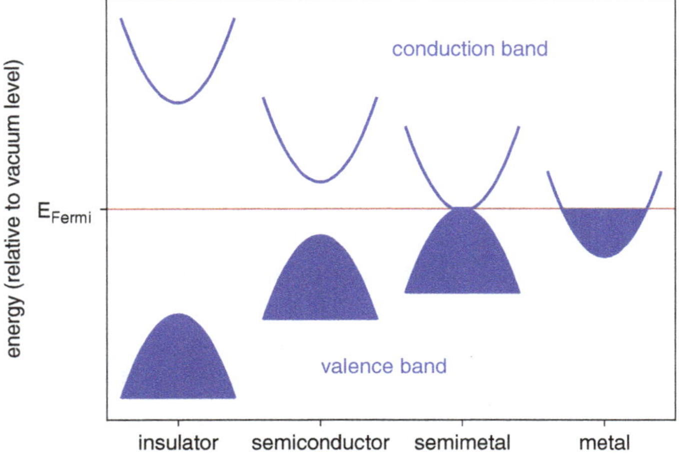

Figure 3.8.1: Band relations around Fermi energy and types of electronic conductivity.

charge displacement D and polarization P, by the so-called relative dielectric constant or relative permittivity ε_r (for the sake of simplicity, we disregard the vector nature of D, P and E here):

$$D = \varepsilon_0 \varepsilon_r E$$
$$P = \varepsilon_0 E(\varepsilon_r - 1)$$

In nonmagnetic materials, there are simple relations between refractive index and absorption coefficient on one side, and the components of the full frequency-dependent dielectric function on the other side (the notation used in the following is the "engineering" convention; in physics, the imaginary component of ε_r usually is defined as a negative quantity). With a dimensionless absorption coefficient κ and n as the real part of the refractive index (and letting $\mu_r = 1$ for a nonmagnetic material):

$$\tilde{n}^2 = \varepsilon_r \mu_r$$
$$\tilde{n} = n + i\kappa$$
$$\varepsilon_r = \varepsilon_r' + i\varepsilon_r''$$
$$\varepsilon_r' = n^2 - \kappa^2$$
$$\varepsilon_r'' = 2\kappa n$$

Obviously, ε_r' can become negative in the neighborhood of strong absorption bands (the system then becomes optically reflective) and will be strongly frequency-dependent! In many cases, only two values are reported in this context: the so-called static dielectric constant $\varepsilon_{r,0}$ coupled to polarizability in a static electrical field, and the high-frequency variant $\varepsilon_{r,\infty}$, where ∞ is not clearly defined but typically means "optical" frequencies. The relations can be qualitatively understood by a simple sum of independent-damped oscillators:

$$\varepsilon = \varepsilon_\infty + \sum_i \frac{S_i\omega_i^2}{\omega_i^2 - \omega^2 - i\gamma_i\omega}$$

$$\omega_i \equiv \textit{resonant angular frequency of oscillator } i$$

$$\gamma_i \equiv \textit{damping factor of oscillator } i$$

$$S_i \equiv \textit{oscillator strength of oscillator } i$$

Figure 3.8.2 shows a sketch of these relations in the neighbourhood of a single absorption band treated as an independent damped oscillator. Dielectric polarizability varies strongly in the neighbourhood of absorption bands; this effect is known as "anomalous dispersion". Figure 3.8.2 also shows that ε_r and $\tilde{n}$ are always highest at low frequencies and tend to decrease step by step whenever the frequency of the polarizing field exceeds the resonance frequency of one of the oscillators involved. This has important consequences for functional dielectric materials: vibrations have resonance frequencies in the THz region, therefore, ε_r in the MHz and GHz regions is always much larger than in the optical region for a given material. These contributions, are often, grouped into "electronic polarization" (observable up to energies of the band gap) and "lattice polarization" (observable up to the onset of vibrations).

The imaginary and real components of ε_r and $\tilde{n}$ are not independent of each other and (at least in principle) can be transformed into each other by *Kramers–Kronig* relations. This offers a way to retrieve frequency-dependent information on the full dielectric

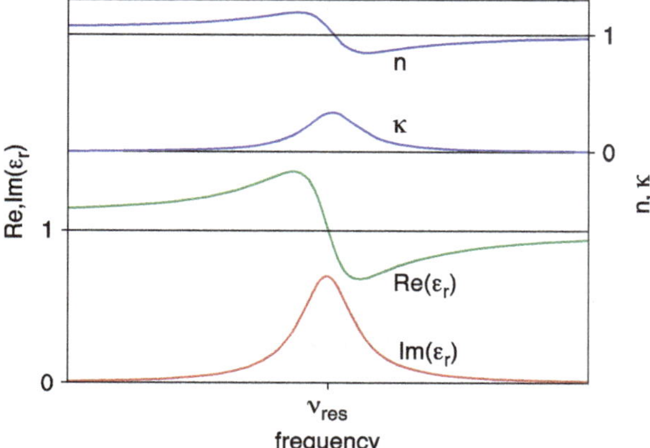

Figure 3.8.2: Relations between dielectric and optical data around resonance in a damped oscillator (n: refractive index, κ: absorption constant).

function from high-quality optical data (absorption or reflectance spectra together with refractive indices determine ε_r''):

$$\varepsilon_r'(\omega) = 1 - \frac{2}{\pi}\int_0^\infty \frac{x\varepsilon_r''(x)}{x^2-\omega^2}dx$$
$$\varepsilon_r''(\omega) = -\frac{2}{\pi}\int_0^\infty \frac{\omega\varepsilon_r'(x)-1}{x^2-\omega^2}dx$$

While, ε_r' is the "ordinary" relative dielectric constant or permittivity, ε_r'' can be linked (in addition to absorption losses) to contributions of electrical conductivity σ (for a systematic description of frequency-dependent contributions see [1]):

$$|\sigma| = \varepsilon_0\varepsilon_r''\omega$$

According to complex algebra, the ratio between imaginary and real components defines a so-called loss angle δ which is often used to define a "quality factor" Q in high frequency applications:

$$\frac{\varepsilon_r''}{\varepsilon_r'} = \tan\delta = \frac{1}{Q}$$

3.8.2 Semiconductors

Most of the binary trivalent rare earth compounds are insulators, but some of the divalent chalcogenides (EuO, EuS, EuSe, EuTe, SmS, SmSe and SmTe) possess interesting semiconducting properties. Since all of them are magnetic, they may be useful for controlling the spin of the mobile electrons. This concept was already investigated back in the 1960 s, when researchers at IBM sandwiched an EuSe layer between metallic electrodes and found large magnetoresistance [2]. More recently, this class of materials was reinvestigated due to the effect of piezoresistance and its potential application in piezoelectronic transistors; films of SmSe as thin as 8 nm were prepared for this purpose [3].

A very peculiar case is SmS [4]: under a pressure of 6.5 kbar at room temperature (which can be reached even by polishing), it transforms discontinuously from a (blackish) n-type semiconductor with 0.15 eV band gap to a metal (golden-yellow). Although there is a slight hysteresis, this effect can be used for solid state pressure sensors. In SmSe and SmTe, however (band gaps of 0.45 and 0.65 eV, respectively), the transition occurs continuously [5]. The effect is explained by a pressure-induced electronic transition from localized 4f electrons to delocalized 5d electrons, changing the valence state effectively from Sm^{2+} to Sm^{3+}. The crystal structure (rocksalt) does not change during transition but the lattice parameter shrinks from 0.597 to 0.570 nm. Other rare earth monosulfides, without electron-configuration-stabilized divalent state, consequently exhibit metallic character and do not show transitions between semiconducting and metallic behavior. A similar trend is observed in rare earth hydrides: dihydrides tend to be metallic conductors (exceptions: EuH_2 and YbH_2), whereas the trihydrides are salt-like insulators.

A rare case of a semiconducting binary compound of a trivalent rare earth is Ce_2S_3 which has even found application as a red pigment. Other semiconducting compounds involving the trivalent state always include the divalent species as well, such as Eu_3S_4 and Sm_3S_4. In contrast, the more complex compounds of trivalent rare earths, especially in perovskites with transition metal oxides (most notably from the Fe, Mn, Co group), are often semiconductors and find interest in sensor applications [6]. (La,Nd)FeO_3, for instance, is a p-semiconductor with a band gap of about 2.3 eV and can be used as electrical sensor. For example, for ethanol, because surface oxide is able to oxidize ethanol, it leaves behind vacancies causing surface conductivity [7].

3.8.3 Superconductors

The discovery of "high-temperature" ceramic superconductors (for a detailed description we refer to Chapter 4.12) is an excellent example of the capability of rare earths to fine-tune material properties. The first such material prepared, back in 1986 [8], contained as superconducting phase $LaBa_2Cu_3O_{7-x}$ with a transition temperature of approximately 35 K. It was soon realized that an exchange of La with other rare earths, especially yttrium, pushes the transition temperature up to about 90 K, above the boiling point of liquid nitrogen [9]. This led to the $YBa_2Cu_3O_{7-x}$-family of so-called high-temperature Y123 superconductors, with optimal properties found around the composition $x = 0.07$. The crystal structure of these materials is a defect perovskite, containing "CuO_2" planes with mixed Cu valency, separated by yttrium ions. Till this day, there is no closed theory explaining this "unconventional" superconductivity in nonmetals, but it appears that the cuprate planes are carriers of the conductivity. The rare earths, with their varying ion diameters, therefore, seem to fine-tune the geometry of the cuprate-based crystal structure.

Due to the involvement of mixed-valence cuprate planes, the electrical properties of all Y123 materials are strongly anisotropic, which makes the reproducible preparation of superconducting polycrystalline films or wires quite demanding. Grain boundaries need to be controlled precisely to avoid breakdown of superconductivity with increasing current or magnetic field. Some of the methods developed for the manufacture of superconducting cables are using Nd123, Sm123 or Gd123 as seeds for grain growth control. Therefore, commercial cables often contain Sm, Nd and Gd as substitutes for Y – partially intentionally doped; partially due to diffusion from seeds.

Rare earths seem to have a similar role in the family of iron-based high-temperature semiconductors ($LaFeAsO_{1-x}F_x$) [10]. Exchanging La with other rare earths pushes the critical temperature up to 55 K when using Sm [11].

3.8.4 Ion conductors and mixed conductors

The most prominent example of a pure rare earth-based material in this class is ceria (CeO_2), which tends to be oxygen-deficient at higher temperatures, leading to oxygen vacancies and oxygen anion conductivity. This property makes it useful as solid electrolyte in high-temperature fuel cells, but does suffer in this context from developing electron conductivity in parallel. Better than pure ceria as oxygen anion conductor are mixed crystals with Gd_2O_3 [12, 13]. The technically most important example for oxygen anion conductivity also uses a rare earth as dopant: yttria-stabilized zirconia, in which, in some cases, the yttria may be exchanged with Sc_2O_3. The simultaneous presence of redox-active defects and resonable conductivity both for ions and for electrons, has also made CeO_2 interesting as basis for electrocatalysts [14].

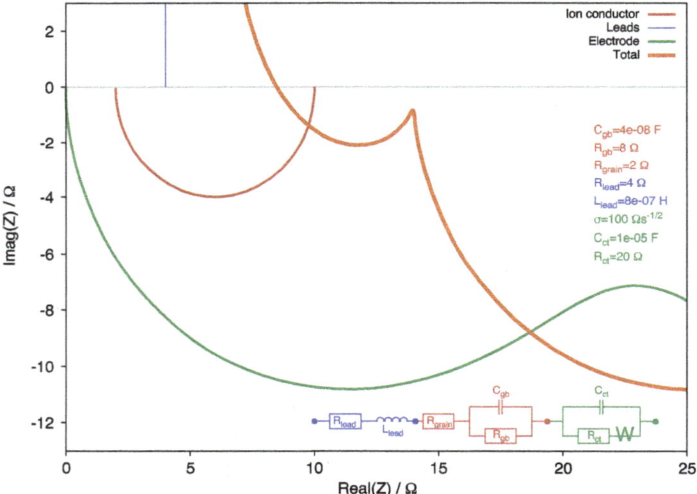

Figure 3.8.3: Nyquist plot of individual impedance contributions from ion conductor (red, with grain and grain boundary contributions), leads (blue) and electrodes (green, as Randles circuit, "W" is a Warburg impedance with parameter σ), together with total system impedance (light red).

The determination of conductivity contributions in materials as these is not always straightforward, since in ceramic objects, contributions of grain volume, grain boundary and electrodes have to be deconvoluted. The standard procedure here is electrochemical impedance spectroscopy (EIS) in which, the impedance of a sample contacted with Pt or Ag leads is measured in a frequency range, typically between

0.1 Hz and 1 MHz, followed by fits to equivalent circuits. A detailed description of the different aspects to be observed here, including artifacts by wiring and instrumentation, can be found in [12]. Figure 3.8.3 displays in a *Nyquist* plot, typical relative contributions by the ceramic conductor at high temperature, by inductivities of leads and by the electrodes (the latter modeled here as a *Randles* circuit, "W" is a *Warburg* impedance). The approximate parameter values were taken from [12], using the equivalent circuit depicted in the inset of Figure 3.8.3. It is obvious from this example that accurate ionic conductivities can only be extracted by careful fitting, not just from qualitative inspection of the comparatively small and distorted central "semicircle." Like in the dielectric function, the real and imaginary components of impedance Z are not independent but related by *Kramers–Kronig* transforms [15, 16]; they can be used to evaluate the statistical validity of experimental data:

$$Z'(\omega) = Z'(\infty) - \frac{2}{\pi}\int_0^\infty \frac{xZ''(x) - \omega Z''(\omega)}{x^2 - \omega^2}\,dx$$

$$Z'(\omega) = Z'(0) - \frac{2\omega}{\pi}\int_0^\infty \frac{\frac{\omega}{x}Z''(x) - Z''(\omega)}{x^2 - \omega^2}\,dx$$

$$Z''(\omega) = \frac{2\omega}{\pi}\int_0^\infty \frac{Z'(x) - Z'(\omega)}{x^2 - \omega^2}\,dx$$

Mixed conductors (oxide anion conduction combined with electronic conduction) may be useful in cathodes of fuel cells; prominent examples again contain lanthanum: LSC ($LaCoO_3$), LSCF ($La_{1-x}Sr_xCo_{1-y}Fe_yO_3$) or LSM ($La_{1-x}Sr_xMnO_3$). A review of the related compounds using various rare earths can be found in [17], [18] and [19]. For the characterization of these materials, EIS is very useful, as demonstrated in a sensor example with LSM [20].

Rare earth cations are comparatively large and, therefore, may also facilitate ionic conductivity for other smaller anions, especially in fluorides [21]. LaF_3, therefore, does find application in fluoride-selective electrodes (doped with EuF_2 to induce vacancies) and may reach a fluoride conductivity of 10^{-4} S m^{-1} at room temperature in single crystals. More recently, $La_{1-x}Ba_xF_{3-x}$ has been of interest as solid electrolyte in fluoride ion batteries [22] which promises high energy density once suitable anode and cathode materials are developed.

For the same reasons, rare earth compounds have been tested as solid state proton conductors, for example, in steam electrolysis using $SrCeO_3$ or $SrCe_{0.95}Yb_{0.05}O_3$ membranes [23, 24], or in high-temperature fuel cells using Y-doped $BaCeO_3$ [25]. For similar purposes, rare earth niobates and tantalates such as Ca-doped $LaNbO_4$, $NdNbO_4$ and their partially Ta-substituted variants also show promise as stable high-temperature proton conductors [26]. There are also some rare earth containing MOFS (metal organic frameworks) that are stable enough to be considered as proton conductors [27].

3.8.5 Thermoelectrics

Thermoelectric materials can convert waste heat to electrical power or cool objects with the help of an electrical current (although in both cases only up to the *Carnot* limit). The thermoelectric effect cannot be observed in a single material because the necessary leads would expose the measurement system to the same temperature gradient as the material itself. A single thermoelectric element, therefore always contains junctions between different materials – metals as well as semiconductors. When using semiconductors, the typical arrangement is the connection between a p- and an n-conductor on one side and terminals on these branches on the other side, positioned in a temperature gradient causing or sustaining heat flow (see Figure 3.8.4 for an illustration of the setup in a generator as well as in a cooler).

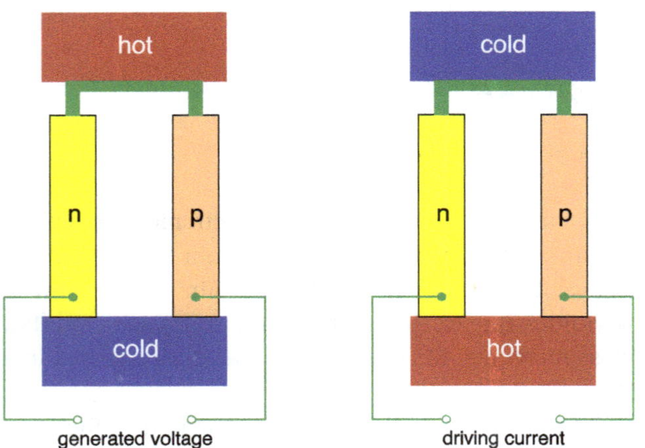

Figure 3.8.4: Single elements of a semiconductor based thermoelectric generator (left) and a thermoelectric cooler (right).

The thermoelectric efficiency factor, $z = S^2\sigma/\kappa^2$, calls for materials with large *Seebeck* coefficient $S = -\Delta V/\Delta T$, high electronic conductivity σ and small thermal conductivity κ. Electron scattering therefore should be as small as possible, whereas phonon scattering, for example at defects or heavy atoms, would be helpful to suppress thermal conductivity. The latter request turned out to be hard to fulfill without compromising electronic conductivity and has triggered research in various directions. The S-deficient SmS as n-semiconductor in this context is a promising single-phase candidate for the n-leg of thermoelectric generators [28]. Strong phonon scattering by "rattling" modes in solids with large cavities has also been achieved with the help of rare earths. Filled skutterudites such as $CeFe_3CoSb_{12}$ or its La-analogue show thermal conductivities much lower than their unfilled parent compound [29], and thus increased z-values. Another approach employing rare earths is the use of composites of semiconductors (typically from the III/V family) and nanoparticulate semimetals such as ErAs or TbAs (in n-conductors) or ErSb (in p-conductors) at concentrations of approximately 1% to reduce thermal conductivity [30, 31].

3.8.6 Interface conduction

In contrast to ordinary conductivity by carriers in the volume of a solid or along grain boundaries, interface conductivity due to a true 2D electron gas (better described as "liquid," due to strong correlation) localized at the interface between two epitaxially grown insulating crystalline layers can occur as well. The first example of this phenomenon was the combination of a $SrTiO_3$ substrate and a $LaAlO_3$ thin film [32]. It turned out that the $LaAlO_3$ layer needs to be La/O terminated for such an effect. Although the detailed mechanism of this effect is as yet unknown, many more rare earth-based (polar) perovskites have shown similar effects (always on $SrTiO_3$ substrate), for example $GdTiO_3$ [33], $LaTiO_3$ [34], $LaVO_3$ [34], $LaGaO_3$ [35], $PrAlO_3$ [36], $NdAlO_3$ [36] or $NdGaO_3$ [36]. There are strong indications that rare earth components are involved by exerting local strain due to their varying ionic sizes [37]. These true interface effects must not be confused with the surface conductivity observed in so-called topological insulators. Rare earths play a minor role there although SmB_6 is a classic example at low temperatures [38].

3.8.7 Dielectric function and polarization

Due to their large electron system, most insulating rare earth compounds (basically all trivalent oxides) possess a large relative permittivity, also at high frequencies. This is reflected in a simple relationship between the atomic number and the dielectric constant, and can be used in the prediction of materials properties [39]. This typical behavior is exploited in glasses containing La_2O_3 to achieve a high refractive index (static $\varepsilon_r \simeq 30$ for La_2O_3). Lanthanides with sharp absorption bands may be additionally used to colorize glass and, sometimes, because of the effects due to anomalous dispersion (polarization), in the neighborhood of absorption lines. The resulting high values of the dielectric constant, for the heavier rare earth oxides, are extensively used in class 1 ceramic capacitors. Nd_2O_3, Sm_2O_3 and other rare earth oxides are found besides various titanates, tantalates and niobates as components of the standardized dielectric ceramic type C0G (previously: NP0), guaranteeing low loss, high stability, low temperature drift and high operating voltages. The same holds for multilayer capacitors based on $BaTiO_3$ ceramic ferroelectric dielectrics. Doping with rare earth oxides such as Yb_2O_3, Sm_2O_3, Ho_2O_3, La_2O_3, or Nd_2O_3 allows to fine-tune grain size and grain boundaries and, thus, the performance of standardized X7R dielectrics for high volume efficiency [40].

La_2O_3 (as well as other rare earth oxides, mostly Sm_2O_3, Nd_2O_3 and Er_2O_3) has also been investigated as a high-k gate dielectric in FETs (field effect transistors) due to its insulating properties coupled with a high dielectric constant. The same applies to some more complex oxides such as $LaAlO_3$ ($\varepsilon_r \simeq 30$) [41, 42]. Such materials are technologically in high demand because, on a silicon base, SiO_2 as a convenient dielectric but with low dielectric constant would need to be applied in ever-shrinking FETs with thicknesses of much less than 1 nm – which is impossible because of the onset of tunnel currents. Still, on a silicon semiconductor base, both band gap edges of suitable replacement materials should be offset relative to silicon by at least 1 eV to avoid leakage currents. Suitable materials, therefore, need to have a band gap of at least 3 eV and should be as defect-free as possible, see Figure 3.8.5. On the other hand, a high dielectric constant in the frequency range of interest for electronic applications is coupled to high electronic and lattice polarizability, which does lead to small band gaps. This conflict can be resolved by the presence of relatively heavy and polarizable lanthanide ions. A compilation of relevant quantities for all rare earth oxides can be found in [43]. Their performance may be further enhanced by interfacial layers based on rare earth silicates [44], also in contact with semiconductors other than silicon (e.g. GaAs).

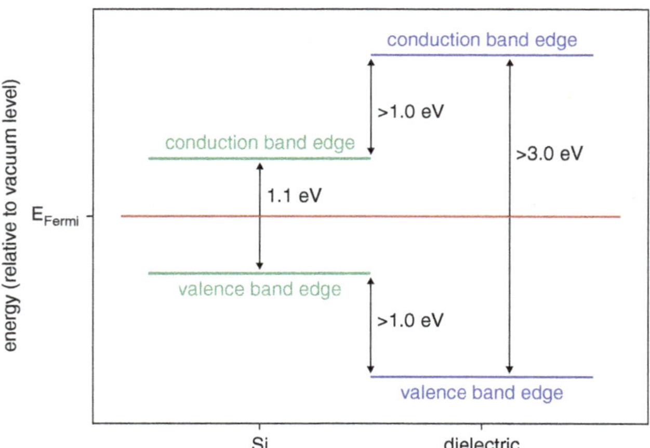

Figure 3.8.5: Necessary band offsets for the dielectric layer in FETs.

A technically important case is dielectrics for electronic high-frequency (MHz–GHz) applications. Without such materials, modern wireless communication would not at all be possible. Rare earth oxides, with their high polarizability and potentially low conductivity, have always played a major role in this context. It was realized already in the 1970 s and subsequently systematically investigated that titanate ceramics (sometimes multiphase) containing BaO and Nd_2O_3 (or further rare earth oxides) are excellent candidates for high frequency dielectrics with high permittivity, albeit with modest quality factor [45–47]. In later developments, these compositions were further augmented with Bi_2O_3, and Ba was replaced by Zn, giving rise to the Nd–zinc titanate family of dielectrics. There are also more simple compositions proposed as dielectric, for example, $NdNbO_4$ or $CeVO_4$.

3.8.8 Piezoelectricity

Piezoelectricity, in general, is a complex phenomenon and describes the relation between mechanical stress σ and induced polarization P. Stress can be compressive, tensile or shearing. The σ tensor, therefore, has six independent components illustrated in Figure 3.8.6 using the conventional notation:

$$P_i = \sum_{j,k} d_{ijk}\sigma_{jk}$$

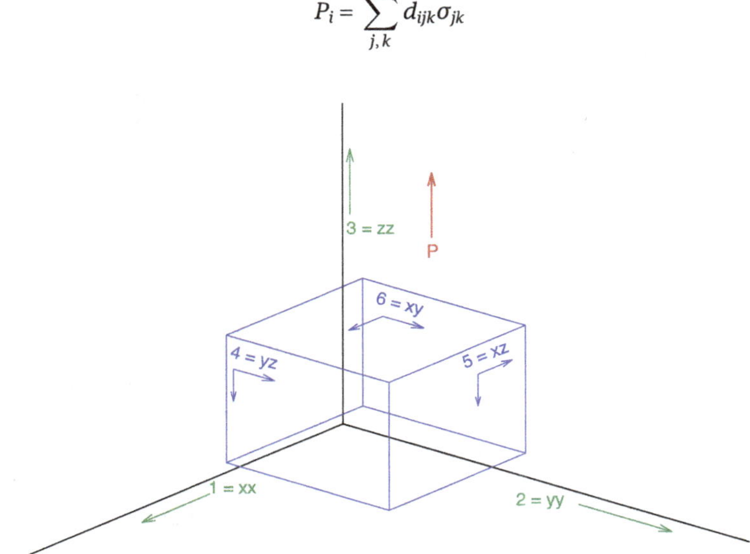

Figure 3.8.6: The six components of stress to induce a polarization P.

The resulting 18 components of the piezoelectric strain coefficient d in a triclinic lattice fortunately, in most cases, reduce to a much smaller number due to symmetry relations. A technically important application is in actuators: mechanical strain caused by an electrical field; the relation between these two quantities is also given by *d*.

Ferroelectric materials can spontaneously polarize and may form polarized domains; these domains can be ordered in an external field: "Poling". Poled ceramics show piezoelectric behavior with respect to their "poled axis". The piezoelectric strain coefficient, *d*, in such a ceramic has only two independent components: d_{33} for longitudinal effects, and d_{31} for transversal effects.

Technically, relevant materials for piezoelectric ceramics are often distorted perovskites on the basis of $Pb(Zr,Ti)O_3$ (PZT), which can be improved by Lanthanum doping (PLZT): $(Pb,La)(Zr,Ti)O_3$. La^{3+} here is a supervalent ion with a large diameter and, therefore, easily substitutes Pb and generates Pb vacancies. Just below the ferroelectric *Curie* temperature (around 600 K in PLZT), such ceramics are poled in DC fields of several MV/m and then cooled to room temperature.

Modern *Diesel* engines need multiple precision fuel injections from a pressure reservoir ("common rail") to stay in line with regulations concerning exhaust gases. Since 2002 (*Bosch, Peugeot*), piezoceramics started to replace mechanical systems. Today, PLZT-based piezoinjectors allow up to eight injections (1–2 ms each) with reaction times of <0.1 ms from a fuel reservoir at approximately 2,500 bar. A stack of approximately 350 layers of piezoceramic (ca. 30 mm total) allows for a total stretch of approximately 40 μm at about 45 V voltage (about 0.11 μm per layer), which is sufficient for the necessary control level.

For some applications, however, piezoelectric crystals are necessary instead of ceramics. Rare earths and their oxyborates ($RECa_4O(BO_3)_3$) have raised some interest for high-temperature sensing [48].

3.8.9 Summary

After walking through the maze of electrical and dielectric properties of rare earth containing functional materials, it is obvious that rare earths are capable of establishing or fine-tuning phenomena in which high polarizability, adjustable ion size and, in some cases, specific redox chemistry with related defects are involved. Functional materials derived from these phenomena, in most cases, are not metallic but insulators or semiconductors with high ε_r, or ion conductors at elevated temperatures.

References

[1] K. Funke, R. D. Banhatti, Solid State Ionics 2005, 176, 1971.

[2] L. Esaki, P. J. Stiles, S. v. Molnar, Phys. Rev. Lett. 1967, 19, 852.

[3] M. Copel, M. A. Kuroda, M. S. Gordon, X.-H. Liu, S. S. Mahajan, G. J. Martyna, N. Moumen, C. Armstrong, S. M. Rossnagel, T. M. Shaw, P. M. Solomon, T. N. Theis, J. J. Yurkas, Y. Zhu, D. M. Newns, Nano Lett. 2013, 13, 4650.
[4] A. Sousanis, P. F. Smet, D. Poelman, Materials 2017, 10, 953.
[5] A. Jayaraman, V. Narayanamurti, E. Bucher, R. G. Maines, Phys. Rev. Lett. 1970, 25, 1430.
[6] T. Arakawa, H. Kurachi, J. Shiokawa, J. Mater. Sci. 1985, 20, 1207.
[7] C. Berger, E. Bucher, A. Windischbacher, A. D. Boese, W. Sitte, J. Solid State Chem. 2018, 259, 57.
[8] J. G. Bednorz, K. A. Müller, Z. Phys. B: Condens. Matter 1986, 64, 189.
[9] P. H. Hor, R. L. Meng, Y. Q. Wang, L. Gao, Z. J. Huang, J. Bechtold, K. Forster, C. W. Chu, Phys. Rev. Lett. 1987, 58, 1891.
[10] H. Takahashi, K. Igawa, K. Arii, Y. Kamihara, M. Hirano, H. Hosono, Nature 2008, 453, 376.
[11] Z.-A. Ren, G.-C. Che, X.-L. Dong, J. Yang, W. Lu, W. Yi, X.-L. Shen, Z.-C. Li, L.-L. Sun, F. Zhou, Z.-X. Zhao, Europhys. Lett. 2008, 83, 17002.
[12] L. Zhang, F. Liu, K. Brinkman, K. L. Reifsnider, A. V. Virkar, J. Power Sources 2014, 247, 947.
[13] A. Maheshwari, H.-D. Wiemhöfer, Ceram. Int. 2015, 41, 9122.
[14] J. Wang, X. Xiao, Y. Liu, K. Pan, H. Pang, S. Wei, J. Mater. Chem. A 2019, 7, 17675.
[15] B. A. Boukamp, Solid State Ionics 1993, 62, 131.
[16] M. Urquidi-Macdonald, S. Real, D. D. Macdonald, Electrochim. Acta 1990, 35, 1559.
[17] P. Shuk, L. Tichonova, U. Guth, Solid State Ionics 1994, 68, 177.
[18] P. Shuk, U. Guth, Ionics 1995, 1, 106.
[19] M. Rahmani, C. Pithan, R. Waser, J. Eur. Ceram. Soc. 2019, 39, 4800.
[20] R. M. L. Werchmeister, K. K. Hansen, M. Mogensen, J. Electrochem. Soc. 2010, 157, P35.
[21] A. Sher, R. Solomon, K. Lee, M. W. Muller, Phys. Rev. 1966, 144, 593.
[22] M. Anji Reddy, M. Fichtner, J. Mater. Chem. 2011, 21, 17059.
[23] H. Iwahara, T. Esaka, H. Uchida, N. Maeda, Solid State Ionics 1981, 3–4, 359.
[24] A. Skodra, M. Stoukides, Solid State Ionics 2009, 180, 1332.
[25] N. Ito, M. Iijima, K. Kimura, S. Iguchi, J. Power Sources, 2005, 152, 200.
[26] R. Haugsrud, T. Norby, Nat. Mater. 2006, 5, 193.
[27] K. Zhang, X. Xie, H. Li, J. Gao, L. Nie, Y. Pan, J. Xie, D. Tian, W. Liu, Q. Fan, H. Su, L. Huang, W. Huang, Adv. Mater. 2017, 29, 1701804.
[28] A. V. Golubkov, M. M. Kazanin, V. V. Kaminskii, V. V. Sokolov, S. M. Solov'ev, L. N. Trushnikova, Inorg. Mater. 2003, 39, 1251.
[29] B. C. Sales, D. Mandrus, B. C. Chakoumakos, V. Keppens, J. R. Thompson, Phys. Rev. B 1997, 56, 15081.
[30] E. Selezneva, L. E. Clinger, A. T. Ramu, G. Pernot, T. E. Buehl, T. Favaloro, J.-H. Bahk, Z. Bian, J. E. Bowers, J. M. O. Zide, A. Shakouri, J. Electron. Mater. 2012, 41, 1820.
[31] G. Zeng, J. E. Bowers, J. M. O. Zide, A. C. Gossard, W. Kim, S. Singer, A. Majumdar, R. Singh, Z. Bian, Y. Zhang, A. Shakouri, Appl. Phys. Lett. 2006, 88, 113502.
[32] M. Breitschaft, V. Tinkl, N. Pavlenko, S. Paetel, C. Richter, J. R. Kirtley, Y. C. Liao, G. Hammerl, V. Eyert, T. Kopp, J. Mannhart, Phys. Rev. B, 2010, 81, 153414.
[33] P. Moetakef, T. A. Cain, D. G. Ouellette, J. Y. Zhang, D. O. Klenov, A. Janotti, C. G. Van de Walle, S. Rajan, S. J. Allen, S. Stemmer, Appl. Phys. Lett. 2011, 99, 232116.
[34] C. He, T. D. Sanders, M. T. Gray, F. J. Wong, V. V. Mehta, Y. Suzuki, Phys. Rev. B 2012, 86, 081401.
[35] P. Perna, D. Maccariello, M. Radovic, U. Scotti di Uccio, I. Pallecchi, M. Codda, D. Marr, C. Cantoni, J. Gazquez, M. Varela, S. J. Pennycook, F. M. Granozio, Appl. Phys. Lett. 2010, 97, 152111.

[36] A. Annadi, A. Putra, Z. Q. Liu, X. Wang, K. Gopinadhan, Z. Huang, S. Dhar, T. Venkatesan, Ariando, Phys. Rev. B 2012, 86, 085450.
[37] F. Schoofs, M. A. Carpenter, M. E. Vickers, M. Egilmez, T. Fix, J. E. Kleibeuker, J. L. MacManus-Driscoll, M. G. Blamire, J. Phys.: Condens. Matter 2013, 25, 175005.
[38] S. Wolgast, Ç. Kurdak, K. Sun, J. W. Allen, D.-J. Kim, Z. Fisk, Phys. Rev. B 2013, 88, 180405.
[39] D. Xue, K. Betzler, H. Hesse, J. Phys.: Condens. Matter 2000, 12, 3113.
[40] S. K. Jo, J. S. Park, Y. H. Han, J. Alloys Compd. 2010, 501, 259.
[41] M. Suzuki, Materials 2012, 5, 443.

[42] J. Robertson, Eur. Phys. J. Appl. Phys. 2004, 28, 265.
[43] K. H. Goh, A. Haseeb, Y. H. Wong, Mater. Sci. Semicond. Process. 2017, 68, 302.
[44] K. Kakushima, K. Okamoto, T. Koyanagi, M. Kouda, K. Tachi, T. Kawanago, J. Song, P. Ahmet, K. Tsutsui, N. Sugii, T. Hattori, H. Iwai, Microelectron. Eng. 2010, 87, 1868.
[45] D. Kolar, S. Z. S. Gaberscek, D. Suvorov, Ber. Dt. Keram. Ges. 1978, 55, 346.
[46] D. Kolar, S. Gaberscek, B. Volavsek, H. Parker, R. Roth, J. Solid State Chem. 1981, 38, 158.
[47] H. Ohsato, M. Imaeda, Mater. Chem. Phys. 2003, 79, 208.
[48] S. Zhang, F. Yu, R. Xia, Y. Fei, E. Frantz, X. Zhao, D. Yuan, B. H. Chai, D. Snyder, T. R. Shrout, J. Cryst. Growth 2011, 318, 884.

4 Materials and applications

Michael Seitz

4.1 Optical materials – molecules

The optical properties in molecular rare earth complexes can be exploited in many different applications, such as in biomedical analyses or as components in various photoactive devices [1]. The overwhelming number of examples in this area are coordination compounds featuring organic ligands with lanthanoid cations in the most stable oxidation state + III exhibiting metal-centered *f–f* luminescence. Almost all lanthanoids show element-specific luminescence bands with emission wavelengths that are only marginally affected by the surrounding ligand sphere. The lanthanoids with the greatest potential for strong luminescence in molecular compounds are the visible emitters Sm, Eu, Tb and Dy (see Figure 4.1.1), as well as the near-infrared (NIR) emitting Yb.

Figure 4.1.2 shows the energy level diagrams for these Ln^{3+} ions and examples for typically observed luminescent transitions with their corresponding emission wavelengths. For the generation of luminescence, electrons have to be first elevated into an excited state (e.g., the $^5D_0$ level for europium in Figure 4.1.2). In trivalent lanthanoid cations, *f–f* transitions are generally forbidden by the Laporte selection rule (transitions between states of the same parity) and in addition, often also by the spin selection rule (change in spin multiplicity, see, e.g., the transition $^5D_0 \rightarrow ^7F_0$ for europium in Figure 4.1.2). Due to the very low molar absorption coefficients ε for direct *f–f* transitions ($\varepsilon < 10\ M^{-1}\ cm^{-1}$), it is very ineffective to directly populate lanthanoid excited states and, therefore, requires very strong excitation sources (e.g., lasers). While this is usually not very problematic for inorganic, solid state materials (see, e.g., Chapter 4.2 "Optical Materials – Microcrystalline Powders" and Chapter 4.3 "Optical Materials – Ceramics") with high thermal and photostability, this approach is usually not desirable for much more sensitive molecular systems in solution. Due to this problem, the common approach for the sensitization of molecular lanthanoid luminescence utilizes the so-called "antenna effect" (Figure 4.1.3) [3]. The strategy involves bringing a chromophoric ligand which is capable of absorbing light efficiently through highly allowed transitions such as $S_0 \rightarrow S_1$ (e.g., π–π*), into close proximity to the lanthanoid. Figure 4.1.4 shows prototypical examples of ligand motifs that can function as antenna units and which are often encountered in successful molecular lanthanoid luminophores. Since the corresponding excited singlet, S_1, is usually rather short-lived, the possibility for energy transfer from this state to the lanthanoid is often limited in efficiency. Therefore, in the best cases, the system undergoes intersystem crossing (ISC: $S_1 \rightarrow T_1$) to the first excited triplet state T_1, which is much more long-lived than S_1, giving energy transfer to the lanthanoid, a much better chance.

In principle, ISC is a spin-forbidden process and, consequently, not always very favorable. In the case of lanthanoid complexes, however, the "heavy atom effect" due to the high atomic number of the metal centers helps lifting the forbiddenness of ISC considerably, making ISC often an efficient process in this context. If the T_1

https://doi.org/10.1515/9783110654929-024

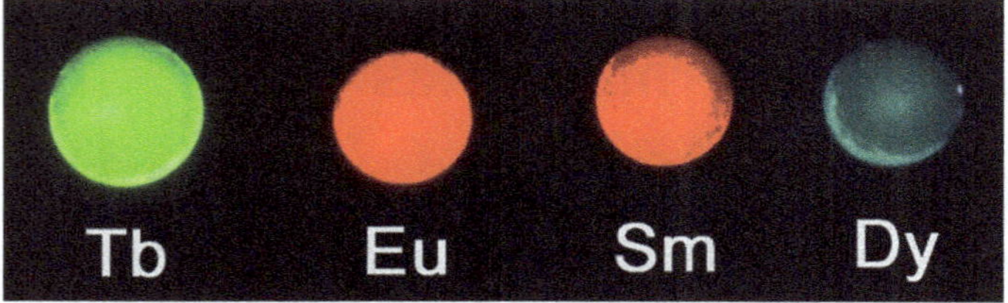

Figure 4.1.1: Typical luminescence colors for the most luminescent lanthanoid ions in molecular complexes (Tb^{3+}, Eu^{3+}, Sm^{3+} and Dy^{3+}).

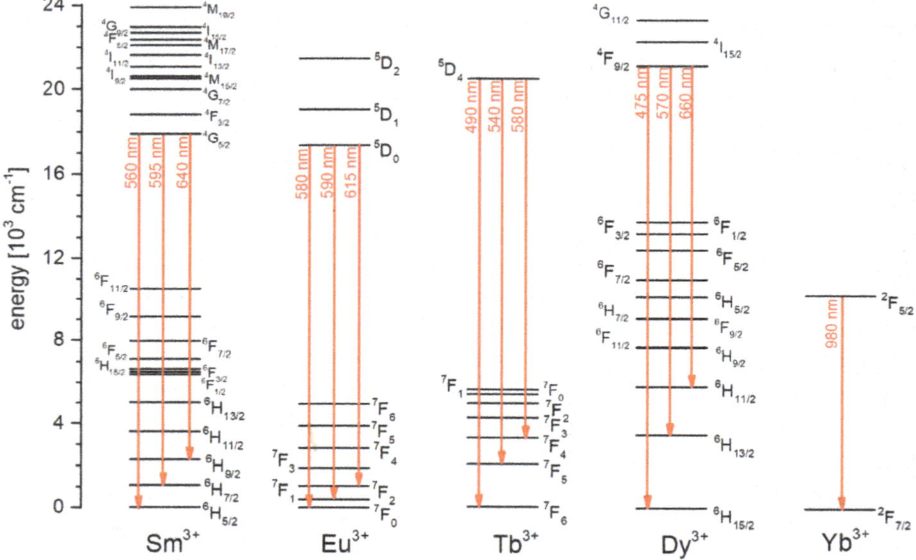

Figure 4.1.2: Partial energy level diagram for the most luminescent trivalent lanthanoids with the most important luminescent transitions and their corresponding approximate wavelengths (energy levels taken from [2]).

state energy is high enough, energy transfer can occur to an excited state of the lanthanoid (see $^{2S'+1}L'_{J'}$ in Figure 4.1.3). As a rule of thumb, $E(T_1)$ should at least be 2,000 cm^{-1} above the emitting lanthanoid state to avoid thermal back transfer from the lanthanoid to the ligand. As each individual lanthanoid has different energies of the relevant emitting states, the antenna moieties must also feature tailor-made T_1 energies to satisfy the requirements outlined above. For example, the emitting states for Eu^{3+} ($^5D_0 \approx 17{,}400$ cm^{-1}) and Tb^{3+} ($^5D_4 \approx 20{,}500$ cm^{-1}) have very different energetic positions (see Figure 4.1.5). Terbium requires an antenna with a relatively high triplet energy $E(T_1)$, which not all ligands can provide. For example, the 2,2′-bipyridine-based cryptand [4a] in Figure 4.1.5 (left side) only has a triplet

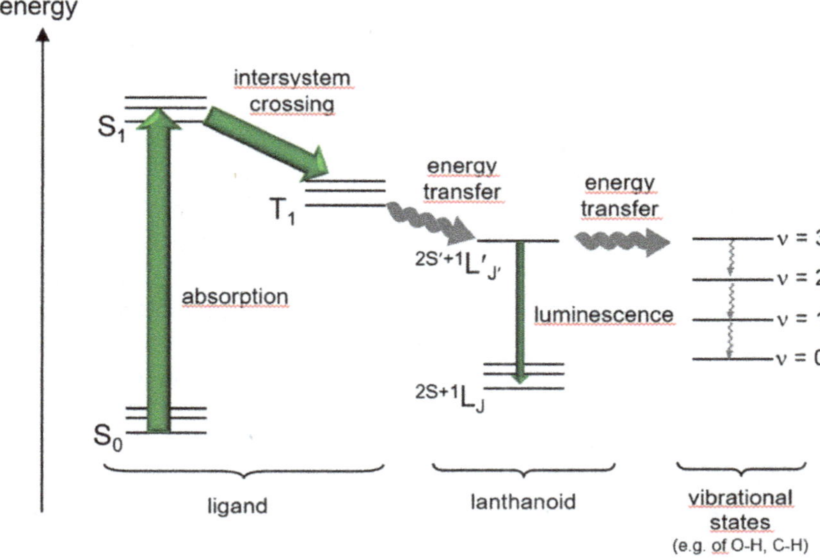

Figure 4.1.3: Jablonski diagram showing the sensitization pathway for lanthanoid luminescence by the antenna effect (ligand and lanthanoid; S_0/S_1: ground/excited singlet states; T_1: excited triplet state) as well as vibrational quenching of lanthanoid excited states by molecular oscillators (lanthanoid and vibrational states with vibrational quantum number v).

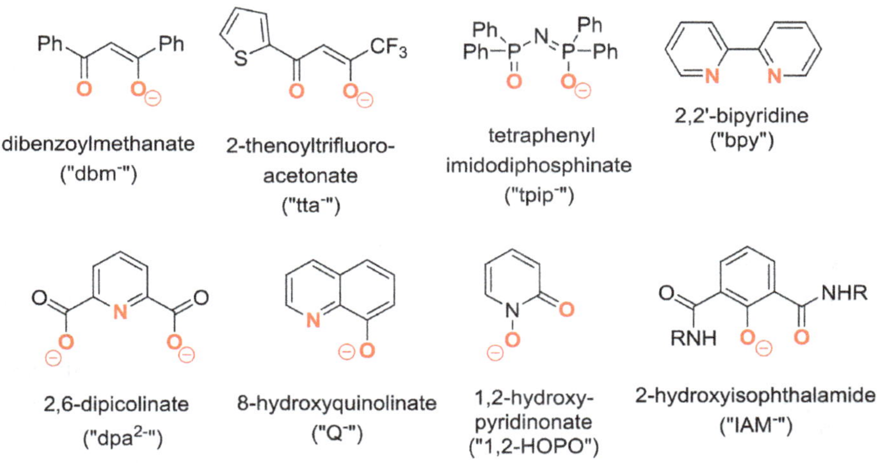

Figure 4.1.4: Selected examples for typical sensitizing chelator motifs ("antenna moieties") in molecular lanthanoid luminophores.

energy of $E(T_1) \approx 22\,100$ cm^{-1} which is only 1600 cm^{-1} above the $^5D_4$ state of Tb^{3+}, but sufficiently high for the effective population of the $^5D_0$ state of Eu^{3+}. Consequently, the cryptand is only a suitable antenna for Eu^{3+} and not for Tb^{3+}. On the other hand, ligand systems with higher antenna triplet levels such as the one shown on the right side of Figure 4.1.5 can provide efficient sensitization for Tb^{3+} [4b].

Another problem for molecular lanthanoid luminophores is that the corresponding metal-centered excited states are subject to non-radiative deactivation ("quenching") by energy transfer to high-energy overtones of vibrational modes in the vicinity of the lanthanoid ion (see Figure 4.1.3). As a dipole-dipole interaction, this effect is strongly dependent on the distance r between Ln and the oscillator moiety (quenching power $\propto r^{-6}$) and is, consequently, mainly operative for vibrational modes in the surrounding ligands and solvent molecules bound in the inner coordination sphere of the metal. The most detrimental oscillators in this respect are the anharmonic stretching modes of O-H-, N-H- and C-H units with vibrational energies > 2,800 cm^{-1} for their fundamental modes [5]. Especially, H_2O molecules are a severe problem in this context because they are very small and good ligands for Ln^{3+} that can bind to the lanthanoids very easily. One of the goals for the development of efficient molecular luminophores is, therefore, the saturation of the entire coordination sphere (usually for coordination numbers of 8 or 9) by multidentate antenna ligands and the concomitant avoidance of free coordination sites for solvent molecules such as H_2O. Another strategy, if the presence of quenching oscillators cannot be avoided, is the isosteric replacement with deuterated or halogenated moieties, for example, the deuteration of O-H to O-D moieties (e.g., by using D_2O instead of H_2O) or the fluorination of C-H to the corresponding C-F units. In both cases, the increase in reduced mass of the diatomic oscillators leads to a considerable lowering of the vibrational energies of the quenching overtones (e.g., fundamental modes for $\nu_{O\text{-}H} \approx 3{,}300$ cm^{-1} vs $\nu_{O\text{-}D} \approx 2{,}500$ cm^{-1} or $\nu_{C\text{-}H}$(aromatic) $\approx 3{,}050$ cm^{-1} vs $\nu_{C\text{-}F} \approx 1{,}250$ cm^{-1}). This in turn necessitates a higher overtone to bridge the energy gap ΔE between the lanthanoid emitting state and the next lower level (see, e.g., the $^5D_0$ and $^7F_6$ states of Eu^{3+} in Figure 4.1.6), leading to reduced efficiency of the detrimental energy transfer from the metal to the vibrational modes and, consequently, improved luminescence efficiency. This difference in quenching efficiency can also be used in a productive way to estimate the number of water molecules (usually called "q") bound to the lanthanoid in a specific ligand environment by measuring the different luminescence lifetimes in H_2O and D_2O and by using empirical equations developed for a number of different lanthanoids [6]. This information is quite useful for the improvement of lanthanoid luminophores but also crucial for the development of gadolinium-based MRI contrast agents (see Chapter 4.5 "Medical applications of rare earth compounds"). The vibrational quenching becomes more and more severe, the smaller the energy gap ΔE turns out to be. The lanthanoids Eu and Tb with large gaps ($\Delta E > 12{,}000$ cm^{-1}, see Figure 4.1.2) are usually only affected by O-H (and for Eu also by N-H) to a relevant extent and show very minor quenching by C-H in

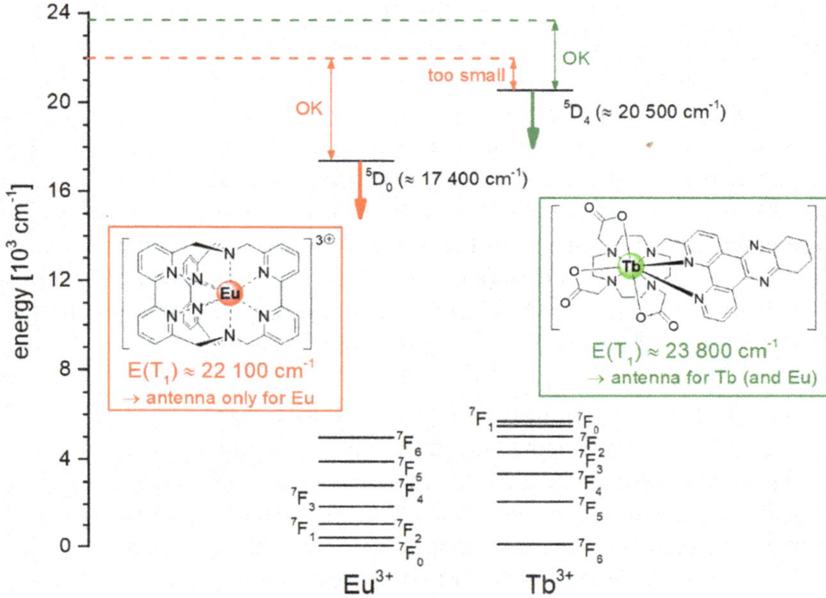

Figure 4.1.5: Importance of the energetic position of the ligand-centered triplet level for the sensitization of different lanthanoids – Left: Cryptate-based europium complex [4a] with relatively small E(T_1), rendering the antenna suitable only for Eu^{3+}; Right: DOTA-based terbium complex [4b] with higher E(T_1) which enables efficient sensitization of Tb^{3+} (the triplet energies E(T_1) were measured in the corresponding Gd^{3+} complexes).

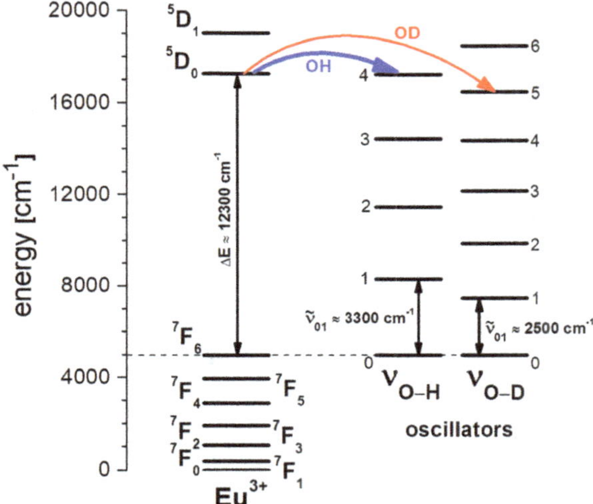

Figure 4.1.6: Vibrational quenching of the emitting $^5D_0$ state of Eu^{3+} by O-H- and O-D-stretching overtones – deuteration of oscillators lowers the vibrational energy and consequently requires a higher overtone for a given energy gap ΔE leading to decreased quenching and increased luminescence.

ligands and/or solvents. Consequently, the best molecular luminophores with these lanthanoids can reach quantum yields well above $\Phi_L > 50\%$ even in aqueous solution [1]. Other lanthanoids, such as Sm, Dy, and Yb with medium energy gaps ($\Delta E \approx 7{,}000$–$10{,}000\ cm^{-1}$, see Figure 4.1.2) are affected by all X-H oscillator classes (X = O, N, C). They can often profit considerably from deuteration/halogenation and in favorable circumstances can achieve quantum efficiencies of $\Phi_L > 10\%$. Figure 4.1.7 shows successful examples of the effect that perdeuteration, combined with fluorination, can have on the luminescence efficiency of the NIR-emitting Yb^{3+} ($\lambda_{em} \approx$ 980 nm) [7, 8]. All other lanthanoids (e.g., Pr, Nd, Ho, Er, Tm) have even smaller gaps ($\Delta E < 7{,}000\ cm^{-1}$) and even the complete avoidance of high-energy oscillators is usually not enough to improve quantum yields Φ_L beyond 1%.

In summary, the design of highly luminescent, molecular lanthanoid luminophores in solution should take into account the following aspects:

- The ligand(s) should provide a thermodynamically and/or kinetically stable lanthanoid complex and saturate the metal coordination sphere as much as possible to avoid binding of solvent molecules (e.g., H_2O) which quench the luminescence. If applications in solution or even in biological media are envisaged, high-denticity, preferably macrocyclic ligands with hard Lewis-basic donors (such as DOTA derivatives, see Chapter 2.6 "Rare Earth Coordination Chemistry"), are required.
- The lanthanoid should be close to a chromophoric antenna moiety in the ligand sphere with high molar absorption coefficient at the desired excitation wavelength and with suitable triplet energy $E(T_1)$ tailor-made for the specific lanthanoid in question.

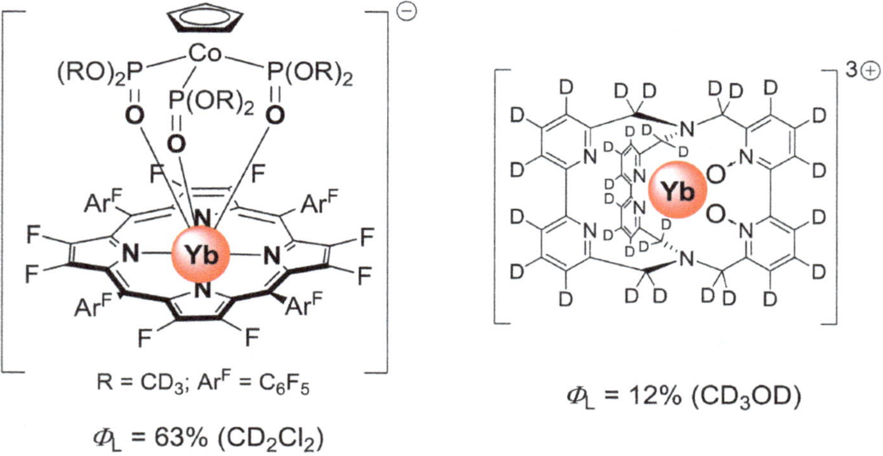

Figure 4.1.7: Selected ytterbium near-infrared luminophores with high luminescence efficiency – Avoidance of detrimental high-energy C-H-oscillators close to the metal center by fluorination and/or deuteration [7].

– High-energy oscillators such as X-H (X = O, N, C) in the vicinity of the metal center should not be present, if possible. If this cannot be avoided, deuteration and/or halogenation can often alleviate the quenching for some lanthanoids, especially the ones which emit in the NIR such as Yb^{3+}.

Using these principles, several very successful molecular lanthanoid luminophores have been developed over the years, especially for the most luminescent lanthanoids Eu^{3+} and Tb^{3+} [4a, 9] (Figure 4.1.8). These complexes are extremely stable and highly luminescent and can even be used in biological media. They allow covalent attachment to bioactive molecules for targeted purposes *via* functional groups (FG in Figure 4.1.8) in the complex periphery.

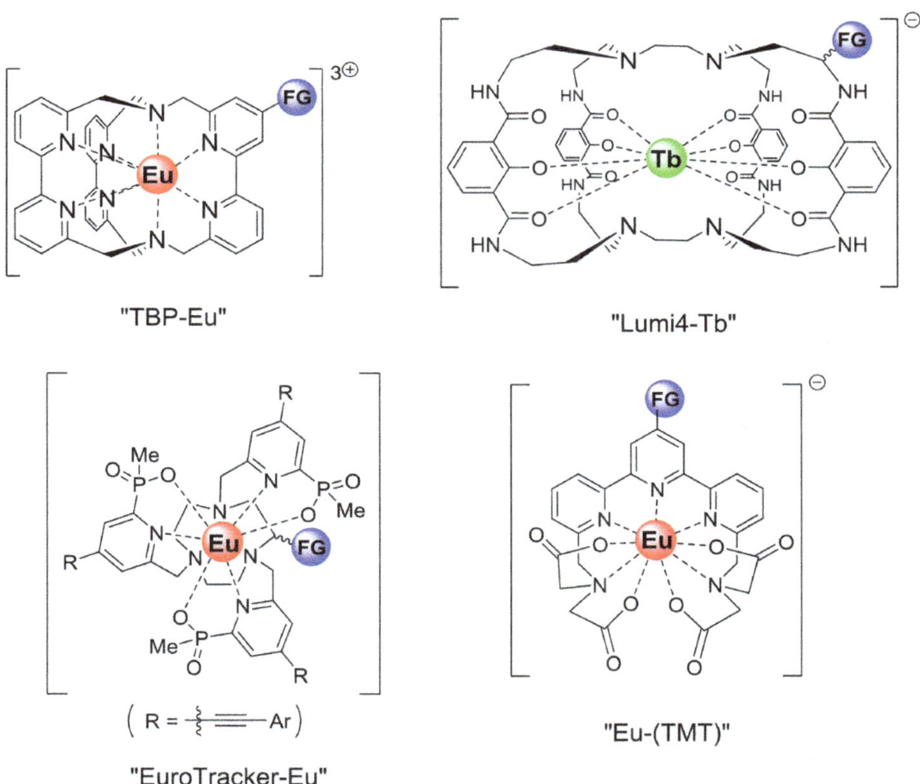

Figure 4.1.8: Prototypical examples for successful europium and terbium luminophores used for the conjugation to biomolecules via functional groups FG [4a, 9].

An aspect that is very unique to molecular lanthanoid luminophores in solution as opposed to solid materials, is the possibility to design responsive systems that allow the modulation of the luminescence properties (e.g., emission intensity, luminescence lifetimes) by a certain chemical or physical stimulus (e.g., temperature, pH, O_2 concentration in solution, the presence of specific analytes) that is of interest for a particular application [10]. For this purpose, a number of structural and electronic parameters important for the luminescence response can be made susceptible to a desired input, for example by modulating the electronic structure of the antenna unit (e.g., changing the energetic position of T_1), by varying the energy transfer efficiency between lanthanoid and antenna through variation of the distance between them, or by changing the number of water molecules bound to Ln^{3+}. Figures 4.1.9 and 4.1.10 show two typical cases.

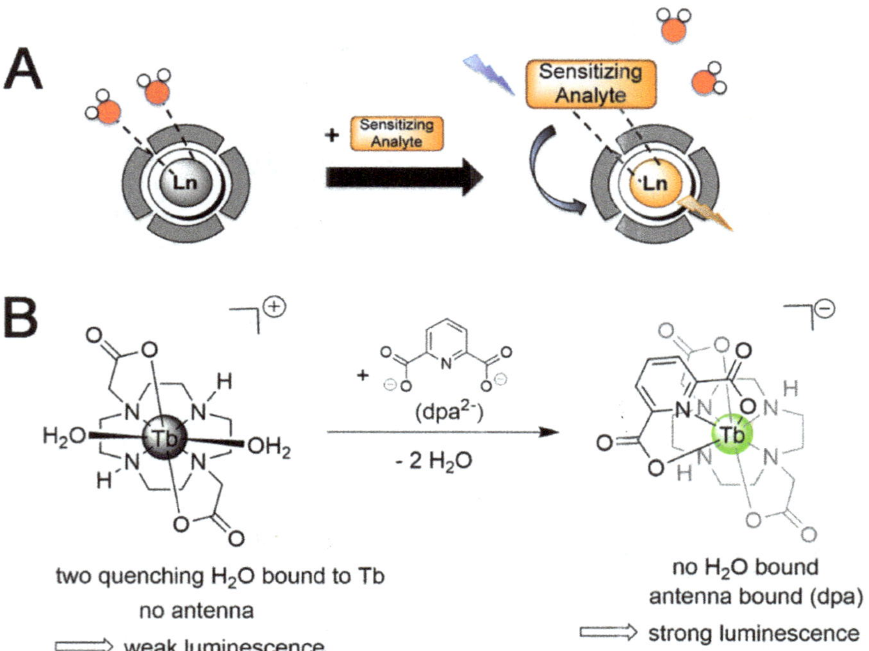

Figure 4.1.9: Molecular terbium complex for the detection of dipicolinate (dpa^{2-}) from bacterial anthrax spores in solution [11] – A: Schematic representation of the functional principle (reprinted with permission from Chem. Rev. 2014, 114, 4496. Copyright 2014 American Chemical Society); B: Complexation process in the presence of dpa^{2-} showing the structural changes and the corresponding effect on luminescence response.

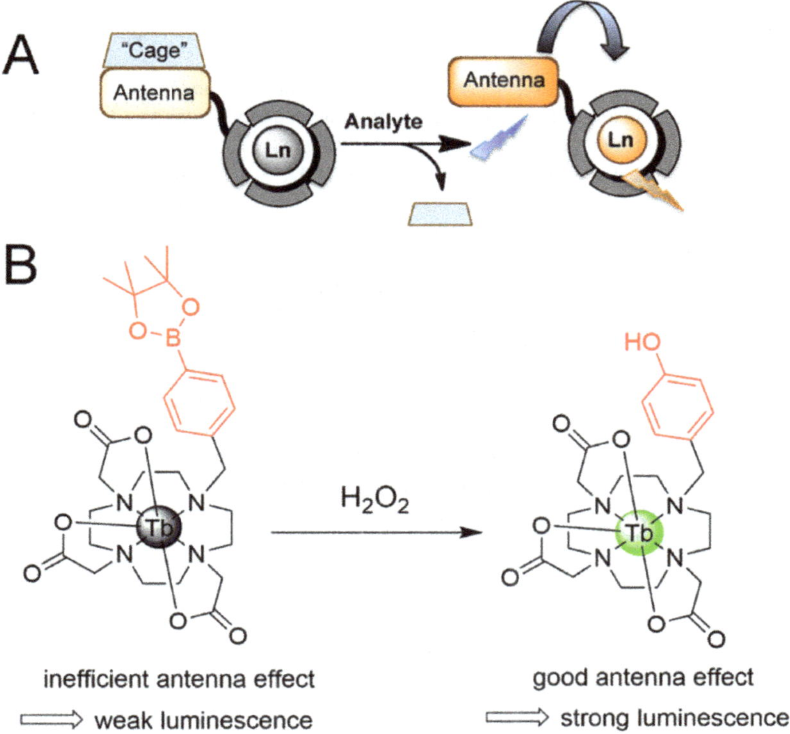

Figure 4.1.10: Molecular terbium complex for the detection of hydrogen peroxide in solution [12] – A: Schematic representation of the functional principle (reprinted with permission from Chem. Rev. 2014, 114, 4496. Copyright 2014 American Chemical Society); B: Chemical and physical changes in the antenna properties upon reaction of the "caged" antenna in the sensor with H_2O_2.

References

[1] a) J.-C. G. Bünzli, S. V. Eliseeva, Photophysics of Lanthanoid Coordination Compounds, in Comprehensive Inorganic Chemistry II (ed. J. Reedijk, K. Poeppelmeier), Vol. 8, Elsevier, Amsterdam, 2013, 339–398; b) J.-C. G. Bünzli, Lanthanide Luminescence: From a Mystery to Rationalization, Understanding, and Applications, in Handbook on the Physics and Chemistry of Rare Earths (ed. J.-C. G. Bünzli, V. K. Pecharsky), Vol. 50, Elsevier, Amsterdam, 2016, 141–176; c) Luminescence of Lanthanide Ions in Coordination Compounds and Nanomaterials (ed. A. de Bettencourt-Dias), Wiley, Chichester, 2014; M. Sy, A. Nonat, N. Hildebrandt, L. J. Charbonniere, Chem. Commun. 2016, 52, 5080; d) J.-C. G. Bünzli, J. Lumin. 2016, 170, 866.

[2] a) W. T. Carnall, P. R. Fields, K. Rajnak, J. Chem. Phys. 1968, 49, 4424; b) W. T. Carnall, P. R. Fields, K. Rajnak, J. Chem. Phys. 1968, 49, 4447; c) W. T. Carnall, P. R. Fields, K. Rajnak, J. Chem. Phys. 1968, 49, 4450; d) W. T. Carnall, G. L. Goodman, K. Rajnak, R. S. Rana, J. Chem. Phys. 1989, 90, 3443.

[3] S. I. Weissman, J. Chem. Phys. 1942, 10, 214.

[4] a) B. Alpha, J.-M. Lehn, G. Mathis, Angew. Chem. Int. Ed. Engl. 1987, 26, 1266; b) C. P. Montgomery, E. J. New, L. O. Palsson, D. Parker, A. S. Batsanov, L. Lamarque, Helv. Chim. Acta 2009, 92, 2186.

[5] E. Kreidt, C. Kruck, M. Seitz, Nonradiative Deactivation of Lanthanoid Luminescence by Multiphonon Relaxation in Molecular Complexes, in Handbook on the Physics and Chemistry of Rare Earths, Vol. 53, Elsevier, Amsterdam, 2018, 35–79.

[6] a) R. M. Supkowski, D. W. Horrocks Jr., Inorg. Chim. Acta. 2002, 340, 44; b) A. Beeby, I. M. Clarkson, R. S. Dickins, S. Faulkner, D. Parker, L. Royle, A. S. de Sousa, J. A. G. Williams, M. Woods, J. Chem. Soc., Perkin Trans. 1999, 2, 493.

[7] a) J.-Y. Hu, Y. Ning, Y.-S. Meng, J. Zhang, Z.-Y. Wu, S. Gao, J.-L. Zhang, Chem. Sci. 2017, 8, 2702; b) C. Doffek, M. Seitz, Angew. Chem. Int. Ed. 2015, 54, 9719.

[8] a) Y. Ning, M. Zhu, J.-L. Zhang, Coord. Chem. Rev. 2019, 399, 213028; b) S. Comby, J.-C. G. Bünzli, Lanthanide Near-Infrared Luminescence in Molecular Probes and Devices, in Handbook on the Physics and Chemistry of Rare Earths (ed. K. A. Gschneidner Jr., J.-C. G. Bünzli, V. K. Pecharsky), Vol. 37, Elsevier, Amsterdam, 2007, 217–470.

[9] a) S. J. Butler, M. Delbianco, L. Lamarque, B. K. McMahon, E. R. Neil, R. Pal, D. Parker, J. W. Walton, J. M. Zwier, Dalton Trans. 2015, 44, 4791; b) J. Xu, T. M. Corneille, E. G. Moore, G.-L. Law, N. G. Butlin, K. N. Raymond, J. Am. Chem. Soc. 2011, 133, 19900; c) A. K. Saha, K. Kross, E. D. Kloszewski, D. A. Upson, J. L. Toner, R. A. Snow, C. D. V. Black, V. C. Desai, J. Am. Chem. Soc. 1993, 115, 11032.

[10] M. C. Heffern, L. M. Matosziuk, T. J. Meade, Chem. Rev. 2014, 114, 4496.

[11] M. L. Cable, J. P. Kirby, D. J. Levine, M. J. Manary, H. B. Gray, A. Ponce, J. Am. Chem. Soc. 2009, 131, 9562.

[12] A. R. Lippert, T. Geschneidtner, C. J. Chang, Chem. Commun. 2010, 46, 7510.

Thomas Jüstel, Florian Baur

4.2 Optical materials – microcrystalline powders

Inorganic luminescent materials such as microcrystalline powders also called phosphors are generally characterized by the emission of photons with energy beyond thermal equilibrium. In other words, the nature of luminescence is different from incandescence or black body radiation, which gives rise to light emission of (halogen) incandescent lamps or celestial objects. Microcrystalline inorganic luminescent pigments thus emit nonthermal radiation after the absorption of energy from an external source. In contrast, color pigments absorb and subsequently convert the up-taken energy into heat (phonons), without the emission of light (Figure 4.2.1).

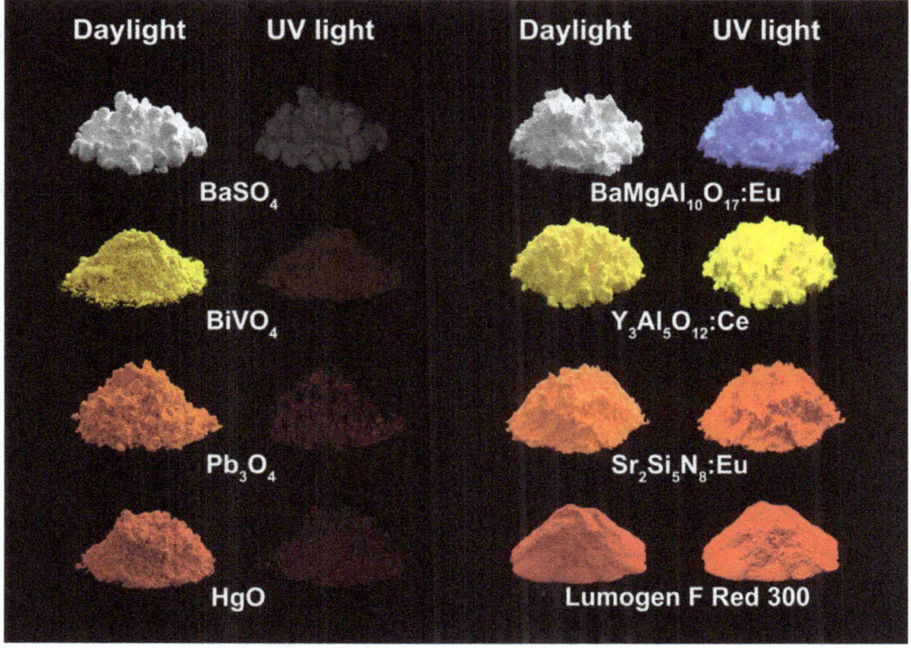

Figure 4.2.1: Appearance of some color pigments (left column) and luminescent pigments (right column) under daylight and upon UV irradiation.

Luminescence can occur as a result of very different types of excitation pathways, which is reflected by established terms such as bio-, cathodo-, chemo-, electro-, photo-, radio-, sono-, thermo- or triboluminescence [1]. In case of microscale materials

https://doi.org/10.1515/9783110654929-025

of practical importance, most commonly, the excitation takes place by X-rays, cathode rays, UV radiation emitted by a gas discharge, near-UV to blue emission of inorganic or organic LEDs. The role of the phosphor is to convert the incident radiation, mostly into visible light, and occasionally, into UV or NIR radiation. In addition to the type of excitation, three other terms are used quite often to classify inorganic luminescent materials. These expressions are related to the decay time (τ): fluorescence ($\tau < 10\ \mu s$), phosphorescence ($\tau > 10\ \mu s$) and afterglow ($\tau > 0.1$ s).

The luminescence of inorganic microcrystalline solids can be described by two mechanisms: luminescence from localized centers, for example, an activator A, or luminescence from semiconductor defects states [1, 2]. For reasons that are discussed below, the activator A is very often a rare earth ion, namely, Ce^{3+}, Pr^{3+}, Nd^{3+}, $Sm^{2+/3+}$, $Eu^{2+/3+}$, Gd^{3+}, Tb^{3+}, Dy^{3+}, Ho^{3+}, Er^{3+}, Tm^{3+} or Yb^{3+}, that means, ions with an electronic configuration $[Xe]4f^n$ with n = 1–13 are used. These ions show either band emission or narrow emission lines. In the latter case, narrow and weak absorption lines are also observed, which gives rise to the need for sensitization by another ion. Such sensitizer ion, S, can either be used to obtain sufficient absorption strength of incident radiation or to ease the energy transfer from host lattice-related energy levels to the activator A (Figure 4.2.2).

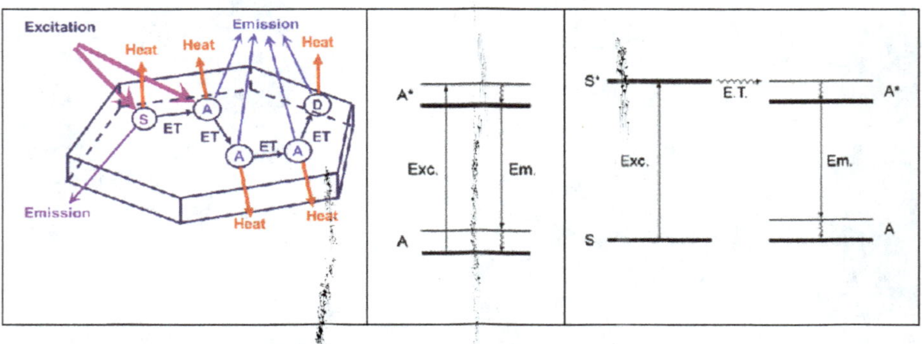

Figure 4.2.2: Sketches illustrating the underlying processes in microcrystalline luminescent pigments (left), on an optical center A (middle), which might be sensitized by S (right); D denotes a defect, which causes quenching.

Luminescence from localized ions is represented by transitions between energy levels of dopant ions, for example, the 4f–4f transitions of Eu^{3+} in Y_2O_3:Eu^{3+} or complex ions, or the ligand-to-metal charge-transfer (LMCT) transition of the $[WO_4]^{2-}$ moiety in $CaWO_4$ or $MgWO_4$. In case of luminescent centers, the transition rate is (more or less strongly) correlated to the relevant quantum-mechanical selection rules, and reflected in intensity as well as decay time of the transition. Excitation and emission can be both (Figure 4.2.2) localized to a single center, for example,

$[WO_4]^{2-}$ in tungstates or separated from each other: excitation on sensitizer, for example, Ce^{3+} in $LaPO_4:Ce^{3+},Tb^{3+}$, emission on activator, for example, Tb^{3+} in $LaPO_4:Ce^{3+},Tb^{3+}$. Consequently, energy transfer between sensitizer and activator, for instance, due to dipole–dipole, multipole–multipole or exchange interaction, is required for the latter.

Luminescence from semiconductor defect states can be observed from any kind of microcrystalline materials with a band gap larger than 0 eV. Such self-trapped exciton luminescence (STE) occurs subsequent to band-to-band excitation, between impurity states localized in the band gap, for example, after VUV or X-ray excitation of $Lu_3Al_5O_{12}$ (LuAG), see Figure 4.2.3. Excess of STE luminescence is dependent on the defect density of a microcrystalline powder and demonstrates the importance of crystallinity and purity of such materials.

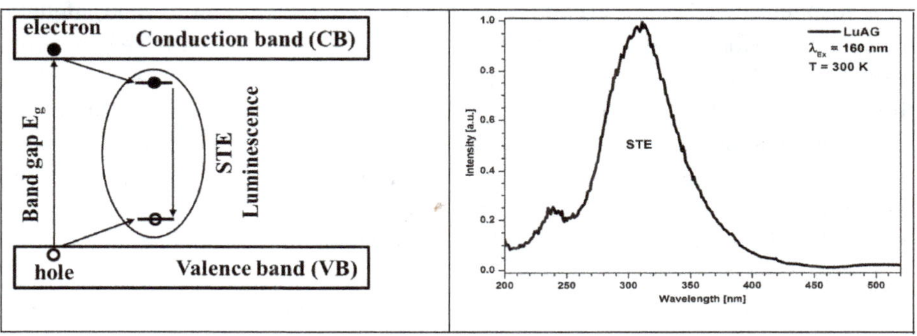

Figure 4.2.3: Sketch illustrating the underlying process causing self-trapped exciton (STE) luminescence (left), and the related STE spectrum of $Lu_3Al_5O_{12}$ (LuAG) after vacuum UV excitation at ambient temperature (right).

Although inorganic luminescent materials have been known since the tenth century in China and Japan as well as since the middle age in Europe [3, 4], the first artificially made microcrystalline luminescent powders just showed up at the end of the nineteenth century. $CaWO_4$ and doped zinc sulfides, for example, $ZnS:Ag^+$, $ZnS:Cu^{2+}$ and $ZnS:Mn^{2+}$ were used by Becquerel, Braun, Edison, and Pupin to produce visible light from Hg discharge lamps, X-ray sources or cathode ray tubes (CRTs). The latter application led to the development of color TVs in 1960, with the need for a large color gamut, that is, saturated red, green and blue primary colors. To this end, efficient, stable and color saturated cathodoluminescent phosphors were required. In view of the efficiency linearity and stability, microcrystalline inorganic materials yielded the best CRT phosphors [5].

Such CRT phosphors with a superior color purity could be obtained by the use of rare earth activators, since the trivalent rare earth ions show characteristic narrow line spectra in the visible range; this has been known since the early twentieth century [6]. In the 1960s the very efficient and temperature-stable microcrystalline phosphors $Y_2O_3:Eu^{3+}$, $Gd_2O_3:Eu^{3+}$, and $YVO_4:Eu^{3+}$ were discovered [7–9], while GTE labs proposed the use of the latter for full color TV sets [10]. The desire to enhance the color gamut and the screen frame frequency resulted in the development of $Y_2O_2S:Eu^{3+}$, some years later [11]. This red emitting phosphor dominated CRTs due to its superior properties until this display technology was replaced by plasma, liquid crystal and OLED displays at the dawn of the twenty-first century.

The most important application of microcrystalline phosphors by scale is the conversion of the emission lines of a low-pressure Hg discharge at 185.0 and 253.7 nm into visible light, as it is done in compact or linear fluorescent lamps since the early twentieth century [12–14]. Rare-earth-activated luminescent microscale materials came into play in the early 1970s and have widely replaced the rare-earth free halophosphates $Ca_5(PO_4)_3(F,Cl):Sb^{3+},Mn^{2+}$ due to the higher quantum yield, lifetime, and luminous efficacy, and to enable fluorescent lamps with an optimized spectrum to enable a high color rendering index [15–17]. As demonstrated in Table 4.2.1, phosphors for fluorescent lamps are activated by few rare earth ions solely, as these ions enable emission over the whole visible range, namely, Eu^{2+} for broad band emission in the blue, Tb^{3+} for line emission in the green and Eu^{3+} for line emission in the red spectral range. Ce^{3+} is used for the sensitization of Tb^{3+}, since in the applied host materials, Tb^{3+} alone does not absorb the 253.7 nm Hg line.

Table 4.2.1: Microscale rare earth activated phosphors used in low-pressure Hg discharge lamps (data partly from [17]).

Phosphor	Structure type of the host material	Emission peak (nm)	Absorption at 254 nm (%)	Quantum yield (%)	Luminous efficacy (lm/W)
$BaMgAl_{10}O_{17}:Eu^{2+}$	ß-Alumina	450	90	90	90
$(Sr,Ba)_5(PO_4)_3Cl:Eu^{2+}$	Apatite	475	95	90	200
$(Ce,Tb)MgAl_{11}O_{19}$	Magnetoplumbite	541	95	90	495
$(Ce,Gd,Tb)MgB_5O_{10}$	–	542	95	90	490
$LaPO_4:Ce^{3+},Tb^{3+}$	Monazite	545	95	93	500
$Y_2O_3:Eu^{3+}$	Bixbyite	611	75	90	280

The dominance of Tb^{3+} and Eu^{3+} as activators for green and red emitting fluorescent lamp phosphors can be explained by the many suitable energy levels (→ Dieke diagram), which is exemplary for Eu^{3+}, as shown by the following figure (Figure 4.2.4). Most important for the distance of the terms involved in the luminescence process, that is, between the ^5D and ^7F terms, is the electron-electron repulsion, which splits the first excited term ^5D from the ground term ^7F by about 17,000 cm^{-1}. However,

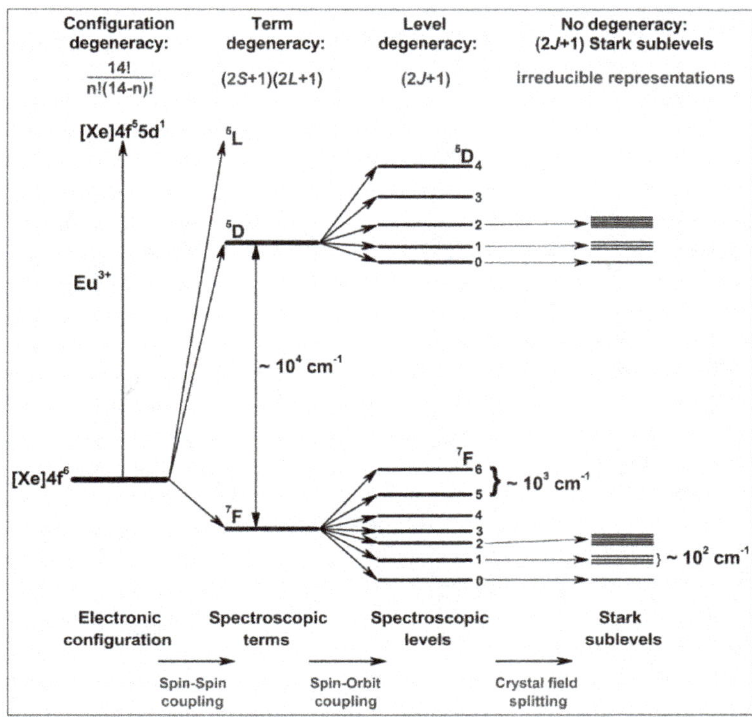

Figure 4.2.4: Sketch of the splitting of the energy levels of the [Xe]$4f^6$ configuration of Eu^{3+} (adapted from [18, 19]).

the details of the emission spectrum of Eu^{3+} activated microcrystalline phosphors are rather complicated, since spin-orbit and crystal-field splitting further separates the ^5D and ^7F terms into sub-levels, while the extent of splitting is a function of size and symmetry of the crystallographic position.

As the emitting state of Eu^{3+} is the $^5D_0$ level, the return to the ground term levels ^7F_J (with J = 0–6) is due to spin- and parity-forbidden intraconfigurational 4f-4f transitions. The majority of these transitions are induced electric dipole transitions (ED transitions), while their intensities can be described by the Judd-Ofelt theory [20, 21]. Some transitions such as the $^5D_0 \rightarrow {}^7F_1$ transition have magnetic dipole character. Magnetic dipole transitions (MD transitions) are allowed by the Laporte selection rule, but their intensities are weak and comparable to those of the induced electric dipole transitions [20]. While the intensity of a magnetic dipole transition is independent of the surrounding, the ED transitions are strongly dependent on the environment. Particularly, the $^5D_0 \rightarrow {}^7F_2$ transition is a so-called hypersensitive transition, which means that its intensity is more strongly governed by the local symmetry of the Eu^{3+} ion and the nature of the surrounding anions, than the

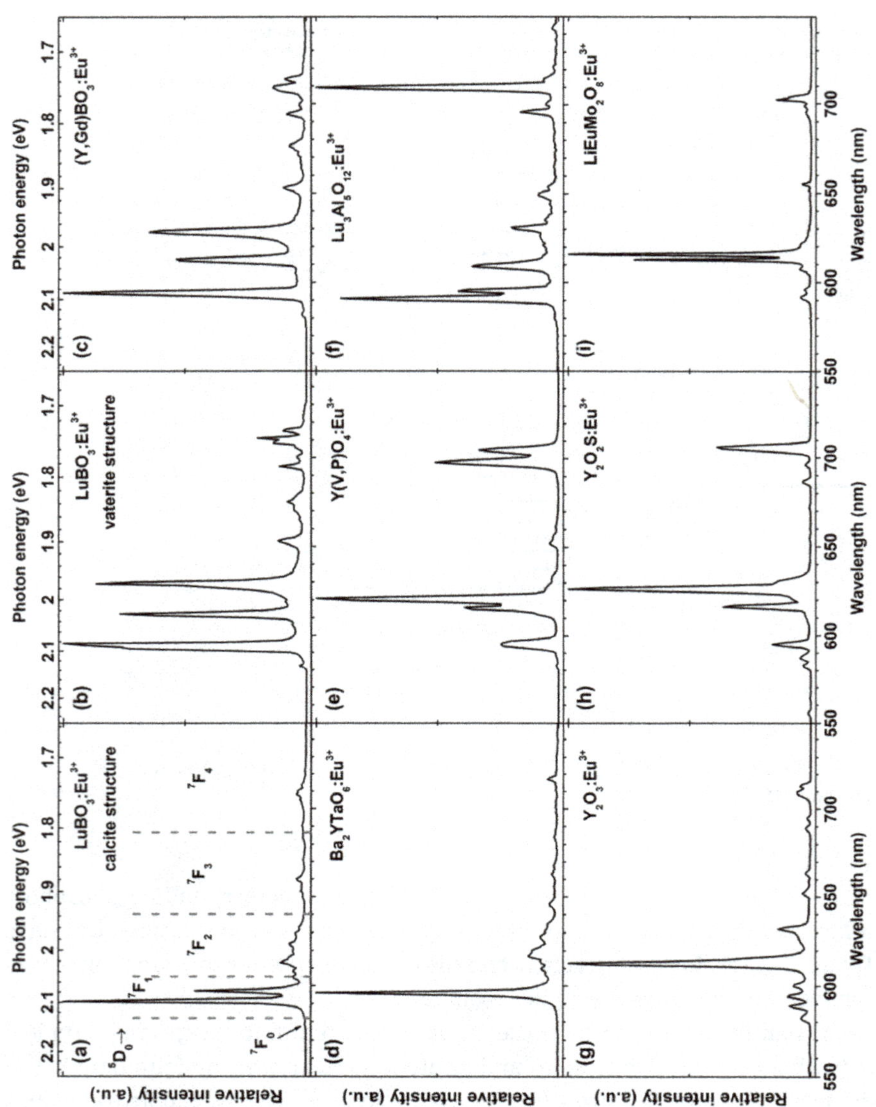

Figure 4.2.5: Emission spectra of selected Eu^{3+} phosphors.

intensities of the other ED transitions. If a Eu^{3+} phosphor with deep red emission is desired, the host materials must offer sites with a low local symmetry for Eu^{3+}, promoting the hypersensitive $^5D_0 \rightarrow {}^7F_2$ transition.

The optimization of the emission spectrum and thus the color point of Eu^{3+} activated microcrystalline powders (Figure 4.2.5) is of tremendous importance in obtaining suitable red emitting phosphors (Table 4.2.2) for emissive displays, such as cathode-ray tubes and plasma panels. A comparison of emission spectra of selected Eu^{3+} phosphors show that the most relevant transitions are located between 580 and 720 nm, but that the intensity pattern of the Eu^{3+} emission lines is a sensitive function of the chemistry and symmetry of the host material [21–26].

Table 4.2.2: Properties of selected Eu^{3+} activated phosphors.

Phosphor	Structure type of the host material	Emission peak (nm)	Decay time [ms]	Luminous efficacy [lm/W]	CIE1931 color point *x*, *y*
$InBO_3:Eu^{3+}$	Calcite	590	10	350	0.61, 0.39
$(Y,Gd)BO_3:Eu^{3+}$	Vaterite	595	8	290	0.65, 0.35
$Y_2O_3:Eu^{3+}$	Bixbyite	611	2.5	280	0.64, 0.34
$Y(V,P)O_4:Eu^{3+}$	Zircon	615	1	220	0.67, 0.33
$Y_2O_2S:Eu^{3+}$	–	627	1	200	0.66, 0.33

Even though Eu^{3+} activated luminescent materials are widely applied in displays and fluorescent lamps, so far, they have not been applied as a microscale conversion screen in light sources with main emission in the near-UV or visible range, since their absorption cross section in this spectral range is much too low.

Since the invention of the blue emitting (In,Ga)N LED [24], inorganic LEDs can be used for the construction of all types of colored and white light sources, while the simplest approach to obtain white light is the use of a single yellow emitting phosphor, like YAG:Ce, on top of a blue emitting LED chip [27–30].

However, this approach results in a high color temperature and rather low color rendering due to the lack of photon flux in the red spectral region. Another way to obtain white light sources is, for example, the construction of a package, which comprises a blue emitting (In,Ga)N chip and a luminescent screen, either yellow and a red, or a green and a red microcrystalline phosphor. Such LEDs comprising a dichromatic phosphor blend are superior than single phosphor LEDs, because the color temperature and the color rendering values can be adjusted according to the requirements in different application areas [31, 32]. A graphical illustration of these three different ways of generating white light by phosphor converted LEDs is depicted in Figure 4.2.6.

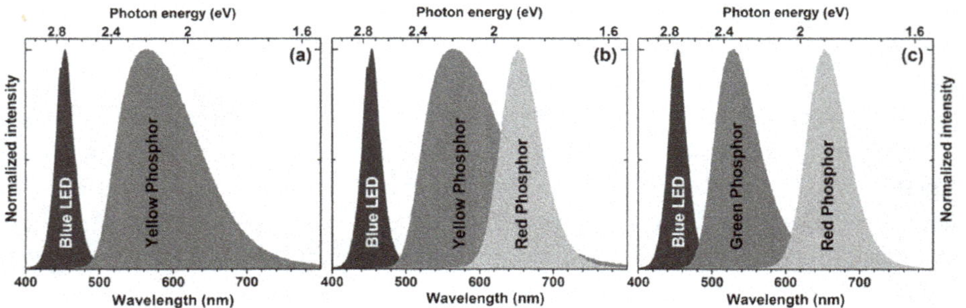

Figure 4.2.6: Normalized emission spectra of the components of phosphor converted LEDs: (a) blue/yellow, (b) blue/yellow/red, and (c) blue/green/red.

The invention and global spread of warm white LEDs initiated the development of luminescent materials being able to down-convert the blue light emitted from the chip to the cyan, green, yellow, orange to red spectral range. Applied converter materials are microcrystalline powders embedded into a polymer resin for the sake of scattering to enable sufficient color mixing and the need to fulfil a range of requirements in order to be applicable in white LEDs. Such requirements are listed as follows [33]:

- An emission spectrum, which in combination with the emission of the other components (chip, other phosphors), leads to white light with a specific color rendering and color temperature
- High absorption cross section strongly overlapping with the emission spectra of the chip
- High thermal quenching temperature
- High quantum yield
- Long-term photo-chemical and thermal stability
- Absence of emission saturation upon a high flux
- Good thermal conductivity

A short overview of widely applied microscale phosphors for blue emitting LEDs is given in Table 4.2.3, whereby the list of microcrystalline LED phosphors under consideration is much longer [34].

The dominance of Ce^{3+} and Eu^{2+} activated luminescent materials is caused by two factors: On the one hand, these rare earth ions show rather strong and broad absorption bands, which are needed for a sufficiently high conversion efficiency onto high brightness or high power LED chips; on the other hand, the underlying 4f-5d states allow strong modulation of the distance between the ground and excited state and thus enable tuning of the emission spectrum, as depicted by the following graph.

Table 4.2.3: Exemplary microcrystalline luminescent materials activated by rare earth ions for application in light emitting diodes.

Phosphor	Density (g cm^{-3})	Crystal system	Space group	Band gap (eV)	Peak wavelength (nm)
$Lu_3Al_5O_{12}$:Ce^{3+}	6.68	Cubic	$Ia\bar{3}d$	7.5–8.0	525
$Y_3Al_5O_{12}$:Ce^{3+}	4.38	Cubic	$Ia\bar{3}d$	6.5–7.0	560
$Tb_3Al_5O_{12}$:Ce^{3+}	5.96	Cubic	$Ia\bar{3}d$	4.5	570
$CaAlSiN_3Eu^{2+}$	3.26	Orthorhombic	$Cmc2_1$	4.3	650
$SrLiAl_3N_4$:Eu^{2+}	3.67	Triclinic	$P\bar{1}$	3.0	650
β-SiAlON:Eu^{2+}	3.09	Hexagonal	$P6_3/m$	6.2	540
$(Ba,Sr)_2Si_5N_8$:Eu^{2+}	4.27	Orthorhombic	$Pmn2_1$	4.2	620
$SrSi_2O_2N_2$:Eu^{2+}	3.77	Triclinic	$P1$	5.6–6.7	540
Ba_2SiO_4:Eu^{2+}	5.48	Orthorhombic	$Pmcn$	~ 8.0	505

Even though the spectral position of the interconfigurational 4f–5d transitions of all rare earth ions could be modulated in this way, it is in just Ce^{3+} and Eu^{2+}, where the 4f–5d distance (ΔE(f,d)) for the free rare earth ion is small enough (Table 4.2.4), that the impact of surrounding ligands by covalent interaction (nephelauxetic effect) and crystal field splitting enables the shift of the absorption and emission band towards the visible range. Therefore, these ions dominate microcrystalline phosphors for solid state lighting and displays.

Among the lanthanide series, the Ce^{3+} ion is the simplest one from a spectroscopic point of view, due to its [Xe]$4f^1$ ground state configuration which is just split into two sub-levels, namely, $^2F_{5/2}$ and $^2F_{7/2,}$ by spin-orbit coupling. The excited configuration [Xe]$5d^1$ is sensitive to the local crystal field and the respective energy level is thus split into two to five crystal-field components [1]. The energy of emitted photons depends on the nephelauxetic effect (covalent character), the crystal field splitting and the Stokes shift, Since the emission spectrum of Ce^{3+} ions is caused by transitions between the lowest crystal field component of the [Xe]$5d^1$ configuration and the two spin-orbit split sub-levels of the ground state configuration, characteristic emission spectra with a double-band are observed [1]. Selected examples of Ce^{3+} emission spectra in inorganic host matrices are given in Figure 4.2.8. It is obvious that in hosts with a rather ionic character, namely, halides, phosphates or borates, the Ce^{3+} emission occurs at high energy due to the small nephelauxetic effect as shown in Figure 4.2.7. The highest energy of the 4f-5d emission bands of Ce^{3+} ions is usually observed in halides, for example, at 280 nm in $LiCaAlF_6$:Ce^{3+} [36]. However, if doped into more covalent hosts, Ce^{3+} emit at lower energy, that ism, from the blue to the deep reed

Table 4.2.4: Energy difference ΔE(f,d) between the ground state and the excited 5d state for free (gaseous) RE^{3+} ions (data from [35]).

Rare earth ion	Ground state	Excited state	ΔE(f,d) (cm^{-1})
Ce^{3+}	[Xe]$4f^1$	[Xe]$5d^1$	49,340
Pr^{3+}	[Xe]$4f^2$	[Xe]$4f^15d^1$	61,580
Nd^{3+}	[Xe]$4f^3$	[Xe]$4f^25d^1$	72,040
Sm^{3+}	[Xe]$4f^5$	[Xe]$4f^45d^1$	75,840
Eu^{3+}	[Xe]$4f^6$	[Xe]$4f^55d^1$	85,240
Eu^{2+}	[Xe]$4f^7$	[Xe]$4f^65d^1$	34,000
Gd^{3+}	[Xe]$4f^7$	[Xe]$4f^65d^1$	95,140
Tb^{3+}	[Xe]$4f^8$	[Xe]$4f^75d^1$	62,540
Dy^{3+}	[Xe]$4f^9$	[Xe]$4f^85d^1$	74,440
Ho^{3+}	[Xe]$4f^{10}$	[Xe]$4f^95d^1$	81,140
Er^{3+}	[Xe]$4f^{11}$	[Xe]$4f^{10}5d^1$	79,340
Tm^{3+}	[Xe]$4f^{12}$	[Xe]$4f^{11}5d^1$	78,640
Yb^{3+}	[Xe]$4f^{13}$	[Xe]$4f^{12}5d^1$	87,340
Lu^{3+}	[Xe]$4f^{14}$	[Xe]$4f^{13}5d^1$	98,510

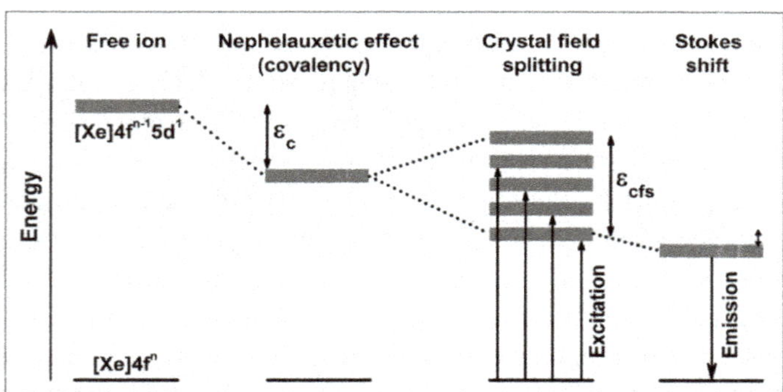

Figure 4.2.7: Simplified energy level diagram of RE^{n+} ions to illustrate the impact of the nephelauxetic effect and the crystal field splitting on the interconfigurational transition between the [Xe]$4f^n$ and [Xe]$4f^{n-1}5d^1$ states.

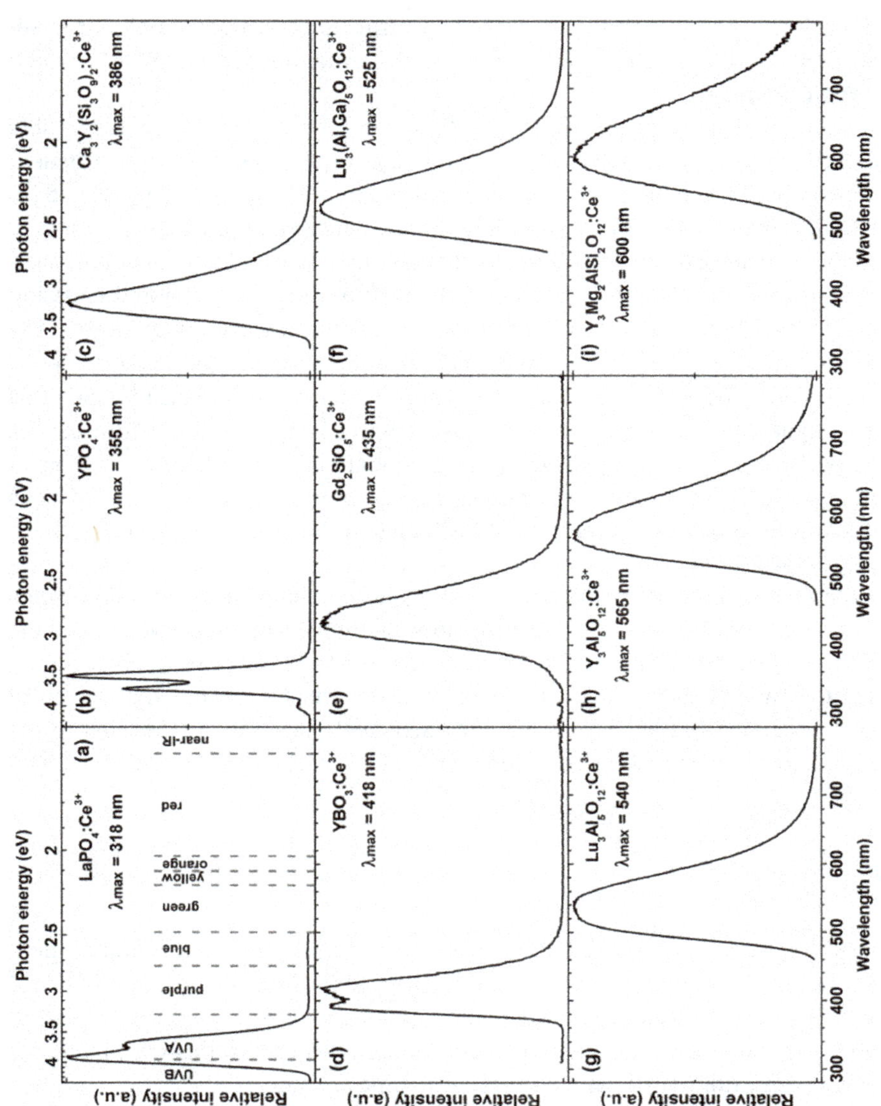

Figure 4.2.8: Emission spectra of Ce^{3+} in some selected host materials.

range. Ce^{3+} phosphors emitting in the red spectral range, unfortunately, are usually accompanied by strong concentration and thermal quenching [37–41], which limits their practical use.

In view of the design of microcrystalline phosphors emitting in the visible range, Eu^{2+} is the most versatile rare earth ion, by far [3]. In contrast to its trivalent counterpart (Eu^{3+}), the luminescence of Eu^{2+} is dominated by the parity-allowed interconfigurational $[Xe]4f^7$-$[Xe]4f^65d^1$ transitions. In theory and practice, all colors of the rainbow are accessible with the help of these transitions, since the rather small 4f–5d splitting of 34,000 cm^{-1} [42] allows the shift of the absorption and emission bands into the whole visible and even towards the near-infrared range. Another important feature of microcrystalline materials considered for a luminescent screen is their emission band width. In most cases, narrow bands are preferred, since this yields a higher luminous efficacy and thus, lamp efficacy. The line width can be controlled, to some extent, by selecting the host material with different numbers of crystallographic sites where Eu^{2+} ions can be located. The emission band thus broadens with an increasing number of Eu^{2+} sites in the structure, as also with increasing Stokes shift.

If the numbers of Eu^{2+} sites and the Stokes shift are both small, emission bands with a rather narrow width are observed, since neither the emitting excited nor the ground state are split. This is, for instance, the case in $Sr[LiAl_3N_4]:Eu^{2+}$, which shows an emission band at 650 nm and a band width of 1180 cm^{-1} (~ 50 nm) [43]. The emission band position and width of a few Eu^{2+} activated materials are depicted in the following figure to demonstrate the strong impact of the host material (Figure 4.2.9).

4.2.1 Outlook

The need for efficient and stable luminescent materials upon high excitation density, for example, as color converter in high power LEDs, laser diodes, or micro LED displays further enhances the demand on the photochemical and thermal stability, as well as on the linearity of such microcrystalline powders. To this end, materials with a high activator density and a short decay time are required.

As a high excitation density also enhances the thermal load on the luminescent screen, materials with superior thermal conductivity are required. This can be achieved by phosphors comprising larger particles and higher crystallinity, which in turn yields less phonon scattering. This development goal might lead to luminescent ceramics or even single crystals, since their thermal conductivity is large and the activator concentration can be high, before concentration quenching sets in [44].

Ceramic bodies can also comprise several luminescent materials to optimize the absorption and emission spectra, which can, for example, be exploited to achieve practical Eu^{3+} comprising luminescent screens in LEDs for illumination purposes [45].

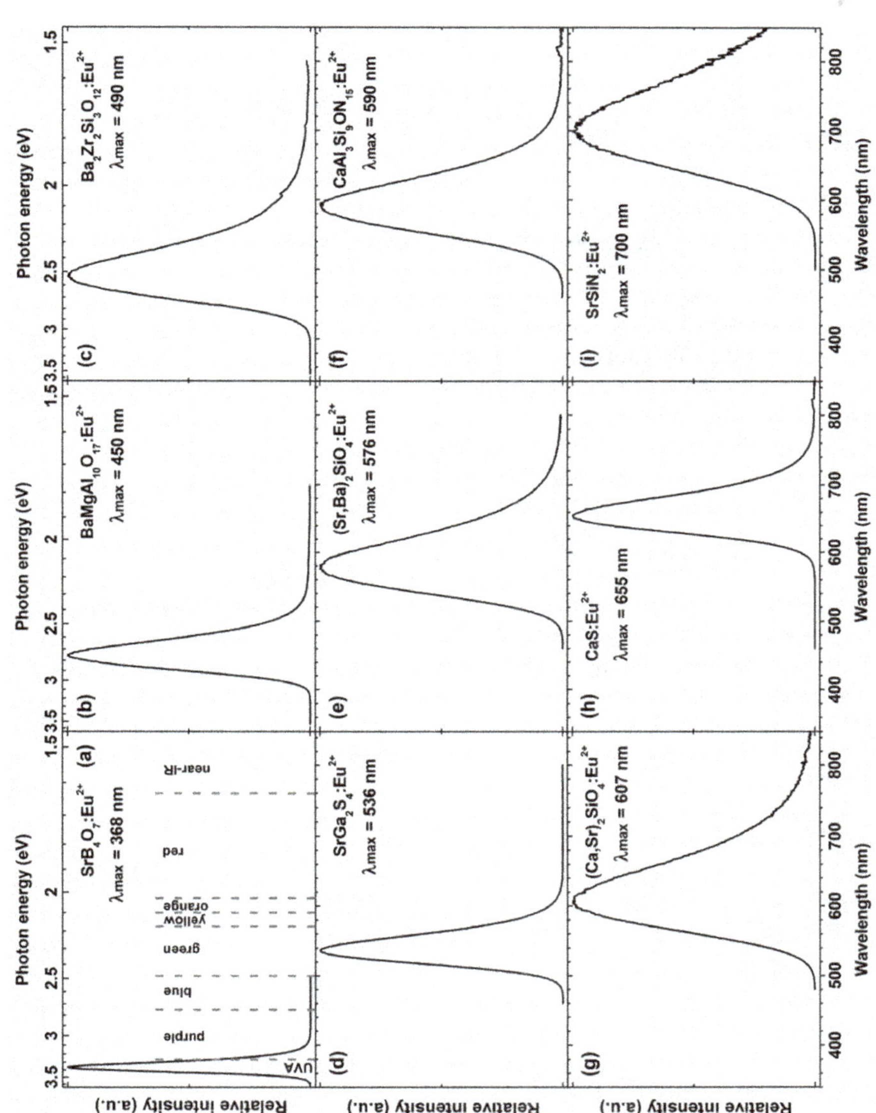

Figure 4.2.9: Emission spectra of Eu^{2+} in some selected host materials.

References

[1] G. Blasse, B. C. Grabmaier, Luminescent Mater., Springer-Verlag, Berlin, 1994.

[2] G. Blasse, A. Bril, Philips Techn. Rundsch. 1970, 10, 320.

[3] S. Shionoya, W. M. Yen (Ed.) Phosphor Handbook, CRC Press, Boca Raton, 1999.

[4] E. N. Harvey, A History of Luminescence, American Philosophical Society, 1957, 18.

[5] L. Ozawa, Cathodoluminescence, VCH, Weinheim, 1990.

[6] P. Goldberg, Luminescence of Inorganic Solids, Academic Press Inc., New York, 1966.

[7] N. C. Chang, J. Opt. Soc. Am. 1963, 53, 3500.

[8] K. A. Wickersheim, R. A. Lefever, J. Electrochem. Soc. 1964, 111, 47.

[9] A. Bril, W. L. Wanmaker, J. Electrochem. Soc. 1964, 111, 1363.

[10] A. K. Levine, R. C. Pallila, Appl. Phys. Lett. 1964, 5, 118.

[11] M. R. Royce, Rare Earth Activated Yttrium and Gadolinium Oxy-chalcogenide Phosphors, Patent US3418246A, 1968.

[12] K. H. Butler, Fluorescent Lamp Phosphors, University Park, 1980.

[13] S. Shinoya, W. M. Yen, Phosphor Handbook, CRC Press, Boca Raton, 1999.

[14] J. R. Coaton, A. M. Marsden, Lamps and Lighting, Arnold, London, 1997.

[15] M. Koedam, J. J. Opstelten, Lighting Res. Technol. 1971, 3, 205.

[16] J. M. P. J. Verstegen, D. Radielovic, L. E. Vrenken, J. Electrochem. Soc. 1974, 96, 1.

[17] T. Jüstel, H. Nikol, C. R. Ronda, Angew. Chem. 1998, 110, 3250.

[18] R. B. Leighton, Principles of Modern Physics, International Series in Pure and Applied Physics, McGraw-Hill, New York, 1959.

[19] B. M. Walsh, Judd-Ofelt Theory: Principles and Practices, in Advances in Spectroscopy for Lasers and Sensing (ed. B. Di Bartolo, O. Forte), Springer, Dordrecht, 2006, 403.

[20] C. Görller-Walrand, K. Binnemans, Spectral Intensities of f-f Transitions, in Handbook on the Physics and Chemistry of Rare Earths (ed. K. A. Gschneidner Jr., L. Eyring), Vol. 25, Elsevier, Amsterdam, 1998, 101.

[21] K. Binnemans, Coord. Chem. Rev. 2015, 295, 1.

[22] C. R. Ronda, Luminescence, Wiley-VCH, Weinheim, 2008.

[23] S. Schwung, S. Rytz, A. Gross, U. Ch. Rodewald, R.-D. Hoffmann, B. Gerke, B. Heying, C. Schwickert, R. Pöttgen, T. Jüstel, Opt. Mater. 2014, 36, 585.

[24] F. Baur, T. Jüstel, Aust. J. Chem. 2015, 68, 1727.

[25] F. Baur, F. Glocker, T. Jüstel, J. Mater. Chem. C 2015, 3, 2054.

[26] F. Baur, T. Jüstel, J. Mater. Chem. C 2018, 6, 6966.

[27] S. Nakamura, G. Fasol, The Blue Laser Diode: GaN Based Light Emitters and Lasers, Springer, Berlin, 1997.

[28] A. Zukauskas, M. S. Shur, R. Caska, Introduction to Solid-State Lighting, John Wiley & Sons, Hoboken, 2002.

[29] R. J. Xie, Y. Q. Li, N. Hirosaki, H. Yamamoto, Nitride Phosphors and Solid State Lighting, CRC Press, Boca Raton, 2011.

[30] K. N. Shinde, S. J. Dhoble, H. C. Swart, K. Park, Phosphate Phosphors for Solid-State Lighting, Springer, Berlin, 2012.

[31] W. M. Yen, M. J. Weber, Inorganic Phosphors, CRC Press, Boca Raton, 2004.

[32] R. B. Mueller-Mach, G. O. Mueller, T. Jüstel, P. J. Schmidt, Red Deficiency Compensating Phosphor Light Emitting Device, Patent US6680569B2, 2004.

[33] P. F. Smet, A. B. Parmentier, D. Poelman, J. Electrochem. Soc. 2011, 158, R37.

[34] C. C. Lin, R.-S. Liu, J. Phys. Chem. 2011, 2, 1268.

[35] P. Dorenbos, J. Lumin. 2000, 91, 155.

[36] V. N. Makhov, J. Lumin. 2012, 132, 418.

[37] M. S. Kishore, N. P. Kumar, R. G. Chandran, A. A. Setlur, Electrochem. Solid-State Lett. 2010, 13, J77.
[38] J. Ziqiang, W. Yuhua, W. Linshui, J. Electrochem. Soc. 2010, 157, J155.
[39] M. C. Maniquiz, K. Y. Jung, S. M. Jeong, J. Electrochem. Soc. 2010, 157, H1135.
[40] M. C. Maniquiz, K. Y. Jung, ECS Trans. 2010, 28, 175.
[41] A. A. Setlur, W. J. Heward, Y. Gao, A. M. Srivastava, R. G. Chandran, M. V. Shankar, Chem. Mater. 2006, 18, 3314.
[42] L. G. Uitert, J. Lumin. 1984, 29, 1.
[43] P. Pust, V. Weiler, C. Hecht, A. Tücks, A. S. Wochnik, A.-K. Henß, D. Wiechert, C. Scheu, P. J. Schmidt, W. Schnick, Nat. Photonics 2014, 13, 891.
[44] P. Pues, S. Schwung, D. Rytz, L. Schubert, S. Klenner, F. Stegemann, R. Pöttgen, T. Jüstel, J. Lumin. 2019, 215, 116653.
[45] M. A van de Haar, J. Werner, N. Kratz, T. Hilgerink, M. Tachikirt, J. Honold, M. R. Krames, Appl. Phys. Lett. 2018, 113, 132101.

Jan Werner

4.3 Optical materials – ceramics

4.3.1 Preliminary note

Although ceramics belong to the oldest man-made materials, they continue to be very important. They not only cover ceramic construction materials in civil engineering (tiles, bricks, sanitary ware) and household products (tableware, porcelain, earthenware) but also as engineered and functional ceramics in a broad variety of barely recognized technical products of our daily life. Some examples of their application include electrical insulators, resistors, actuators, sensors, nearly all kind of electronic devices, computers, telephones, consumer products, lighting, and automobiles. Ceramics also play a key role in many industrial applications, for example as components with high thermal and chemical resistance and mechanical stability, for example in the metal industry, the chemical, pharmaceutical and food industry, as well as in plant and mechanical engineering, transportation and automotive industry. Some ceramics are well established for decades as successful biocompatible materials in modern surgery and dentistry.

Rare earth elements (REE) often play an essential role in modern technical ceramics, for example, $Y_3Al_5O_{12}$ (yttria alumina garnet), YAG (thermal barrier coatings for turbine rotor blades), tetragonal 3-YSZ (yttria stabilized zirconia) ZrO_2 with 3 mol% Y_2O_3 for various technical applications or in environments that demand high mechanical strength and toughness and long term stability even under harsh chemical and thermal conditions. Considering its high inertness, YSZ is also used in surgeries as a biocompatible implant material (as joint replacement component) or in dentistry for root implants and all-ceramic prosthetics. Often, REEs are also essential components of a broad variety of important functional ceramics, for example in the so called high temperature super conductors (i.e., *Curie* temperature T_c above the boiling point of nitrogen) of the $YBa_2Cu_3O_{7-x}$ type.

Last but not least, many transparent or translucent ceramics containing REE are used as passive and active optical components. Their production, characteristic properties and main applications are presented in this chapter. Transparent ceramics that do not include REE (mainly transparent armor for ballistic protection) are not discussed in detail but are only briefly mentioned here for the sake of completeness.

Some of the properties of the ceramic material systems presented here are also characteristic for other material systems, especially for single crystals, glasses and powders or powder-containing composites of the same or similar chemical composition. These are described in detail in the respective chapters of this book insofar as the properties are not exclusively typical for ceramics. On the one hand, the ceramic production process allows an easier, faster, less energy consuming and/or cheaper fabrication and on the other hand, when beneficial effects result from the polycrystalline

https://doi.org/10.1515/9783110654929-026

ceramic microstructure, causing the typical ceramic properties such as high thermal, mechanical and chemical resistance, they will be discussed here. For more general information on all kind of ceramic materials, their properties and manufacturing processes, please refer to the relevant textbooks and technical literature.

4.3.2 Brief introduction to ceramics and their fabrication

4.3.2.1 Definition and explanation of terms

In this chapter, the term "ceramics" is used solely for inorganic, nonmetallic, solid materials with a highly densified polycrystalline structure and produced by sintering, a high temperature treatment of compacted powders. In other contexts, semicrystalline (glass ceramic) and amorphous (glass) materials as well as set inorganic binders (lime, cement, gypsum, and liquid glass) are also referred to as ceramic materials. Thus, ceramics differ from metals and organic solids (e.g. polymeric materials) by their inorganic-nonmetallic character (chemical composition and type of chemical bonding), from setting binders (sequence of thermal processing and shaping) and from single crystals (polycrystalline structure with grain sizes ranging from a few nanometers to a few millimeters, typically in the micron or submicron range).

Ceramics can generally be divided into three main material categories:

- Silicate ceramics, SiO_2-containing oxide ceramics: alkali aluminum silicates (porcelains), magnesium silicates (steatites), magnesium aluminum silicates (cordierites), aluminum silicon oxide (mullites) and other ceramics, produced with naturally occurring raw materials, mainly clay, kaolin, quartz, feldspar and lime (earthenware, stoneware, porcelain, bone china).
- Oxides (except silicates): alumina = Al_2O_3, magnesia = MgO, spinel = $MgAl_2O_4$, yttria = Y_2O_3, yttria alumina garnet (YAG) = $Y_3Al_5O_{12}$, zirconia = ZrO_2.
- Nonoxides: carbides – silicon carbide SiC, tungsten carbide WC, borides – titanium diboride TiB_2, lanthanum hexaboride LaB_6, nitrides – aluminum nitride AlN, silicon nitride Si_3N_4.

In addition these may be part of ceramic composite materials as e.g. particulate reinforced composites – zirconia toughened alumina (ZTA), alumina toughened zirconia (ATZ), fiber reinforced ceramic matrix composites (CMCs) and combinations of different oxides or nonoxides.

It is also useful to distinguish ceramic products by different applications. This allows dividing them into four main types:

- white wares, including tableware, cook- and kitchenware, wall tiles, pottery products and sanitary ware,

- structural ceramics, including bricks, pipes, floor and roof tiles,
- refractories, such as kiln linings, gas fire radiants, steel and glass making crucibles,
- technical ceramics, also known as engineering, advanced, special, or fine ceramics, including gas burner nozzles, ballistic protection, vehicle armor, nuclear fuel pellets, biomedical implants (bone grafts, artificial joints), dental ceramics, coatings of jet engine turbine blades, ceramic disk brakes, missile nose cones, space shuttle heat protection tiles, ball bearings, and a large variety of functional ceramics like piezo-electric, magnetic, ion-conducting, superconductive, high resistivity, radiation converting ceramics, biosensors and many others.

4.3.2.2 Ceramic fabrication process

Apart from the recent techniques of additive manufacturing (AM) and special thermal processing techniques such as spark plasma sintering (SPS) or field assisted sintering technology (FAST), sometimes also called current-activated pressure-assisted densification (CAPAD), the classical ceramic fabrication processes typically include the following steps:

1. (a) Fabrication of good moldable slurries for slip casting by mixing, grinding and homogenization of raw material powders, often referred to as ceramic powders, with water or other solvents, utilizing surface active additives, for example, deflocculants, dispersants, plasticizers, lubricants, defoaming agents and binders; or
1. (b) fabrication of good moldable pasty masses for extrusion forming with a higher solid content, as earlier mentioned molding slurries; or
1. (c) fabrication of good flowing and easy compactable granulates by spray drying or other granulation techniques, for example, pelletizing or special mixing techniques for different press forming techniques, as axial pressing and/or cold isostatic pressing; followed by
2. optional intermitted machining of thus compacted so called "green ceramics" by cutting, punching, drilling, structuring;
3. thermal processing, sometimes subdivided into several subsequent heating processes, like binder-burnout, sintering (in electrically driven furnaces with resistance-heating elements, microwave generators, induction heating coils or gas-fired burners) and finally, optional post treatment like pressure assisted sintering or hot isostatic pressing (HIP);
4. finishing of the so-called as-fired ceramics, meaning optional machining of the parts or some of their functional surfaces, where necessary, for example by grinding and polishing, lapping etc. to a desired/needed surface quality.

Figure 4.3.1 illustrates the entire ceramic production chain and demonstrates the influence of the processing steps on the characteristics of the ceramic microstructure which determine the ceramic properties.

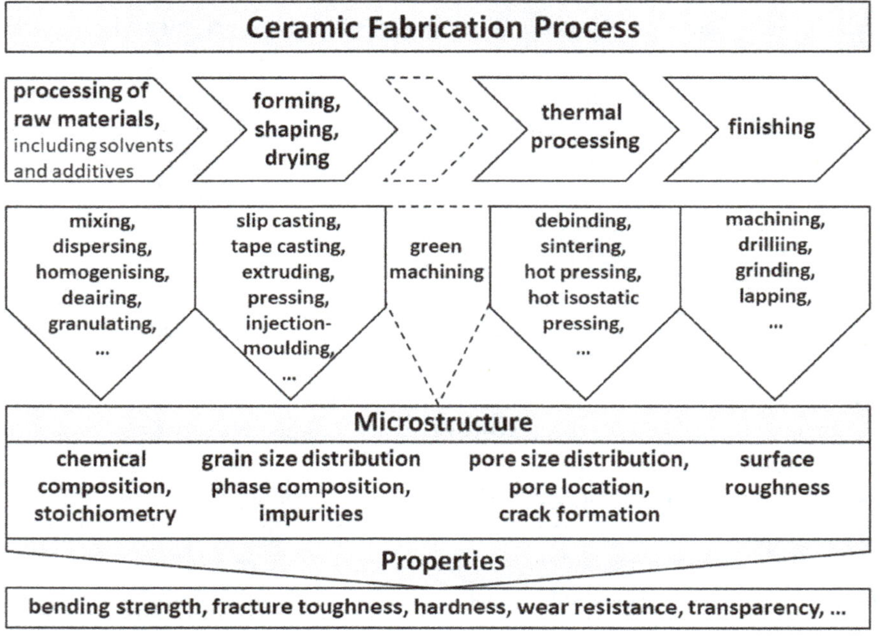

Figure 4.3.1: Flow chart of a conventional ceramic production chain, demonstrating the influence on the microstructural characteristics and on the properties of the produced ceramic components.

Most ceramics are typically opaque. Electro-magnetic radiation, in the range of ultra violet (UV), visible (VIS) or infrared (IR), cannot or can only to a very limited extent pass through these materials. This is due to reflection (also at inner surfaces), absorption and scattering of the incident light. Such phenomena are mainly caused by residual porosity, even in amounts of far less than a few per mille, as well as by incorporating particulate impurities, secondary mineral phases, grain boundary segregations, amorphous components, and deviations from the intended mineralogical composition or chemical stoichiometry, respectively. In noncubic materials, birefringence by optical anisotropy causes further scattering losses (Figure 4.3.2).

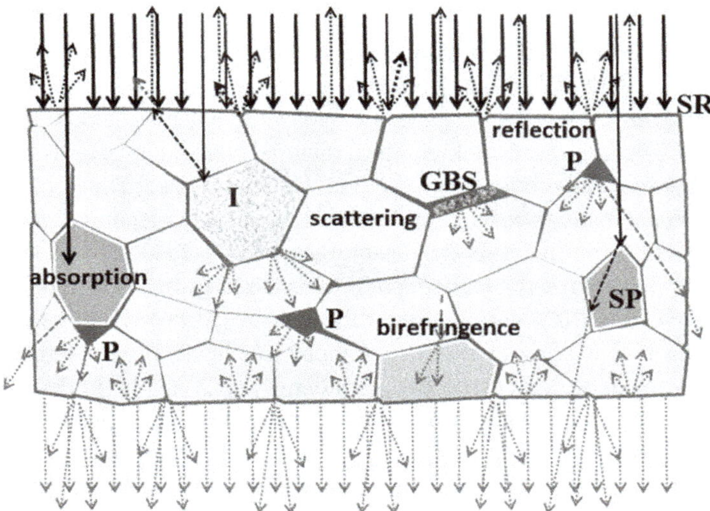

Figure 4.3.2: Schematic cross section illustration of the physical interaction of incident light with a typical polycrystalline microstructure of a translucent ceramic. SR: surface roughness, P: pore, GBS: grain boundary segregation, I: impurity / contaminant, SP: secondary phase.

4.3.2.3 Production of optical ceramics

For the fabrication of optically transparent ceramics, it is necessary to achieve a nearly defect-free microstructure in the sintered body and porosities, in the range of $<0.1\%$ [1], to avoid absorption and scattering of transmitting light. Therefore, often, powders of cubic crystal materials systems (due to their optically isotropic index of refraction) with high chemical and mineralogical purity and very small particle size and, thus, high sintering activity are homogeneously and most densely packed by suitable forming techniques with nearly no abrasion based contamination. This is followed by a subsequent sintering step that leads to homogeneous and almost defect-free microstructures with small and narrow distributed grain sizes and maximum density.

Although, each step in the production process influences the later properties of the ceramic, sintering is the most decisive. In traditional silicate-based ceramics, typically glassy phases occur and they take part as liquid phases during the sintering process. In the so called technical ceramics (often nonsilicate oxide ceramics), sintering most often takes place without the occurrence of significant amounts of viscous liquid phases and the densification process is dominated by solid-state diffusion processes or gas transport reactions. At temperatures below the melting temperature, diffusion processes lead to a further densification of the compacted powder particles and reduction of the voids between the particles. Larger particles grow at the expense of smaller particles which disappear with sintering temperature and dwell time. Under optimal conditions, the number and size of the pores decrease simultaneously. Starting from an optimally compacted green body of homogeneously packed, sintering active particles, a homogeneous and almost pore-free solid body with nearly a maximum of the theoretical materials density is obtained at the end of the sintering process after cooling. The final ceramic part exhibits a more or less densely sintered microstructure with randomly oriented grains separated from each other by grain boundaries. Increasingly, emerging innovative field-assisted sintering technology (FAST) [2] allows the fabrication of transparent ceramics directly in a single and rapid step from the loose powders to the sintered polycrystalline ceramic. During pressing and heating of the thus compacted powders, an electrical field is applied, yielding transparent ceramics of high density with nearly no defects and pores, comprising a fine grained microstructure, even after very short processing times, in the range of only a few minutes. Therefore, FAST is also promising for the fabrication of ceramics with noncubic crystal structure, as the scattering by birefringence of optically nonisotropic grains declines at grain sizes smaller than the incident light. However, FAST is expensive and the process is hard to control. There are also limitations with regard to component size and complexity of the achievable shapes. If the conventional production method is used, sintering according to today's standards for highest optical qualities, sintering in vacuum or under hydrogen atmosphere and/or hot isostatic pressing are indispensable.

4.3.2.4 Application of optical ceramics without Rare Earth Elements Alumina (REE)

Research on translucent ceramics was initiated with the development and invention of transparent polycrystalline alumina (PCA) by General Electrics in the early 1960 s [3]. Small amounts of magnesia have been used to inhibit abnormal grain growth. Due to the optical anisotropy of the trigonal crystal structure of the alumina, in combination with the relatively large average grain size of these ceramics, initially only a translucency could be achieved. However, this translucency proved to be sufficient for the intended use of PCA in the arc tubes of high-pressure sodium lamps, thus paving the way for subsequent developments.

Persistent efforts and the use of special techniques have led to better understanding, new concepts and thus, to great progress in improving the mechanical and optical properties of alumina, opening up new potential applications (Figure 4.3.3) [4–8].

Meanwhile, applying new materials preparation and sintering strategies, highly transparent alumina ceramics, even as composites with REE-oxides or doped with REE for laser applications, have been successfully produced [9–11]. Due to the availability, price of raw materials and to some intrinsic properties of alumina, for

Figure 4.3.3: Different structured polycrystalline alumina (pca) samples after pressing and conventional sintering (background), with a typical opaque appearance, in a heap of raw materials granulate and highly translucent ceramic sample, sintered under hydrogen atmosphere (foreground). (© FGK 2020).

example, its high surface hardness and ballistic performance, the main commercial interest still lies in the use for transparent armor.

Despite the success achieved, the main disadvantage of alumina remains its noncubic structure. Therefore, the most competitive transparent armor materials for ballistic protection application are the polycrystalline cubic alumina magnesia spinel $MgAl_2O_4$ [12] and the ceramic aluminum oxynitride $(Al_2O_3)_{0.67}(AlN)_{0.33}$ with a defect cubic spinel crystal structure [13–14]. There have also been first attempts to produce fluorescent REE-doped spinel ceramics of following compositions, Yb: $MgAl_2O_4$ [15] and Ce^{3+}:MgAlON [16], and to make them available for laser applications. Using FAST, transparent AlN ceramics could also be produced [17] and recently, the high-temperature high-pressure fabrication of small samples of extremely hard cubic Si_3N_4 ceramics has been reported for the first time [18].

4.3.3 Fabrication and application of optical ceramics based on Rare Earth Elements (REE)

Many ceramics containing one or more rare earth elements (REE) are used in numerous technical applications. As the field of ceramics with optical functionalities is still young, many possible future applications are currently being investigated and further new developments can be expected. The following sections provide an overview of the major optical ceramics currently used and studied that contain REE, either as part of the base material or as a functional component or both.

4.3.3.1 Yttria (Y_2O_3), Scandia (Sc_2O_3), and Lutetia (Lu_2O_3)

The so-called sesquioxides of the REE yttrium, scandium, and lutetium, yttria Y_2O_3, scandia Sc_2O_3, and lutetia Lu_2O_3, exhibit refractive indices with $n > 1.9$ [19]. This is remarkably high compared to optical glasses, that are limited to lower values in this respect. They all have a cubic (C-type) structure, which makes them interesting as transparent optical ceramics. Due to their low thermal expansion, high thermal resistance and mechanical strength, they are potential candidates for various applications such as transparent windows in industrial applications with extreme conditions or IR-transparent missile domes. In another potential application area, they could be used as high power solid state laser hosts (see Section 3.6.1), as they can be doped with other laser active trivalent REE^{3+} ions. However, because of their very high melting points of > 2400 °C, reaching highest density and optical quality by sintering is a challenge. Promising results with un-doped Y_2O_3 [20], Yb^{3+}:Y_2O_3 [21, 22], Eu^{3+}:Lu_2O_3 [23, 24], Er^{3+}:Sc_2O_3 and $(Nd, Yb, Tm)^{3+}$:Sc_2O_3 (for IR-pumped up-conversion laser application) [25] could be achieved, using vacuum sintering, sintering in hydrogen atmosphere, hot pressing, hot isostatic pressing, or field assisted sintering. In addition, the successful use of sintering aids (LiF, ZrO_2 and HfO_2) has been described in several papers [26, 27] and in one case even with a slip casting approach and subsequent vacuum sintering [28].

Recent encouraging results indicate that Eu^{3+}:Y_2O_3 ceramics, for example, could also be promising alternatives to single crystals for applications in quantum information technology [29].

4.3.3.2 Yttria-Stabilized Zirconia (YSZ, FSZ, CSZ, Y-ZrO_2)

Due to its exceptional mechanical strength and fracture toughness, its high temperature resistance and good chemical durability, zirconium oxide is one of the most important and widely used technical ceramics and also refractory materials.

It has also been successfully used for many decades as a remarkably inert and biocompatible ceramic material for artificial joint replacements in implantology and in the form of mechanically highly resilient bridge constructions in aesthetic dentistry. Furthermore, there are also applications of highly transparent parts and components made of zirconia in optical technologies.

However, one must distinguish the different modifications of ZrO_2 which determine the respective properties. ZrO_2 is a polymorphic material and occurs in three different crystal structures: monoclinic, tetragonal and cubic. The monoclinic phase is stable at room temperatures of up to 1170 °C, the tetragonal, at temperatures of 1170–2370 °C and the cubic, at over 2370 °C, according to the following scheme:

monoclinic (1170 °C) $\leftrightarrow$ tetragonal (2370 °C) $\leftrightarrow$ cubic (2690 °C) $\leftrightarrow$ melt

Obtaining pure stable zirconia ceramic products is difficult because of a volume change of approx. +5 vol% accompanying the transition from the tetragonal to the monoclinic structure, during the final cooling of the sintering process.

However, by adding divalent magnesia or calcium oxide or trivalent REE oxides such as cerium oxide or yttrium oxide, the tetragonal and/or cubic phases can be stabilized during cooling, depending on the amount of stabilizing agent added. Thus, ceramics with a single- or multi-phase structure can be produced, either as partially stabilized zirconium oxide (PSZ) or as fully stabilized zirconium oxide (FSZ). The addition of approx. 3 mol% yttrium oxide as a stabilizer in zirconium oxide enables the sintering of pure tetragonal fine-grained zirconium oxide ceramics (tetragonal zirconia polycrystals, Y-TZP) [30]. The transformation of the tetragonal into the monoclinic phase caused under external stress leads to an increase in volume of transforming crystals in the microstructure by approx. 5%, and thus to a mechanical reinforcement mechanism called transformation toughening [31, 32]. Y-TZP ceramics thus achieve outstanding mechanical strength and fracture toughness. The addition of higher amounts (usually ≥ 8 mol% yttria) results in complete stabilization of the cubic high-temperature phase. Although this phase has lower mechanical strength and no transformation toughening, its high refractive index of $n > 2$ over the hole visible spectral range and the optical isotropy of the cubic crystal structure makes it particularly attractive for the fabrication of highly transparent ceramics [33, 34], for example, for use in optical applications in the form of windows, prisms or lenses [35]. The following figures illustrate the progress of the optical properties of ceramic samples at different stages of ceramic processing (Figure 4.3.4) and the visual impression of a transparent cubic zirconia ceramic (Figure 4.3.5).

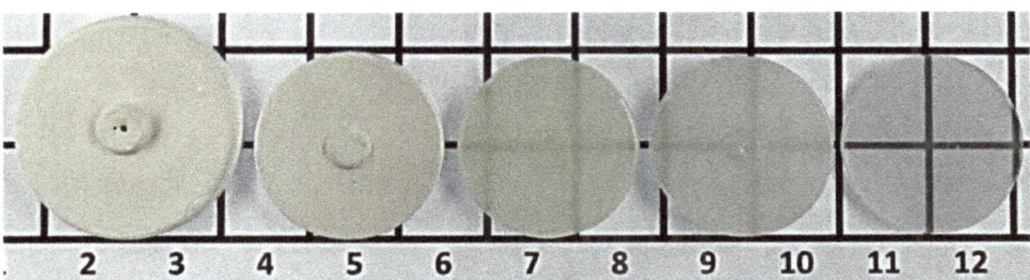

Figure 4.3.4: Ceramic samples of with Y_2O_3 fully stabilized cubic zirconia at different states of ceramic processing: injection molded, sintered, hot isostatically pressed, annealed, polished (from left to right), sample diameter after sintering approx. 20 mm and sample thickness after finishing 0.85 mm. (© FGK 2020).

Miniaturized optical components such as micro lens arrays and micro prisms can be produced, for example, by using specific powder preparation and processing methods and by using near-net-shape thermoplastic molding [36, 37] (see Figure 4.3.6).

Figure 4.3.5: View through a highly transparent ceramic sample of fully with Y_2O_3 stabilized cubic zirconia (© FGK 2020).

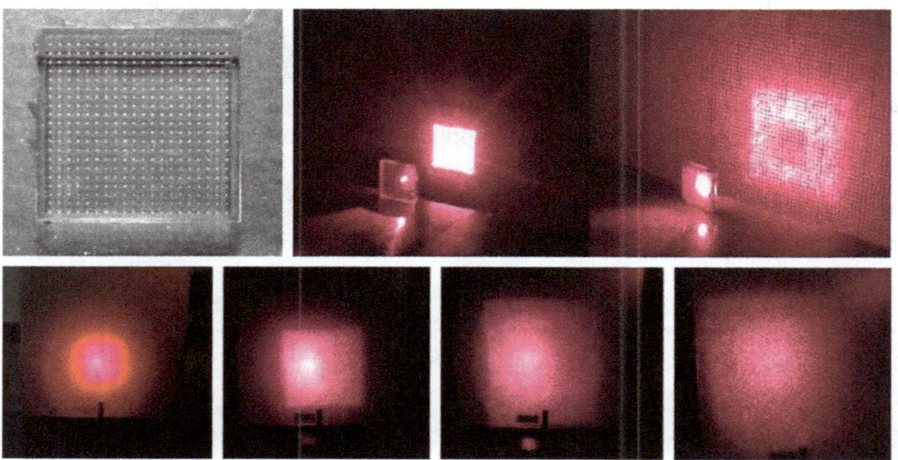

Figure 4.3.6: Highly transparent ceramic sample (0.8 mm x 0.8 mm micro lens array with 20 × 20 lenses) of fully with yttria stabilized cubic zirconia, optical magnification test in comparison to optical glass with 630 nm red laser light). (© FGK 2020).

In modern medicine, 8 mol% stabilized cubic ZrO_2 ceramics are also under investigation for potential use as transparent windows implanted in the skull, for ultrasonic [38] and IR-laser-brain therapy [39, 40].

4.3.3.3 REE pyrochlores ($REE^{3+}{}_2M^{4+}{}_2O_7$ with M^{4+} = Zr, Hf, Ti)

The crystal structure of $A_2B_2O_7$ compounds depends on the ionic radius ratio r_A/r_B, temperature and pressure, whereas the ionic radius ratio is the most important. The most common structures are cubic pyrochlore and cubic defect fluorite phases and sometimes monoclinic phases. At r_A/r_B ratios < 1.46, anion-deficient fluorite with disordered A and B ions is the stable structure, at $r_A/r_B > 1.78$, the monoclinic structure is stable and at ratios $1.46 < r_A/r_B < 1.78$, the cubic pyrochlore structure with ordered distribution of A and B ions is stable. Recently, many transparent ceramics with $A_2B_2O_7$ composition and cubic pyrochlore structures were fabricated and investigated, where A is represented by one or more trivalent REE^{3+} ions and B is represented by tetravalent transition metal ions, especially Zr^{4+}, Hf^{4+} and Ti^{4+}. Due to the great variability of the composition within this material system, the properties of the materials can be particularly well adjusted to the desired requirements.

The versatile potential applications of the transparent polycrystalline pyrochlores include high refractive optical lenses [41] and magneto-optic Faraday rotators [42], based on nondoped systems, as well as scintillators for X-ray imaging or as solid state laser host materials, based on REE^{3+}-doped systems [43–55]. Since pyrochlores have only been studied in more detail as optical materials for 15–20 years, there are currently no commercial products in practical use [56]. However, the spectrum of possible applications continues to grow.

4.3.3.4 Lead lanthanum zirconate titanate ($(Pb_{1-x}La_x)(Zr_{1-y}Ti_y)O_3$, PLZT)

Lead lanthanum zirconate titanates with variable Zr/Ti ratios and with general compositions $Pb_{1-x}La_x(Zr_yTi_{1-y})_{1-0.25x}V^B{}_{0.25x}O_3$, for example $Pb_{0.91}La_{0.09}(Zr_{0.65}Ti_{0.35})O_3$ (PLZT 9/65/35) crystallize in a cubic ABO_3 perovskite structure and exhibit ferro-electric and piezo-electric properties due to their crystallographic structure. PLZT materials have been developed based on ferroelectrics known earlier, such as barium titanate $BaTiO_3$ or lead zirconate titanate $Pb(Zr,Ti)O_3$ (PZT). The partial substitution of the divalent Pb^{2+} ion by the trivalent La^{3+} ion improves the piezoelectric and electro-optical properties on the one hand and favors the densification during sintering to high-density and thus transparent ceramics on the other hand. Well-densified polycrystalline ceramics with high optical transparency can be produced by sintering in oxygen atmosphere, comparable to the fabrication of dense PZT ceramics [57] or by hot pressing [58].

When exposed to an electric field, a polarization of the unit cell and the material occurs due to a shift of the Zr^{4+}/Ti^{4+} ions located in the central BO_6 octahedron. The polarization can cause changes in the refractive index as well as in the macroscopic shape which can be observed and used as piezoelectric effects [59, 60].

Thus, a large number of useful electro-optical components for various technical applications can be produced from PLZT, such as very fast operating electro-optical switches, electro-optical filters and polarizers [61], micro-actuators, wireless actuators in micro-electro-mechanical systems (MEMS) and micro-opto-electro-mechanical systems (MOEMS), which advantageously combine electrical, optical and mechanical properties [62] and electro-optical devices in modern fiber-optic communication [63].

Compositional changes within this ferroelectric system can significantly alter the materials properties and behavior under applied electric fields or temperature variations. This allows such a system to be tailored to a wide range of converter applications. For instance, PLZT ceramics have been suggested for use in optical devices [64–67] because of their good transparency, from the visible to the near-infrared, and for their high refractive index ($n \approx 2.5$), which is advantageous in light wave guiding applications [68, 69]. Using a CO_2-IR laser, the use of a PLZT ceramic as a thermal lens could also be successfully demonstrated for focusing a red HeNe-laser beam [70]. Ferroelectrics will also play a major role in the next generation of optical devices. They can make a valuable contribution to wireless laser communication systems or in ground-based fiber optic networks to transmit large data volumes at very high speed. The miniaturization of practical devices and the advances in nanotechnology have recently made it possible to produce ferroelectric materials with actuator effects in the nanometer range. Beyond their industrial and technological use, ferroelectric materials may also play an important role in consumer goods in the future [70].

4.3.3.5 REE alumina garnets (YAG, $Y_3Al_5O_{12}$, $Y_3ScAl_4O_{12}$ YSAG, $Lu_3Al_5O_{12}$, LuAG)

In the system RE_2O_3-Al_2O_3, namely Y_2O_3-Al_2O_3, there are three phases of yttrium oxide-aluminum oxide compounds, a monoclinic $Y_4Al_2O_9$ phase (YAM), an orthorhombic $YAlO_3$ perovskite phase (YAP) and a cubic $Y_3Al_5O_{12}$ cubic phase with garnet structure (YAG).

However, the reaction of a mixture of Y_2O_3 and Al_2O_3 with stoichiometric YAG composition does not take place directly to the target compound, but proceeds in several temperature-dependent steps. First, the YAM phase forms in the temperature range between 900–1100 °C:

$$2\ Y_2O_3 + Al_2O_3 \rightarrow Y_4Al_2O_9(\text{YAM})\ [900 - 1100\ ^\circ\text{C}]$$

Together with still unreacted aluminum oxide, YAM forms the YAP phase in the temperature range between 1100–1300 °C:

$$Y_4Al_2O_9 + Al_2O_3 \rightarrow 4YAlO_3(\text{YAP})\ [1100 - 1300\ ^\circ\text{C}]$$

Between 1400–1600 °C, the YAP phase finally reacts with previously unreacted aluminum oxide to form the desired YAG phase:

$$3YAlO_3 + Al_2O_3 \rightarrow Y_3Al_5O_{12}(YAG)\ [1400 - 1600\ ^\circ C].$$

Similar processes can also be found when using other, for example, more reactive starting compounds or via synthesis routes with wet chemical precipitation (co-precipitation of poorly soluble hydroxides) and their calcinates. The starting materials are often pre-calcined between 900 and 1000 °C to prevent sintering of loose agglomerates of the nano or submicron scale particles.

Such powders are frequently, mixtures of the still unreacted Al_2O_3 and some of the yttria-alumina phases, YAM, YAP, and YAG. However, these can be further processed by ceramic technology as the complete conversion into YAG usually takes place during the further sintering process. This form of reactive sintering generally also allows the fabrication of defect-free and, therefore, highly transparent ceramics (see Figure 4.3.7).

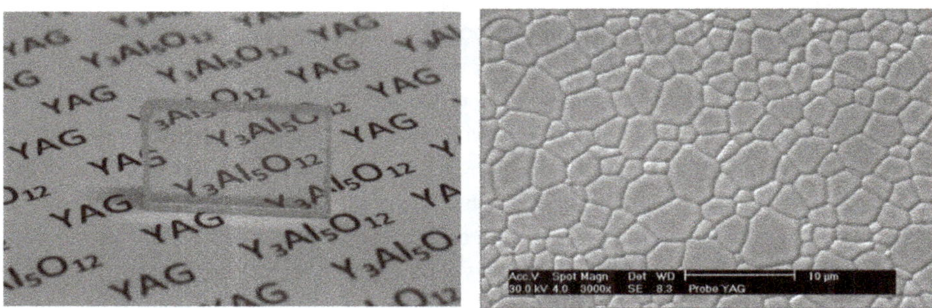

Figure 4.3.7: Highly transparent yttria alumina garnet (YAG) ceramic sample (left) and its corresponding defect free microstructure (right). (© FGK 2020).

In the Y_2O_3-Sc_2O_3-Al_2O_3 system exists a wide range of garnet-structured solid solutions. In some LASER applications, for example, instead of YAG the corresponding $Y_3ScAl_4O_{12}$ (yttria scandia alumina garnet, YSAG) is utilized. Also, a manifold of other garnets like gadolinium gallium garnet (GGG) is well known. However, these garnets are rarely used nondoped, but almost exclusively as host materials for laser or LED converter materials, as described in the following sections.

4.3.3.6 REE doped light emitting ceramics

In addition to the structure-forming properties of rare earths in various compounds described in the previous sections, there are also a number of applications that open up when specific effects can be achieved in existing structures (which may also contain rare earth ions such as cubic sesquioxides, pyrochlores, or garnets) and also by doping, specific ion exchange in these structures by rare earth ions. Such dopants, when excited by electromagnetic radiation, can emit part of the absorbed radiation energy in the form of light in the wavelength range UV–VIS–IR. Such luminescent ceramics which are designed for use in lasers, scintillators and light converters, some of which have already been introduced successfully, are presented in the following sections.

4.3.3.6.1 Solid state lasers

A laser (Light Amplification by Stimulated Emission of Radiation) is a source of coherent, quasi-monochromatic and sharply focused radiation in the visible and adjacent regions of the electromagnetic spectrum (far infrared, infrared, ultraviolet and X-ray). In principle, every laser consists of three components: first, an active laser medium which largely determines the properties of the laser, for example a gas, a solid material (solid state laser, SSL) or a diode; second, a pumping mechanism that supplies energy to the laser medium, for example, a flash lamp or an electrically driven gas discharge emitting at a wavelength shorter than the laser wavelength; and third, a laser resonator, a system of transparent and semitransparent mirrors and other optical elements that provides optical backcoupling and thus, induced emission of the radiation. Depending on the specific design and the choice of components, there are a number of different types of lasers which differ mainly in the attainable power (between a few microwatts and many kilowatts) and frequency characteristics.

For more information on SSL based on single crystals, see Chapter 4.4.

Since the construction of the first functional prototype in 1960 at the Hughes Research Laboratories [71] (a ruby, chromium-doped aluminum oxide ($Cr:Al_2O_3$) laser with a red emission line at 694.3 nm), the laser has become widespread in science, technics, industry, medicine, entertainment and diverse consumer products.

Nowadays, many REE such as neodymium (Nd), ytterbium (Yb), holmium (Ho), thulium (Tm), and erbium (Er) are used as dopants in solid-state-laser materials. Neodymium (Nd) is a common dopant in various solid state laser crystals, including yttrium orthovanadate ($Nd:YVO_4$), and yttrium aluminum garnet (Nd:YAG). These lasers can generate high-power infrared emission at 1064 nm that is used, for example, for cutting, welding and marking metals and other materials. They are also used in spectroscopy, medical surgeries, and in many other applications.

Initially, there were only two types of SSL gain media: single crystals and glasses doped with laser-active ions. Compared to most crystalline materials, ion-doped glasses usually exhibit much broader amplification bandwidths, allowing for large wavelength tuning ranges and the generation of ultrashort pulses. Drawbacks are inferior thermal properties (limiting the achievable output powers) and lower laser cross sections, leading to a higher threshold pump power. As the fabrication of single crystals (e.g., via Czochralski growth from a melt) is time consuming and expensive, the size and geometries with which they can be produced are limited.

The first lasing in a $Dy^{2+}:CaF_2$ ceramic was demonstrated in 1966 [72]. A further breakthrough was the fabrication of a Nd:YAG ceramic in laser quality in 1995 by Ikesue et al. [73]. Figure 4.3.8 shows a nanoscale Nd:YAG starting powder from own wet chemical synthesis after calcination and a ceramic produced from it. In 2002, the next milestone was achieved by demonstrating an output power of > 1 kW using a Nd: YAG ceramic [74]. Great improvements in powder synthesis, forming and sintering led to further remarkable achievements. For example, 105 kW output power was realized from a Nd:YAG ceramic laser system in 2009 [75].

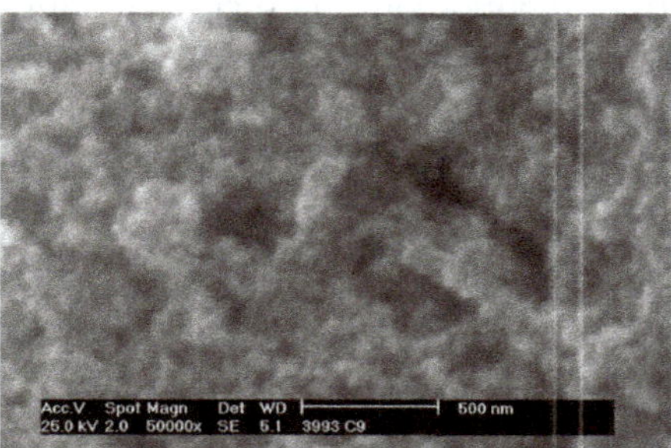

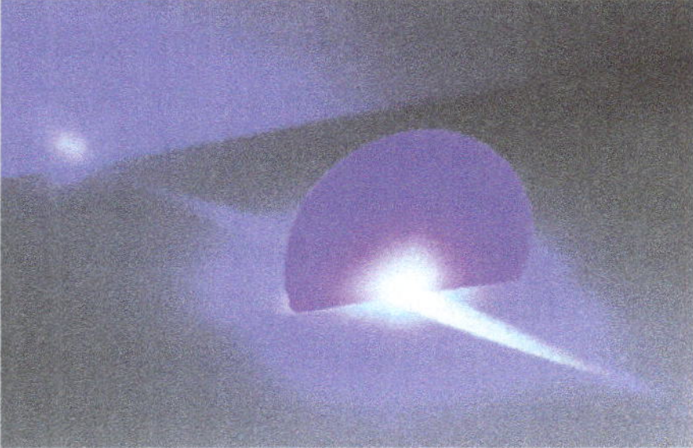

Figure 4.3.8: Nanoscaled Nd:YAG powder (from MJR – micro jet reactor – wet chemical synthesis) after calcination (left) and ceramic sample of the material under UV Laser irradiation (right). (© FGK 2020).

Meanwhile, ceramic laser materials with different composition based on lutetia alumina garnet (e.g., Nd:LuAG) and yttria-scandia alumina garnet (e.g., Yb:YSAG) are under investigation. Compared to glass and single-crystal laser technologies, advanced ceramic lasers are anticipated to be highly attractive alternatives in the future.

Although the first ceramic laser gain medium was based on polycrystalline cubic $Dy:CaF_2$, further development mainly concentrated on doped oxide ceramic lasers. This is due to certain technical advantages such as high transparency in a wide wavelength region (from the VUV to the IR), a decreased probability of non-radiative transition probability (i.e., higher efficiency) and a lower melting point, compared to most oxide based lasers. In the last few years, interest in these materials increased and the use of REE-doped CaF_2 ceramics was successfully reported on $Yb:CaF_2$ [76, 77], $Er:CaF_2$ [78] and $Nd,Y:CaF_2$ [79].

As mentioned already earlier, materials with noncubic crystal structure lack nonisotropic optical properties, leading to light scattering in micro-structured polycrystalline media. For instance, Figure 4.3.9 shows a translucent but (not transparent) sample of a polycrystalline YVO_4 ceramic with tetragonal crystal structure, due to pronounced scattering.

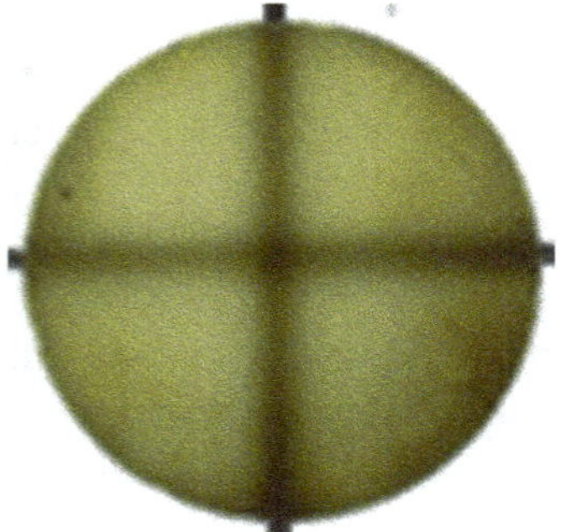

Figure 4.3.9: Translucent sample of a tetragonal YVO_4 ceramic. (© FGK 2020).

Very promising, in this context, is a current report about the successful fabrication of a highly transparent neodymium-doped fluorapatite (Nd:FAP) ceramic with noncubic crystal structure. A sample of 1 mm thickness with a dense microstructure and ultrafine microstructure (mean grain size $d_{50} = 140$ nm) was reported to reach 98.7% of the theoretical transmittance [80]. This indicates the possibility of the development and application of further noncubic laser ceramics in the future.

Ceramic production, especially shaping, joining and bonding techniques, also offers special perspectives for further improving the performance of laser media made of different materials (composite laser ceramics) [81–86].

Further detailed information on laser ceramics with respect to the various innovation steps can be found in the several reviews and the literature cited therein [87–89].

4.3.3.6.2 Scintillators

Since the discovery of ionizing radiation more than 100 years ago, it has been used in a wide variety of technical and medical applications. To make the radiation that is invisible and harmful to the human eye usable, materials are needed that can convert the received radiation into visible radiation. Such materials are called scintillators. Sometimes, the excited state is metastable. So, the relaxation back down from the excited state to the lower states is delayed. Scintillating materials are, currently, widely used in many detection systems, addressing different fields of application such as medical imaging (CT scanners and gamma cameras in medical diagnostics), civil protection, industrial control and oil drilling exploration. Among the desired technical properties of scintillators are their fast operation speed, high density, radiation hardness and durability of operational parameters. Especially because of the latter, stable inorganic scintillators are of great interest. In addition to a multitude of monocrystalline scintillators, there is an interest in polycrystalline ceramic alternatives comparable to the development of the laser materials described above. Although the possibility of the use of ceramics was not considered at the beginning of the 2000s [90], a number of ceramic scintillators have been investigated to-date. These are often REE-doped materials of the pyrochlore structure type such as Eu^{3+}:$(La,Gd)_2Hf_2O_7$ [91] or of the garnet structure type such as cerium-doped lutetia alumina garnet (Ce:LuAG) [92], cerium-doped gadolinium-yttrium gallium-aluminum garnet Ce:$(Gd,Y)_3(Ga,Al)_5O_{12}$ (Ce:GYGAG) [93], cerium-doped yttria alumina garnet (Ce:YAG) [94], cerium-doped gadolinium gallium aluminum garnet Ce:$Gd_3Ga_3Al_2O_{12}$ (Ce:GGAG) [95] which have been covered in the previous sections. They will, therefore, not be treated in detail again here. Instead, reference is made to a comprehensive current overview [96].

4.3.3.6.3 Luminescent converters for LEDs

As mentioned earlier, luminescent inorganic powders (also referred to as "phosphors") play an important role in modern solid state lighting (SSL). In modern white light LEDs, the white color impression is created by mixing several basic colors. This is often achieved by blue or ultra-violet light from a semiconductor chip made of indium-gallium-nitride (In,Ga)N. This partially excites one or more emitter materials, which in turn, emit in the yellow, green or red spectral range. In such so-called phosphor converted light emitting diodes (pc-LEDs), a certain proportion of the exciting blue light is transmitted through the emitter materials so that, in the optimum case, a complete spectrum is generated in the visible range. A large number of luminescent materials, many of them based on REE doped inorganic compounds, have been developed and successfully proven in recent decades as stable, durable and efficient light converters, thus helping modern LED lighting to achieve a breakthrough. And, the search for even better materials, technically and commercially, continues unabated. For a detailed description of the different special concepts, material classes and materials in particular, please refer to the section on converter powders (Chapter 4.2).

With an increasing demand for improved performance and efficiency, powder-based LED converter solutions are reaching their limits. In such white light LEDs, the powders are usually embedded in a polymer matrix, for example, of silicone or epoxy resin. Its inhomogeneity, strong scattering losses and poor heat dissipation can cause problems. At high power levels, operating temperatures are reached which can lead to a damage to the polymer matrix and result in associated negative optical effects such as brow discoloration.

Compared to such luminescent powders that are embedded in polymeric matrices, overall light emission with customized color temperature and high power output can be achieved by the combination of a suitable excitation source, that is, a blue to ultraviolet emitting LED or a laser diode LD with an appropriate ceramic converter, as tiny ceramic plate of approx. 1×1 mm^2 and only a few hundreds of microns thick. This platelet is directly attached to the top of the LD/LED chip or alternatively, at some distance from the LD/LED chip, as the so called "remote phosphor" approach. Thus, on the one hand, luminescent ceramics offer higher efficiency and long-term stability. On the other hand, they permit operation at temperatures much higher than polymer luminescent materials composites, which are severely limited due to their thermal instability at temperatures already below 200 °C.

The ceramic fabrication techniques are the same as for other optical and luminescent ceramics, described (in detail depending on the respective materials, especially with respect to the sintering parameters) in the earlier chapters. The ceramic converters are most often produced by pressing, sintering, polishing and finally slicing to the demanded size, while tape-casting techniques are discussed as alternatives to pressing.

Whenever new luminescent materials are developed and found suitable, it is necessary to determine the best production parameters for the individual processing steps.

Till date, the most common system by far is a combination of cerium-doped YAG [97] with a blue LED. In this simple binary concept, the relatively broadband emission of Ce:YAG with an emission maximum in the yellow region of the spectrum produces a cold white light. By choosing the optimum LED, in combination with a converter with adjusted chemical composition and the right thickness of the ceramic plate and an appropriate amount of scattering (e.g., by tailoring the porosity of the ceramic in the range of up to 3 vol%, see Figures 4.3.10 and 4.3.11), satisfactory results can be achieved [98–100].

Figure 4.3.10: Three ceramic Ce:YAG converter samples with different porosity and hence transparency. (© FGK 2020).

A green shift of the emitted light can, for example, be realized by changing from Ce:YAG to Ce:LuAG, or by using varieties with adjusted yttria and lutetia proportions, that is, Ce:(Y,Lu)AG, or cerium-doped yttrium aluminum gallium garnet Ce:YAGG [101], respectively.

With the desire for better color rendering and a warmer white light, concepts with additional orange to red components were developed. Particular reference is

Figure 4.3.11: White light impression of a highly translucent YAG:Ce converter ceramic with adjusted porosity under blue back light illumination. (© FGK 2020).

made here to the europium-doped nitride compounds of the type (Ca,Sr)$AlSiN_3$:Eu^{2+} and $(Ca,Sr)_2Si_5N_8$:Eu^{2+} [102, 103] and Mn^{4+} activated K_2SiF_6:Mn^{4+} emitters [104]. Due to the challenging and costly production and limited chemical stability, there is still a sustained demand for advanced red emitters.

An alternative concept is based on the use of UV-emitting LEDs whose light is generated entirely by a combination of red, green and blue emitters (instead of the blue LEDs discussed so far in combination with a red and a green converter).

A suitable blue phosphor is Eu^{2+}:$BaMgAl_{10}O_{17}$ (Eu^{2+}:BAM) which can also be produced in the form of monolithic translucent ceramics [105]. Figure 4.3.12 shows the appearance of differently doped ceramic fluorescent converters with different host structures under daylight and under ultraviolet irradiation. By a suitable combination, good color rendering values, a desired color temperature and stable color points can be achieved with comparatively high yields.

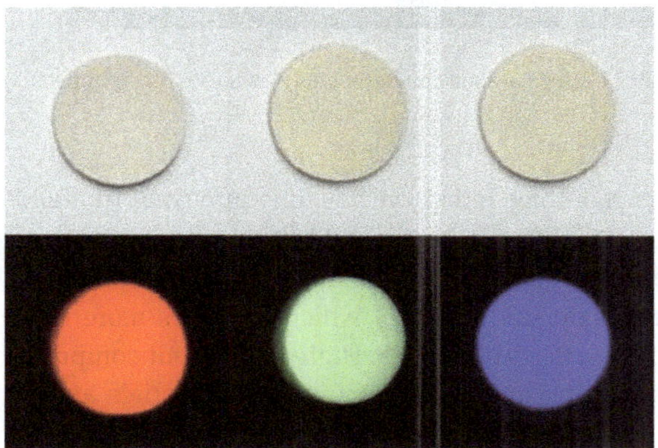

Figure 4.3.12: Red, green and blue ceramic converter samples with different host structures and REE dopants under daylight (above) and UV radiation (below). (© FGK 2020).

While it is relatively easy to create mixtures when using powders, the combination of several ceramics with each other is more elaborate and can be achieved, for example, by a layer-wise assembly of several individual ceramics [106], which may have to be separated from each other by protective layers to prevent reactions during the bonding process at high temperatures, or by coating one phosphor with another [107]. In an alternative conceptual approach, a composite was produced from a low-melting matrix of a europium-doped $Li_3Ba_2La_3(MoO_4)_8$:Eu^{3+} red phosphor [108, 109] sinterable at low temperatures, in which a stable phosphor Ce^{3+}:(Y, $Lu)_3Al_5O_{12}$ was embedded [110]. Combined with an adequate blue emitting LED, color rendering and efficiency could be further improved, compared to conventional warm white LEDs.

Figure 4.3.13 shows different REE-ion doped light-emitting ceramics. Depending on the respective REE-doped ceramic, various emission behaviors can be generated.

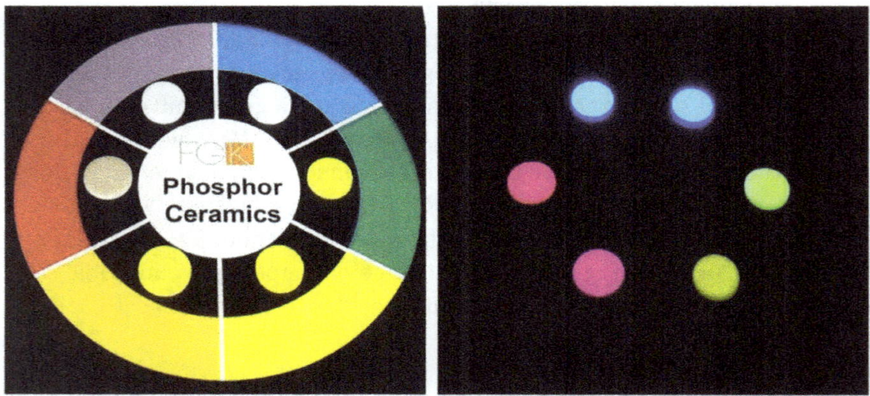

Figure 4.3.13: Comparison of different REE-doped ceramic converter samples under daylight (left) and UV radiation (right). (© FGK 2020).

The discovery of new phosphors has long relied on the time-intensive trial-and-error experimental synthesis based on different successful but strenuous strategies. Other methodologies based on big data and computation are emerging as an important complement to experimental materials research, which promises more rapid discoveries of novel materials with particular interests. High-throughput computing has been used in successfully screening and finding interesting materials and recently, data-driven methodologies have contributed to computational design and experimental identification of novel luminescent materials. For example, density functional theory (DFT) calculations have been used to provide deep insights into composition-structure-property relationships, to screen and discover novel phosphors with desired photo-luminescence properties by using data-driven methods, raising high expectations [111].

4.3.4 Summary and concluding remarks

After almost 60 years of intensive research and development in the field of, initially translucent and later in highly transparent ceramics, a profound understanding of the essential parameters of production has been achieved today. The spectrum of applications ranges from ballistic armor protection over various technical and industrial purposes to modern medicine. Especially, in the fields of optical technologies and artificial lighting, there are a variety of applications. Here in particular, rare earth ions often play a key role, whether as an essential structural component of the ceramic compound itself or as an important part of the host structure of an optically active solid state medium and/or as its light absorbing and/or light emitting component.

Interest in the use of transparent ceramics in new applications is steadily growing and currently reaches into quantum computing. A comprehensive review with numerous references was recently published [112]. With the increasing awareness of the existence and known applications of transparent ceramics, the use of new process technologies (such as additive manufacturing or field-assisted sintering technologies) and the further development of conventional ceramic process technologies, the path of transparent ceramics is brightly lit. With an increasing number of commercially available high-quality powders, the development and use of new synthesis concepts and methods to identify and provide novel innovative materials and the potential offered by the use of composite materials, the path of translucent REE-containing ceramics is leading into a colorful future.

References

[1] A. Krell, T. Hutzler, J. Klimke, J. Eur. Ceram. Soc. 2009, 29, 207.

[2] M. Nygren, Z. Shen, Hot Pressing and Spark Plasma Sintering, in Ceramics Science and Technology, Vol. 3: Synthesis and Processing (ed. R. Riedel, C. I. Wei), Wiley-VCH, Weinheim, 2012.

[3] R. Coble, US Patent Application 3026210 A, 1962.

[4] A. Krell, E. Strassburger, Ceram. Trans. 2001, 134, 463.

[6] R. Apetz, M. van Bruggen, J. Am. Ceram. Soc. 2003, 86, 480.

[7] A. Krell, P. Blank, H. Ma, T. Hutzler, M. van Bruggen, R. Apetz, J. Am. Ceram. Soc. 2003, 86, 12.

[8] A. Krell, J. Klimke, T. Hutzler, J. Eur. Ceram. Soc. 2009, 29, 275.

[9] B.-N. Kim, K. Hiraga, K. Morita, H. Yoshida, J. Eur. Ceram. Soc. 2009, 29, 323.

[10] I. Alvarez-Clemares, G. Mata-Osoro, A. Fernández, S. Lopez-Esteban, C. Pecharromán, J. Palomares, R. Torrecillas, J. Moya, Adv. Eng. Mater. 2010, 12, 1154.

[11] E. Penilla, L. Devia-Cruz, M. Duarte, C. Hardin, Y. Kodera, J. Garay, Light: Sci. Appl. 2018, 7, 33.

[12] M. Rubat du Merac, H.-J. Kleebe, M. Müller, I. Reimanis, J. Am. Ceram. Soc. 2013, 96, 3341.

[13] C. Warner, T. Harnett, D. Fisher, W. Sunne, Proc. SPIE 2005, 5786, 95.

[14] J. McCauley, P. Patel, M. Chen, G. Gilde, E. Strassburger, B. Paliwal, K. Ramesh, D. Dandekar. J. Eur. Ceram. Soc. 2009, 29, 223.

[15] S. Balabanov, A. Belayev, E. Gavrishchuk, I. Mukhin, A. Novikova, O. Palashov, D. Permin, I. Snetkov, Opt. Mater. 2017, 71, 17.
[16] X. Liu, B. Chen, B. Ta, H. Wang, W. Wang, Z. Fu, Materials 2017, 10, 792.
[17] Y. Xiong, Z. Fu, Y. Wang, F. Quan, J. Mater. Sci. Lett. 2006, 41, 2537.
[18] N. Nishiyama, R. Ishikawa, H. Ohfuji, H. Marquardt, A. Kurnosov, T. Taniguchi, B.-N. Kim, H. Yoshida, A. Masuno, J. Bednarcik, E. Kulik, Y. Ikuhara, F. Wakai, T. Irifune, Sci. Rep. 2017, 7, 44755.

[19] CRC Handbook of Chemistry and Physics (ed. D. R. Lide), 90th Edition, CRC Press/Taylor and Francis, Boca Raton, Index of Refraction of Inorganic Crystals, 2010, 10–245–10–248.
[20] J. Mouzon, A. Maitre, L. Frisk, N. Lehto, M. Odén, J. Eur. Ceram. Soc. 2009, 29, 311.
[21] J. Zhang, L. An, M. Liu, S. Shimai, S. Wang, J. Eur. Ceram. Soc. 2009, 29, 305.
[22] J. Wang, J. Ma, J. Zhang, P. Liu, D. Luo, D. Yin, D. Tang, L. Kong, Opt. Mater. 2017, 71, 117.
[23] S. Bagayev, V. Osipov, M. Ivanov, V. Solomonov, V. Platonov. A. Orlov, A. Rasuleva, S. Vatnik, Opt. Mater. 2009, 31, 740.
[24] Y. Shi, Q. Chen, J. Shi, Opt. Mater. 2009, 31, 729.
[25] A. Lupei, V. Lupei, C. Gheorghe, A. Ikesue, E. Osiac, Opt. Mater. 2009, 31, 744.
[26] A. Kruk, A. Wajler, M. Mrózek, L. Zych, W. Gawlik, T. Brylewski, Opt. Appl. 2015, 45, 585.
[27] W. Liu, L. Jin, S. Wang, Mat. Chem. Phys. 2019, 236, 121835.
[28] Y. Xu, X. Mao, J. Fan, X. Li, M. Feng, B. Jiang, F. Lei, L. Zhang, Ceram. Int. 2017, 43, 8839.
[29] J. Karlsson, N. Kunkel, A. Ikesue, A. Ferrier, P. Goldner, J. Phys.: Condens. Matter 2017, 29, 125501.
[30] I. Nettleship, R. Stevens, Int. J. High Tech. Ceram. 1987, 3, 1.

[31] C. Garvie, R. Hannink, R. Pascoe, Nature 1975, 258, 703.
[32] D. Green, M. Swain, R. Hannink, Transformation Toughening of Ceramics, CRC Press, Boca Raton, 1988.
[33] K. Tsukuma, I. Yamashita, T. Kusunose, J. Am. Ceram. Soc. 2008, 91, 813.
[34] I. Yamashita, M. Kudo, T. Koji, Tosoh Res. Technol. Rev. 2012, 56, 11.
[35] U. Peuchert, Y. Okano, Y. Menke, S. Reichel, A. Ikesue, J. Eur. Ceram. Soc. 2009, 29, 283.
[36] M. Zwick, J. Werner, N. Kratz, S. Vetter, DE Patent Application 10 2017 104 168 A1, 2018.
[37] M. Zwick, J. Werner, DE Patent Application 10 2017 104 166 A1, 2018.
[38] M. Gutierrez, E. Penilla, L. Leija, A. Vera, J. Garay, G. Aguilar, Adv. Healthcare Mater. 2017, 1700214.
[39] Y. Damestani, C. L. Reynolds, J. Szu, M. S. Hsu, Y. Kodera, D. K. Binder, B. H. Park, J. E. Garay, M. P. Rao, G. Aguilar, Nanomed.: Nanotechn., Biol. Med. 2013, 9, 1135.
[40] M. Cano-Velazquez, N. Davoodzadeh, D. Halaney, C. Jonak, D. Binder, J. Hernandez-Cordero, G. Aguilar, Biomed. Opt. Expr. 2019, 10, 3369.
[41] U. Peuchert, Y. Menke, US Patent 7,710,656 B2, 2010.
[42] R. Yasuhara, A. Ikesue, Opt. Expr. 2019, 27, 289.
[43] Y. Ji, D. Jiang, T. Fen, Mater. Res. Bull. 2005, 40, 553.
[44] J. Trojan-Piegza, S. Gierlotka, E. Zych, J. Am. Ceram. Soc. 2014, 97, 1595.
[45] H. Yi, X. Zou, Y. Yang, J. Am. Ceram. Soc. 2011, 94, 4120.
[46] Z. Wang, G. Zhou, F. Zhang, J. Lumin. 2016, 169, 612.
[47] L. An, A. Ito, T. Goto, Key Eng. Mater. 2011, 484, 135.
[48] T. Feng, D. Clarke, D. Jiang, Appl. Phys. Lett. 2011, 98, 151105.
[49] Z. Wang, G. Zhou, X. Qin, J. Eur. Ceram. Soc. 2013, 33, 643.
[50] Z. Wang, G. Zhou, X. Qin, J. Alloys Compd. 2014, 585, 497.
[51] L. An, A. Ito, T. Goto, J. Eur. Ceram. Soc. 2011, 31, 237.
[52] Y. Ji, D. Jian, J. Shi, Mater. Lett. 2005, 59, 868.
[53] Y. Ji, D. Jiang, J. Shi, J. Mater. Res. 2005, 20, 567.

[54] Z. Wang, G. Zhou, J. Zhang, J. Am. Ceram. Soc. 2015, 98, 2476.
[55] Z. Wang, G. Zhou, J. Zhang, Opt. Mater. 2017, 71, 5.
[56] Z. Wang, G. Zhou, D. Jiang, S. Wang, J. Adv. Cer. 2018, 7, 289.
[57] T. Murray, R. Dungan, Ceram. Ind. 1964, 82, 74.
[58] G. Haertling, C. Land, J. Am. Ceram. Soc. 1971, 54.
[59] G. Haertling, J. Am. Ceram. Soc. 1971, 54, 303.
[60] G. Haertling, Ferroelectrics 1987, 75, 25.
[61] G. Haertling, J. Am. Ceram. Soc. 1999, 82, 797.
[62] X. Wang, J. Huang, Y. Tang, Adv. Mech. Eng. 2015, 7, 1.
[63] H. Jiang, Y. Zou, Q. Chen, K. Li, R. Zhang, Y. Wang, H. Ming, Z. Zheng, Proc. SPIE 2005, 5644, 598.
[64] A. Glebov, V. Smirnov, M. Le, L. Glebov, A. Sugama, S. Aoki, V. Rotar, IEEE Photon. Technol. Lett. 2007, 19, 701.
[65] G. Liberts, A. Bulanovs, G. Ivanovs, Ferroelectrics 2006, 333, 81.
[66] F. Wei, Y. Sun, D. Chen, G. Xin, Q. Ye, H. Ai, R. Qu, IEEE Photon. Technol. Lett. 2011, 23, 296.
[67] Q. Ye, Z. Dong, Z. Fang, R. Qu, Opt. Expr. 2007, 15, 16933.
[68] T. Kawaguchi, H. Adachi, K. Setsune, O. Yamazaki, K. Wasa, Appl. Opt. 1984, 23, 2187.
[69] R. Thapliya, Y. Okano, S. Nakamura, J. Lightwave Technol. 2003, 21, 1820.
[70] R. Sabat, P. Rochon, Ferroelectrics 2009, 386, 105.

[71] T. Maiman, Nature 1960, 187, 493.
[72] E. Carnall, E. Hatch, W. Parsons, Mater. Sci. Res. 1966, 3, 165.
[73] A. Ikesue, T. Kinoshita, K. Kamata, K. Yoshida, J. Am. Ceram. Soc. 1995, 78, 1033.
[74] J. Lu, K. Ueda, H. Yagi, T. Yanagitani, Y. Akiyama, A. A. Kaminskii, J. Alloys Compd. 2002, 341, 220.
[75] B. Bishop, "Northrop Grumman Scales New Hights in Electric Laser Power, Achieves 100 kW from a Solid-state Laser" (March 18th, 2009), can be found under https://news.northropgrumman.com/news/releases/photo-release-northrop-grumman-scales-new-heights-in-electric-laser-power-achieves-100-kilowatts-from-a-solid-state-laser (accessed Jan 18, 2020).
[76] P. Aubry, A. Bensalah, P. Gredin, G. Patriarche, D. Vivien, M. Mortier, Opt. Mater. 2009, 31, 750.
[77] P. Aballea, A. Suaganuma, F. Druon, J. Hostalrich, P. Georges, P. Gredin, M. Mortier, Optica 2015, 2, 288.
[78] W. Zhou, F. Cai, G. Zhi, B. N. Mei, Mater. Sci. Pol. 2014, 32, 358.
[79] Z. Sun, B. Mei, W. Li, X. Liu, L. Su, Opt. Mater. 2017, 71, 35.
[80] H. Furuse, N. Horiuchi, B.-N. Kim, Sci. Rep. 2019, 9, 10300.
[81] Y. Fu, L. Ge, J. Li, Y. Liu, M. Ivanov, L. Liu, H. Zhao, Y. Pan, J. Guo, Opt. Mater. 2017, 71, 90.
[82] A. Ikesue, Y. Aung, T. Kamimura, S. Honda, Y. Iwamoto, Mater. 2018, 11, 271.
[83] E. Kupp, G. Messing, J. Anderson, V. Gopalan, J. Dumm, C. Kraisinger, G. Quarles, J. Mater. Res. 2010, 25, 476.
[84] J. Sulc, H. Jelínková, V. Kubecek, K. Nejezchleb, K. Blažek, Proc. SPIE 2002, 4630, 128.
[85] D. Kracht, M. Frede, R. Wilhelm, C. Fallnich, Opt. Expr. 2005, 13, 6212.
[86] H. Yagi, K. Takaichi, K. Ueda, Y. Yamasaki, T. Yanagitani, A. Kaminski, Laser Phys. 2005, 15, 1338.
[87] Ikesue, Y. Aung, T. Taira, T. Kamimura, K. Yoshida, G. Messing, Annu. Ref. Mater. Res. 2006, 36, 397.
[88] V. Lupei, Opt. Mater. 2009, 31, 701.
[89] J. Sanghera, W. Kim, G. Villalobos, B. Shaw, C. Baker, J. Frantz, B. Sadowski, I. Aggarwal, Opt. Mater. 2013, 35, 693.

[90] S. Derenzo, M. Weber, E. Bourret-Courchesne, M. K. Klintenberg, Nucl. Instr. Meth. 2003, 505, 111.

[91] Z. Wang, G. Zhou, J. Zhang, X. Qin, S. Wang, Opt. Mater. 2017, 71, 5.

[92] J. Kuntz, J. Roberts, M. Hough, N. Cherepy, Scr. Mater. 2007, 57, 960.

[93] N. Cherepy, Z. Seeley, S. Payne, P. Beck, E. Swanberg, S. Hunter, L. Ahle, S. Fisher, C. Melcher, H. Wei, T. Stefanik, Y.-S. Chung, J. Kindem, Proc. SPIE 2017, 9213, 921302.

[94] V. Osipov, A. Ishchenko, V. Shitov, R. Maksimov, K. Lukyashin, V. Platonov, A. Orlov, S. Osipov, V. Yagodin, L. Viktorov, B. V. Shulgin, Opt. Mater. 2017, 71, 98.

[95] Y. Ye, P. Liu, D. Yan, X. Xiu, J. Zhang, Opt. Mater. 2017, 71, 23.

[96] C. Dujardin, E. Auffray, E. Bourret-Courchesne, P. Dorenbos, P. Lecoq, M. Nikl, A. Vasilev, A. Yoshikawa, R.-Y. Zhu, IEEE Trans. Nucl. Sci. 2018, 65, 1977.

[97] G. Blasse, A. Bril, Appl. Phys. Lett. 1967, 11, 53.

[98] S. Nakamura, G. Fasol, The Blue Laser Diode: GaN Based Light Emitters and Lasers, Springer, Berlin, 1997.

[99] L. Chen, C. C. Lin, C.-W. Yeh, R.-S. Liu, Mater. 2010, 3, 2172.

[100] M. Raukas, J. Kelso, Y. Zheng, K. Bergenek, D. Eisert, A. Linkov, F. Jermann, ECS J. Solid State Sci. Technol. 2013, 2, R3168.

[101] H. Hua, S. Feng, Z. Ouyang, H. Shao, H. Qin, H. Ding, Q. Du, Z. Zhang, J. Jiang, H. Jiang, J. Adv. Eram. 2019, 8, 389.

[102] R. Müller-Mach, G. Müller, M. Krames, O. Shchekin, P. Schmidt, H. Bechtel, C.-H. Chen, O. Steigelmann, Phys. Stat. Solid. RRL 2009, 3, 215.

[103] W. Schnick, Phys. Stat. Solid. RRL 2009, 3, A113.

[104] H. Sijbom, R. Verstraete, J. Joos, D. Poelman, P. Smet, Opt. Mater. Expr. 2017, 7, 3332.

[105] C. Cozzan, M. Brady, N. O´Dea, E. Levin, S. Nakamura, S. DenBaars, R. Seshadri, AIP Adv. 2016, 6, 105005.

[106] I. Pricha, W. Rossner, R. Moos, J. Am. Ceram. Soc. 2015, 99, 211.

[107] A. Schricker, K. Mai, G. Basin, U. Mackens, J. Vogels, A. Wejers, K. Zijtfeld, US Patent 10,205,067 B2, 2016.

[108] A. Katelnikovas, J. Plewa, S. Sakirzanovas, D. Dutczak, D. Enseling, F. Baur, H. Winkler, A. Kareiva, T. Jüstel, J. Mater. Chem. 2012, 22, 22126.

[109] D. Böhnisch, F. Baur, T. Jüstel, Dalton Trans. 2018, 47, 1520.

[110] M. van de Haar, J. Werner, N. Kratz, T. Hilgerink, M. Tachikirt, J. Honold, M. Krames, Appl. Phys. Lett. 2018, 112, 132101.

[111] S. Li, R.-J. Xie, ECS J. Solid State Sci. Techn. 2020, 9, 016013.

[112] Z. Xiao, S. Yu, Y. Li, S. Ruan, L. Bing Kong, Q. Huang, Z. Huang, K. Zhou, H. Su, Z. Yao, W. Que, Y. Liu, T. Zhang, J. Wang, P. Liu, D. Shen, M. Allix, J. Zhang, D. Tang, Mater. Sci. Eng. Rep. 2020, 100518.

Daniel Rytz, Klaus Dupré, Andreas Gross, Christoph Liebald, Mark Peltz, Patrick Pues, Sebastian Schwung, Volker Wesemann

4.4 Crystal growth of rare-earth doped and rare-earth containing materials

4.4.1 Introduction

Rare earth-containing or rare earth-doped single crystals are common in the fields of solid state lasers, scintillators and light converters, among others. Prominent examples of rare earth containing crystals are:

$RE_3Al_5O_{12}$ and $RE_3Ga_5O_{12}$ (garnets)
$REAlO_3$ (perovskites)
RE_2O_3 (sesquioxides)
$CaREAlO_4$ and $SrREGaO_4$
$RE_2Ti_2O_7$ and $RE_2Zr_2O_7$ (pyrochlores)
$LiREF_4$
KRE_3F_{10}
RE_2SiO_5 (orthosilicates)
$KRE(WO_4)_2$ and $KRE(MoO_4)_2$ (double tungstates or molybdates)
$REAl_3(BO_3)_4$ (huntites)
$RECa_4O(BO_3)_3$ (calcium oxoborates)

where RE^{3+} designates a rare earth ion, occupying a lattice site. This list is far from exhaustive; however, it contains crystalline materials of current interest in several technologically relevant areas, which are the basis of numerous ongoing technological and scientific developments. The present chapter will describe four types of materials among the listed members, namely: $RE_3Al_5O_{12}$ (garnets), $CaREAlO_4$, $KRE(WO_4)_2$ (double tungstates) and $REAl_3(BO_3)_4$ (huntites).

Historically, the garnet family was probably the first among these crystals to play an important role. The structures of the rare earth (including yttrium) aluminum and gallium garnets family have many properties that are desirable as solid state laser host materials. They are cubic and thus optically isotropic, stable and mechanically robust, hard and highly suitable for optical polishing. Most importantly and unlike the single crystalline form of Al_2O_3 (which could otherwise be an almost ideal optical material), they can accept trivalent ions of the rare earth group as substitutional dopants. Doped garnet crystals with high structural perfection and low optical losses have been used as hosts for lasers based on Ce^{3+}, Nd^{3+}, Dy^{3+}, Ho^{3+}, Er^{3+}, Tm^{3+} and Yb^{3+}.

To the best of our knowledge, the first publication describing a garnet-based laser was written by Geusic et al. [1] and described a Nd^{3+}:YAG laser, based on

https://doi.org/10.1515/9783110654929-027

garnet crystals grown by various techniques. Today Nd^{3+}:YAG, Yb^{3+}:YAG and related laser hosts play an essential role in the industry (see for example [2]).

Cerium-doped garnets act as a phosphor-emitting yellow light when subjected to blue, ultraviolet or X-ray excitation [3, 4]. Garnets are used as light-converting phosphors on high-brightness blue (In,Ga)N light-emitting diodes, converting part of the blue light into yellow, which, in combination, then appears as white. Light-emitting diodes (LEDs) combined with phosphors such as Ce:YAG or Ce:LuAG are not only the new reference in general lighting, they can also be used as alternatives to halogen or arc lamps for more specific, bright illumination sources. As an example, one can note extremely bright point sources of light, where multiple LEDs are combined to directly pump a light guide made of a phosphor rod [5]. LEDs have also recently shown their potential for lower-cost alternatives as energy sources for laser pumping [6], based on a light conversion scheme in Ce:YAG crystalline light guiding structures.

Scintillators are relatively simple but accurate detectors for high energy radiation such as X-rays, gamma-rays and high energy particles exceeding a few kilo electron Volts (keV) in energy. When high energy radiation strikes on a scintillating crystal, it creates a large number of electron-hole pairs inside the crystal. Recombination of these pairs will release photons with energy in the range of a few eV. The latter can then be detected by a photomultiplier tube or an avalanche photodiode.

In the early 1980s, Ce-doped Gd_2SiO_5 crystals came to be used as a scintillator material. This material has an adequately high density (6.71 g/cm^3) and is non-hygroscopic. In the late 1980s, Ce-doped Lu_2SiO_5 crystals were developed with an even higher density (7.4 g/cm^3), a significantly better light yield (reaching 75% of NaI(Tl), the scintillation reference material) and a faster (42 ns) decay time. Moreover, unlike Gd_2SiO_5 that has a strong tendency to cleave during fabrication of detector devices, Lu_2SiO_5 does not exhibit any cleavage properties. Compared with all the other existing known scintillator crystals, Ce-doped Lu_2SiO_5 seems to have the best combination of all the needed properties for positron emission tomography or other high energy gamma-ray detector applications. For recently developed aspects related to Lu_2SiO_5, the reader is referred to a publication by Yuntao Wu [7].

Most of the crystals mentioned above are produced by the Czochralski technique. This crystal growth technique is suitable for materials with a congruent melting behavior and with melting points up to 2150 °C. A schematic crystal puller for the Czochralski process is represented in Figure 4.4.1. A crucible (1) contains the melt (2) from which a crystal (3) is slowly pulled. At the same time, the pullrod (4) is also rotated. Growth is initiated at the start of the process by dipping a seed (5) into the melt. At the end of the process, the grown crystal will be extracted and separated from the melt.

Adequate thermal conditions are obtained with a heater (6, here an induction coil) and a passive afterheater (7). For more details, the reader is referred to, for example, the books by Nassau [8], or the review by Uecker [9] or by Hongjun Li et al. [10].

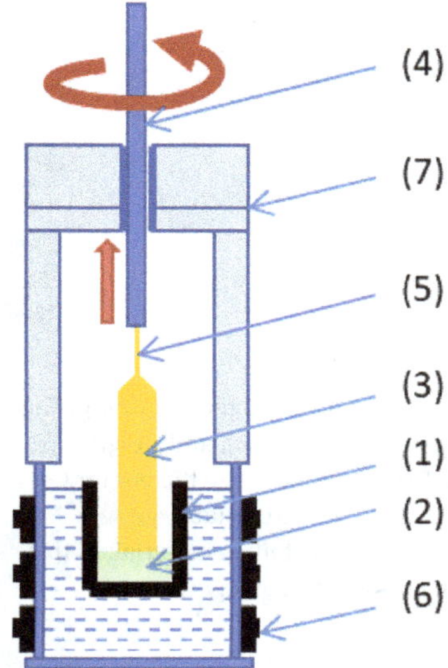

Figure 4.4.1: Schematic crystal puller for the Czochralski process. A crucible (1) contains the melt (2) from which a crystal (3) is slowly pulled. At the same time, the pullrod (4) is also rotated. Growth is initiated by dipping a seed (5) into the melt. Adequate thermal conditions are obtained with a heater (6) and a passive after-heater (7).

4.4.2 Czochralski growth of YAG and other garnets

Single crystalline garnets of the $RE_3Al_5O_{12}$ family are grown by the Czochralski method in a setup described above in Figure 4.4.1. An iridium crucible with a diameter of 80 to 150 mm is used, mounted in a standard thermal insulation setup. The growth process takes place at temperatures of approximately 1950 to 2050 °C, depending on the rare earth *RE*. In the vicinity of the melt level, the temperature gradient is estimated to be near 50 K cm^{-1}. The seeds used in the process can be oriented along <111>, <110> or <001> . The growth atmosphere is generally, nitrogen plus a small fraction of oxygen, at the level of about 0.5%. The seed rotation and pulling parameters are in the ranges 10–18 rpm and 0.3 to 2.0 mm h^{-1}, respectively. The resulting boules can reach masses near 5 kg, corresponding to diameters up to 80 mm and usable cylinder lengths up to 200 mm. Standard growth interfaces appear to be conical, with small {121} facets present near the center of the boule. In such a case, the boule will have a longitudinal core, that is, an axial zone in the center of the boule with strong strain induced birefringence. Depending on the composition and the doping of the crystal, the growth duration for a boule with 50 mm diameter and a usable length of 150 mm will be about 7 days for a Yb:YAG crystal and approximately 21 days for a Nd:YAG crystal. A detailed description of the growth of garnets can be found in [11]. An example for a Ce:YAG crystal is provided in Figure 4.4.2.

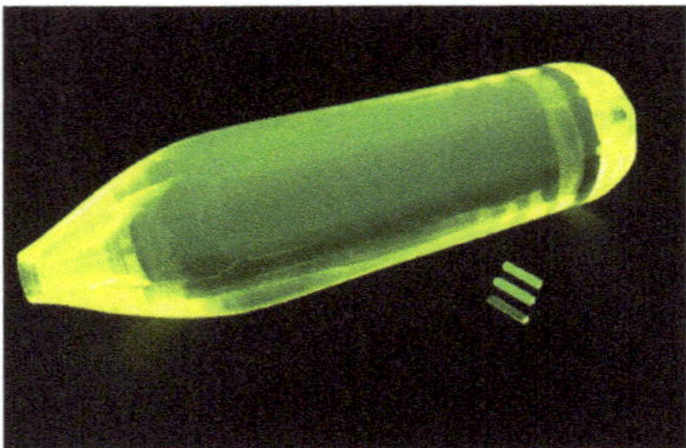

Figure 4.4.2: Ce:YAG boule, grown by the Czochralski technique from a melt weighing 9,800 g. The boule has a diameter of 60 mm, a length (from start of cone to tail end) of 155 mm and weighs 2,488 g. The Ce concentration in the starting melt is 0.4 mol%. The segregation coefficient for Ce in YAG leads to a much lower concentration in the crystal (see below). The three slabs (2 × 2 × 15 mm) on the right are polished parallelepipeds of the same crystalline material. In this image, the crystal and the samples are illuminated by a 365 nm excitation source.

Due to segregation phenomena, doped YAG crystals can exhibit sizeable concentration gradients, especially along the growth axis. If the following quantities are defined:

$k = C_S/C_L$ = effective segregation coefficient = dopant concentration in crystal / concentration in melt,

C_0 = dopant concentration in initial melt,

f_S = mass of crystal / total initial melt mass,

one can use the so-called Scheil's equation:

$$C_S = kC_0(1 - f_S)^{k-1}$$

to calculate the dopant concentration, C_S, as a function of the distance along the growth direction in the crystal. Figure 4.4.3 shows the curve describing the situation for the Ce:YAG boule shown above in Figure 4.4.2.

For this calculation, the segregation coefficient, k, has to be known, for example, from actual measurements of the dopant concentration in the crystal. Here, a value of k = 0.10 was assumed, in agreement with values quoted in the literature [12].

The segregation coefficient is strongly dependent on the dopant. For example, in the case of Yb^{3+} substituting for Y^{3+} ions in YAG, the segregation coefficient is close to one, which implies a very small concentration gradient. An example for such a boule is provided in Figure 4.4.4.

Yb:YAG will display an optical absorption such as described in Figure 4.4.5. Yb^{3+} in YAG exhibits strong absorption peaks at 945 and 970 nm.

In the UV range, Yb-doped garnets as well as undoped crystals will usually show extrinsic absorption features induced by residual impurities, as shown in Figure 4.4.6.

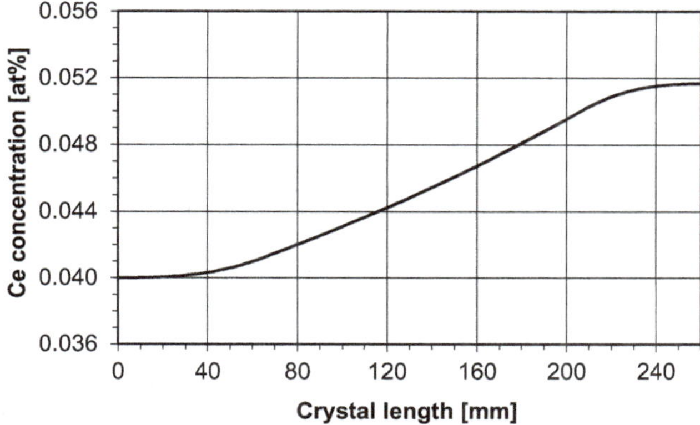

Figure 4.4.3: Calculated concentration profile in the Ce:YAG crystal of Figure 4.4.2. Due to the relatively low segregation coefficient for Ce, the concentration of Ce is expected to vary from 0.040 at% at the seed end to 0.052 at% at the tail end.

Figure 4.4.4: Yb:YAG boule, grown by the Czochralski technique. The boule has the dimensions diameter 53 mm × length (from start of cone to tail end) 174 mm and weighs 1,300 g. The Yb concentration in the starting melt is 3.0 mol% and varies only slightly along the boule axis.

A broad absorption approximately centered at 255 nm appears. It is attributed to Fe^{3+} impurities introduced into the melt based on a combination of contaminants originating from starting materials, crucibles and ceramic insulation materials. Such impurities have been analyzed in [13–15] and are still the topic of ongoing investigations, not only in garnets but also in other crystals as we will see in the following chapters.

4.4.3 Czochralski growth of CALGO and related materials

$ABCO_4$ oxides with K_2NiF_4 structure (space group *I4/mmm*) where $A^{2+} = Ca^{2+}$ or Sr^{2+}, B^{3+} is a rare earth RE^{3+}, and $C = Al^{3+}$ or Ga^{3+}, exhibit a statistical distribution of A^{2+} and B^{3+} ions on a common Wyckoff site. When such crystals are doped with laser active ions such as Yb^{3+}, one can also expect a statistical distribution for the

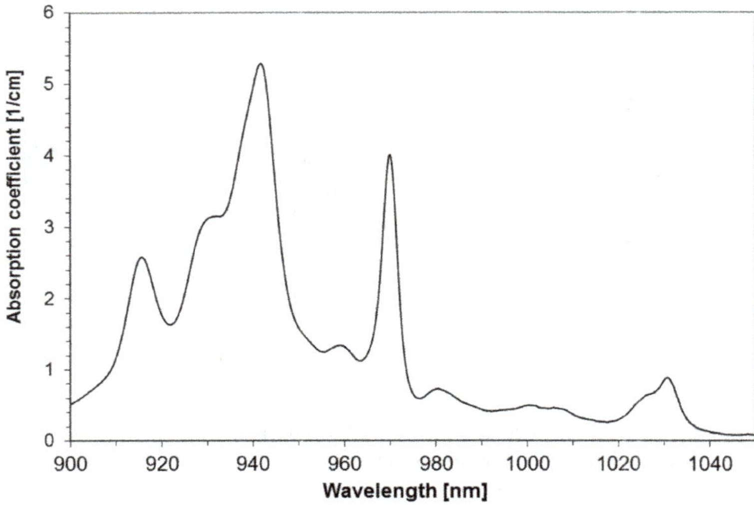

Figure 4.4.5: Extinction coefficient measured for Yb(5%):YAG in the wavelength range between 900 and 1,100 nm.

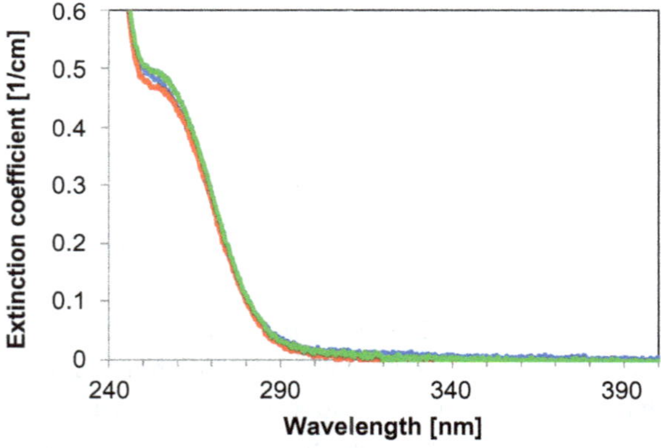

Figure 4.4.6: Extinction coefficients in three different Yb doped LuAG crystals in the wavelength range between 240 and 400 nm. The parasitic absorption near 260 nm is due to residual impurities.

dopant, resulting in a noticeable broadening of the emission linewidth, while, at the same time, the thermal conductivity of the $ABCO_4$ crystals stays at a relatively high level. This unique combination of attributes leads to properties reminiscent of glasses (i.e., broad emission) and crystals (i.e., favorable thermal behavior) in one and the same crystalline material, the most prominent example being $CaGdAlO_4$ (CALGO). Properties of $ABCO_4$ materials have been described by Pajaczkowska and

Gloubokov [16]. These materials were initially used as substrates for epitaxy of high T_c superconductors [17, 18].

Numerous members of the $ABCO_4$ family exhibit noncongruent melting [19–21]. CALGO, however, appears to be congruently melting, although the optimal melt composition for growth may deviate from the stoichiometric composition. Nd^{3+}-doped CALGO as a laser material appears first in the work of Lagatskii et al. [22] and Yb^{3+} doping for lasers is described [23–25]. Based on the broad emission of Yb:CALGO, ultrashort pulses are reported in 2006 (47 fs by Zaouter et al. [26]) and 2012 (40 fs by Agnesi et al. [27]). Beil et al. [28] demonstrated very high slope efficiencies and a large tuning range in Yb:CALGO. As a result of these research activities, Yb:CALGO has become a material used in commercially available ultrafast lasers.

State-of-the-art Yb:CALGO crystals for laser applications are shown in Figure 4.4.7. These crystals have been grown by the Czochralski method in a setup described by Liebald [29]. An Iridium crucible with a diameter of 80 mm is used, mounted in a standard thermal insulation setup. The growth process takes place at a temperature of approximately 1750 °C. In the vicinity of the melt level, the temperature gradient was measured to be near 7 K mm^{-1}. The seeds used in the process were oriented along the *a*-axis. The growth atmosphere was nitrogen. The seed rotation and pulling parameters were 12 rpm and 1 mm h^{-1}, respectively. The resulting boules had masses near 250 g corresponding to diameters near 25 mm and usable cylinder lengths near 60 mm. The growth interfaces appear to be slightly convex, with small {101} facets sometimes present near the center of the boule. When the cone angles on the seed end of the boules are too wide, occasional cleaving of the boules along planes parallel to {001} faces may appear across the entire boule.

Figure 4.4.7: Boules of Yb(1%):CALGO, with diameter 25 mm, cylinder length 60 mm and mass 250 g.

The boules appear to have varying colors (from light amber to colorless), depending on the exact starting composition of the melt and the growth atmosphere. In laser applications, almost colorless boules are preferred, as they appear to exhibit lower densities of scattering centers (see below). Yb substitution levels up to 10% of Yb and higher have been demonstrated.

Yb:CALGO boules exhibit a concentration gradient: between the top and the bottom of a boule, the Yb concentration increases, as can be seen in Figure 4.4.8. The intensities of the main absorption peaks at 981 nm, increase from approximately 16 to 18 cm^{-1} for *c*-polarized light and from 6 to 7 cm^{-1} for *a*-polarized light over a longitudinal (along the growth axis) distance of 70 mm.

Yb:CALGO crystals show scattering centers as depicted in Figure 4.4.9. These defects have varying densities both longitudinally and transversally. Several different types of scattering centers with small (<2 µm) and larger (5 µm) volumes seem to coexist, the larger ones having been identified as $Ca_3Al_2O_6$ [29]. The CALGO crystals with less coloration than the crystal used for Figures 4.4.8 and 4.4.9 also exhibit scattering centers, with a tendency to lower densities when the color is less pronounced.

In a recent development work [30], a new material of the CALGO family, namely CALTO or $CaTbAlO_4$, where the rare earth site is fully substituted by terbium, was also grown by the Czochralski technique. CALTO has the same tetragonal symmetry as CALGO. Figure 4.4.10 displays such a CALTO boule grown under similar conditions as the CALGO described above.

It should be noticed that the CALTO crystals display an intense yellow-orange color. This is due to the addition of 1% of Ce to the melt in order to prevent the incorporation of Tb^{4+} ions into the crystal. Without Ce, the resulting boules are very dark in coloration and thus not usable for optical applications. The compensation mechanism is the subject of ongoing investigations.

CALTO is a potentially interesting material for applications in optical isolators based on the Faraday effect. For such applications, in the wavelength range between 400 and 1,100 nm, the reference crystals are TGG = $Tb_3Ga_5O_{12}$ or KTF = KTb_3F_{10}, with Verdet constants (for 1,064 nm radiation) of 39 Rad T^{-1} m^{-1} and 36 Rad T^{-1} m^{-1}, respectively.

CALTO appears to possess a much higher Verdet constant amounting to about 170% of the value of the constant of TGG, as shown in Figure 4.4.11, where the rotation angle induced by a magnetic field near 0.500 tesla is measured for several Ce-doped CALTO crystals with respect to a TGG standard of the same length [30].

The measurements show no dependence of the rotation angle in Ce:CALTO on the Ce concentration. Ce concentrations of 1%, 2% and 4% corresponding to the value in the melt have been used, the exact Ce concentrations in the crystals not being known at present!

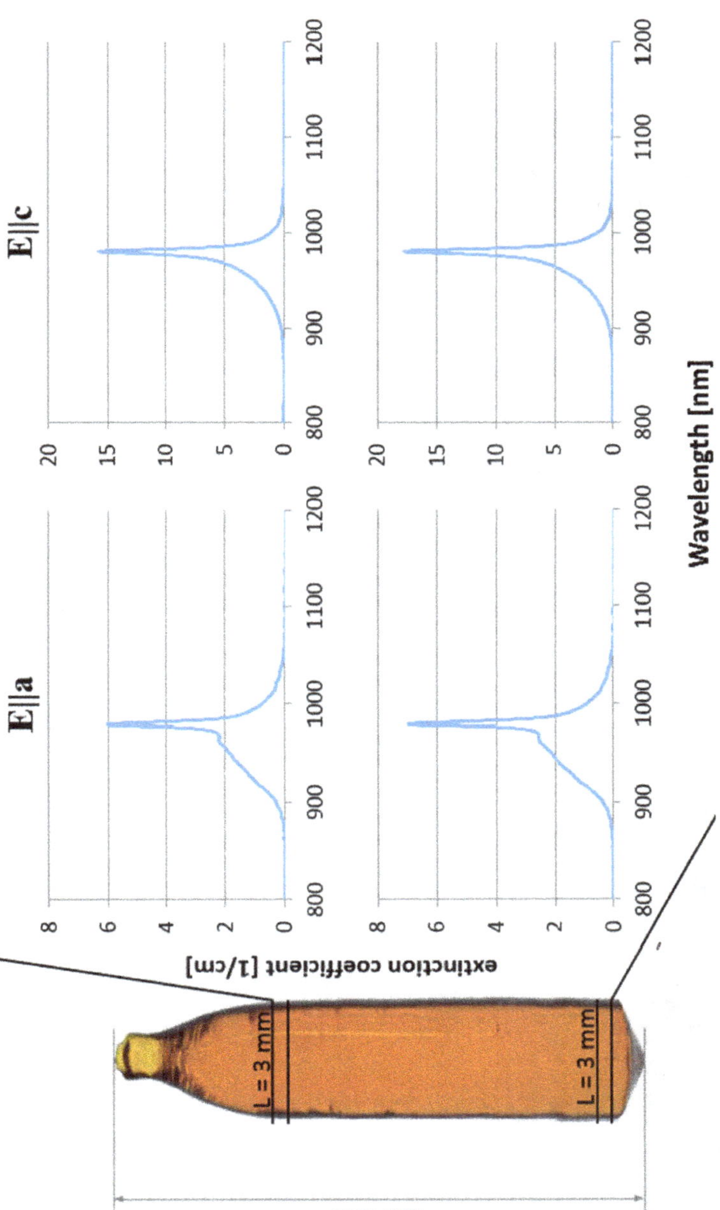

Figure 4.4.8: Picture of a Yb(5%):CALGO boule showing the locations of the two samples separated by 70 mm used for the measurements of the absorption peaks near 981 nm for *c*- and *a*-polarized light in order to illustrate the Yb concentration gradient along the growth direction.

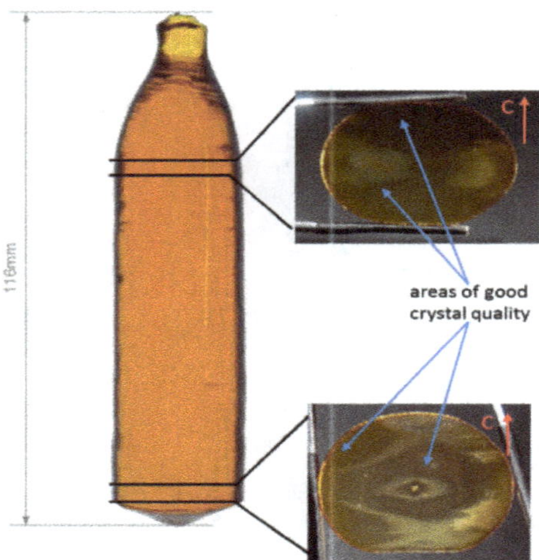

Figure 4.4.9: Same Yb(5%):CALGO as above, showing the transverse distribution of scattering centers in two crystal plates. The foggy areas diffusing light contain scattering centers of 2 to 5 µm size.

Figure 4.4.10: Boule of CALTO, with diameter 35 mm, cylinder length 40 mm and a mass of 180 g.

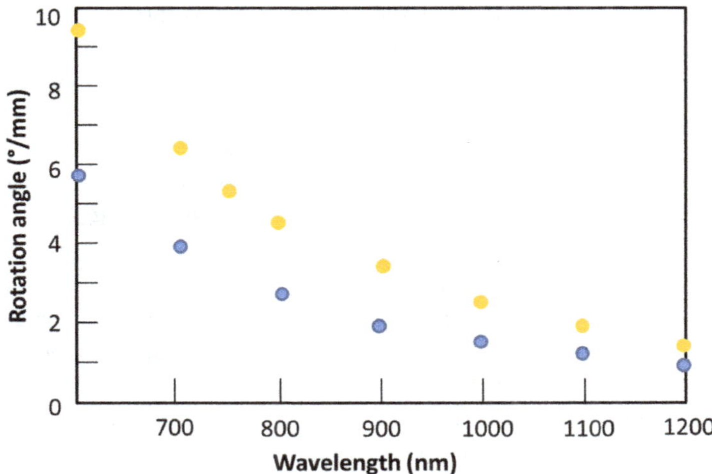

Figure 4.4.11: Measured rotation angles as a function of wavelength for Ce:CALTO crystals (orange dots), as compared to a TGG reference crystal (blue dots).

As a conclusion, it appears that CALTO crystals indeed display an attractively high Verdet constant, a fact that has also been reported in the literature by Man Mei [31]. In addition, we have verified that the optical absorption of CALTO is comparable to the absorption of TGG in the range from 600 to 1100 nm. This implies that the use of CALTO could pave the way for the implementation of smaller geometries and thus for optical isolators with smaller volumes and footprints for laser sources in that range.

4.4.4 TSSG of KYW = $KY(WO_4)_2$ and related materials

Top-seeded solution growth (TSSG) is a high temperature crystal growth technique that is often used in the growth of non-congruently melting materials such as LiB_3O_5 [32] and $KNbO_3$, or for the growth of materials exhibiting phase transitions such as $BaTiO_3$ [33]. The TSSG approach is based on a furnace geometry similar to the one used in the Czochralski technique, that is, a seed which is dipped in a molten solution, often with weight control of the growing crystal. The molten solution is composed of two distinct chemical components, the material to be grown (for example, LiB_3O_5, $KNbO_3$ or $BaTiO_3$) and an adequate high temperature solvent (MoO_3 for LiB_3O_5, K_2O for $KNbO_3$, and TiO_2 for $BaTiO_3$). The resulting solutions allow the growth of the desired crystals by inserting a seed crystal at the appropriate equilibrium temperature and subsequently slowly cooling the solution, in order to obtain very high structural quality crystals. In the case of LiB_3O_5, crystals with more than 1 kg mass can be obtained by cooling a solution of LiB_3O_5+MoO_3 from 700 to 640 °C in approximately 40 days.

$KNbO_3$ crystals of up to 300 g can be obtained from 0.95 $KNbO_3$ + 0.05 K_2O by cooling from 1040 to 1033 °C in approximately 5 days. $BaTiO_3$ crystals of up to 100 g can be obtained from 0.51 $BaTiO_3$ + 0.49 TiO_2 by cooling from 1390 to 1340 °C in 6 days. As can be inferred from the case of LiB_3O_5, scaling the TSSG process to larger batches, yielding larger boules, is possible and TSSG should not be viewed as a process restricted to small boules only. It should be noted that in the two examples above, in the cases of both $KNbO_3$ and $BaTiO_3$, the respective fluxes do not introduce foreign ions into the high temperature solutions suitable for the crystal growth process. Such a situation is, of course, favorable from the point of view of flux-induced impurities. On the other side, when a flux containing foreign atoms with respect to the crystal to be grown has to be used, the flux may then represent a potential source of contamination.

Rare earth double tungstates of the general type $KRE(WO_4)_2$, where *RE* designates a rare earth ion, belong to the class of materials that have to be grown by TSSG, due to solid-solid phase transitions occurring around 1000 °C [34, 35].

The growth process is based on the use of $K_2W_2O_7$ as a flux, as described in the following phase diagram (Figure 4.4.12). This diagram is based on work by Pujol et al. [35] for the growth of $KGd(WO_4)_2$. Herein, it is assumed that a similar diagram can be drawn for KYW = $KY(WO_4)_2$ which is the compound that is discussed in some detail in this chapter.

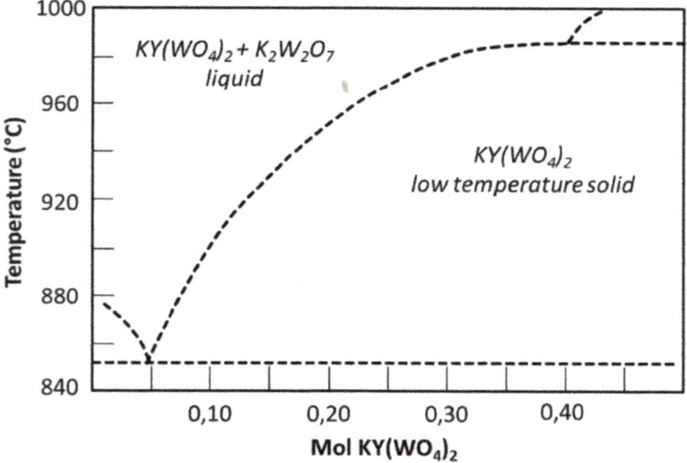

Figure 4.4.12: Qualitative phase diagram $K_2W_2O_7$–KYW illustrating the composition range near 18% to 20% of solute used for flux growth of KYW.

For TSSG growth of KYW with $K_2W_2O_7$ as a solvent, the following reaction equation applies:

$$y/2[K_2CO_3 + Y_2O_3 + 4WO_3] + (1-y)\ [K_2CO_3 + 2WO_3] \rightarrow$$
$$yKY(WO_4)_2 + (1-y)\ K_2W_2O_7 + (1-y/2)CO_2$$

High purity (99.99%) starting materials K_2CO_3, WO_3 and Y_2O_3 are weighed in and mixed to form a 700 g solution with an initial molar ratio between $KY(WO_4)_2$ and $K_2W_2O_7$ close to 1:5. For doped crystals, the appropriate amount of dopant, for example, in the form of Yb_2O_3, is added in replacement of the corresponding molar quantity of Y_2O_3.

The mixture is melted in a platinum crucible at 970 °C and held at that level for 24 h to ensure a homogenous solution. The latter is then cooled to about 955 °C at a rate of 1 K h^{-1} and held for an additional period of 5 h to ensure a homogenous temperature distribution within the solution. The growth then starts at about 955 °C by contacting an oriented [010] $KY(WO_4)_2$ seed crystal with the liquid. After the start of the crystallization process, the growth on the seed is initially carried out without pulling until a mass of 1.2 g has been reached. At that mass level, pulling is initiated with 0.02 mm h^{-1} and increased to 0.15 mm h^{-1} within 13 hours. A slow initial temperature ramp, starting at −0.01 K h^{-1} is started at the beginning of the growth process. As soon as a mass of 1.2 g is reached, the temperature ramp is increased to −0.06 K h^{-1} and accelerated by an additional −0.04 K h^{-1} step, every 24 hours. This adjustment of the temperature ramp is necessary to stabilize the crystallization process at a close-to-optimal volume increase per unit time.

Figure 4.4.13 shows several typical Yb^{3+}-doped and undoped KYW boules grown along the monoclinic *b*-axis. In this case, the growth is terminated by a clearly visible *b*-facet at the bottom of the boule. The edges of this facet correspond to the

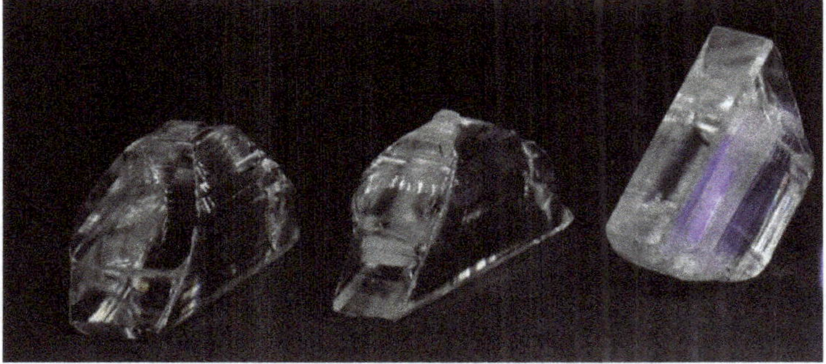

Figure 4.4.13: Undoped and Yb-doped KYW boules with typical dimensions (a = 20 mm, b = 25 mm, c = 35 mm). The *b*-axis corresponds to the growth direction. The two boules on the left rest on their *b*-facets, whereas the boule on the right is lying on a (101) facet, with the smooth *b*-facet terminating the growth clearly observable.

monoclinic *a*- and *c*-axes which form an angle of 94.0° from *c* to *a*. The position of the *a*-axis is clearly recognizable from the position of the (101) facets. Boules grown from the solutions described above reach a mass near 100 g after a growth time of approximately five days.

The TSSG parameters could, in principle, be scaled to achieve much larger boules using larger crucibles and thus, larger melt volumes. Crystals with masses in excess of 500 g have been demonstrated for some compositions in the K*RE*W family of crystals.

Figure 4.4.14 shows a KTbW = $KTb(WO_4)_2$ boule with its (101) facets clearly visible. Its *b*-facet is positioned in the *a*-*c* plane, which also contains the so-called N_m and N_g axes. As is usual for monoclinic crystals, the principal axes labelled N_p, N_m and N_g describing the optical indicatrix of the crystal (and thus forming an orthonormal system) are positioned parallel to the *b*-axis for N_p, and contained in the *a*-*c* plane for N_m and N_g. The dispersion angle φ is wavelength dependent and amounts to about 19.8° for 1000 nm light.

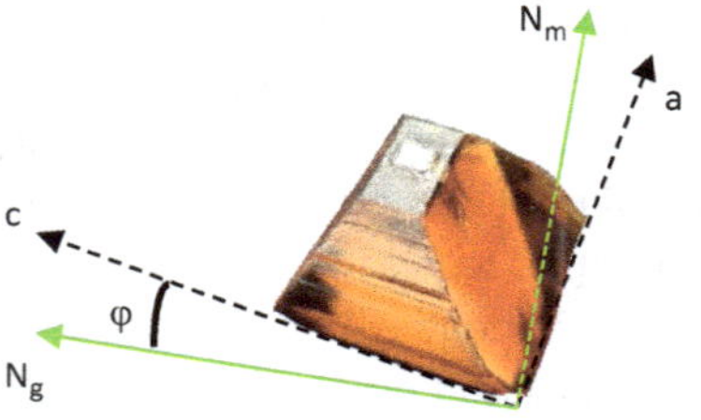

Figure 4.4.14: Position of the monoclinic *a*- and c-axes and of the crystal optical axes N_m and N_g. All four axes are in the same plane. The angle between *c* and *a* amounts to approximately 94.4° in the case of the KTbW crystal shown here. The dispersion angle has a value of approximately 19.8° for a wavelength of 1000 nm. The brown coloration of the boule is likely caused by Tb^{4+} ions present in the crystal due to the mildly oxidizing growth conditions.

The parameters *a*, *b*, *c*, *β*, N_p, N_m, N_g, φ are specific for every member of the K*RE*W family. The following Table 4.4.1 provides the values for KGdW, KYW, and KLuW [36–41].

The optical properties of Yb^{3+}-doped double tungstates make them interesting and useful for laser applications, especially as laser hosts for the generation of ultrashort pulses. A comparison among the most commonly used laser hosts (including Yb:CALGO) is summarized in Table 4.4.2.

It follows from this summary that Yb:KYW noticeably has simultaneously high absorption and emission cross sections, while at the same time, the emission bandwidth is sizeable. On the down side, its thermal conductivity is quite low. From this combination of properties, it follows that Yb:KYW is a highly efficient laser host for ultrafast sources in the low-to-medium average power range.

Table 4.4.1: Detailed crystallographic and crystal optical parameters for three selected members of the K*REW* family.

Symbol	Property	$KGd(WO_4)_2$	$KY(WO_4)_2$	$KLu(WO_4)_2$
	Structure	Monoclinic (*I*2/*c*)	Monoclinic (*I*2/*c*)	Monoclinic (*I*2/*c*)
a (Å)	Lattice parameter	8.08	8.05	8.01
b (Å)	Lattice parameter	7.58	7.54	7.49
c (Å)	Lattice parameter	10.37	10.35	10.21
β (°)	Angle from *c* to *a*	94.4	94.0	94.2
ϕ (°) at 1,000 nm	Angle from N_g to *c*	19.8	19.4	18.9
N_g at 1,000 nm	Refractive index	1.986 (1,067 nm)	1.97	1.995
N_m at 1,000 nm	Refractive index	2.033 (1,067 nm)	2.01	2.030
N_p at 1,000 nm	Refractive index	2.049 (1,067 nm)	2.03	2.084

Table 4.4.2: Summary of relevant parameters for Yb^{3+} doped laser hosts used for the generation of ultrashort laser pulses.

	Yb:YAG	Yb:KYW	$Yb{:}CaGdAlO_4$	$Yb{:}Lu_2O_3$	$Yb{:}YAl_3(BO_3)_4$
Peak absorption (nm)	942	981	979	976	975
Absorption cross section (10^{-20} cm^2)	0.75	13.3	2.8 (pi)	3.0	4.1 (sigma)
Absorption bandwidth (nm)	3	3.5	4 (pi)		20
Peak emission (nm)	1,031	1,025	1,050	1,032	1,039
Emission cross section (10^{-20} cm^2)	2.1	2.5	0.8 (sigma)	1.1	0.47 (sigma) 0.29 (pi)
Emission bandwidth (nm)	10	24	60 (sigma)		40 (sigma)
Lifetime (μs)	951	600	440	820	450
Thermal conductivity (Wm^{-1} K^{-1})	7.1	2.8	6.9	10.8	3 to 4
Experimentally observed pulse duration (fs)	100	71	47		

When using Yb:KYW for such applications, it becomes important to determine the homogeneity of the dopant in such boules. Figure 4.4.15 shows the optical extinction (i.e., the optical absorption if diffusion losses can be neglected, which appears to be the case in Yb:KYW) measured for samples having two different locations in the boule, one near the seed and another near the tail end. The

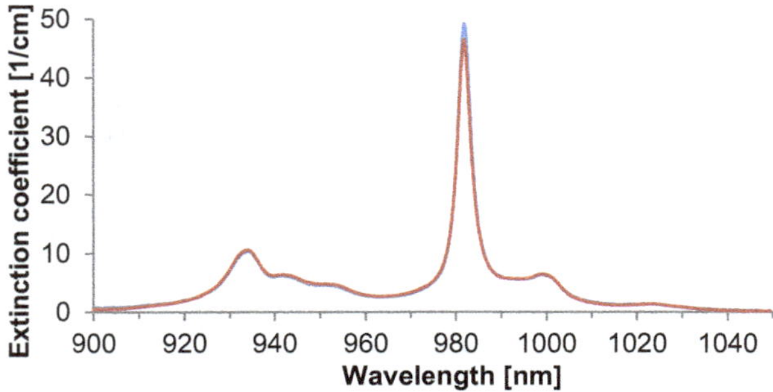

Figure 4.4.15: Optical extinction measurements in the wavelength range between 900 and 1,050 nm for two samples of a Yb(5%):KYW crystal: one cut near the top of the boule near the seed (curve in blue), another cut near its tail end (curve in red), the distance separating the two samples being 24 mm (along the growth axis).

wavelength range of interest is near 981 nm corresponding to the maximum absorption caused by Yb^{3+} for light with nm polarization.

It can be clearly seen that the two curves overlap precisely. Thus, one concludes that the crystal is homogeneous and that the Yb^{3+} concentration does not vary from the top to the bottom of the KYW boule (within experimental uncertainty).

The crystalline perfection of KYW crystals has been tested by Guguschev [42] who determined the widths of rocking curves of a Ho-doped KYW boule segment to be 29 and 30 arcsec for two different geometries. Such relatively narrow widths indicate a high degree of structural perfection of the tungstate crystals.

Work in progress is nevertheless revealing striations in similar crystals, as shown in the micrograph of Figure 4.4.16.

The observed striations were also present in the segment whose mosaic structure has been measured. Can the mosaicity and thus the crystal perfection of KYW crystals be further improved by reducing the striae in the bulk? This is one of the current topics under investigation at our facility, in order to further improve the laser performances achievable with this material.

It should be noted that, due to their relatively large interionic *RE*-*RE* distances, K*RE*W crystals are able to fluoresce with high efficiency and relatively little quenching when they contain active ions as stoichiometric compounds. Two such examples are shown in Figure 4.4.17 for KTbW and KEuW.

Previous work on Eu:KGW or Eu:KYW, and on Tb:KYW by Dashkevich et al. [43] and Schwung et al. [44, 45] dealt with doped crystals. Here, stoichiometric crystals where Gd or Y have been fully replaced by Eu and Tb are demonstrated, and further experimental details can be found for KEuW [46], or will be provided in the near future, in the case of KTbW.

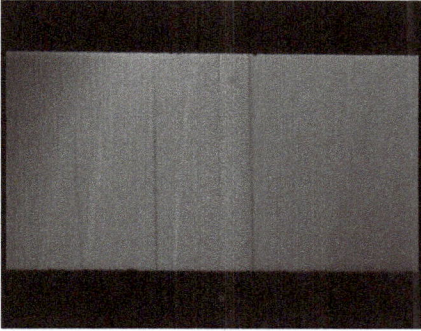

Figure 4.4.16: 2 × 4 mm micrograph revealing striations in a 4 mm long sample with a 2 × 2 mm aperture. The striations appear as lines oriented along the 2 mm long N_m edge. Along the 4 mm propagation direction N_g, one can count approximately 100 striae with an average spacing of 40 µm. Highly pronounced isolated striae appear (some at an angle) in the left half of the picture.

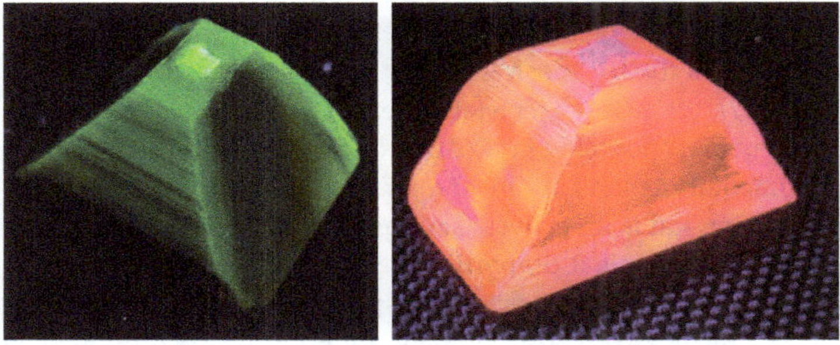

Figure 4.4.17: As-grown KTbW (left) and KEuW (right) crystals under 365 nm irradiation. KTbW fluoresces at 535 nm and KEuW at 614 nm.

It is assumed that such crystals may be key to the implementation of fluorescence standards with a high stability and long lifetimes, which is another topic of current interest related to double tungstate materials.

4.4.5 TSSG of YAB and related materials

Crystals belonging to the family described by the formula $REX_3(BO_3)_4$ (with RE = Y, La, rare earth, X = Al, Ga, Sc) are materials with interesting potential nonlinear optical (NLO) properties for the generation of wavelengths in the visible and in the UV.

$REX_3(BO_3)_4$ crystals belong to the space group $R32$ with a rhombohedral unit cell with lattice parameters in the ranges $a = 0.925–0.979$ and $c = 0.718–0.795$ nm depending on the exact composition. The crystal optical axes of the index ellipsoid are defined by X = *a* and Z = *c* (Y being thus automatically set). The propagation direction chosen for a given nonlinear optical frequency conversion scheme is defined in the XYZ system based on two polar angles Phi (in the XY plane, measured from X towards Y) and Theta (in the plane containing Z and the projection of the propagation direction on the XY plane, measured from Z towards the projected direction).

As a representative member of this crystal family, $YAl_3(BO_3)_4$ (YAB) is transparent down to about 170 nm and shows an acceptable effective nonlinear optical (NLO) coefficient. The refractive indices ($n_o = 1.7553$, $n_e = 1.6869$ at 1064 nm) of YAB permit phase matching for several types of frequency conversion to UV light, including second harmonic generation (SHG) from 532 to 266 nm [50]. In contrast to many other NLO borates, YAB is non-hygroscopic and hard (Mohs hardness 7.5) as compared to β-BaB_2O_4 (BBO), LiB_3O_5 (LBO) or $CsLiB_6O_{10}$ (CLBO). Moreover, YAB is free of toxic elements, such as beryllium, an essential component of other proposed NLO materials including $Sr_2Be_2B_2O_7$ (SBBO) or $KBe_2BO_3F_2$ (KBBF). The combinations of suitable optical, chemical and mechanical properties make YAB a promising crystal for NLO applications in the UV spectral range.

Known since 1962 [47], $REX_3(BO_3)_4$ crystals became attractive following the demonstration of self-frequency doubling in NYAB = $Nd_xY_{1-x}Al_3(BO_3)_4$ by Dorozhkin [48, 49]. In this experimental demonstration, the NYAB crystal plays a dual role: due to the Nd dopant, the crystal is an active laser host emitting at 1062 nm on one hand, and due to its nonlinear optical properties the same crystal can, when oriented for so-called phase matching, generate the second harmonic of the laser radiation at 531 nm. Besides NYAB, self-frequency doubling has been observed in several members of the crystal family, namely $Nd_xGd_{1-x}Al_3(BO_3)_4$ (NGAB) with $x = 3$–10%) and $Yb_xY_{1-x}Al_3(BO_3)_4$ (YbYAB) with $x = 5$–10%, as can be found in the bibliography of [50].

In the latter reference, it could be shown that undoped $REX_3(BO_3)_4$ crystals such as YAB or GdAB could also be used as efficient frequency converters for the generation of UV wavelengths in the range between 360 and 250 nm. The effective nonlinear coefficient is larger than 0.30 pm V^{-1} for all these wavelengths, down to 249 nm. The shortest theoretically achievable wavelength would be 245 nm; however, the nonlinear coefficient would reach zero at that very wavelength.

Crystals of $REX_3(BO_3)_4$ can be grown by the top-seeded solution growth (TSSG) technique from solutions based on the solvent systems $K_2Mo_3O_{12}$ [51], LaB_3O_6 [52, 53], Li_2O–WO_3–B_2O_3 [54] and Li_2O–Al_2O_3–B_2O_3 [55]. Depending on the solvent, the habitus of the YAB crystals may vary between rounded and prismatic, with pyramidal rhombohedral and trigonal prismatic facets. Crystals can be grown in all directions with Z = *c* and Y being the most commonly used ones.

In Figure 4.4.18, an Yb,Er:YAB crystal grown from a solution with the Li_2WO_4 flux in air from a 90 ml platinum crucible is shown. The solution had an approximate

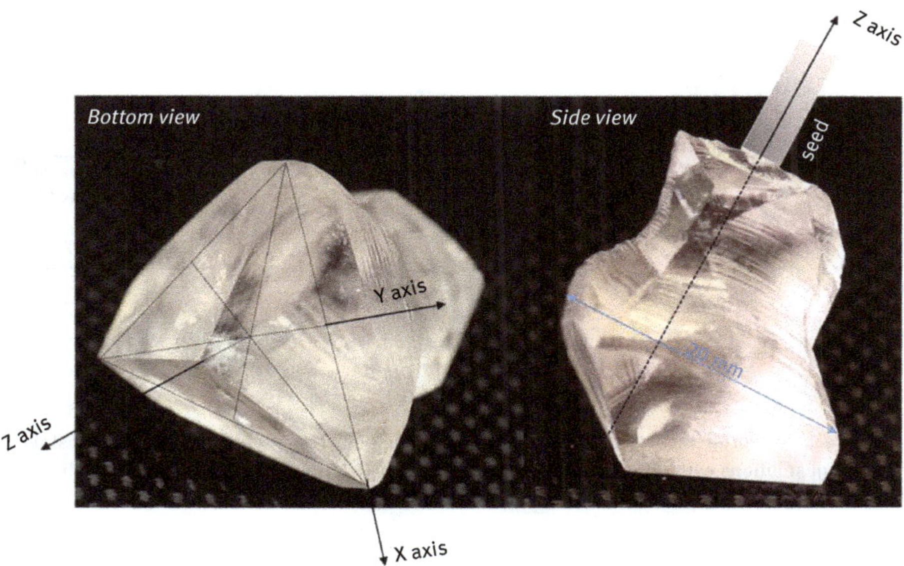

Figure 4.4.18: Bottom and side views of an Er(1.5%),Yb(11%):YAB crystal grown by TSSG from Li_2WO_4 based solution. The orientation of the X, Y, and Z axes in the crystal are shown. The width of the boule is 20 mm.

1:3 composition (solute:flux). The starting growth temperature was close to 960 °C and a cooling interval of 25 °C in 25 days was necessary to grow a boule of about 10 g in mass. As shown below, a Z-axis oriented seed was used, leading to rhombohedral facets when viewing the crystal from below. Increasing the volume of the crucible and thus the melt size, crystals up to 50 g can be grown using a similar approach.

When measuring the optical transmission of undoped YAB or other *RE*AB crystals, an extrinsic and often intense absorption is observed for wavelengths below 350 nm. This residual absorption is shown in Figure 4.4.19 for YAB crystals obtained from different fluxes.

Octahedrally coordinated trivalent iron is suspected to be responsible for the strong UV-absorption in YAB [56]. The variable shapes of the absorption curves are typical of YAB crystals and can occur in samples obtained with the same flux. Reducing the Fe^{3+} contamination is the subject of ongoing efforts to improve crystal growth of YAB and related compounds.

Another typical characteristic of crystals from the YAB family is the twin structure that is shown in Figure 4.4.20.

Both pure and *RE*-doped YAB crystals show this typical lamellar microstructure that has also been revealed by X-ray diffraction topography [57] surface etching patterns [58, 59] and atomic force microscopy [60]. A similar microstructure has also been detected in crystals of $Y_{0.57}La_{0.72}Sc_{2.71}(BO_3)_4$ (YLSB [61, 62]). Based on these

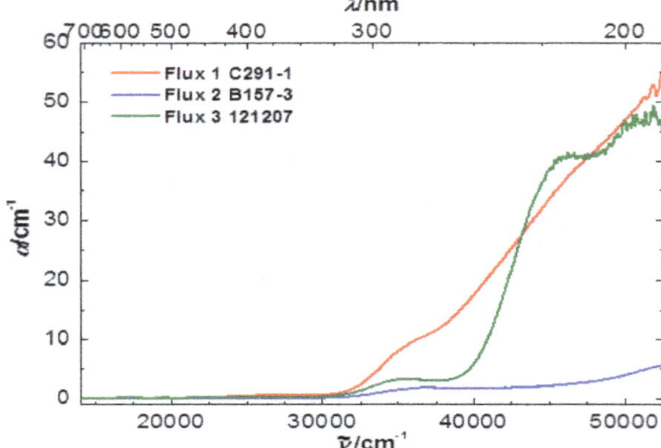

Flux	Crystal nb.	Absorption coeff. (cm^{-1})
1	C292-1	11.7
2	B157-3	1.7
3	121207	2.9

Figure 4.4.19: Absorption coefficient α (cm^{-1}) vs. wavenumber v (cm^{-1}) measured for YAB crystals grown from solutions with different fluxes (1: Li_2WO_4; 2: Li_2O-Al_2O_3-B_2O_3; 3: LaB_3O_6). Note the absorption peaks in the range below 300 nm (=33,000 cm^{-1}).

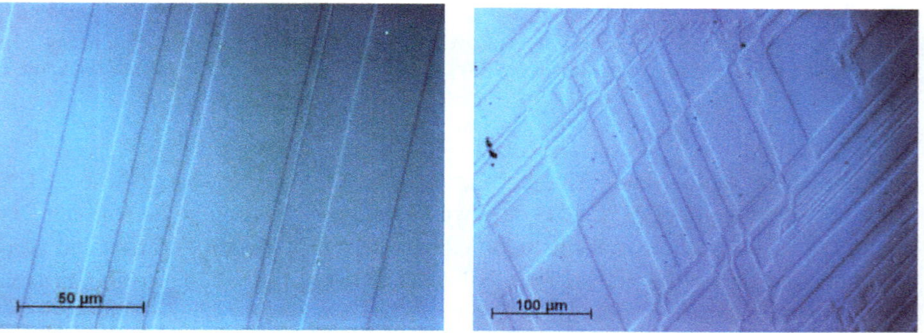

Figure 4.4.20: Micrographs obtained by C-DIC = Circular Polarized Light Differential Interference Contrast Microscopy, showing lamellar structures with widths between 5 and 200 µm. Both samples are viewed along their X-axis. On the right, one of the angles between lines delimiting lamellas is near 109°, which corresponds to the angle obtained under X-axis incidence, when viewing the rhombohedral cell.

observations, the lamellae have been interpreted as polysynthetic inversion twins, with reflection on $\{11\bar{2}0\}$ planes as the twin operation and adjacent twin domains sharing planes parallel to $\{10\bar{1}1\}$ ([57, 61, 63]). This twin law, that is, reflection on $\{11\bar{2}0\}$ with domain boundaries parallel to $\{10\bar{1}1\}$ also describes Brazil twins in low-quartz (α-SiO_2, [64, 65]).

Such twin laws generate merohedral twins [63, 64] that are not readily detected by conventional X-ray diffraction techniques. Reflecting the crystal structure on planes perpendicular to the *a*-axes not only inverts the polar two-fold axes along *a*, but also converts the right-handed polytype of each structure into its left-handed counterpart, and vice versa.

Measuring the rotatory power was applied to determine the optical rotatory dispersion of YAB crystals and to characterize their domain structures [66]. By probing different locations in oriented crystal sections with a He-Ne laser beam, the sign and magnitude of the optical rotation were found to vary within individual YAB crystals. As right and left-handed domains rotate the electric field vector in opposing senses, the variations in apparent rotatory power correlate with the microstructure of the crystals, when viewed between crossed polarizers. To relate an observed optical rotation to inversion twinning, however, the unaffected or intrinsic rotatory power of the crystal species needs to be known. The method, as described by Buchen, allows a direct assessment of the extent of inversion twinning and selection of high-quality (twin-free) crystal specimens is now possible.

As a conclusion for this chapter on YAB crystals, we will emphasize a few relatively new experiments performed with such rare-earth-doped or rare-earth-containing crystals in order to underline the versatility of these crystals. A similar conclusion holds for the other crystal families described previously.

Doped Yb:YAB and Er,Yb:YAB crystals have also shown interesting new developments as laser hosts. Measurements from Junhai Liu et al. [67] performed on a Yb (5,6%):YAB crystal have emphasized favorable emission properties that could be implemented in a thin-disk geometry emitting 109 W in a continuous wave (CW) laser, by Weichelt et al. [68]. Such thin disk elements are likely to benefit from a reduction of the twin density in the crystal, which would then lead to even more efficient laser elements.

Using the Er^{3+} transition $^4I_{13/2}$ to $^4I_{15/2}$, a laser wavelength in the near-infrared between 1.5 and 1.6 µm can be generated in the so-called eye-safe spectral range. The ytterbium co-dopant can be used in order to broaden the absorption band and increase the absorption coefficient of the active medium for the use of common laser diodes working near 970 nm for efficient pumping.

The laser active material Er,Yb:YAB has been described in numerous papers: CW laser with 1 W output power, slope efficiency of 35% [69, 70], wavelength tuning between 1.5 and 1.6 µm [71], passive Q-switching [72] and passive mode-locking [73]. Work in progress is expected to lead to additional developments in this area.

Very recently, fluorescence properties of $TbAB = TbAl_3(BO_3)_4$ crystals were investigated. Such a crystal is shown in its as-grown form in Figure 4.4.21.

This TbAB crystal was grown in air. The boule appears colorless without any reducing post-growth treatment. Nevertheless, no brownish coloration has been detected. Does this imply that TbAB contains almost no Tb^{4+} ions? This will be a subject for future investigations.

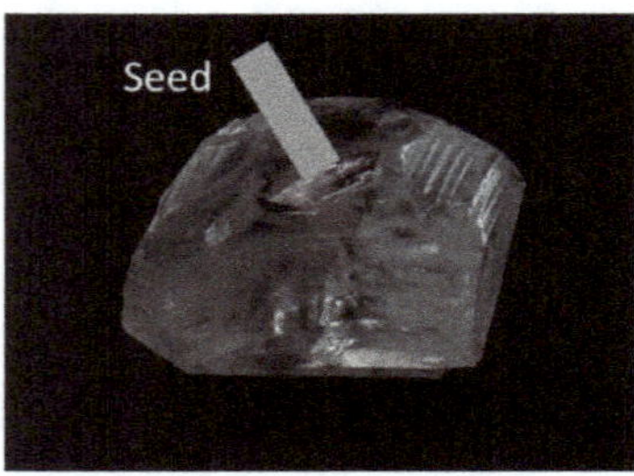

Figure 4.4.21: As-grown TbAB crystal grown by TSSG with a Z-axis oriented seed. On the left, the picture shows the crystal under standard daylight illumination. On the right, the same crystal emits green light under UV excitation at 365 nm.

References

[1] J. E. Geusic, H. M. Marcos, L. G. Van Uitert, Appl. Phys. Lett. 1964, 4, 182.

[2] W. Koechner, Solid State Laser Engineering, 6th Edition, Springer, Berlin, Heidelberg, New York, 2006.

[3] G. Blasse, A. Bril, Appl. Phys. Lett. 1967, 11, 53.

[4] G. Blasse, B. C. Grabmaier, Luminescent Materials, Springer, Berlin, Heidelberg, New York, 1994.

[5] D. K. G. De Boer, D. Bruls, C. Hoelen, H. Jagt, Proc. of SPIE 2017, 10378, 103780M.

[6] T. Gallinelli, A. Barbet, F. Druon, F. Balembois, P. Georges, T. Billeton, S. Chenais, S. Forget, Opt. Express 2019, 27, 11830.

[7] Y. Wu, M. Koschan, C. Foster, C. L. Melcher, Cryst. Growth Des. 2019, 19, 4081.

[8] K. Nassau, Gems Made by Man, Chilton, 1980.

[9] R. Uecker, J. Cryst. Growth 2014, 401, 24.

[10] H. Li, J. Xu, Springer Handbook of Crystal Growth (ed. K. Byrappa, V. Prasad, M. Dudley), Ch. 15, 2010, 479.

[11] D. Mateika, Advanced Crystal Growth (ed. P. M. Dryburg, B. Cockayne, K. G. Barraclough), Prentice Hall, 1987.

[12] Z. Yu, T. Voznyak, V. Gorbenko, E. Zych, S. Nizankovski, A. Dan'ko, V. Puzikov, Radiat. Meas. 2010, 45, 389.

[13] C. Y. Chen, G. J. Pogatshnik, Y. Chen, M. R. Kokta, Phys. Rev. B 1988, 38, 8557.

[14] D. Fagundes-Peters, N. Martynyuk, K. Lünstedt, V. Peters, K. Petermann, G. Huber, S. Basun, V. Laguta, A. Hofstaetter, J. Lumin. 2007, 125, 238.

[15] S. Rydberg, M. Engholm, J. Appl. Phys. 2013, 113, 223510.

[16] A. Pajaczkowska, A. Gloubokov, Progr. Cryst. Growth, Charact. Mater. 1998, 36, 123.

[17] R. Brown, V. Pendrik, D. Kalokitis, B. Chai, Appl. Phys. Lett. 1990, 57, 1351.

[18] H. J. Scheel, M. Berkowski, B. Chabot, Physica C 1991, 185, 2095.

[19] A. Dabkowski, H. A. Dabkowski, J. E. Greedan, J. Cryst. Growth 1993, 132, 205.

[20] R. Uecker, P. Reiche, S. Ganschow, D. C. Uecker, D. Schultze, Acta Phys. Pol. A 1997, 2, 23.

[21] A. E. Becerra-Toledo, L. D. Marks, Surf. Sci. 2010, 604, 1476.

[22] A. A. Lagatskii, N. V. Kuleshov, V. G. Shcherbitskii, V. F. Kleptsyn, V. P. Mikhailov, V. G. Ostroumov, G. Huber, Quantum Electron. 1997, 27, 15.

[23] J. Petit, PhD thesis, Univ. Pierre-et-Marie-Curie Paris VI, 2006.

[24] P.-O. Petit, PhD thesis, Univ. Pierre-et-Marie-Curie Paris VI, 2010.

[25] A. Jaffres, PhD thesis, Univ. Pierre-et-Marie-Curie Paris VI, 2013.
[26] Y. Zaouter, J. Didierjean, F. Balembois, G. Lucas Leclin, F. Druon, P. Georges, J. Petit, P. Goldner, B. Viana, Opt. Lett. 2006, 31, 119.
[27] A. Agnesi, A. Greborio, F. Pirzio, G. Reali, J. Aus der Au, A. Guandalini, Opt. Express 2012, 20, 10077.
[28] K. Beil, B. Deppe, C. Kränkel, Opt. Lett. 2013, 38, 1966.
[29] C. Liebald, PhD thesis, Johannes Gutenberg-Universität Mainz, 2017.
[30] S. Haab, Bachelor of Sciences thesis, Univ. of Applied Sciences Trier, Birkenfeld Campus, 2019.
[31] M. Mei, L. L. Cao, Y. He, R. R. Zhang, F. Y. Guo, N. F. Zhuang, J. Z. Chen, Adv. Mater. Res. 2011, 306, 1722.
[32] A. Kokh, A. N. Kononova, G. Mennerat, Ph. Villeval, S. Durst, D. Lupinski, V. Vlezko, K. Kokh, J. Cryst. Growth 2010, 312, 1774.
[33] D. Rytz, B. A. Wechsler, C. C. Nelson, K. W. Kirby, J. Cryst. Growth 1990, 99, 864.
[34] P. V. Klevtsov, L. P. Kozeeva, Yu. L. Kharchenko, Sov. Phys. Crystallogr. 1975, 20, 732.
[35] P. V. Klevtsov, R. F. Klevtsova, J. Struct. Chem. 1977, 18, 339.
[35] C. Pujol, M. Aguiló, F. Dı́az, C. Zaldo, Opt. Mater. 1999, 13, 33.
[36] A. A. Kaminskii, J. B. Gruber, S. N. Bagaev, K. Ueda, U. Hömmerich, J. T. Seo, D. Temple, B. Zandi, A. A. Kornienko, E. B. Dunina, A. A. Pavlyuk, R. F. Klevtsova, F. A. Kuznetsov, Phys. Rev. B 2002, 65, 125108.
[37] X. Mateos, R. Sole, J. Gavalda, M. Aguilo, J. Massons, F. Díaz, V. Petrov, U. Griebner, Opt. Mater. 2006, 28, 519.
[38] G. Métrat, M. Boudeulle, N. Muhlstein, A. Brenier, J. Cryst. Growth 1999, 197, 883.
[39] Y. E. Romanyuk, École Polytechnique Fédérale de Lausanne, PhD thesis 3390, 2005.
[40] O. Silvestre, A. Aznar, R. Solé, M. C. Pujol, F. Díaz, M. Aguiló, J. Phys.: Condens. Matter 2008, 20, 225004.
[41] P. Loiko, P. Segonds, P. L. Inácio, A. Peña, J. Debray, D. Rytz, V. Filippov, K. Yumashev, M. C. Pujol, X. Mateos, M. Aguiló, F. Díaz, M. Eichhorn, B. Boulanger, Opt. Mater. Express 2016, 6, 2984.
[42] C. Guguschev (unpublished), 2018.
[43] V. I. Dashkevich, S. N. Bagayev, V. A. Orlovich, A. A. Bui, P. A. Loiko, K. V. Yumashev, A. S. Yasukevich, N. V. Kuleshov, S. M. Vatnik, A. A. Pavlyuk, Laser Phys. Lett. 2015, 12, 085001.
[44] S. Schwung, D. Enseling, D. Rytz, B. Heying, U. Ch. Rodewald, B. Gerke, O. Niehaus, R. Pöttgen, T. Jüstel, J. Lumin. 2015, 159, 251.
[45] S. Schwung, D. Rytz, B. Heying, U. Ch. Rodewald, O. Niehaus, D. Enseling, T. Jüstel, R. Pöttgen, J. Lumin. 2015, 166, 289.
[46] P. Pues, S. Schwung, D. Rytz, L. Schubert, S. Klenner, F. Stegemann, R. Pöttgen, T. Jüstel, J. Lumin. 2019, 215, 116653.
[47] A. A. Ballmann, US Patent 3‘057‘677, 1962.
[48] L. M. Dorozhkin, I. I. Kuratev, N. I. Leonyuk, T. I. Timchenko, A. V. Shestakov, Sov. Tech. Phys. Lett. 1981, 7, 555.
[49] L. M. Dorozhkin, I. I. Kuratev, V. A. Zhitnyuk, A. V. Shestakov, V. D. Shigorin, G. P. Shipulo, Sov. J. Quantum Electron. 1983, 13, 978.
[50] D. Rytz, A. Gross, S. Vernay, V. Wesemann, Proc. SPIE 2008, 6998, 6998–38.
[51] N. I. Leonyuk, J. Cryst. Growth 2017, 476, 69.
[52] T. Alekel, N. Ye, D. A. Keszler, US Patent Application 2006/0054864 A1, 2006.
[53] R. Jinlei, PhD thesis, Université Pierre et Marie Curie, Paris VI, 2017.
[54] J. Liu, S. H. Fang, J. Y. Wang, N. Ye, Mater. Res. Innov. 2011, 15, 102.
[55] X. Yu, J. Yu, L. Liu, N. Zhai, X. Zhang, G. Wang, X. Wang, C. Chen, J. Cryst. Growth 2012, 341, 61.

[56] X. Yu, Y. Yue, J. Yao, Z. G. Hu, J. Cryst. Growth 2010, 312, 3029.
[57] X. B. Hu, S. S. Jiang, X. R. Huang, W. J. Liu, C. Z. Ge, J. Y. Wang, H. F. Pan, J. H. Jiang, Z. G. Wang, J. Cryst. Growth 1997, 173, 460.
[58] X. B. Hu, S. S. Jiang, X. R. Huang, W. J. Liu, C. Z. Ge, J. Y. Wang, H. F. Pan, C. Ferrari, S. Gennari, Nuovo Cimento D 1997, 19, 175.
[59] A. Péter, K. Polgár, E. Beregi, J. Cryst. Growth 2000, 209, 10.
[60] S. Zhao, J. Wang, D. Sun, X. Hu, H. Liu, J. Appl. Crystallogr. 2001, 34, 66.
[61] N. Ye, J. L. Stone-Sundberg, M. A. Hruschka, G. Aka, W. Kong, D. A. Keszler, Chem. Mater. 2005, 17, 2687.
[62] M. Bourezzou, A. Maillard, R. Maillard, P. Villeval, G. Aka, J. Lejay, D. Rytz, Opt. Mater. Express 2011, 1, 1570.
[63] S. Ilas, PhD. thesis, Université Pierre et Marie Curie-Paris VI, 2014.
[64] C. Giacovazzo, H. L. Monaco, G. Artioli, D. Viterbo, M. Milanesio, G. Gilli, P. Gilli, G. Zanotti, G. Ferraris, M. Catti, Fundamentals of Crystallography, 3rd Edition, Oxford University Press, Oxford, UK, 2011.
[65] H. H. Schlössin, A. R. Lang, Phil. Mag. 1965, 12, 283.
[66] J. Buchen, V. Wesemann, S. Dehmelt, A. Gross, D. Rytz, Crystals 2019, 9, 8.
[67] J. Liu, X. Mateos, H. Zhang, J. Li, J. Wang, V. Petrov, IEEE J. Quantum El. 2007, 43, 385.
[68] B. Weichelt, M. Rumpel, A. Voss, A. Gross, V. Wesemann, D. Rytz, M. Abdou Ahmed, T. Graf, Opt. Express 2013, 12, 25708.
[69] N. A. Tolstik, S. V. Kurilchik, V. E. Kisel, N. V. Kuleshov, V. V. Maltsev, O. V. Pilipenko, E. V. Koporulina, N. I. Leonyuk, Opt. Lett. 2007, 32, 3233.
[70] N. A. Tolstik, V. E. Kisel, N. V. Kuleshov, V. V. Maltsev, N. I. Leonyuk, Appl. Phys. B 2009, 97, 357.
[71] Y. J. Chen, Y. F. Lin, X. H. Gong, Q. G. Tan, Z. D. Luo, Y. D. Huang, Appl. Phys. Lett. 2006, 89, 24111.
[72] V. E. Kisel, K. N. Gorbachenya, A. S. Yasukevich, A. M. Ivashko, N. V. Kuleshov, V. V. Maltsev, N. I. Leonyuk, Opt. Lett. 2012, 37, 2745.
[73] A. A. Lagatsky, V. E. Kisel, A. E. Troshin, N. A. Tolstik, N. V. Kuleshov, N. I. Leonyuk, A. E. Zhukov, E. U. Rafailov, W. Sibbett, Opt. Lett. 2008, 33, 83.

Andreas Faust

4.5 Medical applications of rare earth compounds

4.5.1 REE in bioimaging

Besides modern technologies with their tremendous need of rare earth elements (REE), their use in medicine has gained prominence in different molecular imaging techniques and radiotherapies [1]. For diagnostics, one widely used technique is magnetic resonance imaging (MRI). Next to the field of nuclear medicine (second part of this chapter), where radioactive isotopes are common, clinical radiology uses rare earth metal complexes such as paramagnetic Gd(III)-complexes, to enhance contrast in MRI.

To understand these effects, first a few basic principles will be given herein. MRI is based on nuclear magnetic resonance (NMR), but instead of providing chemical shifts and coupling constants of hydrogen nuclei, it provides their spatial distribution of the intensity in tissues, for example, in the human body [2]. The spin of the hydrogen nuclei generates a magnetic field and these magnetic moments are normally randomly oriented. An external permanent magnetic field B_0, as used in a human MR system, results in a so-called “Larmor precession” with a frequency (times the protons precess per second; Larmor frequency ω_0) that is proportional to the applied field strength: $\omega_0 = \gamma B_0$. The gyromagnetic ratio (γ) for hydrogen is 42.6 MHz/T and the normally used field strength is 1.5 or 3 T. On- and off-switching of radiofrequency (RF) pulses, which must be in resonance with the Larmor frequency, results in disturbance and the hydrogen nuclei fall out of alignment with B_0. The energy arising from longitudinal (T_1) and transversal (T_2) relaxation of the spins with a characteristic decay time is recorded in the MRI [3]. This time is dependent on the chemical compound and the molecular environment in which the preceding hydrogen nuclei are located. Therefore, the different tissue types differ characteristically in their signal, which leads to different signal strengths (brightness) in the resulting image. To improve image quality contrast, agents are used that can change relaxation times of neighboring water protons at the place of distribution. These agents are grouped into four major classes: paramagnetic, superparamagnetic, chemical exchange saturation transfer (CEST) and direct detection [4].

Clinically, the most common MRI contrast agents are paramagnetic Gd(III)-complexes (for their analytics, we refer to Chapter 3.1). Next to lanthanides, transition elements such as manganese or iron are used. Gd(III), with its half-filled [Xe]f^7 shell, is highly paramagnetic and the unpaired electrons generate their own magnetic field.

https://doi.org/10.1515/9783110654929-028

Thus, they have a dramatic effect on T_1 and T_2 relaxation times of the nearby water protons, increase the signal intensity of T_1-weighted images and reduce the signal intensity of T_2-weighted images [5]. Because of its highly paramagnetic nature, gadolinium is also usable for magnets or superconductors. Other applications are the use for alloys or in nuclear technology as neutron absorber. Because of the similar ionic radius of Gd^{3+} and Ca^{2+}, the human organism cannot distinguish between these cations and therefore, Gd^{3+} can act as an inhibitor of calcium-based processes in the body. The variety of gadolinium(III)-based contrast agents can be categorized into extracellular fluid agents, blood pool contrast agents and organ-specific agents [6]. To reduce toxicity of the metal ions, chelation is used. Many chelators have been developed so far (Figure 4.5.2). They can be divided in to three classes: the ionic and hydrophilic complexes (e.g., Gd(III)DOTA or Gd(III)DTPA), nonionic and hydrophilic complexes (e.g., Gd(III)DTPA-BMA or Gd(III)DO3A-butrol) and ionic and lipophilic complexes such as Gd(III)BOPTA. Mostly, these ligands satisfy not only the high coordination number preference of Gd^{3+} but also represent hard oxygen donor ligands and leave at least one coordination site open for water coordination. One example for the resulting contrast is given in Figure 4.5.1 where, after the administration of gadovist, the meningeom is clearly visible in comparison to the image without contrast agent. It is important to mention that the stability of these gadolinium-based contrast agents may not be sufficient in all cases. The relationship between their use and nephrogenic systemic fibrosis (NSF) cannot be denied anymore [7]. The most frequent cases of NSF are known for Gd-DTPA-BMA. 90% of the documented cases are due to this contrast agent, 10% are related to Gd-DTPA and Gd-DTPA-BMEA [8]. Gd-DTPA-related contrast agents belong to the linear complexes with the lowest stability constants [9]. All Gd-DTPA-related contrast agents have been suspended by the European Medicine Agency in 2017 [10]. On the other hand, no proven cases of NSF have been documented if cyclic Gd-based contrast agents are applied. Nevertheless, recent studies have confirmed gadolinium accumulation in the human brain, after repeated gadolinium-based contrast agent administration. However, the impact of the retained gadolinium in the brain remains unknown and is still part of ongoing investigation [11]. Further, paramagnetic contrast agents are transition metal ions, such as chelated high-spin manganese ions or superparamagnetic iron oxide (SPIO) and ultrasmall superparamagnetic iron oxide (USPIO), which are normally used as nanoparticles. Several, such as Feridex® or Resovist®, have been withdrawn from major markets due to undesired side effects or just due to lack of users [12].

As an alternative, a new method of magnetic resonance spectroscopy has come into focus of research over the last decade: chemical exchange saturation transfer (CEST). This technique is based on the phenomenon of chemical exchange between solutes and water protons in which the exogenous or endogenous compounds containing either exchangeable protons or exchangeable molecules are selectively saturated. After the transfer of this saturation, they are indirectly detected through the water signal with enhanced sensitivity [13]. To date, there are different types of

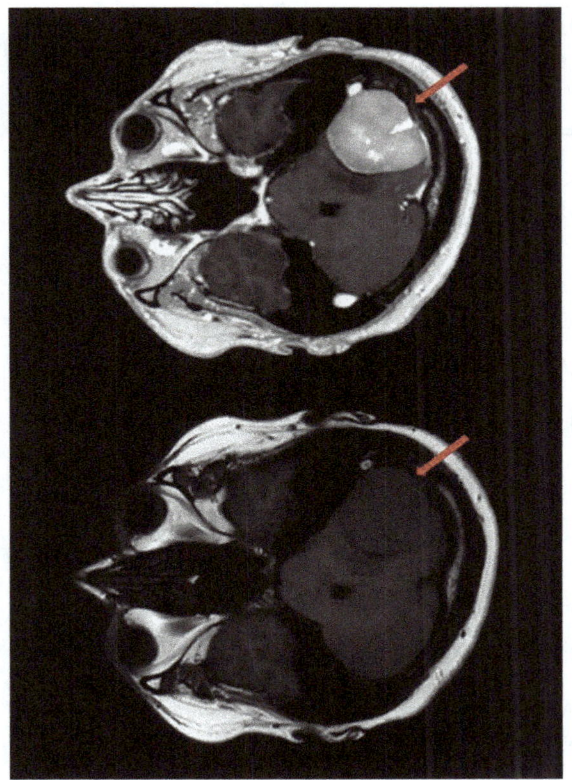

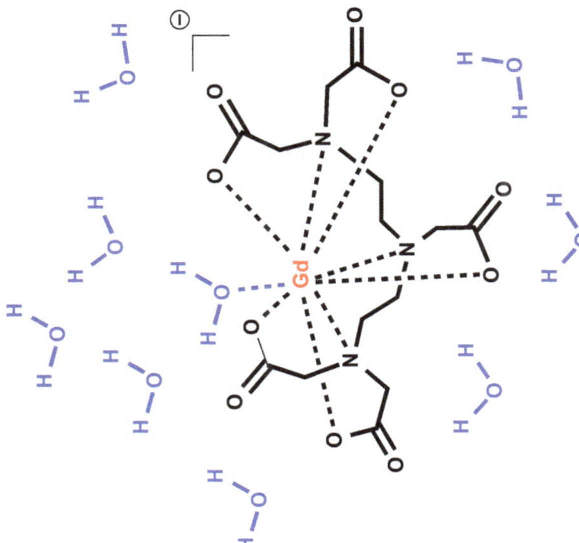

Figure 4.5.1: Left: Effect of Gd-based contrast agents on MR-images: Defect of the blood-brain barrier of a patient with meningeom shown in MRI (T1-weighted images, left image without, right image with Gd-DO3A-butrol (gadovist) administration © St. Franziskus Hospital, Münster, Germany; right: Gd-DTPA (diethylenetriamine pentaacetate) with its influence sphere on the relaxation times (T_1 and T_2) of neighboring water protons.

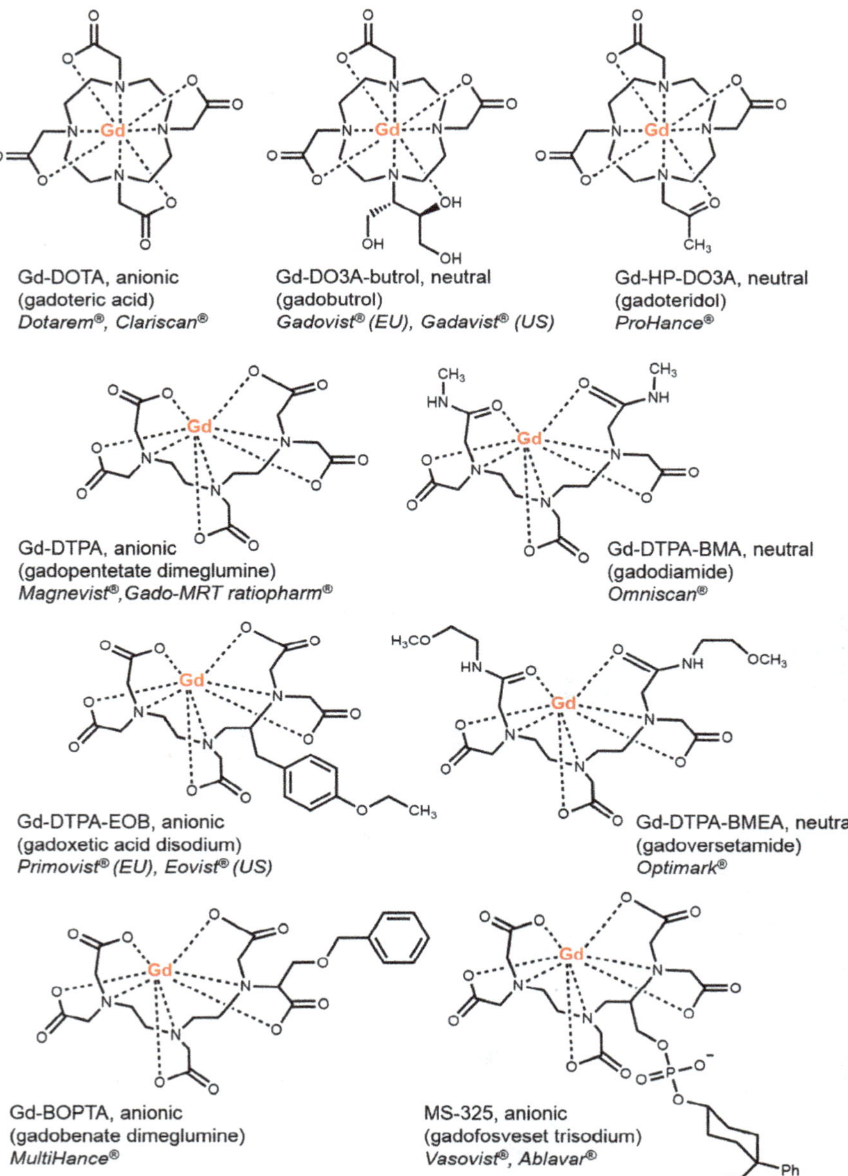

Figure 4.5.2: Clinically used contrast agents based on Gd(III)-complexes. The macrocyclic agents Gd-DOTA, Gd-DO3A-butrol and Gd-HP-DO3A are still available for clinical application, whereas the linear agents Gd-BOPTA and Gd-DTPA-EOB are only available with limited indications because of their lower stability. For the same reason, the agents Gd-DTPA, Gd-DTPA-BMEA and Gd-DTPA-BMA have been suspended by the European Medicine Agency on 20.07.2017 [9a] or withdrawn from different markets in case of MS-325 [9b].

CEST known. Small diamagnetic molecules in the frequency range of 0 ppm to 7 ppm from water and represent compounds found in normal tissues (mostly amide, amine or hydroxyl groups), are summarized as DiaCEST. Paramagnetic complexes of REE containing exchangeable protons on hydroxyl, amide or amino moieties are known as ParaCEST-agents. Here, the molecular exchange rate is much faster and many lanthanide-containing complexes (with Nd^{3+}, Eu^{3+}, Tb^{3+}, Tm^{3+}, Dy^{3+} or Yb^{3+}) are part of recent developments. The third type of CEST contrast-based agents are liposomes filled with high amounts of paramagnetic agents (LipoCEST), where the resonance of the water trapped in the liposome is shifted and shielded from the outside.

For direct detection, it is possible to trace also other nuclei such as ^{31}P, ^{13}C or ^{19}F, but they must be highly concentrated for a measurable and useful MR-signal, typically in the 10–50 mM range. Only ^{19}F-MRI has been popular during the last decade as many magnetic equivalent ^{19}F-nuclei (meaning a high total spin concentration), in combination with the very low background from the normal tissue, make it feasible. In Figure 4.5.3, there are examples of fluorine-containing MR-probes which have

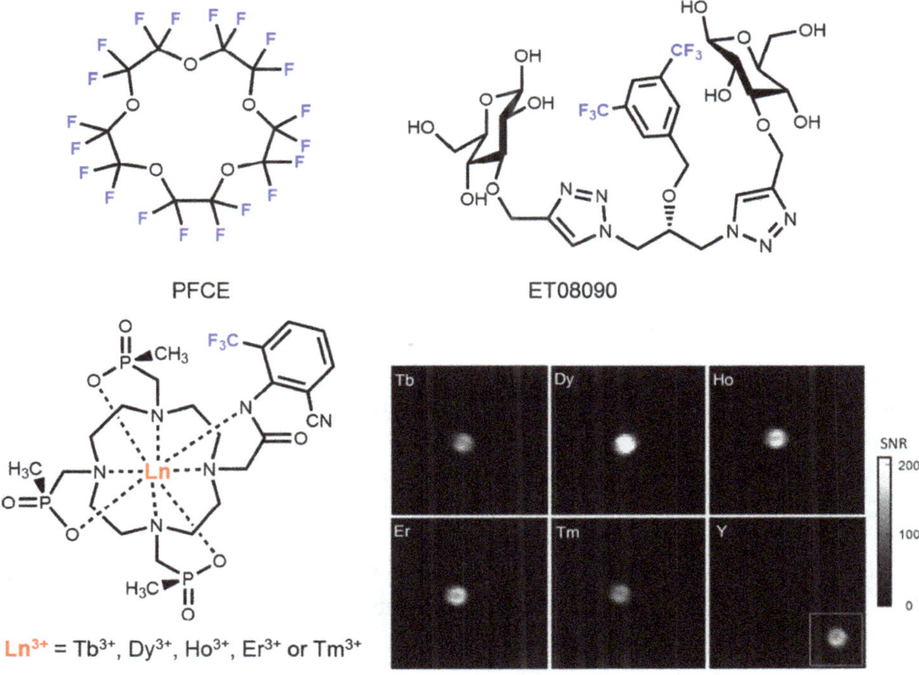

Figure 4.5.3: *Top*: PFCE as an example of perfluorocarbons and ET08090 [16] as hydrophilic fluorinated compounds which are used for direct detection in ^{19}F-MRI and are normally formulated in nanoparticles or liposomes. *Bottom*: To increase the ^{19}F-MRI-signal, different lanthanide complexes are evaluated. Here, phantom imaging data was acquired in each complex at 7.0 T. All images intensity scaled to the same value. Inset Y image scaled in intensity by a factor of 10. (© A. M. Blamire [16]; redrawn with permission).

been evaluated so far. REEs are able to enhance the ^{19}F-signal in MRI. The BLOCH-WANGSNESS-REDFIELD theory [14] accurately predicts how proximate paramagnetic centers of REEs and their overall geometry or the oxidation and spin state of the metal ion could affect the sensitivity of fluorine-based probes. They enhance the ^{19}F-MR-signal by increasing the rate of longitudinal relaxation [15]. One fluorine-containing complex is shown in Figure 4.5.3 and the resulting signal enhancement depending on the proximal metal center can be measured by referencing the signal intensity to an diamagnetic analogue, such as the related yttrium(III)-complex. In the case of Dy^{3+} as the central atom, the sensitivity was improved by a factor of nearly 15 [16].

4.5.2 REE in nuclear medicine

In the field of nuclear medicine, very small amounts of radioactive materials (radiopharmaceuticals) help to examine organ function and structure at a molecular level or to treat diseases like arthritis or cancer. In this area, radioactive isotopes of REEs find use, mainly in therapeutic interventions. To understand this field of nuclear diagnostics and therapy, it is important to understand the physical background and the different resulting techniques. ANTOINE HENRI BECQUEREL discovered radioactivity in 1896 and in honor, the unit for the radiation decay per second is Becquerel (Bq). The released energy (type and spectrum) is typical for each radionuclide and predicts whether it is useful for medical application or not. Unstable atomic nuclei spontaneously emit ionizing radiation upon transformation, including (1) particles (β^--, Auger-e^- or α-emission), (2) they relax from nuclear excited states by releasing energy (γ-rays), or (3) they release positrons (β^+) resulting in two γ-rays (180°, 511 keV) upon annihilation with electrons from the environment (Figure 4.5.4). The low number of useful radionuclides have half-lives from minutes to days and are mainly β^--emitters (radiotherapy) or β^+/γ-emitters for diagnostic techniques like PET (positron emission tomography or SPECT (single photon emission computed tomography). The decision, whether a radionuclide is dangerous or beneficial for the patient, depends on the dose of radiation, measured in Sievert (Sv). Due to the necessity of very low concentrations (picomolar) in the case of diagnostic application, the overall dose is comparable to CT-measurements (computed tomography). In therapeutic interventions, the dose is often higher but focused on the tissue of interest, for example, in the case of cancer, the radionuclide should target the cancer cells with high specificity thus killing them selectively, while avoiding undesired side effects in healthy tissues. Radionuclides of REEs have a wide range of application in nuclear medicine (Table 4.5.1). To bring radionuclides into organic molecules of medical interest, a special field of radiochemistry called radiopharmaceutical chemistry is part of most nuclear medicine centers. This interdisciplinary field between

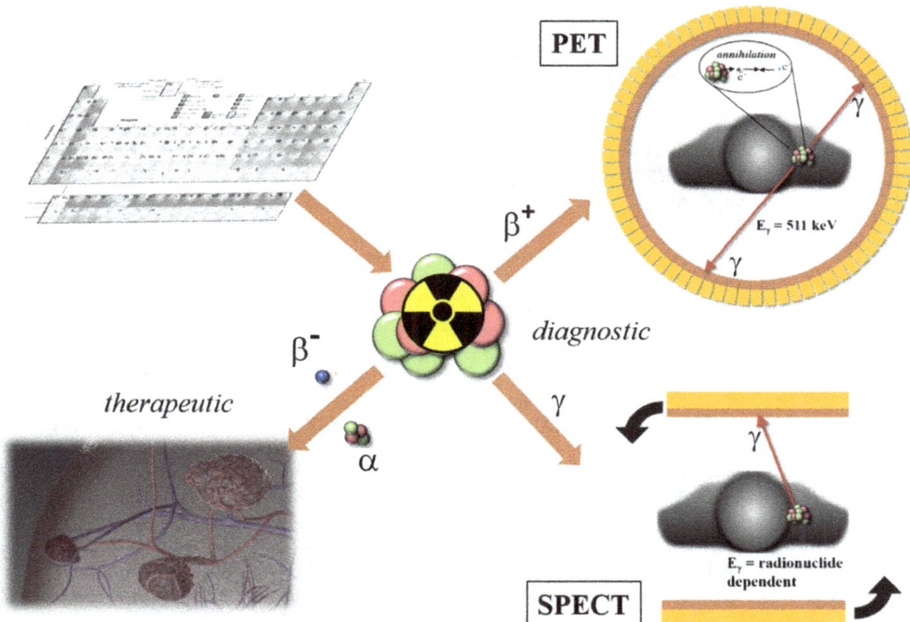

Figure 4.5.4: Different decay-types and their use in medical imaging and therapy (overview).

medicine, pharmacy and labeling chemistry has been growing over the last few decades. There are different ways of radiolabeling: covalent bond formation (labeling with positron emitters like ^{18}F, ^{11}C or ^{13}N) or in the case of the large variety of feasible radiometals, the use of coordination chemistry. The targeting bioactive molecule is often coupled with a bifunctional chelator, serving the possibility of complexation and bioconjugation at the same time. Many different chelators are known to be compatible with the metal of choice. Their most important feature is the prevention of the release of free metal *in vivo* [17].

Scandium, as the lightest element of the rare earth metals, is new to the medical field. The relevant nuclides are ^{43}Sc, ^{44}Sc and ^{47}Sc. The positron emitters (PET-nuclides) ^{43}Sc and ^{44}Sc have longer physical half-lives (~4 h) than the usually used nuclides ^{18}F ($t_{1/2}$ = 1.8 h) or ^{68}Ga ($t_{1/2}$ = 1.1 h), thus allowing imaging in a longer time scale. The chemically identical low-energy $β^-$emitter ^{47}Sc should be suitable for treatment but has an inconvenient production route. Because of the chemical similarities of REEs, often the $β^-$-emitters ^{90}Y and ^{177}Lu are more common for treatment. One example of these theranostic pairs is PSMA-617 labeled with the radionuclides ^{44}Sc or ^{68}Ga (PET) and ^{177}Lu (therapy). The targeting of the prostate-specific membrane antigen (PSMA) is of particular interest for imaging and therapy of prostate cancer and metastasis. Due to the high and specific tumor uptake, these PSMA-derivatives enable the possibility for radionuclide therapy with ^{177}Lu besides

Table 4.5.1: Prominent radionuclides of REEs, their decay parameters and applications [18].

radionuclide	half life (h)	decay mode	energy of the particle (keV)	γ-energy (keV)	application
^{43}Sc	3.89	β^+/EC	1200	373	PET
^{44}Sc	3.92	β^+/EC	1470	1157	PET
^{47}Sc	80.4	β^-	162	159	β^--therapy/ SPECT
^{86}Y	14.7	β^+/EC	664	777, 1077, 1153	PET
^{90}Y	64	β^-	2280	–	β^--therapy
^{141}Ce	780	β^-	580, 435	145	therapy (pain)/ SPECT
^{142}Pr	19.1	β^-	2162	1575	β^--therapy/ SPECT
^{153}Sm	46.3	β^-	808, 704, 634	103, 70	therapy (pain)/ SPECT
^{149}Tb	4.1	α (17%), β^+/EC (83%)	α: 3970 β^+: 728	853, 817, 389, 352, 165	α-therapy/ PET
^{152}Tb	17.5	β^+/EC	1142	344	PET
^{155}Tb	128	EC		105, 87	SPECT
^{161}Tb	165	β^- 12.4 e^-/ decay	154 Auger: 46.5	26 75, 49	β^--therapy Auger-e^--therapy
^{165}Dy	2.33	β^-	454, 415	95, 48	β^--therapy/ SPECT
^{166}Ho	26.8	β^-	694, 651	81	β^--therapy/ SPECT
^{169}Er	226	β^-	101, 98	51	β^--therapy/ SPECT
^{170}Tm	3086	β^-	323, 291	84, 49, 48	β^--therapy/ SPECT
^{177}Lu	159	β^-	134	208, 113	β^--therapy/ SPECT

diagnostic PET-imaging [19]. In many facilities, the fluorine-18-derivative ^{18}F-PSMA-1007 is also used for PET and therapy monitoring of patients with metastasized castration-resistant prostate cancer (Figure 4.5.5). Another prominent ^{68}Ga/^{177}Lu theranostic match pair is the peptide-based radiopharmaceuticals (somatostatin receptor (sstr) agonists and antagonists) for imaging and treatment of neuroendocrine tumors. As sstr agonists, ^{68}Ga/^{177}Lu-DOTATOC and ^{68}Ga/^{177}Lu-DOTATATE

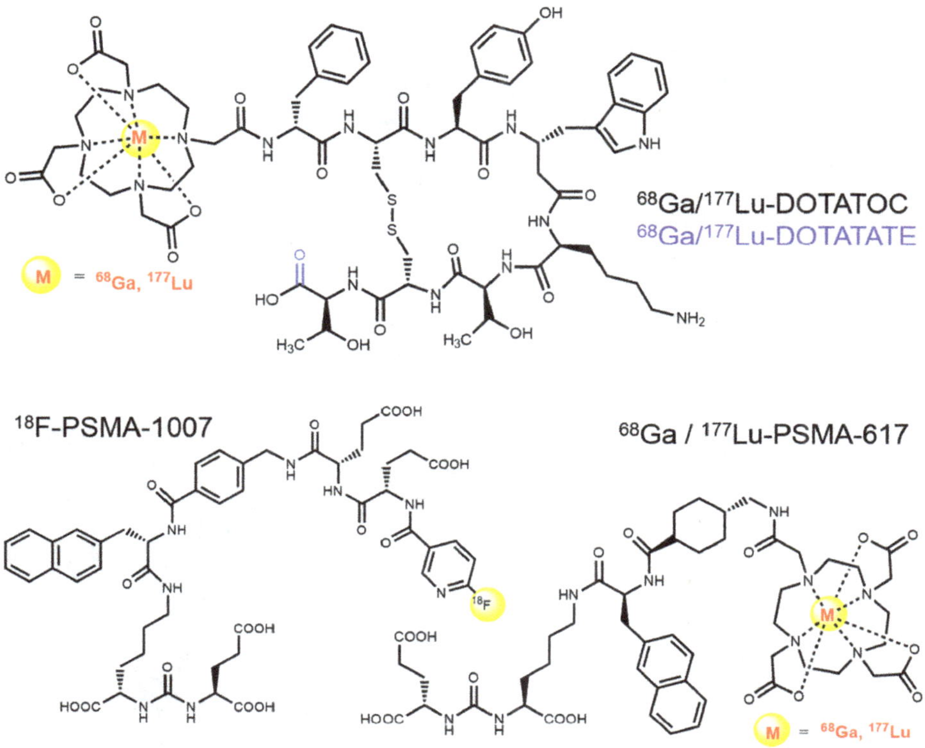

Figure 4.5.5: Top: molecular structure of the sstr-targeting cyclic peptides with the chelator DOTA; bottom: structures of the PSMA-targeting ligands PSMA-617 and PSMA-1007.

are important radiopharmaceuticals in modern nuclear medicine. Recently, radiolabeled sstr-antagonists including $^{68}Ga/^{177}Lu$-DOTA-JR11 came into focus because they have potential to improve imaging and therapy not only for neuroendocrine neoplastic tissues but also of tumors that have not been the focus of sstr targeting [20].

One of the most common therapeutic radionuclides is ^{90}Y. With its nearly three days half-life, pure β^- and high energy-emission resulting in a relatively long tissue penetration of ca. 1 cm, it finds wide use in antibody and peptide-labeling. One example is ^{90}Y-ibritumomab tiuxetan (^{90}Y-Zevalin®) as the first radioimmuno-therapeutic drug approved by the Food and Drug Administration (FDA) in 2002 to treat low-grade or follicular B-cell non-Hodgkin's lymphoma (NHL) [21]. Even though ^{86}Y (as a β^+-emitter) yields a theranostic pair with the chemically identical ^{90}Y, the corresponding ^{111}In-SPECT is usually employed to get diagnostic information before treatment. This is probably due to the low availability of ^{86}Y and more research has to be done for clinical use of this radionuclide [22]. Another very common therapy using ^{90}Y is SIRT (Selective Internal Radiation Therapy), where the radioactive dose (trapped in microspheres) is locally placed via catheter and therefore ideal for the treatment of tumors. Especially, for unresectable liver cancer with no response to systemic chemotherapy, this method is ideal. The liver has a dual blood supply system with a hepatic artery and a portal vein. Most liver tumors get their blood supply from the hepatic artery. Thus, it is possible to deposit radioactive labeled microspheres in the tumor (radioembolization), while avoiding harmful side effects for the healthy tissue (Figure 4.5.6). There are three types of microspheres available. Two of them use ^{90}Y as the radioactive source and are made of glass (TheraSphere®); alternatively, the radioactivity is loaded on small resins (SIR-Spheres®) [23]. The third one is made of poly(L-lactic acid) loaded with the radionuclide ^{166}Ho (QuiremSpheres®) and has a nice side effect, because after the radioactivity decayed, it is possible to see the paramagnetic particles in MRI [24].

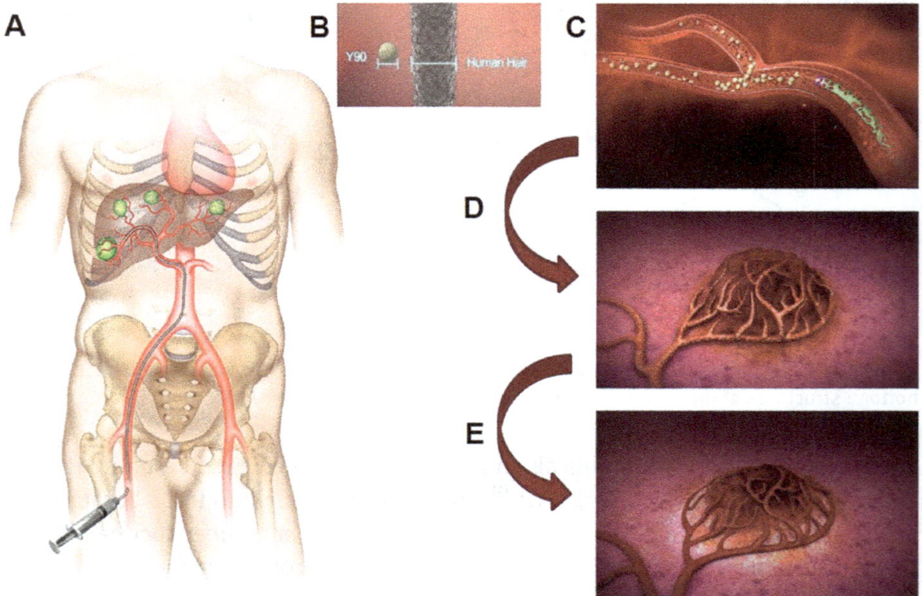

Figure 4.5.6: A: Minimally invasive treatment for liver tumors ^{90}Y-resin microspheres via microcatheter; B: microsphere in comparison to human hair; C: release of the microspheres into the arterial blood supply; D: microspheres are carried directly to the tumor and trapped; E: tumor necrosis upon β^-radiation. Rearranged with permission from Sirtex© Medical Europe GmbH.

β^--emitting radionuclides are used not only in curative therapy but also for palliative pain treatment where metastatic bone pain is a common problem in cancer patients in terminal stages. Often, palliation rather than cure is the goal of treatment, meaning alleviation of pain and avoiding adverse effects by lowering the medication dose. Here, bone-specific radiopharmaceuticals which have a strong affinity for bone metastases have shown favorable results in addressing bone pain. There are different mechanisms of action discussed such as the reduction of cytokines, tumor growth factors or inflammatory cells as well as several others [25]. The most common radiopharmaceutical for this area is ^{153}Sm-EDTMP. Due to the small γ-energy of ^{153}Sm, it is possible to have therapy monitoring via SPECT as shown in Figure 4.5.7. Less common are other radionuclides such as ^{177}Lu-EDTMP [26], ^{170}Tm-EDTMP [27] or ^{141}Ce-DOTMP [28] and they are objects of research due to the possible curative effect. Even ^{175}Yb-labeled polyamino-phosphonates, *bis*-phosphonates or hydroxyapatite were discussed as probes for treatment of bone pain or for use in radiosynovectomy of small joints [29]. The curative treatment of inflamed tissue (radiosynovectomy) has a long tradition in nuclear medicine. It involves an intra-articular injection of small radioactive particles to treat a synovitis. In the early 1960 s, ^{198}Au was first used for knee joints. In the next two decades, the most common radionuclides were ^{90}Y, colloidal

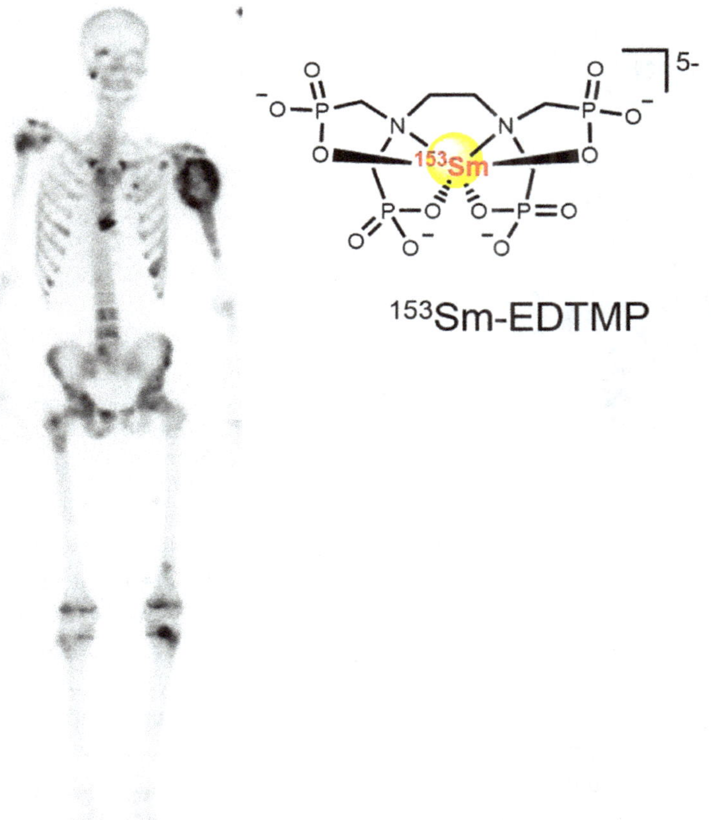

Figure 4.5.7: *Left*: whole body SPECT-Scan of a 14y old patient with osteosarcoma with metastasis in lung and bones receiving palliative pain therapy with 2.1 Gbq ^{153}Sm-EDTMP; *Right*: molecular structure of ^{153}Sm-EDTMP. © Department of Nuclear Medicine, University Hospital Münster, Germany.

chromic phosphate (^{32}P) and ^{186}Re-sufide colloids [30]. Till today, there are several additional radionuclides that are common in radiosynovectomy such as ^{169}Er-citrate and ^{186}Re-citrate or -colloid. Also, more exotic radiopharmaceuticals such as ^{165}Dy-ferric-hydroxide (DFH) are discussed for use in patients with different types of arthritis [31]. Depending on their β^--energies, they have tissue penetrations from the submillimeter range up to several mm for the radiation to be used in small finger or bigger knee joints. Upon β^-radiation, the injected particles cause ultimate damage to the cells of the synovial membrane. Starting with excitation and ionization, a large number of secondary particles like free radicals are created, resulting in subsequent apoptosis and ablation of the inflamed synovial tissue [32].

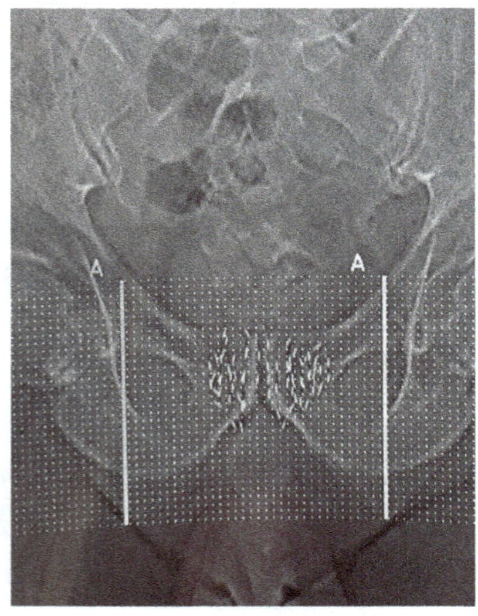

Figure 4.5.8: Brachytherapy of prostate cancer. Implanted radionuclide bearing seeds in the pelvis of a male patient. © Department of Radiation Oncology, University Hospital Münster, Germany.

Another form of radiotherapy is the brachytherapy (βραχύς, gr. – nearby), where in contrast to the previous therapies, sealed radionuclides (seeds) are placed inside (invasive) or next to the area of interest (skin or esophagus, non-invasive) (Figure 4.5.8). Brachytherapy is an effective treatment for prostate, skin or breast cancer, but due to its mostly invasive nature, it is not that common anymore. Due to the rapid development in the ability to focus ionizing radiation with high precision on the area of interest, external beam radiotherapy (EBRT) has been increasingly applied. Nevertheless, next to the usual radionuclides for brachytherapy (e.g., ^{192}Ir), radioisotopes of REEs such as ^{142}Pr are also under consideration. With its energetically high β^-emission and its low γ-emission, it could be suitable not only for therapy, but also for biodistribution tracking via SPECT. It is available as sealed probe and unsealed complex as ^{142}Pr-DTPA for radiosynovectomy [33].

References

[1] T. I. Kostelnik, C. Orvig, Chem. Rev. 2019, 119, 902.
[2] N. Long, W.-T. Wong, The Chemistry of Molecular Imaging, John Wiley & Sons, Hoboken, 2015.
[3] S. Currie, N. Hoggard, I. J. Craven, M. Hadjivassiliou, I. D. Wilkinson, Postgrad. Med. J. 2013, 89, 209.
[4] J. Washner, E. M. Gale, A. Rodriguez-Rodriguez, P. Caravan, Chem. Rev. 2019, 119, 957.
[5] M. C. Heffern, L. M. Matosziuk, T. J. Meade, Chem. Rev. 2014, 114, 4496.
[6] Y.-D. Xiao, R. Paudel, J. Liu, C. Ma, Z.-S. Zhang, S.-K. Zhou, Int. J. Mol. Med. 2016, 38, 1319.
[7] C. Thakral, J. Alhariri, J. L. Abraham, Contrast Media Mol. Imaging 2007, 2, 199.
[8] G. Bongartz, D. Weishaupt, M. Mayr, Schweiz. Med. Forum 2008, 8, 116.
[9] W. P. Cacheris, S. C. Quay, S. M. Rocklage, Magn. Res. Imaging 1990, 8, 467.
[10] a) https://www.ema.europa.eu/en/medicines/human/referrals/gadolinium-containing-contrast-agents.html (accessed Aug 29, 2019); b) https://www.magnetic-resonance.org/ch/13-01.html (accessed Aug 29, 2019).
[11] B. J. Guo, Z. L. Yang, L. J. Zhang, Front. Mol. Neurosci. 2018, 11, 335.
[12] Y. Xiáng, J. Wáng, J.-M. Idée, Quant. Imaging Med. Surg. 2017, 7, 88.
[13] P. C. M. van Zijl, N. N. Yadav, Magn. Res. Med. 2011, 65, 927.
[14] M. Goldman, J. Magn. Res. 2001, 149, 160.
[15] K. L. Peterson, K. Srivastava, V. C. Pierre, Front. Chem. 2018, 6, 160.
[16] a) P. Harvey, I. Kuprov, D. Parker, Eur. J. Inorg. Chem. 2012, 2015; b) K. H. Chalmers, A. M. Kenwright, D. Parker, A. M. Blamire, Magn. Reson. Med. 2011, 66, 931.
[17] E. W. Price, C. Orvig, Chem. Soc. Rev. 2014, 43, 260.
[18] a) https://www-nds.iaea.org/relnsd/vcharthtml/VChartHTML.html (accessed Aug 29, 2019); b) C. Müller, K. A. Domnanich, C. A. Umbricht, N. P. van der Meulen, Br. J. Radiol. 2018, 91, 20180074; c) C. Müller, M. Bunka, S. Haller, U. Köster, V. Groehn, P. Bernhardt, N. van der Meulen, A. Türler, R. Schibli, J. Nucl. Med. 2014, 55, 1658; d) S. Haller, G. Pellegrini, C. Vermeulen, N. P. van der Meulen, U. Köster, P. Bernhardt, R. Schibli, C. Müller, EJNMMI Res. 2016, 6, 13.
[19] C. A. Umbricht, M. Benešová, R. M. Schmid, A. Türler, R. Schibli, N. P. van der Meulen, C. Müller, EJNMMI Res. 2017, 7, 9.
[20] M. Fani, G. P. Nicolas, D. Wild. J. Nucl. Med. 2017, 58, 61S.
[21] A. J. Grillo-López, Expert Rev. Anticancer Ther. 2002, 2, 485.
[22] F. Rösch, H. Herzog, S. Qaim, Pharmaceuticals 2017, 10, 56.
[23] M. A. Westcott, D. M. Coldwell, D. M. Liu, J. F. Zikria, Adv. Radiation Onc. 2016, 1, 351.
[24] M. L. J. Smits, M. Elschot, M. A. A. J. van den Bosch, G. H. van de Maat, A. D. van het Schip, B. A. Zonnenberg, P. R. Seevinck, H. M. Verkooijen, C. J. Bakker, J. Nucl. Med. 2013, 54, 2093.
[25] M. Taheri, Z. Azizmohammadi, M. Ansari, P. Dadkhah, K. Dehghan, R. Valizadeh, M. Assadi, Nuklearmedizin 2018, 57, 174.
[26] M. Alavi, F. Khajeh-Rahimi, H. Yousefnia, M. Mohammadianpanah, S. Zolghadri, A. Bahrami-Samani, M. Ghannadi-Maragheh, Cancer Biother. Radiopharm. 2019, 34, 280.
[27] T. Das, A. Shinto, K. K. Kamaleshwaran, S. Banerjee, Clin. Nucl. Med. 2017, 42, 235.
[28] K. V. Vimalnath, A. Rajeswari, H. D. Sarma, A. Dash, S. Chakraborty, J. Labelled Comp. Radiopharm. 2019, 62, 178.
[29] a) B. Mathew, S. Chakraborty, T. Das, H. D. Sarma, S. Banerjee, G. Samuel, M. Venkatesh, M. R. Pillai, Appl. Radiat. Isot. 2004, 60, 635; b) A. Fakhari, A. R. Jalilian, H. Yousefnia, S. Shanehsazzadeh, A. B. Samani, F. J. Daha, M. S. Ardestani, A. Khalaj, Mol. Imag. Radionucl. Ther. 2015, 24, 110; c) S. Chakraborty, T. Das, S. Banerjee, S. Subramanian, H. D. Sarma, M. Venkatesh, Nucl. Med. Biol. 2006, 33, 585.

[30] K. Liepe, World J. Nucl. Med. 2015, 14, 10.
[31] C. Pirich, A. Pilger, E. Schwameis, D. Germadnik, U. Prüfert, E. Havlik, S. Lang, H. Kvaternik, J. A. Flores, P. Angelberger, A. Wanivenhaus, H. W. Rüdiger, H. Sinzinger, J. Nucl. Med. 2000, 41, 250.
[32] J. Ailland, W. U. Kampen, M. Schünke, J. Trentmann, B. Kurz, Ann. Rheum. Dis. 2003, 62, 1054.
[33] M. K. Bakht, M. Sadeghi, Ann. Nucl. Med. 2011, 25, 529.

Cees Ronda

4.6 Rare earth-based scintillators

4.6.1 Overview

Scintillators are luminescent materials that convert high-energy (ionizing) radiation into photons with considerably lower energy, ranging from UV to IR radiation. An important application field is medical imaging, in which X-ray or γ-radiation is converted into visible light. Scintillators enabled the accidental discovery of X-rays by Wilhelm Conrad Röntgen, in 1895. A barium-cyanido-platinate $Ba[Pt(CN)_4]$-coated screen showed green emission, due to excitation with an, at that time, unknown kind of radiation (X-rays) [1]. The potential of this method for medical applications, especially, was realized very quickly.

$CaWO_4$ has been used to detect X-rays for almost 80 years, a significant improvement of scintillation properties was realized, especially with the advent of halide based materials like CsI:Tl and rare-earth based materials like Gd_2O_2S:Pr.

Luminescent materials, in general, consist of a host structure, with intentionally added impurities (dopants, activators) on which the luminescence takes place. In luminescent materials, for example, used in LEDs or fluorescent lamps, the excitation process can take place on the activator ions directly, or via a so-called sensitizer ion when the activator ion shows an only weak absorption. In such a case, one does not speak of scintillation (see Figure 4.6.1). The emission of most of the luminescent materials is determined by the optical properties of the dopant ion in the specific host lattice.

In case of scintillators, excitation is via the host lattice. The host lattice absorbs the energy of the incident ionizing radiation, ultimately leading to excitations across the band gap (thermalization), followed by transfer of the energy to the dopants that subsequently emit; see Figure 4.6.2 for a schematic picture. This figure also shows that the final process, generating the luminescence, is the same for luminescent materials excited by ionizing radiation and by UV radiation or visible light. In case of scintillators, processes and mechanisms resulting in energy transfer from host lattice states to the localized activator states have to be understood and controlled in great detail, this being a very interesting research challenge.

A number of processes can be involved in energy transfer to the activators; see Figure 4.6.3:

- The electron and hole can couple to form an exciton (as the electron and hole attract each other as they carry opposite charges). The energy of such an exciton is smaller than the band gap energy, because of the attractive Coulomb interaction and also because the exciton polarizes the lattice, which may result in (self) trapping. In principle, the exciton is mobile and its energy can be transferred to an activator ion, followed by emission. Because of self-trapping, the movement

https://doi.org/10.1515/9783110654929-029

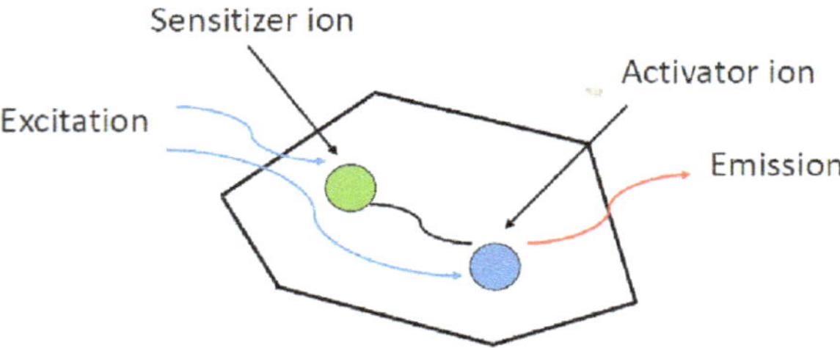

Figure 4.6.1: Luminescent material, consisting of a host lattice containing activator ions. The emission is generated by activator ions. Excitation can be on activator ions or sensitizer ions, see text.

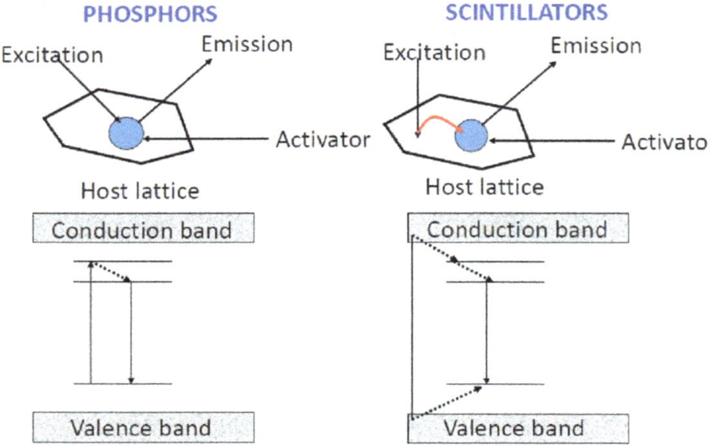

Figure 4.6.2: Luminescent material, consisting of a host lattice containing activator ions. The process via immediate excitation of ions is depicted in the left part of the figure; the scintillation process is depicted in the right part of the figure. The activator ions have energy levels within the band gap of the materials. The emission results from electronic transitions between such states.

of the exciton is temperature- dependent; this may, for example, lead to a clear build-up of the luminescence signal, especially at low temperatures. Please note that energy can be transported in the lattice without charge being transported, as excitons do not carry a charge.

- The activator captures a hole and subsequently an electron or vice versa, followed by emission. This process requires activator ions with more than one valence. Examples are given in Table 4.6.1; also, the nature of the first charge captured is indicated. Metal ions with more than one valence state, in general, are very suitable as activator ions in scintillating material, as they can very easily capture electrons and holes. This process can be very fast, even on the 10^{-12} s (ps) timescale, and no build-up may be seen.

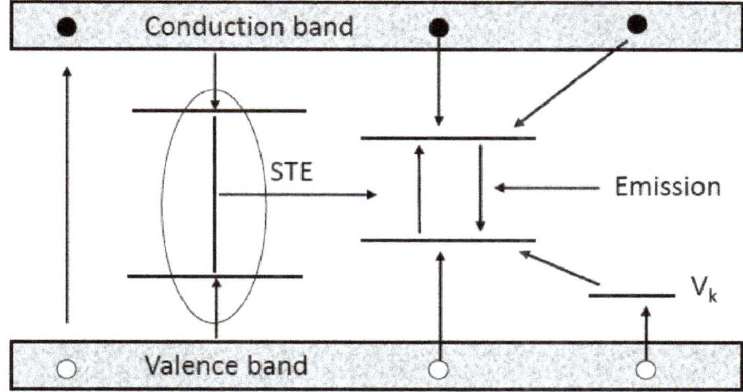

Figure 4.6.3: Processes taking place after thermalization. In the left part of the figure, a Self-Trapped Exciton (STE) is formed, followed by energy transfer to the luminescent center. In the middle part of the figure, the luminescent center is excited immediately by subsequent capture of an electron and a hole (or vice versa), and in the right part of the figure, the hole is first trapped on a vacancy (V_k center), followed by transfer of the hole to the luminescent center.

– A hole is captured on a vacancy, resulting in a self-trapped hole (V_k center), followed by transfer of the hole to an activator ion. Subsequently, an electron is captured and emission follows. This process may lead to considerable afterglow, as the transfer of the hole from the vacancy to the activator ion is thermally activated.

Table 4.6.1: Examples of rare-earth ions leading to efficient scintillator, based on capture of charge.

Emitting ion	First step in scintillation mechanism	Trap type
Ce^{3+} ($4f^1$)	Ce^{3+} ($4f^1$) $+h^+ \rightarrow Ce^{4+}$ ($4f^0$)	Hole trap
Pr^{3+} ($4f^2$)	Pr^{3+} ($4f^2$) $+h^+ \rightarrow Pr^{4+}$ ($4f^1$)	Hole trap
Eu^{2+} ($4f^7$)	Eu^{2+} ($4f^7$) $+h^+ \rightarrow Eu^{3+}$ ($4f^6$)	Hole trap
Eu^{3+} ($4f^6$)	Eu^{3+} ($4f^6$) $+e^- \rightarrow Eu^{2+}$ ($4f^7$)	Electron trap
Tm^{3+} ($4f^{12}$)	Tm^{3+} ($4f^{12}$) $+e^- \rightarrow Tm^{2+}$ ($4f^{13}$)	Electron trap
Yb^{3+} ($4f^{13}$)	Yb^{3+} ($4f^{13}$) $+e^- \rightarrow Yb^{2+}$ ($4f^{14}$)	Electron trap

Scintillating materials are characterized by their light yield, their decay time and their afterglow behavior. One speaks of afterglow when the luminescence intensity decays more slowly than the intrinsic decay time of the activator ion that generates the luminescence.

4.6.2 Light yield

Predicting the maximum light yield of scintillators has been the subject of many papers [2–4]. There is general consensus that the following expression gives a good estimation of the light yield (η) that can be achieved:

$$\eta = \langle h\upsilon \rangle \eta_t \eta_{act} \eta_{esc} / \beta E_g \tag{4.6.1}$$

in which $\langle h\upsilon \rangle$ is the mean energy of the emitted photons, η_t is the transfer efficiency of host lattice energy to the activator ions, η_{act} is the quantum efficiency of the activator ion (ratio of the number of activator ions that emit and that are excited) and η_{esc} is the escape probability (ratio of photons that leave the scintillator and photons generated in the scintillator). β is a pure number and is the ratio of the energy needed to create an electron-hole pair in the scintillation process and the band gap energy (E_g) of the host material. The maximum efficiency (η_{max}) is consequently given by:

$$\eta_{max} = \langle h\upsilon \rangle / \beta E_g \tag{4.6.2}$$

Inspection of this expression shows that the highest light yields can be expected for materials with a small band gap. A related expression is generally used to express the photon yield of scintillators:

$$\eta = 1\text{MeV} / \beta E_g \tag{4.6.3}$$

This expression gives the number of photons that are generated by the scintillator, per MeV of energy impinging on the scintillating material.

In the thermalization processes in which the ionizing radiation is converted into excitation across the bandgap of the host of the scintillating materials, both energy and momentum have to be conserved. Without going into detail, these requirements result in values for β that are significantly larger than 1; in general a value of 2.5 can be used, but this value depends on the nature of the chemical bonding in the host lattice; see [2–5] and references cited therein.

4.6.3 Decay time

The decay time of the emission in scintillating materials is ideally determined by the ion generating the emission in the scintillator host. In many cases, however, there is a build-up in the emission signal, such that the luminescence intensity is not maximal immediately after the excitation has been stopped. This can be due to very shallow traps, or also due to temperature- dependent diffusion of self-trapped excitons to activator ions.

4.6.4 Afterglow

Afterglow requires that energy is stored in the lattice, for example, by trapping charges on vacancies or by ionization of metal ions (including the activator ions) with energy levels in the bandgap of the scintillator host material. To combat afterglow, several strategies have been developed, aiming at keeping the activator ions in the equilibrium charge state. In Gd_2O_2S:Pr, to mention an example, an electron from Pr^{3+} is excited into the conduction band, converting it into Pr^{4+}, followed by capturing the electron by an electron trap. Afterglow occurs due to capturing an electron residing in the conduction band, that was released from an electron trap by Pr^{4+}, converting it into an excited Pr^{3+} ion, followed by emission. Afterglow in Gd_2O_2S:Pr can be reduced significantly by adding a metal ion that has the same electrochemical behavior as Pr^{3+}, like Ce^{3+}. As Ce^{3+} does not emit in Gd_2O_2S, the amount of Ce^{3+} that can be added is limited; otherwise the penalty of light loss is too large (electron-hole recombination may also take place on Ce^{3+}). During irradiation with ionizing radiation, Ce-ions keep Pr ions in their trivalent state. Addition of, for example, Eu^{3+} ions results in excellent electron traps (converting Eu^{3+}-ions into Eu^{2+}-ions), and therefore, to more pronounced afterglow. Another strategy is trap energy level engineering by changing the composition of the host structure. Scintillators in medical imaging equipment are generally operating at temperatures slightly higher than room temperature. Traps that are not active at room temperature (e.g., because the traps are very shallow or very deep) do not, or hardly contribute to afterglow. Garnet materials have a large compositional variety – in this class of materials, trap energy level engineering has been investigated in great detail. An illustration of this approach is given in [6, 7], for the material $Y_3Al_{5-x}Ga_xO_{12}$:Ce that contains Cr^{3+} as impurity, Cr^{3+} ions resulting in afterglow. Using thermoluminescence spectroscopy, the depth of the traps can be calculated, and the depth of the Cr^{3+} related trap turned out to depend significantly on the Al/Ga ratio; see Table 4.6.2.

Table 4.6.2: Cr^{3+}-related trap depth as a function of the Al/Ga ratio in $Y_3Al_{5-x}Ga_xO_{12}$:Ce.

x in $Y_3Al_{5-x}Ga_xO_{12}$:Ce	Cr^{3+}-related trap depth (eV)
0	1.0
1	1.03
2	0.89
3	0.67

Of course, afterglow can also be exploited for commercial use and materials have been developed, that emit light even hours after irradiation with blue light or UV radiation was stopped. An example for such a material is $SrAl_2O_4$:Eu,Dy, used in safety applications or in toys.

4.6.5 Preparational aspects

As is the case of any luminescent material, also in the case of scintillating materials, high purity materials (especially for the host material) have to be used. Underlying reasons are that energy transfer to impurities may take place, and that when excited across the bandgap, impurity ions may also trap charges from the valence or conduction band (holes or electrons), reducing the light yield and potentially leading to afterglow (when the trapped charges are thermally released and recombine on e.g., activator ions) or bright burning. One speaks of bright burning when the intensity increases as a function of time during irradiation. The underlying mechanism is that energy is used to fill traps; once filled, the radiative channel remains.

Scintillators are applied as powder layer in X-ray intensifying screens. To completely absorb the energy of the ionizing radiation impinging on the scintillating materials, in general, relatively thick layers are needed. For example, in the case of CsI:Tl, a material used in flat panel X-ray detectors, the scintillating element is obtained via evaporation in vacuum.

To obtain completely transparent materials, either single crystals are needed or polycrystalline materials that are sintered to transparency. Single crystal growth is expensive, in view of the high melting temperature of many (oxidic) scintillator host structures. Ceramic transparent scintillators require host structures with cubic symmetry. The development of ceramic transparent scintillators was initiated by Greskovich et al. [8]; compounds investigated were, for example, $(Y,Gd)_2O_3$:Eu, Gd_2O_2S:Pr,Ce,F and $Gd_3Ga_5O_{12}$:Cr. In the course of time, the cubic materials $(Y,Gd)_2O_3$:Eu and many cubic garnet materials were also developed in completely transparent form.

4.6.6 Applications

In this section, we focus on scintillators for application in equipment for computer tomography (CT) and positron emission tomography (PET).

In case of CT, the layer thickness needed is in the order of a few mm, and in case of PET (the energy of the impinging radiation is equal to the rest mass energy of an electron: 511 keV), even in the order of a few cm. Therefore, in the case of scintillators for PET, completely transparent materials are needed; in the case of CT, translucent materials can also be used.

In CT machines, both the X-ray generator and the scintillators rotate around the patient. Spatial resolution is obtained by structuring the scintillating ceramics. This is done by intersecting the ceramic layers with reflecting layers. In this way, small, optically-decoupled segments are created and the light generated is coupled into photodiodes; see the inset in Figure 4.6.4. Decay time and afterglow set an upper limit to the rotation speed; for this reason, materials with a short decay time (< 1 ms) and low afterglow values are preferred. Absorption of X-rays is measured, this results in a lower intensity of light generated by the scintillators. CT images, therefore, deliver anatomic information as X-ray absorption is tissue-density-dependent, and more X-ray energy is absorbed in bones than in softer tissue. The decay time requirement for CT scintillators can be fulfilled with activator ions with forbidden transitions; the transitions do not need to be optically allowed.

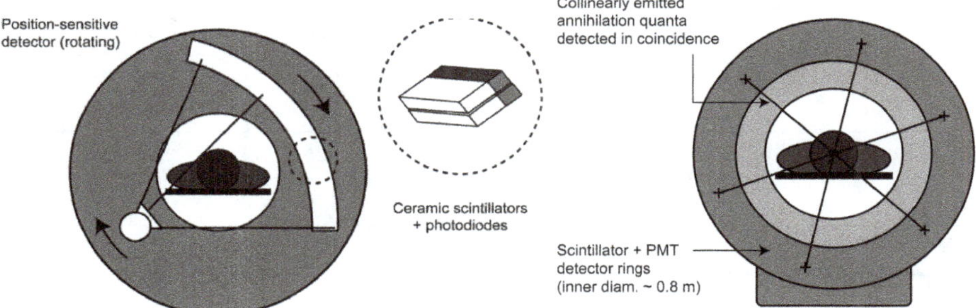

Figure 4.6.4: Left: CT configuration. Both the X-ray source and the scintillator detector array are rotating around the patient. The picture is reproduced from [3, 9]. Right: PET configuration. Picture is taken from [9].

In PET machines, a radioactively-labeled sugar is administered to a patient. This sugar is enriched in tissue with a high metabolic rate, such as cancerous tissue. The radioactive material emits a positron, the positron recombines with an electron, resulting in two γ quanta, both with the energy corresponding to the rest mass energy of an electron (511 keV). The two γ quanta move into exactly opposite directions and build a line of response. Only in this way, both momentum and energy can be conserved in the annihilation process. The direction of the lines of response is not fixed by any physical mechanism; all directions can occur. The tissue with high energy consumption is located where all lines of response cross each other.

For PET scintillators, a very short decay time is preferred, in the order of tens of nanoseconds, as two photons have to be detected in coincidence to determine the line of response. In addition, essentially all photons have to be counted in a short time, preventing dead time in the PET machine, as no second γ quantum can be processed when a previous γ quantum is still being processed in the same detection volume. This means that PET scintillators are based on activator ions with completely allowed transitions. To comply with the spectral sensitivity of Si-based photo detectors, they should not emit in the UV; this excludes very fast emitting ions like Nd^{3+} and Pr^{3+}. Table 4.6.3 gives an overview of some important scintillators.

Table 4.6.3: Some important scintillating materials. $Bi_4Ge_3O_{12}$ and $CaWO_4$ are self-activated materials, not intentionally doped with an activator ion.

Material	Light yield (ph/MeV)	Emission peak (nm)	Decay time (μs)	Application
Gd_2O_2S:Pr	40,000	510	3.4	CT
$(Y,Gd)_2O_3$:Eu	19,000	610	1000	CT
$Bi_4Ge_3O_{12}$	9,000	480	0.3	PET
$(Lu,Y)Si_2O_5$:Ce	33,800	420	0.04	PET
$Gd_{1.5}Y_{1.5}Ga_{2.5}Al_{2.5}O_{12}$:Ce	50,000 [11]	550	0.10	PET/CT
$Lu_3Al_5O_{12}$:Ce	25,000 [12]	525	0.06	
CsI:Tl	66,000	550	0.8	X-ray
$CaWO_4$	15,000	425	8	X-ray

4.6.7 Outlook

The field of scintillators is rapidly developing; see [10–14]. Recent years have seen major new developments, like new scintillator hosts and scintillating nanocrystals. The discovery of very efficiently scintillating $LaBr_3$:Ce in 2001 [14] has resulted in a large research activity dealing with halides, and a number of new efficiently emitting halides have been discovered. Recently developed iodide hosts like CaI_2:Eu, SrI_2:Eu, $CeBr_3$ (a stoichiometric scintillator) and $CsBa_2I_5$:Eu especially show very high light yields in the order of 100,000 photons MeV^{-1} [3, 14]. The high light yield is due to the combination of a relatively small band gap and a relatively small value for the constant β; both factors result in a large value for the maximum light yield (eq. (4.6.2)). Disadvantage of many of these materials, especially for applications in PET, is that the density, and therefore, the stopping power is rather low. In addition, quite a few of these materials are strongly hygroscopic.

As touched upon already above, transparent ceramic scintillators are the subject of a large research activity. There are several reasons for this: when scintillators can be produced in transparent form without tedious single crystal growth procedures, the underlying processes are in general much faster and require less high processing temperatures, and therefore such processes are also less expensive.

In addition, there is no concentration gradient of activator ions for instance, as there is no growth from the melt. Such a gradient occurs when the solubility of activator ions (or other ions) differs in the liquid- and in the solid phase; this, in general, is the case. Moreover, especially in garnets, the so-called anti site defects occur: rare-earth ions take the place of trivalent main group metal ions in the garnet hosts, driven by entropy, and are therefore more pronounced at higher temperature. For these reasons, lower processing temperatures also result in more homogeneous materials. The reader is referred to some overview papers, with many more expected to appear [7–13].

Nanoscintillators are also gaining research interest. See e.g., [15, 16]. Potential use of such scintillators is in medical treatment [15]. On irradiation with ionizing radiation, they may create UV light, or energy may, for example, be transferred to porphyrins that subsequently create $^1O_2$, a cytotoxin. In both cases, irradiation of the nanoscintillators, while in the body, may result for example, in killing of cancerous cells. To this end, the nanoscintillators should have sizes in the order 50–150 nm to enable a prolonged circulation in the blood stream [15].

Nanocrystalline scintillators are prepared similarly to other nanocrystalline materials, for example, using sol-gel methods, the combustion method (in which addition of a fuel results in very rapid heating of the reaction mixture, enabling a very fast synthesis, in the order of minutes), low temperature reactions under high pressure using autoclaves [17] or also laser ablation techniques. Dispersion of the frequently agglomerated powders may be done using ultrasound techniques. In general, the light yield of nanocrystalline scintillators is significantly lower than that of the corresponding microcrystalline powders, scintillator films, ceramics or single crystals of the same material. Their uses, however, are different and the light yield is less important in these applications.

References

[1] a) W. C. Röntgen, Sitz. Ber. Phys. Med. Ges. Würzburg 1895, 9, 132; b) W. C. Röntgen, Science 1896, 3, 227.

[2] R. H. Bartram, A. Lempicki, J. Lumin. 1996, 68, 225.

[3] C. Ronda, H. Wieczorek, V. Khanin, P. Rodnyi, ECS J. Solid State Sci. Technol. 2016, 5, R3121.

[4] D. J. Robbins, J. Electrochem. Soc. 1980, 127, 2694.

[5] P. Dorenbos, Nucl. Instrum. Meth. A 2002, 486, 208.

[6] I. I. Vrubel, R. G. Polozkov, I. A. Shelykh, V. M. Khanin, P. A. Rodnyi, C. R. Ronda, Cryst. Growth Des. 2017, 17, 1863.

[7] H. Wieczorek, V. Khanin, C. R. Ronda, J. Boerekamp, S. Spoor, R. Steadman, I. Venevtsev, K. Chernenko, A. Meijerink, P. Rodnyi, to be published in IEEE-TNS, 2019.

[8] C. Greskovich, S. Duclos, Ann. Rev. Mater. Sci. 1997, 27, 69.
[9] A. M. Srivastava, C. R. Ronda, Luminescence: from Theory to Applications, VCH Press, Weinheim, 2007.

[10] N. J. Cherepy, S. A. Payne, S. J. Asztalos, G. Hull, J. D. Kuntz, T. Niedermayr, S. Pimputkar, J. J. Roberts, R. D. Sanner, T. M. Tillotson, E. van Loef, C. M. Wilson, K. S. Shah, U. N. Roy, R. Hawrami, A. Burger, L. A. Boatner, W.-S. Choong, W. W. Moses, IEEE Trans. Nucl. Sci. 2009, NS-56, 873.
[11] N. J. Cherepy, Z. M. Seeley, S. A. Payne, P. R. Beck, E. L. Swanberg, S. Hunter, L. Ahle, S. E. Fisher, C. Melcher, H. Wei, T. Stefanik, Y.-S. Chung, J. Kindem, Proc. SPIE 2014, 921302.
[12] P. Lecoq, Nucl. Instrum. Meth. A 2016, 809, 130.
[13] M. Nikl, Meas. Sci. Technol. 2006, 17, R37.
[14] E. V. D. van Loef, P. Dorenbos, C. W. E. van Eijk, K. W. Kraemer, H. U. Guedel, Appl. Phys. Lett. 2001, 79, 1573.
[15] J. Y. Jung, G. A. Hirata, G. Gundiah, S. Derenzo, W. Wrasidlo, S. Kesari, M. T. Makale, J. McKittrick, J. Lumin. 2014, 154, 569.
[16] J. M. Nedelec, J. Nanomater. 2007, Article ID 36392.
[17] A. Wiatrowska, W. Keur, C. R. Ronda, J. Lumin. 2017, 189, 9.

Klaus Stöwe, Marc Armbrüster

4.7 Rare earth metals in heterogeneous catalysis

4.7.1 Introduction

Intermetallic compounds have been employed in catalysis since the discovery of the skeletal Ni-catalysts by Raney in 1925 [1]. The number of groups working in this interesting field is strongly increasing in recent years resulting in more than 2,400 publications [2]. Rare-earth-based materials (i.e., intermetallic compounds, hydrides, nitrides and oxides) have been used in heterogeneous catalysis for a long time and date back as far as 1968 [3, 4]. Since then, the chemical properties of rare-earth-based intermetallic compounds have been used to address catalytic challenges through their direct use, employing their hydrides or by decomposing them deliberately to transition metal nanoparticles supported by the corresponding rare earth oxide (Figure 4.7.1). Also, in few cases, the formation of intermetallic compounds by the reaction of noble metals with the partially reduced supporting rare earth oxides has been reported. In total, more than 300 publications focus on the use of intermetallic compounds based on rare earth metals in catalysis. The application of rare earth elements in catalysis is not restricted to intermetallic compounds, but also extends to rare earth oxides and much less commonly to other compounds. This is due to the observation that many rare earth intermetallic compounds form rare earth hydrides, nitrides or oxides under the reaction conditions.

In this overview, the different approaches to use rare-earth-based materials in catalysis as well as the strategies behind them will be elucidated. The overview is organized by increasing oxidation state, thus first focusing on (inter)metallic catalysts, then on altered (inter)metallic materials and finally on the oxides of rare earth elements as functional supports, promoters or even active phases.

4.7.2 Intermetallic compounds

While the use of rare earth metals as such is not described in literature, the rich literature addressing the use of rare-earth-based intermetallic compounds can be grouped into three categories (Figure 4.7.1). First, the intermetallic compounds can be directly employed as the catalytically active component. In the second case, the intermetallic compounds are altered before or during catalysis, thus acting as a precursor of the active species. The third case is the (un)intended formation of intermetallic compounds from rare-earth-oxide-supported transition metal particles and the corresponding rare earth metal, by partial reduction of the support.

https://doi.org/10.1515/9783110654929-030

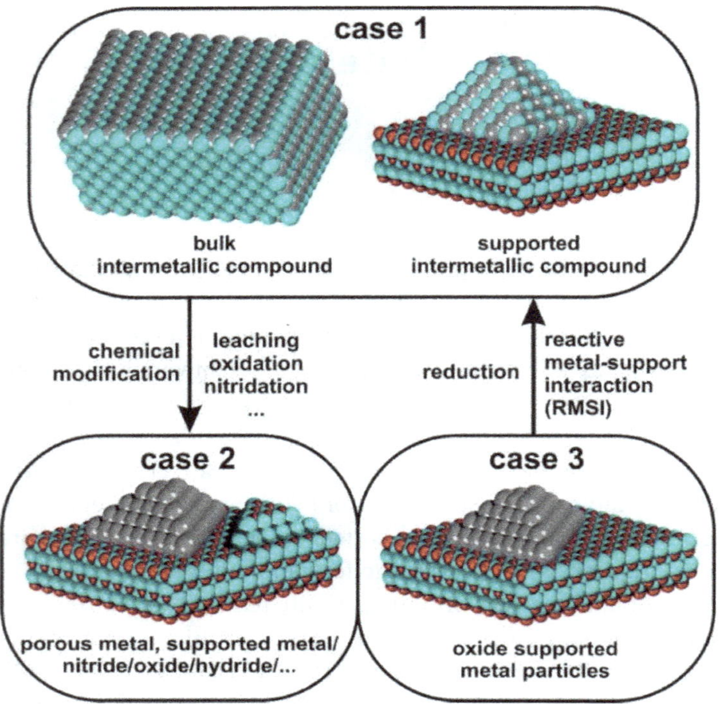

Figure 4.7.1: Different uses and occurrences of intermetallic compounds in catalysis.

This so-called reactive metal-support interaction (RMSI [5]) is the more likely, the easier the support can be reduced. Due to the high stability of the rare earth oxides, the RMSI is not very often observed/used in this class of materials. In this chapter, the principles on which the rare-earth-containing intermetallic compounds are used as catalysts are summarized.

Intermetallic compounds containing rare earth metals have been used in catalysis because of their chemical properties. This comprises the formation of intermetallic hydrides (e.g., $LaNi_5H_6$ from $LaNi_5$), partial oxidation of the intermetallic compound and in few cases, the intrinsic catalytic activity of the intermetallic compound. Many industrially relevant reactions require the activation of hydrogen – a prominent example is the synthesis of ammonia from the elements ($N_2 + 3\ H_2 \leftrightarrow 2\ NH_3$). The activation of hydrogen requires splitting of the H–H bond which also has to take place if (inter)metallic hydrides are formed. Thus, hydride-forming intermetallic compounds based on rare earth metals have been used as hydrogenation and dehydrogenation catalysts. The reasoning behind the latter is that the intermetallic compound easily absorbs the resulting hydrogen thus facilitating the reaction.

4.7.2.1 Ammonia synthesis

In ammonia synthesis, the formation of intermetallic hydrides and the decomposition of the intermetallic compounds into transition metal particles supported by the corresponding rare earth nitride have been reported. Intermetallic compounds in the systems Ce-Co [6, 7], Ce-Fe [6, 7], Ce-Ru [6–8], Dy-Fe [6], Er-Fe [6], Fe-Gd [6], Fe-Ho [6], Fe-Tb [6], Fe-Th [6] and Co-Pr [6] thus result in supported transition metal particles, to which the catalytic activity has to be ascribed. This also holds true for the prominent example of intermetallic hydride formation, $LaNi_5$. Under ammonia synthesis conditions, thin films of Sm_2Fe_{17} form the hydride $Sm_2Fe_{17}H_{0.6}$ [9]. While ammonia formation takes place, no nitride formation was reported in this study, in contrast to later studies [10, 11], thus suggesting the nitride to be the catalytically active phase. Very recently, ammonia synthesis over LaCoSi was reported and the analysis after the reaction showed no additional phases, thus making this compound the first, where the catalytic properties can be assigned to the intermetallic compound itself [12]. The catalytic activity is assigned to "hot electrons" which are present in LaCoSi.

4.7.2.2 Methanol synthesis

Another important process involving the activation of hydrogen is syngas (a mixture of CO, CO_2 and H_2) conversion from which a variety of alcohols, hydrocarbons and oxygenates can be synthesized. Common to all intermetallic compounds based on rare earth metals is the fact that they partially oxidize to the corresponding rare earth oxide and small particles of the corresponding transition metal or its oxide. Intermetallic compounds of copper with La, Ce, Pr or Ho get oxidized in syngas mixtures at elevated temperatures and the resulting materials catalyze the synthesis of methanol and to some part, higher alcohols (C2 to C4) [13]. The materials, based upon intermetallic precursors, are more resistant against sulfur impurities and as long as the materials are not fully oxidized, the underlying intermetallic compound can still accelerate the catalytic reaction by hydride formation, thus activating the hydrogen in the reactant mixture [14]. *In situ* XRD studies on CuNd, Cu_2Nd, Cu_5Nd and Cu_2Ce in syngas revealed that the decomposition occurs via intermediate hydride phases, thus corroborating the earlier findings (Figure 4.7.2) [15–17]. Also, compounds in the systems Co-La [18, 19], La-Ni [18], Cu-Dy [19], Cu-Dy-Zr [19], Cu-Gd [19], Cu-Gd-Ti [19], Cu-Gd-Zr [19], Cu-Mm-Zr [19] (Mm = Misch metal, a mixture of different rare earth metals used to avoid the tedious separation of the rare earth elements), Cu-Nd-Zr [19] and Cu-Sm-Zr [19] have been investigated, always resulting in the decomposition of the intermetallic compound.

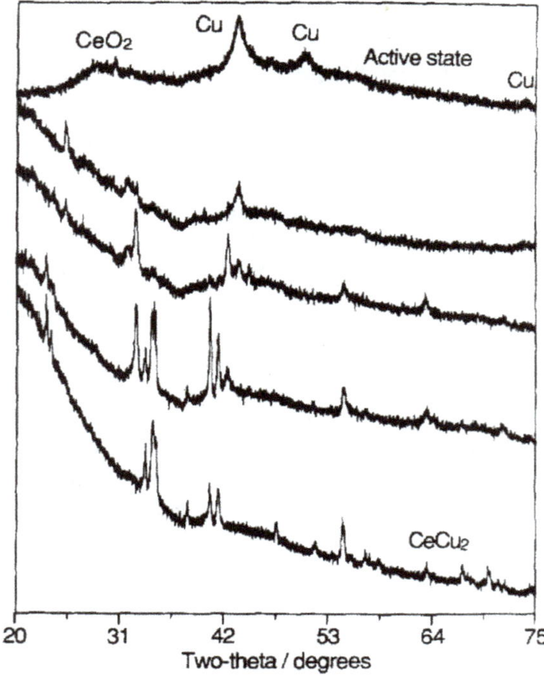

Figure 4.7.2: Conversion of $CeCu_2$ into an active catalyst, through intermetallic hydride phases, in 50 bar CO/H_2 at 100 °C. Figure reproduced from reference [16] with permission from Elsevier.

4.7.2.3 Methanation and Fischer–Tropsch process

Besides methanol, the formation of hydrocarbons as synthetic liquid fuel from syngas generated by the steaming of coal is of high interest (so-called Fischer–Tropsch process). Closely related to this process is the methanation of CO_2/H_2 mixtures which is part of the power-to-gas approach to reduce the CO_2-concentration in our atmosphere and the use of fossil fuels. In these processes, the CO_2 usually acts as an oxidant, making the intermetallic compound in question a precursor for the catalytic material. *RE*Ni_5 (*RE* = La [14, 20], Ce [14, 20], Pr [14], Nd [14], Sm [14], Eu [14], Gd [14, 20], Tb [14, 20], Ho [14, 20], Er [14] or Yb [14, 20]), La_2Ni_7 [21], $LaNi_3$ [21] and LaNi [21] as well as $CeCo_2$ [20] and $ErFe_2$ [20] have been tested. The result is CH_4 and traces of higher hydrocarbons in a $CO/3H_2$ mixture, starting from 225, 300 and 450 °C for Ni-based compounds, $CeCo_2$ and $ErFe_2$, respectively [20]. During the reaction, small particles of the transition metal supported by the corresponding rare earth metal oxide are formed, resulting in an activation period of the materials. Decomposition into nickel supported by the corresponding rare earth oxide also occurs for CeNiAl and compounds of the series *RE*Ni_x (*RE* = La or Ce, $x = 2$ or 5) [22], $CeNi_2$ [23] and Ce_2Ni_{17} [24], while $CeCo_5$ [24] and $LaCo_5$ [24] decompose into supported cobalt particles. Accordingly, the intermetallic compounds *RE*Cu_2 (*RE* = La, Ce, Pr or Ho) [14] partially oxidize into Cu/RE_2O_3. These materials, therefore, show the typical catalysis pattern of the transition metals. Subjecting $LaNi_4Fe$ [25] or $LaNi_4Cr$ [26] to syngas conversion conditions results in supported Fe-Ni and Cr-Ni particles, respectively, where Fe-Ni shows a higher selectivity toward C2, C3 and C4 hydrocarbons compared to nickel [25]. Intermetallic compounds can also be used as precursors by oxidizing them before catalysis. This has been applied to the ternary series $CeNi_{5-x}Co_x$ ($x = 1–5$) which transforms into $Ni_{5-x}Co_x$ particles supported by CeO_2 [27]. Subsequent catalytic testing in the syngas conversion revealed that the use of $CeNi_5$ as precursor resulted in the highest activity, while $CeCo_5$ showed highest selectivity to methane. Besides crystalline intermetallic compounds, the metallic glass $Ni_{78}P_{19}La_3$ has also been tested for its methanation properties [28]. The glassy state did not prevent the oxidation of the compound under reaction conditions, thus resulting in a catalytic behavior similar to nickel.

Among the many intermetallic compounds tested for the conversion of syngas, only $CeAl_2$ [23], $LaNi_4Cu$ [26] and $LaNi_4Al$ [26] seem to be stable. While the latter two were tested to 250 °C, $CeAl_2$ does not decompose even at 700 °C. Thus, these few cases are the only ones where the catalytic properties are intrinsic to the intermetallic compound. While the compounds are very stable against sintering, they show only moderate activity for methanation.

4.7.2.4 Hydrogenation and dehydrogenation

Driven by the hydride-forming abilities, many intermetallic compounds of the rare earths were tested as catalysts for the hydrogenation of unsaturated compounds such as olefins, and also as dehydrogenation catalysts. The idea behind the latter is that the intermetallic compound extracts the hydrogen from the saturated compound, forms the hydride and releases the hydrogen later as molecular hydrogen. Many rare-earth-based materials are catalytically active in (de)hydrogenation reaction due to hydride formation (Table 4.7.1). Unfortunately, only very few studies report activities in relation to commonly used (de)hydrogenation catalysts, thus hindering comparison.

In addition to Table 4.7.1, $LaNi_5$ has been successfully applied to hydrogenate 1-hexene [46], 1-octene [47] and 1-undecene [48] and compounds with the composition $REPd_3$ (RE = La [49], Ce [50], Pr [50], Nd [50] or Sm [50]) to hydrogenate 1-butene. Other functional groups can also be fully hydrogenated by hydride-forming intermetallic compounds: Nitrobenzene is transformed effectively to aniline by hydrogen using $LaNi_5$ [51], $LaNi_{5-x}Cu_x$ [52] or $La_{1-x}Ce_xNi_5$ [52] as catalysts, while $LaNi_3$ and $LaNi_5$ [53] have been tested successfully for the hydrogenation of p-benzoquinone to hydroquinone. $LaNi_5$ is also able to liquefy coal using hydrogen at a pressure of 48.3 bar and temperature of 316 °C [54] and can exchange hydrogen atoms in methane [55]. Enantioselective hydrogenation of ethylacetate [56] and ethylacetoacetate [44, 57] has been attempted by modification of $LaNi_{5-x}M_x$ (M = Al, V, Cr, Mn, Co or Cu), $LaNi_{5-x-y}Co_xM_y$ (M = Cr or Mn) and $Sm_{0.5}Gd_{0.5}Ni_{4.5}Mn_{0.5}$ with R,

Table 4.7.1: Summary of hydrogenation of ethylene and propene on rare-earth-based intermetallic compounds.

Olefin/IMC	Ethylene	Propene
$RECo_2$		La [29], Sm [29]
$RECo_5$	La [30] ([31]*), Ce [32], Pr [30], Sm [30] ([31]*)	La [33]
$RECo_3$	La [34], Ce [34], Pr [34], Gd [34], Tb [34], Dy [34], Ho [34], Er [34]	
$RECu_{6.5}$	Yb [35]	
$RENi_2$	Ce [36]	La [33], Sm [29]
$RENi_3$	La [34], Ce [34], Pr [34], Gd [34], Tb [34], Dy [34], Ho [34], Er [34]	La [33]
$RENi_5$	La [37] ([31]* [38]#), Ce ([38]#), Pr [39] ([38]#)	La [33], $La_{1-x}Pr_x$ [40], Ce[41], Pr [42], Nd [41], Sm [41], Gd [41]
$RENi_{5-x}Al_x$		La [43, 44]
$RENi_{5-x}Co_x$		La [43, 44]
$RENi_{5-x-y}Co_xCr_y$		La [44]
$RENi_{5-x-y}Co_xMn_y$		La [44]
$RENi_{5-x}Cr_x$		La [44]
$RENi_{5-x}Cu_x$		La [44, 45]
$RENi_{5-x}Fe_x$		La [43]
$RENi_{5-x}Mn_x$		La [43, 44], $Sm_{0.5}Gd_{0.5}$ [44]
RE_2Ni_7		La [33]

* Raney-type catalysts obtained by leaching with 1,2-diiodoethane;
# supported catalysts obtained by pre-oxidation of the compounds

R-(+) tartaric acid. Highest enantiomeric excess was obtained for the hydrogenation of ethylacetoacetate over $LaNi_4Co$ (26.6%) [44].

Catalytically more challenging than the above-summarized full hydrogenation reactions is the selective partial hydrogenation of triple bonds or chemically different double bonds. $REPd_3$ (RE = La, Ce, Pr, Nd or Sm) show low activities for the 1-butyne hydrogenation but high selectivity for the semi-hydrogenation, at high conversion [50]. In the semi-hydrogenation of 1,3-butadiene, $LaCo_5$ results in a selectivity toward butenes of 86% and experiments using $LaCo_5D_x$ show that hydrogenation occurs via 1,2- and 1,4-addition [58]. Also, the compounds PrNi, GdNi and TmNi are selective for the semihydrogenation of isoprene and give high selectivities to 2-methyl-2-butene and 2-methyl-1-butene [59]. For the selective hydrogenation of crotonaldehyde to crotyl alcohol, selective catalysts for the hydrogenation of the C=O functionality are needed. Using Pt/CeO_2 as a catalyst in this reaction results in max. 35% selectivity to crotyl alcohol [60]. If the material is strongly reduced at 700 °C, $CePt_5$ is formed by the RMSI effect [5] and the selectivity toward crotyl alcohol increases to 83%.

The hydride formation ability of many of the rare-earth-based intermetallic compounds is exploited for dehydrogenation reactions. The idea is that the hydrogen is taken up by the intermetallic compound by which the equilibrium is shifted to the product side. This approach has been tested for the decomposition of methanol to CO and hydrogen which resulted in a conversion rate of 13.7 mmol $\times$ min^{-1} $\times$ m^{-2}, a CO-selectivity of 98.5 and 1.5% formaldehyde over $LaNi_{4.7}Al_{0.3}$ [61]. Pr_2Co_7, Nd_2Co_7, Sm_2Co_7, $DyFe_2$ as well as $ErFe_2$ are much less active in the methanol decomposition [62]. 2-Propanol is dehydrogenated into acetone over Nd_2Co_7 [63], Sm_2Co_7 [63] and Pr_2Co_7 [64] with nearly 100% selectivity at 44 °C. For the dehydrogenation of cyclohexene to benzene, different hydrogen storage intermetallics have been tested and it was found that $CaNi_5$ is superior to $LaNi_{4.7}Al_{0.3}$ [65]. Deep dehydrogenation of acetylene [66] or acetylene/ammonia mixtures [67] over $LaNi_5$ resulted in carbon nanotubes (CNTs) and nitrogen-doped CNTs. Finally, pre-oxidized PrNi, GdNi and LaNi have been applied to the partial oxidation of methane to hydrogen and carbon monoxide [68].

4.7.2.5 Electrocatalysis

In the last few years, the interest in electrode materials with reduced noble metal content has driven a large number of studies in the field and a recent review on intermetallic compounds in electrocatalysis is available [69]. First activities concerning the use of rare-earth-based intermetallic compounds as electrodes date back to the 1960s and 1970s, where $RENi_5$ (RE = Mm, La, Ce, Pr, Nd, Sm, Eu, Gd, Tb, Dy, Ho, Er, Tm, Yb or Lu) [3] and LaB_6 [70] were investigated as fuel cell electrodes to oxidize hydrogen, hydrazine, ammonia, alcohols, esters, carboxylates, hydrocarbons or carbonyl compounds. This development even included electrodes for borohydride fuel cells (using $NaBH_4$ or KBH_4 as fuel) and $LaNi_{4.5}Al_{0.5}$ [71], PtSm [72], Pt_4Sm_3 [72], PtHo [72], PtCe [72] or Pt_4Ce_3 [72] as electrode. Since the discovery of the advantageous electrocatalytic properties of dealloyed intermetallic compounds by Strasser [73], that is, the formation of a Pt-layer on an underlying intermetallic compound to tune the electronic structure, much research has been driven to identify new and better materials, especially for the hydrogen evolution reaction (HER) and the oxygen reduction reaction (ORR). For the first reaction, a large number of hydride-forming rare-earth-based intermetallic compounds have been tested: $RENi_5$ (RE = Mm [74], La [74, 75] or Ce [76]), $LaNi_{5-x}M$ (M = Al [77, 78], Co [75, 77, 79], Cu [77, 78], Mn [80] or Si [78, 81]), $CeNi_{3-x}Co_x$ [78], $LaFe_5$ [78], $GdNi_4Al$ [82], DyPt

[83], HoPt [83], $Fe_{17}RE_2$ (RE = Ce, Mm) [84], Fe_2Ce [84], $Fe_{12}Sm_2$ [84], Fe_3Sm [84], $MmNi_{3.4}Co_{0.8}Al_{0.8}$ [85] or $MmNi_{3.6}Co_{0.75}Mn_{0.4}Al_{0.27}$ [81]. Unfortunately, in many cases the embrittlement of the electrodes during the cyclic formation and decomposition of the hydride hinders their application [86]. This is not the case for platinum containing intermetallic compounds where the rare earth metal is leached from the first couple of layers. This leaves a strained platinum shell behind which is electronically modified. This shell hinders diffusion of the hydrogen, thus suppressing the hydride formation. Materials tested for the HER comprise PtSm, Pt_4Sm_3, HoPt, PtCe, Pt_4Ce_3 and Ce_3Pt [87], but a real breakthrough could be achieved in the ORR, where studies on Pt_5RE (RE = La [88], Ce [89], Sm [89], Gd [90], Tb [89], Dy [89] and Tm [89]) made full use of the potential of the materials in terms of electronic tuning and corrosion resistance due to the platinum shell (Figure 4.7.3). Besides the HER and ORR, $LaNi_5$ [91], $MmNi_5$ [91], $LaCr_5$ [91], $LaFe_5$ [91], $LaCo_5$ [91], $CeNi_3$ [91], $CeNi_{3-x}Co_x$ [91], $CeCo_3$ [91], $La_{0.8}Nd_{0.2}Ni_{2.5}\,Co_{2.4}Si_{0.1}$ [91] and $LaNi_{5-x}Cu_x$ [92] have been employed to convert nitrobenzene to aniline and $LaNi_5$ [93], to hydrogenate methyl vinyl ketone, 1-decene and 5-hexen-2-one electrochemically.

4.7.2.6 Additional reactions

Besides the reports above, intermetallic compounds based on rare earth metals have been tested as catalysts for the diacetoxylation of 1,3-butadiene (PdTeLa and PdTeCe [94]), methane dry reforming ($LaNi_5$, which shows very low activity [95]) and the hydrodesulfuration of thiophene ($NdNi_5$, decomposition into Ni_3S_2/Nd_2O_2S [96]), none of which revealed an advantage over known catalysts.

4.7.3 Oxides

However, the usage of rare earth elements is not restricted to intermetallic compounds or supports for active metal components. As they show catalytic activity also in pure form, they have additionally been tested as promoters in the form of dopants of classical supports as for instance, alumina, silica or titania or even impregnated onto the surface of these oxidic support materials as an active phase. Especially, rare-earth-oxide-supported metallic catalysts quite commonly reveal strong metal-support interactions (SMSI) so that the role of the rare earth elements is not restricted to a passive carrier for the active phase, but contributes significantly to the catalytic activity of the whole system.

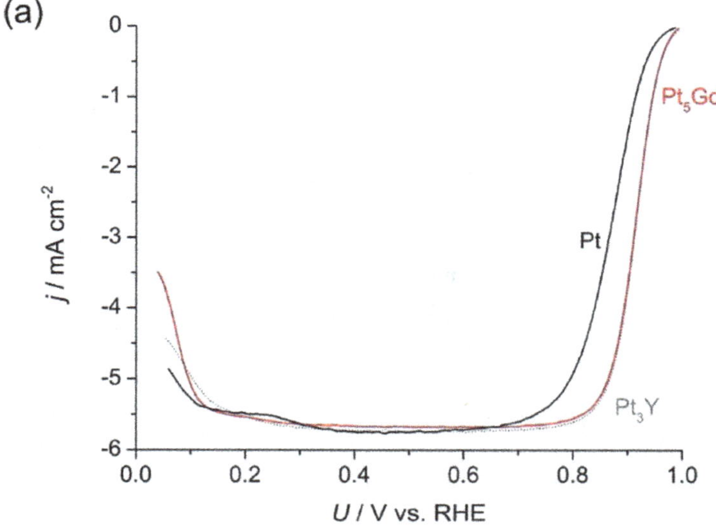

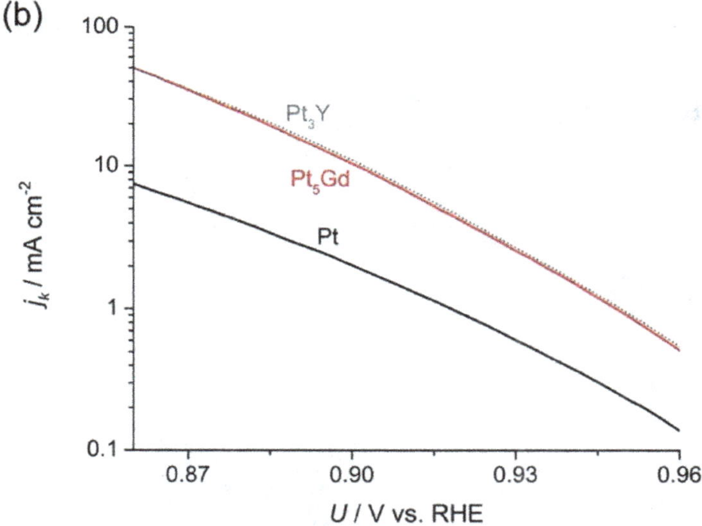

Figure 4.7.3: (a) RRDE polarization curves for the ORR on Pt_5Gd (red curve), Pt (black curve), and Pt_3Y (dotted gray curve) and (b) Tafel plots showing the kinetic current density (j_k) of Pt_5Gd, Pt and Pt_3Y as a function of the potential (U). Reprinted with permission from [90]. Copyright 2012 American Chemical Society.

4.7.3.1 NO decomposition

Relating to the high NO adsorption ability of rare earth oxides, they have been tested by Tsujimoto et al. as direct NO decomposing catalysts for internal combustion engines in automobile applications [97, 98]. Commonly used as NO_x abating catalysts are Cu ion-exchanged zeolites which have a high low-temperature as well as the currently highest mid-temperature activity in the selective catalytic reduction by ammonia (NH_3-SCR) process. However, they lack sufficient hydrothermal stability and deactivate irreversibly above 600 °C. This makes rare earth oxides interesting candidates as alternative catalysts. In particular, a C-type cubic oxide catalyst of the formula $(Y_{0.69}Tb_{0.30}Ba_{0.01})_2O_{2.99+\delta}$ has been found to decompose NO into N_2 and O_2 to 100% at 900 °C [98]. Later, more systematic investigations have been performed by the same group [97]. Rare earth sesquioxides are known to crystallize at temperatures below ca. 2,000 °C and at ambient pressure, in three polymorphic forms, A-type trigonal, B-type monoclinic and C-type cubic, whereas above this temperature, X and H forms are reported [99]. The C-type cubic or bixbyite-type structure derives from the fluorite-type structure by a doubling of unit cell length in all three space directions with one-fourth of the oxygen vacancies regularly ordered in the sesquioxides as shown in Figure 4.7.4. These vacancies, together with basic surface sites, seem to

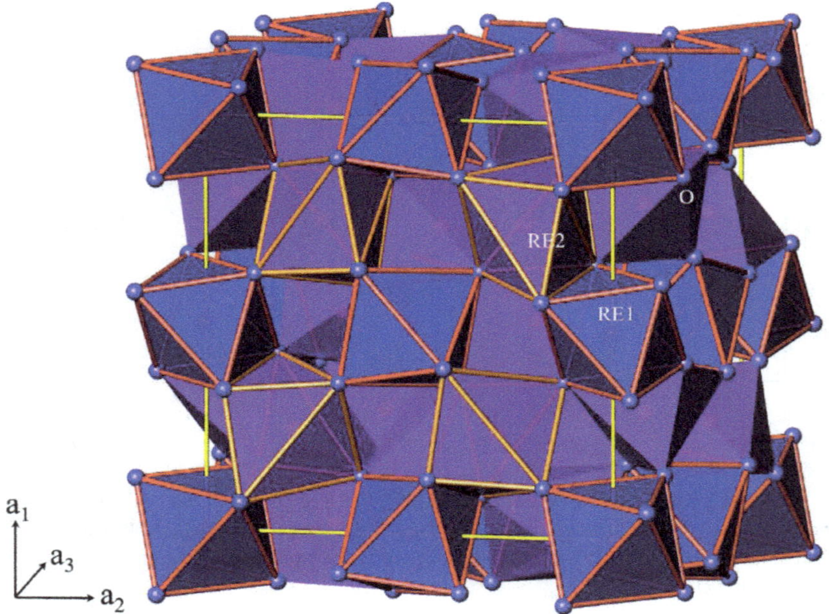

Figure 4.7.4: Crystal structure of C-type cubic rare earth sesquioxides derived from the fluorite-type structure by oxygen vacancy ordering with two crystallographically different rare earth ions (space group $Ia\bar{3}$).

play an important role in the NO decomposition activity of the rare earth oxides crystallizing in this structure type (*RE* = Y, Eu, Gd, Ho, Er, Tm, Yb or Lu). The main drawback for a commercial application of these catalysts might be the high reaction temperature of 900 °C.

4.7.3.2 Reactivity spectrum

Rare earth oxides have also been tested for several different other applications in catalytic reactions, especially dehydrations. For example, the vapor-phase catalytic dehydration of 5-amino-1-pentanol over various rare earth and non-rare-earth oxides has been investigated [100]. While acidic non-rare-earth oxides as Al_2O_3, SiO_2, SiO_2-Al_2O_3, TiO_2 and ZrO_2 resulted in cyclic amines such as piperidine as a major product at temperatures of ≥300 °C, basic rare earth oxides as Tm_2O_3, Yb_2O_3 and Lu_2O_3 with bixbyite-type structure revealed high conversion as well as selectivity of more than 90% to 4-penten-1-amine at 425 °C. Also, the dehydration of 2,3-butanediol and 1,4-butanediol to 1,3-butadiene (one of the most important chemicals for polymer syntheses of styrene-butadiene rubber (SBR), polybutadiene rubber (BR), acrylonitrile-butadiene-styrene resins and adiponitrile) over rare earth oxides have been investigated [101, 102]. In both publications, intermediates and by-products such as 2- or 3-buten-1-ol, butanone, 2-methylpropanal and 2-methyl -1-propanol are described and the various rare earth oxides revealed different selectivity for all these products. Wang et al. stated that Yb_2O_3 inhibits major side reactions, such as the decomposition of 3-buten-1-ol to propene, and provides selective production of 1,3-butadiene from 1,4,butanediol [102]. The transesterification of fats and oils with methanol to biodiesel and glycerol has also been investigated in the presence of various pure rare earth oxides, supported rare earth oxides as well as stoichiometric rare earth mixed oxides of pyrochlore structure (see later) [103]. While the authors found an exceptionally high activity for La_2O_3 among the pure rare earth oxides, rare earth oxides on oxidic supports resulted in the formation of surface mixed oxides, which revealed perovskite and garnet crystal structures and a whole spectrum of different FAME yields in the range of 45-70% at 200 °C.

4.7.3.3 Rare earth oxides as support

The scientific literature on rare earth oxides as supports with active phases on their surfaces formed by impregnation with metal salt solutions and subsequent reduction or as solid solutions prepared by co-precipitation is much more widespread so that only selected examples can be mentioned here. Among Ni catalysts of different loading supported by various rare earth oxides prepared by impregnation, Ni on

Y_2O_3 with an optimal Ni loading of 40 wt% and a high Ni dispersion showed the best performance for ammonia decomposition in terms of catalytic activity for on-site generation of H_2 [104]. Rare earth oxides as supports for precious metals such as Pt have also been tested for lean NO_x reduction by hydrogen in the SCR process [105]. Among the supports tested, CeO_2, Pr_6O_{11}, Eu_2O_3, and Gd_2O_3, Pt/CeO_2 provided good NO conversion of 80% with roughly 40% N_2 selectivity. Ceria plays a special role among rare earth oxides and will be discussed in more detail below. Solid solution formation between SnO_2-based catalysts modified by La, Ce, and Y with a Sn/*RE* atomic ratio of 2:1 prepared by co-precipitation was observed only for the Ce-modified sample. It showed improved catalytic activity for both CO and CH_4 oxidation, compared to the pure support SnO_2 [106]. The Ce-modified sample also showed good reaction durability and thermal stability, all of this was attributed by the authors to the special properties of cerium.

The role of ceria among the rare earth oxides is particularly worth mentioning. Cerium, with its [Xe] $4f^2$ $6s^2$ electron configuration, forms as its most prominent oxide CeO_2 with a cubic fluorite-type crystal structure, but also Ce_2O_3 and other oxides in the compositional range between these two boundary phases are described in the literature [107, 108]. The CeO_{2-x} phase has an extended homogeneity range toward lower oxygen contents, increasing with temperatures from $x = 0.154$ at 449 °C to $x \approx 0.3$ at 1,327 °C and a miscibility gap between [108]. The phase width of ceria with oxygen vacancies induces a high reducibility of the compound and the so-called oxygen storage capacity (OSC) [109, 110] responsible for the high catalytic activity of ceria in partial oxidations, according to a Mars-van-Krevelen [111] mechanism. Charge neutrality in CeO_{2-x} is achieved by Ce^{3+} ions at the Ce^{4+} sites with $4f^1$ instead of $4f^0$ valence electron configuration. In X-ray photoemission spectra (XPS) of the Ce 3d states, final state effects are observed resulting in a spin-orbit splitting into $3d_{3/2}$ and $3d_{5/2}$, all in all, in up to 10 discrete peaks (Ce^{3+}: $v_0 + v' + u_0 + u'$; Ce^{4+}: $v + v'' + v''' + u + u'' + u'''$) [112] in the binding energy range of 880–920 eV [113]. In Figure 4.7.5, a Ce 3d photoemission spectrum of a CeO_2 (111) film deposited at 250 °C on Cu(111) with a composition close to CeO_2 is shown which reveals the final states f^0, f^1 and f^2. Doping of the ceria film with Au resulted in a 4f resonance enhancement of the Ce^{3+} species. In a detailed study which also includes XPS measurements on different lanthanide-doped cerium oxides $Ce_{0.5}RE_{0.5}O_{1.75}$ (*RE* = Gd, La, Pr, Nd, Sm) loaded with 20 wt% Cu as catalysts for the oxidation of ethyl acetate, a common volatile organic compound (VOC), resulted in an activity sequence $CeO_2 \approx Ce_{0.5}Pr_{0.5}O_{1.75} > Ce_{0.5}Sm_{0.5}O_{1.75} > Ce_{0.5}Gd_{0.5}O_{1.75} > Ce_{0.5}Nd_{0.5}O_{1.75} > Ce_{0.5}La_{0.5}O_{1.75}$ and Cu addition improving the catalytic performance [114].

High-quality, single-crystalline films of cerium and praseodymium oxide grown on Si(111) substrates were used as well-defined catalytic model systems to study, among other properties, the oxygen storage and release capability in detail [115]. Ions such as Zr^{4+} substituted into the ceria structure forming a range of solid solutions are known to increase the OSC due to their smaller ionic radius and, as a

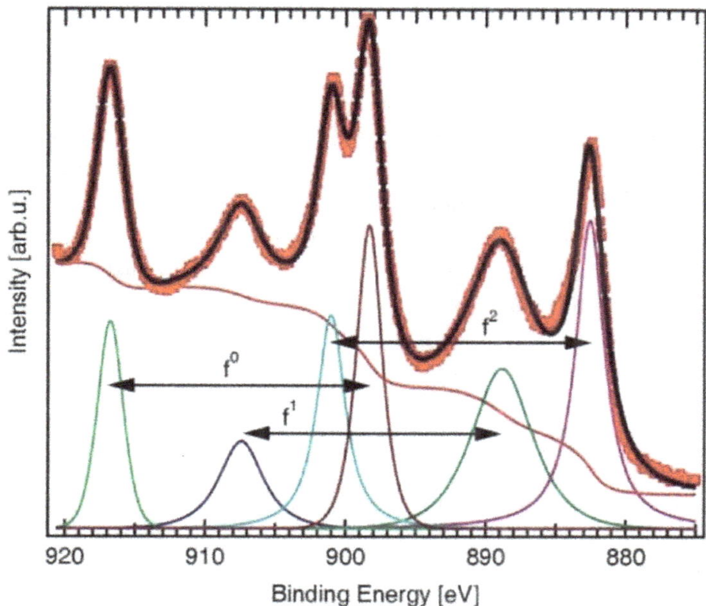

Figure 4.7.5: Ce 3*d* photoemission spectrum of a CeO_2 film epitaxially grown onto Cu(111) at 250 °C revealing the Ce^{4+} components $v + v'' + v''' + u + u'' + u'''$; photon energy 1486.6 eV. Figure reproduced from reference [113] with permission from Elsevier.

consequence thereof, change of cation coordination from CN = 8 (cubic) to CN = 7 (tetragonal, monoclinic). But thermogravimetric studies on $Ce_{0.6}Zr_{0.4-x}Nd_{1.3x}O_2$ ($0 \leq x \leq 0.4$) revealed that the OSC of $Ce_{0.6}Zr_{0.4}O_2$ decreases after substitution of Nd^{3+} due to the larger ionic radius of Nd^{3+} compared to Zr^{4+} [116]. Lattice oxygen ions also play an important role in the oxidative dehydrogenation (ODH) of organic molecules such as ethylbenzene (EB) to styrene, resulting in the formation of water as joint product. A variety of rare-earth-oxide-doped ceria on γ-Al_2O_3 catalysts prepared by a wet impregnation method were tested for ODH of EB using CO_2 as soft oxidant, in vapor-phase atmospheric pressure conditions. The observation of 3 wt% Er_2O_3, Pr_2O_3, or Nd_2O_3 incorporation into 15 wt% CeO_2/γ-Al_2O_3 resulting in unusual catalytic properties have been explained on the basis of active metal-support synergistic interactions involving a superior mobile oxygen storage capacity and a noticeable surface acidic nature [117]. Especially, in the case of doping with reducible ions such as $Pr^{3+/4+}$ together with an increased meso/micro-pore volume and stabilization of external surfaces gave for rare earth modified ceria Ce*RE*O_x (*RE* = La, Pr, Sm, Y) superior soot oxidation activity of $CePrO_x$ compared with $CeSmO_x$, $CeYO_x$, and CeO_2 [118, 119]. Soot oxidation catalysts are of high importance for diesel particle filters (DPF) in automobile applications. For CO instead of soot oxidation, copper-substituted ceria nanoparticles $Cu_{0.1}Ce_{0.9}O_{2-x}$ were also tested and characterized by means of *in-situ* spectroscopic techniques such as X-ray absorption spectroscopy (XAS), atmospheric pressure XPS (AP-XPS) and diffuse reflectance infrared Fourier transform spectroscopy (DRIFTS) [120]. These methods show that CO can be oxidized to CO_3^{2-} even in the absence of O_2 due the OSC of the catalyst, but CO_3^{2-} was found to desorb only under oxygen-rich conditions when the oxygen vacancy is refilled by dissociative adsorption of O_2. Together with DFT calculations, the spectroscopic investigations were the basis for the introduction of the computed oxygen vacancy formation energy as an activity descriptor for substituted ceria materials to successfully rationalize observed activity trends in metal substitutes $M_{0.1}Ce_{0.9}O_2$ with M = Mn, Fe, Co, Ni that span three orders of magnitude.

4.7.3.4 Catalysts with special morphologies

Ceria nanoparticles with a special morphology have been prepared in the form of multi-shelled hollow spheres by a facile coordination polymer (CP) precursor method using 2,5-pyridinedicarboxylic acid [121]. These ceria multi-shelled nanospheres displayed a good photocatalytic activity in the degradation of Rhodamine B under UV light at room temperature, whereas Au nanoparticle loaded multi-shelled ceria nanocomposites were reported to show an excellent catalytic activity for the reduction of *p*-nitrophenol with $NaBH_4$ as reductant. A very prominent problem in heterogeneous catalysis is the deactivation of nanoparticles during the catalytic reactions due to thermally activated sintering phenomena, resulting in a particle size increase and adsorption surface loss. A solution for this problem is to encapsulate nanoparticles into porous nanospheres. Due to their good thermostability and mechanical strength, porous hollow silica nanospheres (PHSNs) encapsulating rare earth oxide nanoparticles were considered for application in catalysis. Xu et al. reported a one-pot synthesis of a series of M_xO_y@PHSNs (M = La, Ce, Eu, Gd) employing charge-driven micelles as templates, which consist of ion pairs of positively charged block copolymer and negatively charged complex chains of multidendate ligand bound rare earth ions [122]. After calcination at 400 °C with removal of the organic precursors, HRTEM images showed the formation of rare earth oxide nanoparticles with porous silica shells, as depicted in Figure 4.7.6. Experiments to degrade methylene blue and new coccine with H_2O_2 over M_xO_y@PHSNs revealed that encapsulated ceria showed the best catalytic properties.

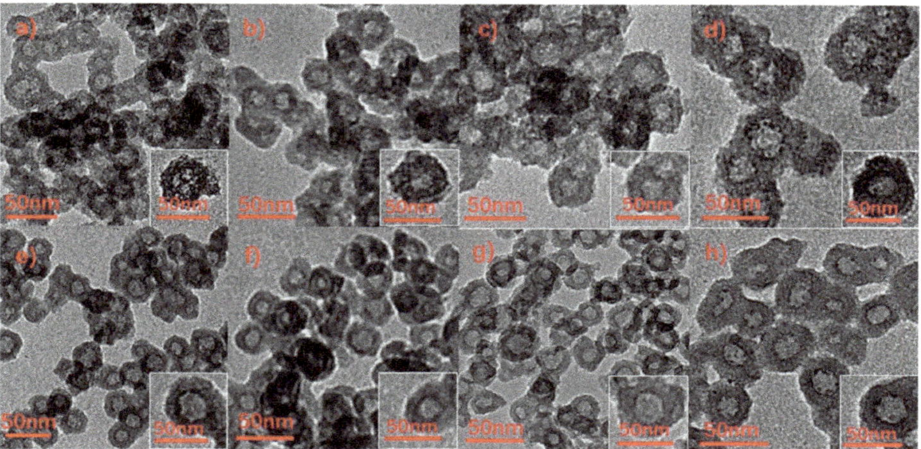

Figure 4.7.6: TEM images showing (a) Eu_xO_y@PHSNs (A); (b) Gd_xO_y@PHSNs (A); (c) La_xO_y@PHSNs (A); (d) Ce_xO_y@PHSNs (A); (e) Eu_xO_y@PHSNs; (f) Gd_xOy@PHSNs (B); (g) La_xO_y@PHSNs (B); (h) Ce_xO_y@PHSNs (B). Insets showing their corresponding HRTEM images. A and B referring to acidic and basic synthesis conditions. Scale bars 50 nm. Reprinted with permission from [123]. Copyright 2019 American Chemical Society.

4.7.3.5 Photocatalysis

Currently, TiO_2 (titania) is the most widely studied photocatalyst among various metal oxide semiconductors for environmental remediation due to a unique photocatalytic activity, low cost, nontoxicity and high stability. But for an efficient use of solar energy as power source, titania reveals one severe obstacle to its effective utilization, the low absorption of less than 5% of the sunlight in the visible range. This obstacle has launched research for oxide semiconductors absorbing in the visible light range of the solar spectrum. Ji et al. studied the adsorption and degradation of the nonbiodegradable azodye acid orange 7 (AO7) on the surface of CeO_2, under visible light irradiation and found an AO7 photodegradation rate by this rare earth oxide to be much faster than that observed for commercial titania (P25 from Evonik) [124]. The authors proposed a possible degradation pathway for this photocatalytic process to determine intermediates and attributed the enhanced photoactivity of ceria to the superior adsorption capacity and special 4f electron configuration. Other examples for potential photocatalysts with visible light activity are doped bismuth-based oxides including bismuth vanadates, molybdates and tungstates [125–127]. Among these Bi-based photocatalysts, Bi_2MoO_6 has attracted enormous interest since it exhibits outstanding visible light photocatalytic activity due to its special geometrical and electronic properties represented by a unique layered structure [128]. But its low efficiency in separating photogenerated electron–hole pairs had to be improved. Mu et al. synthesized a Bi_2MoO_6 catalysts with rare earth dopants by a hydrothermal method resulting with Nd and Sm as dopants in phase mixtures of Bi_2MoO_6 and $Bi_{3.64}Mo_{0.36}O_{6.55}$ [129]. Photodegradation experiments of Rhodamine B with a Xe lamp indicated that rare earth doping improved the photocatalytic activity significantly.

4.7.3.6 Oxidative coupling of methane

Furthermore, in a special reaction considered as a "dream reaction" in catalysis, the oxidative coupling of methane (OCM) to higher hydrocarbons (HC), the oxides or stoichiometric compounds of rare earth elements as, for instance, pyrochlores of the form $RE_2B_2O_7$ (RE = rare earth, B = Ti, Zr, Sn) found entry into research. Research on this topic intensified in the 1980s [130] and 1990s [131, 132]. It was brought into focus recently by the group of Wang et al. [133, 134] and others [135–138]. After identifying $La_2Ce_2O_7$, with defective cubic fluorite-type structure, as promising catalyst for OCM with a C_2 HC yield of 15.3% at 650 °C, a temperature at which the reference catalyst $Mn/Na_2WO_4/SiO_2$ exhibits only 1% C_2 yield [134], their investigations on $RE_2Ce_2O_7$ extended to other rare earth pyrochlores $RE_2Zr_2O_7$ [133]. By powder X-ray diffraction (PXRD) and Raman investigations, they proved that with decreasing RE/Zr radii ratio, the crystal structure changes from an ordered pyrochlore (RE = La) to a less ordered pyrochlore (RE = Pr, Sm) and finally to a defective cubic fluorite-type phase (RE = Y). H_2-TPR, O_2-TPD and XPS measurement data suggested that the amount of surface-active O_2^- species is consistent with the reaction performance, while CO_2-TPD demonstrated the amount of moderate basic sites to be another factor to affect the catalyst performance. The best catalyst in this study, $La_2Zr_2O_7$, showed an improved reaction performance at temperatures below 750 °C compared to the actual state-of-the-art catalyst $Mn/Na_2WO_4/SiO_2$.

4.7.3.7 Fluid catalytic cracking

One of the most important fields of the application of rare earth elements in catalysis is fluid catalytic cracking (FCC), a major conversion technology in crude oil refinery. Currently, using the FCC process, the majority of petrol for the transport sector as well as an important fraction of propene for the polymer industry is produced. In the catalytic cracking process, streams from the vacuum distillation section of the oil refinery are fed into a riser-regenerator reactor unit due to the very high coking rate of the catalyst. In the fluidized bed reactor part, the HCs fed are cracked in an endothermic process via a carbenium-ion-based mechanism, including hydrogen transfer reactions, at temperatures of 493–554 °C into branched lower molecular weight HCs, preferentially of high research octane number (RON). The acidic sites for this process are provided by ion-exchanged zeolites as heterogeneous catalysts, currently in the main rare-earth-stabilized zeolite Y (IUPAC structure code FAU [139]) as well as zeolite ZSM-5 (IUPAC structure code MFI [139]), the latter targeting an increased propene yield. The spent catalyst from the riser is reactivated after residence times in the range of a few seconds in the regenerator at higher temperatures of 650–760 °C through the burning of a small portion of the feedstock with an excess of oxygen and brought back after coke burn-off into the riser. A typical FCC catalyst particle has an average time on stream (TOS) of about 1 month. This, and much more information, can be taken from a very detailed and critical review provided by Vogt and Weckhuysen [140]. Note that reviews on this topic have also been provided by other authors [141–143]. The amount and strength of Brønsted as well as Lewis acid sites of a zeolite are dependent on the structure type as well as on the Si/Al ratio (SAR). The latter can be increased by controlled steaming and washing/leaching cycles, increasing the zeolite stability (ultra stable Y zeolite, USY) and decreasing the number of acidic sites. By ion exchange, for instance, via ammonium ions and subsequent decomposition into ammonia and protons, zeolites can be converted into the acid or H-form, that is, HY or HZSM-5. For an improvement of the effectiveness of the zeolite for the FCC process and for preservation for a longer activity, rare earth ions are commonly ion-exchanged as counter-ions, to a certain fraction. As cationic acids, hydrated rare earth ions $RE(H_2O)_6{}^{3+}$ provide additional Brønsted acid sites. This has been proven by Rabo et al. through assignment of IR vibrations observed for zeolite Y [144]. For rare-earth-exchanged zeolite Y, two IR bands related to hydroxyl groups were described. One band was at 3640 cm^{-1}. This was attributed to bonding with water, benzene and ammonia to Brønsted acid sites exposed within the supercage. The second band was at 3524 cm^{-1}. This does not bind to ammonia and benzene and thus was attributed to an OH-group hidden inside the sodalite cage. The crystal structure of zeolite Y in the form of Si/Al connectivity is shown in Figure 4.7.7. It depicts the secondary building units (SBUs) of truncated octahedron (sodalite cage, dark blue) and hexagonal prism (light blue) thus generating a 3-D pore system, in which pores of a diameter of ca. 7.3 Å are connecting larger cages of 13 Å diameter called supercages (in the center of Figure 4.7.7).

Additionally, the most relevant ion-exchange sites are highlighted as small spheres (U, S-I, S-I′, S-II, S-II′). The exchange of cations in zeolite Y by rare earth ions causes the exothermic peak in differential thermal analysis (DTA), representing the collapse of the framework, to shift toward higher temperature of 800–1,000 °C and thus outside the temperature range relevant for FCC [145]. The investigation of the dealumination, that is, SAR increase of zeolite Y with varying loading of rare earth oxides by IR, PXRD and ^{29}Si MAS NMR, revealed that the unit cell size is not a good indicator for structural stabilization [146]. Instead, it is assumed that RE^{3+} ions in ion exchange positions induce local polarization of the Al atoms in the structure, causing long-range effects in the form of T-O-T angle (T = Si, Al) extensions and thus increases in unit cell sizes [147]. The authors claim that the origin of the enhanced catalytic activity found for the active form of USY and *RE*-Y for acid-catalyzed reactions is identical. All in all, it seems quite clear that rare earth cations

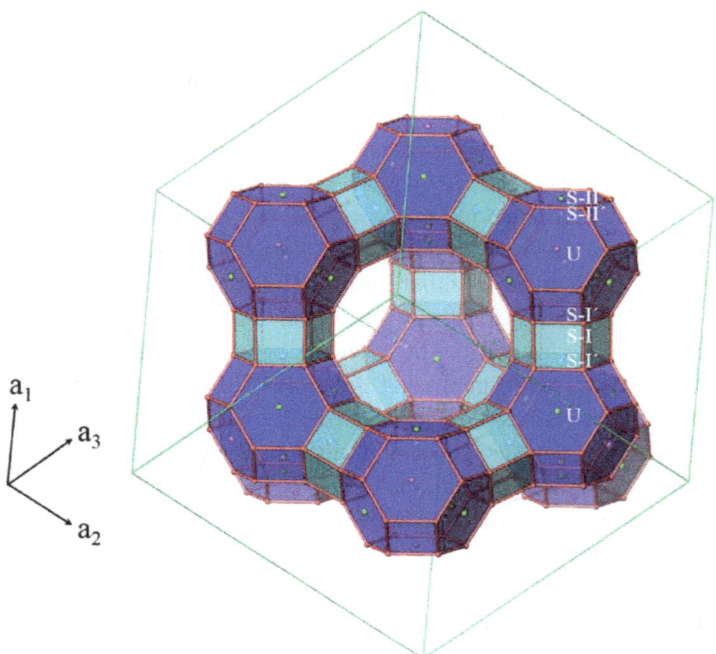

Figure 4.7.7: Framework of zeolite Y (cubic faujasite, FAU) depicted as Si/Al connectivity with structure building units (red lines) and rare earth ion exchange sites U, S-I, S-I´, S-II and S-II´ (small spheres).

in zeolite structures provide some form of stabilization by retaining more Al in the structure as observed by IR and NMR. The rare earth ions stabilize the zeolites by occupying hexagonal prism sites (S-I) as well as supercage sites (S-II), when incorporated in excess, and form strong Brønsted acid sites in connection with framework Al, thus increasing catalytic activity dramatically [140]. With the investigation on this topic already starting in the 1970s and 80s, many groups tried to improve the FCC catalyst materials continuously over the subsequent decades. Among the abundance of papers dealing with this topic, only a recent one investigating the impact of rare earth concentration and matrix modification in FCC catalysts on their performance in a wide array of operational parameters from the view of an industrial application is cited here [148].

4.7.4 Conclusions

Today, rare earth elements play an indispensable role in heterogeneous catalysis. Only a very limited selection of examples could be presented in this chapter. However, many others would have been worth mentioning here. For the interested reader, several reviews worth reading on the catalytic properties of rare-earth-based intermetallic compounds or their decomposition products are available [13, 149–155], as well as of metallic glasses [151, 156].

References

[1] M. Raney, Method of Preparing Catalytic Material, US 1563587, 1925.
[2] M. Armbrüster, Sci. Technol. Adv. Mater. 2020, 21, 303.
[3] L. R. Dilworth, W. J. Wunderlin, Fuel Cell and Fuel Cell Electrode Containing Nickel-Rare Earth Intermetallic Catalysts, US3405008, 1968.
[4] G. V. Samsonov, The Catalytic Properties of Refractory Compounds, and Principles of the Creation of Refractory Compounds with Prescribed Catalytic Properties, Jerusalem, 1968.
[5] S. Penner, M. Armbrüster, ChemCatChem 2015, 7, 374.
[6] T. Takeshita, W. E. Wallace, R. S. Craig, J. Catal. 1976, 44, 236.
[7] A. P. Walker, T. Rayment, R. M. Lambert, J. Catal. 1989, 117, 102.
[8] A. P. Walker, T. Rayment, R. M. Lambert, R. J. Oldman, J. Catal. 1990, 125, 67.
[9] H. Uchida, K. Ishikawa, T. Suzuki, T. Inoue, H. H. Uchida, J. Alloys Compd. 1995, 222, 153.
[10] M. Itoh, K.-I. Machida, H. Nakajima, K. Hirose, G. Adachi, J. Alloys Compd. 1999, 288, 141.
[11] H. Uchida, T. Kawanabe, S. Tachibana, K. Kinoshita, Y. Matsumura, M. Tada, H.-H. Uchida, H. Kaneko, T. Kurino, M. Sato, J. Ceram. Soc. Jpn. 2006, 114, 896.
[12] Y. Gong, J. Wu, M. Kitano, J. Wang, T.-N. Ye, J. Li, Y. Kobayashi, K. Kishida, H. Abe, Y. Niwa, H. Yang, T. Tada, H. Hosono, Nat. Catal. 2018, 1, 178.
[13] W. E. Wallace, J. France, A. Shamsi, Catalysis Using Rare Earth and Actinide Intermetallics Containing Fe, Co, Ni and Cu, Plenum Press, New York, 1982.
[14] V. V. Lunin, O. V. Kryukov, Katal. Fundam. i Prikl. Issled 1987, 86.
[15] R. M. Nix, T. Rayment, R. M. Lambert, J. R. Jennings, G. Owen, J. Catal. 1987, 106, 216.
[16] E. A. Shaw, T. Rayment, A. P. Walker, R. M. Lambert, Appl. Catal. 1990, 67, 151.
[17] E. A. Shaw, T. Rayment, A. P. Walker, R. M. Lambert, T. Gauntlett, R. J. Oldman, A. Dent, Catal. Today 1991, 9, 197.
[18] G. C. Chinchen, P. J. Denny, J. R. Jennings, M. S. Spencer, K. C. Waugh, Appl. Catal. 1988, 36, 1.
[19] J. R. Jennings, R. M. Lambert, R. M. Nix, G. Owen, D. G. Parker, Appl. Catal. 1989, 50, 157.
[20] V. T. Coon, W. E. Wallace, R. S. Craig, Methanation by Rare Earth Intermetallic Catalysts, in the Rare Earths in Science and Technology (ed. G. J. McCarthy, J. J. Rhyne), Plenum Press, New York, 1978, 93.
[21] T. Shirotsuka, K. Onoe, A. Yokoyama, J. Chem. Eng. Jpn. 1986, 19, 375.
[22] V. Paul-Boncour, A. Percheron-Guegan, J. C. Achard, J. Barrault, H. Dexpert, R. C. Karnatak, J. Physique – Suppl. 1986, 47, C8-305.
[23] C. A. Luengo, A. L. Cabrera, H. B. MacKay, M. B. Maple, J. Catal. 1977, 47, 1.
[24] A. Shamsi, W. E. Wallace, Ind. Eng. Chem. Prod. Res. Dev. 1983, 22, 582.
[25] V. Paul-Boncour, A. Percheron-Guegan, J. C. Archard, J. Barrault, G. Jehanno, Eur. J. Solid State Inorg. Chem. 1991, 28, 449.
[26] H. Ando, M. Fujiwara, Y. Matsumura, H. Miyamura, H. Tanaka, Y. Souma, J. Alloys Compd. 1995, 223, 139.
[27] J. E. France, W. E. Wallace, Lanthanide and Actinide Res. 1988, 2, 165.
[28] H. Yamashita, M. Yoshikawa, T. Funabiki, S. Yoshida, J. Catal. 1986, 99, 375.
[29] N. M. Parfenova, I. R. Konenko, A. L. Shilov, M. E. Kost, Izvestiya Akademii Nauk Turkmenskoi SSR, Seriya Fiziko-Tekhnicheskikh, Khimicheskikh i Geologicheskikh 1979, 57.

[30] K. Soga, H. Imamura, S. Ikeda, J. Catal. 1979, 56, 119.
[31] H. Imamura, Y. Kato, S. Tsuchiya, Z. Phys. Chem. – Neue Folge 1984, 141, 129.
[32] K. N. Semenenko, L. A. Petrova, Neftekhimiya 1979, 19, 26.
[33] N. M. Parfenova, I. R. Konenko, A. L. Shilov, A. A. Tolstopyatova, E. I. Klabunovskii, M. E. Kost, Inorg. Mater. 1978, 14, 1333.
[34] L. A. Petrova, K. N. Semenenko, V. V. Burnasheva, Neftekhimiya 1981, 21, 38.
[35] H. Imamura, T. Yoshimura, Y. Sakata, J. Solid State Chem. 2003, 171, 254.
[36] N. Endo, S. Kameoka, A. P. Tsai, T. Hirata, C. Nishimura, Mater. Trans. 2011, 52, 1794.
[37] K. Soga, H. Imamura, S. Ikeda, J. Phys. Chem. 1977, 81, 1762.
[38] H. Imamura, W. E. Wallace, J. Phys. Chem. 1980, 84, 3145.
[39] K. Soga, H. Imamura, S. Ikeda, Nippon Kagaku Kaishi 1977, 1299.
[40] I. R. Konenko, N. M. Parfenova, E. I. Klabunovskii, E. M. Savitskii, V. P. Mordovin, D. E. Bogatin, T. P. Makarochkina, Izv. Akad. Nauk SSSR, Ser. Khim. 1981, 986.
[41] I. R. Konenko, N. M. Parfenova, E. I. Klabunovskii, E. M. Savitskii, V. P. Mordovin, T. P. Makarochkina, Izv. Akad. Nauk SSSR, Ser. Khim. 1981, 981.
[42] I. R. Konenko, N. M. Parfenova, E. I. Klabunovskii, E. M. Savitskii, V. F. Terekhova, I. A. Markova, V. P. Mordovin, D. E. Bogatin, Izv. Akad. Nauk SSSR, Ser. Khim. 1978, 1683.
[43] J. Barrault, A. Guilleminot, A. Percheron-Guegan, V. Paul-Boncour, J. C. Achard, Appl. Catal. 1986, 22, 263.
[44] I. R. Konenko, E. V. Starodubtseva, K. A. Urazbaeva, É. A. Fedorovskaya, E. I. Klabunovskii, A. A. Slinkin, V. P. Mordovin, Kin. Catal. 1988, 29, 858.
[45] I. R. Konenko, E. V. Starodubtseva, Y. P. Stepanov, É. A. Fedorovskaya, A. A. Slinkin, E. I. Klabunovskii, E. M. Savitskii, V. P. Mordovin, T. P. Savost´yanova, Kin. Catal. 1985, 26, 291.
[46] T. A. Solomina, B. B. Bekkulov, R. K. Ibrasheva, K. A. Zhubanov, Kin. Catal. 1986, 27, 1278.
[47] J. R. Johnson, Z. Gavra, J. J. Reilly, Z. Phys. Chem. 1994, 183, 391.
[48] J. R. Johnson, Z. Gavra, P. Chyou, J. J. Reilly, J. Catal. 1992, 137, 102.
[49] K. S. Sim, L. Hilaire, F. Le Normand, R. Touroude, V. Paul-Boncour, A. Percheron-Guegan, Stud. Surf. Sci. Catal. 1989, 48, 863.
[50] K. S. Sim, L. Hilaire, F. LeNormand, R. Touroude, V. Paul-Boncour, A. Percheron-Guegan, J. Chem. Soc. Faraday Trans. 1991, 87, 1453.
[51] R. K. Ibrasheva, T. A. Solomina, B. B. Bekkulov, K. A. Zhubanov, Kin. Katal. 1986, 27, 1278.
[52] R. K. Ibrasheva, T. A. Solomina, G. I. Leonova, V. P. Mordovin, A. E. Bekturov, Int. J. Hydr. Energy 1993, 18, 505.
[53] V. V. Lunin, N. N. Sychev, V. I. Bogdan, Kin. Catal. 1992, 33, 447.
[54] J. E. Smith, S. D. Johnson, Abstr. Pap. Am. Chem. Soc. 1990, 199, 5-FUEL.
[55] Y. M. Baikov, Y. P. Stepanov, A. I. Gorshkov, Kin. Catal. 1981, 22, 795.
[56] I. R. Konenko, L. S. Gorshkova, N. M. Parfenova, E. I. Klabunovskii, D. E. Bogatin, Izv. Akad. Nauk SSSR, Ser. Khim. 1980, 431.
[57] E. V. Starodubtseva, I. R. Konenko, E. I. Klabunovskii, E. M. Savitskii, V. P. Mordovich, T. P. Savost´yanova, Bull. Acad. Sci. USSR Div. Chem. Sci. 1983, 696.
[58] K. Soga, H. Imamura, S. Ikeda, Nippon Kagaku Kaishi 1978, 923.
[59] J. B. Branco, T. A. Gasche, A. P. Gonçalves, A. P. de Matos, J. Alloys Compd. 2001, 323–342, 610.
[60] M. Abid, V. Paul-Boncour, R. Touroude, Appl. Catal. A 2006, 297, 48.
[61] H. Imai, T. Tagawa, K. Nakamura, Appl. Catal. 1990, 62, 348.
[62] H. Imamura, T. Takada, S. Kasahara, S. Tsuchiya, Appl. Catal. 1990, 58, 165.
[63] H. Imamura, H. Yamada, K. Nukui, S. Tsuchiya, J. Chem. Soc. Chem. Comm. 1986, 367.
[64] H. Imamura, K. Nukui, H. Yamada, S. Tsuchiya, T. Sakai, J. Chem. Soc. Faraday Trans. 1 1987, 83, 743.
[65] H. Imai, T. Tagawa, M. Kuraishi, Mater. Res. Bull. 1985, 20, 511.

[66] H. Zhang, Y. Chen, S. Li, X. Fu, Y. Zhu, S. Yi, X. Xue, Y. He, Y. Chen, J. Appl. Phys. 2003, 94, 6417.
[67] J. A. Rajesh, A. Pandurangan, RSC Adv. 2014, 4, 20554.
[68] A. C. Ferreira, A. M. Ferraria, A. M. Botelho do Rego, A. P. Gonçalves, M. R. Correia, T. A. Gasche, J. B. Branco, J. Alloys. Compd. 2010, 489, 316.
[69] L. Rößner, M. Armbrüster, ACS Catal. 2019, 9, 2018.
[70] R. D. Armstrong, A. F. Douglas, D. E. Keene, J. Electrochem. Soc. 1971, 118, 568.
[71] L. Wang, C.-A. Ma, X. Mao, J. Sheng, F. Bai, F. Tang, Electrochem. Comm. 2005, 7, 1477.
[72] D. M. F. Santos, P. G. Saturnino, D. Macciò, A. Saccone, C. A. C. Sequeira, Catal. Tod. 2011, 170, 134.
[73] P. Strasser, S. Kühl, Nano Energy 2016, 29, 166.
[74] T. Kitamura, C. Iwakura, H. Tamura, Chem. Lett. 1981, 965.
[75] S. Trassati, Electrocatalysis of Hydrogen Evolution: Progress in Cathode Activation, VCH, Weinheim, 1992.
[76] F. Rosalbino, G. Borzone, E. Angelini, R. Raggio, Electrochim. Acta 2003, 48, 3939.
[77] D. E. Hall, V. R. Shepard, Int. J. Hydrogen Ener. 1984, 9, 1005.
[78] L. Bing, H. Wei-Kang, J. Appl. Electrochem. 1998, 28, 120.
[79] H. Tamura, C. Iwakura, T. Kitamura, J. Less-Common Met. 1983, 89, 567.
[80] F. Leardini, J. F. Fernández, F. Cuevas, J. Electrochem. Soc. 2007, 154, A507.
[81] W. Hu, Int. J. Hydrogen Ener. 2000, 25, 111.
[82] A. Jukic, M. Metikoš-Hukovic, Electrochim. Acta 2003, 48, 3929.
[83] D. Macciò, F. Rosalbino, A. Saccone, S. Delfino, J. Alloys Compd. 2005, 391, 60.
[84] F. Rosalbino, D. Macciò, E. Angelini, A. Saccone, S. Delfino, J. Alloys Compd. 2005, 403, 275.
[85] R. Bocutti, M. J. Saeki, A. O. Florentino, C. L. F. Oliveira, A. C. D. Angelo, Int. J. Hydrogen Ener. 2000, 25, 1051.
[86] T. Kitamura, C. Iwakura, H. Tamura, Electrochim. Acta 1982, 27, 1723.
[87] D. M. F. Santos, C. A. C. Sequeira, D. Macciò, A. Saccone, J. L. Figueiredo, Int. J. Hydrogen Ener. 2013, 38, 3137.
[88] I. E. L. Stephens, A. S. Bondarenko, U. Grønbjerg, J. Rossmeisl, I. Chorkendorff, Ener. Environm. Sci. 2012, 5, 6744.
[89] M. Escudero-Escribanno, P. Malacrida, M. H. Hansen, U. G. Vej-Hansen, A. Velazquez-Palenzuela, V. Tripkovic, J. Schiøtz, J. Rossmeisl, I. E. L. Stephens, I. Chorkendorff, Science 2016, 352, 73.
[90] M. Escudero-Escribanno, A. Verdaguer-Casadevall, P. Malacrida, U. Grønbjerg, B. P. Knudsen, A. K. Jepsen, J. Rossmeisl, I. E. L. Stephens, I. Chorkendorff, J. Am. Chem. Soc. 2012, 134, 16476.
[91] O. A. Petrii, I. V. Kovrigina, S. Y. Vasina, Mater. Chem. Phys. 1989, 22, 51.
[92] R. K. Ibrasheva, T. A. Solomina, G. I. Leonova, V. V. Kiselev, P. A. Zhdan, G. I. Kaplan, V. P. Mordovin, R. G. Baisheva, Kin. Catal. 1990, 31, 1057.
[93] G. M. R. van Druten, E. Labbé, V. Paul-Boncour, J. Périchon, A. Percheron-Guegan, J. Electroanal. Chem. 2000, 487, 31.
[94] A. V. Devekki, M. I. Yakushkin, T. Y. Kul´chickaya, Kin. Katal. 1988, 29, 1355.
[95] T. Komatsu, T. Uezono, J. Jpn. Petrol. Inst. 2005, 48, 76.
[96] Y.-H. Moon, S.-K. Ihm, Catal. Lett. 1996, 42, 73.
[97] S. Tsujimoto, T. Masui, N. Imanaka, Eur. J. Inorg. Chem. 2015, 1524.
[98] S. Tsujimoto, K. Mima, T. Masui, N. Imanaka, Chem. Lett. 2010, 39, 456.
[99] G.-Y. Adachi, N. Imanaka, Chem. Rev. 1998, 98, 1479.
[100] K. Ohta, Y. Yamada, S. Sato, Appl. Catal. A: Gen. 2016, 517, 73.
[101] H. Duan, Y. Yamada, S. Sato, Appl. Catal. A: Gen. 2015, 491, 163.
[102] Y. Wang, D. Sun, Y. Yamada, S. Sato, Appl. Catal. A: Gen. 2018, 562, 11.

[103] B. M. E. Russbueldt, W. F. Hoelderich, J. Catal. 2010, 271, 290.
[104] K. Okura, T. Okanishi, H. Muroyama, T. Matsui, K. Eguchi, ChemCatChem 2016, 8, 2988.
[105] M. Itoh, K. Motoki, M. Saito, J. Iwamoto, K.-I. Machida, Bull. Chem. Soc. Jpn. 2009, 82, 1197.
[106] X. Xu, R. Zhang, X. Zeng, X. Han, Y. Li, Y. Liu, X. Wang, ChemCatChem 2013, 5, 2025.
[107] B. Predel in Ce-O (Cerium-Oxygen): Datasheet from Landolt-Börnstein – Group IV Physical Chemistry Volume 5C: "Ca-Cd – Co-Zr" in SpringerMaterials (can be found under https://doi.org/10.1007/10086082_837), Springer-Verlag, Berlin, Heidelberg.
[108] H. J. Seifert, P. Nerikar, H. L. Lukas, ECS Trans. 2006, 1, 3.
[109] D. Chen, D. He, J. Lu, L. Zhong, F. Liu, J. Liu, J. Yu, G. Wan, S. He, Y. Luo, Appl. Catal. B: Environ. 2017, 218, 249.
[110] M. Sugiura, Catal. Surv. Asia 2003, 7, 77.
[111] C. Doornkamp, V. Ponec, J. Mol. Catal. A: Chem. 2000, 162, 19.
[112] F. Zhang, P. Wang, J. Koberstein, S. Khalid, S.-W. Chan, Surf. Sci. 2004, 563, 74.
[113] M. Skoda, M. Cabala, I. Matolinova, T. Skala, K. Veltruska, V. Matolin, Vacuum 2009, 84, 8.
[114] S. A. C. Carabineiro, M. Konsolakis, G. E.-N. Marnellos, M. F. Asad, O. S. G. P. Soares, P. B. Tavares, M. F. R. Pereira, J. J. de Meloorfao, J. L. Figueiredo, Molecules 2016, 21, 644/641.
[115] G. Niu, M. H. Zoellner, T. Schroeder, A. Schaefer, J.-H. Jhang, V. Zielasek, M. Baeumer, H. Wilkens, J. Wollschlaeger, R. Olbrich, C. Lammers, M. Reichling, PhysChemChemPhys 2015, 17, 24513.
[116] N. S. Priya, C. Somayaji, S. Kanagaraj, Rare Metals (Beijing, China) 2016, Ahead of Print.
[117] V. R. Madduluri, B. D. Raju, K. S. R. Rao, Res. Chem. Intermed. 2019, 45, 2749.
[118] K. Krishna, A. Bueno-Lopez, M. Makkee, J. A. Moulijn, Appl. Catal. B: Environ. 2007, 75, 210.
[119] K. Krishna, A. Bueno-Lopez, M. Makkee, J. A. Moulijn, Appl. Catal. B: Environ. 2007, 75, 201.
[120] J. S. Elias, K. A. Stoerzinger, W. T. Hong, M. Risch, L. Giordano, A. N. Mansour, Y. Shao-Horn, ACS Catal. 2017, 7, 6843.
[121] Y. Liao, Y. Li, L. Wang, Y. Zhao, D. Ma, B. Wang, Y. Wan, S. Zhong, Dalton Trans. 2017, 46, 1634.
[122] P. Xu, K. Li, H. Yu, M. A. Cohen Stuart, J. Wang, S. Zhou, Ind. Engin. Chem. Res. 2019, 58, 3726.
[123] V. I. Alexiadis, M. Chaar, A. van Veen, M. Muhler, J. W. Thybaut, G. B. Marin, Appl. Catal. B: Environ. 2016, 199, 252.
[124] P. Ji, J. Zhang, F. Chen, M. Anpo, Appl. Catal. B: Environ. 2009, 85, 148.
[125] J. You, Y. Guo, R. Guo, X. Liu, Chem. Engin. J. 2019, 373, 624.
[126] W. Fang, W. Shangguan, Int. J. Hydrogen Ener. 2019, 44, 895.
[127] R. He, D. Xu, B. Cheng, J. Yu, W. Ho, Nanoscale Horiz. 2018, 3, 464.
[128] H. Wang, B. Zhou, X. Zhao, Photoelectrocatalytic Degradation of Organic Contaminants in Nanosemiconductor Film Electrodes under Visible Light Irradiation, McGraw-Hill, 2011, 107.
[129] J. J. Mu, G. H. Zheng, Z. X. Dai, L. Y. Zhang, Z. F. Yao, Y. Q. Ma, J. Mater. Sci. 2017, 28, 14747.
[130] A. T. Ashcroft, A. K. Cheetham, M. L. H. Green, C. P. Grey, P. D. F. Vernon, J. Chem. Soc. Chem. Comm. 1989, 1667.
[131] C. Petit, A. Kaddouri, S. Libs, A. Kiennemann, J. L. Rehspringer, P. Poix, J. Catal. 1993, 140, 328.
[132] C. Petit, J. L. Rehspringer, A. Kaddouri, S. Libs, P. Poix, A. Kiennemann, Catal. Tod. 1992, 13, 409.
[133] X. Fang, L. Xia, L. Peng, Y. Luo, J. Xu, L. Xu, X. Xu, W. Liu, R. Zheng, X. Wang, Chin. Chem. Lett. 2019, 30, 1141.
[134] J. Xu, L. Peng, X. Fang, Z. Fu, W. Liu, X. Xu, H. Peng, R. Zheng, X. Wang, Appl. Catal. A: Gen. 2018, 552, 117.
[135] N. Subramanian, J. J. Spivey, Preprints Am. Chem. Soc., Div. Petrol. Chem. 2012, 57, 182.
[136] T. W. Elkins, B. Neumann, M. Baumer, H. E. Hagelin-Weaver, ACS Catal. 2014, 4, 1972.
[137] T. W. Elkins, S. J. Roberts, H. E. Hagelin-Weaver, Appl. Catal. A: Gen. 2016, 528, 175.
[138] B. Neumann, T. W. Elkins, A. E. Gash, H. Hagelin-Weaver, M. Baeumer, Catal. Lett. 2015, 145, 1251.

[139] W. M. Meier, D. H. Olson, C. Baerlocher, Zeolites 1996, 17, 1.
[140] E. T. C. Vogt, B. M. Weckhuysen, Chem. Soc. Rev. 2015, 44, 7342.
[141] A. Akah, J. Rare Earths 2017, 35, 941.
[142] E. F. Sousa-Aguiar, F. E. Trigueiro, F. M. Z. Zotin, Catal. Tod. 2013, 218-219, 115.
[143] D. Wallenstein, T. Roberie, T. Bruhin, Catal. Today 2007, 127, 54.
[144] J. A. Rabo, C. L. Angell, V. Schomaker, Proc. 4th Int. Congress on Catalysis: Moscow, USSR 23-29 June 1968, Akad. Kiad, 1971.
[145] C.-Y. Li, L. V. C. Rees, Zeolites 1986, 6, 60.
[146] J. W. Roelofsen, H. Mathies, R. L. de Groot, P. C. M. van Woerkom, H. A. Gaur, Effect of Rare Earth Loading in Y-Zeolite on Its Dealumination during Thermal Treatment (ed. Y. Murakami, A. Iijima, J. W. Ward), Vol. 28, Elsevier, 1986, 337.
[147] J. A. van Bokhoven, A. L. Roest, D. C. Koningsberger, J. T. Miller, G. H. Nachtegaal, A. P. M. Kentgens, J. Phys. Chem. B 2000, 104, 6743.
[148] D. Wallenstein, K. Schaefer, R. H. Harding, Appl. Catal. A: Gen. 2015, 502, 27.
[149] W. E. Wallace, A. Elattar, H. Imamura, R. S. Craig, A. G. Moldovan, Intermetallic Compounds: Surface Chemistry, Hydrogen Absorption and Heterogeneous Catalysis, Academic Press, New York, 1980.
[150] F. P. Netzer, E. Bertel, Adsorption and Catalysis on Rare Earth Surfaces, Elsevier, Amsterdam, 1982.
[151] A. Baiker, Faraday Disc. Chem. Soc. 1989, 87, 239.
[152] B. Viswanathan, Catalysts Derived from Hydrogen Absorbing Intermetallics, Springer, Berlin, 1998.
[153] V. Paul-Boncour, L. Hilaire, A. Percheron-Guegan, The Metals and Alloys in Catalysis, Elsevier, Amsterdam, 2000.
[154] W. E. Wallace, Rare Earths and Actinide Intermetallics as Hydrogenation Catalysts, in Hydrides for Energy Storage (ed. A. F. Anderson, A. J. Maeland), Pergamon, Oxford, 1978, 501.
[155] O. V. Chetina, V. V. Lunin, Russ. Chem. Rev. 1994, 63, 483.
[156] A. Molnár, G. V. Smith, M. Bartók, Adv. Catal. 1989, 36, 329.

Yaroslav Mudryk, Vitalij K. Pecharsky

4.8 Materials for solid state cooling

Unlike the modern times, when rare earth materials have become firmly ingrained into many aspects of our lives, the rare earth elements themselves were barely available, considered a scientific curiosity and had no clear path to applications beyond lighter flints at the beginning of the twentieth century. Remarkably, solid state magnetic cooling was one of the first suggested uses that were based on the 4*f*-electron magnetism of lanthanides. Independent of each other, P. Debye in 1926 [1] and W. F. Giauque in 1927 [2] proposed a method of "producing temperatures considerably below 1° absolute" by adiabatic demagnetization, whereas the method itself was derived from "a thermodynamic treatment of certain magnetic effects." The concept was demonstrated experimentally, in 1933 [3], by adiabatically demagnetizating some 60 g of $Gd_2(SO_4)_3 \cdot 8H_2O$ precooled with liquid helium to 1.5–3.4 K in a steady 8 kOe magnetic field; when the magnetic field was reduced to 0, the salt cooled down to a few hundred millikelvin (a low temperature record at the time!) due to disordering of the localized, purely 4*f*-spin magnetic moments of the Gd atoms. W. F. Giauque was awarded the 1949 Nobel Prize in Chemistry, including for his work on low-temperature thermodynamics and adiabatic demagnetization, and P. Debye — the 1936 Nobel Prize in Chemistry for his contributions to the knowledge of molecular structures. With respect to room temperature applications of magnetic field-induced thermal effects, Gd once again played a critical role. The elemental metal was used by G. V. Brown, who, in 1976, showed that it is possible to create temperature differences much greater than the magnetocaloric effect itself [4]. Some 20 years later, the element became an important component of the $Gd_5(Si_{1-x}Ge_x)_4$ family of materials, which both announced the discovery of the giant magnetocaloric effect [5, 6] and opened the door to its room temperature applications [7].

The phenomenon of thermoelectricity is known since the 1820s, but rare earth compounds were brought into the spotlight as potential materials for thermoelectric energy conversion in much more recent times. Traditional narrow band gap thermoelectric materials, such as Si_xGe_{1-x} and Bi_2Te_3, and a variety of chalcogenides, do not rely on lanthanides for their thermoelectric energy conversion properties. However, new discoveries, including filled skutterudites [8–10] and $Yb_{14}MnSb_{11}$ [11], made rare earths relevant and promising for thermoelectric cooling.

4.8.1 Introduction to magnetocaloric effect

To understand why a rare earth compound – $Gd_2(SO_4)_3 \cdot 8H_2O$ – was chosen for the first ever demonstration of magnetocaloric cooling, we need to learn what

https://doi.org/10.1515/9783110654929-031

the magnetocaloric effect is, and what properties are critical to maximize the effect. The magnetocaloric effect (MCE) is a thermal response of a material to a changing magnetic field. It was discovered in 1918, by Weiss and Piccard, who observed a reversible change in the temperature of nickel reaching 0.7 K, when the metal was magnetized and demagnetized with 1.5 Tesla magnetic field in the vicinity of its Curie temperature of 629 K [12]. MCE is one of the fundamental properties of matter, but the effect is difficult to detect since it is weak in most magnetic solids at ambient conditions. Following [13–15], we start with the Gibbs free energy, G, of a solid at constant pressure expressed as a function of internal energy, U, entropy, S, volume, V, and magnetization, M, as shown below:

$$G = U - TS + pV - MH \tag{4.8.1}$$

Then, the response of a system to external stimuli, such as temperature, T, pressure, p, and magnetic field, H, is described by the full differential of the Gibbs free energy as:

$$\mathrm{d}G = V\mathrm{d}p - S\mathrm{d}T - M\mathrm{d}H \tag{4.8.2}$$

The individual internal parameters (S, V, and M) can be determined using the following equations of state:

$$S(T, H, p) = -\left(\frac{\partial G}{\partial T}\right)_{H,p} \tag{4.8.3a}$$

$$M(T, H, p) = -\left(\frac{\partial G}{\partial H}\right)_{T,p} \tag{4.8.3b}$$

$$V(T, H, p) = \left(\frac{\partial G}{\partial p}\right)_{T,H} \tag{4.8.3c}$$

Assuming isobaric process, $\mathrm{d}p = 0$, by coupling eq. (4.8.3a) and (4.8.3b), we obtain one of the following Maxwell relations (which are relations connecting second derivatives of thermodynamic potentials, in our case, Gibbs free energy; not to be confused with the Maxwell equations for electromagnetism):

$$\left(\frac{\partial S(T,\ H)}{\partial H}\right)_T = \left(\frac{\partial M(T,\ H)}{\partial T}\right)_H \tag{4.8.4}$$

This leads to the following expression, which can be used to determine the change in entropy from the change in magnetization:

$$\mathrm{d}S = \left(\frac{\partial M}{\partial T}\right)_T \mathrm{d}H \tag{4.8.5}$$

By integrating eq. (4.8.5) one can see that, for a given magnetic field change $\Delta H = H_\mathrm{F}$ (final field) – H_I (initial field), the change of the system's entropy at constant temperature and pressure can be obtained from the measurements of magnetization as a function of temperature and magnetic field as follows:

$$\Delta S_M(T)_{\Delta H} = \int_{H_I}^{H_F} \left(\frac{\partial M(T,H)}{\partial T}\right)_H dH \tag{4.8.6}$$

Since generation of strong magnetic fields is difficult (*e. g.*, at room temperature the strength of the magnetic field on the surface of the high-grade $Nd_2Fe_{14}B$-based permanent magnets is about 14 kOe and the field decays rapidly away from the surface); H_I is always 0. Finally, we should consider that the heat capacity of a material, C, and its entropy are related as follows:

$$\frac{dS}{dT} = \frac{C_H}{T} \tag{4.8.7}$$

The total differential of the total entropy of the magnetic system (for $dp = 0$) can be written as:

$$dS = \left(\frac{\partial S}{\partial T}\right)_H dT + \left(\frac{\partial S}{\partial H}\right)_T dH \tag{4.8.8}$$

or

$$T\mathrm{d}S = T\left(\frac{\partial S}{\partial T}\right)_H \mathrm{d}T + T\left(\frac{\partial S}{\partial H}\right)_T \mathrm{d}H \tag{4.8.9}$$

In an adiabatic process $TdS = 0$, and by adapting eq. (4.8.4) and (4.8.7) into (4.8.9), we can determine the magnetic field-induced temperature change for the adiabatic-isobaric process as:

$$\mathrm{d}T = -\frac{T}{C_H}\left(\frac{\partial M}{\partial T}\right)_H \mathrm{d}H \tag{4.8.10}$$

Therefore, for a given magnetic field change $\Delta H = H_F - H_I$, the magnetocaloric effect is:

$$\Delta T_{ad}(T)_{\Delta H} = -\int_{H_I}^{H_F} \left(\frac{T}{C(T,H)}\right)_H \left(\frac{\partial M(T,H)}{\partial T}\right)_H dH. \tag{4.8.11}$$

The adiabatic temperature change, ΔT_{ad}, and the isothermal magnetic entropy change, ΔS_M, are the two most important characteristics of MCE. In addition, cooling capacity defined as:

$$q = -\int_{T_1}^{T_2} \Delta S_M(T)_{\Delta H} dT \tag{4.8.12}$$

quantifies the amount of heat that can be transferred by a given material between the temperatures T_2 and T_1, for a given field change ΔH.

The major characteristics of magnetocaloric effect can be determined directly, for example, by measuring temperature change during rapid application and removal of magnetic field to a thermally insulated material [13, 16]. More commonly, however, magnetocaloric effects are determined indirectly, that is, they are computed from magnetization measured as a function of temperature and magnetic field, and then applying eq. (4.8.6) to calculate $\Delta S_M(T)_{\Delta H}$ [17–19], and/or from heat capacity, measured as a function of temperature in different magnetic fields, and then applying eq. (4.8.7) to determine $S(T)_H$, from which both $\Delta S_M(T)_{\Delta H}$ and $\Delta T_{ad}(T)_{\Delta H}$ are easily calculated [18] as:

$$\Delta S_M(T)_{\Delta H} = S(T)_{H_F,T} - S(T)_{H_I,T}, \text{ and} \tag{4.8.13}$$

$$\Delta T_{ad}(T)_{\Delta H} = T(S)_{H_F,S} - T(S)_{H_I,S}. \tag{4.8.14}$$

Figure 4.8.1 shows temperature- and magnetic field-dependent magnetocaloric effects as both the magnetic entropy change and the adiabatic temperature change of elemental Gd – a benchmark magnetocaloric material. The values plotted in Figure 4.8.1 are

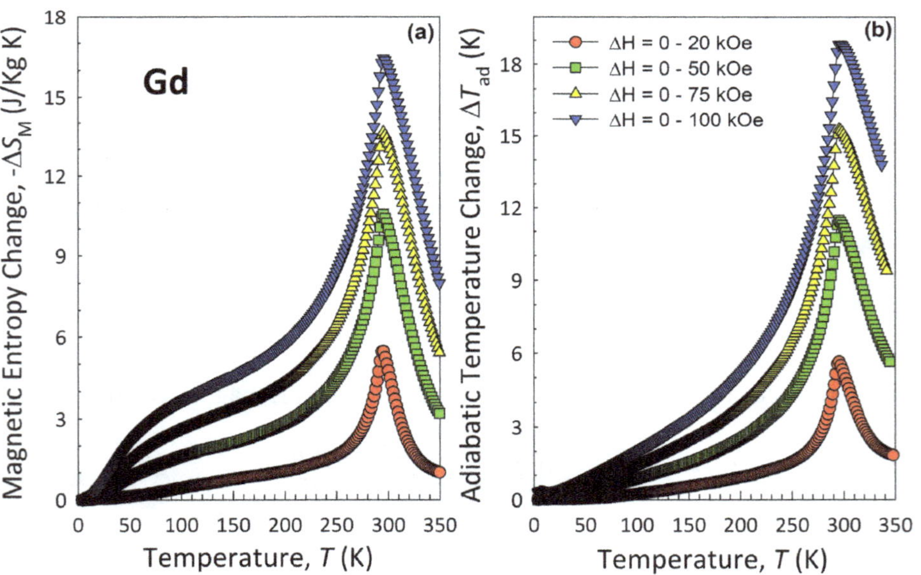

Figure 4.8.1: Temperature dependence of isothermal magnetic entropy change (a) and adiabatic temperature change (b) in Gd metal calculated from heat capacity data measured in applied magnetic fields up to $H = 100$ kOe [20]. ΔH denotes the magnetic field change from $H_I = 0$ to the corresponding nonzero value. Of course, MCE will also occur when the magnetic field changes from, e.g., 20–40 kOe, but, as mentioned above, H_I is always zero for practical reasons.

calculated using eqs. (13) and (14) from heat capacity of the metal measured in a zero magnetic field, and four additional measurements made in constant magnetic fields of 20, 50, 75 and 100 kOe [20]. The MCE of Gd is substantial at just about any temperature, especially so when $H_F \gg H_I = 0$. Clearly, the effect is maximized near the spontaneous ferromagnetic ordering transition temperature of the element, which is $T_C = 294$ K. We note that in very high but impractical magnetic fields, for example, when H_F exceeds some 10^6 Oe (hundreds of Tesla), the conventional caret-like shapes of $\Delta T_{ad}(T)_{\Delta H}$ of Gd seen in Figure 4.8.1 change into step-like anomalies at T_C, with high-temperature tails slowly varying above T_C, and ΔT_{ad} measuring hundreds of Kelvin [21].

As directly follows from Eqs. (4.8.6) and (4.8.11) and illustrated in Figure 4.8.1, besides the obvious, that is, the larger ΔH leads to larger MCE, the rate at which magnetization changes with temperature in the presence of magnetic field, $(dM/dT)_H$, is the most critical intrinsic property of any material with respect to MCE. Not only $|(dM/dT)_H|$ must be large, which is common in weak magnetic fields for many materials in the immediate vicinities of their spontaneous ferro- or ferrimagnetic ordering phase transformations, it must remain large as the magnetic field increases. This explains[1] the selection of a paramagnetic gadolinium salt for the early experiments by W. F. Giauque [3].

Trivalent Gd has the highest localized spin moment, $s = 7/2$, among all stable elements in the periodic table, due to a half-filled $4f$ shell containing seven unpaired $4f$ electrons. The maximum molar magnetic entropy of a solid, S_M, is given [13, 22] as:

$$S_M = R\ln(2J + 1), \tag{4.8.15}$$

where R is the universal gas constant, J is the total angular momentum quantum number, which, for light lanthanides, is $J = L - s$, and it is $J = L + s$ for the heavy lanthanides, and L is the orbital angular momentum quantum number. Large $J = 7/2$ makes Gd more favorable than, for example, Fe^{3+} or Mn^{2+} that may have s as high as 5/2. Further, even though the total angular momenta of other heavy lanthanides can be higher than Gd (*e. g.*, Dy^{3+} has $L = 5$, $s = 5/2$, and $J = 15/2$), gadolinium has no orbital angular momentum ($L = 0$), which makes spin-orbit coupling, magnetocrystalline anisotropy and crystal field influences extremely weak, and does not quench the magnetic entropy available for manipulation by the magnetic field. This factor is particularly important for applications between 2 and 20 K [23].

The entropy and temperature changes due to magnetic field cycling up and down are illustrated schematically in Figure 4.8.2, which describes the magnetocaloric cooling principle in comparison with the conventional gas-compression refrigeration technology. Consider a solid, such as Gd metal or another magnetic

---

1 An ideal paramagnet obeys the Curie law which may be written as $M = cHT^{-1}$, where c is the Curie constant. Hence, for any constant magnetic field, H, magnetization hyperbolically approaches infinity when T approaches 0, thus maintaining large $|(dM/dT)_H|$.

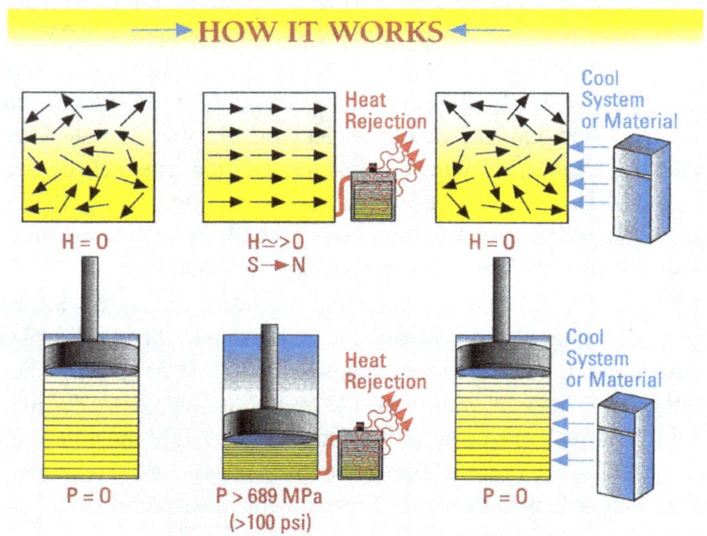

Figure 4.8.2: Principle of operation of a magnetic refrigerator (cartoons on top) compared with conventional gas compression (cartoons on bottom). Left - the initial high-magnetic entropy state of a magnetic material at $H_I = 0$, and the initial high-configurational entropy state of a gas at $P = 0$. Middle – magnetizing with field $H_F > 0$ aligns spins, decreases the magnetic entropy, causing the material to heat up (similarly, gas compression reduces configurational entropy, causing the gas to heat up); heat generated in either case is rejected at the hot end. Right – removal of the magnetic field restores the high-entropy state and the material cools down (gas de-compression restores the high configurational entropy state of the gas allowing the gas to cool down); heat is absorbed at the cold end completing the cycle. Note that adiabatic demagnetization demonstrated by W.F. Giauque [3] is different from continuous magnetocaloric cooling as the former ends at the last step.

material, containing a system of large, weakly interacting magnetic moments in a low-magnetization state at $H_I = 0$. Application of magnetic field $H_F > 0$ forces the moments to align along the field direction, reducing magnetic disorder and, consequently, decreasing the magnetic entropy of the system by $-\Delta S_M$. When the magnetizing is performed adiabatically, that is, without heat exchange with the surroundings, and recalling that changes in p and V are negligible, lattice entropy (and electronic entropy if the solid is a metal) must increase by the equivalent of $+\Delta S_M$ to maintain the total entropy of the magnetic material constant. The result is increased lattice vibrations and rise of the global temperature of the solid. When the magnetic field is removed, the material returns to its original disordered state, magnetic entropy goes up by $+\Delta S_M$, lattice and electronic entropies combined go down by $-\Delta S_M$ and the solid cools down. A single cycle equivalent to that described in this paragraph was demonstrated by W. F. Giauque [3]. For continuous heat pumping, a heat exchange fluid, liquid or gas, is commonly used to transfer the heat from and to the solid material, as was originally demonstrated by G. V. Brown [4].

A number of other important materials characteristics, in addition to the total available magnetic entropy and $|(dM/dT)_H|$, must be considered. Heat capacity of the solid material should be small, because $|\Delta T_{ad}|$ is inversely proportional to C, (eq. 4.8.11). Paramagnets work well for sub-1 K temperatures because C approaches 0 as T approaches 0, but in all other applications spontaneous magnetic ordering is needed to maximize $|(dM/dT)_H|$, $|\Delta S_M|$ and $|\Delta T_{ad}|$. The magnetic ordering temperature must be near, ideally inside, the operating temperature window, defined as $T_{hot} - T_{cold}$, where T_{hot} is the temperature of the hot reservoir and T_{cold} is the temperature of the cold reservoir. Efficient heat exchange with the heat transfer fluid requires that the magnetic solid has reasonably high thermal conductivity. Other desirable characteristics include availability and affordability of raw materials, chemical and mechanical stability, easy synthesis and fabrication into shapes to maximize heat transfer, long-term durability and high gravimetric density (large magnetization per unit volume). The latter is related to the fact that maintaining strong magnetic fields in large volumes is both difficult and expensive.

As mentioned above, solid state magnetocaloric cooling (more generally, magnetocaloric heat pumping) was conceived and initially utilized to achieve temperatures much below 1 K. It remains a useful laboratory tool for cryogenic applications below 20 K. Due to a number of scientific and engineering breakthroughs announced over the past two decades, magnetocaloric cooling is now under development for continuous heat pumping at much higher temperatures. Proof-of-principle magnetocaloric prototypes operating with efficiencies comparable to vapor-compression systems near room temperature demonstrate that magnetic refrigeration is capable of providing cooling powers comparable to those of modern household appliances and air-conditioners [24]. While today heat pumps utilizing the magnetocaloric effect are more expensive than the well-refined vapor-compression counterparts due to the high cost of production of magnetic fields in excess of 10 kOe and lack of magnetic materials that can operate in fields much below 10 kOe, room for improvements is significant. Besides, solid-state magnetocaloric (and, indeed, thermoelectric) heat pumps are expected to have small environmental footprints because they do not use volatile refrigerants common in vapor-compression systems, majority of which rely on high global-warming potential hydrofluorocarbons.

While the principle of operation illustrated in Figure 4.8.2 is the same regardless of the temperature of operation, magnetocaloric materials for use in cryogenic and room temperature heat pumps are different. Magnetocaloric materials for low temperatures are almost exclusively lanthanide-based. Among materials for near-room temperature application, both rare earth-containing and rare earth-free compounds have been proposed. Below, we will discuss some of the compounds that contain rare earths and exhibit the strongest magnetocaloric effects in both cryogenic and near-room temperature regimes.

4.8.2 Rare earth magnetocaloric materials for cryogenic applications

Cryogenic magnetocaloric cooling technologies can be grouped into two major areas. One area is to cool and maintain detectors and instruments at low temperature on board of spacecraft, for example, satellites. To achieve resolution and sensitivity that would allow capturing every single incoming photon, some detector systems must be cooled down to, and maintained at ~0.1 K. Low thermal and mechanical noise profiles, high reliability and efficiency are critical, and magnetocaloric cryocoolers can provide many of these characteristics. Another area of interest is liquefaction of helium (boiling temperature, $T_b = 4.2$ K), hydrogen ($T_b = 20.3$ K) and natural gas (mainly methane with $T_b = 111.7$ K, some propane with $T_b = 231.1$ K, and other gases such as carbon dioxide, nitrogen, and hydrogen sulfide in various concentrations). Lanthanide-based compounds are materials of choice for applications in both areas due to large available magnetic entropies and wide range of magnetic ordering temperatures that are commonly below 300 K.

On the surface, the easiest solution would be to use pure lanthanide metals or intra-lanthanide alloys. Both the metals and alloys are ductile and workable, they exhibit reasonable mechanical stabilities, have broad range of T_C's, the highest local magnetic moments and total angular momenta known in nature and are concentrated magnetic systems, where every atom in their structures carries a magnetic moment. The total available magnetic entropies (eq. (4.8.15) and Figure 4.8.3) are the highest in heavy lanthanides and, consequently, theoretical limits of the magnetic entropy change are the highest in these elements as well [4, 13, 14].[2] Yet, unlike Gd (Figure 4.8.1) which becomes ferromagnetic near room temperature, other lanthanides show complex magnetism and, commonly, more than one magnetic transition. As a result, their magnetocaloric effects in low magnetic fields may change sign and be spread over wide range of temperatures, becoming substantial only in high magnetic fields, because their magnetic structures in low magnetic fields are often non-collinear. All of this leads to rather complex temperature dependencies of the magnetocaloric effects in heavy lanthanides, as illustrated in Figure 4.8.4 for elemental Dy [25]. The situation becomes more complex for intra-lanthanide alloys, where new crystallographic (and magnetic) phases emerge at certain ranges of concentrations, despite similar chemistry and metallurgy of the lanthanides [26].

2 In the absence of crystallographic phase changes in response to varying magnetic field, the maximum possible magnetic entropy change when $\Delta H \to \infty$ cannot exceed the total available magnetic entropy given by eq. (4.8.15). However, the effects of crystalline electric field, potential for intermediate and mixed valence states of some elements, can significantly impact the actual MCE. For reasonable magnetic fields of 100 kOe and below, the actual ΔS_M values do not exceed 10–20% of the theoretical limit [13].

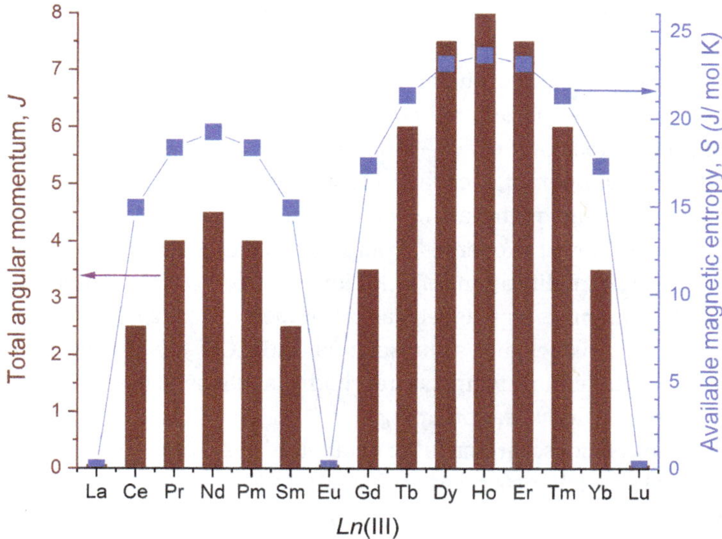

Figure 4.8.3: Correspondence between the available molar magnetic entropies, S (symbols, right hand scale), and the total angular momentum quantum numbers, J (bars, left hand scale), of trivalent lanthanides.

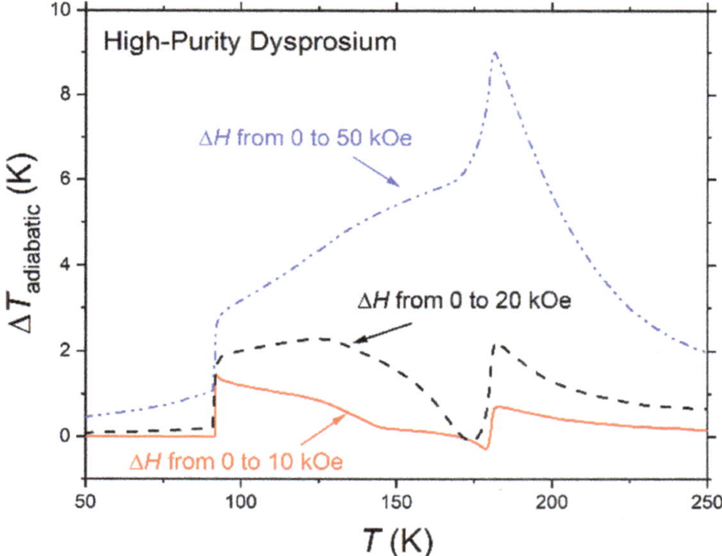

Figure 4.8.4: Magnetocaloric effect of ultra-pure Dy calculated from the heat capacity data [25].

Consequently, with only a few exceptions, Gd-based compounds are considered the consensus materials for cryogenic applications based on the magnetocaloric effect. Due to rather low concentration of the moment-carrying ions and low thermal conductivity, $Gd_2(SO_4)_3 \cdot 8H_2O$ – the original compound employed to demonstrate the usefulness of the magnetocaloric effect – finds no practical use today [23]. In the 1970s, gadolinium gallium garnet, $Gd_3Ga_5O_{12}$ (commonly known as GGG) was suggested and became the lead material for cryogenic magnetocaloric cooling [23, 27], due to its much improved properties, compared to $Gd_2(SO_4)_3 \cdot 8H_2O$. Later studies revealed that chemical modifications of GGG, for example, partial substitution of gallium by iron, further improve its magnetocaloric properties [28, 29]. A number of other complex gadolinium oxides and salts, such as $Gd(HCOO)_3$ [30], have been proposed to achieve and maintain low temperatures, both via adiabatic demagnetization and continuous magnetocaloric heat pumping.

In addition to salts and oxides, intermetallic compounds and alloys can be utilized as working materials for magnetocaloric refrigeration [15]. In an intermetallic solid, reaction of two or more metals or metalloids does not result in the formation of ionic or covalent bonds – the bonding and properties remain typical of a metal – and conduction electrons are present at the Fermi level. Magnetic interactions between lanthanide ions mediated by conduction electrons (also known as indirect Ruderman-Kittel-Kasuya-Yosida, or RKKY [31] interactions) open numerous possibilities for precise tunability of intermetallic magnetism. Thermal conductivity of intermetallics is usually much higher compared to oxides and salts, as well. As a result, a number of rare earth intermetallics with large MCEs at cryogenic temperatures have been characterized. Among them, for example, are binary REM_2 compounds with cubic $MgCu_2$-type crystal structure, commonly known as C15-type or Laves phases, where *RE* is a rare earth and *M* could be a transition metal, such as Ni or Co, an *s*-element such as Mg or a *p*-element such as Al. Laves phases form with a great variety of *M*, hence by varying *RE* and *M*, materials operating at the peak of MCE in a desired temperature range can be engineered. For instance, $ErAl_2$ has a large $|\Delta S_M|$ with a maximum of 36 J/kg K at $T = 13$ K for $\Delta H = 50$ kOe [13, 32], while $HoCo_2$ shows a very respectable $|\Delta S_M|$ peaking at 12.5 J/kg K and $T = 78$ K for $\Delta H = 20$ kOe [33]. Further, many isostructural binary lanthanide intermetallics easily form continuous solid solutions. This makes it possible to fine-tune a material for peak performance at just about any temperature between the Curie temperatures of binary parents. Thus, another Laves phase system, $(Er_{1-x}Dy_x)Al_2$, demonstrates excellent magnetocaloric effects between 13 K (T_C of $ErAl_2$) and 64 K (T_C of $DyAl_2$), controlled by adjusting x between 0 and 1 [14, 32]. The orbital moments of heavy lanthanides in intermetallics may or may not be quenched by crystalline electric field, and in some compounds, their total magnetic moments can reach their corresponding theoretical limits around 10 μ_B/RE atom, providing a sizeable boost to $|(dM/dT)_H|$ and to the magnetocaloric effect.

Gas liquefaction is another potential application for cryogenic magnetocaloric cooling. In order to cool, for example, hydrogen gas down to its boiling temperature of 20.3 K from either room temperature or from liquid nitrogen at 77 K, a magnetic cooler needs to have more than a single stage, and therefore, more than one magnetocaloric material should be used to cascade the temperature down. Lanthanides and their intermetallics, which otherwise are excellent materials for cryogenic cooling, may not be the best choice for this particular purpose, because both the lanthanides and some of their intermetallics [34] can absorb hydrogen. This can change Curie temperatures, and in many cases, compounds decompose, forming stable rare earth hydrides. Thus, direct contact of hydrogen with the cooling materials may not be possible, complicating the system design and lowering cooling efficiency. Oxides and salts mentioned earlier do not react with hydrogen and doped garnets, such as $Gd_3(Ga_{0.5}Fe_{0.5})_5O_{12}$ (GGIG) and $Dy_{2.4}Ga_{0.6}Al_5O_{12}$ (DGAG), are good options to liquefy hydrogen gas at ~20 K, but their cooling performance significantly drops above 25 K [34], requiring additional materials. One of the possible solutions is to use rare earth mononitrides, *RE*N, where *RE* = Gd, Tb, Dy and/or Ho; see Figure 4.8.5. These materials possess high magnetic moment density per unit volume, good thermal conductivity and are inert to hydrogen. Holmium nitride, HoN, has a peak MCE at 18 K, right around the boiling point of hydrogen, while others can be used to pre-cool gaseous H_2 from 77 K [35].

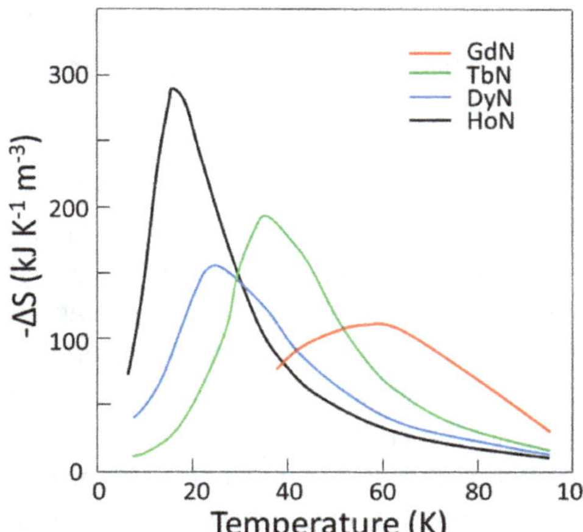

Figure 4.8.5: Isothermal magnetic entropy changes in GdN, TbN, DyN and HoN for $\Delta H = 50$ kOe [35].

4.8.3 Near-room-temperature magnetic refrigeration and giant magnetocaloric effect

As temperature increases above a few Kelvin, the adiabatic temperature change is negatively impacted by the rapidly rising lattice heat capacity of a solid, which increases as a function of $(T/\Theta_D)^3$, where Θ_D is the Debye temperature, saturating at $3R$

above θ_D. If the solid is a metal, its electronic heat capacity also rises as γT, where γ is the electronic specific heat, which is proportional to the electronic density of states at the Fermi level. Maximum magnetization and $|(dM/dT)_H|$, on the other hand, both decrease because rising thermal energy competes with the nearly constant magnetic exchange energy, reducing the strength of field-moment coupling and impeding alignment of magnetic moments. As follows from eq. (4.8.11), high C and low $|(dM/dT)_H|$ inevitably lead to the reduction of ΔT_{ad}. Among conventional ferromagnets, only a handful of highly concentrated magnetic materials such as Gd and $Gd_{1-x}Y_x$ alloys with $x \ll 1$ [13, 14], and Tb and $Gd_{1-x}Tb_x$ alloys [13, 14, 36], whose magnetocaloric effects are similar to those shown in Figure 4.8.1, are suitable near room-temperature magnetocaloric refrigerants. To realize the potential of high efficiency in magnetocaloric heat pumping near room temperature, magnetic fields above 10–15 kOe are highly desirable, but they require either bulky permanent magnet arrays or superconducting magnets, both of which are expensive. An alternative is to employ a different class of magnetocaloric materials, in which conventional magnetic entropy changes may be enhanced by either or both lattice and electronic entropy changes associated with coupled magnetostructural or magnetovolume transformations [37, 38]. Such solids are commonly known as materials exhibiting a giant magnetocaloric effect, and they are briefly described in this section.

The term "giant magnetocaloric effect" was coined in 1997,[3] when a very large magnetocaloric effect reaching $|\Delta S_M| = 14$ J/kg K and $|\Delta T_{ad}| = 7$ K for magnetic field change $\Delta H = 20$ kOe, (Figure 4.8.6), was discovered near room temperature in a ternary intermetallic compound, $Gd_5Si_2Ge_2$ [5]. The isothermal entropy changes of $Gd_5Si_2Ge_2$ are substantially higher than those of the Gd metal (compare Figure 4.8.6 with Figure 4.8.1). The adiabatic temperature changes are greater as well, but the differences in ΔT_{ad} are not as significant because of much higher heat capacity of $Gd_5Si_2Ge_2$. By comparing Figures 4.8.1 and 4.8.6, one also notices a distinct difference in shape: Gd has caret-like temperature dependence of its MCE parameters in any field of 100 kOe or below, while in $Gd_5Si_2Ge_2$, the giant MCE broadens and flattens in high magnetic fields after reaching a steep maximum at T_C, exhibiting weakly decaying plateaus at $T > T_C$, which then abruptly drop at different ΔH-dependent temperatures. The discrepancy comes from the different thermodynamic nature of the corresponding magnetic transitions: the conventional MCE in Gd is associated with a second-order (continuous) paramagnetic-ferromagnetic phase transition, while the anomalous (giant) MCE in $Gd_5Si_2Ge_2$ is related to the paramagnetic-ferromagnetic transition coupled with a crystallographic transformation [39], making the transition at T_C of the first-order type (discontinuous).

3 The giant magnetocaloric effect became mainstream avenue of research after the discovery of first order magnetostructural transformation in $Gd_5Si_2Ge_2$ [5, 39] and related $Gd_5Si_{4-x}Ge_x$ [6], even though some 7 years earlier, a large magnetocaloric effect associated with first order magnetovolume ferromagnetic-antiferromagnetic phase transformation was reported in FeRh [37].

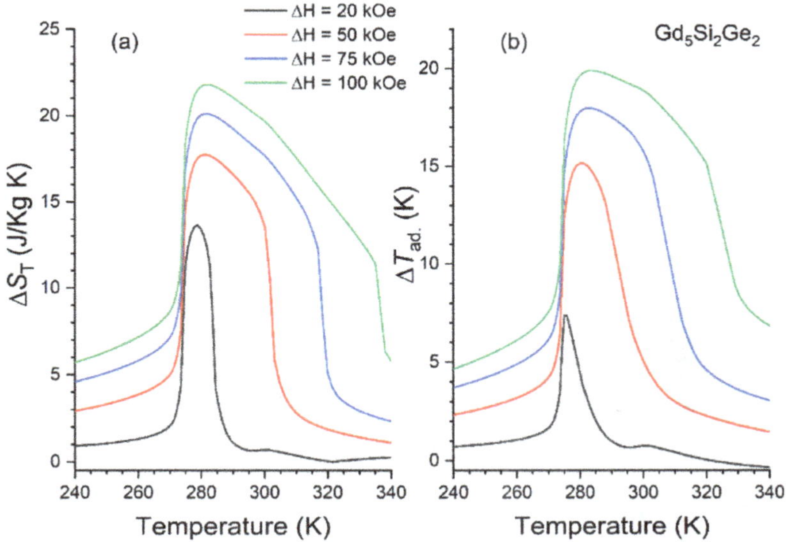

Figure 4.8.6: Magnetocaloric effect in $Gd_5Si_2Ge_2$: (a) magnetic entropy change; (b) adiabatic temperature change in magnetic fields up to 100 kOe.

The coupling means that both the magnetic (Figure 4.8.7a) and structural transformations (Figure 4.8.7b) in $Gd_5Si_2Ge_2$ occur simultaneously. A specific magnetic state of $Gd_5Si_2Ge_2$ is always connected to a specific crystal structure: as long as the compound retains a monoclinic structure it remains paramagnetic, but when it orders ferromagnetically, the crystallography changes from the monoclinic to a closely related orthorhombic phase. When the paramagnetic monoclinic $Gd_5Si_2Ge_2$ phase transforms into the ferromagnetic orthorhombic structure, some of the interatomic bonds that connect pairs of quasi-two-dimensional atomic layers or slabs (Figure 4.8.7b) are created when the slabs shift with respect to one another, which, in turn, is associated with a drastic change in the magnetic exchange in the material, making it strongly ferromagnetic [40, 41]. Both the magnetic and structural transitions are reversible. The same transition may also be triggered by applied hydrostatic pressure and, more importantly with respect to MCE, by applied magnetic field. The first-order nature of the transition makes the change of magnetization with temperature nearly discontinuous (Figure 4.8.7a), and the resulting $|(dM/dT)_H|$ may reach very high values, remaining so in high magnetic fields.[4] High $|(dM/dT)_H|$ enhances

4 Theoretically, when a first-order transition is truly discontinuous, $|(dM/dT)_H| \rightarrow \infty$ (the same is true for $|(dV/dT)_H|$, $|(dS/dT)_H|$ and a number of other properties). However, in real solids, first-order transitions usually occur over a certain range of temperatures (pressures, magnetic fields) while traversing phase-separated states, so $|(dM/dT)_H|$ remains finite.

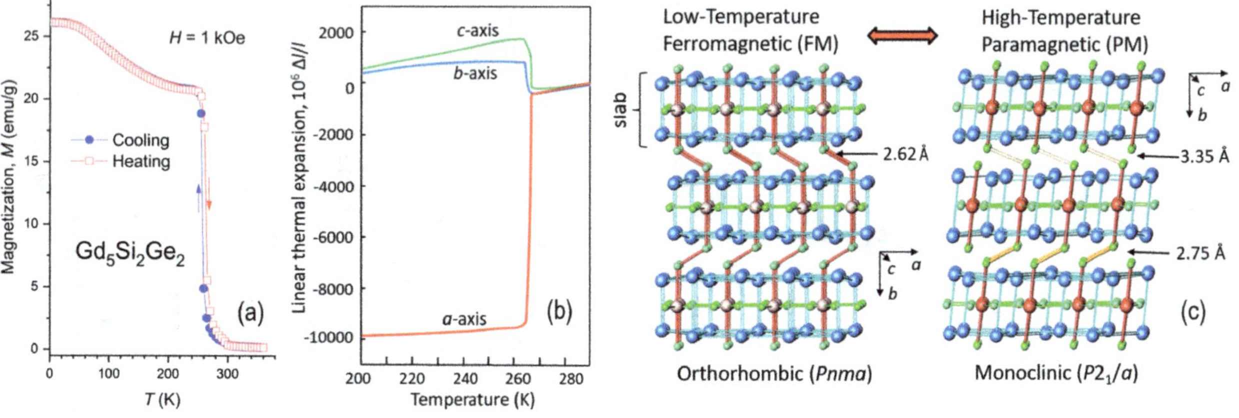

Figure 4.8.7: Magnetic (left) and structural (middle) transformations in $Gd_5Si_2Ge_2$ occurring simultaneously at $T_C = 270$ K, which is also the temperature of the first-order transformation. The right-hand panel illustrates the atomic scale mechanism of the structural phase transition.

both the isothermal entropy change and the adiabatic temperature change, leading to giant MCE.

A characteristic feature of the magnetostructural transformation shown in Figure 4.8.7 is the presence of thermal (and magnetic) irreversibilities, or hysteresis, namely, in the vicinity of the transition, $M(T)$ data recorded during heating and cooling do not match. When looking for the next promising magnetocaloric material, the presence of hysteresis in either or both $M(T)$ and $M(H)$ data should cause both excitement and caution: on the one hand, the materials with first-order transitions are highly desirable, on the other, if hysteresis is too large (some 10–20 K or more, or some 5–10 kOe or more) the material may not be useful for applications that rely solely on actuation of the giant magnetocaloric effect with magnetic field [42]. Hysteresis is generally a detriment, because in the magnetocaloric refrigerator the material must be cycled through the transition many times, but any and all irreversibilities represent energy losses. Many materials with giant MCEs, including $Gd_5Si_2Ge_2$, have at least some irreversibility, and minimization of hysteresis-related losses remains one of the major thrusts in research on giant magnetocaloric materials [43, 44].

Different contributions to giant magnetocaloric effect can be analyzed and quantified by comparing magnetic field-induced isothermal entropy changes of closely related materials, if one exhibits a first- and another a second-order phase transformation. $Gd_5(Si_{1-x}Ge_x)_4$ is an excellent model system for such exercise. Whereas $Gd_5Si_2Ge_2$ has a distinct first-order magnetostructural transition, a minor increase in the concentration of Si, for example, to $Gd_5Si_{2.5}Ge_{1.5}$, makes the transition magnetic-only and second-order. The difference in magnetocaloric effects of these two materials is large: from $|\Delta S|$ of 14 J/kg K in $Gd_5Si_2Ge_2$ (first-order transition) to $|\Delta S|$ of only about 5 J/kg K for $Gd_5Si_{2.5}Ge_{1.5}$, both values are quoted here for the same magnetic field change of 20 kOe. In $Gd_5Si_2Ge_2$, there is an additional contribution from the crystal structure change, $|\Delta S_{st}| > 0$, which in $Gd_5Si_{2.5}Ge_{1.5}$ is reduced to 0. Hence, the total magnetic field-induced entropy change, $|\Delta S_T|$, can be expressed as a sum as follows:

$$\Delta S_T = \Delta S_M + \Delta S_{st}. \tag{4.8.16}$$

Obviously, ΔS_M is magnetic field-dependent and its sign depends on the signs of both $(dM/dT)_H$ and ΔH (eq. (4.8.6)). As follows from the mean field theory, ΔS_M scales with $H^{2/3}$ when $H_I = 0$ [45], but ΔS_{st} is not, provided ΔH is large enough to complete the crystallographic transition. Since ΔS_{st} only depends on the difference of entropies between the two crystallographically different phases, in this case monoclinic and orthorhombic, for any given ΔH, eq. (4.8.16) can be rearranged as:

$$\Delta S_{T,\Delta H} - \Delta S_{M,\Delta H} = \Delta S_{st} = \text{const}, \tag{4.8.17}$$

making it easy to estimate ΔS_{st} by comparing ΔS_T measured for various ΔH [46]. We note that the sign of ΔS_{st} may be either the same or different than the sign of ΔS_M. If the signs are identical, the magnetocaloric effect is enhanced, but it is reduced when the signs of ΔS_{st} and ΔS_M are opposite. The difference between the isothermal magnetic entropy changes of $Gd_5Si_2Ge_2$ and $Gd_5Si_{2.5}Ge_{1.5}$ remains nearly constant, regardless of $|\Delta H|$ between 20 and 100 kOe, yielding $|\Delta S_{st}|$ of approximately 9.8 J/kg K. Importantly, the signs of ΔS_{st} and ΔS_M are identical, and $|\Delta S_{st}|$ exceeds $|\Delta S_M|$ by a factor of nearly 2 in a relatively weak $|\Delta H|$ of 20 kOe. The large $|\Delta S_{st}|$ in $Gd_5Si_2Ge_2$ is due to a major reshuffling of the atomic structure (Figure 4.8.7c), which is also accompanied by more than 1% volume change. Similar considerations and analyzes apply to first-order magnetocaloric materials for cryogenic applications. For example, $DyCo_2$ undergoes a first-order transformation between cubic paramagnetic and tetragonal ferrimagnetic phases at T_C of 140 K. By replacing Co with small amounts of Si or Ge, one can turn this material into the one with a second-order transformation. For $DyCo_2$, phase volume change $\Delta V/V$ is 0.2%, much smaller than that in $Gd_5Si_2Ge_2$, and $|\Delta S_{st}|$ is reduced as well to 4.5 J/kg K [20].

Substantial crystallographic changes altering symmetry of the co-existing phases are not the requirement, and some of the top-performing magnetocaloric materials exhibit isosymmetric volume changes, together with magnetic (dis)ordering. For example, in $LaFe_{13-x}Si_x$, which contains non-magnetic lanthanum, both the magnetism and giant magnetocaloric effect are defined by $3d$ electrons of iron. The compounds crystallize in the cubic $NaZn_{13}$-type crystal structure, which remains the same below and above the transition. Depending on x, $LaFe_{13-x}Si_x$ exhibits either first- or second-order transitions at T_C (Figure 4.8.8).[5] Due to a rather large phase volume change of

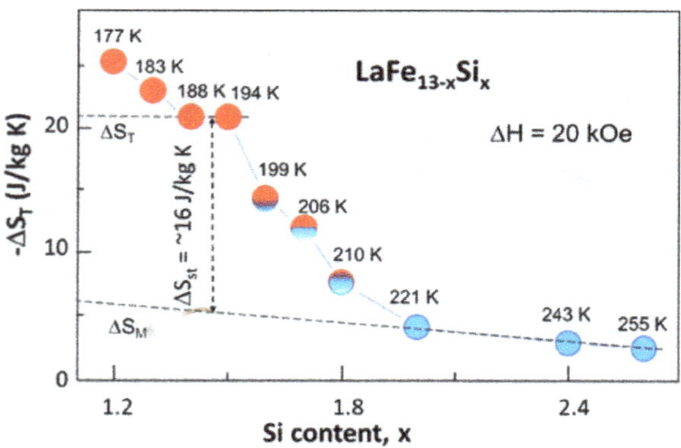

Figure 4.8.8: The total isothermal entropy change, $|\Delta S_T|_{\Delta H}$, in $LaFe_{13-x}Si_x$ as a function of x for $\Delta H = 20$ kOe [20, 47]. The red (blue) circles represent alloys with clearly defined first- (second)-order phase transitions, respectively. The mixed-colored circles show the region where the first-order phase transition gradually weakens and then vanishes, transforming into a second-order phase change as x increases. The numbers near the symbols represent the corresponding T_Cs.

5 Compounds $LaFe_{13-x}Si_x$ with $x < 1.1$ are thermodynamically unstable and binary $LaFe_{13}$ is not known.

1.5%, the structural contribution to the total isothermal entropy change reaches 16 J/kg K as illustrated by the dashed lines in Figure 4.8.8 [20].

As also follows from Figure 4.8.8, the boundary between first- and second-order transitions in real materials is not always clearly defined. Here, the alloys with $1.5 < x \leq 1.8$ exhibit gradual weakening of the first-order nature of the transition at T_C as x increases from 1.5 to 1.8 [47], finally becoming a second-order phase transition when $x = \sim 2$ and beyond, where phase volume change is reduced to 0. Materials that fall into the intermediate region may combine desirable properties, such as large MCE with much reduced hysteresis losses.

Overall, $LaFe_{13-x}Si_x$ compounds are excellent magnetocaloric materials when $x < 1.8$, but their magnetic ordering temperatures, and, consequently, peak MCEs are located at and below 200 K. This makes them attractive candidates for the high temperature range in cryogenic applications, but they are not suitable for use at ambient conditions. There are two known ways to bring their operating temperature range to room temperature: (1) partial substitution of Fe or Si with Co [48]; (2) expansion of the lattice with hydrogen [49]. In the first case, T_C increases because Co has a higher T_C compared to Fe, and, for example, Co-substituted $LaFe_{10.97}Co_{1.03}Si$ has $T_C = \sim 293$ K [48]. However, the transition becomes second order, and its entropy change is less than a half when compared to the parent $LaFe_{11.8}Si_{1.2}$. The second approach, hydrogenation of $LaFe_{13-x}Si_x$ alloys, preserves the first-order nature of the phase transformation and the giant MCE, while raising T_C (which depends on the amount of absorbed hydrogen) up to 340 K [49]. The fully hydrogenated materials contain up to 1.65 hydrogen atoms per formula unit and, in principle, by varying the hydrogen content, one can control the T_C. However, compounds with less than full hydrogen content and T_C at room temperature phase-separate into hydrogen-rich and hydrogen-poor phases, with two different T_Cs when held at room temperature, for a few days to a few weeks [50]. Thus, these magnetocaloric materials must be fully hydrogenated for practical use. The necessary tuning of T_Cs is achieved by minor substitutions of Fe with Mn, which strongly reduces T_C of the $LaFe_{13-x}Si_x$ parent. The Si content is adjusted as well, so the actual alloys can be described by the formula $LaFe_xMn_ySi_zH_{1.65}$, where $11.22 \leq x \leq 11.76$, $0.06 \leq y \leq 0.46$, and $1.18 \leq z \leq 1.32$, making these complex intermetallic hydrides useful over 270–340 K temperature range [51].

4.8.4 Novel thermoelectric materials containing rare earths

The phenomenon of thermoelectricity allows for direct conversion of heat into electricity or, conversely, to cool materials directly by passing electric current. In 1821, T. J. Seebeck discovered that a temperature gradient maintained across a conductor may lead to a notable generation of voltage – he showed that when two junctions in a closed circuit, made of two different conductors, are held at different temperatures, the continuous flow of electric current occurs. For a given pair of conductors the measured electric potential is proportional to the magnitude of the temperature difference between the junctions, as given by:

$$V = -\alpha \Delta T, \tag{4.8.18}$$

where α is Seebeck coefficient, V is voltage and T is temperature. The Seebeck voltage is, essentially, an induced electrostatic potential compensating the change in the chemical potential, which occurs due to difference in the carrier concentrations between the hot and cold ends of the paired conductors.[6] Conversely, application of electric current to a similar system generates temperature difference between the junctions, and is known as Peltier effect, named after J. C. A. Peltier, who co-discovered thermoelectricity independently from T. J. Seebeck, and is shown by:

$$\Pi = \dot{Q}/I, \tag{4.8.19}$$

where Π is the Peltier coefficient, $\dot{Q}$ is the heat flow, and I the electric current. Both Peltier and Seebeck effects are closely related to the Thompson effect, which describes thermal (caloric) effect in a current carrying material subjected to a temperature gradient. The relationship between Peltier and Seebeck coefficients is as follows:

$$\Pi = \alpha T \tag{4.8.20}$$

Equation (4.8.20) holds true for magnetically disordered materials and in the absence of external magnetic field. One must remember that, in practice, the Seebeck coefficient, or coefficient of thermoelectric power, α, is measured in paired circuits. This way it is determined with respect to the other conductor and, technically, has to be noted as such. At the same time, α for actual thermoelectric materials is several orders of magnitude higher than for good metallic conductors such as Cu. It is possible to determine the absolute value of α, but it has more fundamental than applied value.

The sign of the Seebeck coefficient (direction of the current flow with respect to the direction of the heat flow) depends on the type of the prevailing charge carriers. In metals, these are electrons, so α is negative, but semiconductors can be both n-type (charge carriers are electrons) and p-type (charge carriers are holes), so there are two types of thermoelectric materials, that is, n-type and p-type.

Suitability of materials for thermoelectric applications depends on α (the higher the better), but it is only one among several important parameters of materials to be considered. The other two are electrical conductivity, σ (inverse of electrical resistivity, ρ) and thermal conductivity, κ. In 1958, z, the material's figure of merit, was

6 When the generated voltage is measured and calibrated, such a pair may serve as a "thermocouple" – a crucial part of digital thermometry.

introduced as a measure of efficiency for the thermoelectric energy conversion [52] and is given by:

$$z = \frac{\alpha^2 \sigma}{\kappa}. \tag{4.8.21}$$

This quantity is universally adopted as a measure of materials performance simply because it is related to the maximum (a.k.a., Carnot) efficiency, η_{max}, of a thermoelectric device determined as follows:

$$\eta_{max} = \frac{T_h - T_c}{T_h} = \frac{\sqrt{1+zT}-1}{\sqrt{1+zT}+T_c/T_h}, \tag{4.8.22}$$

where T_h and T_c are, respectively, temperatures of the hot and cold junctions, and the product, zT, is the dimensionless figure of merit in common use. Both z and zT are temperature dependent, so different materials show peak efficiency at different temperatures. A thermoelectric material is considered suitable for application if its zT is about 1 or higher, although for some applications (such as waste heat utilization) lower values of zT (e.g., 0.3–0.8) may become acceptable, if that material is considerably cheaper.

Small band-gap semiconductors are good candidates as thermoelectric materials because they combine a high Seebeck coefficient, typical of semiconductors, with the low electrical resistivity of metals. In that regard, plenty of promising thermoelectrics are found among Zintl phases, whose electronic properties are intermediate between those of intermetallic and insulating compounds. Zintl phases are charge-neutral, with a high degree of covalent bonding, yet they offer wide-ranging opportunities for chemical substitutions, and, consequently, tuning of the band gap. There are many known Zintl phase rare earth compounds. Adjustment of the band gap and doping levels optimizes, that is, maximizes the numerator of eq. (4.8.21), which is termed a thermoelectric power factor $P = \alpha^2\sigma$.

Obviously, the best thermoelectric materials must have large α and σ, but at the same time their thermal conductivity should be kept as small as possible, if the goal is high efficiency. The total thermal conductivity combines electronic κ_{el} and phononic κ_{ph} parts, $\kappa = \kappa_{el} + \kappa_{ph}$. The immediate problem here is that the κ_{el} is directly proportional to σ according to the Wiedemann-Franz law [22] and is given by:

$$\kappa_{el} = LT\sigma, \tag{4.8.23}$$

where L is a proportionality constant named Lorenz number. Therefore, reducing κ_{el} will proportionally reduce σ, without changing zT. The focus of recent research on thermoelectric materials lies in the minimization of κ_{ph} while maintaining good κ_{el}. It is known that amorphous, glass-like materials poorly conduct both heat and electricity, yet a periodic arrangement of atoms is best to maintain high electrical conductivity. In 1995, G. A. Slack [53] proposed a concept of phonon-glass electron-crystal (PGEC) – a solid combining poor lattice thermal conductivity of a glass with high electrical conductivity of a crystalline conductive solid. Even though the approach appears to be counterintuitive, such materials were indeed discovered among crystalline compounds that contain large voids or "cages" that can be filled by heavy atoms, for example, lanthanides. When diameters of lanthanides and chemically similar alkali metal atoms are smaller than the diameters of the voids, chemical bonds between the filler atoms and the host structure are weak. In this case, the heavy atoms vibrate with large amplitudes – in other words, rattle. The rattling disrupts coherent propagation of phonons through the lattice and strongly reduces phonon-based thermal conductivity κ_{ph}. At the same time, the electrical conductivity is determined by the conduction bands of the host atoms and is practically unaffected by the rattling. Examples of successful thermoelectric materials designed using the PGEC concept are filled skutterudites and intermetallic clathrates.

Skutterudites is a family of stable compounds with rather large voids in their unit cells. The host structure is cubic with 32 atoms/unit cell and MPn_3 stoichiometry, where M is transition metal, typically Group 9 element (Co, Rh, Ir), and Pn is pnictogen (Group 15 element), such as P, As or Sb. The voids that exist naturally in skutterudites can be partially filled by heavy atoms, including lanthanides, which allows scientists to study the effects of heavy atom doping by comparing properties of both filled and unfilled host structures. In addition, both M and Pn elements can be substituted by their neighbors in the periodic table (for example, Co for Fe), opening additional options for property tuning.

Unfilled binary skutterudites are semiconductors with good α and σ, but the thermal conductivity is also quite high. Doping the atomic cages with rattler atoms reduces $\kappa_{ph,}$ in accordance with the PGEC concept. The actual occupancy by the heavy atoms is typically at or below 25%, so the compositions of filled skutterudites are commonly represented as REM_4Pn_{12}. The concentrations of RE can be as low as a few per cent atomic. For example, the $CoSb_3$ host can be filled by rather heavy Yb atoms to form $Yb_xCo_4Sb_{12}$ with $x = 0$, 0.066, and 0.19 [9]. As shown in Figure 4.8.9, the reduction of the lattice thermal conductivity is, indeed, very significant, even for the heavily underdoped $Yb_{0.066}Co_4Sb_{12}$. The dimensionless figure of merit zT of $Yb_{0.19}Co_4Sb_{12}$ is 0.3 near room temperature, and it reaches zT of approximately 1 at 600 K, becoming comparable with that of the best thermoelectric materials, such as Bi_2Te_3. Other skutterudite materials, such as $CeFe_4Sb_{12}$ or $CsFe_{4-x}Co_xSb_{12}$ show even higher zTs, which approach 1.3.

Clathrates. A term "clathrate" covers a large variety of compounds, both organic and inorganic. In general, it means an inclusion complex, where atoms or molecules of one compound (guest) are completely enclosed inside cavities present in the molecular or crystalline structure of the other (host). The "complete enclosure" of the guest differentiates clathrates from chelates, in which the "guest" is trapped but is not fully encapsulated. The hydrate clathrates, in particular, methane clathrate ($4CH_4 \cdot 23H_2O$) have been found in large quantities under sediments

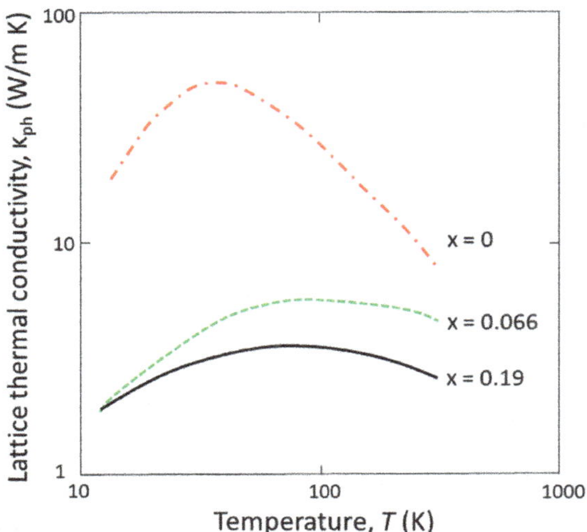

Figure 4.8.9: Thermal conductivity as a function of temperature for the unfilled ($x = 0$) and filled ($x = 0.066$, 0.19) skutterudites $Yb_xCo_4Sb_{12}$ [9].

of the ocean floor. In the 1960s, clathrate-like structures were found among intermetallic compounds crystallizing in two structure types: type I (M_8T_{46}, illustrated in Figure 4.8.10) and type II (M_xT_{136}), where M is a large guest atom, usually alkali or alkaline earth metal, and T is p-element (most commonly Si, Ge, or Sn). Other p-elements, as well as late transition metals can become part of the network. Two characteristic features of clathrate structures are: 1) presence of large cages formed by a network of tetrahedrally coordinated atoms, which are commonly filled by guest atoms, fully or partially, and 2) pentagondodecahedron polyhedra as the base structural units. Another principal difference between clathrates and skutterudites is that the latter are fully stable without the guest atoms, while clathrate compounds without guest atoms are very difficult to obtain [54].

The basic clathrate crystallography fits the PGEC model very well. The host network of intermetallic clathrates is commonly formed by p-elements from Groups 13 to 15, but doping by late d-elements, such as Cu or Zn, is possible, and opens a lot of possibilities for tuning the electronic transport properties. At the same time, presence of large guest atoms lowers the lattice thermal conductivity, increasing zT.

The voids in the clathrate structures are actually too large to host the normally trivalent lanthanide ions. However, one particular lanthanide, europium, in its divalent state, is large enough to stabilize the clathrate structure. Both full Eu clathrates, such as $Eu_8Ga_{16}Ge_{30}$ [55] and mixed clathrates, such as $Eu_{0.27}Ba_{7.22}Al_{13}Si_{29}$ are known [56]. These compounds exhibit long range magnetic ordering at low temperatures with T_C of about 40 K, however, thermoelectric performance of Eu-containing

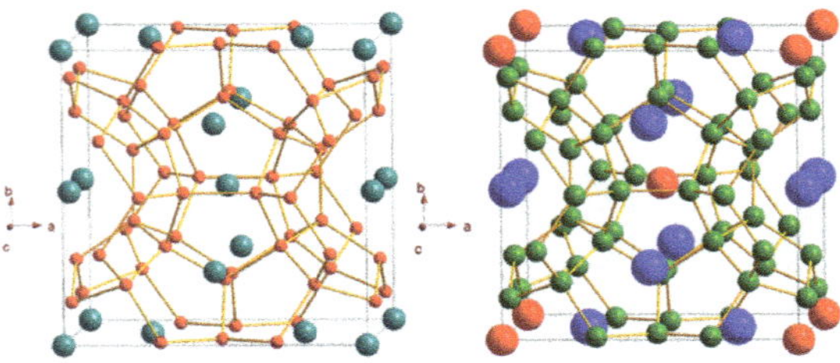

Figure 4.8.10: Comparison of type I hydrate (left) and intermetallic clathrate (right) structures. In the hydrate structure, only oxygen atoms (red) from water molecules are shown for clarity.

clathrates is not very high, with zT less than 0.1 due to low α. For comparison, zT of $Ba_8Ga_{16}Ge_{30}$ approaches 1.3 at 1,000 K.

$Yb_{14}MnSb_{11}$ Zintl phases. Even though $Yb_{14}MnSb_{11}$ is considered a Zintl phase, it is not a typical semiconductor [11]. It crystallizes in the $Ca_{14}AlSb_{11}$ type of crystal structure. Mimicking the 2+ state of Ca in $Ca_{14}AlSb_{11}$, ytterbium in $Yb_{14}MnSb_{11}$ is divalent, but Mn is also in a 2+ state, contributing one electron per formula unit less than Al^{3+}. The shortage of one electron makes this compound a *p*-type material. $Yb_{14}MnSb_{11}$ has the best thermoelectric performance for high-temperature applications, reaching zT of 1 above ~1,000 K. The reason for this is that even though its $\alpha < 200$ μV/K is rather modest, and its electrical resistivity increases with temperature (metallic type), its thermal conductivity is very low and comparable to that of common glass ($\kappa = 7$–9 mW/cm K). Both the complexity of the crystal structure (208 atoms/unit cell) and presence of strongly vibrating atoms are responsible for low κ_{ph}. A disadvantage of $Yb_{14}MnSb_{11}$ lies in its chemical instability in the presence of water. This is a common problem of many other lanthanide metallic and semimetallic compounds (including skutterudites and clathrates), which, in addition to cost, somewhat limits their application perspectives despite very unusual and exciting chemistry and physics.

References

[1] P. Debye, Ann. Phys. 1926, 81, 1154.
[2] W. F. Giauque, J. Am. Chem. Soc. 1927, 49, 1864.
[3] W. F. Giauque, D. P. MacDougall, Phys. Rev. 1933, 43, 768.
[4] G. V. Brown, J. Appl. Phys. 1976, 47, 3673.
[5] V. K. Pecharsky, K. A. Gschneidner Jr., Phys. Rev. Lett. 1997, 78, 4494.
[6] V. K. Pecharsky, K. A. Gschneidner Jr., Appl. Phys. Lett. 1997, 70, 3299.
[7] K. A. Gschneidner Jr, V. K. Pecharsky, Int. J. Refrig. 2008, 31, 945–961.
[8] G. S. Nolas, G. A. Slack, D. T. Morelli, T. M. Tritt, A. C. Ehrlich, J. Appl. Phys. 1996, 79, 4002.
[9] G. S. Nolas, M. Kaeser, R. T. Littleton IV, T. M. Tritt, Appl. Phys. Lett. 2000, 77, 1855.
[10] G. S. Nolas, D. T. Morelli, T. M. Tritt, Annu. Rev. Mater. Sci. 1999, 29, 89.

[11] S. R. Brown, S. M. Kauzlarich, F. Gascoin, G. J. Snyder, Chem. Mater. 2006, 18, 1873.
[12] P. Weiss, A. Piccard, Comptes Rendus 1918, 166, 352.
[13] A. M. Tishin, Y. I. Spichkin, The Magnetocaloric Effect and Its Applications, IoP Publishing, Bristol and Philadelphia, 2003.
[14] K. A. Gschneidner Jr, V. K. Pecharsky, Annu. Rev. Mater. Sci. 2000, 30, 387.

[15] K. A. Gschneidner Jr., V. K. Pecharsky, A. O. Tsokol, Rep. Prog. Phys. 2005, 68, 1479.
[16] A. E. Clark, E. Callen, Phys. Rev. Lett. 1969, 23, 307.
[17] M. Foldeaki, R. Chahine, T. K. Bose, J. Appl. Phys. 1995, 77, 3528.
[18] V. K. Pecharsky, K. A. Gschneidner Jr., J. Appl. Phys. 1999, 86, 565.
[19] H. Nevez Bez, H. Yibole, A. Pathak, Y. Mudryk, V. K. Pecharsky, J. Magn. Magn. Mater. 2018, 458, 301.
[20] K. A. Gschneidner Jr., Y. Mudryk, V. K. Pecharsky, Scr. Mater. 2012, 67, 572.
[21] T. Gottschall, M. D. Kuz'min, K. P. Skokov, Y. Skourski, M. Fries, O. Gutfleisch, M. Ghorbani Zavareh, D. L. Schlagel, Y. Mudryk, V. Pecharsky, J. Wosnitza, Phys. Rev. B 2019, 99, 134429.
[22] C. Kittel, Introduction to Solid State Physics, 2nd Edition, John Wiley & Sons, New York, 1956.
[23] J. A. Barclay, W. A. Steyert, Cryogenics 1982, 73.
[24] S. Jacobs, J. Auringer, A. Boeder, J. Chell, L. Komorowski, J. Leonard, S. Russek, C. Zimm, Inter. J. Magn. Refrig. 2014, 37, 84.
[25] A. M. Tishin, K. A. Gschneidner Jr., V. K. Pecharsky, Phys. Rev. B 1999, 59, 503.
[26] K. A. Gschneidner Jr., J. Less-Common Met. 1985, 114, 29.
[27] G. E. Brodale, E. W. Hornung, R. A. Fischer, W. F. Giauque, J. Chem. Phys. 1975, 62, 4041.
[28] R. D. McMichael, J. J. Ritter, R. D. Shull, J. Appl. Phys. 1993, 73, 6946.
[29] K. Matsumoto, A. Matsuzaki, K. Kamiya, T. Numazawa, Jpn. J. Appl. Phys. 2009, 48, 113002.
[30] G. Lorusso, J. W. Sharples, E. Palacios, O. Roubeau, E. K. Brechin, R. Sessoli, A. Rossin, F. Tuna, E. J. L. McInnes, D. Collison, M. Evangelisti, Adv. Mater. 2013, 25, 4653.
[31] a) M. A. Ruderman, C. Kittel, Phys. Rev. 1954, 96, 99; b) T. Kasuya, Prog. Theor. Phys. 1956, 16, 45; c) K. Yosida, Phys. Rev. 1957, 106, 893.

[32] a) K. A. Gschneidner Jr., V. K. Pecharsky, M. J. Gailloux, H. Takeya, Adv. Cryog. Eng. 1996, 42A, 465; b) K. A. Gschneidner, V. K. Pecharsky, S. K. Malik, Adv. Cryog. Eng. 1996, 42A, 475.
[33] Y. Mudryk, V. K. Pecharsky, K. A. Gschneidner Jr., Proceedings of the 3rd IIF-IIR International Conference on Magnetic Refrigeration at Room Temperature, Des Moines, 2009, 127.
[34] T. Numazawa, K. Kamiya, T. Utaki, K. Matsumoto, Cryogenics 2014, 62, 185.
[35] T. A. Yamamoto, T. Nakagawa, K. Sako, T. Arakawa, H. Nitani, J. Alloys Compd. 2004, 376, 17.
[36] S. A. Nikitin, A. M. Tishin, S. V. Redko, Phys. Met. Metallogr. 1988, 66, 77.
[37] S. A. Nikitin, G. Myalikgulyev, A. M. Tishin, M. P. Annaorazov, K. A. Asatryan, A. L. Tyurin, Phys. Lett. A 1990, 148, 363.
[38] V. K. Pecharsky, A. P. Holm, K. A. Gschneidner Jr., R. Rink, Phys. Rev. Lett. 2003, 91, 197204.
[39] L. Morellon, P. A. Algarabel, M. R. Ibarra, J. Blasco, B. García-Landa, Z. Arnold, F. Albertini, Phys. Rev. B 1998, 58, R14721.
[40] G. D. Samolyuk, V. P. Antropov, J. Appl. Phys. 2003, 93, 6882.
[41] V. K. Pecharsky, K. A. Gschneidner Jr., Adv. Mater. 2001, 13, 683.
[42] V. K. Pecharsky, J. Cui, D. D. Johnson, Phil. Trans. A 2016, 374, 20150305.
[43] F. Guillou, G. Porcari, H. Yibole, N. van Dijk, E. Bruck, Adv. Mater 2014, 26, 2671.
[44] E. Lovell, A. M. Pereira, A. D. Caplin, J. Lyubina, L. F. Cohen, Adv. Energy Mater. 2015, 5, 1401639.
[45] H. Oesterreicher, F. T. Parker, J. Appl. Phys. 1984, 55, 4334.
[46] V. K. Pecharsky, K. A. Gschneidner Jr., Magnetocaloric Effect Associated with Magneto-Structural Transitions, in Magnetism and Structure in Functional Materials. Springer Series in

Materials Science (ed. A. Planes, L. Mañosa, A. Saxena), Vol. 79, Ch. 11, Springer, Heidelberg, 2005, 199.
[47] B. G. Shen, J. R. Sun, F. X. Hu, H. W. Zhang, Z. H. Cheng, Adv. Mater. 2009, 21, 4545.
[48] J. D. Moore, K. Morrison, K. G. Sandeman, M. Katter, L. F. Cohen, Appl. Phys. Lett. 2009, 95, 252504.
[49] A. Fujita, S. Fujieda, Y. Hasegawa, K. Fukamichi, Phys. Rev. B 2003, 67, 104416.
[50] C. B. Zimm, S. A. Jacobs, J. Appl. Phys. 2013, 113, 17A908.
[51] V. Basso, M. Küpferling, C. Curcio, C. Bennati, A. Barzca, M. Katter, M. Bratko, E. Lovell, J. Turcaud, L. F. Cohen, J. Appl. Phys. 2015, 118, 053907.
[52] T. Takabatake, K. Suekuni, T. Nakayama, E. Kaneshita, Rev. Mod. Phys. 2014, 86, 669.
[53] G. A. Slack, CRC Handbook of Thermoelectrics (ed. D. M. Rowe), CRC Press, Boca Raton, 1995, 407.
[54] A. M. Guloy, R. Ramlau, Z. Tang, W. Schnelle, M. Baitinger, Yu. Grin, Nature 2006, 443, 320.
[55] S. Paschen, W. Carrillo-Cabrera, A. Bentien, V. H. Tran, M. Baenitz, Yu. Grin, F. Steglich, Phys. Rev. B 2001, 64, 214404.
[56] C. L. Condron, R. Porter, T. Guo, S. M. Kauzlarich, Inorg. Chem. 2005, 44, 9185.

Holger Kohlmann

4.9 Rare earth metal-based hydride materials

Rare earth metals and many of their intermetallic compounds react readily in an exothermic reaction with hydrogen gas to form stable binary hydrides. These compounds are of fundamental interest, since they exhibit chemical bonding and physical properties intermediate between salt-like, semiconducting and metallic. Therefore, they bridge the gap between alkali and alkaline earth metal hydrides (such as NaH or CaH_2) on one hand and transition metal hydrides (such as $PdH_{0.7}$ or $VH_{0.5}$) on the other hand. While the former represent typical stoichiometric, ionic compounds and the latter are metallic and nonstoichiometric with respect to hydrogen, rare earth metals may form both types of hydrides, depending on the rare earth metal and the conditions (T, $p(H_2)$). In all those types, however, hydrogen atoms are more or less negatively polarized, that is, hydridic in nature. Ternary and multinary rare earth hydrides also range from salt-like to metallic and are important for various types of applications such as hydrogen storage or the production of magnetic materials. The present chapter gives a general overview of rare earth metal hydrides with a focus on the characteristics of binary and ternary hydrides as well as some applications. More in-depth coverage may be found at [1, 2].

4.9.1 Binary rare earth metal hydrides and their characterization

All rare earth metals react with gaseous hydrogen to form binary hydrides at room temperature, even at hydrogen pressures below 1 bar. However, because of the low reaction rates, slightly elevated temperatures of, typically, 400 K $\leq T \leq$ 600 K are applied usually. With the exception of the red colored EuH_2, rare earth metal hydrides are grey to black in color or show metallic appearance. Hydrogenation of the metals yields fine powders that are easily ground due to their brittle nature. Hydrides of the type $RE\text{H}_{2\leq x\leq 3}$ are nonstoichiometric with respect to hydrogen and exhibit densities that are 12%–18% lower than those of the hydrogen free elements. Dihydrides of the type $RE\text{H}_2$ have a residual electric resistivity of at least one order of magnitude lower than the respective hydrogen-free rare earth metals. This might be related to reduced scattering because of enlarged cell volumes and they are paramagnetic at room temperature [2]. Some of them order ferromagnetically (T_C in K: NdH_2 (10), EuH_2 (24)) or antiferromagnetically, (T_N in K: GdH_2 (21), TbH_2 (40), DyH_2 (8), HoH_2 (8)) with magnetic moments coupled via conduction electrons, that is, by the RKKY

https://doi.org/10.1515/9783110654929-032

(Ruderman–Kittel–Kasuya–Yosida) mechanism. Since the conduction band is depleted of electrons upon evolution of the trihydrides due to hydride ion formation, the RKKY mechanism is no longer viable and the ordering temperatures are considerably lower in $RE\text{H}_3$. The loss of conduction electrons in the reaction 2 $RE\text{H}_2 + \text{H}_2 \rightarrow$ 2 $RE\text{H}_3$, is also responsible for a metal–semiconductor transition, with an appropriate change in optical properties. A more detailed description of this phenomenon and its use in switchable mirrors is given below. The hydrides of the type $RE\text{H}_{2\leq x\leq 3}$ are not stable in air. Red (sometimes dark violet) colored EuH_2 reacts very quickly to form a yellow powder, probably $\text{Eu(OH)}_2 \times \text{H}_2\text{O}$ via EuOOH [3]. For most other binary rare earth metal hydrides, stability in air depends on the particle size and ranges from being pyrophoric to slow reaction, over several weeks.

In addition, there are hydrides, often called α phases, with lower and strongly temperature-dependent hydrogen concentration. As an example, the phase diagram La-H is discussed here (Figure 4.9.1). The α phase retains the *hc* packing of lanthanum metal (Chapter 1.7) and exhibits low hydrogen concentration of $x \leq 0.01$. Above 600 K, β-LaH_x ($0 \leq x \leq 0.27$) with cubic close packing of lanthanum atoms and higher hydrogen content, probably in tetrahedral voids, appears. Yet, at higher temperatures, γ-LaH_x ($0 \leq x \leq 0.85$) crystallizes with a W-type structure and hydrogen, probably in

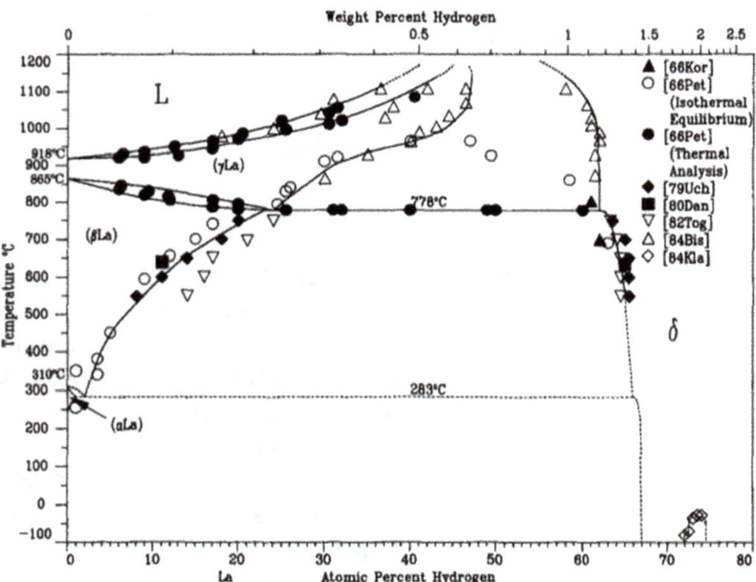

Figure 4.9.1: The La-H phase diagram (Reprinted by permission from Springer Nature, D. Khatamian, F. D. Manchester, The H-La (Hydrogen-Lanthanum) System, Bull. Alloy Phase Diagrams 1990, 11, 90–99).

pseudo-tetrahedral voids. δ-LaH_x ($1.38 \leq x \leq 3.00$) is separated by a temperature-dependent miscibility gap (at room temperature $0.01 \leq x \leq 1.94$; Figure 4.9.1). Its crystal structure is of the (defect) BiF_3 type, that is, lanthanum atoms form a cubic close packing and hydrogen atoms in tetrahedral and octahedral voids (Figure 4.9.2). Some occupation of the octahedral voids is already seen for $x < 2$, but the greater part is successively filled for increasing x in the range $2 < x \leq 3$, accompanied by a 1.2% decrease in unit cell volume. This is due to a change from metallic to ionic bonding, accompanied by a metal-semiconductor transition. This can be well illustrated by following the optical properties, which change from a dark color to metallic reflectance.

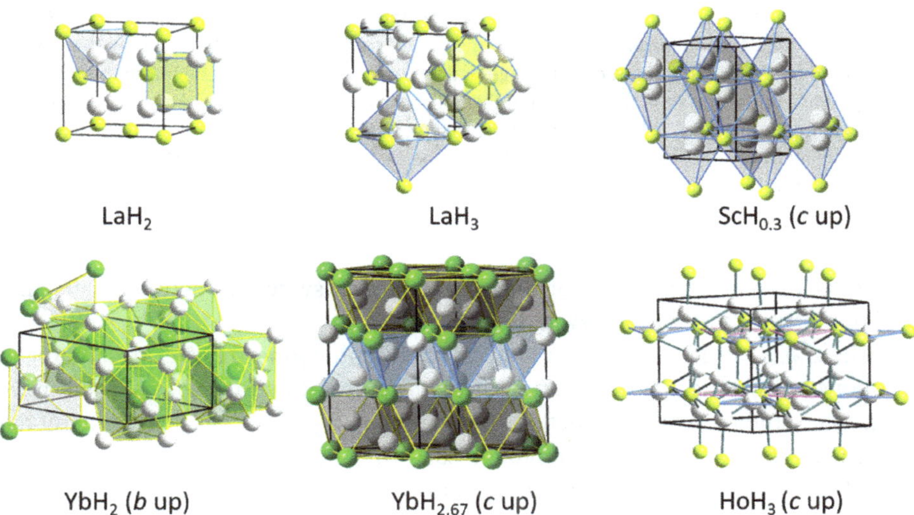

Figure 4.9.2: Crystal structures of binary rare earth metal hydrides with ytterbium atoms shown as green; other rare earth metal atoms shown as yellow and hydrogen atoms shown as white spheres; polyhedra: hydrogen occupies La_4 tetrahedra and lanthanum occupies H_8 cubes in LaH_2, hydrogen occupies La_4 tetrahedra and La_6 octahedra and lanthanum occupies H_{12} rhombic dodecahedra in LaH_3, hydrogen occupies (50%) distorted Sc_4 tetrahedra in $ScH_{0.3}$, distorted hydrogen filled Yb_4 tetrahedra and Yb_5 square pyramids and YbH_9 tricapped trigonal prisms in YbH_2, hydrogen filled distorted Yb_4 tetrahedra (blue) and Yb_6 octahedra (grey) in $YbH_{2.67}$, hydrogen centered Ho_3 triangles (blue), hydrogen atoms slightly above Ho_3 triangles (purple) and distorted Ho_4 tetrahedra (bonds drawn out) in HoH_3.

This behavior can be exploited in switchable mirrors which consist of thin metal films in hydrogen. By varying the partial pressure of the hydrogen gas, the thin film can be switched between a reflecting (REH_2) and transparent (REH_3) state. For example, α-$YH_{x<0.23}$ has a shiny silvery appearance. β-$YH_{\approx 2}$ is also reflecting but with a blue tint and a small transparency window in the red, thus acting as a band-pass filter. γ-$YH_{2.85<x}$ is a yellowish transparent semiconductor (Figure 4.9.3). By

Figure 4.9.3: Switchable mirror, based on hydride formation in a 500-nm-thick yttrium film covered with a 20-nm-thick palladium protection layer. Left: Metallic YH_2 after 1 min exposure to 90 kPa hydrogen acting as a bandpass filter. Right: Transparent, semiconducting YH_3 revealing the chessboard pattern behind (faint mirror image of the knight caused by reflection from the Pd layer; photographs courtesy of Prof. R. Griessen, University of Amsterdam; reprinted from J. N. Huiberts, R. Griessen, J. H. Rector, R. J. Wijngaarden, J. P. Dekker, D. G. Groot, N. J. Koeman, Yttrium and lanthanum hydride films with switchable optical properties, Nature 1996, 380, 231–234, with permission from Nature).

changing the partial pressure of the hydrogen gas, switching between reflecting and transparent states is achieved.

The crystal structures discussed so far exhibit pronounced defects for hydrogen positions with statistical distribution at room temperature and above. There have been some studies on superstructures of the cubic BiF_3 type and the ordering of hydrogen atoms, for example, tetragonal $LaH_{2.28 \le x \le 2.41}$, space group type $I4_1md$ at room temperature or tetragonal $TbH_{2.25}$, space group type $I4/mmm$ [4]. A peculiar feature of the trihydride LaH_3 is the displacement of the hydrogen atoms along the body diagonal of the cubic unit cell by about 22 pm, creating local disorder [5]. Neutron diffraction and NMR spectroscopy suggest a dynamic origin associated with large anharmonic thermal vibration. Upon cooling to 230 K a phase transition occurs, whose origin is not yet clear but might be due to the positional ordering of the displaced hydrogen atoms (Figure 4.9.1). Such order–disorder phase transitions in metal hydrides are often driven by repulsive interaction between hydrogen atoms. H–H interactions are present in the disordered phases, as proved by isotope dilution neutron spectroscopy [6].

The crystal structures of metal hydrides, as discussed above, are usually determined by neutron diffraction. X-ray diffraction, in general, does not contain enough information on light elements when much heavier atoms are also present. Since the ^1H isotope has a huge incoherent neutron scattering cross section, resulting in very high unwanted background in a diffraction experiment, deuterides are used most often. The structural isotope effect is usually very small and manifests in slightly smaller unit cell volumes of the deuterides, as compared to the hydrides [7].

Therefore, for reasons of simplicity, even if it is determined on deuterides, only the formulae for hydrides are given throughout this text.

Diffraction experiments give information on the long-range structure, but average local ordering, for which spectroscopic techniques are better probes. Inelastic neutron scattering (INS) is particularly sensitive to hydrogen due to the large incoherent scattering cross section of the ^1H nucleus, making it an ideal technique for hydrogen poor metal hydrides. Local phenomena such as hydrogen atom vibrations, metal-hydrogen and hydrogen-hydrogen interactions can be easily studied by INS [6]. Unlike collective vibrations of hydrogen atoms in hydrides with high concentrations, the situation for low concentrations is best described by independent local Einstein oscillators. The resulting so-called localized modes depend on the potential formed by the surrounding metal atoms and are thus, a useful probe for hydrogen atom positions (e.g., tetrahedral vs. octahedral voids) and symmetry. This is particularly appealing for compounds with low hydrogen concentration where diffraction methods reach their detection limits. Hydrogen, in octahedral voids, usually has an optical vibration frequency around 80 meV, lower than for hydrogen in the smaller tetrahedral voids, with typical values of 160 meV. INS, quasi-elastic neutron scattering (QENS) and nuclear magnetic resonance (NMR) spectroscopy suggest local order in the hydrogen poor α phases. Sc, Y and Lu show some unusual features, for example, high hydrogen concentration in α phases and peculiar hydrogen diffusion mechanisms [8]. Hydrogen occupies tetrahedral sites in a hexagonal close packing of scandium atoms in α-$ScH_{0.3}$ (Figure 4.9.2). The most probable jump is between the nearest neighbor sites along [001] with about 100 pm distance and shows rates which decrease with decreasing temperatures (10^8 s^{-1} at 100 K). An even faster motion (10^{12} s^{-1}) is seen for a fraction of atoms. This bimodal distribution is probably associated with local order. Slower hydrogen atoms arrange into pairs whereas faster hydrogen atoms do not participate in this local order. There is no long-range order of hydrogen atoms in these α phases of Sc, Y and Lu, that is, no superstructure by diffraction is visible [8]. Due to the quantum origin of the localized hydrogen motion, there is a large isotope effect (H, D).

For the rare earth metals heavier than praseodymium, for Sc and for Y, solid solution ranges of the dihydrides REH_{2+x} are much smaller than for the lighter lanthanides with x_{max} values of 0.077(4) for Y, 0.87(2) for Nd, 0.218(8) for Tb, 0.182(5) for Dy, 0.120(3) for Ho, 0.096(7) for Er, 0.058(6) for Tm and 0.010(3) for Lu (Figure 4.9.4) [9]. There is a change from the cubic BiF_3 to the trigonal HoH_3 type due to the lanthanide contraction. Octahedral voids are too large for smaller lanthanides which would lead to close anion-anion contacts. In the trigonal HoH_3 type, this is avoided by more degrees of freedom (enlarged c/a ratio). The hydrides REH_3 (RE = Y, Sm, Nd, Tb, Dy, Ho, Er, Tm, Lu) are centrosymmetric ($P\bar{3}c1$, HoH_3 structure), which may actually be a superposition of very small noncentrosymmetric twin domains ($P3c1$) [10]. *RE* metal atoms form a hexagonal close packing, and hydrogen atoms occupy tetrahedral and trigonal sites. Hydrogen sites are usually described by split atom positions but essentially display trigonal metal configurations (trigonal LaF_3-type structure). Reports on phase widths differ strongly in literature, with the most probable values being $x < 0.01$ in REH_{3-x} (RE = Y, Sm, Nd, Tb, Dy, Ho, Er, Tm, Lu; Figure 4.9.4) [9].

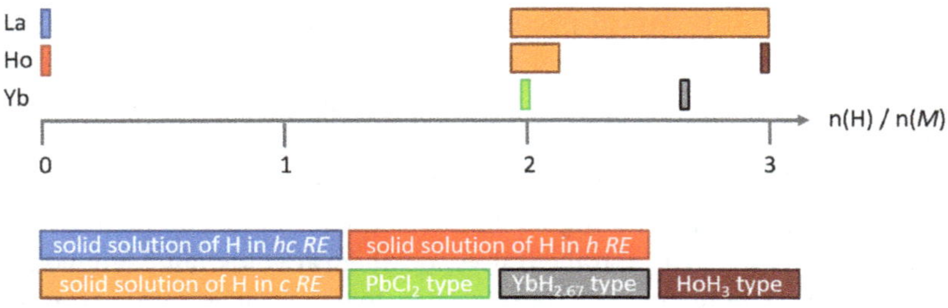

Figure 4.9.4: Three main types of rare earth hydride sequences from RE to REH_3, including stoichiometric ranges at room temperature: La, Ho, Yb.

The solid solution series REH_{2+x} and REH_{3-y} are fascinating due to the complex interplay of several interactions. In the nonstoichiometric regions of the phase diagram, hydrogen atom ordering occurs. For the magnetic *RE*s cooperative magnetism is common and upon increasing the hydrogen content, a metal–semiconductor transition takes place. Magnetic anisotropy, which might be coupled to structural distortions, and the RKKY coupling mechanism are in competition, which gives rise to a plentitude of different magnetically ordered phases [11]. Usually these structural, magnetic and electronic effects are somehow coupled, yielding a more complicated picture than what is outlined so far. This will be briefly explained by two examples. $TbH_{2.00}$ exhibits two sinusoidally modulated antiferromagnetic (AF) configurations. One, a low-T phase, stable below 16 K commensurate with the lattice, and the other between 15 K and 19 K (also called 'intermediate' structure) that is incommensurate with a slightly T-dependent propagation vector k ≈ 1/8(116). Additionally, a broad hump in the background of neutron powder diffraction on the deuterides reveals magnetic short-range order [11]. Upon slightly increasing the hydrogen concentration to $TbH_{2.18}$, thereby occupying octahedral voids, drastic changes occur. Hydrogen atoms order at 200–250 K, causing a tetragonal distortion in a structure with the ideal composition $TbH_{2.25}$. This also increases both magnetic ordering temperatures to 42 and 32.5 K and changes the direction of the propagation vectors [11]. Exchanging the rare earth element will again alter the picture. $DyD_{2.00}$ shows behavior similar to $TbD_{2.00}$, but both ordered structures are incommensurate. In contrast, $DyD_{2.135}$ does not show any long-range but only weak short-range order. These examples show that the interplay of electronic and structural effects strongly influence magnetic phenomena. For increasing the hydrogen content, we see a loss of RKKY coupling and an increased chance for structural ordering, and both effects compete for cooperative magnetism. This leads to antiferromagnetically ordered REH_{2+x} (*RE* = Tb, Dy, Ho) with critical temperatures that are strongly dependent on *x* and *RE* and propagation vectors that are dependent on *RE*, *x* and *T* [11]. This competition between electronic (RKKY) and magnetoelastic interactions triggers a subtle game of opposing effects, resulting in complex structural and magnetic ordering phenomena. A further parameter for tuning these properties is hydrostatic pressure [12].

Europium and ytterbium behave differently due to their particular electronic situations, which favor the divalent oxidation state. This causes analogies to alkaline earth metals with europium resembling strontium and ytterbium resembling calcium, due to similar ionic radii. Europium forms a stoichiometric, salt-like hydride EuH_2 with red and sometimes dark violet color. It crystallizes in the $PbCl_2$ type with ninefold coordination of europium by hydrogen atoms and four- and fivefold coordination of hydrogen by europium atoms. Composition, crystal structure, magnetism and Mößbauer spectroscopy unequivocally show the presence of divalent europium [13]. EuH_3 forms [14] only under extreme pressures in diamond anvil cells. The divalent oxidation state is also favored for ytterbium but less so, as compared to europium. Thus, the hydride YbH_2 (isotypic to EuH_2) is more prone to further hydrogen uptake. At hydrogen pressures above 100 bar and at 600 K, the mixed-valent hydride $YbH_{2.67}$ forms, which exhibits a hydrogen-filled, trigonally distorted cubic close ytterbium atom packing [15]. Hydrogen occupies all tetrahedral and 2/3 of the octahedral voids, according to A β γ α B γ □ β C α β γ (Figure 4.9.2); however, some hydrogen atoms are severely displaced from the center of the octahedra.

4.9.2 Ternary and multinary rare earth metal hydrides

Partial substitution of the rare earth element by other metal atoms or hydrogen by other anions, leads to ternary or multinary compounds. The latter is seen in a variety of mixed anionic rare earth metal hydrides. Hydride halides of trivalent rare earth elements $REH_{1-x}X$ ($0 \le x < 1$) are black metallic compounds with hydrogen in tetrahedral voids of a close *RE* atom packing. They reversibly incorporate additional hydrogen atoms in trigonal voids to form semiconducting and optically transparent valence compounds REH_2X. The low-valent hydride halides may be described as cluster compounds and are covered in more detail in Chapter 2.3 of this book. Again, europium and ytterbium behave differently due to a strong preference for the divalent state. Mixed hydride halides are known for compositions $EuHX$ and $YbHX$ [16], Eu_2H_3X, $Eu_7H_{12}X_2$ [17], as well as fluorite-type and $PbCl_2$-type solid solutions $EuH_{2-x}F_x$ [18].

Also known are multinary hydride oxide halides such as $LiEu_2HOCl_2$ [19]. Pure hydride conduction has been discovered in K_2NiF_4-type hydride oxides $La_{2-x-y}Sr_{x+y}LiH_{1-x+y}O_{3-y}$ [20] which are based on Ln_2LiHO_3 (Ln = La, Ce, Pr and Nd) compounds [21]. Most rare earth elements form hydride oxides *RE*HO with disordered fluorite type or ordered variants of the fluorite type such as the anti-LiMgN type in YHO [22]. Hydride oxides REFeAsOH$_x$ have attracted a lot of interest lately due to the discovery of high temperature superconductivity in this class of long known 1:1:1:1 compounds (see also Chapter 4.12) [23, 24]. Hydride sulfides, selenides and tellurides are known as well, for example, *RE*HSe, *RE*HTe, Pr_2H_2S and RE_2H_2Se [25, 26]. Rare earth metal carbides and carbide halides may take up hydrogen to form the corresponding hydrides such as $La_2C_3H_{1.5}$ [27], Gd_2CXH [27], and Y_2CH [28]. In the latter, carbon atoms orderly occupy Y_6 octahedra, hydrogen atoms occupy Y_4 tetrahedra and Y_6 octahedra. Upon replacement of carbon by silicon, Zintl-like phases are observed, for example, CrB-type or FeB-type compounds $LnTt$ (Ln = La, Nd; Tt = Si, Ge, Sn). They are metallic, may be described by a limiting ionic formula $Ln^{3+}Tt^{2-}e^{-}$ and react with hydrogen to electron-precise hydrides $LnTt$H (= $Ln^{3+}Tt^{2-}H^{-}$) [29]. $Yb_5Bi_3H_x$ and further, $M^{II}{}_5Pn_3H_x$ (Pn = pentel), are classic examples of compounds believed to be unusual, electron-imprecise Zintl phases, which later turned out to be hydrides, instead [30].

Numerous ternary and multinary rare earth metal hydrides with more than one kind of metal atom are known and only a short glimpse of the multifaceted world of mixed cationic hydrides can be given here. Laves phases are an important class of materials for hydride research as many of them reversibly take up hydrogen, making them storage materials in the early days. Moreover, Laves phase hydrides show many intriguing fundamental chemical and physical properties such as order–disorder transitions, lattice gas and liquid behavior, fast hydrogen diffusion and the changes in magnetic order and electronic properties due to hydrogen insertion. Many Laves phases REM_2 take up hydrogen and form hydrides, ranging from typical metallic interstitial type (M = transition metal, Al) to ionic (Mg). The latter form semiconducting, sometimes colored (Ln = La (brownish grey), Ce (brownish grey), Pr (colorless), Nd (colorless), Sm (olive green)), mostly salt-like hydrides $LnMg_2H_7$ with a tetragonally distorted cubic Laves phase-type structure ($LaMg_2H_7$ type, Figure 4.9.5) and hydrogen in trigonal and tetrahedral voids [31]. The hexagonal Laves phase $EuMg_2$ takes up hydrogen to form $EuMg_2H_6$ as a red powder. Its structure is derived from a cubic perovskite-type structure by omitting every other layer of europium. According to the formula $Eu_{1/2}\square_{1/2}MgH_3 = EuMg_2H_6$ ($EuMg_2H_6$ type, also called $Th_{0.5}Nb_2O_6$ type, anti-$Na_5[NiO_2][CO_3]$ type, Figure 4.9.5) [13]. $GdMg_2$ and $TbMg_2$ form hydrides, for which the parent hexagonal Laves phase structure is orthorhombically distorted and, presumably, has lower hydrogen content than those mentioned above. They are probably metallic and resemble hydrides of REM_2 (M = transition metal, Al) rather than $LnMg_2H_7$ (Ln = La, Ce, Pr, Nd, Sm) and $EuMg_2H_6$. For the remaining $LnMg_2$, decomposition to binary hydrides or intermetallic compounds was found [31]. Laves phases such as $REFe_2$ and $REMn_2$ incorporate up to $x = 6$ hydrogen atoms per formula unit. For $0 < x \leq 4.5$, hydrogen

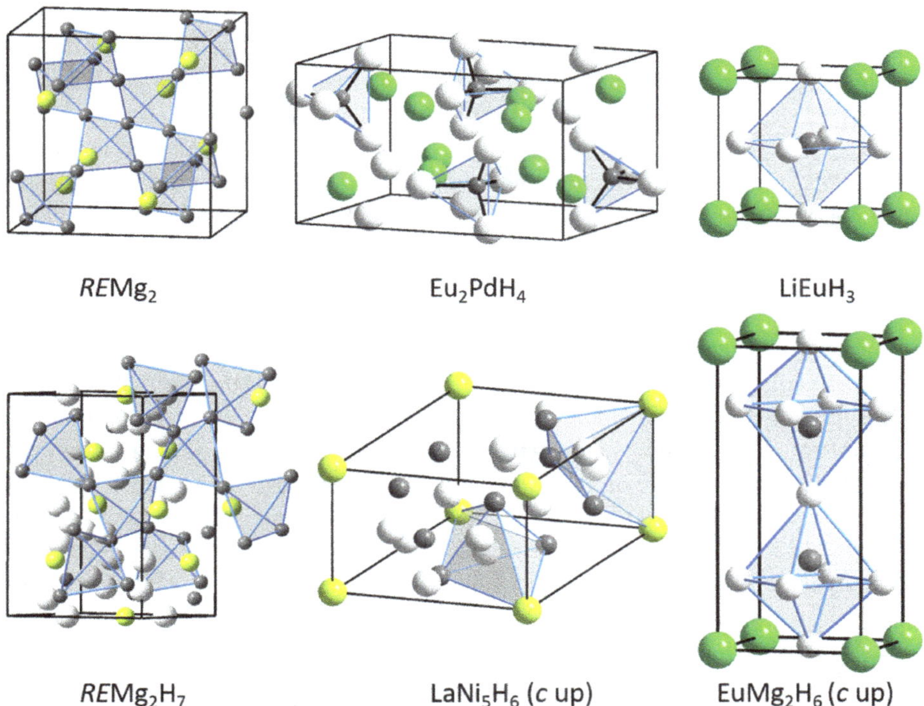

Figure 4.9.5: Crystal structures of selected ternary rare earth metal hydrides with europium atoms shown as green, other rare earth metal atoms shown as yellow and hydrogen atoms shown as white spheres; polyhedra: empty (distorted) Mg_4 tetrahedra in $REMg_2$ and $REMg_2H_7$, distorted tetrahedral $[PdH_4]^{4-}$ complexes in Eu_2PdH_4, hydrogen occupied distorted La_2Ni_3 trigonal bipyramids in $LaNi_5H_6$ (simplified 2-site model for hydrogen [33]), LiH_6 octahedra in $LiEuH_3$ and distorted MgH_6 octahedra in $EuMg_2H_6$.

resides in tetrahedral voids in the parent Laves phase-type structure. At higher pressures ($p(H_2) > 100$ MPa), an orthorhombic distortion occurs for $x = 5$. For $x = 6$, a complete reorganization to a K_2PtCl_6-type related structure takes place. These hydrides show a strong interplay between hydrogen and magnetic order, depending on the hydrogen content [32]. Compounds of rare earth metals and transition metals, with and without hydrogen, offer the possibility to study many fundamental magnetic properties. These are often related to and may be controlled by the coupling between the localized 4f magnetism of the rare earth atoms and the itinerant 3d electrons. Direct 3d-3d interactions are often stronger than RKKY coupled 4f-4f and 4f-3d interactions, which determine the coupling between rare earth and transition metal atoms, and both can be strongly influenced by incorporating hydrogen. Hydrogen insertion often results in a unit cell volume increase and consequently, a weakening of the rare earth transition metal interaction, for example, in $Tm_2Fe_{17}H_5$ [32].

REM_x phases with many other compositions have been investigated with respect to their hydrogen uptake and influence on magnetic behavior. Because of the complex interplay of the variation of interatomic distances by volume expansion due to hydrogen incorporation and the DOS at the Fermi level (Stoner criterion), the changes of magnetic bulk properties upon hydrogen incorporation are manifold. Cerium often changes its valence from +IV to +III upon hydrogen uptake, seen for example, by the appearance of ferromagnetism in Pauli-paramagnetic $CeNi_3$ by hydrogenation. The weak itinerant ferromagnetism in YNi_3 disappears upon hydrogen uptake, due to a decrease in the density of states at the Fermi level [1]. $RECo_5$ and RE_2Fe_{17} comprise important magnetic materials, with $SmCo_5$ being the one of the strongest permanent magnets. Upon hydrogen uptake, $RECo_5H_3$ is formed and magnetic moments of the cobalt atoms are reduced. In contrast, Fe atom magnetic moments in $REFe_2$ increase upon hydrogenation. In the case of iron compounds, ^{57}Fe Mößbauer spectroscopy can be used to not only track the hyperfine splitting as an indication of magnetic ordering, but line widths can also be used as a measure for atomic disorder and symmetry consideration. These may give hints for the preference of hydrogen atoms for certain voids inside the crystal structure. Some rare earth elements have also Mößbauer active isotopes, such as ^{151}Eu, ^{155}Gd, ^{161}Dy or ^{169}Tm. However, most are less convenient to handle and are therefore less available (see Chapter 3.6). Only ^{151}Eu was used extensively, for example, to show that europium is always divalent in hydrides at least at ambient conditions. Even the few examples given here clearly show that the effects of hydrogenation on magnetic behavior are complex. The effects depend on the changes of interatomic distances and of the electronic structure by partial localization of electrons at the hydrogen atoms (hydride anions, H^-, or at least negatively polarized hydrogen atoms, $H^{\delta-}$) as well as on the nature of the rare earth metal and of the transition metal.

Equiatomic phases $REMX$ (M = transition metal; X = Mg, Zn, Cd, Hg, Al, Ga, In, Tl, Si, Ge, Sn, Pb, P, As, Sb and Bi) often crystallize in the hexagonal ZrNiAl-type structure, the orthorhombic TiNiSi-type structure or its distortion (coloring) variants (Chapter 2.4, Figure 2.4.7). Because of the huge combinatorial opportunities for M and X, they offer a nice playground for synthesis and tuning of chemical and physical properties. This is particularly true for cerium compounds in which two valence states, Ce^{III} and Ce^{IV}, may be easily realized and electrons are often strongly correlated. These fascinating physical phenomena (Kondo effect, heavy fermions, magnetic ordering, mixed or fluctuating valency) may be altered by the insertion of hydrogen into the crystal structure. Typical hydrogen atom sites are distorted Ce_4 and Ce_3M tetrahedra or Ce_3M_2 and Ce_3X_2 bipyramids. Often, a strong lattice anisotropy is observed upon hydrogenation (shrinkage in a and expansion in c). The occupation of the Ce_4 voids by hydrogen atoms in the antiferromagnet CeMnGe suppresses the cerium ordering and forces manganese magnetic moments to parallel order (ferromagnet, ferrimagnet or canted spin system) [34]. Another example of drastic changes in magnetic and electronic properties is the nonmagnetic heavy fermion compound CeRuGe. This property is lost in forming CeRuGeH in the ZrCuSiAs type, which is an antiferromagnet ($T_N = 4$ K) instead [34]. Again, the presence of hydrogen triggers a subtle interplay of competing electronic and magnetic coupling in the solid and no simple rules exist to predict the influence of hydrogen on ground state electronic and magnetic properties of such *REMX* phases.

Cubic (anti) perovskite-type structures are found for metallic phases ($CeRh_3H_{0.84}$, $EuPdH_3$) and ionic compounds ($LiEuH_3$) alike. Defect variants like $Eu_{1/2}\square_{1/2}MgH_3 = EuMg_2H_6$ are also known (Figure 4.9.5). *RE*Pd_3 phases ($AuCu_3$ type) are of interest for catalytic hydrogenation reactions (Chapters 2.4 and 4.7). Many *RETM*$_3$ phases with $PuNi_3$-type structure and ordering variants such as $LaMg_2Ni_9$, reversibly take up hydrogen [35]. They crystallize in distorted variants of the hydrogen-free parent intermetallic structures which may be described as an intergrowth of Laves phase-like slabs with $CaCu_5$ like slabs ($PuNi_3$ type). Hydrogen preferentially occupies the interstices in the Laves phase structure-like slabs. Complex tetrahedral NiH_4^{4-} moieties were suggested from crystal structures [36], which would make these compounds interesting border cases between interstitial and covalent hydrides. However, further experimental and theoretical investigations are needed to properly explain chemical bonding in these hydrides.

The number of compounds containing hydrido complexes of transition metals with trivalent rare earth metals are limited: $LaMg_2NiH_7$ with $[NiH_4]^{4-}$ [37], isolated $[Ni_2H_7]^{7-}$ and $[Ni_4H_{12}]^{12-}$ ions in $La_2MgNi_2H_8$ [38], and $LaMg_2PdH_7$ with $[PdH_4]^{4-}$ anions [39]. Divalent europium and ytterbium on the other hand, may easily replace Sr or Ca cations in complex hydrides, for example, $RE^{II}FeH_6$, $RE^{II}CoH_5$, $RE^{II}NiH_4$, $RE^{II}RuH_6$, $RE^{II}IrH_5$ $RE^{II}PdH_4$ (Figure 4.9.5), $RE^{II}Mg_2FeH_8$, $RE^{II}MgNiH_4$, $(RE^{II})_4Mg_4Fe_3H_{22}$, and $(RE^{II})_4Mg_4Co_3H_{19}$ [40].

4.9.3 Applications

$LaNi_5$ hydrides are well-known for their use in electrochemical hydrogen storage in nickel-metal-hydride (Ni-*M*H) rechargeable batteries with storage capacities of typically 80 Wh/kg, 250 Wh/L. The cell reaction involves the reversible intercalation (charging) and deintercalation (discharging) of hydrogen in a storage material *M*, in an aqueous potassium hydroxide solution as electrolyte. This may be schematically written as $MH + NiOOH = M + Ni(OH)_2$. Here, $M = LaNi_5$ and *M*H is in reality $LaNi_5H_6$ (Figure 4.9.5). Lanthanum may be replaced by the cheaper *Misch metal* (*Mm*), a mixture of rare earth metals, and nickel is substituted for improving corrosion stability such that a typical composition is $MmNi_{3.55}Mn_{0.4}Al_{0.3}Co_{0.75}$. In recent years, other materials such as *RE*-Mg-Ni compounds [41] have also been employed in Ni–MH batteries. Ni-MH batteries have been widely used since the 1990s in portable devices but have been largely succeeded by lithium ion batteries and are nowadays mainly found in small appliances. $LaNi_5$ is also used for hydrogen gas storage because of low desorption pressure, slightly above 1 bar, fast absorption and desorption kinetics, easy activation and good cycling stability. Its low gravimetric storage capacity limits its use to stationary stores. Further uses include hydrogen compression, hydrogen separation and purification, and in heat exchangers.

Switchable mirrors based on thin films of rare earth metals have been discussed (*vide supra*). Modern devices use Mg-*RE* compounds, Mg_2Ni or Mg_3Ir and have switching times as low as 20 ms at room temperature. The easy change between reflecting and transparent state makes them interesting devices for a wide range of optical applications, including smart windows for blocking of sunlight or even as electrically switchable light filter for preterm infant incubators [42].

Rare earth metal hydrides have been used for nuclear reactions, for example, as moderator materials in nuclear reactors because of the rather high hydrogen density. They were also used as target materials for neutron production, especially deuterides, for example, $D(d,n)^3He$. Lanthanides are common dopants in luminescent materials. Metal hydrides have been recently investigated in this regard because the strong nephelauxetic effect of hydrogen can be used to shift the emission wavelength of Eu^{II} considerably [43]. Therefore, emission wavelengths covering the whole range of visible light are accessible. Rare earth magnets are often processed by the hydrogenation-disproportionation-desorption-recombination (HDDR) process in which a material such as $SmCo_5$ [44] or $Nd_2Fe_{14}B$ [45] takes up hydrogen, disproportionates into binary rare earth hydrides and transition metal (borides), and finally releases hydrogen to yield the elements in a very fine-grained homogeneous mixture. The recombination step recovers the starting material but with improved material properties, especially coercivity, since small crystallites can be easily aligned in an external magnetic field. HDDR is thus aiming at optimizing magnetic properties by controlling the material's microstructure. The formation of metal hydrides may also be an undesired reaction, for example, in hydrogen embrittlement. This causes deterioration in mechanical properties and plays a role in light-weight magnesium alloys and high-strength steel with rare earth metal addition.

References

[1] J.-C. G. Bünzli, V. K. Pecharsky (Eds.), Handbook on the Physics and Chemistry of Rare Earths, Vols. 1–54, North-Holland, Elsevier, Amsterdam, 1978–2019.

[2] H. Bergmann, Hydride von Scandium, Yttrium, Lanthan und den Lanthaniden sowie Deuteride und Tritide, Gmelin Handbuch der Anorganischen Chemie, Seltenerdmetalle Teil C1, Sc, Y, La und Lanthanide, Springer-Verlag, Berlin, 1974.

[3] H. Bärnighausen, Z. Anorg. Allg. Chem. 1966, 342, 233–239.

[4] Q. Huang, T. J. Udovic, J. J. Rush, J. Schefer, I. S. Anderson, J. Alloys Compd. 1995, 231, 95–98.

[5] G. Renaudin, K. Yvon, W. Wolf, P. Herzig, J. Alloys Compd. 2005, 404–406, 55–59.

[6] I. Anderson, The Dynamics of Hydrogen in Metals Studied by Inelastic Neutron Scattering, in Neutron Scattering from Hydrogen in Materials (ed. A. Furrer), World Scientific, Singapore, 1994, 142–167.
[7] V. P. Ting, P. F. Henry, H. Kohlmann, C. C. Wilson, M. T. Weller, Phys. Chem. Chem. Phys. 2010, 12, 2083–2088.
[8] R. Hempelmann, A. Skripov, Hydrogen Motion in Metals, in Hydrogen Transfer Reactions (ed. J. T. Hynes, J. P. Klinman, H. H. Limbach, R. L. Schowen), Wiley-VCH, Weinheim, 2007, 787–829.
[9] T. J. Udovic, Q. Huang, J. J. Rush, J. Alloys Compd. 2003, 356–357, 41–44.
[10] T. J. Udovic, Q. Huang, A. Santoro, J. J. Rush, Z. Kristallogr. 2008, 223, 697–705.
[11] P. Vajda, G. André, J. Alloys Compd. 2001, 326, 151–156.
[12] T. Palasyuk, M. Tkacz, Solid State Commun. 2007, 141, 354–358.
[13] H. Kohlmann, Eur. J. Inorg. Chem. 2010, 2582–2593.
[14] H. Saitoh, A. Machida, T. Matsuoka, K. Aoki, Solid State Commun. 2015, 205, 24–27.
[15] G. Auffermann, Z. Anorg. Allg. Chem. 2002, 627, 1615–1618.
[16] H. P. Beck, A. Limmer, Z. Naturforsch. 1982, 37b, 574–578.
[17] O. Reckeweg, F. J. Di Salvo, S. Wolf, T. Schleid, Z. Anorg. Allg. Chem. 2014, 640, 1254–1259.
[18] N. Kunkel, A. Meijerink, H. Kohlmann, Inorg. Chem. 2014, 53, 4800–4802.
[19] D. Rudolph, D. Enseling, T. Jüstel, T. Schleid, Z. Anorg. Allg. Chem. 2017, 643, 1525–1530.
[20] G. Kobayashi, Y. Hinuma, S. Matsuoka, A. Watanabe, M. Iqbal, M. Hirayama, M. Yonemura, T. Kamiyama, I. Tanaka, R. Kanno, Science (Washington, DC, U. S.) 2016, 351, 1314–1317.
[21] H. Schwarz, Neuartige Hydrid-Oxide der Seltenen Erden: Ln_2LiHO_3 mit Ln = La, Ce, Pr und Nd, Dissertation, Universität Karlsruhe, 1991.
[22] N. Zapp, H. Kohlmann, Inorg. Chem. 2019, 58, 14635.
[23] S. Iimura, S. Matsuishi, H. Hosono, Phys. Rev. B 2016, 94, 024512.
[24] R. Pöttgen, D. Johrendt, Z. Naturforsch. 2008, 63b, 1135–1148.
[25] M. Folchnandt, D. Rudolph, J.-L. Hoslauer, T. Schleid, Z. Naturforsch. 2019, 74b, 417–422.
[26] T. Schleid, M. Folchnandt, Z. Anorg. Allg. Chem. 1996, 622, 455–461.
[27] A. Simon, Angew. Chem. 2012, 124, 4354–4361; Angew. Chem. Int. Ed. 2012, 51, 4280–4286.
[28] J. P. Maehlen, V. A. Yartys, B. C. Hauback, J. Alloys Compd. 2003, 351, 151–157.
[29] A. Werwein, H. Auer, L. Kuske, H. Kohlmann, Z. Anorg. Allg. Chem. 2018, 644, 1532–1539.
[30] E. A. Leon-Escamilla, P. Dervenagas, C. Stassis, J. D. Corbett, J. Solid State Chem. 2010, 183, 114–119.
[31] A. Werwein, F. Maaß, L. Y. Dorsch, O. Janka, R. Pöttgen, T. C. Hansen, J. Kimpton, H. Kohlmann, Inorg. Chem. 2017, 56, 15006–15014.
[32] V. Paul-Boncour, O. Isnard, M. Guillot, A. Hoser, J. Magn. Magn. Mater. 2019, 477, 356–365.
[33] V. A. Yartys, V. V. Burnasheva, K. N. Semenenko, N. V. Fadeeva, S. P. Solov'ev, Int. J. Hydrogen Energy 1982, 7, 957–965.
[34] R. Pöttgen, O. Janka, B. Chevalier, Z. Naturforsch. 2016, 71b, 165–191.
[35] M. Latroche, J. Alloys Compd. 2003, 356–357, 461–468.
[36] Y. E. Filinchuk, K. Yvon, J. Solid State Chem. 2006, 179, 1041–1052.
[37] G. Renaudin, L. Guénée, K. Yvon, J. Alloys Compd. 2003, 350, 145–150.
[38] J.-N. Chotard, Y. Filinchuk, B. Revaz, K.Yvon, Angew. Chem. 2006, 118, 7934–7937; Angew. Chem., Int. Ed. 2006, 45, 7770–7772.
[39] K. Yvon, J.-P. Rapin, N. Penin, Z. Ma, M. Y. Chou, J. Alloys Compd. 2007, 446–447, 34–38.
[40] K. Yvon, G. Renaudin, Hydrides: Solid State Transition Metal Complexes, in Encyclopedia of Inorganic Chemistry (ed. R. B. King), Vol. 3, Wiley, New York, 2005, 1814–1846.
[41] S. Han, Y. Li, B. Liu, Hydrogen Storage Alloys, De Gruyter, Berlin, 2017.

[42] K. Tajima, M. Shimoike, H. Li, M. Inagaki, H. Izumi, M. Akiyama, Y. Matsushima, H. Ohta, Appl. Phys. Lett. 2013, 102, 161913.

[43] N. Kunkel, T. Wylezich, Z. Anorg. Allg. Chem. 2019, 645, 137–145.
[44] I. I. Bulyk, V. V. Panasyuk, A. M. Trostianchyn, J. Alloys Compd. 2004, 379, 154–160.
[45] O. Gutfleisch, K. Khlopkov, A. Teresiak, K.-H. Müller, G. Drazic, C. Mishima, Y. Honkura, IEEE Trans. Magn. 2003, 39, 2926–2931.

Niko Hildebrandt, Loïc J. Charbonnière

4.10 Lanthanide nanoparticles and their biological applications

The unique luminescence properties of lanthanides (Ln) can become even more interesting when they are combined with nanoparticles (NPs). Nanoparticles mainly add the features of high surface-to-volume ratios and quantum confinement to the lanthanide properties. LnNPs have been mainly developed for biological applications, including biosensing, bioimaging, drug delivery, theranostics and forensics, and for light-emitting diodes (LEDs), all of which has been recently reviewed in detail by Bünzli et al. [1]. In this chapter, we will explain the various principles of photoluminescent LnNPs, as well as their synthesis in view of biological applications, and provide some representative examples concerning biosensing (spectroscopy), bioimaging (microscopy) and theranostics (combination of therapy and diagnostics). Taking into account that LnNPs can also provide other properties than photoluminescence, we will also discuss other biological applications of LnNPs in computer tomography (CT), single photon emission tomography (SPECT), positron emission tomography (PET), radioluminescence and Cerenkov photosensitization and magnetic resonance imaging (MRI).

4.10.1 Principles of Ln nanoparticles (LnNPs)

Although there exists no official classification of LnNPs, they can be defined by various properties [1–10], which are summarized in Figure 4.10.1. The classical photoluminescence case, in which one excitation photon leads to one emission photon with the same or longer wavelength (lower energy), for example, green excitation (~500 nm) leads to red emission (~600 nm), is termed *downshifting*. The emission is Stokes shifted compared to the excitation, and the maximum theoretical quantum yield is 100%. If two or more excitation photons result in one emitted photon with shorter wavelength (higher energy), the term that is used is *upconversion*. The emission is anti-Stokes shifted compared to excitation, and the maximum theoretical quantum yield is 50%, though the practical quantum yields are much lower (below 1%) in most cases. The inverse case, for which one excitation photon leads to two or more emission photons with longer wavelength (lower energy), is referred to as *downconversion* (or quantum cutting). The emission is Stokes shifted compared to excitation, and the maximum theoretical quantum yield can be larger than 100%.

Luminescence itself is an important property for the desired application of LnNPs. It can be divided into two characteristics, namely temporally- (defined by the luminescence lifetime) and spectrally- (defined by the luminescence wavelength) resolved emission. Temporal and spectral measurements are usually referred to as time-resolved (or

https://doi.org/10.1515/9783110654929-033

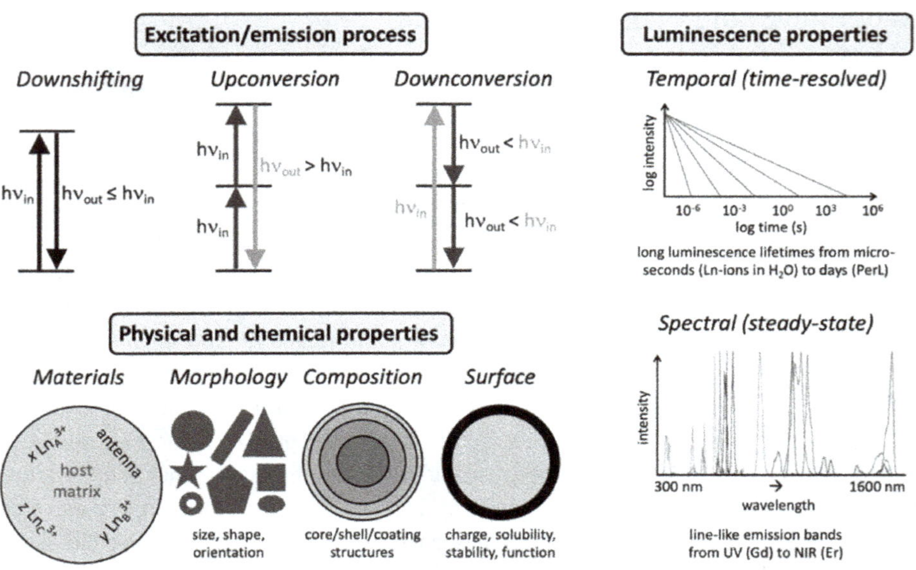

Figure 4.10.1: Schematic presentations of possible classifications of LnLPs by their excitation/emission properties, luminescence properties, or physical and chemical properties.

time-gated) and steady-state (unless time-resolved spectra are recorded), respectively. Luminescence lifetimes of LnNPs can range from the nano- to microsecond (e.g., Ln ions in contact with water), micro- to millisecond (e.g., Ln ions protected by ligands), or millisecond to days (persistent luminescence, PerL, or afterglow) scale. The luminescence spectra of LnNPs are usually composed of several narrow emission bands that can range from the UV (for Gd) to the NIR (for Er).

Another way of describing LnNPs is by their physical and chemical properties or the way they have been synthesized. These properties do not only define the photophysical and photochemical characteristics of LnNPs, but also their stability (during storage and during application), their biocompatibility (interactions with the biological system) and bioapplicability (e.g., conjugation with biologicals or interaction with, and penetration into cells). In addition to one or several types of Ln ions, the NP material will be composed of one or several organic and/or inorganic host materials that can also have the function of protecting the Ln ions from the surrounding environment, and light harvesting and transferring the harvested energy to the emitting Ln ions (which can also transfer the energy among them, so-called energy migration). Morphology (e.g., size and shape) and composition (e.g., core/shell or layer-by-layer structures) are further important features that define a LnNP. One of the most important advantages of NPs is their large surface-to-volume ratio and therefore, the surface properties (e.g., charge, functionality and stability) are of paramount importance, in particular, when the LnNPs are used for biosensing, where the surface is the interface between LnNP and biological system.

4.10.2 Synthesis and characterization

Before planning any synthesis of an LnNP for bio-analytical applications, a crucial point to be determined is the choice of the matrix. Importantly, some Ln based solids may be partly soluble in aqueous media, thus leading to a leaching of the particles. This is typically the case of Ln_2O_3 based materials which can be fully dissolved in neutral and acidic aqueous solution. Alternatively, this property can be applied to use the soluble NPs as a template for the preorganization of other nanosized structures, as it was smartly done for the synthesis of ultrasmall rigid particles for theranostic applications [11]. Particular care should then be taken in the choice of matrix, as even fluoride-based lanthanide materials, which display very low solubility product constants [12–14] and which are commonly used for the preparation of LnNPs have been shown to be subject to leaching, with loss of luminescence properties in aqueous solutions [15].

LnNPs can be obtained in a large range of forms and sizes [16]. Their shape varies from spheres [17], rods [18], spindles [19], cubes [20], stars [21], wires [22], plates [23], octahedra [24] to even polyhedra [25]. The size and shapes are generally deeply dependent on the synthetic conditions used to obtain the NPs, among which the solvent, the starting materials, the temperature, the pressure, the reaction time and the use of possible additives are the main variables. One may distinguish between two major types of synthetic protocols. Those obtained under ambient pressure, possibly at very high temperature, and the ones realized under solvothermal conditions at high pressure, either using autoclave ovens or closed vessel microwave irradiation [21]. As a general rule, the higher the temperature, the better the crystallinity of the material, but LnF_3 NPs can eventually be obtained in water at room temperature, in the presence of an additive such as aminoethylphosphonate [26].

At ambient pressure, the synthesis is generally performed at the boiling point of the solvent, and the use of high boiling point organic solvents, such as octadecene (315°C) or trioctylphosphine oxide (411°C) allows to reach very high reaction temperatures (provided that an inert atmosphere is used during synthesis), resulting in highly monodisperse NPs [27]. An interesting strategy lies in the use of Ln salts that can be degraded during the synthesis, resulting in the slow delivery of anions along the process. This is typically the case in the use of $Ln(CF_3CO_2)_3$, the triflate slowly decomposing during the heating, resulting in the delivery of fluoride anions [25]. The strategy can further be used for the growing of multifunctional core/shell nanoparticles with the core and the shell having various compositions, depending on the expected properties of the NPs [28]. The use of organic solvents for producing LnNPs in apolar media requires the transfer of the NPs into an aqueous medium, when biological applications are envisioned. This is generally obtained by biphasic exchange of the ligands at the capping layer, potentially playing with the pH of the aqueous solution to increase the kinetic of the ligand exchange [29]. Alternatively, the synthesis can be performed directly in aqueous media, potentially under hydrothermal conditions in autoclaves [30] or under microwave irradiation [31].

The classical characterizations of NPs concern their composition and their morphology. Transmission electron microscopy (TEM), possibly at high resolution (HR-TEM) and scanning electron microscopy (SEM) are ideally suited to reveal the size and shape of the NPs. When coupled with energy dispersive X-ray spectroscopy (EDX), information of the elemental composition of the sample can be obtained at least for heavy elements. Dynamic light scattering (DLS) is a relatively simple way to obtain the hydrodynamic size of the particles in solution, but it refers to a spherical model and only gives an average radius or diameter. In combination, measurement of the zeta potential (ζ-potential), which can be viewed as the charge of the surface per area unit gives important data on the colloidal stability of the NPs, the stability being assumed to be good enough – above or below values of plus or minus 30 mV, respectively. Concerning the elemental composition of the NPs, the elemental analysis can be obtained by using inductively coupled plasma associated to detection by mass spectrometry (ICP-MS) or emission spectroscopy. In the presence of an organic shell, further elemental analysis, such as CHNS, and thermogravimetric methods may be of interest. The optical properties of LnNPs are usually characterized by absorption, luminescence spectroscopy and fluorescence microscopy. Regarding biological applications, storage conditions, stability (in storage and in the biological media), and toxicity (including phototoxicity) may be other properties to be characterized.

4.10.3 Application in biosensing, bioimaging, and theranostics

The various types of LnNPs described above have been used in different biological applications. Here, we will discuss some representative recent examples of luminescent LnNPs used in biosensing, imaging, and theranostics. Other biological applications of LnNPs can be found at the end of this chapter. The principal steps necessary for applying LnNPs in biological research are shown in Figure 4.10.2. In all applications, the first step consists of conjugating the LnNPs with biological recognition molecules (e.g., antibodies, DNA, peptides) for targeting a biomarker of interest or with drugs for disease treatment. While the general types (sensing, imaging, theranostics) all have common features, many different approaches and methods have been applied, and all need a certain amount of specialization and adaption of the general procedures. For a larger overview of LnNP bioapplications, we refer the interested reader to recent reviews concerning LnNPs in general [1, 4, 8], NIR-emitting LnNPs [2, 10], PerL-LnNPs [3, 5, 7], and upconversion LnNPs (UCNPs) [6, 9].

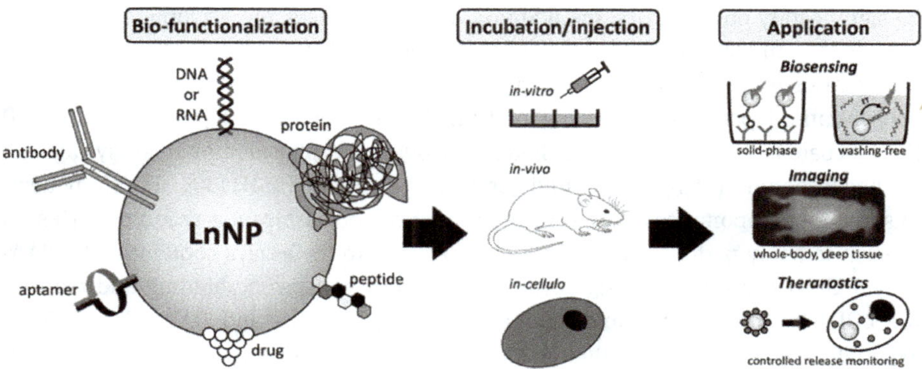

Figure 4.10.2: Schematic representation of the various steps of LnNP-based biological applications (biosensing, imaging, and theranostics).

4.10.3.1 LnNPs for biosensing

Biosensing is usually referred to as the identification and/or quantification of biomarkers (e.g., disease-relevant biomolecules, such as proteins or nucleic acids, for clinical diagnostics) in vitro and in solution. The sample (e.g., plasma, serum, or lysed cells) is mixed and incubated with the LnNP bioconjugates and other assay components, for example, in microtiter plates, cuvettes, or microfluidic setups. Specific binding of the LnNP bioconjugates with the biological target (biomarker) leads to LnNP-target complexes that can be separated physically, for example, by washing/separation steps from unbound components in a solid-phase assay, for which secondary recognition molecules are attached to the surface of the microwell (or cuvette, microfluidic channel etc.), or optically (e.g., in an energy transfer (ET)-based liquid-phase assay, for which the LnNP-target complexes provide a specific luminescence signal distinct from all other components). Solid-phase (or heterogeneous) assays are more time- and reagent-consuming because they necessitate several physical separation steps, whereas liquid-phase (or homogeneous) assays are quicker because binding occurs in the liquid phase and separation steps are not required. On the other hand, heterogeneous assays can provide better (lower) limits of detection (LODs) because physical separation (and eventually signal amplification – e.g., in enzyme-linked immunosorbent assays, ELISAs) can result in higher signal-to-noise (or signal-to-background) ratios.

Different LnNPs have been developed for the detection and quantification of biomarkers for various diagnostic applications [32]. In one example, simultaneous Tb and Eu downshifting luminescence was used for a ratiometric sensor for dipicolinic acid (DPA) [33]. Tb/DPA-based SiO_2 NPs were mixed with guanosine monophosphate (GMP) and $Eu(NO_3)_3$. In the resulting Tb/Eu NPs, the Eu luminescence could be significantly enhanced by the addition of DPA, whereas the Tb luminescence was not significantly affected. This LnNP-based ratiometric DPA sensor could be applied to a time-resolved DPA assay and colorimetric (from green for low to red for high DPA concentration) paper strip sensing of DPA. In another example, poly(acrylic acid) (PAA-)-coated UCNPs were used in heterogeneous (solid-phase) immunoassays against cardiac troponin I (cTnI) and thyroid stimulating hormone (TSH) [34]. This study demonstrated the importance of the surface properties (and surface reactivity) of LnNPs, in relation to assay performance. Addition of free PAA to the assays containing the PAA-coated UCNPs resulted in a significant reduction of nonspecific binding and reduced the LODs by more than 3-fold, down to 0.48 ng L^{-1} cTnI and 0.02 mIU L^{-1} TSH. Recently, it was also shown that adequate preparation and separation of UCNPs can lead to a significant improvement of both conventional (analogue) bioaffinity assays and single-UCNP counting (digital) assays. In addition to the improvement of assay performance, the UCNP-bioconjugates could be purified (with agarose-gel electrophoresis) in milligram amounts, which could also be highly beneficial for reproducibility and costs of UCNP-based diagnostic assays. Homogeneous (liquid-phase) bioassays without separation steps were also developed by exploiting ET from UCNPs to dyes [35]. ET or FRET (Förster resonance energy transfer) between Ln complexes and quantum dots (QDs) was also applied for multiplexed diagnostic assays [36]. However, in this case, the Ln complexes were attached to the surfaces of distinct NPs (QDs), and therefore, these QD-Ln nanohybrids cannot be classified as typical (core) LnNPs.

4.10.3.2 LnNPs for luminescence imaging

Imaging is used to visualize (qualitatively or quantitatively) cells, subcellular compartments, biomolecules, or drugs in cells or tissues and can be performed in vitro (e.g., cultured cells) or in vivo (e.g., live animals). The experimental setup can vary significantly depending on the system to be analyzed. For in vitro imaging (inside, outside, or whole cells), LnNP bioconjugates can be simply incubated with the cells and binding to surface receptors or passive (e.g., diffusion across the cell membrane) or active cellular uptake (e.g., endocytosis) can occur. In case the LnNPs do not enter into the cells, external means of cellular delivery, such as microinjection or cell permeabilization, can be applied. For in vivo imaging (mainly mice are used), LnNP bioconjugates are often intravenously injected (e.g., via the tail veins) to specifically bind to or enter their targets (e.g., cancer cells) *via* the blood circulation. However, for research purposes, direct injection below the skin or into the organ of interest can be applied too. The imaging setup will also depend on the sample. For cells or tissues, conventional fluorescence microscopes (often inverted microscopes) are used, and the cells can be alive or fixed (and permeabilized), whereas tissues are usually frozen or paraffin embedded. For whole animals, specific imaging setups (boxes) in which the animals can be fixed and anesthetized, are used. LnNPs that can be excited by (and/or luminesce in the) NIR (e.g., UCNPs) are advantageous, as they allow for deeper light penetration into cells and tissues, because of reduced water and tissue absorption in the NIR. Such LnNPs can be used for imaging below the skin, but penetration is still limited to a few millimeters. PerL LnNPs are also beneficial for deep tissue imaging, because their long luminescence (up to days) allows for an excitation before injection and therefore, no excitation light needs to enter into the tissue. However, the emitted light still needs to exit the tissue for imaging.

Although UCNPs are the most often applied as LnNP optical imaging agents, a large variety of LnNPs and applications have been developed for bioimaging [2–5, 8, 10]. In a recent example for in vivo imaging, PerL LnNPs (Cr^{3+} and Sm^{3+} doped $LaAlO_3$ NPs) were coated with PEG to accomplish a longer circulation time in the blood [37]. The PerL NPs were excited at 254 nm ex vivo for 1 min. and then injected into living mice. In vivo imaging at 734 nm showed improved biodistribution of the PEGylated PerL NPs compared to hydroxylated ones. UCNPs ($NaYF_4$:Yb,Tm) were used for super-resolution imaging (stimulated emission depletion, STED) [38]. Combined excitation at 980 nm and depletion at 808 nm resulted in upconversion imaging with a sub-diffraction resolution of 28 nm. Developments of downshifting LnNPs include extremely bright ($>10^6$ $M^{-1}cm^{-1}$) particles, for which Eu-complexes were loaded into polymer LnNPs for background-free single-particle and live-cell imaging [39] or antenna ligands were coated to the surface of $La_{0.9}Tb_{0.1}F_3$ LnNPs to yield strongly enhanced Tb emission from the LnNP surfaces for confocal imaging of cells [31].

4.10.3.3 LnNPs for theranostic applications

In addition to their numerous and versatile applications in the fields of biosensing (diagnostics) and imaging, LnNPs offer a large potential for therapeutic applications, thereby becoming theranostic agents. Examples of therapeutic applications concerning core-based LnNPs are still scarce, probably due to the fact that the toxicity of NPs in general [40], and of LnNPs, in particular [41], are still under debate. Of particular interest, is the size of NPs associated with a specific phenomenon called EPR for enhanced permeability and retention [42]. The phenomenon is characterized by an increased permeability of the blood vasculature for macromolecular objects near tumor tissues or inflamed regions. Thus, large molecules (with a molecular weight of more than 40 kDa) [41] can escape from the blood flow in these specific regions and accumulate there. Provided LnNPs are designed with a therapeutic property (photothermal effect, presence of a radioelement, etc.), one might find interesting therapeutic applications, for example, in cancer treatment.

Nanospheres of composition $Na_9[Gd(W_5O_{18})_2]$ conjugated to chitosan have been developed and used as radiosensitizers for radiotherapy [43]. Radiotherapy is based on the irradiation with high-intensity ionizing radiation, to produce reactive oxygen species (ROS) that degrade DNA and impede the proliferation of cells, ideally cancer-related ones, selectively. In addition to their radiosensitizing ability, the Gd NPs have been shown to present a synergistic effect on hypoxic tumor cells. The rapid growth of cancer cells results in low oxygen content and subsequently low production of ROS. Once produced, ROS are normally reduced in the cells by glutathione (GSH), to avoid damage to DNA strands. For the described NPs, the irradiation also led to *in situ* oxidation of GSH, which resulted in an increased efficiency of the produced ROS for DNA damage. UCNPs are also of potential interest in the field of theranostics, as they produce light upon NIR excitation, where the penetration depth of the excitation beam into tissues is at its maximum. It is then possible to phototrigger some chemical process deeper in the affected areas. This strategy also allows for luminescence imaging of the processes, typically resulting in a theranostic approach. Advances in the field are summarized in a recent review on the topic [44]. Another example of therapeutic applications of LnNPs concerns the synthesis of multifunctional nanostructures, based on Eu-doped $CaMoO_4$ NPs protected by a silica shell, decorated with silver nanorods and labeled with Prussian blue, and an antibody directed against epidermal growth factor receptors (overexpressed at the surface of cancer cells, e.g., for breast cancer). When excited for 5 min with a 0.8 W cm^{-2} laser at 808 nm, the NPs displayed a strong and specific photothermal effect, which resulted in the death of the cancer cells [45]. Although not *stricto sensu* core LnNPs, AGuIX (commercial name, *cf.* nhtheraguix.com) [46], NPs are based on a siloxane skeleton decorated with nearly 10 Gd complexes at their surfaces [11]. Such NPs have been shown to display an interesting radiosensitizing effect when injected into mice bearing cancerous tumors, which improved the efficiency of the radiotherapy alone, and AGuIX NPs have now reached the step of clinical trials [46].

4.10.4 Other biological applications

4.10.4.1 Ln NPs for computer tomography (CT), single photon emission tomography (SPECT) and positron emission tomography (PET) imaging

Although luminescence is generally the main focus of interest for LnNPs, other physicochemical properties of LnNPs can be used for bioanalytical applications. Because Ln are heavy elements, they are targets of choice for use as contrast agents in X-ray CT. This imaging technique is based on the attenuation of X-ray through tissues and allows for 3D mapping of large volumes, which is of interest. As the absorption cross-sections of X-rays by an element is proportional to the fourth power of the atomic number Z, heavy elements such as Ln ($Z = 57$ to 71 from La to Lu) are potentially good contrast agents for CT to replace clinically used iodinated compounds or $BaSO_4$ [47]. As an example, $BaYbF_5$ NPs were developed combining the heavy Ba (Z = 56) and Yb (Z = 70) elements in a single NP, and proved to be highly efficient for high resolution CT of mice, when coated by a SiO_2 shell with polyethylene glycol chains to improve their biocompatibility [48]. Other imaging modalities, like SPECT or PET, make use of a radioactive element, which is injected into the body, and for which one can follow the biodistribution by recording the radiation emitted by the radio-isotope. In SPECT, the radio-isotope is a gamma emitter. In addition to their potential use as CT contrast agents, LnNPs can also be doped with a radioactive Ln element. This was, for example, achieved by the synthesis of oxyfluoride EuOF NPs doped with ^{153}Sm [49], which was used for the dual mode CT/SPECT imaging of mice. For PET imaging, the radioelement is a positron emitter. A positron is the antiparticle of an electron with the same mass, but a positive charge. After the radioelement decays, the emitted positron will travel through the matter before encountering an electron, resulting in its annihilation and the emission of two γ rays of almost opposite direction. The γ rays are collected by cameras, allowing for the positioning of the annihilation event in the sample with a precision down to few mm [50]. ^{18}F is a radioelement of choice for PET imaging, because it can be produced on sites with small cyclotrons. As fluoride anions are hard Lewis bases and Ln^{3+} hard Lewis acids, they produce strong electrostatic interactions which can be used for fluoride sensing, for example [51]. This strong interaction was used to fix ^{18}F fluoride anions at the surface of UCNPs, which were injected intravenously into mice, allowing for whole body PET imaging of the mice and observation of the rapid clearance of the NPs into the liver and spleen, as soon as 5 minutes after injection [52].

4.10.4.2 LnNPs for radioluminescence and Cerenkov photosensitization

Interestingly, when LnNPs are doped with luminescent Ln ions, one can also take advantage of the luminescence properties of the ion, which can be excited by X-ray beams, or potentially through the presence of some positron emitters in the medium. As an example, NPs of $Ba_{0.55}Y_{0.3}F_2$ composition doped with approximately 5% of Eu were shown to display radioluminescence upon exposure to X-rays, with the characteristic emission spectrum of europium in the red to NIR region [53]. The luminescence can also be observed when such NPs are placed in the presence of a compound containing ^{18}F. In practice, the emission of the positron is accompanied by the generation of photons in the UV-blue domain, a phenomenon called Cerenkov radiation. Such radiation can indirectly excite atoms such as Eu in Eu_2O_3 NPs, to generate its luminescence in the red region of the spectrum [54], which can then be used for in vivo imaging.

4.10.4.3 LnNPs for magnetic resonance imaging (MRI)

Because of their intrinsic paramagnetic properties (except for La and Lu), Ln cations have long been known to be excellent contrast agents for MRI, with a particular emphasis on Gd and its seven unpaired electrons [55]. Medical MRI is also a 3D rendering technique based on the detection of H atoms in the body. When placed in a static high magnetic field, the sum of the nuclear dipoles of H atoms result in a magnetization of the sample. Superimposition of a pulsed radiofrequency magnetic field perpendicular to the static field results in a perturbation of the magnetization, which returns to equilibrium with two characteristic kinetic parameters, the longitudinal (T_1) and transversal (T_2) relaxation times. For ^1H nuclear magnetic resonance spectroscopy, T_1 of diamagnetic compounds can be as long as tens of seconds, requiring long periods between successive pulses while allowing relaxation of the system. The presence of paramagnetic compounds with unpaired electrons can drastically reduce the relaxation times allowing for faster acquisition and/or better signal-to-noise ratios. Gd-based NPs have been found to be interesting candidates for such applications, and ultrasmall $NaGdF_4$ NPs of approximately 2 nm size have, for example, been used as T_1 contrast agents, with a longitudinal relaxivity (r_1) enhanced by a factor of 2.4, with respect to clinically used contrast agents [56]. Considering that the dominant factors for efficient T_1 contrast agents is mainly governed by interactions of the water molecules with the paramagnetic cations, the size of the surface is important, and some general trends have been observed between the measured longitudinal relaxivity and the surface-to-volume ratio of the NPs, with this ratio decreasing along the increasing size and a concomitant decrease of the relaxivity per Gd atom in $Eu_{0.2}Gd_{0.8}PO_4 \cdot H_2O$ NPs [57]. For T_2 contrast agents, $NaHoF_4$ and $NaDyF_4$ NPs have been shown to be potential candidates of interest [58].

4.10.5 Conclusions and perspectives

The combination of Ln (photophysical properties) and NPs (nano-sized properties) into LnNPs provides many benefits for bioapplications, with a large impact on both lanthanide and nano research. With the field of LnNPs being relatively young, there is still a long way to go before assessing if LnNPs will find their application beyond fundamental research, where they have contributed to significant advances in previously-unaccomplished approaches, methods or technologies. Their application requires additional research; many photophysical properties on the nanoscale require further investigation for a full understanding. UCNPs are, arguably, the most frequently used LnNPs, with many new studies published in a large variety of scientific journals, on a daily basis. However, conventional downshifting LnNPs and PerL LnNPs are gaining significant interest and have found their place in biological applications, including sensing, imaging, and theranostics. This chapter provided only an introductory overview of LnNP-based technologies and applications, with the aim of motivating more young (and old) researchers to contribute new ideas and approaches to the field, which can be expected to strongly and rapidly develop, within the near future.

References

[1] J. Zhou, J. L. Leaño, Z. Liu, D. Jin, K.-L. Wong, R.-S. Liu, J.-C. G. Bünzli, Small 2018, 14, e1801882.

[2] Y. Fan, F. Zhang, Adv. Opt. Mater. 2019, 7, 1801417.

[3] S.-K. Sun, H.-F. Wang, X.-P. Yan, Acc. Chem. Res. 2018, 51, 1131.

[4] P. Llano Suarez, M. Garcia-Cortes, M. Teresa Fernandez-Arguelles, J. Ruiz Encinar, M. Valledor, F. Javier Ferrero, J. Carlos Campo, J. Manuel Costa-Fernandez, Anal. Chim. Acta 2019, 1046, 16.

[5] J. Liu, T. Lecuyer, J. Seguin, N. Mignet, D. Scherman, B. Viana, C. Richard, Adv. Drug Deliv. Rev. 2019, 138, 193.

[6] H. H. Gorris, U. Resch-Genger, Anal. Bioanal. Chem. 2017, 409, 5875.

[7] J. Wang, Q. Ma, Y. Wang, H. Shen, Q. Yuan, Nanoscale 2017, 9, 6204.

[8] Q. Ma, J. Wang, Z. Li, X. Lv, L. Liang, Q. Yuan, Small 2019, 15, 1804969.

[9] C. Duan, L. Liang, L. Li, R. Zhang, Z. P. Xu, J. Mater. Chem. B 2018, 6, 192.

[10] L. Liang, N. Chen, Y. Jia, Q. Ma, J. Wang, Q. Yuan, W. Tan, Nano Res. 2019, 12, 1279.

[11] F. Lux, A. Mignot, P. Mowat, C. Louis, S. Dufort, C. Bernhard, F. Denat, F. Boschetti, C. Brunet, R. Antoine, P. Dugourd, S. Laurent, L. Vander Elst, R. Muller, L. Sancey, V. Josserand, J.-L. Coll, V. Stupar, E. Barbier, C. Rémy, A. Broisat, C. Ghezzi, G. Le Duc, S. Roux, P. Perriat, O. Tillement, Angew. Chem. Int. Ed. 2011, 50, 12299.

[12] T. Mioduski, C. Guminski, D. Zeng, J. Phys. Chem. Ref. Data 2014, 43, 013105.

[13] T. Mioduski, C. Guminski, D. Zeng, J. Phys. Chem. Ref. Data 2015, 44, 013102.

[14] T. Mioduski, C. Guminski, D. Zeng, J. Phys. Chem. Ref. Data 2015, 44, 023102.

[15] O. Dukhno, F. Przybilla, V. Muhr, M. Buchner, T. Hirsch, Y. Mély, Nanoscale 2018, 10, 15904.

[16] F. Wang, X. Liu, Chem. Soc. Rev. 2009, 38, 976.

[17] S. Zeng, M.-K. Tsang, C.-F. Chan, K.-L. Wong, B. Fei, J. Hao, Nanoscale 2012, 4, 5118.

[18] Y. Wu, C. Li, D. Yang, J. Lin, J. Colloid Interface Sci. 2011, 354, 429.

[19] D. Yang, X. Kang, M. Shang, G. Li, C. Peng, C. Li, J. Lin, Nanoscale 2011, 3, 2589.

[20] S. Rodriguez-Liviano, A. I. Becerro, D. Alcántara, V. Grazú, J. M. de la Fuente, M. Ocaña, Inorg. Chem. 2013, 52, 647.

[21] A. Escudero, A. I. Becerro, C. Carrillo-Carrión, N. O. Núñez, M. V. Zyuzin, M. Laguna, D. González-Mancebo, M. Ocaña, W. J. Parak, Nanophotonics 2017, 6, 881.

[22] Q. Li, V. W.-W. Yam, Angew. Chem. Int. Ed. 2007, 46, 3486.

[23] Z. Li, Y. Zhang, Nanotechnology 2008, 19, 345606.

[24] X. Jiang, C. Cao, W. Feng, F. Li, J. Mater. Chem. B 2015, 4, 87.

[25] H.-X. Mai, Y.-W. Zhang, R. Si, Z.-G. Yan, L. Sun, L.-P. You, C.-H. Yan, J. Am. Chem. Soc. 2006, 128, 6426.

[26] P. R. Diamente, F. C. J. M. van Veggel, J. Fluoresc. 2005, 15, 543.

[27] H.-T. Wong, F. Vetrone, R. Naccache, H. L. W. Chan, J. Hao, J. A. Capobianco, J. Mater. Chem. 2011, 21, 16589.

[28] S. Bhuckory, E. Hemmer, Y.-T. Wu, A. Yahia-Ammar, F. Vetrone, N. Hildebrandt, Eur. J. Inorg. Chem. 2017, 5186.
[29] Y. Huang, E. Hemmer, F. Rosei, F. Vetrone, J. Phys. Chem. B 2016, 120, 4992.
[30] X. Wang, J. Zhuang, Q. Peng, Y. Li, Inorg. Chem. 2006, 45, 6661.
[31] J. Goetz, A. Nonat, A. Diallo, M. Sy, I. Sera, A. Lecointre, C. Lefevre, C. F. Chan, K.-L. Wong, L. J. Charbonnière, ChemPlusChem 2016, 81, 526.
[32] S. Y. Lee, M. Lin, A. Lee, Y. I. Park, Nanomaterials 2017, 7, 411.
[33] Q.-X. Wang, S.-F. Xue, Z.-H. Chen, S.-H. Ma, S. Zhang, G. Shi, M. Zhang, Biosens. Bioelectron. 2017, 94, 388.
[34] S. Lahtinen, A. Lyytikainen, N. Sirkka, H. Pakkila, T. Soukka, Microchim. Acta 2018, 185, 220.
[35] L. Mattsson, K. D. Wegner, N. Hildebrandt, T. Soukka, RSC Adv. 2015, 5, 13270.
[36] M. C. Dos Santos, N. Hildebrandt, TrAC – Trends Anal. Chem. 2016, 84, 60.
[37] M. Pellerin, E. Glais, T. Lecuyer, J. Xu, J. Seguin, S. Tanabe, C. Chaneac, B. Viana, C. Richard, J. Lumin. 2018, 202, 83.
[38] Y. Liu, Y. Lu, X. Yang, X. Zheng, S. Wen, F. Wang, X. Vidal, J. Zhao, D. Liu, Z. Zhou, C. Ma, J. Zhou, J. A. Piper, P. Xi, D. Jin, Nature 2017, 543, 229.
[39] M. Cardoso Dos Santos, A. Runser, H. Bartenlian, A. M. Nonat, L. J. Charbonnière, A. S. Klymchenko, N. Hildebrandt, A. Reisch, Chem. Mater. 2019, 31, 4034.
[40] E. Fröhlich, J. Nanobiotechnol. 2017, 15, 84.
[41] A. Gnach, T. Lipinski, A. Bednarkiewicz, J. Rybka, J. A. Capobianco, Chem. Soc. Rev. 2015, 44, 1561.
[42] H. Maeda, Bioconjugate Chem. 2010, 21, 797.
[43] Y. Yong, C. Zhang, Z. Gu, J. Du, Z. Guo, X. Dong, J. Xie, G. Zhang, X. Liu, Y. Zhao, ACS Nano 2017, 11, 7164.
[44] Y. Wang, S. Song, S. Zhang, H. Zhang, Nano Today 2019, 25, 38.
[45] A. K. Parchur, Q. Li, A. Zhou, Biomater. Sci. 2016, 4, 1781.
[46] L. Sancey, F. Lux, S. Kotb, S. Roux, S. Dufort, A. Bianchi, Y. Crémillieux, P. Fries, J.-L. Coll, C. Rodriguez-Lafrasse, M. Janier, M. Dutreix, M. Barberi-Heyob, F. Boschetti, F. Denat, C. Louis, E. Porcel, S. Lacombe, G. Le Duc, E. Deutsch, J.-L. Perfettini, A. Detappe, C. Verry, R. Berbeco, K. T. Butterworth, S. J. McMahon, K. M. Prise, P. Perriat, O. Tillement, Br. J. Radiol. 2014, 87, 20140134.
[47] Y. Liu, K. Ai, L. Lu, Acc. Chem. Res. 2012, 45, 1817.
[48] Y. Liu, K. Ai, J. Liu, Q. Yuan, Y. He, L. Lu, Adv. Healthc. Mater. 2012, 1, 461.
[49] Y. Wu, Y. Sun, X. Zhu, Q. Liu, T. Cao, J. Peng, Y. Yang, W. Feng, F. Li, Biomaterials 2014, 35, 4699.
[50] D. Brasse, A. Nonat, Dalton Trans. 2015, 44, 4845.
[51] T. Liu, A. Nonat, M. Beyler, M. Regueiro-Figueroa, K. Nchimi Nono, O. Jeannin, F. Camerel, F. Debaene, S. Cianférani-Sanglier, R. Tripier, C. Platas Iglesias, L. J. Charbonnière, Angew. Chem. Int. Ed. 2014, 53, 7259.
[52] Q. Liu, M. Chen, Y. Sun, G. Chen, T. Yang, Y. Gao, X. Zhang, F. Li, Biomaterials 2011, 32, 8243.
[53] C. Sun, G. Pratx, C. M. Carpenter, H. Liu, Z. Cheng, S. S. Gambhir, L. Xing, Adv. Mater. 2011, 23, H195.
[54] E. C. Pratt, T. M. Shaffer, Q. Zhang, C. M. Drain, J. Grimm, Nat. Nanotechnol. 2018, 13, 418.
[55] E. Boros, E. M. Gale, P. Caravan, Dalton Trans. 2015, 44, 4804.
[56] H. Xing, S. Zhang, W. Bu, X. Zheng, L. Wang, Q. Xiao, D. Ni, J. Zhang, L. Zhou, W. Peng, K. Zhao, Y. Hua, J. Shi, Adv. Mater. 2014, 26, 3867.
[57] Y. Li, T. Chen, W. Tan, D. R. Talham, Langmuir 2014, 30, 5873.
[58] X. Zhang, B. Blasiak, A. J. Marenco, S. Trudel, B. Tomanek, F. C. J. M. van Veggel, Chem. Mater. 2016, 28, 3060.

Marina M. Lezhnina, Ulrich Kynast

4.11 Rare-earth-based hybrid materials

4.11.1 Survey

This chapter covers somewhat deliberately selected rare earth hybrid systems; a concise treatment truly matching the size inferred by the title would, by far, exceed the frame of this survey. Far more comprehensive reviews on the subject have been published by K. Binnemans in 2009 and a recent book by B. Yan in 2017 [1, 2]. Hence, the present content reflects the author's view of sometimes surprising and promising, mostly optical properties of selected matrix materials. However, valuable, polymeric matrices, pure silica based materials, MOFs (metal organic frameworks) and macroporous materials are not considered. Much of the work discussed still covers fundamental aspects; however, where appropriate, application potential will be pointed out. The first series of hybrid materials deals with 3D confinements (0D micropores of up to 2 nm diameter in zeolites), a second set with 2D confinement (1D channel type of mesopores ranging in diameters from 2 to 50 nm as in MCM materials) and finally, selected materials with 1D confinement will be inspected (2D gaps of 1 to app. 40 nm thickness in layered structures).

4.11.2 Hybrid materials

4.11.2.1 Micropores / Zeolites

The materials class of zeolites reviewed in this overview is structurally based on sodalite cages and their spatial alignment. These provide a variety of rigid, near spherical cavities, with sizes ranging from approximately 660 pm (sodalite, Figure 4.11.1) to 1,200 pm (faujasites, see Figure 4.11.2), the access to which is restricted by entrance windows of 260 and 740 pm, respectively. For other rare earth functionalized zeolites not considered here due to accessibility, availability or stability constraints, for example, channel structures or chemical compositions other than Si, Al and O, we would like to refer to recent reviews [3–5] and an excellent site maintained by the International Zeolite Association (IZA) [6]. Chemically, the zeolites under discussion here are composed of Si^{4+} and Al^{3+} ions located at the corners of truncated octahedra ("sodalite cages" or "β-cages"), each of which is tetrahedrally coordinated by oxygen ions at the edges of the soadalite cages. The sodalite cages may be linked via quadratic prisms (with oxygen atoms residing on the edges) or hexagonal prisms (again, oxygen atoms on the edges). In the most simplified view, they derive from a Si_2O_4 unit, in which one Si^{4+} ion is replaced by an Al^{3+} ion, hence yielding the formula $SiAlO_4^-$, that is, each Al^{3+} gives rise to a negative charge,

https://doi.org/10.1515/9783110654929-034

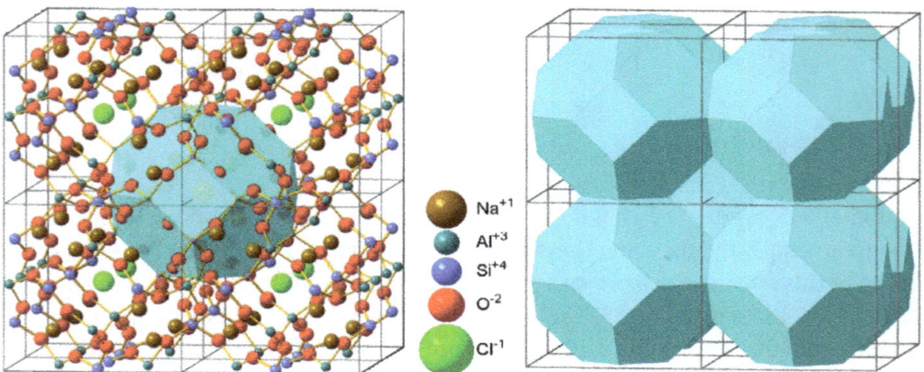

Figure 4.11.1: Chlorosodalite ($Na_8[(AlSiO_4)_6Cl_2]$). The atoms of eight unit cells are reproduced, one sodalite cage is inscribed (left). As zeolite structures can become rather complicated, a sketched description is often used to illustrate the arrangement of the cages and entrance windows (right).

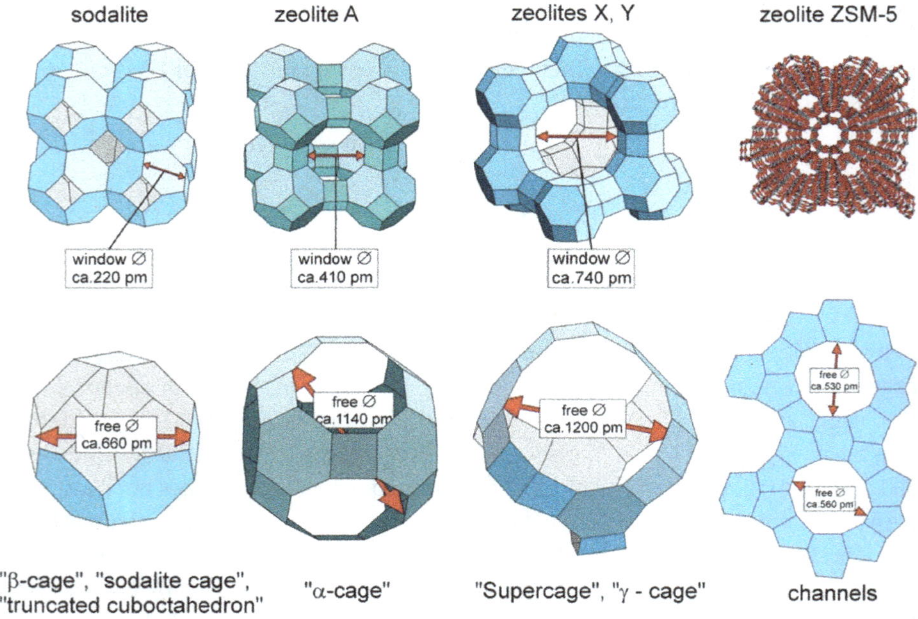

Figure 4.11.2: Sketches of typical zeolite structures. Zeolites X and Y are also referred to as Faujasites, the main difference between them being the Si:Al ratio.

which is compensated by a cation residing in the cages (e.g., Na^+); furthermore, remaining void volume is filled up by water molecules. In real zeolites, the Si/Al ratio can vary between 20 to 1; the sodalite structure is also realized solely with Al^{3+} ions [7].

The uses of zeolites are manifold, major applications of for example, faujasites include ion exchangers, as used in detergents, sorbents, industrial catalysis and dehydrating agents, their production scale exceeding 2×10^5 t/year. During the last two decades, considerable effort has been devoted to size control in zeolite syntheses, to the effect that crystalline materials ranging from a few nanometers to several microns are available, which is not only important for the abovementioned "classical" applications, but also paves the way for innovative uses, for example, as carriers for luminescent markers in biomedical assays.

4.11.2.2 Zeolites incorporating inorganic emitters

Due to the structural features and charging of the zeolite structures described, they constitute promising playground for aqueous rare earth ion exchange (one trivalent RE^{3+} ion readily releases and replaces three Na^+) and chemistry, based on smaller molecular and complex species. Numerous attempts have thus been undertaken to accommodate RE^{3+} ions by the mentioned ion exchange. Next to fundamental questions, these were initially of interest in technical applications (fluorescent lamps, LEDs, lasers), because the cheap zeolite hosts were thought to enable the replacement of "classical" phosphors, which frequently comprise costly, high purity host structures like Y_2O_3 and $Y_3Al_5O_{12}$ [8]. Of the RE^{3+} ions, only Ce^{3+} produces luminescent materials potentially applicable in these fields (almost unity quantum yields could be realized for UV allocated emission of Ce^{3+}-doped zeolites [9]), as practically all other ions useful for lighting only exhibit quantum mechanically-forbidden f→f-transitions in the spectral excitation range of interest ($\lambda > 250$ nm), others allowed transitions of the rare earth ions (such as d→d-transitions of for example, Tb^{3+} and Eu^{2+}, charge transfer (CT)-transitions of Eu^{3+}), proved to be inefficient or yielded redox-unstable materials. However, exploiting the efficient energy transfer of $Ce^{3+} \rightarrow Tb^{3+}$ enabled the sensitization of Tb^{3+} emission. Unfortunately, it soon became evident that the system was efficient, but the required concentration balance between sensitizer (Ce^{3+}) and emitter (Tb^{3+}) to accomplish high absorptivity and high quantum yield of the Tb^{3+}-emission simultaneously, could not be established within the zeolites due to stoichiometric restraints (Figure 4.11.3) [10].

Subsequently, systems containing sensitizers other than Ce^{3+} were investigated. In combining rare earth ions and transition metalates in the sodalite structure ("*RE*TMO$_4$-Sodalite"; *RE* = La, Nd, Eu, Gd, Tb, Ho, Er, Yb; TM = W, Mo) (Figure 4.11.4), tungstate or molybdate ions, instead of Ce^{3+} can be utilized as the absorber for UV-radiation. In these, the allowed O→TM^{+VI} CT feeds the RE^{3+}-excitation. Using Eu^{3+}/MoO_4^{2-} in $[Eu_4(Al_8Si_4O_{24})(MoO_4)_2]$, quantum efficiencies of 60% could be obtained, whereas Tb^{3+}/WO_4^{2-} yielded only 35% [11]. Interestingly, the absence of occluded water also allowed emissions from low energy states, for example, in the case of Nd^{3+} and WO_4^{2-} in $[Nd_{0.2}Gd_{1.8}La_2(Al_8Si_4O_{24})(WO_4)_2]$; infrared emission at 1064 nm ($^4F_{3/2} \rightarrow {}^4F_{7/2}$)

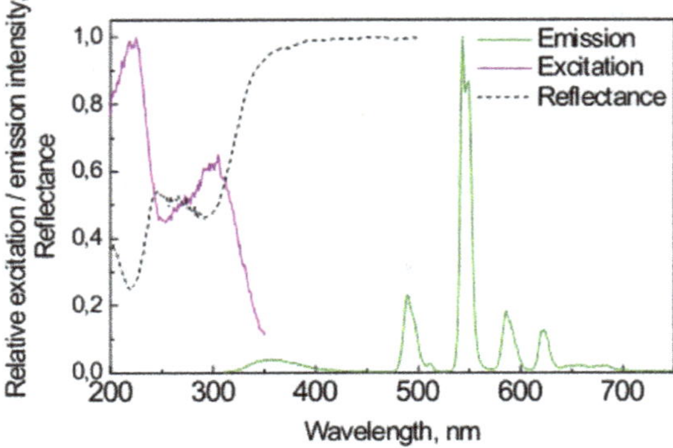

Figure 4.11.3: Emission, excitation and reflectance spectra of zeolite X loaded with 20 Tb^{3+} and 4 Ce^{3+} per unit cell. The quantum yield at 320 nm excitation amounted to 85%; however, the absorption is low. Increasing the Ce^{3+} content happens at the cost of Tb^{3+} and increases the undesired UV emission at 360 nm. Figure reproduced from reference [10] with permission from Wiley.

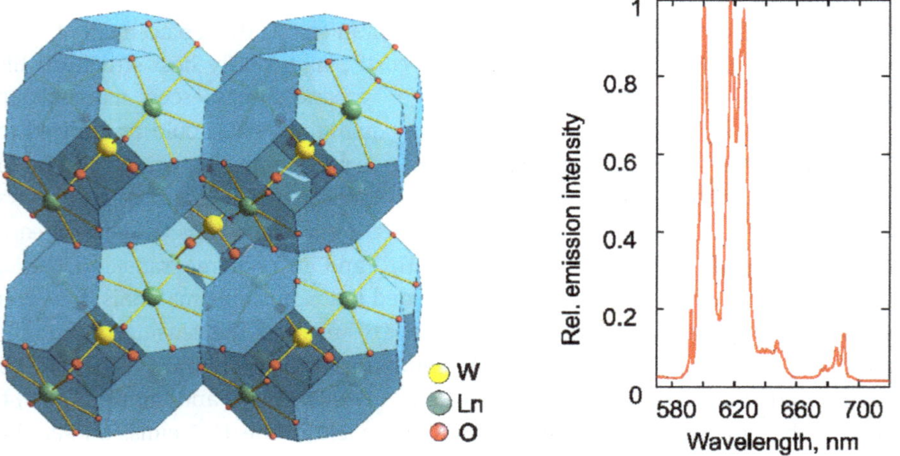

Figure 4.11.4: Transition metalate-sensitized sodalites $RE_4Si_4Al_8O_{24}(TMO_4)_2$ (TM = Mo, W; *RE* = La, Eu, Tb, Gd, Nd, Er, Ho or Yb) (left). Emission spectrum of $Eu_4Si_4Al_8O_{24}(MoO_4)_2$; the strong magnetic dipole emission at 590 nm ($^5D_0 \rightarrow {}^7F_1$) is due to the high symmetry site occupied by the Eu^{3+} ion. Figure reproduced from reference [11] with permission from Wiley.

comparable in efficiency to commercial Nd-glasses can thus be realized [12]. Even up-conversion, that is, the conversion of two near infrared photons (of $\lambda = 980$ nm) into a visible photon is possible in the sodalite system, as was demonstrated for example for $[(ErYbLa_2)(Al_8Si_4O_{24})]$, although, up to now, well below the efficiencies obtained for the well-known hexagonal Yb^{3+}, Er^{3+}-doped $NaYF_4$. However, while the abovementioned, purely inorganic zeolite materials are interesting from a scientific point of view and have promising optical and structural properties, up to date, classical phosphors are not outperformed and have hitherto not truly entered commercial exploitation.

4.11.2.3 Zeolites incorporating metallo-organic emitter complexes

Another approach towards zeolite-based rare earth emission is the incorporation of metallo-organic rare earth complexes serving as suitable ligands here. Obviously, their use will be restricted to applications beyond pure lighting purposes, as their photostability is too poor to compete with purely inorganic materials, that is, aggressive ambience or high power conditions, as in fluorescent lamps and LEDs can be excluded. In the metallo-organic systems, the sensitization to overcome the bottleneck of forbidden f→f transitions is accomplished via the UV-absorption of ligand-centered, allowed $\pi \rightarrow \pi^*$-transitions. The initially excited singlet ($^1S^*$) state subsequently undergoes intersystem crossing (ISC), populating the ligands' long-lived triplet state (^3T) that is greatly enhanced by spin-orbit coupling, effected by the central rare earth ion. The ligand ^3T-state, then, intramolecularly transfers its energy to suitable rare earth states, from which the characteristic emission occurs eventually (see Figure 4.11.5).

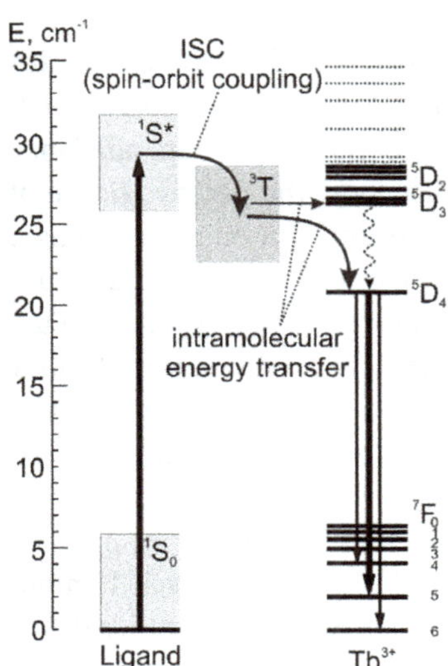

Figure 4.11.5: Principle mechanism of rare earth sensitization by organic ligands ("antenna effect") of Tb^{3+} with a deliberately chosen ligand. In the depicted case, luminescence may be expected, because the emitting Tb^{3+} state ($^5D_4$), from which several emissions are possible, clearly locates below the ^3T state. Typically $^5D_4 \rightarrow {}^7F_5$ is the most intense. The dotted arrow indicates vibrational relaxation of higher states to $^5D_4$, which may occur, if the intramolecular energy transfer can also reach the $^5D_3$ level.

In this scheme, the organic ligand is frequently referred to as an "antenna", collecting the incident radiation. In metallo-organic complexes, some essential boundary conditions have to be fulfilled for high (quantum) efficiencies: (1) the emissive rare earth level has to be allocated below the ligands' ^3T state energetically, (2) the separation between ^3T and the lowest excited rare earth state (i.e., $^5D_0$ for Eu^{3+} and $^5D_4$ for Tb^{3+}) should amount to at least 2,000 cm^{-1} to prevent energy back transfer and (3), additional coordination of ligands possessing vibrational modes of high energy such as H_2O, R-OH, NH_3, R-NH_2 should be circumvented; excited Eu^{3+}-states, in particular, suffer from radiationless deactivation via vibrations. In several instances, the attachment of ancillary ligands like chelating 1,10-phenanthroline or 2,2′-bipyridine can efficiently screen the rare earth ions from high energy vibrations. Ligands that have been used to serve as antennae in zeolites range from chelating organic nitrogen bases (phenanthroline, bipyridine) to carboxylates (benzoates and derivatives, picolinic and dipicolinic acids and derivatives) and diketonates (acetylacetonates and a multitude of derivatives) [13–18], first reports dating back more than 25 years. We will, here, discuss two selected examples for Eu^{3+} and Tb^{3+} coordinated by Htta (2-thenoyltrifluoroacetone, or more precisely, 4,4,4-trifluoro-1-(2-thienyl)-1,3-butanedione) and Hsal (salicylic acid, 2-hydroxybenzoic acid) within the supercages of faujasites, that is, zeolites X and Y. Thus, after rare earth ion exchange, the zeolites were moderately dried in vacuum (250–300 °C) to remove intra-zeolite water and combined with the ligands in an ampoule, which was sealed under vacuum, and heated at 120°C, overnight. After breaking the ampoule and thorough washing with ethanol, the materials were mildly dried again and exposed to the ambient atmosphere (days) to allow rehydration, before the product was exposed to a base (NH_3 vapor, or to aminopropyltriethylsilane vapors in a dynamic vacuum) eventually to deprotonate the ligands and support intra-zeolite complex formation. The gas phase loading was found to be advantageous, if high loading levels for strong UV absorption are desired for high brightness, that is, the product of quantum efficiency and absorption constant, which is, in numerous applications, the decisive figure of merit. Overall, the efficiency of the materials is strongly dependent on ion exchange, ligand loading levels and the ion to ligand ratios; in case of Eu^{3+} and Httfa, the optimum composition amounted to four complex $Eu(tta)_4^-$ ions per unit cell, that is, Eu^{3+} occupation of half of the available supercages. Spectra, analyses and molecular modelling corroborate the presentation depicted in Figure 4.11.6: the complex is probably too large to be confined within one supercage and partially extends into neighboring cages.

The experimentally determined quantum efficiencies well in excess of 45%, at 95% absorption are very promising. As the quantum efficiency of a given luminophore is proportional to its emission decay time, the latter can, among other determinable information, qualitatively support the experimentally determined quantum efficiency. Thus, the measured decay time of >1,200 μs is also unusually high for such complexes. For comparison, a composition of $[Eu(tta)_2]_8$-zeolite X, utilized all of the eight supercages and demanded the formation of $Eu(tta)_2^+$ cations, and either the additional coordination of lattice oxygen atoms or water molecules – at the cost of efficiency and brightness (Figure 4.11.6; see also the appearance of cationic species

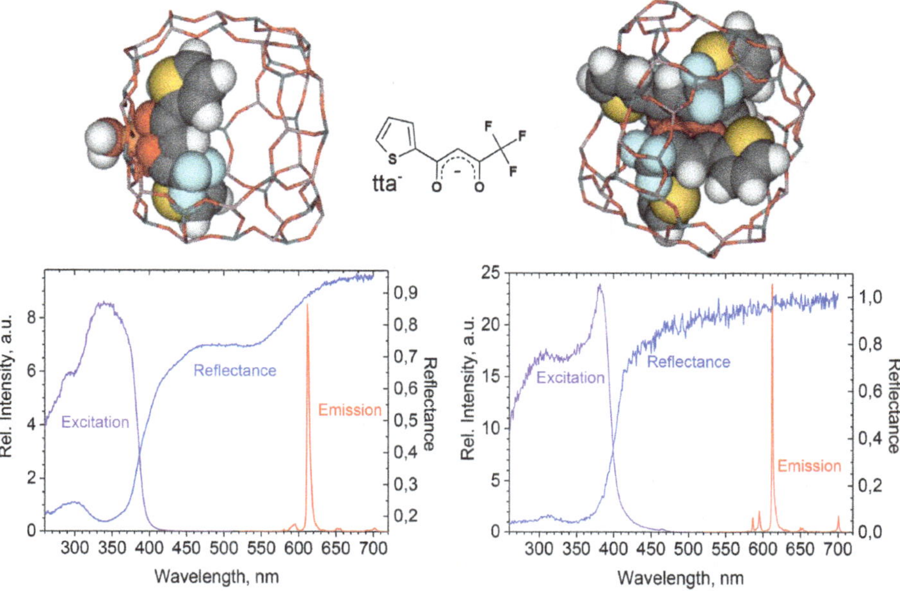

Figure 4.11.6: Tentative structures of encaged Eu(tta)-species in zeolite X (results from molecular modelling with Hyperchem; the encaging supercage was taken from crystal structure data and kept rigid); reflectance, excitation and emission spectra of $[Eu(tta)_2]_8$-zeolite X (left, all supercages occupied) and $[Eu(tta)_4]_4$-zeolite X (only half of the supercages filled, right).

in Tb-salicylates in the next paragraph). The quantum yield is now found at well below 35%, and consequently the decay time is reduced to 550 µs.

Analogous Tb^{3+} salicylate complexes, for example, the up to now most efficient hybrid $[Tb(sal)_{1.65}]_{16}$-zeolite X, are less sensitive to lattice or water co-coordination, as expected. In view of the spatial constraints, this analytically verified stoichiometry suggests that eight $Tb(sal)_3$ complex species are accommodated in the supercages, while another eight Tb^{3+} ions reside in sodalite cages or in six-ring windows respectively, the latter additionally coordinating water. Whether or not the occluded $Tb(sal)_3$ complexes additionally coordinate water or lattice oxygen atoms is not known. The arrangement yielded materials with quantum yields > 60% at > 90% absorption, accompanied by decay times as large as 1900 µs. Nevertheless, the Tb^{3+}-sal-zeolite X system has its complications: the Tb^{3+}/sal ratio never exceeded 0.5, which means that at medium loading levels as in $[Tb_8(sal)_{16}]$-zeolite X (all supercages occupied), the most likely species is a $Tb(sal)_2^+$ cation. Furthermore, at comparably low Tb^{3+} contents ($[Tb(sal)_{2.4}]_2$-zeolite X, Tb^{3+}/sal = 0.42, $[Tb(sal)_{2.05}]_4$-zeolite X, Tb^{3+}/sal = 0.48), the appearance of a strong phosphorescence band from free salicylate next to the Tb^{3+} emission lines implies the presence of a di-cationic $Tb(sal)^{2+}$ species, at least as a by-product (Figure 4.11.7).

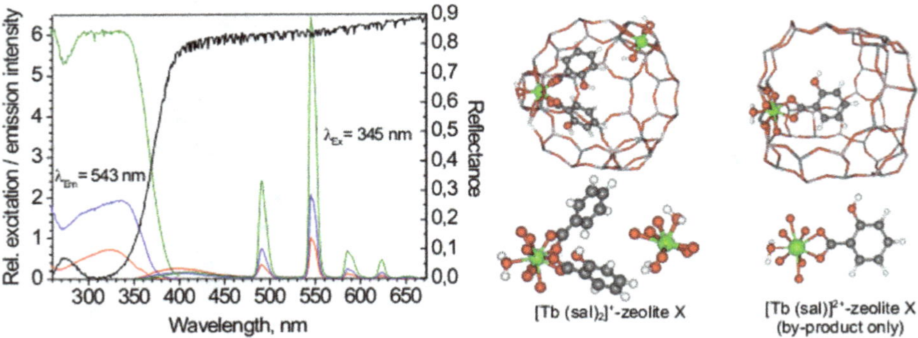

Figure 4.11.7: Luminescence spectroscopy of Tb-salicylates in zeolite X and tentative structures from elementary analyses and molecular models (Hyperchem). The most efficient material, $[Tb_{16}(sal)_{24}]$-zeolite X, green curves, contains $Tb(sal)_3(H_2O)_n$ units (structure not shown here) next to zeolite lattice and water-coordinated Tb^{3+}. At lower Tb^{3+}-contents (≤8 per unit cell) $Tb(sal)_2^+$ ions prevail, at very low Tb^{3+} contents ($[Tb_2(sal)_4]$-zeolite X, red and $[Tb_4(sal)_8]$-zeolite X, blue) the emission of the salicylate ions indicates $Tb(sal)^{2+}$ ions.

Most of the given metallo-organic examples were based on zeolite X, with a particle diameter of approximately 200 nm, which after suitable surface modification for protein linkage, makes them promising candidates for highly sensitive lateral flow analyses for example, point-of-care test strips (analogous to “pregnancy tests”). Other bio-assay applications are currently tested or anticipated in ELISA-type and DELFIA-based determinations of (bio-)medically relevant parameters (proteins, antibodies, toxins etc.; immunobiology). The exploitation of the full particle size regime between 20 nm and microns will, foreseeably, even extend this scope.

4.11.2.4 Mesopores

Mesoporous hosts are materials greatly extending the pore size range over the microporous zeolites (≤ 2 nm); by definition (IUPAC), mesopores exhibit diameters ranging from 2 to 50 nm, while beyond that, the term macropores applies. Today, numerous important mesoporous materials are known – for example, active carbon and numerous oxides of transition metals among them. However, in the present context, the focus shall be set on silica-based mesoporous materials. These have first been reported in the 1970’s, MCM-41 (Mobil Composition of Matter No. 41) being among the first materials. They exhibit a hexagonal alignment of parallel channels with well-defined diameters of 2 to ≤ 20 nm (see sketch, Figure 4.11.8), depending on the surfactant used in the synthesis. Subsequently, probably a publication in Nature [19], triggered an intense search for similar materials and their technical exploitation, to the avail that SBA-15 (Santa Barbara Amorphous type material) was discovered a few

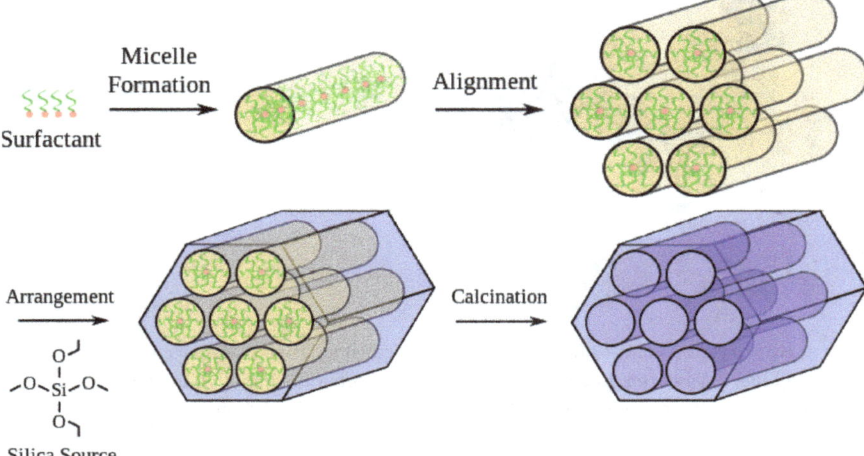

Figure 4.11.8: Synthesis of mesoporous MCM-41 and SBA-15. The channels form after self-assembly of the surfactant tubes to a hexagonal array. (Work from Hermann Luyken after dissertation of Markus Reichinger, Ruhr-Universität Bochum, 2007; licensed under CC0 1.0 Universal Public Domain Dedication 2014).

years later; it displays yet larger channel diameters of up to 30 nm [20] and beyond, and wall thicknesses of up to 3 nm. These, and additionally, cubic MCM-48 (having intersecting channels), are the most intensely studied materials, not the least because their synthesis is fairly straightforward and simple. Chemically, they essentially consist of nanoscaled, amorphous SiO_2, which is assembled in the structures described by the use of amphiphilic templates like cetylammonium bromide (CTAB), around which hydrolysable silica precursors like tetramethylorthosilicate are deposited, in the presence of tetramethyl ammonium hydroxide (TMAOH). The initial product thus still contains the surfactant, which may eventually be removed by extraction, or, more completely, by calcination at 550 °C (Figure 4.11.8).

After removal of the organic template, they display enormous internal surfaces well beyond 1,000 m^2/g due to micropores arising between the silica nanoparticles formed, which makes them ideal candidates for heterogeneous catalysis and sorption; recent trends have been described by Suib [21]. Due to the inevitable presence of surface Si–OH groups (see Figure 4.11.9 for a quantum mechanical model of hydroxylated MCM-41 [22]), the materials are inviting subsequent grafting of further extended functions. The most convenient method is the condensation of aminoalkylalkoxysilanes (e.g. $H_2N\text{-}CH_2\text{-}CH_2\text{-}CH_2\text{-}Si(OC_2H_5)_3$, “APTES”) onto the inner and outer surface, the terminal $-NH_2$ – group opening a realm of possibilities for further covalent or ionic linkages. Two examples are discussed below.

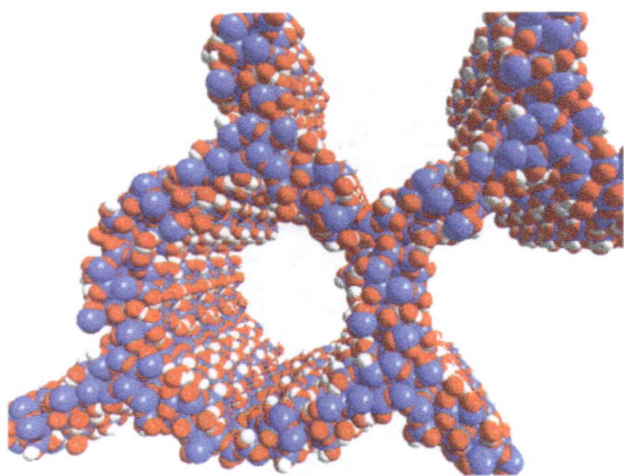

Figure 4.11.9: Hydroxylated MCM-41 from quantum chemical simulation (Large-scale Periodic B3LYP); silicon blue, oxygen red, hydrogen white. Crystal information file by P. Ugliengo, Univ. Torino. The apparent order of hydrogen is an artefact from the multiplication of the unit cell.

Applications of the modified materials are manifold, and now include drug delivery, specific ion sensors, and numerous bioassays [23]. However, even "native" MCM-41 and SBA-15 are instructive and rewarding hosts for luminescence, as demonstrated in the following two examples. One particular advantage of MCM-41 (and SBA-15) is the fact that the accessible sizes for guests by far exceed the zeolites of the previous chapter, where the entrance to the supercages of zeolite X limits the guest size to roughly 7.4 Å (Figure 4.11.10). Molecular complexes beyond 12 Å in size can only be encaged if a "ship-in-the-bottle" strategy is available. Numerous molecules and complexes with bulky ligands or co-ligands, which are known to exhibit superior luminescence, are thus not expected as possible in zeolite X. Among

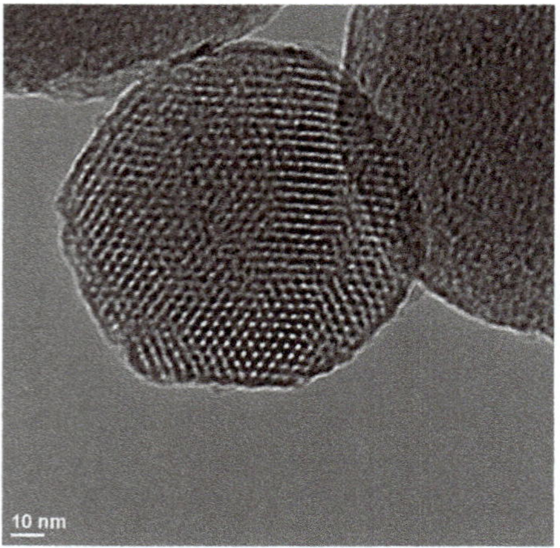

Figure 4.11.10: TEM-Image of MCM-41 displaying the hexagonal arrangement of channels. Courtesy, Dr. Victor Lin group at Iowa State University (licensed under Wikimedia commons).

these are trioctylphosphineoxide (TOPO) and triphenylphosphineoxide (TPPO); both, carboxylate and diketonate complexes of Tb^{3+} and Eu^{3+}, are known to be efficient green and red emitters, respectively, if equipped with ancillary TOPO or TPPO.

The results of loading MCM-41 with $Tb(sal)_3TOPO_2$ and $Eu(tta)_3TOPO_2$ are reproduced in Figure 4.11.11. Loading can be performed by simply stirring the complexes, which are readily soluble in toluene with the powderous MCM-41, followed by filtration, washing with toluene three times and drying; the loading in both cases amounted to approximately 10% w/w, initially. Both hybrids are very interesting in their own right: $Tb(sal)_3TOPO_2$-MCM-41 shows a decay time of 1240 µs, which is somewhat lower than the pure complex with 1412 µs, but is in agreement with a lower quantum efficiency Θ of $Tb(sal)_3TOPO_2$-MCM-41 (30%), as compared to the pure complex with Θ = 48%. In case of $Eu(tta)_3TOPO_2$, the final washing procedures led to a severe loss of the complex, such that its absorption A is unfortunately reduced to 57%, which comes at the expense of brightness. Nevertheless, $Eu(tta)_3TOPO_2$-MCM-41 is at the verge of ultimate efficiency (Θ > 90%); correspondingly, the decay time of 770 µs is the longest ever measured in our lab for this complex. The weaker leaching of $Tb(sal)_3TOPO_2$ as compared to $Eu(tta)_3TOPO_2$ from MCM-41 can be attributed to a somewhat stronger adherence of the more polar salicylate to the MCM-41 walls, this, in turn, implying that for truly satisfying materials, a covalent attachment of the less polar $Eu(tta)_3TOPO_2$ may be mandatory.

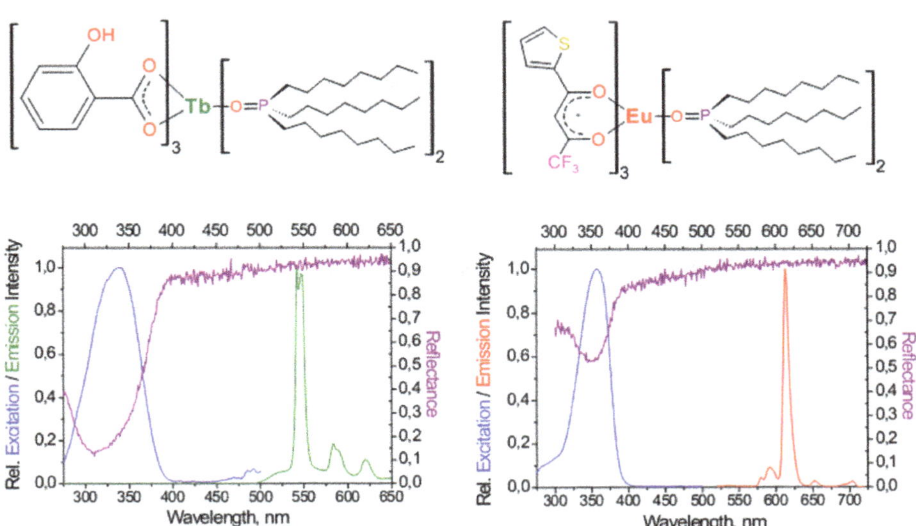

Figure 4.11.11: Structures and optical spectra of $Tb(sal)_3TOPO_2$ (τ = 1,240 µs, Θ = 30%, R = 17%; left) and $Eu(tta)_3TOPO_2$ (τ = 770 µs, Θ > 90%, R = 57%; right). The products were extensively extracted with toluene.

As pointed out, grafting aminoalkylalkoxysilanes to the MCM-41 or SBA-15 walls can be used to covalently bind rare earth guests to the hosts. Numerous schemes have been reported [23], two of which, also featuring the tta ligand to Eu^{3+}, shall be mentioned here. Guo et al. went through appreciable and successful effort to anchor the complex via a silane-modified phenanthroline co-ligand (see Figure 4.11.12, left) in SBA-15. However, the effort did not lead to what might have been expected: the efficiencies did not exceed $\Phi_L = 23\%$ at decay times of 520 μs [24]. Another approach was undertaken to exploit ionic interactions and provide an organic moiety by the incorporation of the efficient $[Eu(tta)_4]^-$ ion together with the large organic *N*-hexadecyl pyridinium cation by Xu et al. [25]. Prior to the actual optical activation, the MCM-41 matrix was pretreated with N-(3-trimethoxysilyl)ethylethylenediamine to remove (acidic) Si-OH groups. Efficiencies and absorption were not given, but remarkably, the decay time rose by 260% on encapsulation; hence, efficient hybrid materials may again be expected in usage of the $[Eu(tta)_4]^-$.

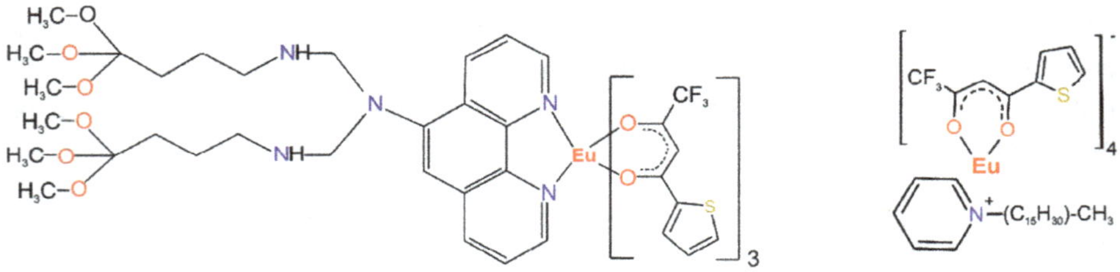

Figure 4.11.12: Examples for methods to anchor rare earth complexes within mesoporous hosts. The compound to the left was used to covalently bond $Eu(tta)_3$ to the walls of SBA-15. The *N*-hexadecyl pyridinium salt of the tetrakis diketonate $[Eu(tta)_4]^-$ to the right was realized in MCM-41; it relies on ionic forces to retain the complex within the host.

In summary, up to now it has been demonstrated that mesoporous hosts and rare earth complexes can very efficiently cooperate with respect to optical performance. Clear advantages are their degrees of spatial freedom and the ease of preparation, and since no principle bottlenecks are in sight, further efforts towards employment, in the biomedical field in particular, are more than justified; in this area, beneficial breakthroughs seem to be at hand.

4.11.3 Layered structures

As opposed to previous chapters, in layered structures, the confinement is now only one-dimensional, although the third dimension is also accessible to some degree in swellable materials as the layer thicknesses may adjust to the intercalated species. As in the previous sections, a multitude of layered structures involving rare earths calls for a rigorous focus on selected or particularly exciting material classes and optical properties. The somewhat deliberate choice will address examples of Layered Double Hydroxides (LDHs, for an extensive review and recent update see [26, 27]), but mainly embrace clay type materials with high silica contents. In these structures, the accommodation of rare earth guest species typically takes place in interlayers between two "hard" layers; the focus is set on luminescent complexes. In LDHs, the layers are cationic in nature, hence they are suited for the intercalation of anions, while silicate type clays, composed of anionic sheets, enable the insertion of cationic guests.

4.11.3.1 Layered double hydroxides

LDHs may be perceived as derivatives of brucite that display octahedral layers of Mg $(OH)_2$, in which a Mg^{2+} ion is isomorphously replaced by Al^{3+} resulting in $[Mg_9Al(OH)_{18}]^+$, obviously to be charge compensated by an anion, that is, an intercalated anionic (rare earth) guest species. The $Mg^{2+} \leftrightarrow Al^{3+}$ substitution is not restricted to one Mg^{2+}: Mg^{2+}/Al^{3+} ratios of 2 can readily be obtained. The corresponding anions are organized in the interlayers between octahedral sheets. The interlayer thickness corresponds to the difference between the so-called basal spacing (determined from $(hkl) = (0\ 0\ 3)$ in XRD) and the thickness of a single $Mg,Al(OH)_2$ sheet of 480 pm. An illustrative example is the envelopment of derivatives of $[Tb(C_5H_4N{-}COO)_4]^-$ ("picolinates", pyridine-2-carboxylates) by the $Mg,Al(OH)_2$ sheets (Figure 4.11.13). Quantum yields Φ_L of approximately 50% could be obtained on increasing the Al^{3+} content. Analysis of the interlayer separations, chemical composition and optical spectra surprisingly revealed that not the tetra-picolinate $[Tb(C_5H_4N{-}COO)_4]^-$ formed, as typically obtained for the free complex, but rather a tris-picolinate $[Tb(C_5H_4N{-}COO)_3]$ and even a formally cationic species $[Tb(C_5H_4N{-}COO)_2]^+$ were intercalated. To compensate for the overall charges, it has to be assumed that Tb-OH bonds evolve in case of the tris-picolinates, that is, water molecules indicated in the molecular sketch (Figure 4.11.13) of the solution species are replaced by Mg,Al-OH, eventually leading to an oxygen bridge between the sheet atoms and Tb. At low Mg^{2+}/Al^{3+} ratios < 5, a high lattice charge results, allowing a large amount of anionic complex to be accommodated, at the same time causing the larger interlayer spacing of near 900 pm, while at lower Al-contents, the sheets can come closer and an interlayer spacing of approximately 300 pm is monitored. For this to be feasible, the complex has to become confined in the dimension perpendicular to the layers. For the resulting linear / planar species, at least two lattice oxygen atoms are assumed to coordinate to the lanthanide ion (Figure 4.11.13, right), resulting in a notably reduced quantum efficiency ($\Phi_L < 30\%$). The fairly high quantum efficiencies obtained at $Mg^{2+}/Al^{3+} = 2$ are somewhat surprising, as the LDHs constitute an ambience with high frequency vibrations, usually disadvantageous with respect to efficiency. Following this argument, it

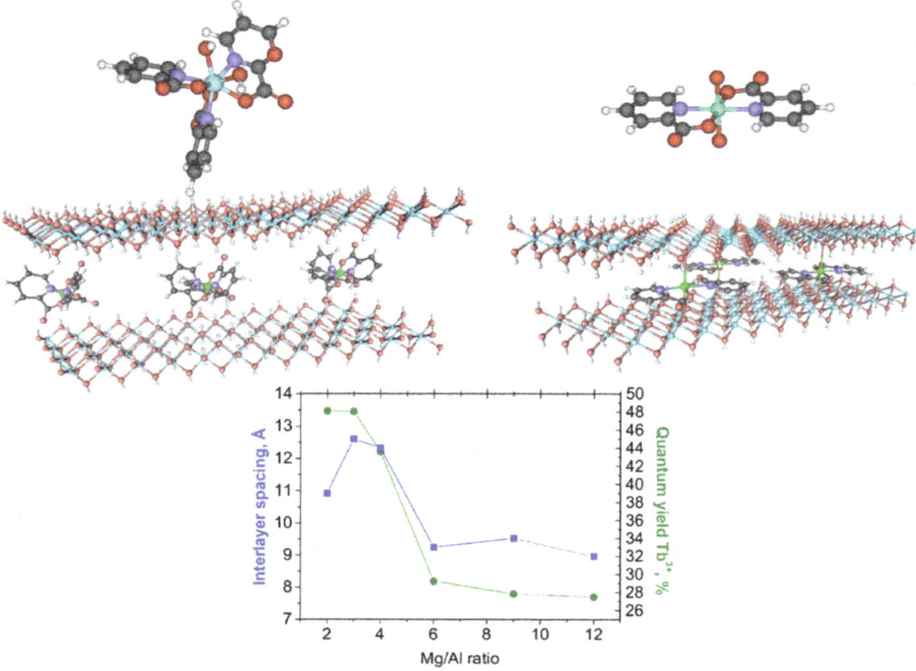

Figure 4.11.13: Tb(picolinate) species enclosed between Double Layer Hydroxide layers (Mg,Al $(OH)_2$). Top left: at high Al^{3+} content, hydrated or hydroxlated $Tb(pic)_3$, is formed. Top right: lower Al^{3+} contents lead to (hydroxylated) $Tb(pic)_2$ – species. Figure reproduced from reference [29] with permission from Elsevier.

is worth noting that analogous Eu-picolinates, as expected, never exceeded a 10% margin in efficiency [28, 29]. Hence, LDHs, despite the respectable efficiency of Tb^{3+}, seem to be less suited as photonic hybrids for the visible Eu^{3+} emission, in particular.

4.11.3.2 Clays

The second prominent type of layered structures are historically among the oldest construction materials and the overwhelming amount of literature on clays is therefore not surprising. An excellent overview, although by no means comprehensive, was edited by Bergaya, Theng and Lagaly already in 2006, but still serves as a reference work [30]. This article shall focus on the smectite family, members of which are montmorillonite, beidellite, hectorite and saponite, all of them swellable clays. Hectorite and its synthetic nanoscaled relative will serve as the relevant examples. Structurally, these may be perceived as derived from talc, with octahedral MgO,OH sandwiched and linked to tetrahedral SiO_4 layers on top and bottom. The resulting structure is made up from TOT sheets (tetrahedral–octahedral–tetrahedral), with an idealized composition of $[Mg_6Si_8O_{20}(OH)_4]$ for talc. If some Li^+ is isomorphously placed into Mg^{2+} sites, a negative charge is caused within the TOT layers. The relation of hectorite may then already be seen from its idealized stoichiometry: $[Mg_5LiSi_8O_{20}(OH)_4]^-$. Cations to counterbalance negative charges within the TOT sheets lead to a lamination of the individual sheets by attractive Coulomb forces in solid clays; Figure 4.11.14 reproduces the structural essentials (the spacing data given are taken from an analysis of Cs $[Mg_5LiSi_8O_{20}F_4]$ [31]). Typically, the interlayer guests are alkali cations and water. As in the case of zeolites, the alkali ions are also exchangeable by rare earth ions and can thus activate the hectorites; efficient sensitization of Tb^{3+} by Ce^{3+} (see first paragraph, Figure 4.11.3) is also possible. Moreover, similar to the situation in zeolites, Tb^{3+} is coordinated by some water within the hectorite interlayer to give an overall composition, $Tb_{0.4}Mg_{5.4}Si_8O_{20}(OH)_4(bipy)_{0.96}(H_2O)_3$. The efficiency of this hectorite-complex hybrid is greatly enhanced in comparison to the naked Tb^{3+} ion, but is still restricted to a quantum efficiency of approximately 20% [32].

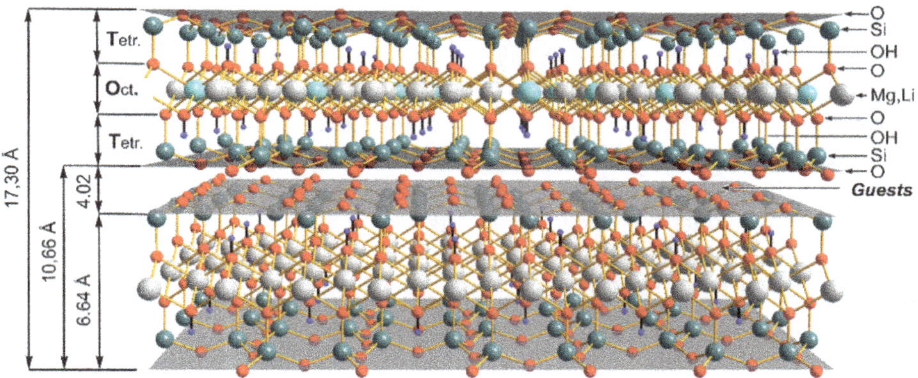

Figure 4.11.14: Structure of hectorite/laponite clays. The given interlayer spacing varies with the size of the intercalated guests. Data were reconstructed from a crystal structure analysis of Cs $[Mg_5LiSi_8O_{20}F_4]$. Figure reproduced from reference [29] with permission from Elsevier.

A nanoscaled variant of hectorite clays named “Laponite”[33] exhibits particular charm in the context of photonic materials, its outstanding solution behavior being the driving force behind the interest. The Laponite’s composition is best given as $[Na_{0.7}\{Li_{0.3}Mg_{5.5}Si_8O_{20}(OH)_4)\}(H_2O)_n]$ (the apparent stoichiometric mismatch is probably due to foreign clay impurities [34]). Such nanoclays are readily available commercially at competitive prices (e.g., “Laponite RD©” [35]) and are extensively used as laquer and polymer additives for the adjustment of rheological and mechanical properties. Their unusual properties arise from the anisotropic dimensions of single particles with diameters of 25–30 nm at a thickness of only 1 nm and a high degree of surface ionization (see Figure 4.11.15), which can yield glass clear, aqueous solutions at pH approximately ≥ 9 at concentrations even in excess of 1% w/w due to the highly charged surfaces: the natively white powders then exfoliate, as the sodium ions migrate into the aqueous surrounding and water hydrates the individual particles. Depending on pH, the rims, at which Mg-OH sites are inevitably present (Mg-O-Mg + H_2O → 2Mg-OH) are ionized, such that the edges of the platelets become positively charged (Mg-OH + H_2O ⇔ $MgOH_2^+$ + OH^-) (Figure 4.15.15, right). At higher concentrations or in the presence of surplus foreign cations (including protons), negatively charged faces and positive rims first form a "house-of-cards"-structure, accompanied by a dramatic increase in viscosity (Figure 4.11.15, center). Eventually, the precipitation of more or less regularly stacked particles occurs. The most exciting feature of the Laponites is the capability of the exfoliated, dispersed particles to adsorb and mobilize even extremely hydrophobic molecules, natively not found in water, into the aqueous phase.

Numerous cases with organic dyes have recently been reported, in which they even profit from the nanoclay platform with regard to their optical properties or emission efficiency, among them phthalocyanines, for example [36].

However, rare earths, complexes in particular, can also benefit greatly on forming hybrids with Laponites. While the interaction of cationic species like the trivalent rare earth ions or cationic $Tb(bipy)_2^{3+}$-complexes with the negatively-charged nanoclay particle surfaces is no more surprisingly an option to accomplish fairly efficient dispersions [37], the full potential of rare earth complex-nanoclay hybrids is revealed on studying non-charged complexes that are usually not encountered in aqueous ambience. This approach has also been pursued by several other scientists, see for example, the numerous efforts reported by H. Li and coworkers and references cited therein [38]. With regard to insolubility in water, the complexes Tb $(sal)_3TOPO_2$ and $Eu(tta)_3TOPO_2$ (see Figure 4.11.11 for ligand structures), also discussed in the context with microporous materials above, assume an outstanding rank. The corresponding Laponite hybrids can be obtained from solutions of the complexes in toluene, by stirring with the solid powders, filtration and re-dispersion in water. Although restricted in efficiency, the salicylates are fruitful objects for this study, as the occurrence of free salicylate emission serves as an indicator for the stability of the hybrid in aqueous dispersion (Figure 4.11.16).

The broad emission band at 420 nm is due to the free salicylate anion, which is most likely still adsorbed, presumably at the positively charged Laponite's rim, where it can coordinate the protruding Mg ions. As this emission is notably reduced at high overall complex concentrations (nominally 100 molecules per Laponite disk), a competition of the rims with $Tb(sal)_3TOPO_2$ for salicylate seems to occur.

The solubilization of Eu-TOPO complexes can proceed analogously, as examples, $Eu(tta)_3TOPO_2$ and $Eu(btfa)_3TOPO_2$ are represented in Figure 4.11.17 (btfa = benzoyltrifluor-acetylacetonate). The difference in emission intensity is brought about by the differing absorption and excitation maxima, but their efficiencies are roughly comparable. The determination of exact quantum efficiencies is somewhat hampered by increasing

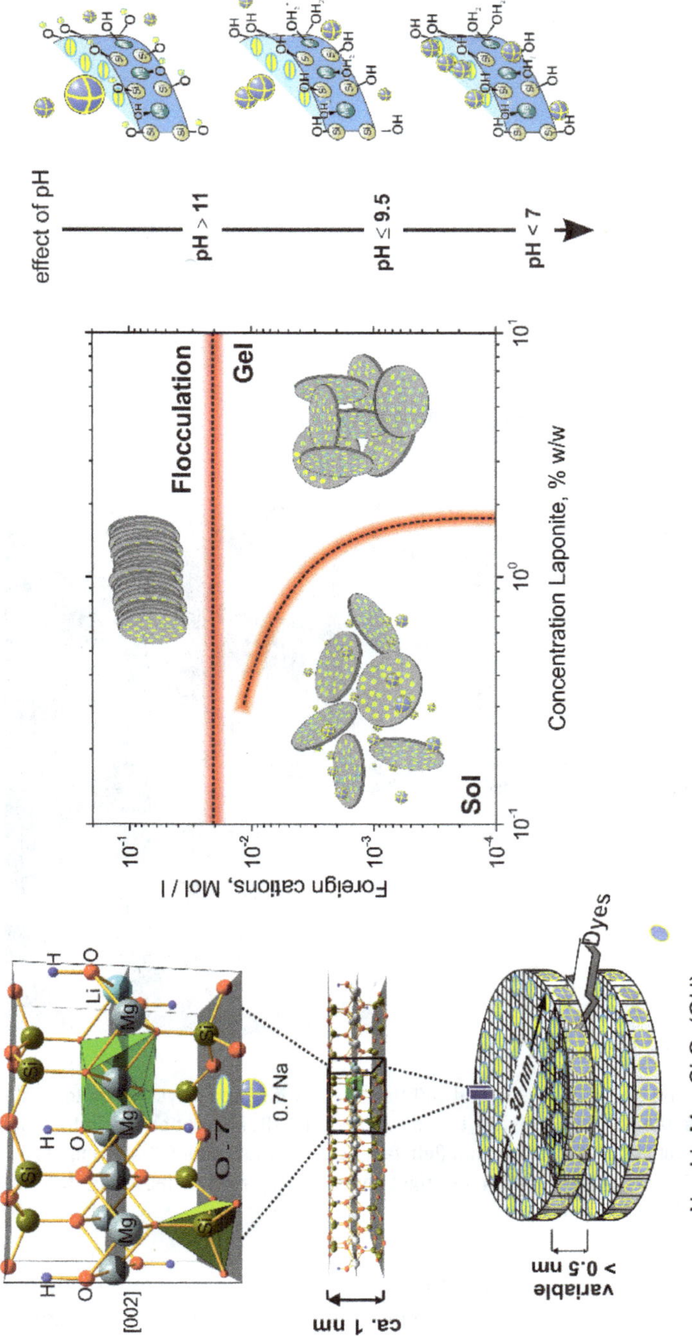

Figure 4.11.15: Unit cell, morphology and guest allocation (left), lamination behavior (center) and surface ionization of Laponite (right), cations are symbolized by blue and yellow spheres. Figure reproduced from reference [29] with permission from Elsevier.

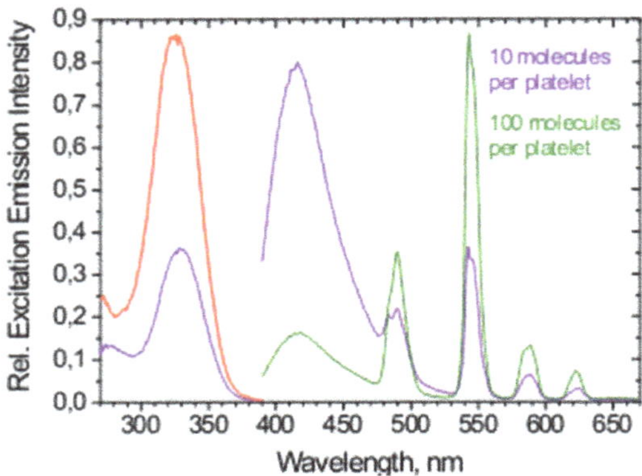

Figure 4.11.16: Solubilization of $Tb(sal)_3TOPO_2$ with nano-clays (Laponite).

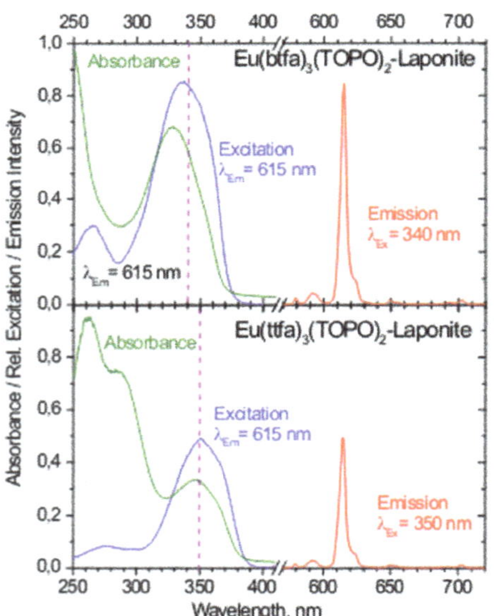

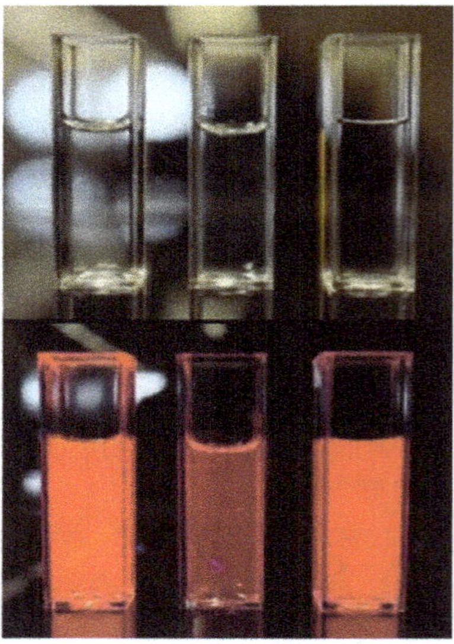

Figure 4.11.17: Solubilization of $Eu(tta)_3TOPO_2$ and $Eu(btfa)_3TOPO_2$ with nano-clays (Laponite, in 1% w/w dispersion; samples had 40 complexes per Laponite platelet). Left: absorption, excitation and emission spectra; Right: $Eu(ttfa)_3TOPO_2$ in ethanol (left samples), ethanol water (10:1) (samples in middle), and $Eu(ttfa)_3TOPO_2$-Laponite in pure water (right samples). Top row: under daylight, bottom row: broadband UV-excitation.

scatter in the UV at the high, but desirable loading levels, which amounted to 40 complexes per Laponite platelet, and time-related stability issues. However, in the solids, quantum efficiencies amounted to $\Phi_L = 18\%$, almost identical to the Laponite hybrid with $Eu(tta)_3phen$ (bidentate 1,10-phenanthroline coordinating Eu^{3+} instead of TOPO, not depicted), which had a solid state efficiency of 20%. In aqueous solution, a roughly 20-fold emission advantage of $Eu(tta)_3TOPO_2$ over $Eu(tta)_3phen$ emerges, which can be ascribed to the much better screening of the TOPO-complexes towards attack by water molecules. The images of pure $Eu(tta)_3TOPO_2$ in ethanol, a 10:1 water/ethanol mixture and the Laponite- $Eu(tta)_3TOPO_2$-hybrid shown in Figure 4.11.17 illustrate this vividly.

In brief, rare earth clay materials proved to be worth looking into and provided numerous unexpected, pleasant surprises. However, ultimate quantum efficiencies and brightnesses seem to be difficult to achieve; furthermore, in aqueous dispersions, the complex hybrids suffered from substantial depreciation of the emission (sometimes within hours). Hence, their employment is useful in applications where water solubility and the long decay times of the complexes are required, and ultimate brightness and long-term stability are of secondary importance. Typically, such applications may be found in bio-medical analysis or in microbiological contexts, but may have to compete with a large number of matured organic dyes.

References

[1] K. Binnemans, Chem. Rev. 2009, 109, 4283–4374.

[2] B. Yan, Photofunctional Rare Earth Hybrid Materials, Springer Singapore, Singapore, 2017.

[3] Y. Wang, H. Li, CrystEngComm. 2014, 16, 9764–9778.

[4] H. Li, P. Li, Chem. Comm. 2018, 54, 13884–13893.

[5] B. Yan, B. Yan, Photofunctional Rare Earth Hybrid Materials, Springer Singapore, Singapore, 2017, 83–106.

[6] http://www.iza-online.org

[7] W. Depmeier, Acta Crystallogr. C 1984, 40, 226–231.

[8] C. R. Ronda, J. Alloys Compd. 1995, 225, 534–538.

[9] U. Kynast, V. Weiler, Adv. Mater. 1994, 6, 937–941.

[10] T. Jüstel, D. U. Wiechert, C. Lau, D. Sendor, U. Kynast, Adv. Funct. Mater. 2001, 11, 105–110.

[11] C. Borgmann, J. Sauer, T. Jüstel, U. Kynast, F. Schüth, Adv. Mater. 1999, 11, 45–49.

[12] M. Lezhnina, F. Laeri, L. Benmouhadi, U. Kynast, Adv. Mater. 2006, 18, 280–283.

[13] M. Bredol, U. Kynast, C. Ronda, Adv. Mater. 1991, 3, 361–367.

[14] I. L. V. Rosa, O. A. Serra, E. J. Nassar, J. Lumin. 1997, 72–74, 532–534.

[15] M. Alvaro, V. Fornés, S. García, H. García, J. C. Scaiano, J. Phys. Chem B 1998, 102, 8744–8750.

[16] J. Dexpert-Ghys, C. Picard, A. Taurines, J. Inclusion Phenom. 2001, 39, 261–267.

[17] D. Sendor, U. Kynast, Host-Guest-Systems Based on Nanoporous Crystals, Wiley-VCH Verlag GmbH & Co. KGaA, 2005, 558–583.

[18] Y. Wang, H. Li, L. Gu, Q. Gan, Y. Li, G. Calzaferri, Microp. Mesop. Mater. 2009, 121, 1–6.

[19] C. T. Kresge, M. E. Leonowicz, W. J. Roth, J. C. Vartuli, J. S. Beck, Nature 1992, 359, 710–712.

[20] D. Zhao, J. Feng, Q. Huo, N. Melosh, G. H. Fredrickson, B. F. Chmelka, G. D. Stucky, Science 1998, 279, 548–552.

[21] S. L. Suib, Chem. Rec. 2017, 17, 1169–1183.

[22] P. Ugliengo, M. Sodupe, F. Musso, I. J. Bush, R. Orlando, R. Dovesi, Adv. Mater. 2008, 20, 4579–4583.

[23] B. Yan, Photofunctional Rare Earth Hybrid Materials, Springer Singapore, Singapore, 2017, 57–82.
[24] X. Guo, H. Guo, L. Fu, R. Deng, W. Chen, J. Feng, S. Dang, H. Zhang, J. Phys. Chem. C 2009, 113, 2603–2610.
[25] Q. Xu, L. Li, B. Li, J. Yu, R. Xu, Microp. Mesop. Mater. 2000, 38, 351–358.
[26] X. Duan, D. Evans (Vol. Eds.), Layered Double Hydroxides, in Struct. Bonding, Vol. 119, Springer, Berlin, Heidelberg, 2006.
[27] G. Arrabito, A. Bonasera, G. Prestopino, A. Orsini, A. Mattoccia, E. Martinelli, B. Pignataro, P. G. Medaglia, Crystals 2019, 9, 361.
[28] N. G. Zhuravleva, A. A. Eliseev, A. V. Lukashin, U. Kynast, Y. D. Tretyakov, Mendeleev Commun. 2004, 14, 176–178.
[29] M. M. Lezhnina, U. H. Kynast, Opt. Mater. 2010, 33, 4–13.
[30] F. Bergay, B. K. G. Theng, G. Lagaly, Handbook of Clay Science, Elsevier, Amsterdam, 2006.
[31] J. Breu, W. Seidl, A. Stoll, Z. Anorg. Allg. Chem. 2003, 629, 503–515.
[32] M. Lezhnina, E. Benavente, M. Bentlage, Y. Echevarria, E. Klumpp, U. Kynast, Chem. Mater. 2007, 19, 1098–1102.
[33] B. S. Neumann, Rheol. Acta 1965, 4, 250–255.
[34] G. E. Christidis, C. Aldana, G. D. Chryssikos, V. Gionis, H. Kalo, M. Stöter, J. Breu, J.-L. Robert, Minerals 2018, 8, 314.
[35] BYK-Chemie GmbH, Germany.
[36] M. C. Staniford, M. M. Lezhnina, M. Grüner, L. Stegemann, R. Kuczius, V. Bleicher, C. A. Strassert, U. H. Kynast, Chem. Commun. 2015, 51, 13534–13537.
[37] M. M. Lezhnina, M. Bentlage, U. H. Kynast, Opt. Mater. 2011, 33, 1471–1475.
[38] X. Chen, Y. Wang, R. Chai, Y. Xu, H. Li, B. Liu, ACS Appl. Mater. Interfaces 2017, 9, 13554–13563.

Dirk Johrendt

4.12 Rare earth based superconducting materials

Superconductivity is a macroscopic quantum phenomenon occurring in metallic matter, characterized by zero electrical resistivity and perfect diamagnetism below a critical temperature (T_c). Superconducting cables can carry a thousand times more current as a copper cable without any loss or heat generation. Magnets with superconducting windings produce extremely high and constant magnetic fields which allow for techniques like magnetic resonance tomography in hospitals, magnetic levitation trains, or high-power wind generators [1].

It was the Dutch physicist Heike Kammerlingh-Onnes, who discovered the abrupt loss of resistivity in mercury metal at 4.2 K in 1911, and coined the term superconductivity. More than 20 years later, the German physicists Walther Meissner and Robert Ochsenfeld found that superconductors completely expel magnetic fields from their interior, which means perfect diamagnetism. It took again more than two decades until the formulation of a physical explanation, the famous BCS-theory, named after their creators Bardeen, Cooper, Schrieffer, who won the 1972 Nobel Prize in Physics. Meanwhile, more than thousand superconducting materials had been discovered, but one has still thought that the phenomenon is limited to metals and alloys at extremely low temperatures. This changed with the discovery of the first copper oxide superconductors in 1986 by Georg Bednorz and Karl Alexander Müller, who won the Nobel Prize in 1987. These ceramic compounds are superconducting at temperatures as high as 138 K, which no one has ever thought possible. The term high-temperature superconductivity (high-T_c) was born, which mesmerizes material scientists till today. It took another two decades until a second class of high-T_c materials emerged, the iron based superconductors. Iron was for a long time neglected in the search for new superconductors because of its magnetic properties and the paradigm that magnetism generally destroys superconductivity. Amazingly enough, it turned out that superconductivity in iron compounds occurs just in the proximity of magnetic ordering, and theory presumes that magnetic spin-fluctuation is one ingredient of the still incompletely understood mechanism of high-T_c superconductivity.

Nevertheless, it is true that localized magnetic spins and especially ferromagnetism often destroy superconductivity, which is the reason why rare earth elements with unpaired electrons in partially filled $4f$-shells appear rather detrimental to the superconducting state. One may therefore expect that superconductivity occurs in the nonmagnetic elements lanthanum, yttrium, ytterbium, and lutetium, at least at very low temperatures. Indeed, only lanthanum is superconducting at ambient pressure, while the others require high-pressure conditions. The outstanding superconducting

https://doi.org/10.1515/9783110654929-035

properties of lanthanum, among the nonmagnetic rare earth elements, are still not fully understood, and often referred to as the "lanthanum problem" [2]. However, a remarkable number of rare earth compounds are superconductors, including several high-T_c cuprates and some of the iron based materials. Rare earth based superconductors have intensively been studied for the intertwining and co-existence of magnetic order with the superconducting state. They have significantly contributed to the understanding of both actually antagonistic phenomena.

4.12.1 Rare earth metals and alloys

Mendelssohn and Daunt discovered superconductivity in lanthanum metal in 1937 [3]. The initial transition temperature of 4.7 K was later determined to 5.0 K for the hexagonal and 5.95 K for the cubic form of lanthanum [4]. The pressure dependence of T_c in lanthanum has been intensively studied and displayed increasing T_c up to 13 K at 20 GPa. An anomaly in the $T_c(P)$ curve at 5.8 GPa was attributed to an isostructural phase transformation [2]. Alloys of lanthanum with nonmagnetic yttrium or lutetium show deceasing transition temperatures [2]. Alloying with magnetic rare earth metals have much stronger effects. For instance, 1 % gadolinium in lanthanum reduces the transition temperature from 6 K to 0.5 K. Alloys with 2.5 % gadolinium are no longer superconducting but ferromagnetic. It turned out that the T_c depression in La-*RE* alloys correlates with the number of unpaired spins and not with the *RE* effective magnetic moment [4]. The detrimental interaction between unpaired spins and the cooper-pairs (the decisive quasiparticles for superconductivity) turned out to be temperature-dependent at small concentrations. As an example, 0.6 % Ce in $La_{0.094}Ce_{0.006}Al_2$ does not suppress the superconducting transition at 1.2 K, but on further cooling, the pair-breaking strength of Ce^{3+} ($4f^1$) increases until it destroys superconductivity below 0.15 K, as shown in resistivity and susceptibility data (Figure 4.12.1) [5].

Yttrium is not superconducting above 6 mK at ambient pressure. A first superconducting transition at $T_c \approx 1.4$ K was found at 11 GPa, which increases to remarkable 19.5 K at 115 GPa [6]. Lutetium has a very low transition temperature of 22 mK at a pressure of 4.5 GPa [7] and reaches not more than 1 K at 18 GPa [2]. Ytterbium is likewise nonmagnetic due the $4f^{14}$ configuration and becomes superconducting under 86 GPa pressure at $T_c = 1.4$ K, which increases to 4.6 K at 179 GPa [8]. Cerium becomes superconducting under pressure in spite of the formally magnetic $4f^1$ configuration. An isostructural phase transition under pressure (γ-Ce → α-Ce) delocalizes the *f*-electron into the conduction band and generates a nonmagnetic "quenched moment state" in α-Ce [9]. Superconductivity of α-Ce remains in the mK-range up to 4 GPa, where it jumps to 1.9 K. Higher pressures reduce the critical temperature down to 1.2 K at 20 GPa [2].

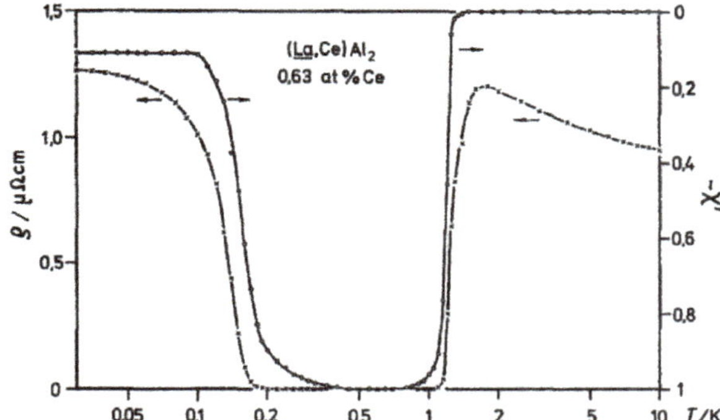

Figure 4.12.1: Electrical resistivity and magnetic susceptibility of $La_{0.094}Ce_{0.006}Al_2$. Reproduced with permission from [5].

4.12.2 Intermetallic compounds

While superconductivity in rare earth elements is largely confined to lanthanum, an enormous number of rare earth intermetallic compounds have been identified as superconductors. A comprehensive discussion is beyond the scope of this book; therefore, we give selected examples and refer to the literature for more complete studies [10, 11].

The binary intermetallic lanthanum compounds $LaSn_3$, $LaPb_3$ and $LaTl_3$ with the Cu_3Au-type structure are superconducting at 6.45 K, 4.05 K and 1.57 K, respectively [12]. This is also the case for Cu_3Au-type compounds such as La_3Tr (Tr = Al-Tl) with higher critical temperatures of 6.16 K (La_3Al), 5.84 K (La_3Ga), 10.4 K (La_3In) and 8.95 K (La_3Tl) [13]. Among the large family of Laves phases, one finds $LaOs_2$ with a relatively high transition temperature of 8.9 K [14] and $LaRu_2$ which is superconducting only below 1.63 K [15]. Further superconducting Laves phases are the $REAl_2$ compounds (RE = La, Lu, Y, Sc), where $LaAl_2$ has the highest T_c with 3.26 K [2]. YGe_3 and $LuGe_3$ are superconductors with the orthorhombic $DyGe_3$-type structure, which contains layers of Ge zigzag-chains and Ge cubes, each separated by the Lu atoms. The transition temperatures are ≈ 2 K in YGe_3 and 3.3 K in $LuGe_3$ [16, 17]. Further superconducting binary rare earth germanides are YGe_2 ($ThSi_2$-type, T_c = 3.8 K), $LaGe_2$ (distorted $ThSi_2$-type, T_c = 1.49 K), and $LuGe_2$ ($ZrSi_2$-type, T_c = 2.6 K) [18]. Among ternary germanides, one finds the superconductors Y_2PdGe_3 (AlB_2-type, T_c = 3 K) [19] and Y_2AlGe_3 with a covalently bonded Al-Ge network and T_c = 4.5 K [20]. An interesting case are the rare earth iron silicides $RE_2Fe_3Si_5$ with the tetragonal $U_2Mn_3Si_5$-type structure, representing the first superconducting ternary iron compounds with critical temperatures up to 6.8 K in $Lu_2Fe_3Si_5$ [21].

4.12.3 Superconductivity and magnetic ordering

One of the most intriguing properties of the rare earth elements is their magnetism emerging from the unpaired electrons in strongly localized (spatially contracted) $4f$ orbitals. Given the actually antagonistic relationship of superconductivity and magnetism, rare earth elements appear to be not promising components for superconductors. Nevertheless, a huge number of rare earth compounds are superconductors, thus the magnetic properties are not detrimental in all cases. The possible co-existence of superconductivity and magnetic ordering has been a problem in solid-state physics for decades. It was V. L. Ginzburg who pointed out in 1957 that both phenomena exclude each other and unlikely occur in the same material [22]. Indeed, it turned out that not only ferromagnetic ordering but also small amounts of randomly distributed paramagnetic impurities can destroy superconductivity, for instance a few percent of magnetic gadolinium in nonmagnetic lanthanum metal, as described above. However, several families of superconducting compounds with magnetic rare earth elements on regular lattices have been discovered where superconductivity is not destroyed by the local spin moments, and may even coexist with magnetic ordering. The most prominent classes in this field are the Chevrel phases $REMo_6X_8$ (X = S, Se, Te), the rhodium borides $RERh_4B_4$, and the borocarbides $RENi_2B_2C$, as well as the iron arsenide $EuFe_2(As_{1-x}P_x)_2$. In these materials, the magnetic ions and the superconducting electrons are separated and therefore the conventional pair-breaking mechanism by local unpaired spins does not apply. Furthermore, the magnetic ordering temperatures are mostly very low (≈ 1–19 K) and one assumes rather weak dipolar interactions between the *RE* ions. Antiferromagnetic ordering has weak influence on superconductivity because the magnetization averages to zero within the unit cell. Ferromagnetic ordering interacts much stronger with the superconducting state. It turned out that coexistence occurs only with sinusoidal modulated ferromagnetism in very narrow temperature ranges, while long range ordering mostly destroys superconductivity due to the internally generated field.

4.12.4 Rare earth Chevrel phases

The ternary molybdenum sulfides MMo_6S_8 (M = $3d$ metal, In, Sn, Pb, alkali-/alkaline-earth elements) were first reported by R. Chevrel in 1971 [23]. The crystal structure contains Mo_6 octahedra with sulfur atoms located above the triangular faces (Figure 4.12.2 left). In 1972, Matthias et al. discovered that some of these compounds are "high-temperature" superconductors up to T_c = 13 K in $PbMo_6S_8$ [24]. Although this critical temperature appears not very high from a present-day perspective, it is the very large upper critical field $H_{c2}(0)$ up to 60 Tesla [25] which makes these materials outstanding. The rare earth Chevrel phases $REMo_6S_8$ (RE = La-Lu) were reported in 1975 and

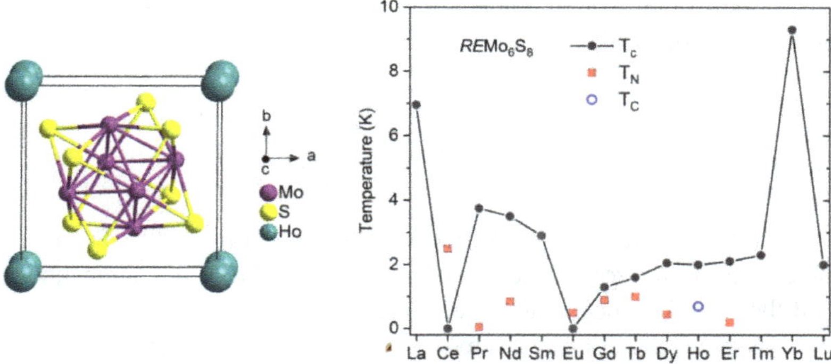

Figure 4.12.2: Left: Crystal structure of the $REMo_6S_8$ compounds. Right: Superconducting (T_c) and magnetic (T_N, T_C) transition temperatures. Data from [26].

surprisingly, among them are superconductors with magnetic *RE* ions except for *RE* = Ce, Eu. Moreover, it turned out that in most cases magnetic ordering emerges at temperatures below T_c, which coexists with the superconducting state. Figure 4.12.2 shows the crystal structure of the $REMo_6S_8$ compounds, together with a plot of the superconducting and magnetic transition temperatures [26].

The magnetic ordering patterns are mostly antiferromagnetic (*RE* = Gd, Tb, Dy, Er) with the exception of $HoMo_6S_8$, where for the first time ferromagnetic order co-existing with superconductivity has been observed, though in a narrow temperature range. $HoMo_6S_8$ is the most extensively investigated compound in the $REMo_6S_8$ series, and comprehensive neutron diffraction studies revealed a complex phase behavior at very low temperatures. $HoMo_6S_8$ becomes superconducting at $T_{c1} = 1.8$ K and re-enters the normal state at $T_{c2} = 685$ mK. The effect is referred to as re-entrant superconductivity. A sinusoidal modulated magnetic ordering state emerges at 750 mK and coexists in a small 15 mK range with superconductivity between T_{c2} and the ferromagnetic Curie-temperature T_C.

Figure 4.12.3 shows this situation on the temperature scale. It is important to note that these subtle low-temperature properties depend on the sample preparation (polycrystalline or single crystal, homogeneity, imperfections etc.), and are more

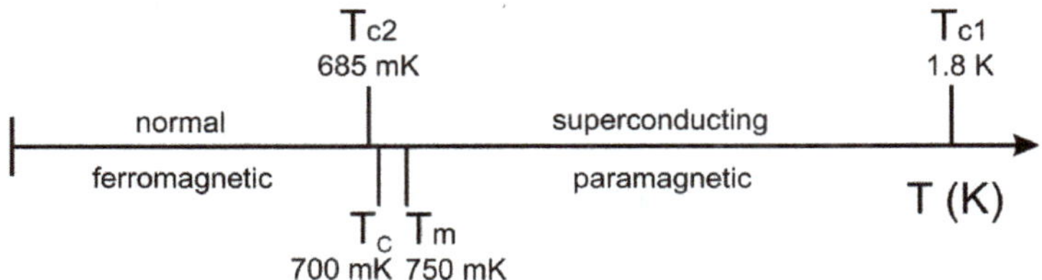

Figure 4.12.3: Superconducting and magnetic phases of $HoMo_6S_8$ on the temperature scale.

complicated when external magnetic fields are applied. For details, we refer to the review by Peña and Sergent [26].

4.12.5 Rare earth rhodium borides

Superconductivity in the ternary rare earth rhodium borides $RERh_4B_4$ (RE = Sc, Y, Pr, Nd, Sm, Gd-Tm, Lu) has been discovered by Matthias et al. in 1977 [27]. The nonmagnetic members exhibit the highest critical temperatures around 12 K (YRh_4B_4 and $LuRh_4B_4$), but even the strongly magnetic $ErRh_4B_4$ is superconducting at 8.7 K, much higher than the Chevrel phases with magnetic rare earth elements. The crystal structure of $ErRh_4B_4$ with space group $P4_2/nmc$ is shown in Figure 4.12.4 (left). Distorted Rh_4B_4 cubes are connected via B–B and Rh–B bonds and form a 3D network with erbium atoms in the voids. Figure 4.12.4 (right) shows the re-entrant behavior. It should not be confused with co-existence of superconductivity and magnetic ordering. The resistivity drops to zero at T_c = 8.7 K and returns to normal at the ferromagnetic ordering temperature $T_C \approx 0.9$ K. Thus, the ferromagnetic order destroys superconductivity in $ErRh_4B_4$ [28]. However, neutron diffraction experiments on $ErRh_4B_4$ single crystals have shown that a sinusoidal modulated ferromagnetic state with a wavelength of ~ 100 Å coexists with superconductivity between 1.2 K and 0.71 K. Below this temperature, long range ferromagnetic order emerges and destroys superconductivity [29].

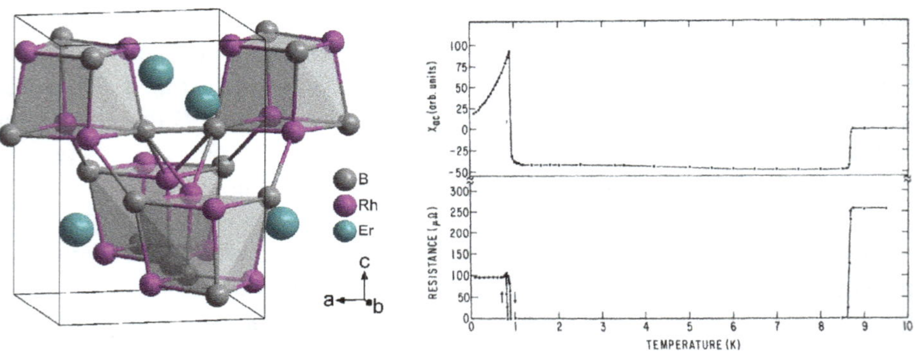

Figure 4.12.4: Left: Crystal structure of $ErRh_4B_4$ Right: Resistivity and AC-susceptibility of $ErRh_4B_4$ showing the re-entrant behavior. Reproduced with permission from [28].

4.12.6 Rare earth borocarbides

First reports about possible superconductivity in the system Y-Ni-B at 12 K appeared in 1993 [30], but the superconducting fraction was only 2 % of an unidentified phase with the nominal composition "YNi_4B". Shortly afterwards it turned out that adding carbon drastically increases the superconducting fraction [31, 32], and it became clear that the unidentified compound is the borocarbide YNi_2B_2C [33, 34]. The record transition temperature of 23 K among the borocarbides occurs in the palladium-compound YPd_2B_2C, which was also published [34] before the correct composition and structure was known [35]. The crystal structure is a filled variant of the tetragonal $ThCr_2Si_2$-type structure (space group *I4/mmm*) with an additional carbon atom located between the layers (Figure 4.12.5 left).

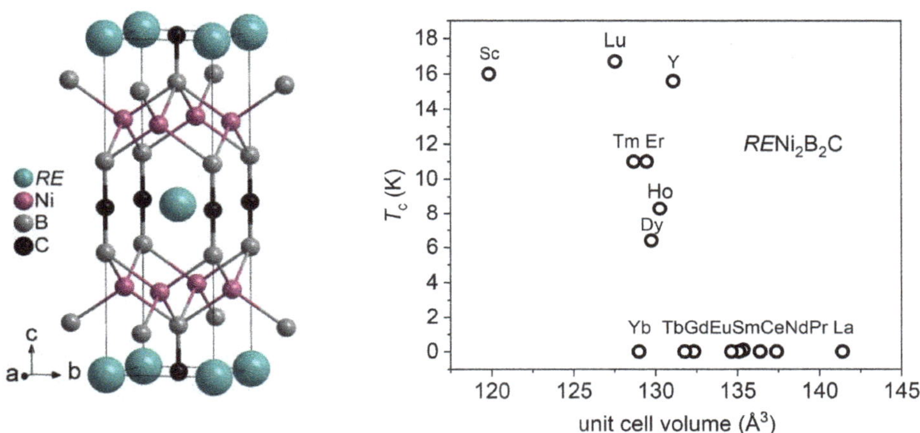

Figure 4.12.5: Left: Crystal Structure of the $RENi_2B_2C$ compounds (*RE* = Sc, Y, La-Lu). Right: Plot of the critical temperature (T_c) against the unit cell volume.

The compounds $RENi_2B_2C$ exist with all rare earth elements [36]. The highest critical temperatures around 16–17 K occur with the smaller nonmagnetic elements Lu, Y, and Sc. A plot of the critical temperature (T_c) against the unit cell volume reveals a remarkable sharp drop of T_c near 130 Å^3 (Figure 4.12.5, right). The compounds at this border (*RE* = Dy, Ho, Er, Tm) are exceptional in another context: Obviously, superconductivity is not destroyed by the magnetism of the *RE* ions. Moreover, it has been shown that in these compounds magnetic ordering can coexist with the superconducting state. $RENi_2B_2C$ compounds haven been intensively studied by neutron diffraction at low temperatures to figure out the magnetic structures [37]. $ErNi_2B_2C$ and $TmNi_2B_2C$ are superconducting below 11 K and enter antiferromagnetically ordered states at 6.8 K and 1.1 K, respectively. Thus, the antiferromagnetic order develops in the superconducting regime. Conversely, $DyNi_2B_2C$ first becomes an antiferromagnet at 10.6 K and enters the superconducting state at 6 K. This means the superconducting state develops during magnetic order. A special case is $HoNi_2B_2C$, which is another example of re-entrant superconductivity: Antiferromagnetism and superconductivity appear almost at the same temperatures ($T_N = 8.5$ K, $T_c = 8$ K), but on further cooling to 6.3 K, the spin structure changes from *c*-axis modulated incommensurate to *a*-axis modulated incommensurate, accompanied by loss of zero resistivity, that is, loss of superconductivity. At even lower temperatures below 5 K, the magnetic structure again changes commensurately and the resistivity drops back to zero (Figure 4.12.6).

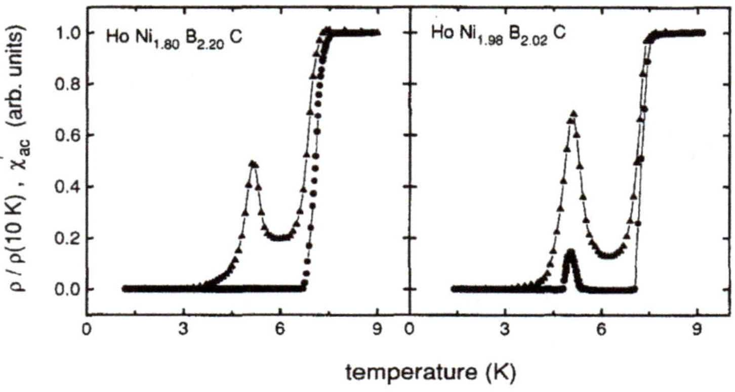

Figure 4.12.6: Re-entrant superconductivity. Electric resistance and AC susceptibility for two samples of $HoNi_{2-x}B_{2+x}C$. Reproduced with permission from [38].

4.12.7 Rare earth carbides and carbide halides

The dicarbides REC_2 of the nonmagnetic rare earth elements Y, La, and Lu crystallize in the tetragonal CaC_2-type with C_2 dimers and are superconducting at 3.88 K, 1.61 K and 3.33 K, respectively [39]. Higher critical temperatures occur in the sesquicarbides RE_2C_3 with the cubic Pu_2C_3-type, where it turned out that T_c strongly depends on the synthesis method [2]. Superconductivity in Y_2C_3 was first reported with $T_c = 11$ K. Later reports give T_c values up to 18 K. The critical temperatures of La_2C_3 and Lu_2C_3 are 11 K and 15 K, respectively [10]. An interesting class of carbide halide superconductors were discovered by A. Simon and co-workers (see also Chapter 2.3) [40]. $Y_2Br_2C_2$ has a layered structure with closed packed double sheets of yttrium and C_2 dimers in the octahedral interstices (Figure 4.12.7). These slabs are sandwiched between layers of bromine. The critical temperature of $Y_2Br_2C_2$ is 5 K and increases to 10 K in the isostructural $Y_2I_2C_2$ [41, 42]. $RE_2X_2C_2$ compounds with $RE =$ La and Lu are likewise superconductors, while the compounds with magnetic rare earth elements are not, as expected.

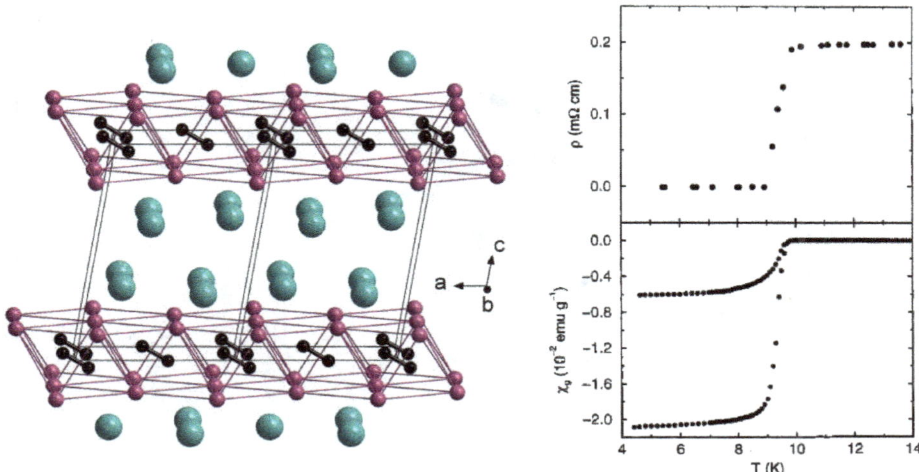

Figure 4.12.7: Left: Crystal Structure of $Y_2X_2C_2$ (X = Cl, Br, I). Carbon: black; yttrium: red; X: green. Right: Superconducting transition of $Y_2I_2C_2$ showing the resistivity and magnetization in a 10 Oe field. Reproduced with permission from [42].

4.12.8 Rare earth copper oxides

The lanthanum copper oxide La_2CuO_4 is the parent compound of the cuprate superconductors, which were discovered by Bednorz and Müller in 1986 [43, 44], who won the Nobel Prize in Physics in the following year. La_2CuO_4 is an antiferromagnetic Mott-insulator ($T_N = 325$ K) and crystallizes above 520 K in the tetragonal K_2NiF_4-type structure [45], the so-called "T-type" (Figure 4.12.8a). Corner-sharing $CuO_{4/2}O_2$ octahedra form layers that are separated by La^{3+} cations. The Jahn-Teller effect of the Cu^{2+} (d^9) ions distorts the octahedra in a way that the two apical Cu–O bonds are significantly longer (222 pm) than the four equatorial ones (186 pm). Below 520 K, the structure transforms to an orthorhombic variant with tilted and even stronger distorted octahedra (242/190 pm), as depicted in Figure 4.12.8b. Partial replacement of La^{3+} by Ba^{2+} or Sr^{2+} removes electrons from the CuO_4-layer (hole-doping) and gradually suppresses the antiferromagnetic ordering as well as the orthorhombic distortion, as shown in the phase diagram (Figure 4.12.8c) [46]. Superconductivity in $La_{2-x}Sr_xCuO_4$ ("LSCO", "214") emerges from $x \approx 0.05$ in the orthorhombic phase and reaches a maximum critical temperature of 38 K near $x \approx 0.2$ [47].

La_2CuO_4 is the only RE_2CuO_4 cuprate which forms the K_2NiF_4-type structure with octahedral copper-coordination, while all other members (RE = Pr-Gd; RE = Tb-Tm, Y as high-pressure phases [48]) crystallize in the likewise tetragonal Nd_2CuO_4-type structure (T´-type) with square coordination in planar CuO_2 layers, as depicted in

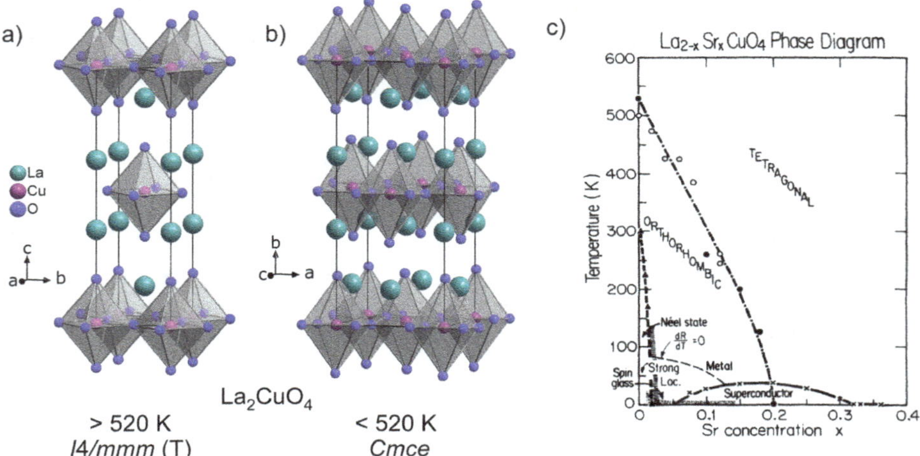

Figure 4.12.8: Crystal structures of La_2CuO_4 (a,b) and the phase diagram of $La_{2-x}Sr_xCuO_4$ (c). Phase diagram reproduced from [46] with permission.

Figure 4.12.9a. Partial substitution of Nd^{3+} by Ce^{4+} adds electrons to the CuO_2-layers (electron-doping) and superconductivity emerges up to 24 K in $Nd_{2-x}Ce_xCuO_4$ (NCCO). The antiferromagnetic order in the CuO_2 layers persists to much higher electron-doping concentrations ($x \approx 0.15$), in comparison to hole-doping ($x \approx 0.02$), and the superconducting dome in electron-doped 214 materials is much smaller ($x \approx 0.15$–0.175) than in hole-doped ones ($x \approx 0.05$–0.32, see Figure 4.12.8c) [49]. For a long time, it was believed that hole-doped RE_2CuO_4 superconductors generally have the T-type and the electron-doped have always the T´-type, until the hole-doped T´-type compounds $La_{1.8-x}Eu^{III}_{0.2}Sr_xCuO_4$ were found [50]. The formation of the T- and T´-type structures depends on the radius of the RE^{3+}-ions, respectively, and only La^{3+} is big enough to stabilize the T-type. Substitution of even small amounts of La^{3+} in La_2CuO_4 by *RE*-ions with smaller ionic radius stabilizes the T´-type. As an example, bulk samples of the T´-type electron-coped compounds $La_{2-x}Ce^{IV}{}_xCuO_4$ exist in the narrow range $x = 0.065$–0.08 and are superconducting up to 30 K [51]. Higher Ce-concentrations, up to $x = 0.2$, are possible in thin films grown by molecular beam epitaxy. For more details about the RE_2CuO_4 superconductors, we refer to the comprehensive reviews by Raveau [52] and Maple [53].

The discovery by Bednorz and Müller immediately triggered an unprecedented worldwide "gold rush" for new superconducting cuprates, and the next breakthrough was quick. On March 2, 1987, the seminal publication titled "Superconductivity at 93 K in a New Mixed-Phase Y-Ba-Cu-O Compound System at Ambient Pressure" appeared in the *Physical Review Letters* [54]. This was the first observation of superconductivity at temperatures above the boiling point of liquid nitrogen (77 K, Figure 4.12.9c). No one ever had thought earlier that this would be possible given that superconductivity was

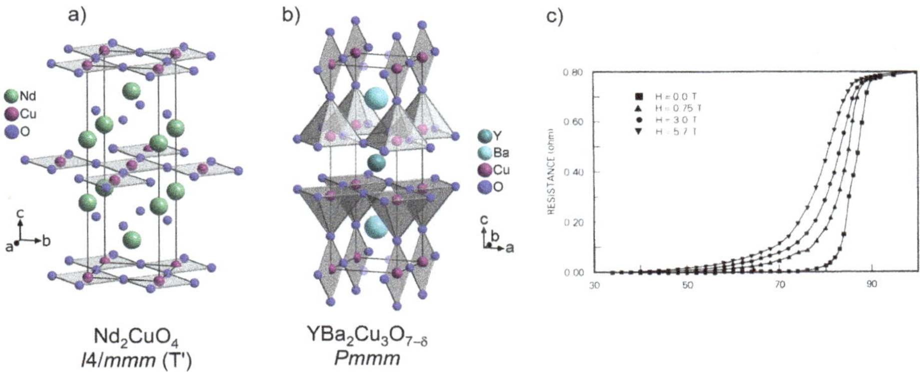

Figure 4.12.9: Crystal structures of Nd_2CuO_4 (a) and $YBa_2Cu_3O_{7-\delta}$ (b). (c) Shows the first superconducting transition above 77 K of a sample, which later turned out to contain $YBa_2Cu_3O_{7-\delta}$. Diagram (c) reproduced from [54] with permission.

considered as an ultra-low temperature phenomenon. However, the composition and structure of the 93 K superconductor was still unclear. One month later, in April 1987, a publication appeared which identified the superconducting phase as "$Ba_2YCu_3O_{9-\delta}$ ($\delta \approx 2.1$)" [55]. The complete structure was published in 1987, and it turned out that the superconducting phase is orthorhombic (space group *Pmmm*) with similar *a*- and *b*-lattice parameters, which simulates tetragonal symmetry through twinning. The composition is $YBa_2Cu_3O_{7-\delta}$ with variable oxygen contents (YBCO, "123-type"). Figure 4.12.9b shows the crystal structure. It contains layers of five-fold coordinated copper in square pyramids and layers of copper which are four-fold coordinated by oxygen. $YBa_2Cu_3O_7$ ($\delta = 0$) is orthorhombic (space group *Pmmm*) with square planar copper coordination (Figure 4.12.9b). However, the oxygen atoms of the CuO_4 squares may be disordered over the positions 1*e* (0 1/2 0) and 1*b* (1/2 0 0), as indicated by the translucent O-atoms in Figure 4.12.9b, which makes the average structure tetragonal (space group *P*4/*mmm*).

The oxygen content is crucial for superconductivity: The highest critical temperatures occur near $\delta = 0$ in orthorhombic $YBa_2Cu_3O_7$, while superconductivity disappears at $\delta \geq 0.5$, where the structure becomes tetragonal [56, 57]. $YBa_2Cu_3O_{6.5}$ is a special case because all copper atoms are formally Cu^{2+} (d^9) and the compound is an antiferromagnetic Mott-insulator. As in the RE_2CuO_4-type materials described above, magnetic ordering and orthorhombic distortion become suppressed by removing electrons from the Cu-O planes (hole doping), in this case by increasing the oxygen content to $\delta < 0.5$ in $YBa_2Cu_3O_{7-\delta}$. The effects of doping on the structure and superconducting transition temperatures have been intensively studied by numerous X-ray and neutron diffraction experiments [57].

Another surprise with respect to the high-T_c cuprates was the insensitivity of the critical temperature of $YBa_2Cu_3O_{7-\delta}$ to the substitution of magnetic rare earths for yttrium. The series $REBa_2Cu_3O_{7-\delta}$ (*RE*BCO) exists for all *RE* elements, except Ce and Tb, which form no single *RE*BCO phases [53]. Superconductivity above 90 K occurs in all *RE*BCO compounds except for $PrBa_2Cu_3O_{7-\delta}$ in which the properties strongly depend on the sample preparation method [58] and $LaBa_2Cu_3O_{7-\delta}$, where mixing of lanthanum and barium on both sites strongly reduces T_c. The effective magnetic moments of the RE^{3+} ions in the *RE*BCO compounds matches the values of the free ions [59]. Thus, seven unpaired spins of Gd^{3+} do not affect the superconducting state in $GdBa_2Cu_3O_{7-\delta}$ and even the antiferromagnetic ordering at ~2 K has no effect. It is widely accepted that the reason for this insensitivity is essentially the spatial separation of the superconducting CuO_2-layers and the magnetic *RE*-layers. Thus, superconductivity and magnetism should coexist perfectly in these compounds, though no ferromagnetic ordering has been observed so far. The superconducting copper oxides are among the most intensively studied materials ever. For further reading, we refer to reviews [52, 53, 60].

Immediately after the discovery of superconductivity at temperatures above the boiling point of liquid nitrogen, an enormous enthusiasm emerged about the possible applications. Loss-free power transmission, powerful superconducting magnets for magnetic levitation trains (MAGLEV) and magnetic resonance imaging (MRI), extremely sensitive detectors, motors and high-power wind generators are only some examples. However, the technological development of superconductivity with cuprates followed the typical path of innovations: Discovery, euphoria, disillusion, reorientation and ascendancy. It turned out that the two-dimensional character of the cuprate materials is detrimental for the production of superconducting wires. Not only the crystal structure, but also the symmetry of the superconducting order parameter ("*d*-wave") is highly anisotropic. This means, crystallites that form superconducting wires needs to be aligned in parallel. The critical current is so sensitive to this texture that orientation deviations by 3°–5° reduces the flowing current by 90 %. Thus, the production of wires with almost perfectly aligned material over kilometer lengths is an enormous challenge and quite cost-intensive. It is therefore not surprising that power transmission with copper oxide based (mostly YBCO and GdBCO) high-T_c superconducting wires (HTS) could hardly compete economically with conventional copper wires for decades. But times changed with the development of the 2G-HTC tapes based on $REBa_2Cu_3O_7$ (*RE*BCO; *RE* = Y, Gd). Thin superconducting films (1–4 µm thickness) are deposited on metal tapes by physical vapor deposition (PVD), chemical solution deposition (CSD), ion-beam assisted deposition (IBAD) or related methods [61]. These coated conductors allow enormous current densities of 1–5 kA/mm^2, making electrical machinery compact and lightweight. Figure 4.12.10a shows the layers of a coated conductor and Figure 4.12.10b compares a coated superconductor tape with a copper conductor. Both can carry the same amount of current. Note that the superconducting layer of this tape is only a few microns thick. A recent application example of such coated superconductors is the successful operation of a multi-megawatt direct drive HTS (*RE*BCO-based) wind generator at a coastal site in western Denmark (EcoSwing project [62]). For almost one year, many Danish households could claim: "powered

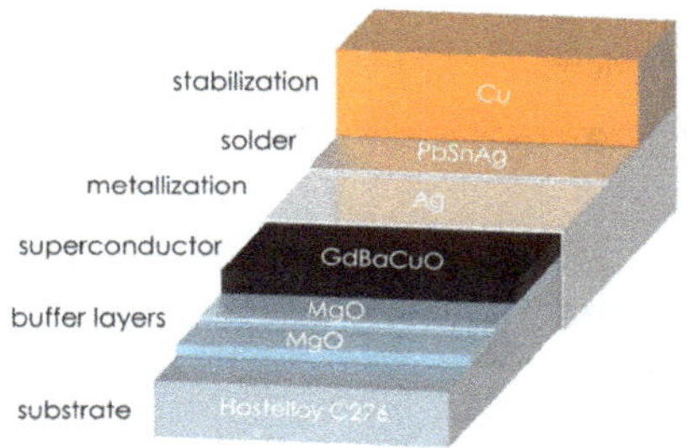

Figure 4.12.10: Left: Layers of the coated conductor tape based on $GdBa_2Cu_3O_7$ as manufactured by THEVA (Ismaning, Germany). The thickness is about 200 μm, of which the superconductor amounts 3.5 μm. Right: Comparison of a superconducting tape with a copper conductor needed to carry similar amounts of current. Reproduced from [62] and [61] (open access).

by superconductivity". This is only one example showing that HTS technology is in the ascendancy phase and may reach the final "diffusion" step of innovation in the near future.

4.12.9 Rare earth iron arsenide superconductors

Exactly 20 years after the discovery of the copper oxide superconductors appeared an article about superconductivity of the iron phosphide oxide LaOFeP [63]. The transition temperature was near 4 K and it therefore attracted not much attention. This rapidly changed in 2008 when superconductivity in the isotypic iron arsenide $LaO_{1-x}F_xFeAs$ at 26 K was discovered [64]. There was tremendous international interest and it sparked a second "gold-rush" for new iron arsenide superconductors, comparable with the discovery of the cuprates in 1986. It took only a few weeks for the critical temperature to be raised to 55 K in $SmO_{1-x}F_xFeAs$ [65], and it was clear that the iron arsenides represent the second class of high-T_c superconductors [66]. Figure 4.12.11 shows the crystal structure of the parent compound LaOFeAs. It is a filled variant of the PbFCl-type structure (space group *P*4/*nmm*) and consists of alternately stacked layers of edge-sharing $OLa_{4/4}$- and $FeAs_{4/4}$ tetrahedra. According to the equiatomic stoichiometry, these materials are referred to as the 1111-type.

Undoped LaOFeAs is a metallic conductor and passes through a structural transition to orthorhombic symmetry at $T_O \approx 160$ K (*P*4/*nmm* → *Cmma*) followed by antiferromagnetic order of the iron moments below $T_N \approx 150$ K with a weak ordered moment around 0.3–0.5 μ_B/Fe. The magnetic ordering pattern is stripe-type with the moments parallel to the *ab*-plane, as shown in Figure 4.12.12 [67].

Superconductivity is achieved by doping of either electrons or holes to the $(FeAs)^-$ layers. The first is achieved by substitution of fluoride for oxygen ($LaO_{1-x}F_xFeAs$) or by

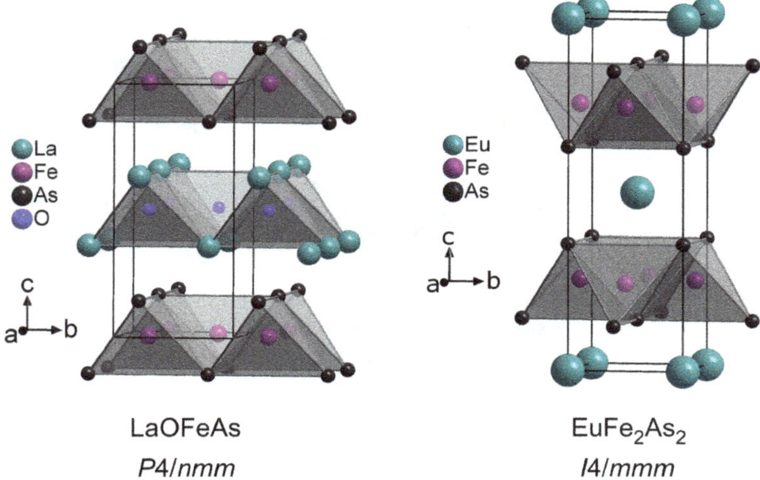

Figure 4.12.11: Crystal structures of LaOFeAs (1111-type) and $EuFe_2As_2$ (122-type).

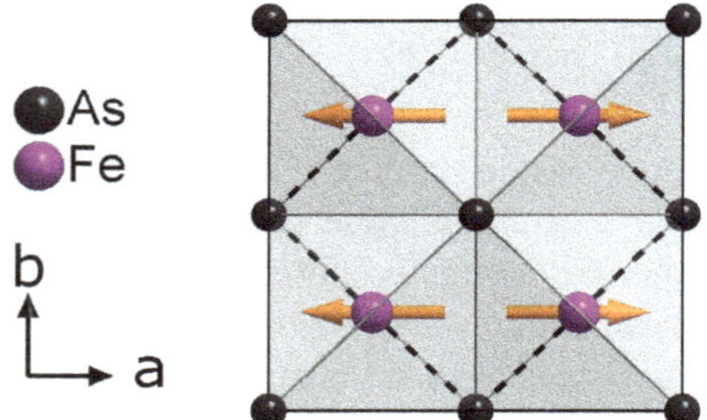

Figure 4.12.12: Crystal and magnetic structure of the FeAs layer in LaOFeAs below $T_N \approx 150$ K. The stripe-type magnetic ordering is antiferromagnetic along the longer axis *a*, and ferromagnetic along the shorter axis *b*. The dashed line indicates the unit cell of the tetragonal phase.

oxygen-vacancies in $LaO_{1-x}FeAs$, and the latter by the substitution of divalent Sr^{2+} for trivalent La^{3+} ($La_{1-x}Sr_xOFeAs$). The antiferromagnetic ordering becomes successively suppressed with doping. Thus, the overall scenario is remarkably similar to the cuprate superconductors. This is evident from a simplified generic phase diagram for cuprate and iron arsenide superconductors as depicted in Figure 4.12.13. In both classes, high-T_c superconductivity emerges once antiferromagnetic ordering becomes suppressed by doping. This inspired the theory that magnetic spin-fluctuations may be involved in the still not fully understood mechanism of superconductivity at high temperatures [68].

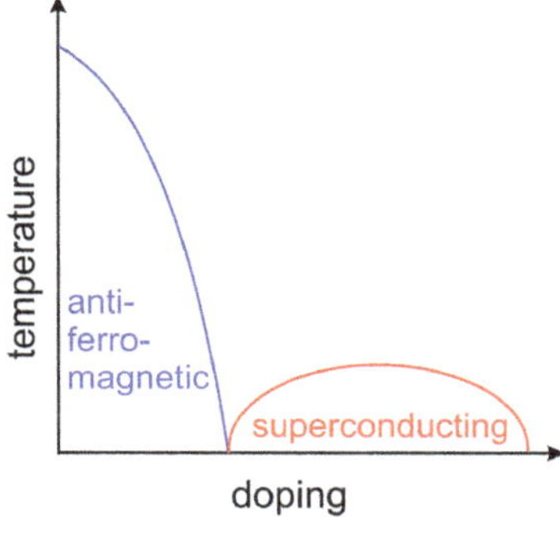

Figure 4.12.13: Simplified generic phase diagram of rare earth cuprate and iron arsenide superconductors emphasizing the proximity of superconductivity and magnetism.

The critical temperatures of hole-doped $LaO_{1-x}F_xFeAs$ is 26 K at $x \approx 0.16$, and increases along the series $\mathit{RE}O_{1-x}F_xFeAs$ to a maximum of 55 K in $SmO_{1-x}F_xFeAs$ [69]. $\mathit{RE}O_{1-x}F_xFeAs$ with *RE* = Gd, Tb as well as the oxygen deficient $\mathit{RE}O_{1-x}FeAs$ with *RE* = Dy, Ho and Er require high-pressure/high-temperature synthesis. While T_c quickly increases from La (26 K) to Pr (52 K) due to the effect of chemical pressure (volume shrinking), further decrease of the RE^{3+} ionic radii has no strong effect on T_c, which remains around 50 ± 5 K for *RE* = Nd-Er [69, 70]. The rare earth moments of the *RE*OFeAs compounds order antiferromagnetically at low temperatures ($T_N \approx 4–12$ K) [71, 72] and do not interfere with superconductivity. As in the cuprates, the magnetic (*RE*) and superconducting (FeAs) layers are spatially well separated and the unpaired spins can co-exist with superconductivity.

Another important family within the iron arsenide superconductors are the compounds $\mathit{AE}Fe_2As_2$ (*AE* = Ca, Sr, Ba, Eu) with the tetragonal $ThCr_2Si_2$-type structure [73, 74], referred to as 122-materials (Figure 4.12.11). Here, the same scenario as described for the *RE*OFeAs phases with antiferromagnetic order and suppression by doping is applicable. The highest critical temperature among these materials is 38 K in $Ba_{1-x}K_xFe_2As_2$ at $x \approx 0.4$ [74]. The only known 122-type iron arsenides with rare earth elements are $EuFe_2As_2$ [75] with divalent europium and $LaFe_2As_2$ [76], which can be synthesized under high-pressure conditions only. $EuFe_2As_2$ shows two antiferromagnetic transitions, one at 200 K by the iron moments according to Figure 4.12.12, and a second at 12 K by the europium moments [77]. Figure 4.12.14a shows the complete magnetic structure. Superconductivity at 35 K was achieved by substitution in $Eu_{1-x}K_xFe_2As_2$ at $x \approx 0.5$ [78] and by chemical pressure through substitution of smaller phosphorus atoms for arsenic. The critical temperature is about 22–28 K in $EuFe_2(As_{1-x}P_x)_2$ at $0.13 \leq x \leq 0.22$ [79]. It turns out that the magnetic structure of the Eu^{2+} moments changes to ferromagnetic on P-doping and coexists with superconductivity [80, 81]. Figure 4.12.14a–c illustrate the magnetic ordering patterns in $EuFe_2As_2$ and $EuFe_2(As_{0.85}P_{0.15})_2$, together with the magnetic susceptibility plot. The latter shows the diamagnetic signal at the onset of superconductivity at 26 K. The diamagnetism weakens below 19 K and competes with the emerging internal

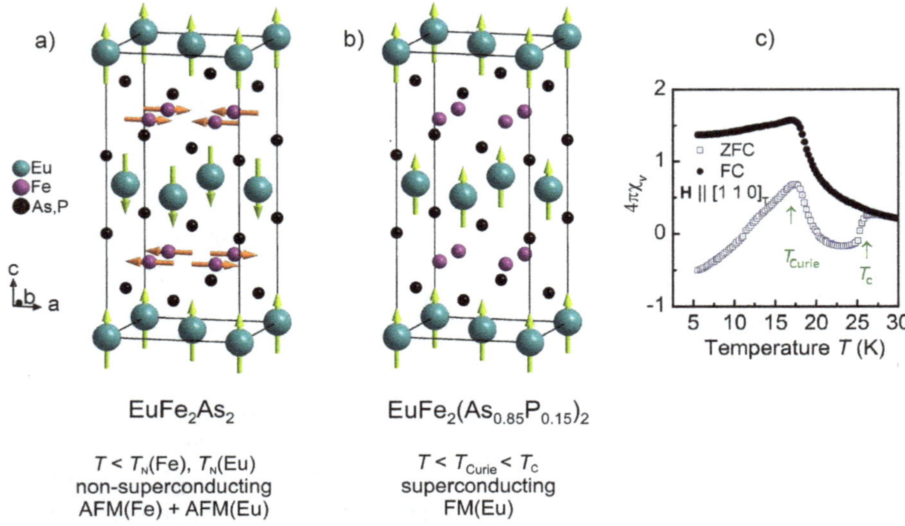

Figure 4.12.14: Magnetic ordering in $EuFe_2As_2$ and coexistence with superconductivity in $EuFe_2(As_{0.85}P_{0.15})_2$. a) Antiferromagnetic order of the europium and iron moments in $EuFe_2As_2$. b) Superconducting $EuFe_2(As_{0.85}P_{0.15})_2$ ($T_c \cong 26$ K) with ferromagnetic order of the europium moments ($T_{Curie} \cong 19$ K). c) Magnetic susceptibility of $EuFe_2(As_{0.85}P_{0.15})_2$. Figure 4.12.14c reproduced from [81] with permission.

magnetic field generated by the ferromagnetic ordering of the europium moments (Figure 4.12.14c). The ferromagnetic transition is complete below ≈ 17 K and the increasing diamagnetism prevails over the internal field.

Similar co-existence of superconductivity and ferromagnetism has been found in $CeOFe(As_{1-x}P_x)$ [82] and $RbEuFe_4As_4$ [83]. However, the detailed nature of these competing orders is not fully understood and subject of current research. One fascinating approach is the possible formation of a spontaneous vortex state [84], where the internal field of the ferromagnetic state penetrates the superconductor and forms the Shubnikov phase, that is, a vortice lattice in the absence of an external magnetic field.

Iron based superconductors have large critical fields comparable with cuprate materials and a significantly lower anisotropy. This makes them good candidates for applications in superconducting wires, especially for high-field magnets. However, the wire production with iron-base superconductors is so far not developed to such an extent as for the cuprates, which are 20 years in advance. The most promising candidates for applications in wires are currently not the rare earth iron arsenides, therefore we refer to recent reviews for further reading [85, 86].

References

[1] R. Kleiner, W. Buckel, Superconductivity – An Introduction, 3rd Edition, Wiley-VCH Verlag GmbH & Co. KGaA, 2016.

[2] C. Probst, J. Wittig, Handbook on the Physics and Chemistry of Rare Earths, Vol. 1, Elsevier, 1978, 749–795.

[3] K. Mendelssohn, J. G. Daunt, Nature 1937, 139, 473–474.

[4] B. T. Matthias, H. Suhl, E. Corenzwit, Phys. Rev. Lett. 1958, 1, 92–94.

[5] K. Winzer, Z. Phys. A: Hadrons Nucl. 1973, 265, 139–164.

[6] J. J. Hamlin, V. G. Tissen, J. S. Schilling, Physica C 2007, 451, 82–85.

[7] The_Editors_of_Encyclopaedia_Britannica, Vol. 2019, Encyclopædia Britannica, Inc., 2017.

[8] J. Song, G. Fabbris, W. Bi, D. Haskel, J. S. Schilling, Phys. Rev. Lett. 2018, 121, 037004.

[9] D. C. Koskenmaki, K. A. Gschneidner, Handbook on the Physics and Chemistry of Rare Earths, Vol. 1, Elsevier, 1978, 337–377.

[10] B. W. Roberts, J. Phys. Chem. Ref. Data 1976, 5, 581–822.

[11] C. P. Poole, H. A. Farach, J. Supercond. 2000, 13, 47–60.

[12] R. J. Gambino, N. R. Stemple, A. M. Toxen, J. Phys. Chem. Solids 1968, 29, 295–302.

[13] F. Heiniger, E. Bucher, J. P. Maita, P. Descouts, Phys. Rev. B 1973, 8, 3194–3205.

[14] A. C. Lawson, J. Less-Common Met. 1973, 32, 173–174.

[15] V. B. Compton, B. T. Matthias, Acta Crystallogr. 1959, 12, 651–654.

[16] M. S. Wei, H. H. Sung, W. H. Lee, Physica C 2005, 424, 25–28.

[17] J.-M. Hübner, M. Bobnar, L. Akselrud, Y. Prots, Y. Grin, U. Schwarz, Inorg. Chem. 2018, 57, 10295–10302.

[18] Y. R. Chung, H. H. Sung, W. H. Lee, Phys. Rev. B 2004, 70, 052511.

[19] S. Majumdar, E. V. Sampathkumaran, Phys. Rev. B 2001, 63, 172407.

[20] D. Johrendt, A. Mewis, K. Drescher, S. Wasser, G. Michels, Z. Anorg. Allg. Chem. 1996, 622, 589–592.

[21] H. F. Braun, Phys. Lett. A 1980, 75, 386–388.

[22] V. L. Ginzburg, Sov. Phys. JETP-USSR 1957, 4, 153–160.

[23] R. Chevrel, M. Sergent, J. Prigent, J. Solid State Chem. 1971, 3, 515–519.

[24] B. T. Matthias, M. Marezio, E. Corenzwit, A. S. Cooper, H. E. Barz, Science 1972, 175, 1465–1466.

[25] R. Odermatt, O. Fischer, H. Jones, G. Bongi, J. Phys. C Solid Sate 1974, 7, L13–L15.

[26] O. Peña, M. Sergent, Prog. Solid State Chem. 1989, 19, 165–281.

[27] B. T. Matthias, E. Corenzwit, J. M. Vandenberg, H. E. Barz, Proc. Natl. Acad. Sci. U. S. A. 1977, 74, 1334–1335.

[28] W. A. Fertig, D. C. Johnston, L. E. DeLong, R. W. McCallum, M. B. Maple, B. T. Matthias, Phys. Rev. Lett. 1977, 38, 987–990.

[29] S. K. Sinha, G. W. Crabtree, D. G. Hinks, H. Mook, Physica B+C 1982, 109–110, 1693–1698.

[30] C. Mazumdar, R. Nagarajan, C. Godart, L. C. Gupta, M. Latroche, S. K. Dhar, C. Levy-Clement, B. D. Padalia, R. Vijayaraghavan, Solid State Commun. 1993, 87, 413–416.

[31] L. C. Gupta, R. Nagarajan, C. Godart, S. K. Dhar, C. Mazumdar, Z. Hossain, C. Levy-Clement, B. d. Padalia, R. Vijayaraghavan, Physica C 1994, 235–240, 150–153.

[32] R. Nagarajan, C. Mazumdar, Z. Hossain, S. K. Dhar, K. V. Gopalakrishnan, L. C. Gupta, C. Godart, B. D. Padalia, R. Vijayaraghavan, Phys. Rev. Lett. 1994, 72, 274–277.

[33] R. J. Cava, H. Takagi, H. W. Zandbergen, J. J. Krajewski, W. F. Peck, T. Siegrist, B. Batlogg, R. B. van Dover, R. J. Felder, K. Mizuhashi, J. O. Lee, H. Eisaki, S. Uchida, Nature 1994, 367, 252–253.

[34] R. J. Cava, H. Takagi, B. Batlogg, H. W. Zandbergen, J. J. Krajewski, W. F. Peck, R. B. van Dover, R. J. Felder, T. Siegrist, K. Mizuhashi, J. O. Lee, H. Eisaki, S. A. Carter, S. Uchida, Nature 1994, 367, 146–148.
[35] Y. Y. Sun, I. Rusakova, R. L. Meng, Y. Cao, P. Gautier-Picard, C. W. Chu, Physica C 1994, 230, 435–442.

[36] R. Niewa, L. Shlyk, B. Blaschkowski, Z. Kristallogr. 2011, 226, 352.
[37] J. W. Lynn, S. Skanthakumar, Q. Huang, S. K. Sinha, Z. Hossain, L. C. Gupta, R. Nagarajan, C. Godart, Phys. Rev. B 1997, 55, 6584–6598.
[38] H. Schmidt, M. Weber, H. F. Braun, Physica C 1995, 246, 177–185.
[39] R. W. Green, E. O. Thorland, J. Croat, S. Legvold, J. Appl. Phys. 1969, 40, 3161–3162.

[40] A. Simon, H. Mattausch, R. Eger, R. K. Kremer, Angew. Chem. Int. Ed. 1991, 30, 1188–1189.
[41] R. W. Henn, W. Schnelle, R. K. Kremer, A. Simon, Phys. Rev. Lett. 1996, 77, 374–377.
[42] A. Simon, M. Bäcker, R. W. Henn, C. Felser, R. K. Kremer, H. Mattausch, A. Yoshiasa, Z. Anorg. Allg. Chem. 1996, 622, 123–137.

[43] J. G. Bednorz, K. A. Müller, Z. Phys. B: Condens. Matter 1986, 64, 189–193.
[44] J. G. Bednorz, M. Takashige, K. A. Müller, Europhys. Lett. 1987, 3, 379–386.
[45] J. M. Longo, P. M. Raccah, J. Solid State Chem. 1973, 6, 526–531.
[46] B. Keimer, N. Belk, R. J. Birgeneau, A. Cassanho, C. Y. Chen, M. Greven, M. A. Kastner, A. Aharony, Y. Endoh, R. W. Erwin, G. Shirane, Phys. Rev. B 1992, 46, 14034–14053.
[47] R. J. Cava, R. B. van Dover, B. Batlogg, E. A. Rietman, Phys. Rev. Lett. 1987, 58, 408–410.
[48] P. Bordet, J. J. Capponi, C. Chaillout, D. Chateigner, J. Chenavas, T. Fournier, J. L. Hodeau, M. Marezio, M. Perroux, G. Thomas, A. Varela, Physica C 1992, 193, 178–188.
[49] G. M. Luke, L. P. Le, B. J. Sternlieb, Y. J. Uemura, J. H. Brewer, R. Kadono, R. F. Kiefl, S. R. Kreitzman, T. M. Riseman, C. E. Stronach, M. R. Davis, S. Uchida, H. Takagi, Y. Tokura, Y. Hidaka, T. Murakami, J. Gopalakrishnan, A. W. Sleight, M. A. Subramanian, E. A. Early, J. T. Markert, M. B. Maple, C. L. Seaman, Phys. Rev. B 1990, 42, 7981–7988.
[50] T. Takamatsu, M. Kato, T. Noji, Y. Koike, Appl. Phys. Express 2012, 5, 073101.
[51] T. Yamada, K. Kinoshita, H. Shibata, Jpn. J. Appl. Phys. 1994, 33, L168–L169.

[52] B. Raveau, C. Michel, M. Hervieu, Handbook on the Physics and Chemistry of Rare Earths, Vol. 30, Elsevier, 2000, 31–65.

[53] M. B. Maple, Handbook on the Physics and Chemistry of Rare Earths, Vol. 30, Elsevier, 2000, 1–30.
[54] M. K. Wu, J. R. Ashburn, C. J. Torng, P. H. Hor, R. L. Meng, L. Gao, Z. J. Huang, Y. Q. Wang, C. W. Chu, Phys. Rev. Lett. 1987, 58, 908–910.
[55] R. J. Cava, B. Batlogg, R. B. Vandover, D. W. Murphy, S. Sunshine, T. Siegrist, J. P. Remeika, E. A. Rietman, S. Zahurak, G. P. Espinosa, Phys. Rev. Lett. 1987, 58, 1676–1679.
[56] R. J. Cava, B. Batlogg, C. H. Chen, E. A. Rietman, S. M. Zahurak, D. Werder, Nature 1987, 329, 423–425.
[57] R. J. Cava, A. W. Hewat, E. A. Hewat, B. Batlogg, M. Marezio, K. M. Rabe, J. J. Krajewski, W. F. Peck, L. W. Rupp, Physica C 1990, 165, 419–433.
[58] Y. Nishihara, Z. Zou, J. Ye, K. Oka, T. Minawa, H. Kawanaka, H. Bando, Bull. Mater. Sci. 1999, 22, 257–263.
[59] M. B. Maple, Y. Dalichaouch, J. M. Ferreira, R. R. Hake, B. W. Lee, J. J. Neumeier, M. S. Torikachvili, K. N. Yang, H. Zhou, R. P. Guertin, M. V. Kuric, Physica B+C 1987, 148, 155–162.

[60] R. Hackl, Z. Kristallogr. 2011, 226, 323–342.
[61] M. Bäcker, Z. Kristallogr. 2011, 226, 343–351.
[62] A. Bergen, R. Andersen, M. Bauer, H. Boy, M. t. Brake, P. Brutsaert, C. Bührer, M. Dhallé, J. Hansen, H. ten Kate, J. Kellers, J. Krause, E. Krooshoop, C. Kruse, H. Kylling, M. Pilas,

H. Pütz, A. Rebsdorf, M. Reckhard, E. Seitz, H. Springer, X. Song, N. Tzabar, S. Wessel, J. Wiezoreck, T. Winkler, K. Yagotyntsev, Supercond. Sci. Technol. 2019, 32, 125006.
[63] Y. Kamihara, H. Hiramatsu, M. Hirano, R. Kawamura, H. Yanagi, T. Kamiya, H. Hosono, J. Am. Chem. Soc. 2006, 128, 10012, 10013.
[64] Y. Kamihara, T. Watanabe, M. Hirano, H. Hosono, J. Am. Chem. Soc. 2008, 130, 3296, 3297.
[65] Z.-A. Ren, W. Lu, J. Yang, W. Yi, X.-L. Shen, Z.-C. Li, G.-C. Che, X.-L. Dong, L.-L. Sun, F. Zhou, Z.-X. Zhao, Chin. Phys. Lett. 2008, 25, 2215–2216.

[66] D. Johrendt, R. Pöttgen, Angew. Chem. Int. Ed. 2008, 47, 4782.
[67] D. Johrendt, J. Mater. Chem. 2011, 21, 13726–13736.
[68] D. J. Scalapino, Rev. Mod. Phys. 2012, 84, 1383–1417.

[69] D. C. Johnston, Adv. Phys. 2010, 59, 803–1061.
[70] P. M. Shirage, K. Miyazawa, K. Kihou, C.-H. Lee, H. Kito, K. Tokiwa, Y. Tanaka, H. Eisaki, A. Iyo, Europhys. Lett. 2010, 92, 57011.
[71] M. A. McGuire, R. P. Hermann, A. S. Sefat, B. C. Sales, R. Jin, D. Mandrus, F. Grandjean, G. J. Long, New J. Phys. 2009, 11, 025011.
[72] N. R. Lee-Hone, D. H. Ryan, J. M. Cadogan, Y. L. Sun, G. H. Cao, J. Appl. Phys. 2014, 115, 17D705.
[73] M. Rotter, M. Tegel, I. Schellenberg, W. Hermes, R. Pöttgen, D. Johrendt, Phys. Rev. B 2008, 78, 020503.
[74] M. Rotter, M. Tegel, D. Johrendt, Phys. Rev. Lett. 2008, 101, 107006.
[75] M. Tegel, M. Rotter, V. Weiss, F. Schappacher, R. Pöttgen, D. Johrendt, J. Phys.: Condens. Matter 2008, 20, 452201.
[76] A. Iyo, S. Ishida, H. Fujihisa, Y. Gotoh, I. Hase, Y. Yoshida, H. Eisaki, K. Kawashima, J. Phys. Chem. Lett. 2019, 10, 1018–1023.
[77] H. Raffius, E. Morsen, B. D. Mosel, W. Müller-Warmuth, W. Jeitschko, L. Terbüchte, T. Vomhof, J. Phys. Chem. Solids 1993, 54, 135–144.
[78] H. S. Jeevan, Z. Hossain, D. Kasinathan, H. Rosner, C. Geibel, P. Gegenwart, Phys. Rev. B 2008, 78, 092406.
[79] H. S. Jeevan, D. Kasinathan, H. Rosner, P. Gegenwart, Phys. Rev. B 2011, 83, 054511.
[80] G. Cao, S. Xu, Z. Ren, S. Jiang, C. Feng, Z. A. Xu, J. Phys.: Condens. Matter 2011, 23, 464204.

[81] S. Nandi, W. T. Jin, Y. Xiao, Y. Su, S. Price, D. K. Shukla, J. Strempfer, H. S. Jeevan, P. Gegenwart, T. Brückel, Phys. Rev. B 2014, 89, 014512.
[82] A. Jesche, T. Förster, J. Spehling, M. Nicklas, M. de Souza, R. Gumeniuk, H. Luetkens, T. Goltz, C. Krellner, M. Lang, J. Sichelschmidt, H.-H. Klauss, C. Geibel, Phys. Rev. B 2012, 86, 020501.
[83] Y. Liu, Y.-B. Liu, Z.-T. Tang, H. Jiang, Z.-C. Wang, A. Ablimit, W.-H. Jiao, Q. Tao, C.-M. Feng, Z.-A. Xu, G.-H. Cao, Phys. Rev. B 2016, 93, 214503.
[84] W.-H. Jiao, Q. Tao, Z. Ren, Y. Liu, G.-H. Cao, npj Quantum Mater. 2017, 2, 50.
[85] I. Pallecchi, M. Eisterer, A. Malagoli, M. Putti, Supercond. Sci. Technol. 2015, 28, 114005.
[86] H. Hosono, A. Yamamoto, H. Hiramatsu, Y. Ma, Mater. Today 2018, 31, 278–302.

Hellmut Eckert

4.13 Rare-earth-containing glasses

4.13.1 Introduction

For several centuries, glasses have been the *eyes of science* empowering us and guiding our insight into both the microscopic world and the macroscopic universe. Today, with a plethora of uses, ranging from the domestic realm to high-end technologies, they not only pervade all aspects of our everyday life, but are at the forefront of fundamental and applied research and development. Glassy optical fibers have become the highways of the modern information age. Fast ion-conducting glasses provide the high power densities required for batteries used in electromotion. Ultra-stable and impact-resistant glass ceramics ensure personal safety. Bioactive glasses and scaffolds activating osteoproduction genes promote bone and tissue healing. As the physical and chemical properties of glasses can be continuously varied by their compositions and their preparation and processing conditions, glasses offer a vast parameter space for designing the most suitable material for its desired application, which usually requires a combination of multiple properties. In entering this new technological era, appropriately termed the ***glass age***, the rare-earth elements have been and continue to be an important part of this journey. While rare-earth oxides themselves cannot be prepared in the glassy state, their incorporation into multicomponent glasses is widely used for imparting crucial optical, magnetic and medical functions to them [1–8]. The present chapter will describe the most important features and aspects of rare-earth containing glasses, their key physical and technologically relevant properties, and the functional materials resulting from them. As these properties are critically controlled by the local environment and spatial distribution of the rare-earth element, this review will also cover some key insights from diffraction and spectroscopic experiments. The structure/function relations uncovered by these structural studies keep guiding our path towards improved materials.

4.13.2 Physical–chemical criteria of the glassy state

Numerous definitions of “glass” can be found in the literature, and efforts of improving the definition are continuing [9]. Glasses are noncrystalline solids, originating from the freezing of supercooled liquids. This classifies them as nonequilibrium, nonstationary states [10–12]. The situation is best illustrated on the basis of a state diagram, in which a thermodynamic state function (molar volume, enthalpy or entropy) is plotted as a function of temperature (see Figure 4.13.1). If crystallization is kinetically inhibited, the

https://doi.org/10.1515/9783110654929-036

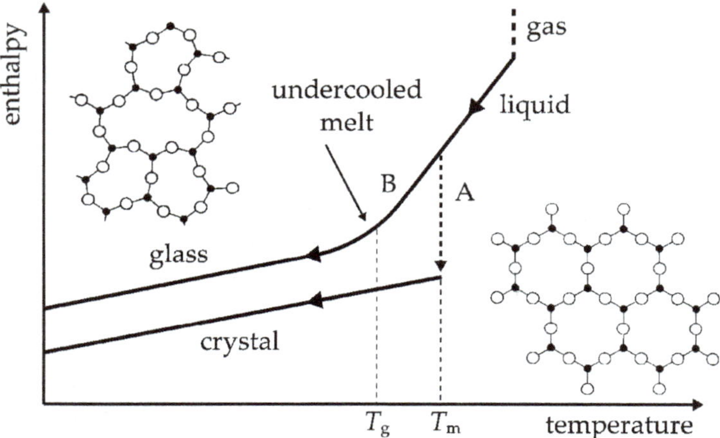

Figure 4.13.1: The glassy state illustrated in terms of the molar enthalpy. Trajectory A (crystallization at T_m) is followed under thermodynamic equilibrium conditions, whereas along trajectory B, the liquid is supercooled at T_m, corresponding to a nonequilibrium state, which becomes nonstationary at the glass-transition temperature T_g. The structural sketches illustrate the translational symmetry of the crystalline state, which is missing in the glass.

supercooled liquid state persists below the freezing point T_m, and state function values continue to decrease with decreasing temperature, as long as molecular relaxation is fast in relation to the cooling rate. However, as the temperature keeps decreasing further, relaxation keeps slowing down, until the system reaches the characteristic *glass transition region* where relaxation towards the stationary supercooled liquid or, ultimately, the thermodynamically favored crystalline state can no longer occur. The temperature where this is observed, known as the *glass transition temperature*, T_g, manifests itself in a change of slope, dX/dT, displayed by numerous macroscopic observables, X. Most commonly, these observables are the enthalpy, the thermal expansion coefficient or the viscosity. The actual value of T_g detected in this way depends on the observable and the rate with which it is being monitored, as well as the thermal history of the glass under investigation. Most importantly, however, T_g depends on the structural organization of the glass, which is determined by its chemical composition. Broadly speaking, T_g increases with increasing strength of the chemical bonds and their overall connectivity.

While a great variety of compounds and compositions can be prepared in the glassy state, the most important rare-earth containing systems are the so-called *covalent network glasses*, based on main group oxides, chalcogenides (sulfides, selenides, tellurides), and halides (fluorides, chlorides). These species, called ***network formers*** establish an aperiodic network of randomly-connected coordination polyhedra, according to the *continuous random network* hypothesis originally formulated by Zachariasen [13]. The structure is further transformed by the incorporation of

network modifiers, which are generally oxides, chalcogenides or halides of mono- and divalent elements, as well as of the trivalent rare earths. The network modifier cations are generally surrounded by bridging and nonbridging O, S, Se, Te, F or Cl species and have network former units and/or other modifier cations, in their second coordination spheres. Some elements can act both as network former and network modifier species. The most prominent examples of these *intermediate oxides* are aluminum and gallium, which in many glass systems, are surrounded by four ligands, forming anionic tetrahedron-like units, and, as such, acting as network formers. They can also occur in the five- and six-coordinated states, and are then considered modifiers. Likewise, the rare-earths are usually considered modifiers, although some exceptions may occur in fluoride glasses. More recent evidence has shown that the concept of random connectivity is not truly realistic, and the situation is better described by distribution functions of angles linking the different polyhedra [14].

Glass structure can thus be discussed in terms of different length scales (Figure 4.13.2). Like liquids they possess a significant amount of ***short-range order***, defining the types, numbers of ligands and coordination geometries in the first atomic coordination spheres; in terms of ***intermediate-range order***, they involve spatial correlations in the second coordination sphere. This includes (a) the connectivity between various network former species, distance correlations between network formers and network modifiers, or the overall spatial distribution of the network modifiers [14]. The

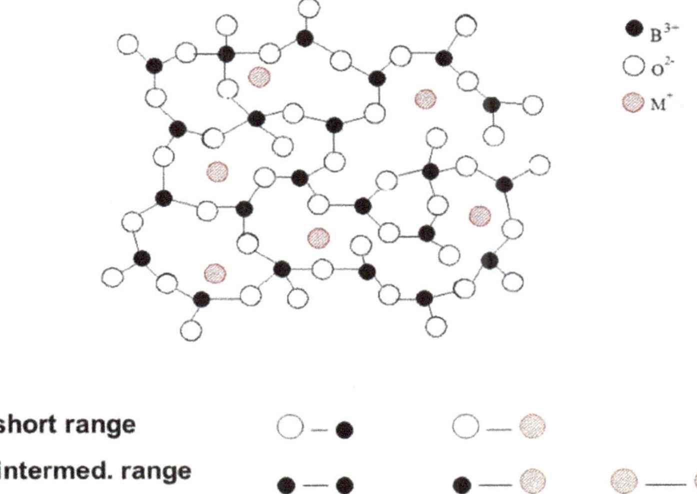

Figure 4.13.2: Aspects of short and intermediate-range order illustrated at the example of oxide glasses involving network former connectivities, spatial correlations between network formers and network modifiers and spatial correlations among network modifiers. Black circles denote network formers, red circles denote network modifiers and open circles denote oxygen atoms.

latter aspect is of great concern in rare-earth containing glasses, whose optical and magnetic properties depend, to a great extent, on how these ions are distributed in space. Ordering processes beyond the second or third coordination spheres are labeled in terms of ***medium-range order***, which encompasses superstructural units, clusters, chains or rings. Pronounced medium range order may be the consequence of incipient (and potentially kinetically suppressed) nanosegregation or phase separation tendencies, leading to nonrandom atomic distributions (clustering). In rare-earth doped glasses such nonrandom distributions are often undesired, if they tend to interfere with their desired material function.

4.13.3 Glass functionalization by rare-earth ions

Owing to their excellent optical transparency, their ability to accommodate and disperse rare-earth ions in high concentrations and their ease of fabricating flexible sizes and shapes, glasses are particularly well-suited host matrices for rare-earth ions, resulting in a plethora of optical and photonic devices. Common formulations are summarized in Figure 4.13.3. Unlike polycrystalline materials that have excessive optical losses due to light scattering from grain boundaries, glasses can be manufactured without grain boundaries, thus minimizing scattering losses. By varying the chemical composition, their physical properties may be tailored to the specific requirements of the

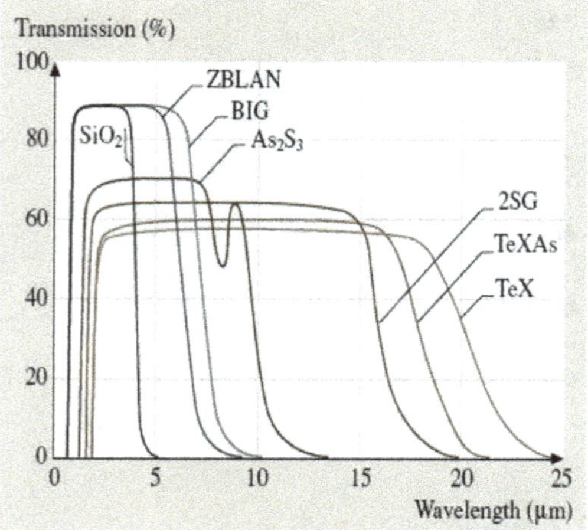

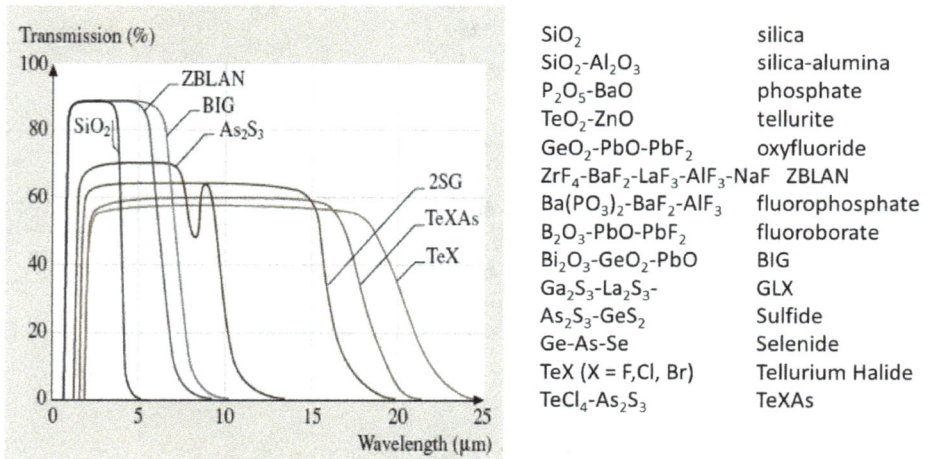

SiO_2	silica
SiO_2-Al_2O_3	silica-alumina
P_2O_5-BaO	phosphate
TeO_2-ZnO	tellurite
GeO_2-PbO-PbF_2	oxyfluoride
ZrF_4-BaF_2-LaF_3-AlF_3-NaF	ZBLAN
$Ba(PO_3)_2$-BaF_2-AlF_3	fluorophosphate
B_2O_3-PbO-PbF_2	fluoroborate
Bi_2O_3-GeO_2-PbO	BIG
Ga_2S_3-La_2S_3-	GLX
As_2S_3-GeS_2	Sulfide
Ge-As-Se	Selenide
TeX (X = F,Cl, Br)	Tellurium Halide
$TeCl_4$-As_2S_3	TeXAs

Figure 4.13.3: Common glass matrices used for *RE*-containing formulations and typical optical transmission windows. Reproduced from R. DeCorby, “Glasses for Photonic Integration,” in Springer Handbook of Electronic and Photonic Materials, S. Kasap and P. Capper, Ed. Springer Science + Business Media Inc., New York, 1041–1061, 2006, with permission from Springer.

photonic application envisioned. The type of glass matrix to be chosen depends on the desired wavelength of operation, as the material must be transparent in this range. To this end, a wide variety of oxide, halide or chalcogenide matrices are chosen. At the high-frequency end, the transmission window is limited by electronic transitions of the matrix which generally lie in the ultraviolet region. The low-frequency end of the transmission window is defined by infrared absorption of vibrational modes, which limit the upper working wavelength in oxide glasses to 3–4 μm. The transmission window can be significantly extended by using heavy metal oxide (Bi_2O_3, GeO_2, TeO_2), fluoride or chalcogenide formulations (see Figure 4.13.3).

For many optical materials and photonic devices that are only dependent on luminescence upon optical excitation, glass functionalization requires the incorporation of only minor concentrations of the rare-earth element (0.1–5 mol%). This can often be achieved by straightforward modifications of the glass preparation procedure. In many cases, however, the solubilities of the rare-earth ions in common glass-forming melts are rather limited at the melting temperature of the host glass, requiring increased melting and homogenization temperatures [1]. In other cases, the rare earths have been noted to act as crystallization catalysts. For the design of high-power lasers or magneto-optical devices (Faraday rotators), incorporation of rare-earth oxides at high concentration levels (> 10 mol%) is necessary. In general, it is desired that the rare-earth ions are homogeneously dispersed to avoid compromising optical transparency and/or producing spatially inhomogeneous refraction caused by clustering or incipient phase separation. The preparation of such glasses is generally challenging. For silicate and aluminosilicate-based systems, the high melting temperatures required pose severe restrictions on glass compositions, suitable crucible materials and effective melt cooling strategies to avoid phase separation and crystallization. Usually, glasses can only be formed in certain composition regions, which can be identified in triangular diagrams as the one shown in Figure 4.13.4. The extension of these glass-forming regions also depends on melting and cooling conditions.

4.13.4 Rare-earth-containing glasses for passive optical device applications

4.13.4.1 High refractive index glasses

Owing to their large number of electrons, rare-earth atoms in glasses tend to increase the refractive indexes, n, typically in a linear fashion, with increasing concentration and atomic number of the rare-earth ion utilized. Of course, there are many formulation alternatives for high-refractive index glasses such as the inclusion of other heavy elements (Sn, Pb and Te). Rare-earth oxide ingredients are the choice selection, however, if any combination with high hardness, E-module or compressive strength is

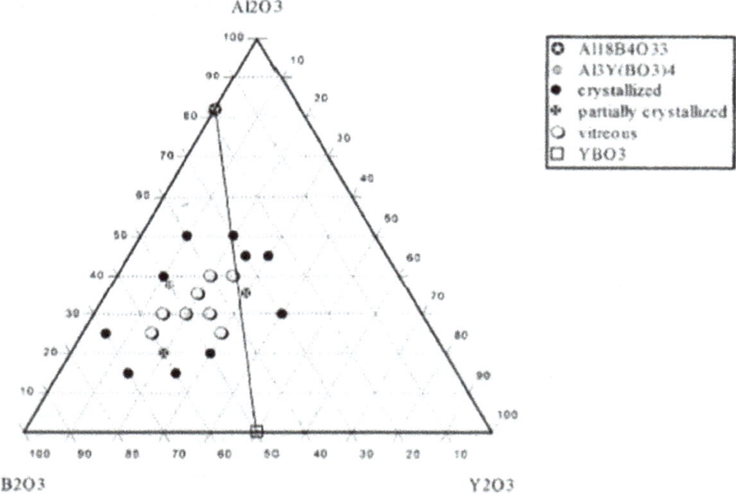

Figure 4.13.4: Glass-forming region in the system Y_2O_3-Al_2O_3-B_2O_3 [15]. Compositions forming glasses are given by open circles. Taken from [15] with permission from Springer.

desired. In going from La to Lu, the concomitant increase of ionic potential as a result of the lanthanide contraction always results in increased bonding strengths and packing densities, enhancing mechanical, thermal and chemical stability [1].

4.13.4.2 Glass coloring and de-coloring

Another rather straightforward application of rare-earth ions in glasses is coloration, based on 4f → 4f transitions effected by the absorption of visible light. Pr^{3+} (green), Nd^{3+} (purple) and Er^{3+} (pink) added at the level of 2–6 mol% of the rare-earth oxides are the most widely used additives. As the 4f → 4f transitions are disallowed by the electric dipole moment selection rules, coloration tends to be not very intense. The color imparted to the glass by Nd^{3+} additives is complementary to the yellow tinge frequently observed in glasses containing small iron impurities. For this reason, neodymium can also be used as an effective glass *decolorization* agent, bringing the spectrum back into balance [16]. This technique was already practiced by the famous Venetian glass makers.

4.13.4.3 Photochromic and photothermal refractive glasses

Finally, Ce^{3+} and Eu^{2+} dopants were shown to impart the *photochromic* effect upon glasses, darkening their color upon light exposure. The process is reversible and involves light absorption in the ultraviolet range and transfer of photoelectrons to nearby traps which absorb in the visible region [17]. The effect is related to reversible photo-oxidation to Ce^{4+} and Eu^{3+}, resulting in intervalence transfer absorption bands in the visible region. The photosensitivity of the cerium valence state is also key to the behavior of *photothermal refractive (PTR) glasses*, whose optical properties can be modified in a controlled fashion, by UV-exposure and subsequent annealing above the glass transition temperature, T_g [18]. The photo-thermo-induced refractive index decrement can be related to the formation of Ag clusters from Ag^+ dopants, via the process $Ce^{3+} + Ag^+ \rightarrow Ce^{4+} + Ag$. By applying holographic techniques, optical elements with different spatial profiles of refractive index can be created, including volume Bragg gratings [19], phase masks [20] and complex holograms, and lenses [21]. Such holographic elements are widely used for narrow band optical filtering [22] and laser design [23].

4.13.4.4 Magneto-optic glasses

Another important optical property utilizing rare-earth ions in glasses is the *Faraday effect*, which describes the rotation of the plane of polarization of an electromagnetic wave propagating in a medium, under the influence of an externally applied magnetic field parallel to the propagation direction [24]. The effect is caused by a small difference in refractive indexes (and hence propagation velocities) of the two circular components associated with a linearly polarized wave. This difference arises from the interaction of the magnetic induction vector of the electromagnetic wave with the electronic shell magnetization that builds up in the applied magnetic field. It is quantified by the *Verdet constant V*, which characterizes the change in optical polarization direction (rotation angle α) under the influence of the magnetic field (magnetic induction *B*) in a material of thickness *d* and given by:

$$\alpha = V\,d\,B$$

Figure 4.13.5 shows a simple experimental setup concept for the measurement of *V*. Diamagnetic and paramagnetic contributions produce rotations in opposite directions and are associated with positive and negative signs of *V*. For rare-earth ions, the strong paramagnetism of the open-shell 4f electronic configurations gives rise to large negative Verdet constants, which find use in optical insulators or modulators and magnetic field sensors in telecommunication systems. While many applications use rare-earth garnet crystals, glasses containing high concentrations (10^{21}–10^{22} ions/cm^3) of rare-earth ions are also very attractive materials, owing to their optical transparency, flexibility of shaping and adjustability of their other physical properties.

The theory of the Faraday effect was developed by van Vleck and Webb, who predict the Verdet constant in quantitative terms from the relation [24-28]:

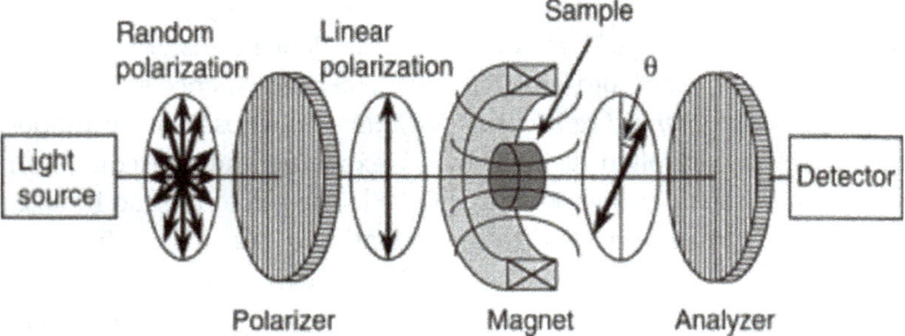

Figure 4.13.5: Measurement of the Verdet constant. The sample is exposed to linearly polarized light within a magnet with its field vector along the propagation direction. The tilt in the plane of polarization is measured by a second polarizer (analyzer), whose plane of polarization can be adjusted for extinction if both polarization planes are perpendicular. Reproduced with permission from N. Miura, Magneto–Spectroscopy of Semiconductors, in P. Bhattacharya, R. Fornari, H. Kimura, eds, Comprehensive Semiconductor Science and Technology, ISBN 978-0-444-53153-7, Elsevier Science Amsterdam, 2011, 256-342.

$$V = \frac{4\pi^2\chi}{g\mu_B ch}\sum_i C_i\left(1-\frac{\lambda^2}{\lambda_i^2}\right)^{-1} \tag{4.13.1}$$

In this expression, χ is the magnetic susceptibility, g the electronic Landé factor and μ_B, c and h are the universal constants Bohr magneton, speed of light and Planck`s constant, respectively. The summation extends over all optical $4f^n \rightarrow 4f^{n-1}5d^1$ transitions i, occurring at the wavelength λ_i (usually in the ultraviolet), with the probability, C_i. This expression is frequently approximated by replacing the summation with an effective transition probability, C_t relating to the wavelength of the closest optical transition, λ_t. Typical values are summarized in Table 4.13.2. Equation (4.13.1) further expresses the dependence of V on the wavelength of the electromagnetic wave, which arises from the refractive index dispersion. While according to Table 4.13.1, glasses containing trivalent Gd, Tb, Dy, Ho and Er appear as the most promising candidates because of their large magnetic moments, applications with Ce^{3+} and Pr^{3+} benefit from particularly high transition probabilities, while consideration of the factor $(1-\lambda^2/\lambda_t^2)^{-1}$ would favor the use of Eu^{2+}. Furthermore, the optimum choice of the rare-earth ion also depends on the wavelength at which one wishes to operate. The exact value of λ_t also depends on glass composition and ion concentration, as the $4f^{n-1}5d^1$ excited state is greatly influenced by the ligand field. For example, in the case of Tb_2O_3-containing boroaluminogalliogermanate glasses, values between 230 and 300 nm (instead of the free-ion value of 215 nm) have been reported.

Glassy Faraday rotators containing all of the elements listed in Table 4.13.2 have been reported; however, to date, the majority of studies have focused on glasses containing Tb_2O_3. Up to 60 mol% can be incorporated successfully into various

Table 4.13.1: Properties of rare-earth ions relevant for their magneto-optic properties in glasses.

Ion	C_t	λ_t (nm)	$(1-\lambda^2/\lambda_t^2)^{-1}$	μ_{theor} [a]	μ_{eff}
Ce^{3+}	28	289	−0.264	2.54	2.4
Pr^{3+}	28	210	−0.124	3.58	3.5
Eu^{2+}	3.0	384	−0.583	7.94	8.0
Tb^{3+}	7.8	215	−0.131	9.72	9.5
Dy^{3+}	8.5	175	−0.083	10.63	10.6

[a] Defined as $g_J\mu_B\{J(J+1)\}^{1/2}$ listed as multiple of the Bohr magneton.

Table 4.13.2: Verdet constants (in units of rad/Tm) at 632.8 nm of typical magneto-optic glass formulations.

$30Tb_2O_3$-$70B_2O_3$	−103 [30]
$25EuO$-$15Al_2O_3$-$60SiO_2$	−102 [29]
$25Tb_2O_3$-$16.5GeO_2$-$21.5B_2O_3$-$37Al_2O_3$	−119 [33]
$33Tb_2O_3$-$25GeO_2$-$25B_2O_3$-5-SiO_2-$12Al_2O_3$	−119 [32]
$35Tb_2O_3$-(Al_2O_3-SiO_2-B_2O_3-P_2O_5-ZnO-ZrO_2)*	−137 [34]
$40Tb_2O_3$-$10Dy_2O_3$-$16.7B_2O_3$-$10Ga_2O_3$-$10SiO_2$-$3.3P_2O_5$*	−185 [31]
$20Tb_2O_3$-$25Ga_2O_3$-$35B_2O_3$-$20GeO_2$	−236 [35]
$60Tb_2O_3$-(B_2O_3-SiO_2-Al_2O_3)*	−234 [37]
$58EuO$-$12Al_2O_3$-$20B_2O_3$-$10SiO_2$	−300 [37]

*Exact composition not revealed.

multiple network former glass formulations [30–36], and some examples are listed in Table 4.13.2. Aside from aluminosilicate glass matrices, more recent compositions include B_2O_3 and GeO_2 as network former components. As a figure of merit, one specifies V at room temperature measured at 632.8 nm, the wavelength of the He-Ne laser. For optimum performance, a homogeneous dispersion of the rare-earth ions is desired, avoiding any clustering and concomitant antiferromagnetic interactions. For such homogeneous glasses, the large body of data already amassed on Tb^{3+}-containing systems indicates that V is strictly proportional to the Tb^{3+} ion concentration (ions per cm^3). Thus, increasing the contribution of boron oxide to the network former inventory at fixed molar rare-earth ion content is a good strategy for increasing V, as incorporation of this light constituent has the effect of decreasing molar volume [35, 36]. The linear relation between *V* and ion concentrations breaks down in glasses containing Tb in both the trivalent and the tetravalent state. As Tb^{4+} ions have some stability due to their half-filled 4f shells, part of the Tb^{3+} ions are oxidized under standard melting conditions. The valence ratio can be estimated from X-ray photoelectron spectroscopy. It has been argued that the presence of Tb in two oxidation states produces additional disorder, thereby stabilizing the glassy state against crystallization [35]. The same caveat applies to EuO-based magneto-optic glasses, whose performance can be decreased, if part of the rare-earth is oxidized to the essentially inactive trivalent state. In this case, ^{151}Eu Mößbauer spectroscopy (see chapter 3.6) is a powerful tool for quantifying the valence distribution [36]. Recently, a record value of −300 rad/Tm was detected in an EuO-based system, where the magnetization was enhanced by ferromagnetic Eu^{2+}–Eu^{2+} superexchange interactions. These were detected by *T*-dependent magnetic susceptibility data, indicating a positive Weiss temperature [37]. Finally, it should be noted that Faraday rotation effects of technological interest are not limited to visible light operation. For applications at shorter wavelengths and in the ultraviolet region, Pr_2O_3-containing oxide [38] and Dy_2O_3 containing fluoride phosphate glasses are the choice systems [39], whereas for operation in the infrared, chalcogenide or fluoride glasses can be used.

4.13.5 Luminescent devices based on rare-earth containing glasses

4.13.5.1 Electronic structure of the rare-earth ions

By far, the largest body of literature dealing with glasses functionalized by rare-earth ions is driven by their luminescent properties. Applications are nearly infinite and have been summarized in a large number of reviews [2-7, 40-46]. The *RE* ions' emission properties are based on the element-specific $4f^n$ electronic energy levels, which are populated by optical absorption under resonance conditions, followed by radiative and nonradiative transfer to intermediate levels and eventual re-emission. Excitation wavelengths are typically in the near UV/visible/near IR region, depending on the element. Usually, multiple-peak emission spectra are observed in the near- to mid-infrared region, which are assigned on the basis of the energy level diagrams summarized in Figure 4.13.6 [47].

The energies of the various different electronic states in the order of decreasing importance are: (1) inter-electronic Coulomb repulsion effects between the f-electrons of the ion considered, (2) vectorial coupling of orbital and spin angular momenta, (3) Coulombic repulsion effects between the f-electrons and their surrounding ligands (*crystal field splitting*) and (4) the Stark effect that removes the degeneracy of the electronic states owing to electric fields caused by the ligand environment. The electronic

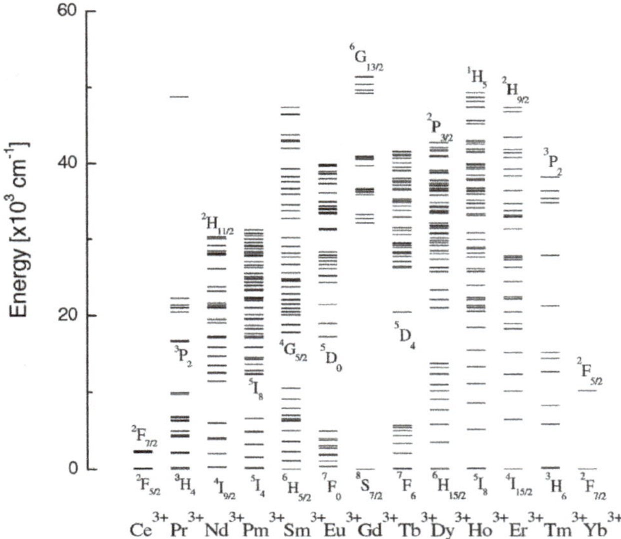

Figure 4.13.6: Electronic structure of the rare-earth ions. Reproduced from [45], with permission from Elsevier.

states are customarily described by the Russell-Saunders scheme $^{2S+1}\mathfrak{L}_J$, where the symbol $\mathfrak{L}$ is represented by a capital letter **S, P, D, F, G, H,** or **I** to label the quantum number denoting the total orbital angular momentum quantum number (L = 0, 1, 2, 3, 4, 5, and 6, respectively), $2S$+1 denotes the spin multiplicity, and J is the quantum number of the total angular momentum arising from the vector addition of the total orbital and spin angular momenta. Following Hund's rules, the electronic ground state configuration corresponds to the one with maximum spin multiplicity, and (in case of the existence of various states with the same S), maximum orbital angular momentum. With regard to J, the electronic ground state corresponds to the minimum total angular momentum quantum number, $|L\text{-}S|$, for less than half-filled shells, whereas, it corresponds to the maximum total angular momentum quantum number, $L+S$, in case the shell is more than half-filled. With a few exceptions, the magnetic properties of rare-earth compounds are dominated by the electronic ground state configuration, but in the case of Eu^{3+} and Sm^{3+} the experimentally-observed magnetic moments defined by $g_J \mu_B \{J(J+1)\}^{1/2}$ are significantly influenced by low-lying excited electronic states. The optical properties of the rare-earth ions are based on the information given in Figure 4.13.6, which also lists the electronic ground state term symbols [47]. Absorption and emission intensities are crucially influenced by the optical transition probabilities, which reflect on the efficiency of the interaction of the electric and magnetic dipoles of the corresponding electronic state to interact with the electromagnetic wave, as expressed by orbital and spin selection rules. Allowed $4f^n \rightarrow 4f^{n-1}5d^1$ transitions usually occur in the ultraviolet region and are relevant in defining the strength of the above-discussed magneto-optical effects. The transitions most relevant for optical devices are the $4f^n \rightarrow 4f^n$ transitions, which occur in the infrared to visible range. While these transitions are not electric-dipole allowed, this selection rule is partially relaxed by crystal field effects and noncentrosymmetric vibrations, when the rare-earth ions are incorporated into solid matrices. As the 4f orbitals are significantly shielded by the outer occupied 5s and 5p shells, rather narrow transitions (with line widths on the order of 1–10 nm) are observed in crystals, where the ions occupy well-defined sites. In glasses, these transitions are somewhat broadened (50–200 nm), as the local environments are less regular and are subject to local variations in ligand field parameters. Owing to this *inhomogeneous broadening* effect, Stark splittings are generally not resolved in the luminescence spectra of glasses.

4.13.5.2 Rare-earth lasing transitions in glass matrices

One important application of the luminescent properties of *RE* containing glasses is the design of lasers. Since the first discovery of Nd^{3+} lasing in a glass fiber matrix in 1961 [48], this field has grown to an immense size. Table 4.13.3 summarizes the currently known laser transitions for the various rare-earth ions, their emission wavelengths and the glass medium in which lasing was observed. Aside from building lasers based on transitions within bulk glassy materials, the construction of optical fiber lasers is important for designing optical amplifiers in telecommunications. As this application normally utilizes the 1.2–1.5 µm transmission window of glassy silica, the design of Nd- and Er-based lasers is of particular importance. In more recent years, eyesafe lasers operating at wavelengths exceeding 2,000 nm have come into focus for many new applications in the medical field, as recently summarized in a comprehensive review [43]. Currently the longest lasing wavelength realized in an optical fiber is 3900 nm, using the $^5I_5 \rightarrow {}^5I_6$ transition of Ho^{3+} in fluorozirconate glass. Various types of lasing schemes are known, some of which are illustrated in Figure 4.13.7.

The simplest case, involving *direct emission* from the excited state populated by pumping, is realized in the case of the $^2F_{5/2} \rightarrow {}^2F_{7/2}$ transition of Yb^{3+} ions. In the more common *three-level lasers*, the system is pumped by a laser from the ground state E_1 to an excited state E_3, from which it relaxes via radiationless transition to a long-lived electronic state E_2, eventually resulting in population inversion, and intense emission back to the optical ground state. In *four-level lasers*, the emitting state E_3 is reached via radiationless transfer from the pumped state E_4, the stimulated emission from E_3 does not reach the original ground state, but rather another excited state E_2. In this case, population inversion between the latter two levels is much easier to accomplish. In the quasi-three-level systems, E_3 and E_4 are connected by excited-state absorption, and E_2 is re-populated by significant re-absorption of the emitted laser radiation. In addition,

Table 4.13.3: Common rare-earth laser transitions, their wavelengths and matrices.

Ion	Laser transition	Λ [nm]	Type	Glass matrix
Pr^{3+}	$^1G_4 \rightarrow {}^3H_5$	1,260–1,350		SiO_2 fiber
	$^3P_0 \rightarrow {}^3F_2, {}^3H_6, {}^3H_4$	635,605,492	UC	FZ [51]
	$^1D_2 \rightarrow {}^3H_6$	890		SiO_2 fiber
	$^1D_2 \rightarrow {}^3H_4$	1,080		SiO_2 fiber
Nd^{3+}	$^4F_{3/2} \rightarrow {}^4I_{13/2}$	1,320–1,400	4-level	Various oxide hosts, FZ, FP
	$^4F_{3/2} \rightarrow {}^4I_{11/2}$	1,060	4-level	Various oxide hosts, FZ, FP
	$^4F_{3/2} \rightarrow {}^4I_{9/2}$	930	3-level	Various oxide hosts, FZ, FP
	$^4G_{7/2} \rightarrow {}^4I_{13/2}, {}^4I_{11/2}, {}^4I_{9/2}$	668, 600, 536	UC	$80TeO_2–5TiO_2–15Nb_2O_5$ [52]
Sm^{3+}	$^4G_{5/2} \rightarrow {}^6H_{9/2}$	651	4-level	SiO_2 fiber [49]
Tb^{3+}	$^5D_4 \rightarrow {}^7F_5$	540	4-level	Borate glass
Dy^{3+}	$^6H_{9/2}+{}^6F_{11/2} \rightarrow {}^6H_{15/2}$	1,300		
	$^6H_{13/2} \rightarrow {}^6H_{15/2}$	2,950–3,350	3-level	FZ
	$^6H_{11/2} \rightarrow {}^6H_{13/2}$	4,300	2-level	Chalcogenide [50]
	$^4F_{9/2} \rightarrow {}^6H_{11/2,13/2,15/2}$	662,576,482	UC	FZ
Ho^{3+}	$^5I_7 \rightarrow {}^5I_8$	2,040–2,080	3-level	FZ fiber, SiO_2 fiber
	$^5S_2 \rightarrow {}^5I_5$	1,380		FZ
	$^5I_6 \rightarrow {}^5I_7$	2,700	4-level	FZ
	$^5I_5 \rightarrow {}^5I_6$	3,900	4-level	
Er^{3+}	$^4I_{13/2} \rightarrow {}^4I_{15/2}$	1,550	3-level	Various oxide hosts, FZ
	$^4I_{11/2} \rightarrow {}^4I_{13/2}$	2,700	4-level	FZ
	$^4F_{9/2} \rightarrow {}^4I_{9/2}$	3,500		FZ
	$^4I_{11/2} \rightarrow {}^4I_{15/2}$	980–1,000	3-level	FZ
	$^4S_{3/2} \rightarrow {}^4I_{13/2}$	850		FZ
	$^4S_{3/2} \rightarrow {}^4I_{13/2}$	550	UC	Various oxide hosts, FZ, FP
Tm^{3+}	$^3F_4 \rightarrow {}^3H_6$	1,700–2,015		Silicate, FZ, Si/GeO_2 fiber
	$^3H_4 \rightarrow {}^3H_5$	2,250		
	$^1D_2 \rightarrow {}^3F_4$	450	UC	FZ
	$^1G_4 \rightarrow {}^3H_6$	480	UC	FZ
Yb^{3+}	$^2F_{5/2} \rightarrow {}^2F_{7/2}$	980–1,064	2-level	Various oxide hosts, SiO_2 fiber

*FZ = fluorozirconate glass, FP= fluorophosphate glass.

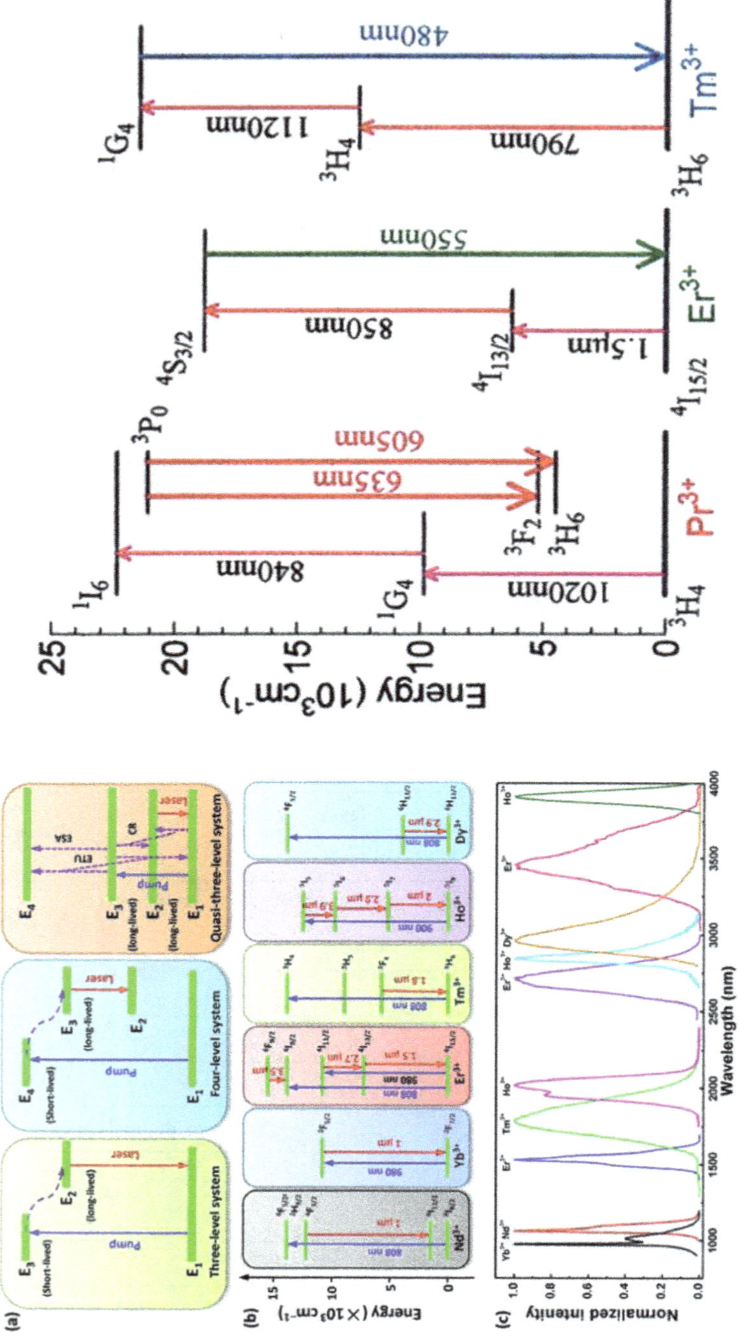

Figure 4.13.7: Common laser schemes involving rare-earth ions in glasses. Left: schemes illustrating three-level, four-level, and quasi-three level systems (top), key energy levels involving some *RE* ion lasers in glasses (middle), and corresponding laser emission spectra (bottom) [43]. Right: Relevant energy level schemes for some upconversion lasers. Taken from [43], with permission from Elsevier and from [3], with permission from Wiley.

other rare-earth (or transition metal) ion *sensitizers* can be present, increasing the pumping efficiency through energy transfer mediated by ion–ion interactions. An important example is the enhancement of the $^4I_{11/2}$ excited state of Er^{3+} by pumping the $^2F_{7/2} \rightarrow {}^2F_{5/2}$ transition of Yb^{3+} ions in glasses co-doped with Er^{3+} and Yb^{3+}. Furthermore, numerous rare-earth elements show the phenomenon of *excited-state absorption* (ESA), leading to a population of higher excited states, producing stimulated emission in the visible region, a process known as *upconversion*. Three recent examples of such upconversion lasers are shown in Figure 4.13.7, right.

4.13.5.3 The influence of the glass matrix

In choosing the most suitable glass matrix for a given application in rare-earth luminescence, one needs, of course, to take into consideration the transmission window (see Figure 4.13.3), which must include both the excitation as well as the emission wavelengths. Aside from this obvious fact, the glass matrix influences (1) the exact emission wavelength, (2) the line width of the transition and (3) its intensity. Figure 4.13.8 shows the effect of the glass matrix upon the position and width of the $^4F_{3/2} \rightarrow {}^4I_{11/2}$ transition of Nd^{3+}, illustrating the influence of the ligand field on the energies of the 4f orbitals. While this influence is relatively small due to the shielding from the occupied outer 5s and 5p orbitals, Figure 4.13.8 illustrates that the emission wavelength can sometimes be changed over a range of hundreds of nm by compositional adjustment of the glass matrix. In addition, local variations of the ligand field caused by the disorder in the glass can produce emission line broadening of the order of 100 nm. Regarding the emission intensity, the composition of the glass matrix chosen will have a decisive influence. The relevant figure of merit is the *stimulated emission cross-section,* which is influenced by multiple parameters relating to the glass composition, structure, and vibrational characteristics in the *RE* emitter's immediate local environment [43-46]. Important factors are the refractive index, the emission wavelength and the line width of the transition. In addition, it is of great importance to evaluate radiative and nonradiative decay processes of related 4f → 4f transitions. The Judd-Ofelt theory [53, 54] is usually adopted to obtain the transition probabilities including radiative decay rates, by utilizing the data of absorption cross sections of several f-f electric dipole transitions. Local ligand field properties are estimated from three parameters, Ω_2, Ω_4, and Ω_6. The nonradiative decay rate is evaluated, combined with the lifetime data obtained by luminescence decay measurements, which include the contributions of multiphonon decay, cross-relaxation and energy transfers. The effect of the glass matrix manifests itself in the way the average excited state lifetime is limited by radiationless de-excitation due to (1) energy migration via ion–ion interactions and (2) interactions with phonons associated with the ions' first coordination spheres. With regard to (1), ion clustering in the glassy matrix produces luminescence quenching; thus, it is desired that the ions are distributed as

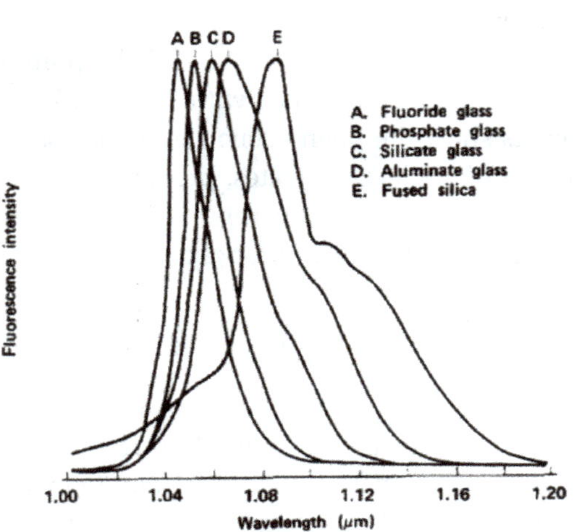

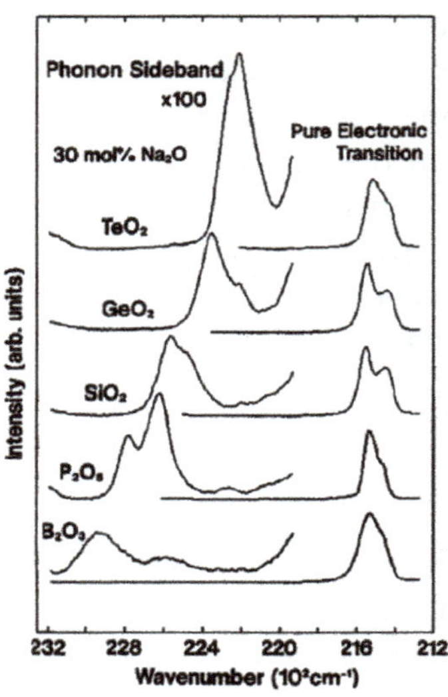

Figure 4.13.8: Influence of the glass matrix on rare-earth emission. Left: emission wavelengths of the $^4F_{3/2}$>$^4I_{11/2}$ transition of Nd^{3+}. Right: Phonon Sideband Spectra in various oxidic glass matrices containing 30 mol% sodium oxide. Taken from [40] with permission from Elsevier and from [3] with permission from Wiley.

homogeneously as possible. The ability of the glass matrices and preparation procedures to disperse the ions homogeneously can be comparatively assessed by measuring the luminescence intensity as a function of ion concentration. Clear deviations from the expected linear relationship will be observed as average ion–ion distances decrease with increasing ion contents, facilitating energy migration. With regard to (2), depopulation of the emitting state can occur via radiationless transitions, which are assisted by the vibrational dynamics occurring in the local environment of the emitting ion. The effect becomes increasingly severe with increasing phonon energy: emission efficiencies of (high-phonon) oxide glasses tend to be significantly lower than those of (low phonon) fluoride and chalcogenide glasses. The phonon energies in the local environment of the *RE* ions can be probed directly via *excitation spectroscopy*, in which one monitors the emission intensity of a selected transition as a function of excitation wavelength. The interaction with the local phonons manifests itself in phonon-assisted transitions, the so-called *phonon sidebands*, at $\Delta E_{opt} + E_{phonon}$, thus appearing at lower wavelengths than the expected optical transition. Figure 4.13.8 shows typical phonon sideband spectra observed in different kinds of matrices. Note that the vibrational frequencies that can be extracted from these sidebands also serve to identify the types of ligands attached to the emitting ions.

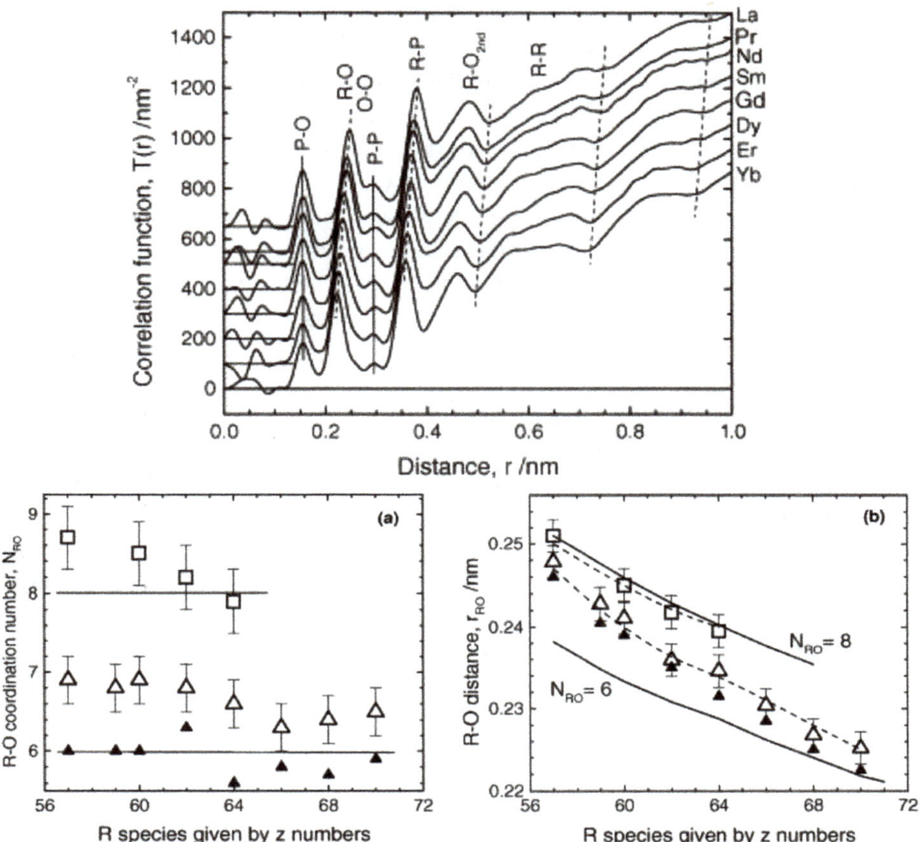

Figure 4.13.9: Results from synchrotron XRD experiments on *RE*-ultraphosphate glasses containing 15 mol% RE_2O_3. Top: Distance correlation functions, bottom: influence of the *RE* species on average *RE*-O coordination number (left), and average *RE*-O distances (right). From reference [55], with permission from Elsevier.

4.13.5.4 Structural characterization of rare-earth local environments in glasses

As discussed in the foregoing paragraphs, the emission properties of the *RE* ions are significantly influenced by their local environments. Thus, the experimental characterization of the latter is of key importance for a rational design of luminescent devices. The most basic information regarding *RE*-O distances and coordination numbers has come from synchrotron [55] or neutron diffraction [56] and **E**xtended **X**-ray **A**bsorption **F**ine **S**tructure (EXAFS) experiments [57-62]. Figure 4.13.9 shows a collection of radial distribution functions for different *RE* ultraphosphate glasses, having a modifier content of 15 mol% [55]. Note the successive shift towards shorter *RE*-O distances as a function of rare-earth atomic number, which corresponds to the expected effect of the lanthanide contraction. Average *RE*-O coordination numbers deduced from these data tend to decrease from values near 8 (for large cations such as La^{3+}, Ce^{3+} and Pr^{3+}) towards 6 for the smallest ions (Tm^{3+}, Yb^{3+} and Lu^{3+}). Analogous results have been obtained from EXAFS experiments at the respective L_{III} rare-earth edges [57–62]. In phosphate glasses, both XRD and EXAFS indicate that the *RE* coordination number tends to decrease with increasing network modifier/network ratio [58]. EXAFS studies of other glass matrices have been reviewed by Ciccone et al. [62]. The evolution of *RE*-O distances and coordination numbers as a function of *RE* atomic number appears to be a general feature in oxide, fluoride and chalcogenide glasses.

While XRD and EXAFS are suitable methods for characterizing the local *RE* environments in pure oxide or pure fluoride glasses, they do not allow for an effective characterization of mixed ligand environments, as they may occur in oxyfluoride and fluorophosphate glasses [63, 64]. For example, systematic changes in the emission intensities, excited state lifetimes and normalized phonon sideband intensities are observed in a suite of Eu^{3+}-doped alkaline earth aluminum fluoride phosphate laser glasses, as a function of fluoride/phosphate ratio [65-73]. The situation is illustrated in Figure 4.13.10. The intensity of the electric dipole allowed (*hypersensitive*)

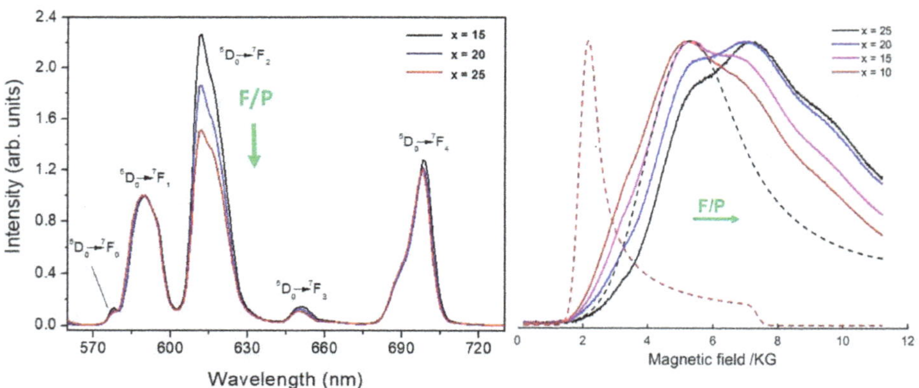

Figure 4.13.10: Left: Emission spectra of Eu-doped fluoride phosphate glasses with different F/P ratios showing the systematic dependence of the hypersensitive $^5D_0 \rightarrow {}^7F_2$ transition on the P/F ratio of the glass composition. Right: solid state X-band EPR spectra, recorded via the echo-detected field sweep of a series of fluoride phosphate glasses with composition $50(Sr/BaF_2)$-$20YF_3$-$(30{-}x)Al(PO_3)_3$-$xAlF_3$, doped with Yb^{3+} for EPR detection. The dashed red curve denotes a typical line shape expected in an oxide glass, the dashed black curve denotes a line shape expected for a pure fluoride-free phosphate glass.

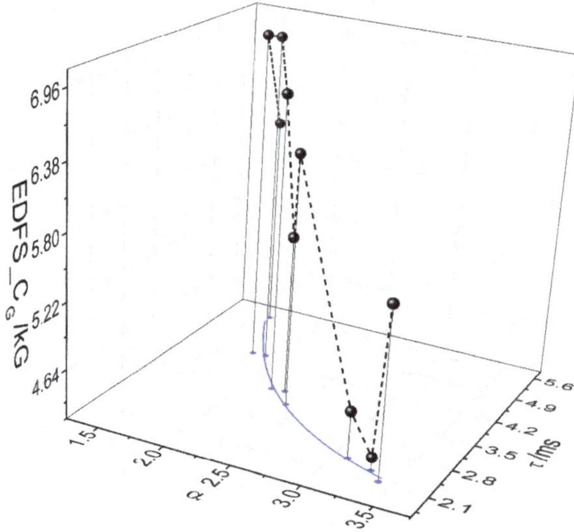

Figure 4.13.11: 3D correlation between the observables $\alpha = I(^5D_0 \rightarrow {}^7F_2)/I(^5D_0 \rightarrow {}^7F_1)$, excited state lifetime τ and the center of gravity C_G, of the X-band EPR signal Yb^{3+} dopant of glasses in the systems 0.5 $(Ba/Sr)F_2$– 0.3–x $(AlPO_3)_3$ – x AlF_3 – 0.19 YF_3 – 0.01 EuF_3 and 0.5 $(Ba/Sr)F_2$– 0.3–x $(AlPO_3)_3$–x AlF_3 – 0.19 ScF_3 – 0.01 EuF_3.

$^5D_0 \rightarrow {}^7F_2$ *transition* is systematically influenced by the phosphate-to-fluoride ratio in the glass formulation, whereas the intensities of the other (predominantly magnetic-dipole allowed) $^5D_0 \rightarrow {}^7F_J$ (J = 1, 3, 4) transitions remain largely unaffected by the glass composition. These results suggest continuous changes in the fluoride/phosphate ratio in the first coordination sphere of the rare-earth ions, as a function of composition. This suggestion is re-enforced by X-band EPR experiments on Yb^{3+} dopants of these samples [70-72]. The center of gravity expressed in units of magnetic field strength measured at the X-band (9.5 GHz microwave frequency), via echo-detected field-sweep detection in these glasses also depends systematically on the F/P ratio in the glass composition, and correlates very well the emission intensity ratio $\alpha = I(^5D_0 \rightarrow {}^7F_2)/I(^5D_0 \rightarrow {}^7F_1)$ and the excited state lifetimes measured by pulsed optical experiments (Figure 4.13.11). As discussed in further detail in the literature, a correlation is also observed with the F/P ligand ratio deduced from ^{45}Sc NMR spectroscopic experiments, in glasses containing scandium as a diamagnetic substitute for the *RE* ions [70, 73]. These results illustrate that the EPR spectrum of Yb^{3+} dopants can be used as an alternative characterization method for the average ligand distribution around the luminescent rare-earth ions in fluoride phosphate glasses. As, in principle, the EPR method is applicable to all of the RE ions with an odd number of unpaired electrons – $4f^1$ (Ce^{3+}), $4f^3$ (Nd^{3+}), $4f^5$ (Sm^{3+}), $4f^7$ (Gd^{3+}), $4f^9$

(Dy^{3+}), $4f^{11}$ (Er^{3+}) and $4f^{13}$ (Yb^{3+}) – this method could also be used to study the distinct dependence on the *RE* ion type.

4.13.6 Rare-earth-doped glasses for in vivo radiation delivery

Finally, rare-earth-containing glasses are used for delivering high-radiation doses in important biomedical applications. For this application, they are prepared in the form of microspheres and subsequently converted to high-intensity β-emitters by neutron activation. The materials in use are, typically, ternary RE_2O_3-Al_2O_3-SiO_2 glasses, which have a significant region of glass formation, extending up to 50 mol% rare-earth oxide [74–80]. They are biocompatible and not corroded by body fluids. Typical SiO_2:Al_2O_3 ratios are about 2:1, and the rare-earth oxide contents range from 15 to 40 mol%. Figure 4.13.12 shows the microsphere preparation scheme. Following standard glass preparation and crushing to a fine powder, these particles are introduced as an aerosol into an acetylene oxygen flame, where they are melted forming gas-suspended spherical droplets, minimizing surface to volume ratios. As they are exiting the flame, these droplets solidify into a soot of microspheres, which are collected from the surface of the reaction vessel.

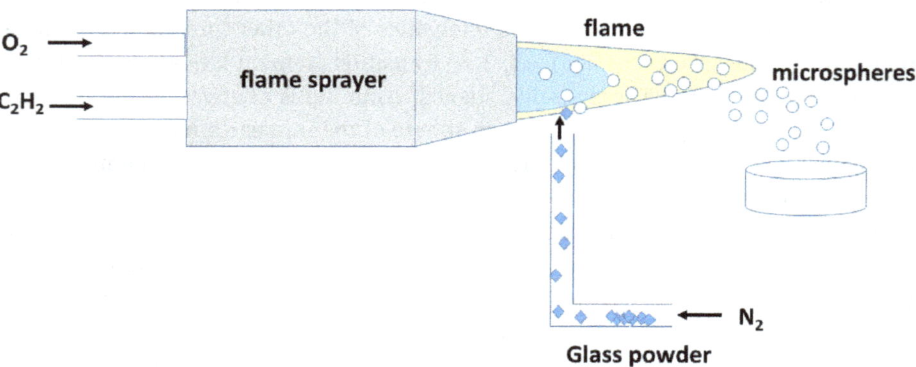

Figure 4.13.12: Generation of rare-earth aluminosilicate (*RE*AS) microspheres as described in [75].

Subsequently, these microspheres are treated by a flux of neutrons to yield radioactive glasses, produced by the reaction $^{m}RE + ^{1}n \rightarrow ^{m+1}RE + \gamma$, under emission of gamma-radiation. Up to 70% of the rare-earth nuclei can be made radioactive in this way, producing beta-emitters that can be utilized for localized radiotherapy of cancerous tissue, particularly in the liver and prostate (*SIRT* = selective internal radiation

Table 4.13.4: Radioactive rare-earth isotopes suitable for radiation therapy and their radiative properties: half life $t_{1/2}$, ß-ray energy, maximum distance range, and maximum activity.

Isotope	$t_{1/2}$/h	E_β/MeV	d_{max}/mm	Max. activity/ (Cig^{-1})
^{90}Y	64	2.27	10.3	2.4
^{153}Sm	46.8	0.80	3.3	6.1
^{166}Ho	26.9	1.84	8.7	6.3
^{165}Dy	2.3	1.29	5.8	230

therapy). A comprehensive recent review is given in reference [76], and an update can be found in [78]. Suitable rare-earth candidates include Y, Sm, Ho and Dy. Their radiative properties, isotopic half life $t_{1/2}$, β-energy, E_β, maximum range, d_{max} and maximum usable activity are summarized in Table 4.13.4. The distance range of activity increases with increasing energy and is maximal for the ^{90}Y isotope. The latter isotope has indeed received, by far, the largest attention, mostly for invasive therapy of advanced stage unresectable liver cancer. The microspheres are introduced into the liver by injection of a suspension into the hepatic artery. They localize directly in the organ, whose healthy tissue is usually impacted by the radiation only in a minor way. In this way, the tumor can be exposed to radiation doses about a factor 10 higher (typically 150 Gy) than by external radiotherapy. Clinical trials have shown great promise in the reduction of the tumor size. As opposed to previously used polymeric resins, there is no leakage into other body parts, as the radioactive emitters are quite tightly bound in the glass structure. Expert recommendations have been formulated by the *Radioembolization Brachytherapy Oncology Consortium*, regarding standardized methods of indications, techniques, multimodality treatments and dosimetry [77]. The current state of the art and various new treatment concepts of this method has been summarized in a recent review article [78]. An important current application is downsizing the tumor for making it operable. Other therapeutic applications have been developed. Biodegradable lithium borate glass doped with dysprosium can be used for localized radiation delivery to joints infected with rheumatoid arthritis [79]. The borate matrix is rapidly dissolved by body fluids, while the dysprosium is metabolized into $DyPO_4$, which can be eliminated from the body at a later stage using chelation therapy. In recent years, other isotopes such as ^{142}Pr (2.16 MeV, $t_{1/2}$ 19.12 h) for treatment of prostate cancer [80] have been investigated. Other useful isotopes include ^{146}Nd, ^{168}Yb, and ^{177}Lu [76], but their potential has not yet been explored.

Acknowledgments: This work was supported by the Center of Research, Technology, and Education in Vitreous Materials (FAPESP 2013/07793-6).

References

[1] J. E. Shelby, Key Engineer. Mater. 1994, 94–95, 1.

[2] F. Gan, Photonic Glasses (ed. F. Gan, L. Xu), Ch. 3, World Scientific Publishing Company, Singapore, 2006, 77.

[3] S. Tanabe, Int. J. Appl. Glass Sci. 2015, 6, 305.

[4] M. Laczka, L. Stoch, J. Less-Common. Met. 1990, 166, 163.

[5] S. Yonezawa, J. Kim, M. Takashima, Functionalized Inorganic Fluorides: Synthesis, Characterization and Properties of Nanostructured Solids (ed. A. Tressaud), J. Wiley & Sons, New York, 2010, 545.

[6] G. C. Rhigini, M. Ferrari, Rivista del Nuovo Cimento 2006, 28, 2005.

[7] J. Lucas, J. Less-Common Met. 1985, 112, 27.

[8] N. Sooraj Hussain, J. D. Da Silva Santos, Physics and Chemistry of Rare Earth Ion Doped Glasses, TransTech. Publ., Zürich, 2008.

[9] E. D. Zanotto, J. C. Mauro, J. Non-Cryst. Solids 2017, 471, 490 and references therein.

[10] W. Vogel, Glass Chemistry, Springer Verlag, Berlin, 1994.

[11] A. Feltz, Amorphous Inorganic Materials and Glasses, VCH, Weinheim, 1993.

[12] R. Zallen, The Physics of Amorphous Solids, John Wiley& Sons, New York, 1983.

[13] W. H. Zachariasen, J. Am. Chem. Soc. 1932, 54, 3841.

[14] H. Eckert, Handbook of Solid State Chemistry (ed. R. Dronskowski, S. Kikkawa, A. Stein), Vol. 1, Wiley-VCH, Weinheim, 2017, 93.

[15] J. Rocherulle, J. Mater. Sci. Lett. 2003, 22, 1127.

[16] B. Locardi, G. Guadagnino, Mater. Chem. Phys. 1992, 31, 45.

[17] A. J. Cohen, H. L. Smith, Science 1962, 137, 981.

[18] T. Cardinal, O. M. Efimov, H. G. Francois-Saint-Cyr, L. B. Glebov, L. N. Glebova, V. I. Smirnov, J. Non-Cryst. Solids 2003, 325, 275.

[19] O. M. Efimov, L. B. Glebov, L. N. Glebova, K. C. Richardson, V. Smirnov, Appl. Optics 1999, 38, 619.

[20] M. SeGall, I. Divliansky, C. Jollivet, A. Schülzgen, L. B. Glebov, Opt. Eng. 2015, 54, 076104.

[21] F. Kompan, I. Divliansky, V. Smirnov, L. B. Glebov, Components and Packaging for Laser Systems III (ed. A. L. Glebov, P. O. Leisher), Proc. SPIE 2017, 100850P 1.

[22] A. L. Glebov, O. Mokhun, A. Rapaport, S. Vergnole, V. I. Smirnov, L. B. Glebov, Micro-Optics, H. Thienpont, J. Mohr, H. Zappe, H. Nakajima (Eds.) Proc. SPIE 2012, 8428 1.

[23] L. Glebov, Rev. Laser Eng. 2013, 41, 684–690.

[24] J. H. Van Vleck, M. N. Hebb, Phys. Rev. 1934, 46, 17.

[25] N. F. Borelli, J. Chem. Phys. 1964, 41, 3289.

[26] C. B. Rubinstein, S. B. Berger, L. G. Van Uitert, W. A. Bonner, J. Appl. Phys. 1964, 35, 2338.

[27] J. T. Kohli, Key Engineer. Mater. 1994, 94-95, 125.

[28] S. Murai, K. Fujita, K. Tanaka, Handbook of Advanced Ceramics, Ch. 5.2, Elsevier Inc., 2013, 383.

[29] M. W. Shafer, J. W. Suits, J. Am. Ceram. Soc. 1966, 49, 261.

[30] K. Tanaka, K. Hirao, N. Soga, Jpn. J. Appl. Phys. 1995, 34, 4825.

[31] T. Hayakawa, M. Nogami, N. Nishi, N. Sawanobori, Chem. Mater. 2002, 14, 3223.

[32] V. I. Savinkov, V. N. Sigaev, N. V. Golubev, P. D. Sarkisov, A. V. Masalov, A. P. Sergeev, J. Non-Cryst. Solids 2010, 356, 1655.

[33] G. Gao, A. Winterstein-Beckmann, O. Shrzhenko, C. Dubs, J. Dellith, M. A. Schmidt, L. Wondraczek, Sci. Rep. 2015, 5, 8942.

[34] J. Ding, P. Man, Q. Chen, L. Guo, X. Hu, Y. Xiao, L. Su, A. Wu, Y. Zhou, F. Zheng, Opt. Mater. 2017, 69, 202.

[35] H. Yin, Y. Gao, H. Guo, C. Wang, C. Yang, J. Phys. Chem. C 2018, 122, 16894.

[36] H. Akamatsu, K. Fujita, Y. Nakatsuka, S. Murai, K. Tanaka, Opt. Mater. 2013, 35, 1997.

[37] F. Suzuki, F. Sato, H. Oshita, S. Yao, K. Nakatsuka, K. Tanaka, Opt. Mater. 2018, 78, 174.

[38] G. T. Petrovskii, I. S. Edelman, T. V. Zarubina, A. C. Malkovskii, V. N. Zabluda, M. Yu. Ivanov, J. Non-Cryst. Solids 1991, 130, 35.

[39] V. Letellier, A. Seignac, A. Le Floch, J. Non-Cryst. Solids 1989, 111, 55.

[40] S. Tanabe, J. Non-Cryst. Solids 1999, 259, 1.

[41] A. G. Clare, Key Engin. Mater. 1994, 94-95, 161.

[42] L. Zhang, L. Hu, S. Jung, Int. J. Appl. Glass Sci. 2018, 9, 90.

[43] W. C. Wang, B. Zhou, S. H. Xu, Y. M. Yang, Q. Y. Zhang, Prog. Mater. Sci 2019, 101, 90.

[44] S. Buddhudu, Trans. Ind. Ceram. Soc. 2015, 64, 69.

[45] M. J. Weber, J. Non-Cryst. Solids 1990, 123, 208.

[46] A. Jha, B. Richards, G. Jose, T. Teddy-Fernandez, P. Joshi, X. Jiang, J. Lousteau, Prog. Mater. Sci. 2012, 57, 1426.

[47] G. H. Dieke, H. M. Crosswhite, Appl. Optics 1963, 2, 675.

[48] E. Snitzer, Phys. Rev. Lett. 1961, 7, 444.

[49] M. C. Farries, P. R. Morkel, P. E. Townsend, Electronics Lett. 1988, 24, 709.

[50] M. R. Majewski, S. D. Jackson, Optics Lett. 2016, 41, 4496.

[51] D. M. Baney, G. Ranking, K. W. Chang, Optics Lett. 1996, 21, 1372.

[52] I. Iparraguirre, J. Azkargorta, J. M. Fernández-Navarro, M. Al-Saleh, J. Fernández, R. Balda, J. Non-Cryst. Solids 2005, 353, 990.

[53] B. R. Judd, Phys. Rev. 1962, 127, 750.

[54] G. S. Ofelt, J. Chem. Phys. 1962, 37, 511.

[55] U. Hoppe, R. K. Brow, D. Ilieva, P. Jovari, A. C. Hannon, J. Non-Cryst. Solids 2005, 351, 3179.

[56] A. G. Clare, A. C. Wright, Key Eng. Mater. 1994, 94-95, 141.

[57] G. Mountjoy, J. Non-Cryst. Solids 2007, 353, 2029.

[58] R. Anderson, T. Brennan, G. Mountjoy, R. J. Newport, G. A. Saunders, J. Non-Cryst. Solids 1998, 232–234, 286.

[59] R. K. Brow, A. K. Wittenauer, Phosphorus Res. Bull. 2002, 13, 95.

[60] J. M. Cole, R. J. Newport, D. T. Bowron, R. F. Pettifer, G. Mountjoy, T. Brennan, G. A. Saunders J. Phys.: Condens. Matter 2001, 13, 6659.

[61] M. Karabulut, E. Metwalli, A. K. Wittenauer, R. K. Brow, G. K. Marasinghe, C. H. Booth, J. J. Bucher, D. K. Shuh, J. Non-Cryst. Solids 2005, 351, 795.

[62] M. R. Cicconi, G. Giuli, E. Paris, P. Courtial, D. B. Dingwell, J. Non-Cryst. Solids 2013, 362, 162.

[63] F. Philipps, T. Töpfer, H. Ebendorff-Heidepriem, D. Ehrt, R. Sauerbrey, Appl. Phys. B 2001, 72, 399.

[64] P. F. Wang, B. Peng, W. N. Li, C. Q. Hou, J. B. She, H. T. Guo, M. Lu, Solid State Sci. 2012, 14, 550.

[65] T. Y. Bocharova, G. Karapetyan, A. Mironov, N. Y. Tagil'Tseva, Phosphorus Res. Bull. 2002, 13, 87.

[66] T. V. Bocharova, J. L. Adam, G. O. Karapetyan, F. Smektala, A. M. Mironov, N. O. Tagil'tseva, J. Phys. Chem. Solids 2007, 68, 978.

[67] S. Tanabe, K. Hirao, N. Soga, J. Non. Cryst. Solids 1992, 142, 148.

[68] S. S. Babu, P. Babu, C. K. Jayasankar, W. Sievers, T. Tröster, G. Wortmann, J. Lumin. 2007, 126, 109.
[69] P. Babu, K. H. Jang, E. S. Kim, R. Vijaya, C. K. Jayasankar, V. Lavín, H. J. Seo, J. Non-Cryst. Solids 2011, 357, 2139.
[70] M. de Oliveira, T. S. Goncalves, C. Ferrari, C. J. Magon, P. S. Pizani, A. S. S. de Camargo, H. Eckert, J. Phys. Chem. C 2017, 121, 2968.

[71] M. de Oliveira, T. Uesbeck, T. S. Gonçalves, C. J. Magon, P. S. Pizani, A. S. S. de Camargo, H. Eckert, J. Phys. Chem. C 2015, 119, 24574.
[72] G. Galleani, S. H. Santagneli, Y. Messaddeq, M. de Oliveira, H. Eckert, Phys. Chem. Chem. Phys. 2017, 19, 21612.
[73] G. Galleani, S. H. Santagneli, Y. Ledemi, Y. Messaddeq, O. Janka, R. Pöttgen, H. Eckert, J. Phys. Chem. C 2018, 122, 2275.
[74] E. M. Erbe, D. E. Day, J. Biomed. Mater. Res. 1993, 27, 1301–1308.

[75] J. E. White, D. E. Day, Key Eng. Mater. 1994, 94-95, 181.
[76] D. Day, Bioglasses, an Introduction (ed. J. R. Jones, A. G. Clare), Ch. 13, J. Wiley & Sons, 2012, 203.
[77] A. Kennedy, S. Nag, R. Salem, R. Murthy, A. J. McEwan, C. Nutting, J. Espat, J. T. Bilbao, R. A. Sharma, J. P. Thoma, D. Coldwell, Int. J. Radiation Oncol. Biol. Phys. 2007, 68, 13.
[78] E. W. Lee, L. Alanis, S. K. Cho, S. Saab, Korean J. Radiol. 2016, 17, 472.
[79] S. D. Conzone, W. W. Hall, D. E. Day, R. F. Brown, J. Biomed. Mater. Res. 2004, 70A, 256.
[80] J. W. Jung, W. D. Reece, Appl. Radiat. Isot. 2008, 66, 441.

Lena J. Daumann

4.14 Bioinorganic chemistry of the elements

That rare earth elements (REE) are essential for our modern technologies is now well established. It is less known that they have also been pivotal in bioinorganic chemistry (the study of the role of metals in biology), in agriculture and in a range of medical applications. And, remarkably, these elements are now also considered to be essential for many bacteria. Table 4.14.1 shows selected applications of REE in bioinorganic chemistry [1].

In medicine, REE are useful both in diagnostics and therapy. One of the most widely applied REEs is gadolinium, that is used as a magnetic resonance imaging (MRI) contrast agent (see also Chapter 4.5 "Medical applications of rare earth compounds"). The seven unpaired electrons of Gd^{3+} in the electronic $[Xe]4f^7$ configuration render it highly paramagnetic. As the free Gd^{3+} aquo complex is toxic, it is administered to patients intravenously bound to a chelator such as DOTA or DTPA. These carboxylate-rich ligands (1) satisfy the high coordination number preference of Gd^{3+}, (2) satisfy the preference for hard oxygen donor ligands and (3) leave at least one coordination site open for water coordination, thus enhancing contrast. One emerging problem with the widespread use of Gd-contrast agents is the concomitant rising levels of this REE in aquatic systems near highly developed healthcare regions [2]. Newer developments in the field of medical imaging include the use of the paramagnetic REEs as chemical shift saturation transfer (PARACEST) agents [3]. Many radionuclides of REEs (^{140}La, ^{141}Ce, ^{153}Sm, ^{160}Tb, ^{170}Tm, ^{169}Yb and ^{177}Lu) find use in scintigraphy, computed tomography (CT) and in therapy [1b, 4]. For example, the beta-emitter ^{153}Sm-EDTMP accumulates in bone metastases and has been used to alleviate pain. A dose of 2–3 g of $La_2(CO_3)_3$ is given to patients with chronic kidney diseases. These patients often struggle with hyperphosphatemia and need to limit their phosphate intake from food. As lanthanum(III) forms poorly-soluble $LaPO_4$ with the phosphates present in food, it allows for excretion via the gastrointestinal tract. The low solubility of $LaPO_4$ is used in an La-containing clay in Australia (Phoslock) that is employed as waterway and wastewater treatment to avoid algal blooms caused by high phosphate contents. More soluble and bioavailable forms of REE salts and complexes find application in agriculture. Lanthanum and cerium citrate additives have long been used in China as livestock feed additives [1a]. In western studies, the effect of improved feed conversion and enhanced growth was also demonstrated. This effect is currently not well understood, and at the time of writing, there is no evidence that these elements are essential for mammals.

However, since 2011, these elements are known to be essential for many bacteria [5]. Why this was discovered only recently has several reasons. The common misconception that REE are rare did not favor the discovery of REE-dependent metalloenzymes and proteins. And while they are abundant in nature, these elements possess low

https://doi.org/10.1515/9783110654929-037

Table 4.14.1: Selected applications of some REE in bioinorganic chemistry.

	Application	Example	Reason for REE use/Notes
Y	Glass spheres for treatment of hepatic neoplasia	Thera-Sphere®	^{90}Y beta-emitter with half-life of 64.1 h
La	Feed additive in agriculture	Lancer®	Lanthanum and cerium citrate shown to improve feed conversion rates in livestock
	Phosphate binder in hyperphosphatemia	Fosrenol®	Lanthanum carbonate, La^{3+} binds and precipitates dietary phosphate and prevents thus phosphate uptake in kidney disease patients
Ce	Treatment for nausea	Peremesin®	Cerium oxalate, historical use
	Used together with La, Nd, Pr as fertilizer	Changle	Mode of action not well understood. Widely used in China
Nd	Anticoagulant	Thrombodym	Administration of Nd salts and complexes for prevention of thrombosis. Replaced by heparin
Sm	Ointment for burns	Phogosam®	Antibacterial properties of samarium complexes. Used in Hungary
	Pain treatment for bone cancer patients	Quadramet	The beta-emitter ^{153}Sm-EDTMP accumulates in bone metastases and has a half-life of 46.3 h
Eu, Tb	Luminescent bioassays	Lumi4-Tb	Long-lived and bright luminescence, possibility of multiplexing
	NMR shift reagent of proteins, Lanthanide Binding Tags (LBT)	EF hand motif peptide	Investigating function and dynamics of proteins, x-ray crystallography
Gd	Contrast agent for MRI	Cyclolux, Dotarem, Magnevist	$S = 7/2$ ground state of Gd^{3+} makes for an excellent T1-contrast agent
Lu	Photodynamic therapy for cancer patients	Lutrim, Lu-Tex	Was previously in clinical trials
	Peptide receptor radionuclide therapy for neuroendocrine tumors	Lutathera	^{177}Lu emitter of low-energy gamma photons (imaging) and medium energy beta particles (therapy)

bioavailability (as them being not available in high concentrations to organisms in soluble form) since they are known to form poorly soluble compounds (e.g., phosphates). It was a group of Japanese microbiologists that first reported, in 2011, that a methanol dehydrogenase (MDH)-type protein was produced in the bacterium *M. radiotolerans* upon addition of La^{3+} to the culture medium [6]. The same authors also reported shortly after that the bacterial strain *M. extorquens* AM1 could grow with La^{3+} instead of Ca^{2+} and isolated a La-MDH from that bacterium [7]. The function of methanol dehydrogenase enzymes in methylotrophic (using C1 carbon compounds such as methanol or methane) bacteria is to oxidize methanol to formaldehyde and as such, they play an integral part in bacterial metabolism. MDH enzymes bear the redox cofactor PQQ (pyrroloquinoline quinone, methoxatin) in addition to a Lewis acid (Ca^{2+} or lanthanides, Ln^{3+}) in the active site (Figure 4.14.1). While REE have coordination chemistry properties similar to those of calcium, being hard Lewis acids, preferring hard (oxygen) donor ligands and showing high flexibility in the surrounding ligand geometry, REE often exhibit higher coordination numbers (See Chapter 2.6 "Rare earth coordination chemistry"). All these coordination chemistry preferences are reflected in the active sites of methanol dehydrogenase enzymes. Figure 4.14.1 shows the active sites of Eu-MDH and Ca-MDH side by side.

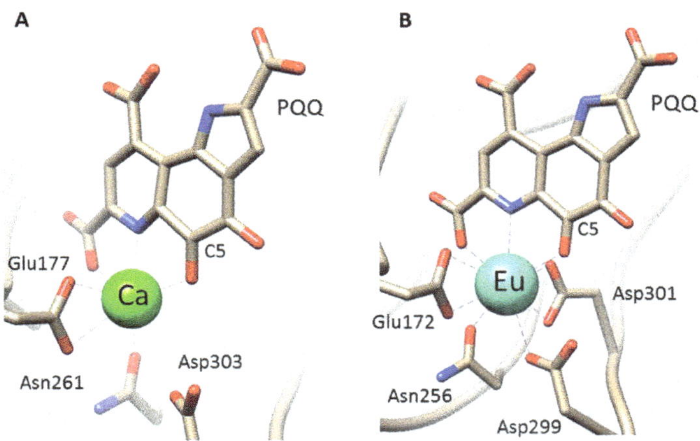

Figure 4.14.1: **A** Active site of Ca-MDH (MxaF type) isolated from *M. extorquens* (PDB: 1W6S). **B** Active site of Eu-MDH (XoxF type) that was isolated from *M. fumariolicum* SolV (PDB: 6FKW).

In addition to the tridentate redox cofactor (PQQ), the metal ions are coordinated bidentately by glutamate and aspartate residues, as well as a monodentately bound asparagine residue. The active site of lanthanide-dependent MDH features an additional bidentate aspartate residue, reflecting the higher coordination number preferences of lanthanides. The Ca-MDH was discovered many decades ago, and was long thought to be the only pathway for methanol oxidation in methylotrophic bacteria. This pathway is encoded by *mxaF* (this denotes a gene; bacterial gene symbols are composed of three lowercase, italicized letters followed by an uppercase letter, the respective protein symbol here then is MxaF). After the initial report of a La-MDH in 2011, the field of lanthanide-dependent bacterial metabolism developed rapidly when the first crystal structure of a Ce-MDH was published in 2014 [8]. Ln-MDH (XoxF-MDH) enzymes are encoded by *xoxF*. The first x in this gene stands for the long-unknown function that puzzled researchers for many decades, until the biological role of lanthanides had been accepted. It has now been established that: (1) A vast number of bacterial strains in different ecosystems – such as in the phyllospheres of plants, in soil and marine environments – carry the gene that encodes for lanthanide-dependent enzymes. (2) There are two major types of lanthanide-utilizing bacteria, those that possess only *xoxF* – and are thus strictly dependent on lanthanides like *M. fumariolicum* SolV – and those that possess both *mxaF* and *xoxF* like *M. extorquens* AM1. (3) The latter bacteria will produce Ln-MDH if concentrations of lanthanides are sufficient; here, concentrations of as low as 2.5 nM La^{3+} were shown to be enough to trigger *xoxF* expression. This flexibility to switch to a different metabolic pathway has been termed the "lanthanide-switch". (4) Other dehydrogenases such as ethanol dehydrogenases have been identified. However, the methanol dehydrogenases are, so far, the best-characterized species of Ln-dependent enzymes. Three crystal structures were available at the time of writing: in addition to the Eu-MDH as well as Ce-MDH from *M. fumariolicum* SolV, an La-MDH from *M. buryatense* have been reported [8–9]. (5) Only the more abundant, early lanthanides (La to Sm) have been shown to stimulate growth of Ln-utilizing bacteria and to promote MDH activity. Efficient uptake into cells is also likely limited to early lanthanides. This demonstrates that, even if lanthanides are often said to be elements *whose properties resemble each other like two drops of water* [10], the differences in ionic radii (due to the lanthanide contraction) of the Ln and thus subtle differences in their coordination chemistries do have an impact on bacterial metabolism.

Y and Sc are often not included in these investigations, but the few studies available showed no significant impact on bacterial growth. Lanthanide-dependent bacteria such as *M. extorquens* AM1 have potential applications in biomining, as they can use Ln-containing scrap metal (e.g., Nd from hard-drive magnets) for growth [11]. Remarkably, this bacterium was shown to store cytoplasmic lanthanum in mineral form (as phosphate) [12]. In addition to methanol-oxidizing enzymes, a high-affinity lanthanide protein, lanmodulin, LanM, has been isolated from this organism [13]. This protein is similar to the well-known Ca^{2+} calmodulin protein family, with a typical "EF-hand" structure (a special structural motif that is found in these calcium binding proteins, Figure 4.14.2). In LanM, three carboxylate-rich metal-binding sites bind Ln^{3+} with pM affinity, and one site with μM affinity and, remarkably, a hundred-million-fold selectivity for Ln^{3+} over Ca^{2+}.[13] Demonstrated here, is how similar the coordination chemistries of Ca^{2+} and Ln^{3+} can be in biology, as both are coordinated almost exclusively by hard O-donors with high coordination numbers. Then again, subtle differences in the peptide sequence – LanM contains a proline residue – can make a big difference and favor Ln^{3+} over Ca^{2+}. Calmodulin and lanmodulin undergo a strong conformational change to a well-folded arrangement upon binding Ca^{2+} and Ln^{3+}, respectively.

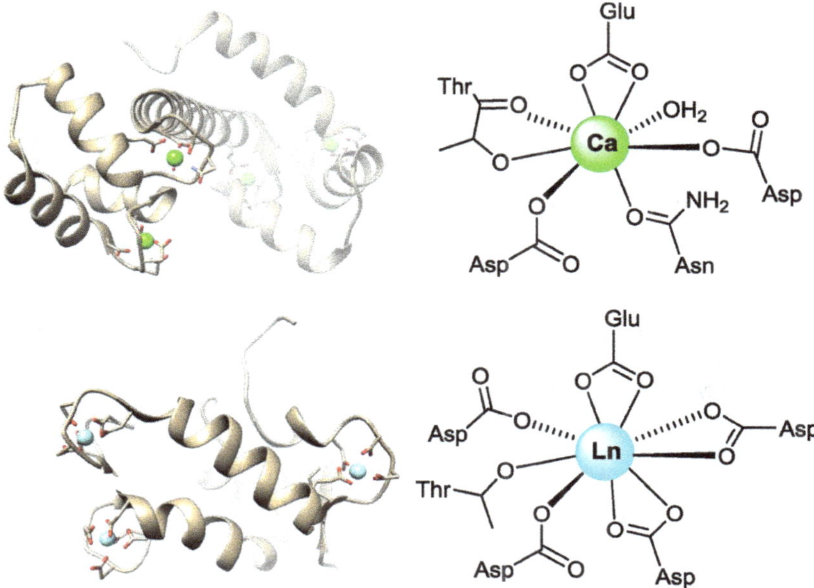

Figure 4.14.2: Chicken calmodulin X-ray structure (PDB: 1UP5) and close up view of one EF-binding motif. Below reported NMR-determined structure [14] of lanmodulin (PDB: 6MI5) and close-up view of the EF1 site.

A fluorescent sensor for lanthanides was developed based on Lanmodulin, and relies on the conformational change of LanM upon Ln^{3+} binding [15]. Here, the lanthanide-based luminescence is not used, but a truncated enhanced cyan fluorescent protein (ECFP), together with a yellow fluorescent protein (citrine, Figure 4.14.3).

As LanM binds to Ln^{3+}, it folds in such way that ECFP and citrine are brought in close proximity, enabling a Förster resonance energy transfer (FRET). This fluorescent sensor was used to track lanthanide distribution in vivo but could also be used as high-affinity sensor for a range of other applications. The luminescent properties of REE themselves have also been exploited extensively to yield responsive probes for biological assays (see Chapter 4.1 “Optical materials – molecules”) [16]. Due to their Laporte-forbidden *f*–*f*-transitions, high-energy lasers have to be used for direct excitation, or more conveniently, the antenna effect is used. A potassium sensor was reported based on the complex shown in Figure 4.14.4. The azacrown-ether moiety binds K^+ and the attached ligands undergo a conformational rearrangement that brings the azaxanthone antenna close enough to sensitize Tb^{3+} [17]. Sensitive and selective potassium detection is of immense importance in diagnosis, as potassium concentrations have been linked to a variety of disorders. This ligand can recognize K^+ with a 93-fold selectivity over Na^+.

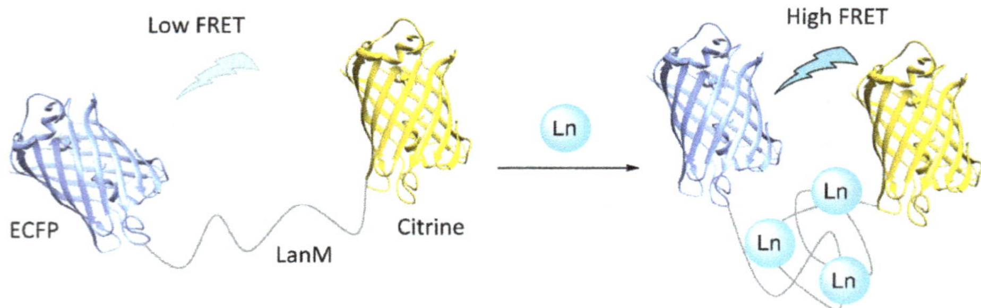

Figure 4.14.3: Fluorescent sensor with pM affinity for rare earth elements based on lanmodulin [15]. As LanM binds lanthanides, the concomitant conformational change of the protein brings a FRET pair into close proximity with each other. Adapted from Cotruvo et al. [15].

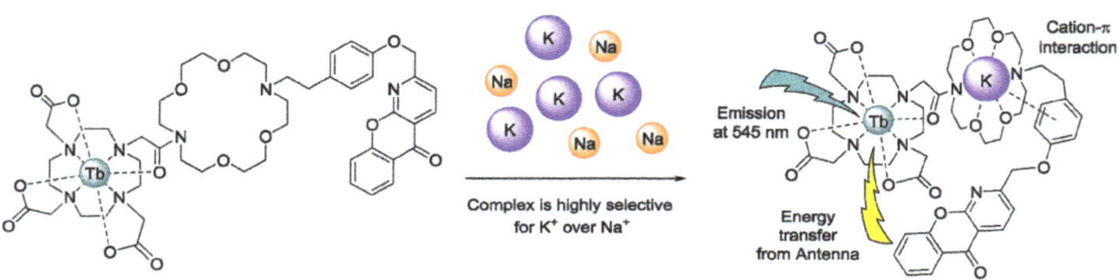

Figure 4.14.4: K^+-sensor based on terbium(III) luminescence [17].

Due to their exceptional photophysical properties, REE have been used as spectroscopically more informative substitutes for calcium(II) and other metal ions, in biochemistry research. The affinity of the apo-form of the aminopeptidase from *E. coli* for metal ions was investigated by europium(III)-luminescence titrations (Figure 4.14.5) [18]. Binding of Eu^{3+} to the active site of the protein reduces the number of bound water molecules in the first coordination sphere, and the protein can simultaneously serve as an antenna, increasing luminescence upon binding. If there is no suitable REE binding site available, the option to add a lanthanide binding tag (LBT) to a protein of interest has been extensively explored. Here, the luminescence can, for example, be used to quickly identify proteins of interest in mixtures or during purification. A tryptophan residue needs to be included in the LBT to enable sensitization via this antenna [19]. LBTs can also be used to help solve phase problems in protein X-ray crystallography and are applied in NMR spectroscopy, where paramagnetic LBTs are used to gain structural insight into proteins and peptides.

REE are also often used as surrogates for actinides in bioinorganic chemistry. It is of great interest to understand how actinides are metabolized and mobilized and how to decontaminate and detoxify organisms and environments after exposure. One

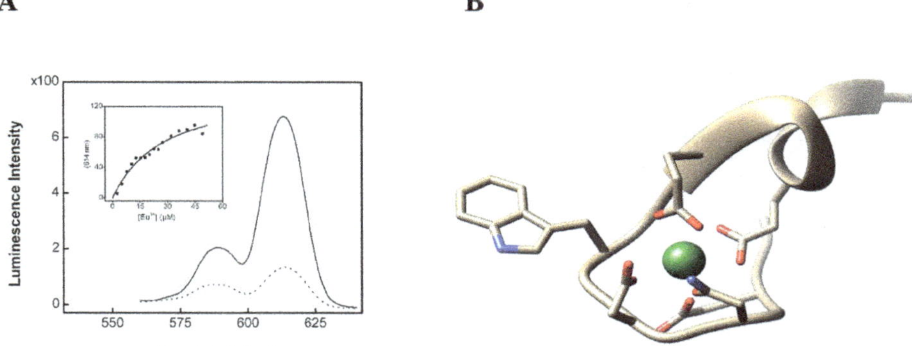

Figure 4.14.5: **A** Binding of free europium(III) (dotted line) to methionine aminopeptidase (solid line) enhances Eu^{3+} luminescence. From the data, affinity constants can be deducted. Redrawn from [18] with permission from Elsevier. **B** Lanthanide Binding Tag with terbium(III) [19].

pathway is possibly, the interaction of actinides with bacterial chelators for iron(III), the siderophores. For example, the bacterial hydroxamate-type siderophore desferrioxamin B (DFO B) can bind thorium(IV), curium(III) and other actinides, while the catecholate-type siderophore enterobactin is also well known to bind to actinides like Pu^{4+} (Figure 4.14.6). In vivo, iron(III)-siderophore complexes are scavenged by the protein siderocalin. Actinide and lanthanide complexes of siderophores and siderophore-like ligands (3,4,3-LI-1,2-HOPO) can also bind to siderocalin (Figure 4.14.6). Remarkably, the protein itself can be used as an antenna for the ligand chromophore (here 1,2-hydroxypyridonate) which then, in turn, excites the metal ion (Eu^{3+}, Sm^{3+}, Cm^{3+} and others) [20].

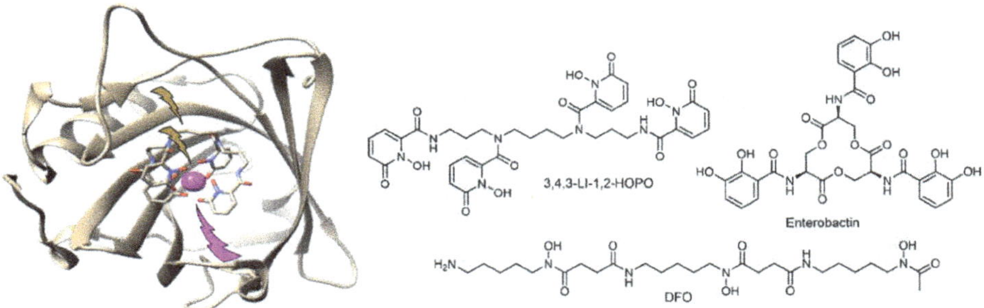

Figure 4.14.6: Siderocalin with a samarium(III) complex of siderophore-inspired chelator and antenna 3,4,3-LI-1,2-HOPO [20] and the two siderophores enterobactin and DFO, which are known to complex lanthanides.

Further, it has been shown that REE from volcanic ash can be mobilized with bacterial chelators such as the siderophore, DFO [21]. Many organisms have been shown to bio-leach and bio-accumulate REE, and it has been further shown that functional groups on bacterial surfaces can be used to separate the heavy REE, Tm, Yb and Lu, from their other family members [22].

References

[1] a) K. Redling, Rare Earth Elements in Agriculture with Emphasis on Animal Husbandry, PhD thesis, Ludwig-Maximilians-Universität München, 2006; b) C. H. Evans, Biochemistry of the Lanthanides, Vol. 8, Springer Science & Business Media, 2013; c) A. Dash, M. R. A. Pillai, F. F. Knapp Jr., Nucl. Med. Mol. Imaging 2015, 49, 85; d) K. N. Allen, B. Imperiali, Curr. Opin. Chem. Biol. 2010, 14, 247; e) S. A. Cotton, J. M. Harrowfield, The Rare Earth Elements: Fundamentals and Applications, ed. DA Atwood, John Wiley & Sons, Chichester, 2012, 65.
[2] P. Ebrahimi, M. Barbieri, Geosci. 2019, 9, 93.
[3] M. Woods, D. E. Woessner, A. D. Sherry, Chem. Soc. Rev. 2006, 35, 500.
[4] N. A. Thiele, J. J. Woods, J. J. Wilson, Inorg. Chem. 2019, 58, 10483.
[5] a) L. J. Daumann, Angew. Chem. Int. Ed. 2019, 58, 12795; b) E. Skovran, N. C. Martinez-Gomez, Science 2015, 348, 862.
[6] Y. Hibi, K. Asai, H. Arafuka, M. Hamajima, T. Iwama, K. Kawai, J. Biosci. Bioeng. 2011, 111, 547.
[7] T. Nakagawa, R. Mitsui, A. Tani, K. Sasa, S. Tashiro, T. Iwama, T. Hayakawa, K. Kawai, PLoS One 2012, 7, e50480.
[8] A. Pol, T. R. M. Barends, A. Dietl, A. F. Khadem, J. Eygensteyn, M. S. M. Jetten, H. J. M. Op den Camp, Environ. Microbiol. 2014, 16, 255.
[9] a) Y. W. Deng, S. Y. Ro, A. C. Rosenzweig, J. Biol. Inorg. Chem. 2018, 23, 1037; b) B. Jahn, A. Pol, H. Lumpe, T. Barends, A. Dietl, C. Hogendoorn, H. Op den Camp, L. Daumann, ChemBioChem 2018, 19, 1147.
[10] D. N. Trifonov, The Rare-earth Elements, Pergamon Press [distributed in the Western Hemisphere by Macmillan, New York], 1963.
[11] N. C. Martinez-Gomez, H. N. Vu, E. Skovran, Inorg. Chem. 2016, 55, 10083.
[12] P. Roszczenko-Jasińska, H. N. Vu, G. A. Subuyuj, R. V. Crisostomo, J. Cai, N. F. Lien, E. J. Clippard, E. M. Ayala, R. T. Ngo, F. Yarza, J. P. Wingett, C. Raghuraman, C. A. Hoeber, N. C. Martinez-Gomez, E. Skovran, Sci. Rep. 2020, 10, 12663.
[13] J. A. Cotruvo, E. R. Featherston, J. A. Mattocks, J. V. Ho, T. N. Laremore, J. Am. Chem. Soc. 2018, 140, 15056.
[14] E. C. Cook, E. R. Featherston, S. A. Showalter, J. A. Cotruvo, Biochemistry 2019, 58, 120.
[15] J. A. Mattocks, J. V. Ho, J. A. Cotruvo, J. Am. Chem. Soc. 2019, 141, 2857.
[16] M. C. Heffern, L. M. Matosziuk, T. J. Meade, Chem. Rev. 2014, 114, 4496.
[17] A. Thibon, V. C. Pierre, J. Am. Chem. Soc. 2008, 131, 434.
[18] N. Sule, R. K. Singh, P. Zhao, D. K. Srivastava, J. Inorg. Biochem. 2012, 106, 84.
[19] M. Nitz, M. Sherawat, K. J. Franz, E. Peisach, K. N. Allen, B. Imperiali, Angew. Chem. Int. Ed. 2004, 43, 3682.

[20] B. E. Allred, P. B. Rupert, S. S. Gauny, D. D. An, C. Y. Ralston, M. Sturzbecher-Hoehne, R. K. Strong, R. J. Abergel, Proc. Nat. Acad. Sci. 2015, 112, 10342.
[21] M. Bau, N. Tepe, D. Mohwinkel, Earth Planet. Sci. Lett. 2013, 364, 30.
[22] a) W. D. Bonificio, D. R. Clarke, Environ. Sci. Technol. Lett. 2016, 3, 180; b) M. Maleke, A. Valverde, J.-G. Vermeulen, E. Cason, A. Gomez-Arias, K. Moloantoa, L. Coetsee-Hugo, H. Swart, E. van Heerden, J. Castillo, Front. Microbiol. 2019, 10, 81.

Marc Wentker, Thomas Jüstel, Jens Leker

4.15 Economic aspects

4.15.1 Criticality overview

Rare earth elements (REEs), platinum-group metals, as well as gallium and indium are most commonly identified as a critical resource in supply criticality studies [1]. Criticality is defined as the risk of a possible supply reduction that is especially important from the perspective of a resource assessor [2]. An example for the change in dependency, and thus in the supply situation for certain commodities, is given by the situation of the US mining industry since 1954, where successively domestic production got replaced by foreign imports [3]. Imports and thus the increasing dependence on foreign sources may lead to restrictions in the availability of raw materials. In the past, such trade relations for technologically highly important raw materials have led to political crises, as was the case between Japan and China with regard to REEs in September 2010 [4, 5]. REEs are usually collectively evaluated as the lanthanum series of elements with atomic numbers from 57 to 71 (La-Lu). Scandium and yttrium are often excluded and evaluated separately due to their different use in functional materials and their specific geological behavior. Mineral raw materials are classified as primary products, coproducts or by-products based on their economic importance to mining companies. Since the economic importance can vary according to the number of deposits, their production as well as the demand reflected by the price, such a classification represents a strong simplification of reality. REEs are obtained as a by-product from the extraction of other raw materials. In particular, one source is iron mining. The term "criticality" reflects the current and potentially expected future demand in high technology manufacturing. These needs strongly depend on the evaluator's perspective and therefore, vary by geographic region. Innovation, ageing or substitution can similarly influence the demand for and importance of particular resources and thus influence their criticality [1].

4.15.2 Historical development and present status

From the mid-1960s to early 1990s, the US was the largest producer of REEs. Production took place mainly at the Mountain Pass mine in south-eastern California. The carbonatite deposits that were mined there primarily consisted of light rare earth elements (LREEs). In the early 1980s, China began mining its REE deposits, processing its ore, extracting, and separating the individual REEs for domestic manufacturing. China started dominating the global REE production with covering constantly above 80% of the global processed REE market since the early 2000s. As presented in

https://doi.org/10.1515/9783110654929-038

Figure 4.15.1, from 2010, China controlled approximately more than 85% of the global production, followed by Australia with 7% and a relaunched production in the USA with about 3%. To preserve its limited resources for electrical and electronic equipment to meet domestic demand and to address concerns about the environmental impact of mining, China began to limit the supply of REEs through the introduction of quotas, licenses and taxes in 2010. As a result, industry responded by increasing its inventory of rare earths. In addition, the search for new deposits outside China and attempts at substitution and recycling are moving into focus. These led to initiatives, for example, from the governments of the US and a joint effort by the European Union to find respective domestic economical deposits [8, 9].

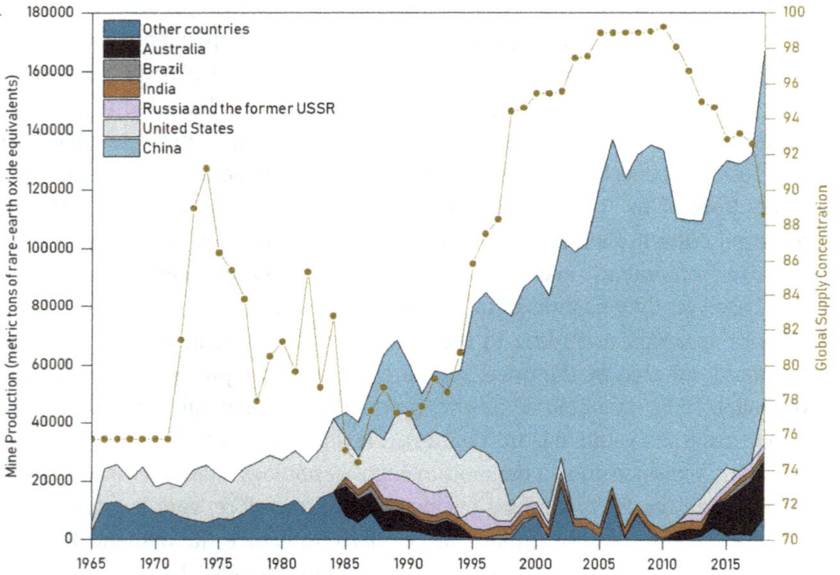

Figure 4.15.1: Mine production of rare-earth oxide equivalents by selected countries and year, starting from 1965 up to today. Left x-axis representing cumulative production data with right x-axis presenting the GSC per year. Data based on [6, 7].

Unfortunately, identifying economically viable heavy rare-earth element (HREEs) deposits outside of China has been without great success. An example of success is the production at the Mount Weld mine in Western Australia which accounted for 11% of the world production in 2018. To meet the constantly growing demand in technical products for these HREEs, a global search for comparable deposits is in progress. Due to the relatively inexpensive mining, efficient chemical extraction processes, and low environmental regulations, the most important sources of HREEs currently located globally are in clay deposits in southern China. Countries in the European Union as well as Japan and the USA with high dependence and economical interest in computer, electronic and green technologies show increasing concerns due to their high dependence on a single resource-supplying source, in this case China. For this reason, the above-mentioned countries classify the current situation with limited sources of supply in connection with the important role of REEs in modern technologies and the economy as "critical" [10–13]. The major surge of applications and the demand for REEs has created an increased interest in identifying new sources, including extraction from former not yet utilized mineral deposits, extraction from coal-based raw materials, and recycling of light bulbs and electronics containing REEs [14]. The Global Supply Concentration (GSC) indicator, also displayed in Figure 4.15.1, which is calculated using the Herfindahl-Hirschman-Index (HHI) of country-based production shares, can be used to show the market power of producing countries, for example to reflect the strong position that China currently possesses. The GSC [2] utilized here is used by the U.S. Department of Justice to identify a highly concentrated market place. A distinction is made at or above a HHI value of 1800, which translates into a GSC value of 70 [15]. Consecutively, the worldwide market for REEs can be described as a highly concentrated marketplace with monopolistic structures that was first dominated by the production in the USA, from the 1960s to the 1990s, and second from China, from around 1995. Interestingly, the period between 1985 and 1995 shows a low in the GSC data due to the fact that at that time, both dominant producing countries were present on the world market with nearly equal volume shares. Table 4.15.1 gives an overview of all the important currently producing countries with quantities per year and the underlying reserves, which are of great importance for further economic analysis.

Table 4.15.1: World reserves and mine production according to the USGS [16].

Country	Estimated reserves [t]	Production [t/a]
China	44,000,000	120,000
Australia	3,400,000	20,000
USA	1,400,000	15,000
Burma (Myanmar)	Not available	5,000
Russia	12,000,000	2,600
India	6,900,000	1,800
Brazil	22,000,000	1,000
Burundi	Not available	1,000
Thailand	Not available	1,000
Vietnam	22,000,000	400
Malaysia	30,000	200
Rest of the world	4,400,000	2,000
Total	**120,000,000**	**170,000**

4.15.3 Usage of rare earth elements

Historical development already revealed that the People's Republic of China is the single biggest producer of LREEs and HREEs currently with an average world production share of over 90% for the last decade. The most important sources of imports of all REEs to the European Union (EU) are China with 40%, the US with 34%, and Russia with 25%. With having 0% domestic production, the EU is an interesting case as it is entirely dependent on foreign sources till date, resulting in an import dependency ratio of 100%. Discussing the topic of usage of a raw material for an economy and therefore for the production of goods is typically not only a matter of primary supply but raw mining production. Furthermore, the point of raw material supply is often also a question if a material is probably substitutable in its functionality in terms of its material properties or if there are sufficient secondary sources to meet the demand, for example, recycling. Therefore, criticality studies often refer to the term "substitutability" which is a measure of the difficulty of replacing a material, evaluated and weighted over all applications. The EU recently introduced substitutability values that range from 0 to 1, where 1 refers to lowest substitutability. The substitutability index additionally gets corrected for the Economic Importance (EI) of a commodity and for the Supply Risk (SR). Presently, REEs are rated at an EI of 0.90 which equals an SR of 0.93 points. This reflects very low substitutability over all REE applications rated by the EU and its executive organs. The very low substitutability is also complemented by an overall low ratio of recycling for the REE group of elements with around 3% of recycled content re-entering the production cycle [17]. Older data even indicated an End-of-Life-Recycling-Rate of under 1% for yttrium, scandium and the lanthanides, although there is up to 10% of scrap metal as input for La, Ce, Pr, Nd, Gd and Dy. The problem with REEs as specialty metals is that they most often get used in very small quantities in complex element compositions in high technology products (e.g. computer chips or multi-component alloys). Most technological products get recycled later but the REEs are often lost because they get integrated in the metal recyclate, leading to the loss of its individual functionality as it becomes part of a compound material [18]. Consequently, the use of materials is further structured based on the applications themselves. REEs have the broadest application range across all consumer products just as any group of commodities usually get clustered as non-fuel mineral raw materials [19]. Due to their similar chemical nature, the various REEs have related or complementary applications. Most of the elements possess strong paramagnetism. One of the most important application fields is the use of REEs in magnets. These magnets are used in engines, hard disc drives, MRI, power generation, microphones, speakers, and magnetic refrigeration. One of the most growing application fields of REE is the use of alloys in magnets ($Nd_2Fe_{14}B$, Sm_5Co and Sm_2Co_{17}) as they are the strongest known magnetic materials in terms of weight and volume. In addition to the use in magnets, the glass industry is one of the leading consumers of REEs, utilizing their

unique material properties for glass polishing, as additives for color, and as additives for special optical properties. All important fields of application and their uses are summarized in the following list [20]:
- Functional ceramics: capacitors, sensors, colorants, scintillators, refractories
- Magnets: engines, hard disk drives, MRI, power generation, microphones and speakers, magnetic refrigeration
- Metallurgical alloys: NiMH-batteries (lanthanum-based alloys), fuel cells, steel, lighter flints, super alloys, aluminum/magnesium
- Phosphors: displays (CRT, LPD, LCD, PDP), fluorescent lighting, LEDs, medical imaging, lasers, fiber optics
- Glass: polishing components, decolorizers/colorizers, UV-resistant glass, X-ray imaging
- Catalysts: petroleum refining (lanthanum), automotive catalytic converters (cerium oxide), diesel additives, chemical processing, industrial pollution scrubber
- Others: nuclear, defense, water and air treatment, pigments, fertilizers

The German "Bundesanstalt für Geowissenschaften und Rohstoffe" has benchmarked the REEs, considering their use for important future technologies. The demand for future technology links the forecasted demand for elements to the current global production of these elements to provide an indicator of the potential need for additional extraction. The technology forecast especially focuses on Sc, Nd, Pr, La, Ce, Y, Dy and Tb. Scandium is currently used in aluminum alloys in the lighting industry in the form of scandium iodide (ScI_3), in mercury high-pressure lamps, and in solid oxide fuel cells (SOFC). Future demand is expected to rise from 17% of mine production in 2013 to over 138% in 2035. This growth will mainly be driven by the metal's demand for use in SOFCs, as scandium aluminum alloys are not expected to play an important role in the aerospace industry in the future [21].

REEs are reported to be especially important metals for use in future technologies such as wind turbines, electric traction motors for hybrid-, electric- and fuel-cell-vehicles, as well as in high-performance permanent magnets. The demand of LREEs neodymium and praseodymium in the production of important future technologies is expected to increase from 79% of mine production in 2013 to 174% in 2035. In addition to the use in magnets, the demand originates from use in solid-state lasers, micro-energy harvesting and in technologies for autonomous driving of automobiles. Furthermore, the demand of lanthanum and cerium for use in SOFCs is not likely to increase over the same period. Demand for heavy rare earth metals, especially yttrium, is expected to increase from 0.3% to 19% in 2035. The demand for yttrium is mainly generated by its use in autonomous driving technology, solid-state lasers, superconductors and SOFCs. Demand for dysprosium and terbium is expected to increase from 85% of mine production to around 313% in 2035, driven by its increased use in micro-energy harvesting and high performance permanent magnet technology [21]. Permanent magnet technology will most likely be the biggest growth driver for both LREEs and HREEs until 2035. The demand for these raw materials is thus very likely to exceed the current production in the future.

The distribution of the various applications of rare earth metals can be seen in detail in Table 4.15.2 for HREEs and in Table 4.15.3 for LREEs. The displayed recycling data was gathered from the United Nations Environment Program report on

Table 4.15.2: Distribution of applications, recycling rates, by-product dependence, substitutability, and Future Technology Demand per element of the HREE group. Partly based on data from [23].

HREE	Applications	End-of-Life-Recycling-Rate	By-Product Dependence (Host metals)	Substitutability	Future Technology Demand: Increase in demand from mine production
Y	7% batteries, 4% glass, 46% displays and lightings, 35% ceramics, 8% others	31%	29% (13% Sm, 9% Ce, 5% La, 3% Fe, <1% Ti, <1% Sn)	20%	6333%
Eu	96% displays and lighting, 4% others	8%	100% (35% Fe, 24% Ce, 19% Y, 15% Sm, 7% La, 1% Ti, <1% Sn)	31%	–
Gd	97% magnets, 3% others	8%	100% (38% Y, 21% Fe, 16% Sm, 14% Ce, 10% La, 1% Ti, <1% Sn)	31%	–
Tb	32% magnets, 68% displays and lighting	8%	100% (44% Y, 19% Fe, 16% Sm, 13% Ce, 7% La, <1% Ti, <1% Sn)	30%	368%
Dy	100% magnets	8%	100% (61% Y, 16% Sm, 12% Ce, 6% La, 4% Fe, 1% ti, < 1% Sn)	30%	368%
Er	74% glass, 26% displays and lighting	8%	100% (66% Y, 16% Sm, 10% Ce, 6% La, 2% Ti, < 1% Sn)	8%	–
Ho, Tm, Yb, Lu	100% glass	8%	100% (<68% Y, <18% Sm, <13% Ce, <7% La, <6% Ti, <1% Sn)	0%	–

Table 4.15.3: Distribution of applications, recycling rates, by-product dependence, substitutability, and Future Technology Demand per element of the LREE group. Partly based on data from [23].

LREE	Application	End-of-Life-Recycling-Rate	By-Product Dependence	Substitutability	Future Technology Demand: Increase in demand from mine production
La	3% metal, 10% batteries, 67% fluid cracking catalysts, 5% polishing, 10% glass, 2% displays and lighting, 2% ceramics	3%	93% (50% Fe, 27% Ce, 7% Sm, 6% Y, 2% Ti, <1% Sn)	12%	0%
Ce	6% metal, 6% batteries, 8% fluid cracking catalysts, 35% automotive catalysts, 11% polishing, 31% glass, 1% displays and lighting, 2% ceramics	3%	73% (67% Fe, 3% Ti, 2% Y, 1% Sm, <1% La, <1% Sn)	23%	0%
Pr	24% magnets, 11% metal, 12% batteries, 10% automotive catalysts, 10% polishing, 8% glass, 15% ceramics, 10% others	3%	100% (61% Fe, 17% Ce, 8% Sm, 7% Y, 6% La, 2% Ti, <1% Sn)	34%	220%
Nd	37% magnets, 12% metal, 13% batteries, 6% automotive catalysts, 8% glass, 11% ceramics, 10% others	3%	100% (57% Fe, 18% Ce, 9% Sm, 7% La, 7% Y, 2% Ti, <1% Sn)	38%	220%

Table 4.15.3 (continued)

LREE	Application	End-of-Life-Recycling-Rate	By-Product Dependence	Substitutability	Future Technology Demand: Increase in demand from mine production
Sm	97% magnets, 3% others	3%	82% (24% Fe, 23% Ce, 22% Y, 12% La, 1% Ti, <1% Sn)	31%	–

recycling rates of metals [18]. The already introduced by-product dependency is a measure for the dependency of the supply of an element from the processing of a different element. A high value indicates that it is not economically justifiable to extract a commodity as a primary product. The companion numbers presented reflect the percentage of global production [22].

4.15.4 Rare earth elements as innovation drivers

REEs are, in general, a group of elements with the broadest range of applications. This broad range of applications is reflected in the market for industrial property rights by a very broad distribution of patents across various International Patent Classification (IPC) classes, as listed in Table 4.15.4. The table differentiates between the status of an application, where a patent request is still pending at the respective patent office, and

Table 4.15.4: Top 10 International Patent Classification fields with the number of REE-related patent applications and grants.

International Patent Classification	Applications	Grants
Materials, metallurgy	51,065	30,311
Electrical machinery, apparatus, energy	39,412	23,087
Chemical engineering	27,923	16,103
Basic materials chemistry	22,503	12,782
Semiconductors	16,968	9,485
Surface technology, coating	16,110	8,793
Optics	14,979	8,710
Organic fine chemistry	14,693	8,353
Environmental technology	11,499	6,466
Audiovisual technology	7,788	4,837

the granted patent that represents fully by the law-protected intellectual property right. Patent owners and/or licensing partners can use, create and sell this work or process using the rights offered by the granted patent or license. Granted patents primarily provide the inventors and manufacturers of a protected work, a monopoly on that work for a certain period of time.

Parallel to the economic development and sharp increase in production and importance of REEs since 1965, as described earlier and shown in Figure 4.15.1, the publication of literature, patent applications and granted patents have likewise increased. The development can be seen in Figure 4.15.2.

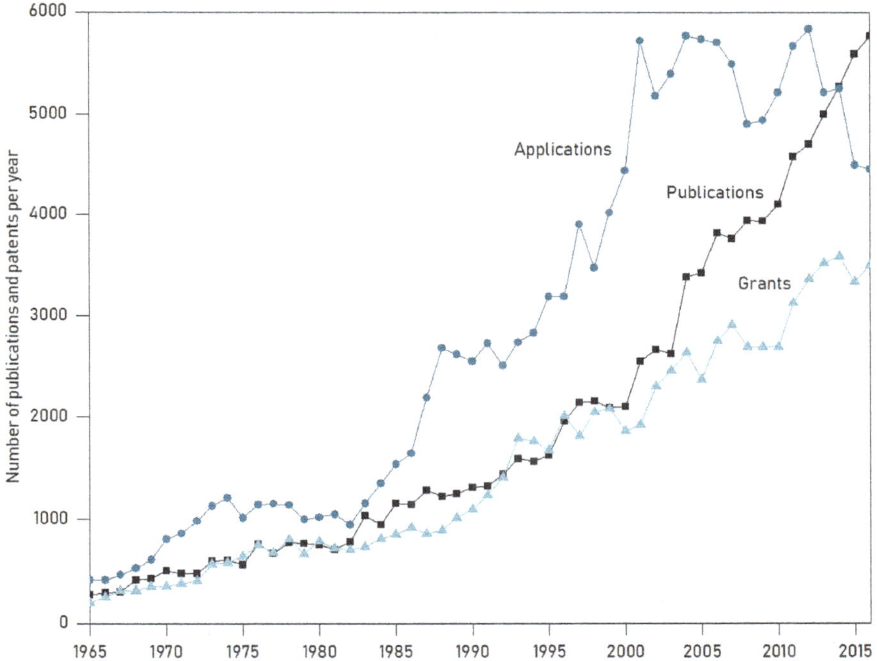

Figure 4.15.2: Growth of publications, granted patents and patent applications over the period 1965 to 2016. The patent data comes from the US Patent Office. The US is the largest and most important consumer market and is therefore a suitable benchmark for collecting this data.

In addition to the general development characterized by strong growth in both economic and intellectual property rights, the relevance of REEs as drivers of innovation can be particularly illustrated by using selected industry cases. An interesting and special case for this purpose is the one for white light-emitting diodes (LEDs). LEDs, as a broad technology, are the most efficient source of emitted visible light today. White LEDs, in comparison to traditional ones, emit polychromatic instead of monochromatic light. In terms of market success, these LEDs are often referred to as solid-state lighting to indicate that they replace conventional lighting sources (incandescent and fluorescent lamps) in many applications. LEDs are usually substitutes to light bulbs in housing, office lighting, advertisement, automotive, street lighting, as well as in liquid crystal displays (LCDs). Together with (Al,In,Ga)P (red, orange, and yellow diodes) and (In,Ga)N LEDs (violet, blue and green diodes) developed in the 80 s and 90 s, it is currently possible to cover the entire CIE1931 color triangle using the LED technology. Usually, the generation of white light is achieved through the combination of three primary colors (red, green and blue), or violet with respect to UV radiation that is absorbed by a phosphor or, and that's the most successful approach, a blue LED coated by a green to yellow and red light emitting phosphor. One of the most important patents that came from Shimizu et al. in the summer of 1996 was by introducing a garnet type phosphor derived from $Y_3Al_5O_{12}$ (yttrium–aluminum–garnet, YAG) doped with trivalent cerium [24].

Since 2008, it is widely accepted that the technology and, especially, white LEDs are revolutionizing the general lighting industry [25]. The 2014 Nobel Prize in Physics was given to Akasaki, Amano, and Nakamura "for the invention of efficient blue light-emitting diodes which has enabled bright and energy-saving white light sources" [26]. The Nobel Committee further acknowledged especially the market penetration of the technology. This is of particular interest in this chapter due to its connection to yttrium and cerium usage as key rare earths. The committee mentioned: "These high-quality LEDs with their very long lifetime (up to 100,000 h) are getting cheaper, and the market is currently exploding" [25]. Comparing 2012 and 2018, the whole LED market grew approximately to about 20% with revenues growing up to 60%. In the same period, manufacturing costs reduced by approximately 50% over all components of LED packages [27]. According to current estimates from 2016, the global sales of LED market is even expected to increase from just under USD 20 million to around USD 46 million in 2022 [28]. The aforementioned development would have never been possible without breakthroughs in the field of phosphors such as the cerium doped $Y_3Al_5O_{12}$ and thus highlighting how REEs are actively shaping the innovation landscape by enabling certain new technologies with unique and valuable properties.

4.15.5 Future supply and demand challenges

India and Brazil were the primary source countries till the 1940s. Between 1940s and 1960s, Australia and Malaysia started the production of monazite from placer deposits. The period up to the 1960s is thus often referred to as the monazite era. From the 1960s to the 1980s, production was dominant from the Mountain Pass mine in the USA. Production was mainly from bastnaesite carbonatite deposits and

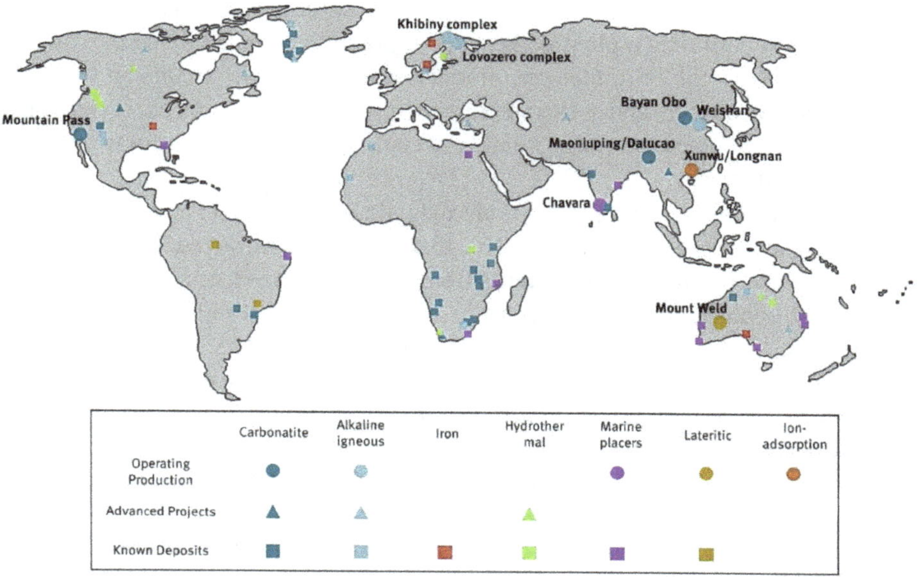

Figure 4.15.3: Global REE production and deposits. Data has been adapted from [20]. For further information please refer to [20].

smaller but growing production was from Australia. The current situation can be summarized as dominant mining production from China, especially from Bayan Obo mine, as a by-product of iron ore mining as well as from the Sichuan province and from South China. Contrary to the earlier production from the USA and Australia, these deposits in China also contain economically feasible amounts of HREE. HREE mining remains as one of the problematic supply fields as it is almost entirely controlled by one mining country. In contrast to the very strong dominance of China as a country in REE production until recently, since 2010, more and more countries, most notably Australia followed by the US, have started to enter the rare earth market once again. As a result, an increasing number of internationally operating companies are now entering the rare earths mining and processing business. Recently, mining started in Russia, again centered in the Kola Peninsula, where the *Lovozerska Mining Company* gathers loparite ore that is further refined to chlorides and carbonates by the *Solikamsk Magnesium Works*. The mining of heavy mineral sands in Brazil is led by *Industrias Nucleares do Brasil* as a by-product from the Taboca Pitinga tin mine and further processed by *Neo Material Technologies Inc.* and *Mitsubishi Corporation*. In Malaysia, monazite is also gathered as a by-product of tin extraction. Australia, historically produced xenotime. The *Lynas Corporation* (Mount Weld mine Australia) recently built concentration and advanced materials plants for the processing of the mined ore. In India, the *Indian Rare Earth Ltd.* (*Nuclear Power Corporation of India*) produces monazite from sand deposits (uranium and thorium main product) for REE chlorides as a by-product. Recently, the Japanese company *Toyota Tsusho Corp* built a plant in the state of Orissa to process rare earth oxide by-products, while in the USA, *Molycorp* (former owner was *Chevron Mining*) took over the Mountain Pass mine and started production there again. The current scenario can be said to be one of an increasing number of companies in an increasingly diversifying world market.

Potential threats leading to a supply reduction include reserve depletion time. Reserves refer to raw material deposits that are being extracted from an active mine, or which are not currently mined but could be mined in a technically and economically viable manner. Reserves are acknowledged through the use of the Static Reach Reserves (SRR) indicator presented in Table 4.15.5. The SRR indicates the number of years till the currently estimated reserves can be exploited, at the current rate of ore production. From the data, it is noticeable that the SRR, since 2006, was never below a value of 706 years of secured production, thus indicating no problem to secure the support of downstream supply of REE ore. Market concentration is reflected through global supply concentration (GSC) which is implemented using the HHI. The GSC is used by the U.S. Department of Justice to identify a highly concentrated marketplace. This distinction is made at or above an HHI value of 1800, which translates into a GSC value of 70 [15]. Both the GSC values in Figure 4.15.1 as well as the HHI presented in Table 4.15.5 indicate a highly concentrated marketplace environment. Political conditions are reflected by political stability, policy perception, and overall political regulation in each elemental mining country as perceived by mining corporations.

Table 4.15.5: Static Reach Reserves, Hirschman-Herfindahl-Index, and political indicators for REEs between 2006 and 2018.

Criterion	2006	2007	2008	2009	2010	2011	2012	2013	2014	2015	2016	2017	2018
SRR [a]	715	710	798	827	827	991	1,273	1,182	1,057	923	930	909	706
HHI	9,365	9,370	9,370	9,412	9,559	8,959	8,285	7,500	7,359	6,637	6,773	6,544	5,213
WGI-PV	72	72	71	69	75	71	71	70	67	67	66	58	–
PPI	59	72	67	55	55	69	57	52	47	53	51	39	–
HDI	65	67	68	69	70	70	72	74	77	77	78	79	–

Political stability is monitored using the Worldwide Governance Indicator – Political Stability and Absence of Violence (WGI-PV), incorporating factors regarding governmental policies and actions that affect the potential to obtain mineral resources [29]. Policy perception is depicted by the Policy Perception Index (PPI), formerly known as the Policy Potential Index, as a single score. It provides "a comprehensive assessment of the attractiveness of mining policies in a jurisdiction, and can serve as a report card to governments on how attractive their policies are from the point of view of an exploration manager" [30]. The risk of political regulation assesses the possibility of restrictions to mining and trade conditions in a country and is indicated by the Human Development Index (HDI), an indicator developed by the United Nations Development Program [31]. A higher score indicates a higher general quality of human life and social progress. This could indicate a more ethical supply chain but also a higher supply risk, as countries with high levels of development are likely to consider quality of life more important than industrial development and are thus more likely to impose restrictions and regulations that could limit material extraction. Table 4.15.5 represents political indicators that are transformed into a score from 0 to 100 points, where 0 indicates no risk and 100, the highest possible risk. The indicators for REE elements are calculated by combining the scores for each source country, weighting each according to their percentage of total production. Accordingly, all three political indicators show medium to high-risk numbers for the period between 2006 and 2017 as they are mainly dependent on the conditions in the dominant sourcing country, China.

4.15.6 The challenge of heavy rare earth elements

Rare earth metals occur only as a group and can be mined only together. Most rare earths are not rare in the earth's crust. On the other hand, they are, unfortunately, also not often enriched in economically degradable quantities. The frequency of their occurrence in the earth's crust is very different and ranges from cerium, with over 60 ppm, to thulium and lutetium, which are less than 0.3 ppm in the earth's crust. Typically, they occur in rock-forming minerals as trivalent cations (exceptions are Sm^{2+}, Eu^{2+}, Ce^{4+}, and Yb^{2+}) in compounds with carbonates, oxides, phosphates and silicates. LREE occur much more frequently than the heavier ones. In most deposits, lanthanum, cerium, praseodymium and neodymium make up more than 90% of the total rare earth deposits [32]. LREE and yttrium are thus cheaper, produced in higher quantities and are generally used more often than the HREE group elements. Most expensive and least used elements are the HREE elements from holmium to lutetium [33]. In the Bayan Obo mine in North China, REEs are mainly extracted from the bastnaesite, monazite and xenotime minerals as well as from ionic adsorption of lateritic weathering crusts. In monazite and bastnaesite, however, the LREE from lanthanum to Neodymium are mainly present. Bastnaesite is also the main product of Australian and US production. In the listed deposits, however, the HREEs account for only a very small fraction. For example, only 2.3% of the rare earth oxides extracted from the Bayan Obo deposit are members of the HREE group. In summary, the situation is that HREE group elements are currently almost exclusively extracted from Chinese mines, which has lead to increased search efforts for deposits outside China.

4.15.7 Financial perspective

The first price data of traded LREEs can be tracked back to the 50 s and 60 s in the United States. Prices ranged between about USD 3.000 per kilogram for CeO_2 and lanthanum to almost USD 6.000 per kilogram for Gd_2O_3. The commencement of the Mountain Pass mine operations at the in 1952 led to a substantial decrease in LREE prices, a maximum of under USD 2,500 USD per kilogram for gadolinium and even lower prices for the rest of the, at these times, processed LREE elements. Prices for HREEs in the same period were between almost 6,000 in 1960 and near to 1,000 USD per kilogram in 1975 for dysprosium, between 30,000 in 1960 and around 10,000 USD per kilogram in 1975 for terbium, and between 90,000 and 20,000 USD per kilogram for europium [16, 34].

Since the 1970s, the prices are based on the price trends for bastnaesite and monazite ores. According to USGS data, the price for 60% bastnaesite rare earth oxide concentrate increased from around USD 4 per kilogram in 1970 to 6–10 USD per kilogram in the 2000s. On the other hand, prices for 55% monazite rare earth oxide concentrate decreased from around USD 2 per kilogram in 1970 to under USD 1 per kilogram since the 1990s. A similar decreasing price trend can be observed for the separated LREE and HREE metals with prices being under USD 200 per kilogram for LREEs in 2010 and around USD 2.000 per kilogram for HREEs. The general trends from the early years remained stable. HREEs such as Tb and Eu traded at more expensive prices compared to the metals from the LREE group by a factor of 10. After the steady decline up to the year 2010, prices suffered a severe increase during the period between February and June of 2011. Prices for LREEs such as neodymium jumped up to a temporary high of around USD 300 per kilogram, while HREEs such as europium traded at almost USD 3,500 per kilogram by the end of the 1980s. Especially for HREEs, the starting of mining in China in the early 1990s led to a substantial price decrease, as mentioned earlier. The reason behind the sharp increase during that period was a result of China's export policy between 2009 and 2012. Chinese export quotas, as well as the supply to domestic consumers and joint ventures with foreign companies in China, decreased by over 80% and around 30–40% during the said period. Prices stabilized again after 2013 till today. It is noteworthy that during these times, many countries reported higher imports from China. These differences are explained by the fact that significant quantities of REEs were exported as a mixture with iron ore and thus as steel composites and other commodities to bypass regulatory requirements [5]. To sum up, especially after the strong temporary price increase between 2011 and 2012, prices are currently back at the level of 2010 or even lower and thus, significantly lower than in the 1970s and 1980s. This applies, in particular, to the HREEs, whereas the prices for the LREEs praseodymium and neodymium could not quite cope with the temporary increase and have indeed become permanently more expensive. Although the overall development of REE volatility since the 1970s can be characterized by increasing stability (especially prices for monazite ore are becoming more stable), the annual price volatility across all REEs continued to be between 40% and 80% in 2015, showing a rising trend since the 2011 crisis and in relation to selected elements such as cerium, which is currently suffering the highest volatility [34]. To give an overview of currently operating REE companies, all leading companies with a brief description and their market capitalization for the year 2016 are shown in Table 4.15.6.

Table 4.15.6: Companies actively operating on the REE market sorted by highest market capitalization [34].

Company	Market capitalization (million USD)	Business	Main operations
Aluminum Corp of China (CHALCO)	8,325	Investment and processing of products.	China
China Molybdenum Co. Ltd.	7,934	Offer molybdenum rare-earth material and other rare-earth products.	China
China Northern Rare Earth Gp	6,449	Offer REE oxides, metals, salt products and magnetic material (both raw and processed materials).	China
Xiamen Tungsten Co.	3,445	Battery materials, nonferrous metal products.	China
Iluka Resources	2,236	Mineral sands exploration, management of mining projects.	Australia
China Minmetals Rare Earth	1,920	REE oxides, aluminum products, luminescent material and lighting products.	China
Rising Nonferrous Metals Share	1,855	Tungsten, REEs and related products. Mining, processing, and distribution of nonferrous metals.	China
Galaxy Resources Ltd.	712	Lithium carbonate and exploration business. Operating in Australia, Argentina, and Canada.	Australia
Lynas Corp.	189	Mining and processing of REE minerals. Operating in Australia and Malaysia. Exploration and development of deposits.	Australia
Kidman Resources	137	Exploration of possible REE deposits in Australia.	Australia
Alkane Resources Ltd.	126	REE and gold metal exploration, development, and mining.	Australia
Wealth Minerals Ltd.	68	REE, lithium, precious metals exploration, and development.	Canada
Ucore Rare Metals	64	REE metal exploration and development in Canada and the USA	USA

Table 4.15.6 (continued)

Company	Market capitalization (million USD)	Business	Main operations
Critical Elements Corporation	55	REE exploration and development. Operates in British Columbia.	Canada
Greenland Minerals	51	Exploration and operating of Kvanefjeld-Rare earth project.	Greenland
Avalon Rare Metals	21	Exploration and development operations involving lithium, gallium, germanium, REEs and other metals.	Canada
Arafura Resources Ltd.	19	Exploration, development, and REE production.	Australia
Tasman Metals	13	REE and tungsten exploration in Scandinavia.	Sweden
Molycorp Inc.	8	REE resources, chemicals, oxides, magnetic material, alloys, and rare metals. Mountain Pass mine California.	USA
Rare Elements Resources	3	Sundance Gold and Bear Lodge REE projects.	USA
Great Western Minerals	0.4	REE metal alloys for magnet, battery, and aerospace industry.	Canada

References

[1] S. M. Hayes, E. A. McCullough, Resour. Policy 2018, 59, 192.
[2] T. E. Graedel, R. Barr, C. Chandler, T. Chase, J. Choi, L. Christoffersen, E. Friedlander, Environ. Sci. Technol. 2012, 46, 1063.

[3] S. M. Fortier, J. H. DeYoung, E. S. Sangine, E. K. Schnebele, Comparison of U.S. Net Import Reliance for Nonfuel Mineral Commodities – A 60-Year Retrospective (1954–1984–2014), U.S. Geological Survey Numbered Series, Reston, 2015.
[4] M. H. Ting, J. Seaman, Asian Stud. Rev. 2013, 37, 234.
[5] N. A. Mancheri, Resour. Policy 2015, 46, 262.
[6] T. D. Kelly, G. R. Matos, D. A. Buckingham, C. A. DiFrancesco, K. E. Porter, Historical Statistics for Mineral and Material Commodities in the United States, 2014.
[7] U.S. Geological Survey, Rare Earths Statistics and Information, 2019.
[8] J. M. Hammarstrom, C. L. Dicken, Focus Areas for Data Acquisition for Potential Domestic Sources of Critical Minerals – Rare Earth Elements, U.S. Geological Survey Numbered Series, Reston, 2019.
[9] E. Machecek, P. Kalvig, GEUS, D'Appolonia, Development of a Sustainable Exploitation Scheme for Europes REE Ore Deposits: European REE Market Survey, 2017.
[10] U.S. Department of Energy, Critical Materials Strategy 2011, 2011.
[11] European Commission, Critical Raw Materials, 2019.
[12] European Commission, Report on Critical Raw Materials and the Circular Economy, 2018.

[13] B. S. Van Gosen, P. L. Verplanck, R. R. Seal II, K. R. Long, J. Gambogi, Rare-Earth-Elements, in Critical Mineral Resources of the United States – Economic and Environmental Geology and Prospects for Future Supply (ed. K. J. Schulz, J. H. DeYoung, R. R. Seal II, D. C. Bradley), 2017.
[14] B. S. Van Gosen, P. L. Verplanck, P. Emsbo, Rare Earth Element Mineral Deposits in the United States, 2019.
[15] U.S. Department of Justice, and Federal Trade Commission, Rev. Ind. Organ. 1993, 8, 231.
[16] U.S. Geological Survey, Rare Earths, 2019.
[17] European Commission, On the 2017 List of Critical Raw Materials for the EU, 2017.
[18] T. E. Graedel, J. Allwood, J.-P. Birat, M. Buchert, C. Hagelüken, B. K. Reck, S. F. Sibley, G. Sonnemann, UNEP Recycling Rates of Metals – A Status Report, 2011.
[19] S. B. Castor, J. B. Hedrick, Rare Earth Elements, in Industrial Minerals and Rocks: Commodities, Markets, and Uses (ed. J. E. Kogel, N. C. Trivedi, J. M. Baker, S. T. Krukowski), 2006.
[20] A. Walters, P. Lusty, A. Hill, Rare Earth Elements, Nottingham, 2011.
[21] F. Marscheider-Weidemann, S. Langkau, T. Hummen, L. Erdmann, L. T. Espinoza, Rohstoffe Für Zukunftstechnologien 2016, 2016.
[22] N. T. Nassar, T. E. Graedel, E. M. Harper, Sci. Adv. 2015, 1.
[23] Deloitte Sustainability, British Geological Survey, Bureau de Recherches Geologiques et Minieres, Netherlands Organisation for Applied Scientific Research, Study on the Review of the List of Critical Raw Materials – Critical Raw Materials Factsheets, 2017.
[24] Y. Shimizu, K. Sakano, Y. Noguchi, T. Moriguchi, Light-emitting Device Having a Nitride Compound Semiconductor and a Phosphor Containing a Garnet Fluorescent Material, 1996.
[25] J. Cho, J. H. Park, J. K. Kim, E. F. Schubert, Laser Photonics Rev. 2017, 11(2).
[26] S. Nakamura, Background Story of the Invention of Efficient Blue InGaN Light Emitting Diodes, Santa Barbara, 2014.
[27] O. Daniel, LED and Technology Transfer at CNRS, Paris, 2014.
[28] P. Smallwood, Lighting, LEDs and Smart Lighting Market Overview, 2016.
[29] The World Bank Group, Worldwide Governance Indicators, 2019.
[30] A. Stedman, K. P. Green, Fraser Institute Annual Survey of Mining Companies 2017, 2018.
[31] United Nations Development Programme, Human Development Data, 2018.
[32] H. Elsner, Commod. Top News 2011, 36, 1.
[33] T. G. Goonan, Rare Earth Elements-End Use and Recyclability, in Rare Earth Elements: Supply, Trade and Use Dynamics, 2012.
[34] V. Fernandez, Resour. Policy 2017, 53, 26.

Subject Index

https://doi.org/10.1515/9783110654929-039

Formula Index

https://doi.org/10.1515/9783110654929-040

www.ingramcontent.com/pod-product-compliance
Lightning Source LLC
Chambersburg PA
CBHW080346220326
41598CB00030B/4617

* 9 7 8 3 1 1 0 6 5 3 6 0 1 *